D0203641

Conversion Between English and SI Units

Quantity	English Unit[a]	SI Unit[a]	Ratio, SI/English: Approximate	Ratio, SI/English: Accurate	Ratio, English/SI: Approximate	Ratio, English/SI: Accurate
Length	inch (in. or ″)	meter (m)	0.0254	2.540 000 E − 2	39.4	3.937 008 E + 1
	foot (ft or ′)	meter (m)	0.305	3.048 000 E − 1	3.28	3.280 840 E + 0
	mile (mi U.S. Statute)	kilometer (km)	1.61	1.609 344 E + 0	0.621	6.213 712 E − 1
Volume	gallon (gal U.S.)	meter^3 (m^3)	0.00379	3.785 412 E − 3	264.	2.641 720 E + 2
Force, (weight)	pound (lb)	newton[d] (N)	4.45	4.448 222 E + 0	0.225	2.248 089 E − 1
Torque	pound-foot (lb-ft)	newton-meter (N·m)	1.36	1.355 818 E + 0	0.737	7.375 620 E − 1
Work, energy	foot-pound (ft-lb)	joule[c] (J)	1.36	1.355 818 E + 0	0.737	7.375 620 E − 1
Power	foot-pound/second (ft-lb/s)	watt[g] (W)	1.36	1.355 818 E + 0	0.737	7.375 620 E − 1
	horsepower[b] (hp)	kilowatt (kW)	0.747	7.456 999 E − 1	1.34	1.341 022 E + 0
Stress, pressure	pounds/in.^{2} (psi)	pascal[e] (Pa)	6890.	6.894 757 E + 3	0.000145	1.450 370 E − 4
	thousand pounds/in.^{2} (ksi)	megapascal (MPa)	6.890	6.894 757 E + 0	0.145	1.450 370 E − 1
Mass	slug[f]	kilogram (kg)	14.6	1.459 390 E + 1	0.0685	6.852 178 E − 2

Examples:
A door 36 in. wide is 36 in. (0.0254 m/in.) = 0.914 m wide;
a door 1 m wide is 1 m (39.4 in./m) = 39.4 in. wide.

[a]The *larger* unit is underlined.
[b]1 hp = 550 ft·lb/s [c]1 J = 1 N·m [d]1 N = 1 kg·m/s^2 [e]1 Pa = 1 N/m^2 [f]1 slug = 1 lb·s^2/ft [g]1 W = 1 J/s

Suggested SI Prefixes for Angular Deflection Calculations

$\theta = TL/K'G^a,\ ML/IE^a$				
θ	T,M	L	K',I	E,G
rad	N·m	m	m^4	Pa
μrad	N·m	m	m^4	MPa
mrad	N·mm	mm	mm^4	GPa
μrad	kN·m	m	m^4	GPa

[a]Illustrated in Table 5.1.

Suggested SI Prefixes for Stress Calculations

$\sigma = P/A,\ Mc/I,\ M/Z;\ \ \tau = P/A,\ V/A,\ Tr/J,\ T/Z',\ (V/Ib)\int y\,dA$						
σ,τ	P,V	M,T	A	I,J	c,r,b,y	Z,Z'
Pa	N	N·m	m^2	m^4	m	m^3
kPa	kN	kN·m	m^2	m^4	m	m^3
MPa	N	N·mm	mm^2	mm^4	mm	mm^3
GPa	kN	N·m	mm^2	mm^4	mm	mm^3

Suggested SI Prefixes for Linear Deflection Calculations

$\delta = PL/AE^a;\ \delta \propto PL^3/EI^a,\ wL^4/EI,\ ML^2/EI^a$							
δ	P	w	M	L	A	E	I
μm	N	N/m	N·m	m	m^2	MPa	m^4
μm	N	N/mm	N·mm	mm	mm^2	GPa	mm^4
μm	kN	N/m	kN·m	m	m^2	GPa	m^4

[a]Illustrated in Table 5.1.

Fundamentals of Machine Component Design

Robert C. Juvinall
Professor of Mechanical Engineering
University of Michigan

John Wiley & Sons
New York *Chichester* *Brisbane* *Toronto* *Singapore*

To Arleene

Copyright © 1983, by John Wiley & Sons, Inc.

All rights reserved. Published simultaneously in Canada.

Reproduction or translation of any part of
this work beyond that permitted by Sections
107 and 108 of the 1976 United States Copyright
Act without the permission of the copyright
owner is unlawful. Requests for permission
or further information should be addressed to
the Permissions Department, John Wiley & Sons.

Library of Congresss Cataloging in Publication Data

Juvinall, Robert C.
Fundamentals of machine component design.

Includes bibliographical references and index.
1. Machinery–Design. I. Title.
TJ230.J88 1983 621.8′15 82–21952
ISBN 0–471–06485–8

Printed in the United States of America

10 9 8 7

PREFACE

This book is intended as a text for first courses in Mechanical Engineering Design and as a reference for practicing engineers. It is assumed that the user has had basic courses in Mechanics, Strength of Materials, and Materials Properties. However, the first nine chapters of the book (Part I) serve to review as well as extend this basic background. The remaining chapters (Part II) deal with the application of these fundamentals to specific machine components, and, in the case of the final chapter, to a complete machine.

Although much of Part I is a review of earlier courses, I would like to call attention to several particular parts: (1) Sections 1.2, 1.3, and 1.4 deal with three of the broadest aspects of engineering—safety, ecology, and social significance. These are concerns to which today's students are particularly responsive. My own classes work Problems 1.1–1.6 (which relate to these concerns) as "extra" assignments, spread throughout the term. (2) Sections 1.7 and 1.8 review fundamental energy relationships. Most students at this level need to gain insight and understanding concerning such basic matters as the relationship between work input to a rotating camshaft and work output at a translating follower, and the relationship between work output of a machine, motor input, and stored energy in the rotating and translating parts. (3) Most teachers of Mechanical Engineering Design lament the weakness of their students in the area of free-body diagram analysis of loads. Unless the loading on a machine component is properly established, subsequent design or analysis is of little value. Section 2.2 and the associated problems are directed toward helping relieve this common deficiency. (4) An elementary treatment of residual stresses is included in Chapter 4. An understanding of the basic concepts involved is vital to modern stress analysis, particularly when fatigue is involved. (5) Castigliano's method for determining elastic deflections and redundant reactions is included in Chapter 5. This method permits ready solution of many problems not amenable to traditional elementary methods. (6) Chapter 6 on Failure Theories, Safety Factors, and Reliability includes introductory treatments of fracture mechanics and of the interference theory of statistical reliability prediction. Even if the chapter is not assigned, the instructor can call attention to these important newer disciplines. (7) Chapter 8 contains a simplified, condensed, and updated version of the material on metal fatigue, which is contained in my earlier book.* This chapter is particularly important, and represents primarily new material for most students. (8) Chapter 9 deals with the various kinds of surface deterioration experienced by machine components. This is of great importance because more machine parts "fail" (cease to be suitable for performing their intended function) because of surface damage than from actual breakage.

Part II is concerned with the application of the fundamentals to specific machine components. In engineering practice, problems involving the design, analysis, or application of machine members can seldom be solved by applying the fundamentals alone. As critically important as a knowledge of the underlying sciences is, it is seldom sufficient. Almost always some empirical information must be used, and good "engineering judgment" brought to bear. Actual engineering design problems seldom have only one correct answer. For example, engineering staffs of competing companies arrive at different product designs as "solutions" to the same problem. And these solutions change as new technology, new materials, new manufacturing methods, and new marketing conditions arise. For many students, the course based on this text will provide their first experience in dealing with these kinds of professional engineering problems.

Most engineers find that the above aspect of engineering adds to the interest and excitement of their profession. There is a close parallel between engineers and medical doctors in this respect: both must solve real-life problems *now,* making full use of the best available scientific information. Engineers must design engines and build electronic apparatus even though scientists are still seeking a more complete knowledge of combustion and electricity. Likewise, medical doctors cannot tell their patients to await treatment until more research has been completed.

Even though the fundamentals treated in Part I are seldom *sufficient* for solving engineering problems relating to machine components, it is important that they be applied fully and consistently. In particular, a special effort has been made in Part II to deal with fatigue and surface considerations in a manner consistent with the treatment given in Chapters 8 and 9. This sometimes results in the development of procedures that vary in detail from those given in the specialized literature, but this is not of major importance. What *is* of major importance is helping the student learn to approach engineering problems by applying the fundamentals and other scientific knowledge as far as possible, and then supplementing these with empirical data and judgment as required to get good solutions within available time limitations.

Few engineering schools allot sufficient time to cover all the machine components treated in Part II. Also, there are many components not treated in the book, and even more that are not yet in existence. For these reasons, each component is treated not only as an end in itself, but also as a representative example of applying basic fundamentals and necessary empirical information to solve practical engineering problems.

Throughout Part II the reader will find numerous instances where ingenuity, insight, and imagination are called for to deal effectively with engineering problems

*Engineering Considerations of Stress, Strain, and Strength, McGraw-Hill, 1967

associated with an individual machine component. The next step in the study of Mechanical Engineering Design usually involves the conception and design of a complete machine. As an introduction to this "next step," the final chapter of the book (Chapter 20) presents a "case study" of the design of the first commercially successful automotive automatic transmission. Here, as with numerous other designs of complete machines, one cannot help being impressed and inspired by the insight, ingenuity, and imagination (as well as prolonged diligent effort) displayed by the engineer(s) involved. Also illustrated in this case study is the way the design of any one component is often influenced by the design of related parts.

Because the present generation of students will inevitably need to deal with both SI and conventional English units, both systems are used in the text and in the problems.

In some cases, this text has retained graphical procedures (like *S-N* curves and mean stress-alternating stress diagrams for fatigue analyses) rather than using equivalent mathematical expressions more quickly handled with calculators and computers. This is done where the graphical procedure helps the student to understand and "visualize" what is going on, develop added insight about the significance of the results, and see how the design might be improved. In actual practice, whenever such procedures are called for on a repetitive basis, the competent engineer will obviously employ modern computing facilities to full advantage.

Robert C. Juvinall

Acknowledgments

It is impossible to give adequate recognition to the many individuals who have contributed substantially to my own professional thinking reflected in this book. Two of the earliest of this distinguished group are Professor Robert R. Slaymaker and Professor Daniel K. Wright of Case Western Reserve University, under whom I studied over 40 years ago, and who remain active at that institution today. I have often wondered how strongly my gravitating to the area of mechanical engineering design was influenced by the fact that I first studied this field under Bob Slaymaker, an outstanding engineer, superb teacher, and gentleman whom I admired greatly. (Those of us in engineering education easily forget how much students are influenced by the character and the professional attitudes and practices of their instructors.)

I would like to recognize with sincere thanks the several engineering authorities who reviewed individual chapters of the book and offered valuable suggestions. Among these are Joseph Datsko (University of Michigan), Robert J. Finkelston (Standard Pressed Steel Co.), Robert Frayer (Federal Mogul Corp.), Alex Gomza (Grumman Aerospace Corp.), Evan L. Jones (Chrysler Corp.), Vern A. Phelps (University of Michigan), Robert R. Slaymaker (Case Western Reserve University), Gus S. Tayeh (New Departure Hyatt Bearings), Paul R. Trumpler (Trumpler Associates), Lew Wallace (Gleason Machine Div.), James E. West (FAG Bearings Corp.), Charles Williams (Federal Mogul Corp.), Ward O. Winer (Georgia Institute of Technology), and William Wood (Associated Spring Barnes Group). In addition to expressing my deep gratitude to these individuals, I would like to state clearly that the responsibility for each chapter is solely mine. If the reader finds errors, or points of view with which he or she disagrees, there should be no inference that these are due to anyone except the author. Also, I would like to state that while every effort has been made to insure the accuracy and the conformity with good engineering practice of all of the material contained in this book, there is no guarantee, stated or implied, that mechanical components designed on the basis of this text will in all cases be proper and safe. Mechanical engineering design is sufficiently complex that its actual practice should always take advantage of the specialized literature in the area involved, the backround of experience with related components, and, most important, appropriate tests to establish proper and safe performance in critical cases.

I would also like to express appreciation to Professors James Barber, Panos Papalambros, and Mohammed Zarrugh at the University of Michigan who made valuable suggestions as a result of teaching from preliminary versions of this text. My thanks go as well to their students and to my own students who have suffered the inconveniences of studying from an incomplete text and who made valuable suggestions for its improvement. I would like to express particular thanks to Professor Emeritus Herbert H. Alvord of the University of Michigan who generously permitted me to use any of the extensive collection of problems he developed for his own classes.

Finally, I would like to express special thanks to my wife, Arleene, for her understanding and encouragement during the years when the preparation of this book preempted time belonging, by all reasonable standards, to important family and social activities.

R. C. J.

CONTENTS

PART 1 FUNDAMENTALS

Chapter 1 Mechanical Engineering Design in Broad Perspective

1.1 An Overview of the Subject 3
1.2 Safety Considerations 5
1.3 Ecological Considerations 8
1.4 Societal Considerations 10
1.5 Overall Design Considerations 13
1.6 Systems of Units 14
1.7 Work, Energy 17
1.8 Power 18
References 24

Chapter 2 Load Analysis

2.1 Introduction 25
2.2 Equilibrium Equations and Free-Body Diagrams 25
2.3 Beam Loading 34
2.4 Locating Critical Sections—Force-Flow Concept 37
2.5 Load Division Between Redundant Supports 40
2.6 Force-Flow Concept Applied to Redundant Ductile Structures 43
References 45

Chapter 3 Materials

3.1 Introduction 46
3.2 The Static Tensile Test—"Engineering" Stress-Strain Relationships 47
3.3 Implications of the "Engineering" Stress-Strain Curve 48
3.4 The Static Tensile Test—"True" Stress-Strain Relationships 51
3.5 Energy Absorbing Capacity 53
3.6 Estimating Strength Properties From Penetration Hardness Tests 54
3.7 Use of "Handbook" Data for Material Strength Properties 56
3.8 Machinability 57
3.9 Cast Iron 57
3.10 Steel 58

3.11 Non-Ferrous Alloys 61
3.12 Plastics 63
References 68

Chapter 4 Static Body Stresses

4.1 Introduction 69
4.2 Axial Loading 69
4.3 Direct Shear Loading 72
4.4 Torsional Loading 73
4.5 Pure Bending Loading, Straight Beams 76
4.6 Pure Bending Loading, Curved Beams 77
4.7 Transverse Shear Loading in Beams 82
4.8 Induced Stresses, Mohr-Circle Representation 88
4.9 Combined Stresses—Mohr-Circle Representation 91
4.10 Stress Equations Related to Mohr's Circle 95
4.11 Three-Dimensional Stresses 96
4.12 Stress Concentration Factors (K_t) 98
4.13 Importance of Stress Concentration 102
4.14 Residual Stresses Due to Yielding—Axial Loading 106
4.15 Residual Stresses Due to Yielding—Bending and Torsional Loading 108
4.16 Thermal Stresses 111
4.17 Importance of Residual Stresses 113
References 114

Chapter 5 Elastic Strain, Deflection, and Stability

5.1 Introduction 115
5.2 Strain Definition, Measurement, and Mohr-Circle Representation 116
5.3 Analysis of Strain—Equiangular Rosettes 119
5.4 Analysis of Strain—Rectangular Rosettes 121
5.5 Elastic Stress-Strain Relationships and Three-Dimensional Mohr Circles 124
5.6 Deflection and Spring Rate—Simple Cases 125
5.7 Beam Deflection 126
5.8 Elastic Deflections by Castigliano's Method 132
5.9 Redundant Reactions by Castigliano's Method 140
5.10 Euler Column Buckling—Elastic Instability 145
5.11 Effective Column Length for Various End Conditions 148
5.12 Column Design Equations—J. B. Johnson Parabola 149
5.13 Eccentric Column Loading—the Secant Formula 152
5.14 Equivalent Column Stresses 154
5.15 Other Types of Buckling 155
References 156

Chapter 6 Failure Theories, Safety Factors, and Reliability

6.1 Introduction 157
6.2 Types of Failure 159

6.3 Fracture Mechanics—Basic Concepts 160
6.4 Fracture Mechanics—Applications 162
6.5 The "Theory" of Static Failure Theories 164
6.6 Maximum-Normal-Stress Theory 165
6.7 Maximum-Shear-Stress Theory 166
6.8 Maximum-Distortion Energy Theory (Maximum-Octahedral-Shear-Stress Theory) 167
6.9 Modified Mohr Theory 169
6.10 Selection and Use of Failure Theories 171
6.11 Safety Factors—Concept and Definition 173
6.12 Safety Factors—Selection of a Numerical Value 174
6.13 Reliability 177
6.14 Normal Distributions 178
6.15 Interference Theory of Reliability Prediction 179
References 181

Chapter 7 Impact

7.1 Introduction 182
7.2 Stress and Deflection due to Linear and Bending Impact 184
7.3 Stress and Deflection due to Torsional Impact 192
7.4 Effect of Stress Raisers on Impact Strength 194
References 199

Chapter 8 Fatigue

8.1 Introduction 200
8.2 Basic Concepts 201
8.3 Standard Rotating-Bending Fatigue Strengths (S'_n) 203
8.4 Fatigue Strengths for Reversed Bending and Reversed Axial Loading 207
8.5 Fatigue Strength for Reversed Torsional Loading 209
8.6 Fatigue Strength for Reversed Biaxial Loading 210
8.7 Influence of Surface and Size on Fatigue Strength 211
8.8 Summary of Estimated Fatigue Strengths for Completely Reversed Loading 214
8.9 Influence of Mean Stress on Fatigue Strength 215
8.10 Influence of Stress Concentration with Completely Reversed Fatigue Loading 222
8.11 Influence of Stress Concentration with Means Plus Alternating Loads 225
8.12 Fatigue Life Prediction with Randomly Varying Loads 232
8.13 Influence of Surface Treatments on the Fatigue Strength of a Part 236
8.14 Mechanical Surface Treatments (Shot Peening, etc.) 238
8.15 Thermal and Chemical Surface-Hardening Treatments (Induction Hardening, Carburizing, etc.) 239
References 239

Chapter 9 Surface Damage

9.1 Introduction 241
9.2 Corrosion—Fundamentals 242
9.3 Corrosion—Electrode and Electrolyte Heterogeneity 244
9.4 Design for Corrosion Control 246
9.5 Corrosion Plus Static Stress 249
9.6 Corrosion Plus Cyclic Stress 251
9.7 Cavitation Damage 252
9.8 Types of Wear 252
9.9 Adhesive Wear 253
9.10 Abrasive Wear 255
9.11 Fretting 256
9.12 Analytical Approach to Wear 257
9.13 Curved Surface Contact Stresses 259
9.14 Surface Fatigue Failures 263
9.15 Closure 267
References 268

PART 2 APPLICATIONS

Chapter 10 Threaded Fasteners and Power Screws

10.1 Introduction 273
10.2 Thread Forms, Terminology, and Standards 274
10.3 Power Screws 278
10.4 Static Screw Stresses 288
10.5 Threaded Fastener Types 294
10.6 Fastener Materials and Methods of Manufacture 296
10.7 Bolt Tightening and Initial Tension 299
10.8 Thread Loosening and Thread Locking 303
10.9 Bolt Tension with External Joint-Separating Force 305
10.10 Bolt (or Screw) Selection for Static Loading 310
10.11 Bolt (or Screw) Selection for Fatigue Loading—Fundamentals 318
10.12 Bolt (or Screw) Selection for Fatigue Loading—Using Special Test Data 324
10.13 Increasing Bolted Joint Fatigue Strength 327
References 328

Chapter 11 Rivets, Welding, and Bonding

11.1 Introduction 330
11.2 Rivets 330
11.3 Welding Processes 333
11.4 Welded Joints Subjected to Static Axial and Direct Shear Loading 336
11.5 Welded Joints Subjected to Static Torsional and Bending Loading 340
11.6 Fatigue Considerations in Welded Joints 346

11.7 Brazing and Soldering 347
11.8 Adhesives 348
References 350

Chapter 12 Springs

12.1 Introduction 351
12.2 Torsion-Bar Springs 352
12.3 Coil Spring Stress and Deflection Equations 353
12.4 Stress and Strength Analysis for Helical Compression Springs—Static Loading 359
12.5 End Designs of Helical Springs 362
12.6 Buckling Analysis of Helical Compression Springs 363
12.7 Design Procedure for Helical Compression Springs—Static Loading 364
12.8 Design of Helical Compression Springs for Fatigue Loading 368
12.9 Helical Extension Springs 375
12.10 Beam Springs (Including Leaf Springs) 377
12.11 Torsion Springs 382
12.12 Miscellaneous Springs 384
References 386

Chapter 13 Lubrication and Sliding Bearings

13.1 Types of Lubricants 387
13.2 Types of Sliding Bearings 388
13.3 Types of Lubrication 389
13.4 Basic Concepts of Hydrodynamic Lubrication 390
13.5 Viscosity 392
13.6 Temperature and Pressure Effects on Viscosity 396
13.7 Petroff's Equation for Bearing Friction 397
13.8 Hydrodynamic Lubrication Theory 398
13.9 Design Charts for Hydrodynamic Bearings 402
13.10 Lubricant Supply 410
13.11 Heat Dissipation, and Equilibrium Oil Film Temperature 413
13.12 Bearing Materials 414
13.13 Hydrodynamic Bearing Design 415
13.14 Boundary and Mixed-Film Lubrication 422
13.15 Thrust Bearings 424
13.16 Elastohydrodynamic Lubrication 425
References 425

Chapter 14 Rolling-Element Bearings

14.1 Comparison of Alternative Means for Supporting Rotating Shafts 427
14.2 History of Rolling-Element Bearings 429
14.3 Rolling-Element Bearing Types 430
14.4 Design of Rolling-Element Bearings 437
14.5 Fitting of Rolling-Element Bearings 441

14.6 "Catalog Information" for Rolling-Element Bearings 442
14.7 Bearing Selection 442
14.8 Bearing Mounting for Proper Thrust-Load Provision 454
References 456

Chapter 15 Spur Gears

15.1 Introduction and History 457
15.2 Geometry and Nomenclature 458
15.3 Interference and Contact Ratio 466
15.4 Gear Force Analysis 470
15.5 Gear-Tooth Strength 473
15.6 Basic Analysis of Gear-Tooth Bending Stress (Lewis Equation) 474
15.7 Refined Analysis of Gear-Tooth Bending Strength—Basic Concepts 476
15.8 Refined Analysis of Gear-Tooth Bending Strength—Recommended Procedure 478
15.9 Gear-Tooth Surface Durability—Basic Concepts 484
15.10 Gear-Tooth Surface Fatigue Analysis—Recommended Procedure 487
15.11 Spur Gear Design Procedures 491
15.12 Gear Materials 495
15.13 Gear Trains 495
References 499

Chapter 16 Helical, Bevel, and Worm Gears

16.1 Introduction 501
16.2 Helical Gear Geometry and Nomenclature 505
16.3 Helical Gear Force Analysis 507
16.4 Helical Gear-Tooth Bending and Surface Fatigue Strengths 510
16.5 Crossed Helical Gears 512
16.6 Bevel Gear Geometry and Nomenclature 512
16.7 Bevel Gear Force Analysis 514
16.8 Bevel Gear-Tooth Bending and Surface Fatigue Strengths 516
16.9 Bevel Gear Trains; Differential Gears 520
16.10 Worm Gear Geometry and Nomenclature 521
16.11 Worm Gear Force and Efficiency Analysis 523
16.12 Worm Gear Bending and Surface Fatigue Strengths 528
16.25 Worm Gear Thermal Capacity 530
References 534

Chapter 17 Shafts and Associated Parts

17.1 Introduction 535
17.2 Provision for Shaft Bearings 536
17.3 Mounting Parts onto Rotating Shafts 536
17.4 Rotating Shaft Dynamics 539
17.5 Overall Shaft Design 542

17.6 Keys, Pins, and Splines 546
17.7 Couplings and Universal Joints 549

Chapter 18 Clutches and Brakes

18.1 Introduction 553
18.2 Disk Clutches 553
18.3 Disk Brakes 559
18.4 Energy Absorbtion and Cooling 560
18.5 Cone Clutches and Brakes 561
18.6 Short-Shoe Drum Brakes 563
18.7 Long-Shoe Drum Brakes With External Shoes 567
18.8 Long-Shoe Drum Brakes With Internal Shoes 573
18.9 Band Brakes 575
18.10 Materials 578

Chapter 19 Miscellaneous Power Transmission Components

19.1 Introduction 580
19.2 Flat Belts 581
19.3 V-Belts 583
19.4 Toothed Belts 586
19.5 Roller Chains 586
19.6 Inverted-Tooth Chains 589
19.7 History of Hydrodynamic Drives 590
19.8 Fluid Couplings 591
19.9 Hydrodynamic Torque Converters 596

Chapter 20 Machine Component Interrelationships—A Case Study

20.1 Introduction 598
20.2 Description of Original Hydra-Matic Transmission 599
20.3 Free-Body Diagram Determination of Gear Ratios and Component Loads 601
20.4 Gear Design Considerations 605
20.5 Brake and Clutch Design Considerations 607
20.6 Miscellaneous Design Considerations 608

Appendix A Units

A-1 Conversion Between English and SI Units 612
A-2 Standard SI Prefixes 613
A-3 Suggested SI Prefixes for Stress Calculations 613
A-4 Suggested SI Prefixes for Linear Deflection Calculations 613
A-5 Suggested SI Prefixes for Angular Deflection Calculations 614

Appendix B Properties of Sections and Solids

B-1 Properties of Sections 616
B-2 Mass and Mass Moments of Inertia of Homogeneous Solids 617

Appendix C Material Properties and Uses

C-1	Physical Properties of Common Metals	619
C-2	Tensile Properties of Some Metals	620
C-3a	Typical Mechanical Properties and Uses of Gray Cast Iron	621
C-3b	Mechanical Properties and Typical Uses of Malleable Cast Iron	622
C-3c	Average Mechanical Properties and Typical Uses of Ductile (Nodular) Iron	623
C-4a	Mechanical Properties of Selected Carbon and Alloy Steels	624
C-4b	Typical Uses of Plain Carbon Steels	626
C-5a	Properties of Some Water-Quenched and Tempered Steels	627
C-5b	Properties of Some Oil-Quenched and Tempered Carbon Steels	628
C-5c	Properties of Some Oil-Quenched and Tempered Alloy Steels	629
C-6	Effect of Mass on Strength Properties of Steel	630
C-7	Mechanical Properties of Some Carburizing Steels	631
C-8	Mechanical Properties of Some Wrought Stainless Steels	632
C-9	Mechanical Properties of Some Iron-Based Superalloys	633
C-10	Mechanical Properties, Characteristics, and Typical Uses of Some Wrought Aluminum Alloys	634
C-11	Tensile Properties, Characteristics, and Typical Uses of Some Cast Aluminum Alloys	635
C-12	Temper Designations for Aluminum and Magnesium Alloys	636
C-13	Mechanical Properties of Some Copper Alloys	637
C-14	Mechanical Properties of Some Magnesium Alloys	638
C-15	Mechanical Properties of Some Nickel Alloys	639
C-16	Mechanical Properties of Some Wrought Titanium Alloys	640
C-17	Mechanical Properties of Some Zinc Casting Alloys	641
C-18a	Representative Mechanical Properties of a Few Plastics	642
C-18b	Properties of a Few Unreinforced vs. Glass-Reinforced Thermoplastic Resins	643
C-18c	Typical Applications of Common Plastics	644

Appendix D Shear, Moment, and Deflection Equations for Beams

D-1	Cantilever Beams	646
D-2	Simply Supported Beams	648
D-3	Beams With Fixed Ends	651

Appendix E Fits and Tolerances 652

Problems 655

Index 751

Fundamentals of Machine Component Design

PART ONE

Fundamentals

CHAPTER ONE

Mechanical Engineering Design in Broad Perspective

1.1 AN OVERVIEW OF THE SUBJECT

The essence of engineering is the utilization of the resources and laws of nature to benefit humanity. Mechanical engineering design is a major segment of engineering; it deals with the conception, design, development, refinement, and application of machines and mechanical apparatus of all kinds. Our basic concern is with the creative and intelligent *application* of solid mechanics and related sciences to "real life" situations. Thus, the material treated in this book is "down the mainstream" of modern mechanical engineering.

For many students, mechanical engineering design is one of their first *professional engineering* courses—as distinguished from background courses in science and mathematics. Professional engineering is concerned with obtaining *solutions* to practical problems. These solutions must reflect an understanding of the underlying sciences, but usually this understanding is not enough; empirical knowledge and "engineering judgment" are also involved. For example, scientists do not completely understand electricity, but this does not prevent electrical engineers from developing highly useful electrical devices. Similarly, scientists do not completely understand combustion processes or metal fatigue, but mechanical engineers use the understanding available to develop highly useful combustion engines. As more scientific understanding becomes available, engineers are able to devise better solutions to practical problems. Moreover, the engineering process of solving problems often highlights areas particularly appropriate for more intensive scientific research. There is a strong analogy between the engineer and the physician. Neither is a scientist whose primary concern is with uncovering basic knowledge, but both *use* scientific knowledge—supplemented by empirical information and professional judgment—in solving immediate and pressing problems.

Because of the professional nature of the subject, most problems in mechanical engineering design do not have a *single* right answer. Consider, for example, the problem of designing a household refrigerator. There is a nearly endless number of workable designs, none of which could be called an "incorrect" answer. But of the "correct" answers, some are obviously *better* than others because they reflect a more sophisticated knowledge of the underlying technology, a more ingenious concept of basic design, a more effective and economical utilization of existing production technology, a more pleasing esthetic appearance, and so on. It is precisely at this point, of course, that one finds the challenge and excitement of modern engineering. Engineers today are concerned with the design and development of products for a different society than any which existed previously, and they have more knowledge available to them than did engineers in the past. Hence, they are able to produce distinctly *better* solutions to meet today's needs. How much better depends upon their ingenuity, imagination, depth of understanding of the need involved, and of the technology that bears on the solutions, etc.

This book is primarily concerned with the design of specific *components* of machines or mechanical systems. Competence in this area is basic to the consideration and synthesis of complete machines and systems in subsequent courses and in professional practice. It will be seen that even in the design of a single bolt or spring, the engineer must use the best available scientific understanding together with empirical information, good judgment, and often a degree of ingenuity, in order to produce the best product for today's society.

The technical considerations of mechanical component design are largely centered around two main areas of concern: (1) stress-strain-strength relationships involving the *bulk* of a solid member, and (2) surface phenomena including friction, lubrication, wear, and environmental deterioration. Part One of the book is concerned with the fundamentals involved, and Part Two with applications to specific machine components. The components chosen are widely used and will be somewhat familiar to the student. It is not feasible or desirable for the student to study the detailed design considerations associated with *all* machine elements. Hence, the emphasis in treating those selected here is on the *methods* and *procedures* used so that the student will gain competence in applying these methods and procedures to mechanical components in general.

When considering a complete machine, the engineer invariably finds that the requirements and constraints of the various components are interrelated. The design of an automotive engine valve spring, for example, depends upon the space available for the spring. This, in turn, represents a compromise with the space requirements for the valve ports, coolant passages, spark plug clearance, and so on. This situation adds a whole new dimension to the imagination and ingenuity required of engineers as they seek to determine an optimum design for a combination of related components. This aspect of mechanical engineering design is illustrated by a "case study" in the final chapter of the book.

In addition to the traditional technological and economic considerations fundamental to the design and development of mechanical components and systems, the modern engineer has become increasingly concerned with the broader considerations of safety, ecology, and overall "quality of life." These topics are discussed briefly in the following sections.

1.2 SAFETY CONSIDERATIONS

It is natural that in the past engineers gave first consideration to the functional and economic aspects of new devices. After all, unless devices can be made to function usefully, they are of no further engineering interest. Also, if a new device cannot be produced for a cost that is affordable by contemporary society, it is a waste of engineering time to pursue it further. But the engineers who have gone before us have succeeded in developing a multitude of products that do function usefully, and that can be produced economically. Partly because of this, increasing engineering effort is now being devoted to broader considerations relating to the influence of engineered products on people and on the environment.

Personnel safety is a consideration engineers have always kept in mind, but that now demands increasing emphasis. In comparison with such relatively straightforward computations as stress and deflection, determination of safety is apt to be an elusive and indefinite matter, complicated by psychological and sociological factors. But this should only add to the appeal of the task for an engineer. It challenges him or her to assemble all pertinent facts, and then to make good decisions reflecting understanding, imagination, ingenuity, and judgment.

The important first step in developing engineering competence in the safety area is cultivating an *awareness* of its importance. Product safety is of great concern now to legislators, attorneys, judges, jurors, insurance executives, and so forth. But none of these people can contribute directly to the safety of a product; they can only underscore the urgency of giving appropriate emphasis to safety in the *engineering development* of a product. It is the *engineer* who must carry out the development of safe products.

Safety is inherently a *relative* matter, and value judgments must be made regarding trade-offs between safety, cost, weight, and so on. Some years ago the author was associated with a particularly safety-conscious company and was in the position of frequently admonishing the staff safety engineer to reduce further the inevitable hazards associated with the company's equipment. When pushed a little too far one day, this engineer responded, "Look, I have made this model fool-proof, but I can never make it *damn*-fool-proof! If someone tries hard enough, he can hurt himself with this machine!" The next day this gentleman inadvertently proved his point when he accidentally dropped the new model prototype on his foot and broke a toe! But the point to be made here is that when society makes decisions relative to safety requirements, engineers should contribute important input.

Following awareness, the second main point of safety engineering is *ingenuity*. The engineer must be imaginative and ingenious enough to *anticipate* potentially hazardous situations relating to a product. The old maxim that anything that *can* happen probably *will* happen sooner or later is relevant. The following are four cases in point, all involving costly liability suits:

1. A large open area with a high ceiling was to be heated and cooled with three cubical units, each suspended from the ceiling by long steel rods at four corners. The cubicles were being fitted with heat exchangers, blowers, and filters by workers inside and on top of the enclosures. The flexibility of the long support rods permitted the cubicles to swing back and forth, and the workers sometimes enjoyed getting their cubicle swinging with considerable amplitude. Fatigue failure of a support rod caused the death of one worker. Since large steam pipes (not yet installed at the time of the accident) prevented significant sway of the completed units, and the rods were designed with a safety factor of 17 (based on static weight of the completed cubicles), no further thought was given to safety. No one responsible for the design and installation of the units had reviewed the installation sequence with the imagination and ingenuity needed to foresee this hazard.
2. A boy was seriously injured by collision with a car when the brakes on his new bicycle failed to respond in an emergency. The cause was discovered to be interference between a fitting on the three-speed shift mechanism and a sharp edge on the caliper brake handle. Both the shift control mechanism and the brake handle were of unusual design. Both were safe within themselves and were safe when used in combination with a conventional design of the other member. But when these two unusual members were used together, it was easy for them to be mounted on the handlebar in such a position that the travel of the brake handle was limited, thereby preventing full application of the brake. Again, no one responsible for the overall design of the bicycle foresaw this hazardous situation.
3. A worker lost a hand in a 400-ton punch press despite wearing safety cuffs that were cam-actuated to pull the hands out of the danger zone before the ram came down. The cause was a loosened setscrew that permitted the cam to rotate with respect to its supporting shaft, thereby delaying hand retraction until *after* the ram came down. This case illustrates the old adage that "a chain is no stronger than its weakest link." Here, an otherwise very positive and strong safety device was nullified because of the inexcusably weak link of the setscrew. A very little imagination and ingenuity on the part of the engineer responsible for this design would have brought this hazard to light before the unit was released for production.
4. A crawling infant lost the ends of three fingers as he attempted to climb up an "exercycle" being ridden by an older sister. When placed on the bottom chain, the infant's hand was immediately drawn into the crank sprocket. In order to minimize cost, the exercycle was very properly designed to take advantage of many high-production, low-cost parts used on a standard bicycle. Unfortunately, however, the chain guard, which provides adequate protection for a bicycle, is totally inadequate for the exercycle. Was it too much to expect that the engineer responsible for this design would have enough imagination to foresee this hazard? Should he or she not have been sufficiently ingenious to devise an alternative guard design that would be economically and otherwise feasible? Should it be necessary for this kind of imag-

ination and ingenuity to be forced upon the engineer by legislation devised and enacted by nonengineers?

Once the engineer is sufficiently *aware* of safety considerations, and accepts this challenge to his or her *imagination and ingenuity,* there are certain techniques and guidelines that are often helpful. Six of these are suggested below.

1. *Review the total life-cycle* of the product from initial production to final disposal, with an eye toward uncovering significant hazards. Ask yourself what kinds of situations can reasonably arise during the various stages of manufacturing, transporting, storing, installing, using, servicing, and so on.
2. Be sure that the safety provisions represent a *balanced approach.* Do not accept a dollar penalty to eliminate one hazard and overlook a twenty-cent possibility for eliminating an equal hazard. And, like the punch-press example above, do not focus attention on how strong the wrist cuffs are while overlooking how weak the cam attachment is.
3. *Make safety an integral feature* of the basic design wherever possible, rather than "adding on" safety devices after the basic design has been completed. An example of this was the development of an electrostatic hand-operated paint gun. Earlier stationary-mounted electrostatic paint guns had metal atomizing heads operating at 100,000 volts. A hand-gun version, incorporating elaborate guards and shields, was quickly recognized as impractical. Instead, a fundamentally new electric circuit design combined with a nonmetallic head was developed so that even if the operator came in contact with the high-voltage head, he or she would receive no shock; the voltage automatically dropped as a hand approached the head, and the head itself had a low enough capacitance to avoid significant discharge to the operator.
4. Use a *"fail-safe" design* where feasible. The philosophy here is that precaution is taken to avoid failure, but if failure *does* occur, the design is such that the product is still "safe"; that is, the failure will not be catastrophic. For example, the first commercial jet aircraft were the British Comets. Some of these experienced catastrophic failure when fatigue cracks started in the outer aluminum "skin" at the corners of the windows (due to alternately pressurizing the cabin at high altitude and relieving the pressurizing stresses at ground level). Soon after the cracks were initiated, the fuselage skin ripped disastrously (somewhat like a toy rubber balloon). After the cause of the crashes was determined, subsequent commercial jet aircraft incorporated the "fail-safe" feature of bonding the outer panels to the longitudinal and circumferential frame members of the fuselage. Thus, even if a crack does start, it can propagate only to the nearest bonded seam. The relatively short cracks in no way impair the safety of the aircraft. (This particular "fail-safe" feature can be illustrated by ripping an old shirt. Once a tear has been started, it is easily propagated to a seam, but it is extremely difficult to propagate the tear through the seam, or "tear-stopper.") Fail-safe designs often incorporate *redundant* members so that if one load-carrying member fails, a second member is able to assume the full load. This is sometimes known as the "belt *and* suspenders" design philosophy. (In extreme cases, a "safety-pin" may be added.)

5. Check government standards (as OSHA and ANSI) and the pertinent technical literature to be sure that legal requirements are complied with, and that advantage is taken of the relevant safety experience of others.
6. Provide warnings of all significant hazards that remain after the design has been made as safe as reasonably possible. The engineers who developed the product are in the best position to identify these hazards. The warnings should be designed to bring the information to the attention of the persons in jeopardy in the most positive manner feasible. Conspicuous warning signs attached permanently to the machine itself are usually best. There are OSHA and ANSI standards pertaining to warning signs. More complete warning information is often appropriately included in an instruction or operating manual that accompanies the machine.

Safety engineering inherently includes important *nontechnical aspects* that are related to the *people* involved. Engineers must be aware of these if their safety-related efforts are to be effective. Three specific points within this category are suggested below.

1. *Capabilities* and *characteristics* of people, both physiological and psychological. When the device is used or serviced, the strength, reach, and endurance requirements must be well within the physiological limitations of the personnel involved. The arrangement of instruments and controls, and the nature of the mental operating requirements, must be compatible with psychological factors. Where the possibility of accident cannot be eliminated, the design should be geared to limiting personnel accident-imposed loads to values minimizing the severity of injury.
2. *Communication.* Engineers must communicate to others the rationale and operation of the safety provisions incorporated in their designs, and in many cases involve themselves in *"selling"* the proper use of these safety provisions. What good does it do, for example, to develop an effective motorcycle helmet if it isn't used? Or to provide a punch press with safety switches for both hands if the operator blocks one of the switches closed in order to have a hand free for smoking? Unfortunately, even the most effective communication does not always guarantee intelligent operator usage. This situation gives rise to controversies like those surrounding the requirement of air bags in cars, because a significant segment of the public cannot be persuaded to use seat belts voluntarily. The best resolution of such controversies requires intelligent input from many quarters, one of which is certainly the engineering profession.
3. *Cooperation.* The controversy mentioned above illustrates the need for engineers to cooperate effectively with members of other disciplines—government, management, sales, service, legal, and so on—in order that joint safety-directed efforts may prove effective.

1.3 ECOLOGICAL CONSIDERATIONS

People inherently depend on their environment for air, water, food, and materials for clothing and shelter. In primitive society, man-made wastes were naturally recycled for repeated use. When open sewers and dumps were introduced, nature became unable to reclaim and recycle these wastes within normal time periods, thus interrupting natural ecological cycles. Traditional economic systems enable products to be mass-produced and sold at prices that often do not reflect the true cost to society in terms of resource consumption and ecological damage. Now that society is becoming more generally aware of this problem, legislative requirements and more realistic "total" cost provisions are having increasing impact upon engineering design. Certainly, it is important for the best available engineering input to go into societal decisions involving these matters.

One can perhaps state the basic ecological objectives of mechanical engineering design rather simply: (1) to utilize materials so that they are economically recycleable within reasonable time periods without causing objectionable air and water pollution, and (2) to minimize the rate of consumption of nonrecycled energy sources (as fossil fuels) both to conserve these resources and to minimize thermal pollution. In some cases, the minimization of noise pollution is also a factor to be considered.

As with safety considerations, ecological factors are much more difficult for the engineer to tie down than are such matters as stress and deflection. The following is a suggested list of points to be considered:

1. Consider all aspects of the *basic design objective* involved, to be sure that it is sound. For example, questions are now being raised as to the overall merits of some major dam constructions. Are there ecological side-effects that might make it preferable to follow an alternative approach? Before undertaking the design of a further expanded highway system or a specific mass-transit system, a determination must be made as to whether the best available knowledge and judgment indicate that the proposed project represents the best alternative.
2. After accepting the basic design objective, the next step is a review of the *overall concepts* to be embodied into the proposed design. For example, a modular concept may be appropriate, wherein specific components or modules most likely to wear out or become obsolete can be replaced with updated modules that are interchangeable with the originals. The motor and transmission assembly of a domestic automatic washing machine might be an example where this approach would be appropriate. Another example is the provision for replaceable exterior trim panels on major kitchen appliances that permit the exterior surfaces to be changed to match a new decorating scheme without replacement of the entire appliance.
3. An important consideration is *designing for recycling.* At the outset of a new design, it is becoming increasingly important that the engineer consider the full ecological cycle including the disposal and reuse of the entire device and its components. Consider an automobile. Parts appropriate for reuse (either with or without rebuilding) should be made so that they can be easily removed from a "junk" car. Dismantling and sorting of parts by material should be made as easy and economical as possible. It has been somewhat facetiously suggested that cars be made so that all fasteners break when dropping a junk car from, say, a height of 30 ft. Automatic devices would then sort the pieces by material for reprocessing. A more realistic proposal is that of attaching the wiring harness so that it can be quickly ripped out in one piece for easy salvaging of the copper.

In developing recycling procedures along these lines it is obviously desirable that the costs to a company for recycling vs. costs for abandoning the old parts and using virgin materials reflect total real costs to society. No individual company could stay in business if it magnanimously undertook a costly recycling program in order to conserve virgin materials and reduce processing pollution if its competitors could utilize inexpensive new materials obtained at a price that did not reflect these total costs.

4. Select *materials* with ecological factors in mind. Of importance here are the known availability in nature of the required raw materials, processing energy requirements, processing pollution problems (air, water, land, thermal, and noise), and recyclability. Ideally, all of these factors would be appropriately reflected within the pricing structure, and this will likely be more nearly the case in the future than it has in the past.

   Another factor to be considered is the relative durability of alternative materials for use in a perishable part. For example, consider the great reduction in razor blade requirements (and in number of scrap razor blades) made by changing the material to stainless steel. (But would it be better, overall, to devise a convenient and effective way to resharpen the blades rather than throwing them away?)

   The engineer should also consider the *compatibility* of materials with respect to recycling. For example, zinc die castings deteriorate the quality of the scrap obtained when present junked cars are melted.

5. Consider ecological factors when specifying *processing*. Important here are pollution of all kinds, energy consumption, and efficiency of material usage. For example, forming operations such as rolling and forging use less material (and generate less scrap) than cutting operations. There may also be important differences in energy consumption.

6. *Packaging* is an important area for resource conservation and pollution reduction. Reusable cartons, and the use of recycled materials for packaging, are two areas receiving increasing attention. Perhaps the ultimate in ecologically desirable packaging is that commonly used ice cream container, the cone.

As a concluding example of the importance of introducing sound engineering thinking into societal ecological decisions, consider the suggestion made by a highly vocal (nonengineering) student that power-plant pollution be virtually eliminated by requiring the power companies to drive their generators with electric motors! But the matter of protecting our environment is a deadly serious one. As the late Adlai Stevenson once said, "We travel together, passengers on a little space ship, dependent on its vulnerable supplies of air and soil . . . preserved from annihilation only by the care, the work, and I will say the love, we give our fragile craft."

1.4 SOCIETAL CONSIDERATIONS

As the reader well knows, the solution to any engineering problem begins with its clear definition. Accordingly, let us define, in the broadest terms, the problem to be addressed

TABLE 1.1
Preliminary List of Factors Constituting the LQI

1. *Physical Health*
2. *Material Well-Being*
3. *Safety* (crime and accident rates)
4. *Environment* (air, water, land, and natural resource management)
5. *Cultural–Educational* (literacy rate, public school quality, college attendance among those qualified, adult educational opportunities, library and museum facilities, etc.)
6. *Treatment of Disadvantaged Groups* (physically and mentally handicapped, aged, etc.)
7. *Equality of Opportunity* (and stimulation of initiative to use opportunities)
8. *Personal Freedom*
9. *Population Control*

when undertaking mechanical engineering design. The opening sentence in this chapter suggests a definition: The basic objective of any engineering design is to provide a machine or device that will benefit humanity. In order to apply this definition it is necessary to think in more specific terms: (1) just how does one benefit humanity, and (2) what "yardstick" (meter stick?) can be used to measure such benefits? The formulation of precise definitions of problem objectives, and the devising of means for measuring results, *fall within the special province of the engineer.*

The writer has suggested [1][1] that the basic objective of engineering design as well as other human pursuits is to improve the quality of life within our society, and that this might be measured in terms of a Life Quality Index (LQI). This index is in some ways similar to the familiar "gross national product," but very much broader. Judgments as to the proper composition of the LQI would, of course, vary somewhat among the many segments of society and also with time.

To illustrate the LQI concept, Table 1.1 lists some of the important factors most people would agree should be included. Perhaps one might arbitrarily assign a value of 100 to the factor deemed most important, with other factors being weighed accordingly. Each factor might then be multiplied by the same fraction so that the total would add up to 100.

The list in the table is admittedly a very rough and oversimplified indication of the direction of thought that would be involved in arriving at an LQI for a given segment of society at a given time. But this *kind* of thinking must be done in order to provide a sound basis for judgment with respect to the fulfillment of the engineering mission of service to humanity.

The professional contribution of engineers engaged in the broad area of engineering design and development plays a major role in determining the LQI. *All* persons are dependent upon the products resulting from this activity for the material needs and comforts of life. Most persons are profoundly affected by the fact that they spend much of their lifetime working in some way to provide these products. Figure 1.1 suggests a diagrammatic representation of several basic relationships. A major segment of the pop-

[1]Bracketed numbers in the text correspond to numbered references at the end of the chapter.

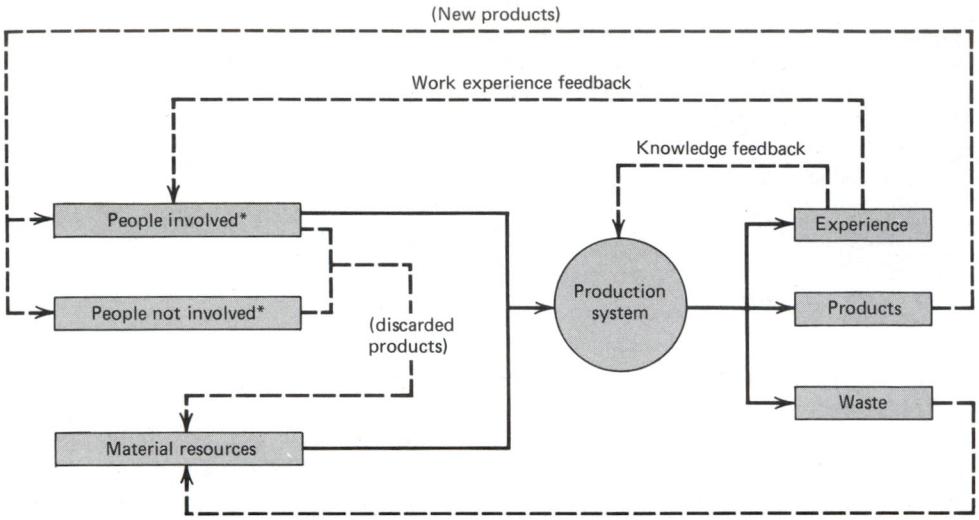

*Involved in research, design, development, production, merchandising, servicing, and recycling of engineered products.

FIGURE 1.1.
Societal relationships involving engineered products.

ulation works within organizations whose function is to do one or more of the following: research, design, develope, produce, merchandise and service engineered products. The efforts of these people, together with appropriate natural resources, go into production systems that yield useful products, waste materials, and experience. The experience is of two kinds: (1) direct working experience of the individuals, which is hopefully constructive and satisfying, and (2) empirical knowledge gained as to the effectiveness of the overall system, with implications for its future improvement. The products made serve all people until discarded, when they constitute long- and short-term recyclable material resources, and possibly pollution.

A valid Life Quality Index must take into account psychological factors. A book of this kind can include only the briefest introduction to this vast subject. But hopefully this will aid in stimulating the student toward a life-long interest and concern with this fundamental area.

We know that people exhibit an infinitely varying and often baffling set of characteristics. But we also know that there are certain inherent human characteristics and needs that remain fixed—for all races and presumably for all time. These have been expressed as proposed *levels of human need* by the late Abraham Maslow, a psychologist at Brandeis University [3,4]. As an aid to memory, these are expressed below in terms of five key words beginning with "S" [2].

Maslow's Hierarchy of Needs

1. Survival
2. Security
3. Social acceptance
4. Status
5. Self-fulfillment

The first level, obviously, is the need for immediate *survival*—food, shelter, clothing, and rest—here and now.

The second level involves *security*—insuring one's safety and future survival.

The third level is *social acceptance*. People need to belong to and interact with a family, clan, or other group; they need love and acceptance.

The fourth level is *status* or recognition—a need not only to fit into a social group but also to earn and receive peer respect and admiration.

The highest level is *self-fulfillment*—growth toward reaching one's full potential, and achievement of the resulting inner satisfaction.

At any given time, both people and nations operate on more than one of these levels; yet the levels define a general path or ladder of advancement that leads from primitive existence to a mature, rich quality of life.

Historically, engineering effort has been directed primarily toward satisfying needs 1 and 2. More recently, an increased percentage of the production systems have been designed to provide society with products going beyond the basic survival and security needs, presumably contributing to satisfying the legitimate higher needs of the consumer. As far as the workers are concerned, it is interesting to note that recent "job enlargement" and "job enrichment" programs are directed toward the workers' higher needs, 3, 4, and 5.

A basic ingredient of human society is *change*. Engineers must seek to understand not only the needs of society today, but also the direction and rapidity of societal changes that are occurring. Moreover, we must seek to understand the influence of technology—and of mechanical products and associated production systems in particular—on these changes. Perhaps the engineering profession's most important objective is to make its input to society such as to *promote changes in the direction of increasing the Life Quality Index*.

1.5 OVERALL DESIGN CONSIDERATIONS

Most engineering designs involve a multitude of considerations, and it is a challenge to the engineer to recognize all of them in proper proportion. Although no simple check list given here can be adequate or complete, it may be helpful to list in some organized fashion the major categories involved.

A. *"Traditional" Considerations*

1. For the bulk or body of the component:
   - **a.** strength
   - **b.** deflection
   - **c.** weight
   - **d.** size and shape
2. For the surfaces of the component:
   - **a.** wear
   - **b.** lubrication
   - **c.** corrosion
   - **d.** frictional forces
   - **e.** frictional heat generated
3. Cost

B. *"Modern" Considerations*

1. Safety
2. Ecology (land, air, water, thermal pollution; resource conservation; noise)
3. Quality of Life

C. *Miscellaneous Considerations*

1. Reliability and Maintainability (This is becoming increasingly recognized as a factor requiring *substantially* increased emphasis.)
2. Aesthetics

Often various design considerations are seemingly incompatible until the engineer devises a sufficiently imaginative and ingenious solution. The design of the lift truck pictured in Figure 1.2 provides a simple example. Here, the objective of achieving a desired aesthetic appearance was seemingly incompatible with cost limitations. Matched metal forming dies were too costly, and inexpensive tooling resulted in unattractive mismatches of mating parts. The solution here was to work deliberate mismatches into the design and eliminate the need for precision fitting. The mismatches were used to give the truck a rugged look. The gap under the hood, for example, creates a strong horizontal line while disguising the fit-up of several frame and fender weldments. It also provides a handhold for lifting the hood. Another gap makes the instrument pod appear to "float" from a steel pylon, again disguising a large tolerance. The large hood simplifies maintenance by providing wide-open access to the engine. By serving also as the seat support, it further reduces cost, while adding to the clean, uncluttered appearance.

1.6 SYSTEM OF UNITS

Because the present generation of American engineers seems destined to suffer the inconvenience of having to deal with both the English and SI systems of units, both are used in this book. Appendix A-1 lists units associated with both systems, conversion factors relating them, and their standard abbreviations.

FIGURE 1.2.
Lift truck designed for functional, attractive, and unique appearance together with low cost. (Courtesy Clark Equipment Co.)

Four fundamental quantities in any system of units are *force, mass, length,* and *time.* Let us designate these by F, m, L, and T, respectively, where the symbols can stand for any units. Thus, L could be meters, inches, miles, and so on. In the seventeenth century Sir Isaac Newton discovered the basic relationship

$$F = ma \tag{1.1}$$

where $a = LT^{-2}$ (i.e., a could be in in./sec^2, etc.). This has the important implication that if three of the four quantities F, m, L, and T are arbitrarily defined, the fourth must be defined so as to satisfy Eq. (1.1). The three arbitrarily defined units are called *basic units* and the fourth one a *derived unit.*

The SI, English, and metric systems all define the unit of time as the second (abbreviated s). All three systems also define an arbitrary unit of length: the inch in

System of units	"Standard object"	Weight, or earth gravitational force (of standard object)	Mass (of standard object)
English		1 lb	$\frac{1\ \text{lb s}^2}{32.2\ \text{ft}} = 0.031$ slug, or $\frac{1\ \text{lb s}^2}{386\ \text{in.}}$
Metric		1 kg	$\frac{1\ \text{kg s}^2}{9.81\ \text{m}}$
SI		$(1\ \text{kg})(9.81\ \text{m/s}^2) = 9.81\ \text{N}$	1 kg

FIGURE 1.3.
Comparison of units for force (or weight) and mass. Note that (1) numerical constants are valid *only* for the standard earth gravitational field. (2) The 1 kg "standard object" has 2.2046 times the mass and weight of the 1 lb "standard object."

the English system and the meter in the metric system and in SI. Having at this point specified units for T and L, each system is free to define the unit of force *or* mass. The final unit must then be derived so as to satisfy Eq. (1.1).

Figure 1.3 shows the choices made in each system. Because the English and metric systems arbitrarily define *force* (in terms of the earth's gravitational field) as a basic unit, and then derive the unit for mass, these are called *gravitational systems of units.* In SI, the unit for *mass* is defined and then the force unit (newton, abbreviated N) is derived. SI is said to be an *absolute system of units* because SI units are defined independently of any gravitational field.

Sometimes the English and metric systems are modified to become absolute systems through the introduction of the *poundal,* the *pound of mass* (abbreviated lbm to distinguish it from the pound of force, abbreviated lb or lbf), or the *dyne.* No use will be made of these three units in this book.

The unit of force is of particular significance in mechanical engineering design and analysis because it is involved in calculations of force, torque, stress (and pressure), work (and energy), power, and elastic moduli. In adopting SI units, engineers accustomed to using either the English or metric systems encountered the unfamiliar unit of force, *newton.* It is interesting to note that a newton comes out to be about the weight of (or earth's gravitational force on) an average apple, such as the one that may have fallen from the tree Sir Isaac was sitting under when his thoughts reportedly first turned to formulating his law of universal gravitation. (It has been reported that some individuals having difficulty distinguishing the units of "long ton" and "short ton" were particularly dismayed upon learning that they must now deal with what they understood to be the "new ton"!)

Appendix A-2 lists standard prefixes for SI units. Appendixes A-3, A-4, and A-5 list compatible combinations of SI prefixes that will be found convenient in solving stress and deflection equations.

1.7 WORK, ENERGY

All mechanical apparatus involves *loads* and *motion,* which, in combination, represent *work,* or *energy.* Thus, it is appropriate to review these basic concepts in the context of representative mechanical devices.

Figure 1.4 shows a wheel being turned by the application of tangential force F acting at radius R. In one revolution the work done (W) is equal to force times the distance through which the force moves. Thus,

$$W = F(2\pi R) \tag{a}$$

If the force is 1 N and the radius 1 m, the work is 2π N·m.

Work is also equal to torque times angular distance measured in radians. Thus, if the wheel in Fig. 1.4 is turned by applying a torque T (equal to the product F times R) to the shaft rather than by the tangential force, the work in one revolution would be

$$\text{Work} = 2\pi T \tag{b}$$

SAMPLE PROBLEM 1.1 Camshaft Torque Requirement

Figure 1.5 shows a rotating *cam* that causes a *follower* to move vertically. For the position shown, the follower is being moved upward with a force of 1 N. Also, for this

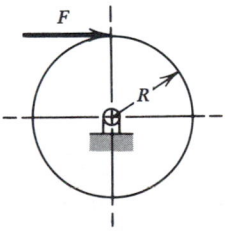

FIGURE 1.4.
Wheel being turned by a tangential force.

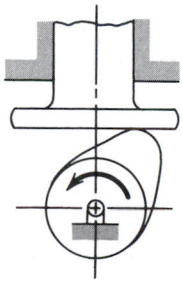

FIGURE 1.5.
Cam and follower.

position it has been determined that a rotation of 0.1 rad (5.73°) corresponds to a follower motion of 1 mm. What is the average torque required to turn the camshaft during this interval?

Solution

If friction can be neglected, the work put into the camshaft is equal to the work done by the follower. Work in is equal to torque T times 0.1 rad; work out is equal to 1 N times 0.001 m. Equating the two and solving for T gives

$$T = \frac{1\ \text{N} \times 0.001\ \text{m}}{0.1\ \text{rad}} = 0.01\ \text{N}\cdot\text{m, or } 10\ \text{N}\cdot\text{mm}$$ ■

1.8 POWER

The work, or energy, involved with mechanical devices is invariably associated with a *time rate.* Thus, *power,* or time-rate of doing work, enters the picture.

In SI units, the Watt (W) is defined as 1 J/s, which is the same as 1 N·m/s. Hence, 1 kW = 60,000 N·m/min. To determine power in kilowatts (kW) corresponding to torque T (measured in Newton meters) and rotating speed n (measured in revolutions per minute), we combine this definition of a kilowatt with Eq. (b):

$$\text{kW} = \frac{2\pi nT\ \text{N}\cdot\text{m}}{60{,}000} = \frac{nT\ \text{N}\cdot\text{m}}{9549} \tag{1.2}$$

In English units, the horsepower is defined as a work-rate of 33,000 ft lb/min.

$$\text{hp} = \frac{2\pi nT\ \text{lb ft}}{33{,}000} = \frac{nT\ \text{lb ft}}{5252} \tag{1.3}$$

There are several facets associated with these simple relationships, some of which are illustrated in the following examples.

SAMPLE PROBLEM 1.2 Camshaft Power Requirement

If the camshaft in the previous example rotates at a uniform rate of 1000 rpm, what is the average power requirement during the time interval involved?

Solution

The rotating speed corresponds to 2000π rad/min, or 33.3π rad/s. Thus, a rotation of 0.1 rad requires $0.1/(33.3\pi)$ sec. During this time interval the work put into the shaft is 0.001 N·m. Power, therefore, is 0.001 N·m per $[0.1/(33.3\pi)]$ sec, or 1.05 N·m/s. This is the same as 1.05 W. ■

The horsepower equivalent (see Appendix A-1) is

$$1.05 \text{ W} \times 0.00134 \text{ hp/W, or } 0.0014 \text{ hp}$$ ■

SAMPLE PROBLEM 1.3 Punch Press Motor Power Requirement Without Flywheel

The crankshaft of a punch press rotates 60 rpm, causing holes to be punched in a steel part at the rate of 60 punches per minute. The crankshaft torque requirement is shown in Fig. 1.6. The press is driven (through suitable speed reducers) by a 1200 rpm motor. Neglecting any "flywheel effect," what motor power is required to accommodate the peak crankshaft torque?

Solution

1. Neglecting friction losses, the 20:1 speed reduction (1200 rpm/60 punches/min.) is associated with a 20:1 torque increase. Hence, the motor must provide a torque of 10 kN·m/20, or 500 N·m.
2. Based on the motor having capacity to deliver this torque *continuously,* the work capacity corresponding to 1 revolution of the shaft is $2\pi(500 \text{ N}\cdot\text{m}) = 1000\pi$ J.
3. In 1 second, during which time the shaft turns 20 revolutions, the work capacity is 20π kJ. This is a work rate (power) of 20π kW, or 62.8 kW. The horsepower equivalent (see Appendix A-1) is 62.8 kW × 1.34 hp/kW, or 84.2 hp. ■

It is obviously wasteful to provide such a large motor when its full capacity is needed but a small fraction of the time. By providing a suitable *flywheel,* a much smaller motor can be used. During the actual punch stroke, energy will be taken from the flywheel, slowing it down. During the relatively long period of time between punch strokes the motor will accelerate the flywheel back to its original speed. This is illustrated in the next sample problem.

SAMPLE PROBLEM 1.4 Punch Press Motor Power Requirement with Flywheel

For the punch press in Sample Problem 1.3 determine the motor power capacity required if use is made of a flywheel.

Solution

1. Figure 1.6 shows that the *average* torque requirement during the actual punch stroke is 6 kN·m, and that the punch stroke lasts for $\pi/3$ radians. (The energy

involved is the area under the curve; 2π kN·m, or 6283 J.) By using a flywheel permitting the motor to deliver a constant torque over the entire 2π radians, the torque requirement would be reduced to 1 kN·m. This is shown in Fig. 1.6 as "uniform torque supplying equal energy."

2. Since, at the same shaft speed, motor torque and motor power are proportional, the 10:1 reduction in motor torque requirement (1 kN·m with flywheel; 10 without) corresponds to a like reduction in power rating required. Hence, the answer, rounded off to two significant figures, is 6.3 kW or 8.4 hp. ■

The next question that arises is how large a flywheel is required. If the flywheel is too small (or, more precisely, if the flywheel has too small a polar moment of inertia), the speed fluctuation will be excessive. If the flywheel is too large, it will involve an excess of weight, bulk, and cost, and there may be problems in getting it up to speed when starting. The next sample problem illustrates a typical flywheel calculation.

SAMPLE PROBLEM 1.5 Design of Punch Press Flywheel

Continuing with the previous problem, assume that the flywheel rotates at ⅓ motor speed and is to limit motor speed fluctuation to the range of 900–1200 rpm. The flywheel is to be made of steel and have the geometric proportions shown in Fig. 1.7. To simplify the calculation, assume that the inertia contributed by the hub and arms is negligible. Determine the required flywheel polar moment of inertia, I, and diameter, d.

Solution

1. Figure 1.6 shows that during the *actual punch stroke,* energy provided by the *motor* is represented by an approximate rectangle 1 kN·m high and $\pi/3$ rad wide. Thus,

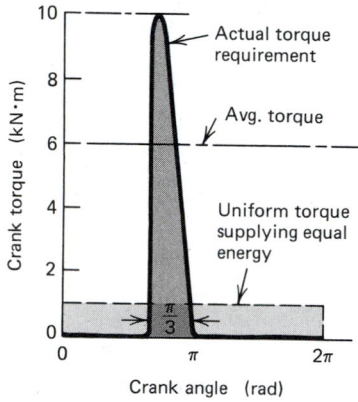

FIGURE 1.6.
Punch press torque requirement.

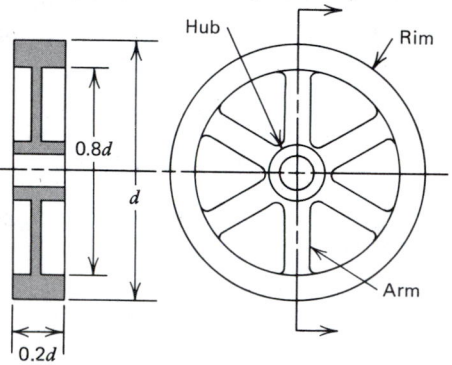

FIGURE 1.7.
Punch press flywheel proportions.

the motor provides 1047 J of the total of 6283 J required. The flywheel must provide the remaining 5236 J.

2. Recalling that kinetic energy is $\frac{1}{2}mv^2$ for linear motion, and $\frac{1}{2}I\omega^2$ for rotational motion, it is evident that the flywheel inertia must be such that

$$5236 = \tfrac{1}{2}I(\omega_{max}^2 - \omega_{min}^2)$$

(Units: I expressed in $kg \cdot m^2$ and ω_{max} and ω_{min} expressed in rad/s.[2])

$$5236 = \tfrac{1}{2}I\,[(13.3\pi)^2 - (10\pi)^2], \text{ or } I = 13.80 \text{ kg} \cdot \text{m}^2$$ ■

3. The moment of inertia for a solid cylinder is

$$I = \frac{md^2}{8} = \pi d^4 L\rho/32 \text{ (see Appendix B-2)}$$

where ρ = mass density = 7700 kg/m^3 for steel (Appendix C-1). Then for a hollow cylinder,

$$I = \pi(d_o^4 - d_i^4)L\rho/32$$

(Units: $I = m^5 \cdot kg/m^3 = kg \cdot m^2$.) Substituting values for this problem:

$$13.80 = \pi[(d)^4 - (0.8d)^4]\,(0.2d)(7700)/32$$

from which d = 0.688 m. ■

SAMPLE PROBLEM 1.6 Automotive Performance Analysis

Figure 1.8 shows a representative engine power requirement curve for constant speed, level road operation of a 4000 pound automobile. Figure 1.9 shows the wide-open-throt-

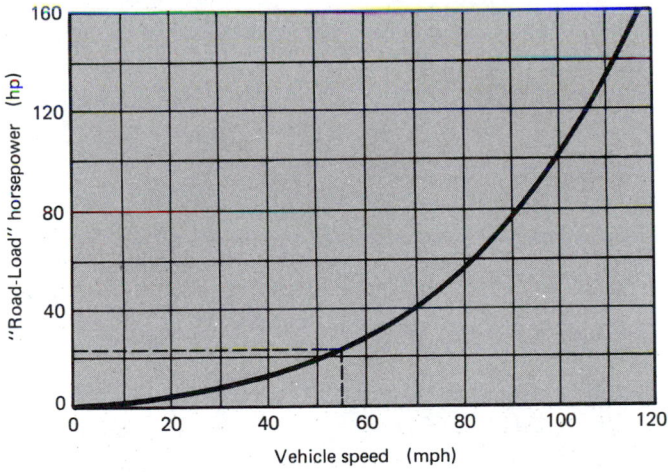

FIGURE 1.8.
Horsepower requirement. Typical 4000-lb sedan (level road, constant speed, no wind).

[2]Recall that 1 J = 1 N·m = 1 $(kg \cdot m/s^2)m$.

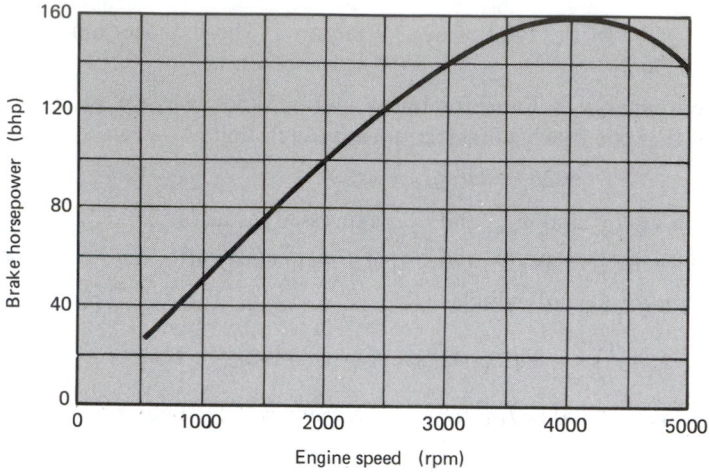

FIGURE 1.9.
Horsepower output curve. Typical 350 cu. in. V-8 engine.

tle horsepower curve of its 350 cubic inch V-8 engine. Figure 1.10 gives specific fuel consumption curves for the engine. (The extreme right-hand point of each curve represents wide-open-throttle operation; points to the left represent increasing throttling.) The rolling radius of the wheels varies a little with speed, but can be taken as 13 in. The transmission provides direct drive in high gear.

1. What axle ratio would result in the highest top speed, and what is that speed?
2. Estimate the gasoline mileage at a constant 55 mph, using this axle ratio.
3. Describe briefly the nature of a theoretically "ideal" automatic transmission. How would it change the 55 mph fuel consumption and the vehicle performance (i.e., acceleration and hill climbing ability)?

Solution

1. Figure 1.9 shows the maximum engine power to be 160 hp at 4000 rpm. Figure 1.8 shows that 160 hp will drive the car 117 mph. The axle ratio must be such that the engine rotates 4000 rpm when the vehicle speed is 117 mph. At 117 mph, wheel speed is

$$\frac{5280 \text{ ft/mi} \times 117 \text{ mi/hr}}{60 \text{ min/hr} \times 2\pi(13/12) \text{ ft/rev}} = 1513 \text{ rpm}$$

The axle ratio required is

$$\frac{4000 \text{ rpm (engine)}}{1513 \text{ rpm (wheels)}} = 2.64$$

■

2. At 55 mph, engine speed is

$$4000 \text{ rpm } (55 \text{ mph}/117 \text{ mph}) = 1880 \text{ rpm}$$

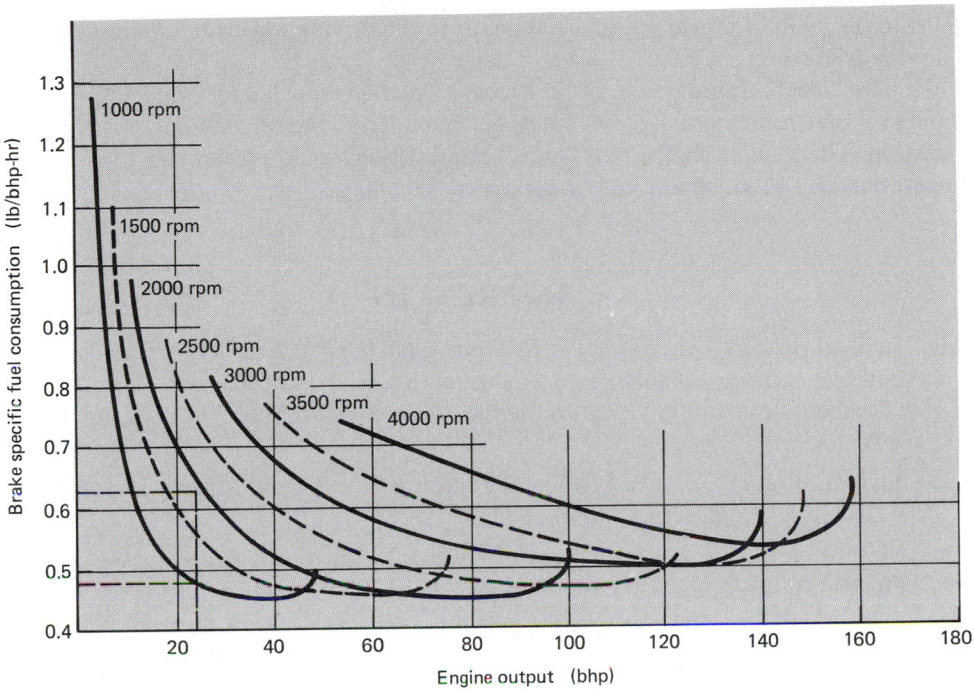

FIGURE 1.10.
Specific fuel consumption, pounds per brake horsepower-hour. Typical 350 cu. in. V-8 engine.

From Fig. 1.8, the 55 mph horsepower requirement is 23 hp. From Fig. 1.10, the specific fuel consumption is about 0.63 lb/bhp-hr, and the hourly fuel consumption is $0.63 \times 23 = 14.5$ lb/hr.

Taking 5.8 lb/gal as the specific weight of gasoline, the fuel mileage is

$$\frac{55 \text{ mi/hr} \times 5.8 \text{ lb/gal}}{14.5 \text{ lb/hr}} = 22 \text{ mi/gal}$$ ■

3. An "ideal" automatic transmission would permit the engine to slow down until either minimum specific fuel consumption (about 0.46 lb/hp-hr) or minimum satisfactory operating engine speed was reached. To provide 23 bhp at 0.46 lb/bhp-hr would require the engine to operate below 1000 rpm. Let us conservatively assume that 1000 rpm is the slowest satisfactory engine speed. At 1000 rpm, 23 bhp can be had with a fuel consumption of 0.48 lb/bhp-hr. In comparison with getting the 23 hp at 1880 rpm, gasoline mileage is increased by the ratio of $0.63/0.48 = 1.31$. Thus, with the "ideal" transmission, 55 mph fuel mileage is 22 mpg $\times$ 1.31 = *28.9 mpg*.

Regarding vehicle performance, an "ideal" transmission would, with fully depressed accelerator, allow the engine to speed up to 4000 rpm and deliver its full 160 hp at *all vehicle speeds* and under all road conditions where this did not result in wheel "spin" and loss of traction. Under the latter conditions, engine speed would be increased just to the point of providing sufficient power to the driving wheels to *almost* overcome driving friction.

The "ideal" transmission would enable a *smaller* (and lighter) engine to be used and still match the normal-speed-range performance of the original engine and transmission. The smaller engine plus "ideal" transmission would presumably give 55 mph operation at 0.46 lb/bhp-hr and an estimated *30.1 mpg.*

REFERENCES

1. Juvinall, Robert C., *Production Research—Basic Objectives and Guidelines,* Second International Conference on Production Research, Copenhagen, August, 1973. (Reproduced in full in *Congressional Record—Senate,* May 29, 1974, pp S9168-S9172.)
2. Juvinall, Robert C., "The Mission of Tomorrow's Engineer: Mission Impossible?", Agricultural Engineering (April 1973).
3. Maslow, Abraham H., "A Theory of Human Motivation," *Psychological Review,* **50** (1943).
4. Maslow, Abraham H., *Motivation and Personality,* Harper, New York, 1954.
5. U.S. Dept. of Commerce, National Bureau of Standards, "The International System of Units (SI)," Special Publication 330, 1980.

CHAPTER TWO

Load Analysis

2.1 INTRODUCTION

This book is concerned with the design and analysis of machine and structural components. Since these are *load carrying* members, an analysis of loads is of fundamental importance. A sophisticated stress or deflection analysis is of little value if it is based on incorrect loads. A mechanical component cannot be satisfactory unless its design is based on realistic operating loads.

Sometimes the service or operating loads can be readily determined, as with some engines, compressors, and electric generators that operate at known torques and speeds. Often the loads are difficult to determine, as with automotive chassis components (which depend upon road surfaces and driving practices) or the structure of an airplane (which depends on air turbulence and pilot decisions). Sometimes experimental methods are used to obtain a statistical definition of applied loads. In other cases, engineers use records of service failures together with analyses of strength in order to infer reasonable estimates of loads encountered in service. The determination of appropriate loads is often a difficult and challenging initial step in the design of a machine or structural component.

2.2 EQUILIBRIUM EQUATIONS AND FREE-BODY DIAGRAMS

After certain initial applied loads have been determined or estimated, the basic equations of equilibrium enable loads at other points to be determined. For a nonaccelerating body, these equations can be simply expressed as

$$\Sigma F = 0 \quad \text{and} \quad \Sigma M = 0 \tag{2.1}$$

For an accelerating body they are

$$\Sigma F = ma \quad \text{and} \quad \Sigma M = I\alpha \tag{2.2}$$

These equations apply with respect to each of any three mutually perpendicular axes (commonly designated X, Y, and Z), although in many problems forces and moments are present with respect to only one or two of these axes.

The importance of equilibrium analysis as a means of load determination can hardly be over-emphasized. The student is urged to study each of the following examples carefully.

SAMPLE PROBLEM 2.1 Automobile Traveling Straight Ahead at Constant Speed on Smooth, Level Road

The 3000 lb (loaded weight) car shown in Fig. 2.1 is going 60 mph and at this speed the aerodynamic drag is 16 hp. The center of gravity (CG) and the center of aerodynamic pressure (CP), are located as shown. Determine the ground reaction forces on the front and rear wheels.

Solution

1. Power is force times velocity; one hp is 33,000 ft lb/min, and 60 mph corresponds to 5280 ft/min; hence:

$$\text{hp} = \frac{\text{drag force (lb)} \cdot \text{velocity (ft/min)}}{33{,}000}$$

$$16 = \frac{(F_d)(5280)}{33{,}000}$$

$$F_d = 100 \text{ lb}$$

■

2. Summation of forces in the direction of motion is zero (no acceleration forces exist at constant velocity), hence thrust force F_t is 100 lb in the forward direction. This is the force applied *by* the road surface *to* the tires. (The force applied by the tires

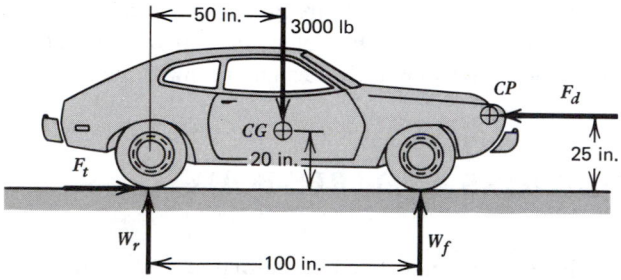

FIGURE 2.1.
Free-body diagram of auto traveling at constant speed.

to the road is equal but opposite in direction.) This force is divided equally between the rear wheels for the rear-wheel-drive car shown; it could be applied to the front tires for a front-wheel-drive car without altering any other forces.

3. Applying the moment equilibrium equation with respect to moments about a transverse axis passing through the points of rear tire road contact, we have:

$$\Sigma M = (3000 \text{ lb})(50 \text{ in.}) - (100 \text{ lb})(25 \text{ in.}) - (W_f)(100 \text{ in.}) = 0$$

from which $W_f = 1475$ lb. ■

4. Finally, from the summation of vertical forces equals zero, we have

$$W_r = 3000 \text{ lb} - 1475 \text{ lb}$$
$$= 1525 \text{ lb}$$

Before leaving this problem we note two further points of interest:

1. Due to symmetry, the weight of the vehicle *when parked* is carried equally by the front and rear wheels—that is, $W_f = W_r = 1500$ lb. When traveling at 60 mph, forces F_d and F_t introduce a front-lifting couple about the lateral axis (any axis perpendicular to the paper in Fig. 2.1) of 100 lb times 25 in. This is balanced by an opposing couple created by the added weight of 25 lb carried by the rear wheels and reduced weight of 25 lb carried by the front wheels. (Note: This simplified analysis neglects *vertical* aerodynamic forces, which can be important at high speeds; hence, the use of "spoilers" and "wings" on race cars.)
2. The thrust force is not, in general, equal to the weight on the driving wheels times the coefficient of friction, but it *cannot exceed* this value. In this problem the wheels will maintain traction as long as the coefficient of friction is equal to or above the extremely small value of 100 lb/1525 lb, or 0.066.

SAMPLE PROBLEM 2.2 Automobile Undergoing Acceleration

The car in Fig. 2.1, traveling 60 mph, is suddenly given full throttle. A curve similar to Fig. 1.9 shows the corresponding engine power to be 96 hp. Estimate the ground reaction forces on the front and rear wheels, and the acceleration of the vehicle.

Solution

1. Before proceeding with the solution, the influence of the *rotating* inertia of the car wheels, engine flywheel, and other rotating members should be considered. When the car accelerates, power is consumed in *angularly* accelerating these members. Detailed calculations typically indicate that in "high" gear the rotational inertia effect is equivalent to adding about 7 percent to the weight of the car. This means that only 100/107 of the power available for acceleration goes to *linearly* accelerating the car mass.

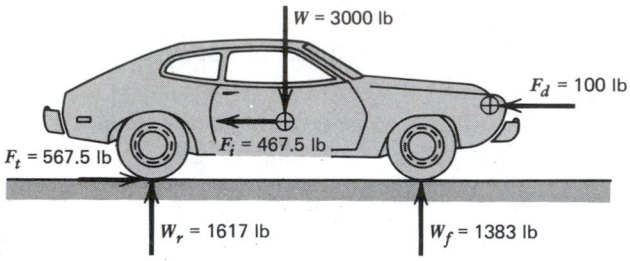

FIGURE 2.2.
Free-body diagram of auto undergoing forward acceleration.

2. In this problem, 16 hp gives the forward wheel thrust of 100 lb needed to maintain constant speed. With total horsepower increased to 96, 80 hp produces acceleration, of which 80(100/107) or 74.8 hp causes linear acceleration. If 16 hp produces 100 lb thrust, then, by proportion, 74.8 hp will increase the thrust by 467.5 lb.
3. From Eq. (2.2):

$$a = \frac{F}{m} = \frac{Fg}{W} = \frac{(467.5 \text{ lb})(32.2 \text{ ft/sec}^2)}{3000 \text{ lb}} = 5.0 \text{ ft/sec}^2$$

4. Figure 2.2 shows the car in equilibrium. The 467.5 lb inertia force acts toward the rear, and causes an additional shift of 92 lb from the front to the rear wheels (calculation details are left to the reader). ■

SAMPLE PROBLEM 2.3 Automotive Power Train Components

Figure 2.3 shows an exploded drawing of the engine, transmission, and propeller shaft of the car in Figs. 2.1 and 2.2. The engine delivers torque T to the transmission, and the transmission speed ratio (ω_{in}/ω_{out}) is R. Determine the loads, exclusive of gravity, acting on these three members.

Solution

1. Consider first the transmission. This member receives torque T from the engine and delivers torque RT to the propeller shaft[1] (through a universal joint, not shown). The propeller shaft applies equal and opposite reaction torque RT to the transmission, as shown. For equilibrium, torque $RT - T$ *must* be applied *to* the transmission housing *by* the engine structure to which it is attached.
2. The engine receives torques T and $RT - T$ from the transmission (action-reaction principle). Moment RT must be applied by the frame (through the engine mounts), as shown.
3. The propeller shaft is in equilibrium under the action of equal and opposite torques applied at its two ends.

[1]Neglecting transmission friction losses

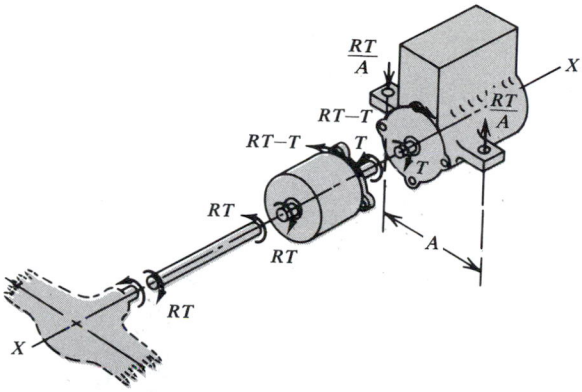

FIGURE 2.3.
Equilibrium of moments about the X-axis for engine, transmission, and propeller shaft of a front engine, rear wheel drive automobile. (T = engine torque, R = transmission torque ratio; engine rotates counterclockwise viewed from transmission.)

SAMPLE PROBLEM 2.4 Automotive Transmission Components

Figure 2.4*a* shows a simplified version of the transmission represented in Fig. 2.3. The engine is delivering a torque, T, of 3000 lb in. to the transmission, and the transmission is in low gear with a ratio, R, of 2.778. (For this problem consider R to be a torque ratio, T_{out}/T_{in}. To the degree that friction losses are present, the speed ratio, ω_{in}/ω_{out}, would have to be slightly greater.) The other three portions of the figure show the major parts of the transmission. Gear diameters are given in the figure. Input gear A rotates at engine speed, and drives countershaft gear B. Countershaft gear C meshes with mainshaft output gear D. (The construction of the mainshaft is such that the input and output ends rotate about a common axis, but the two halves are free to rotate at different speeds.) The mainshaft is supported in the housing by bearings I and II. Similarly, the countershaft is supported by bearings III and IV. Determine all loads acting on the components shown in Figs. 2.4*b*–2.4*d*, thus representing them as free bodies in equilibrium. Assume that the forces acting between mating gear teeth are tangential. (This amounts to neglecting the radial and axial components of load. These load components are discussed in Chapters 15 and 16, dealing with gears.)

Solution

1. A very important initial observation is that equilibrium of the *total* transmission (Fig. 2.4*a*) is *independent* of anything inside the housing. This free-body diagram would pertain equally well to transmissions of $R = 2.778$ having *no* gears inside—such as hydraulic or electric transmissions. In order for the transmission to work, whatever parts are inside *must* provide for the *torque of 5333 lb in. to be reacted by the housing.* (A striking example of this concept came to the author's attention after World War II, when many "inventors" sent the major automobile companies

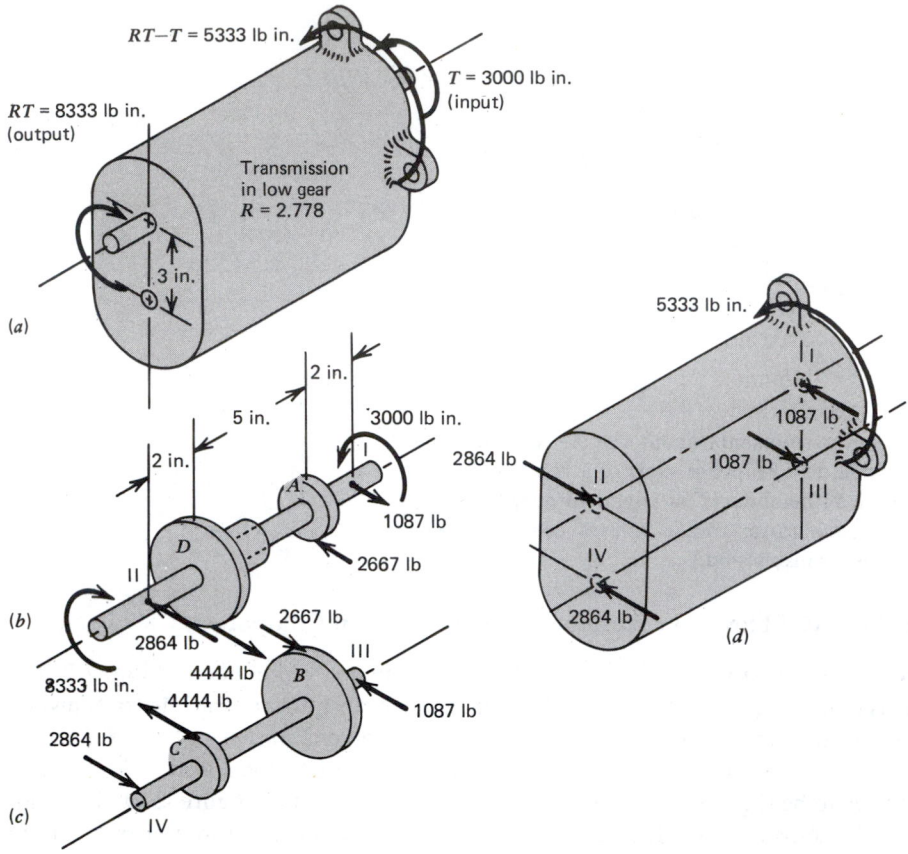

FIGURE 2.4.
Free-body diagram of transmission and major components: (a) Complete transmission assembly. (b) Mainshaft (front and rear halves rotate freely with respect to each other). (c) Countershaft. (d) Housing. Note: Diameters of gears *A* and *C* = 2¼ in. Diameters of gears *B* and *D* = 3¾ in.

voluminous material pertaining to automatic transmission designs they wished to sell. To study all of the drawings, analyses, descriptions, and so on would have required numerous hours. In many cases, however, it could be quickly determined that there was no provision for a torque reaction to be transmitted to the housing, and that therefore the transmission could not possibly work.)

2. The input portion of the mainshaft (Fig. 2.4*b*) requires the tangential force of 2667 lb to balance the input torque of 3000 lb in., thus satisfying $\Sigma M = 0$ about the axis of rotation. This force is applied *to* gear A *by* gear B. Gear A applies an equal and opposite force to gear B, as shown in Fig. 2.4*c*. Since there are no torques applied to the countershaft except by the two gears, it follows that gear C must receive a 4444 lb force from gear D. An opposite 4444 lb force is applied by gear C to gear D. Equilibrium of moments about the axis of the output half of the mainshaft requires that a torque of 8333 lb in. be applied to the output shaft by the

propeller shaft as shown. (Note that the output torque can also be obtained by multiplying the input torque by the gear diameter ratios, B/A and D/C. Thus:

$$3000 \text{ lb in.} \times \frac{3\frac{3}{4} \text{ in.}}{2\frac{1}{4} \text{ in.}} \times \frac{3\frac{3}{4} \text{ in.}}{2\frac{1}{4} \text{ in.}} = 8333 \text{ lb in.})$$

3. The force applied to the mainshaft by bearing II is found by taking moments about bearing I. Thus:

$$\Sigma M = 0: \quad (2667 \text{ lb})\,(2 \text{ in.}) - (4444 \text{ lb})\,(7 \text{ in.}) + (F_{\text{II}})\,(9 \text{ in.}) = 0, \text{ or}$$
$$F_{\text{II}} = 2864 \text{ lb}$$

The force at bearing *I* is found by $\Sigma F = 0$ (or by $\Sigma M_{\text{II}} = 0$). Countershaft bearing reactions are found in the same way.

4. Figures 2.4*b* and 2.4*c* show bearing forces applied *to* the shafts, through the bearings, and *by* the housing. Figure 2.4*d* shows the corresponding forces applied *to* the housing, through the bearings, and *by* the shafts. The *only* members in contact with the housing are the four bearings and the bolts that connect it to the engine structure. Figure 2.4*d* shows that the housing is indeed a free body in equilibrium, as both forces and moments are in balance.

The foregoing examples have illustrated how the powerful free-body-diagram method can be used to determine loads at various levels—that is, loads acting on a complex total device (as an automobile), loads acting on a complex unit within the total device (as the automotive transmission), and loads acting on one part of a complex unit (transmission countershaft). Free-body-equilibrium concepts are equally effective and valuable in determining *internal* loads, as illustrated below in Sample Problem 2.5. This is also true of internal loads in components like the transmission countershaft in Fig. 2.4*c*, as will be seen in the next section.

SAMPLE PROBLEM 2.5 Determination of Internal Loads

Two examples of load carrying members are shown in Figs. 2.5*a* and 2.6*a*. Using free-body diagrams, determine and show the loads existing at cross section *AA* of each member.

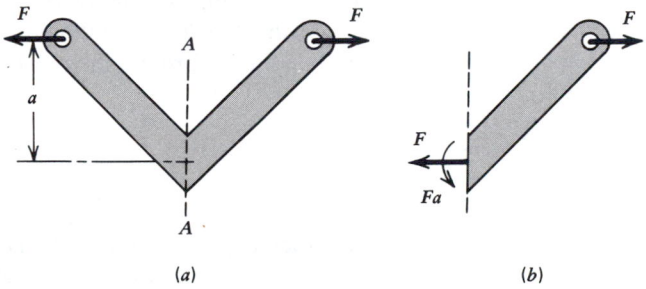

FIGURE 2.5.
Loads acting on an internal section as determined from a free-body diagram.

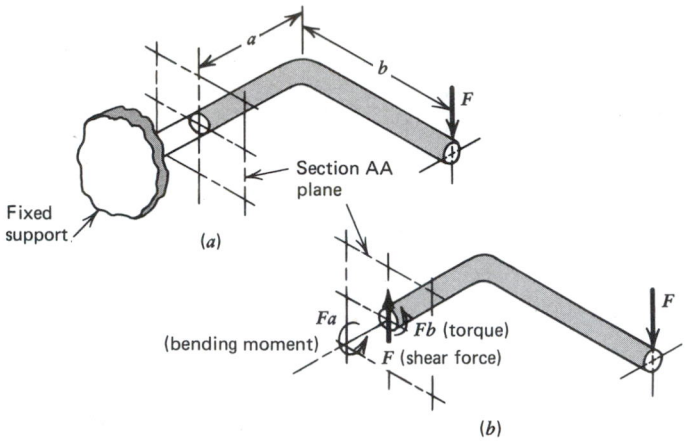

FIGURE 2.6.
Loads acting on an internal section as determined from a free-body diagram.

Solution

Figures 2.5*b* and 2.6*b* show segments on one side of section *AA* as free bodies in equilibrium. The forces and moments acting at the section are determined from the equations of equilibrium.

The next two examples illustrate the determination of loads acting on three-force members where only one of the three forces is completely known and a second is known in direction only.

SAMPLE PROBLEM 2.6 Three-Force Member

Figure 2.7 shows a bell-crank (link 2) that pivots freely with respect to the fixed frame (link 1). A horizontal rod (link 3, not shown) attaches at the top, exerting a force of 40 lb, as shown. (Note the subscript notation: F_{32} is a force applied *by* link 3 *to* link 2.) A rod 30° from horizontal (link 4, not shown) attaches to the bottom, exerting force F_{42} of unknown magnitude. Determine the magnitude of F_{42}, and also the direction and magnitude of force F_{12} (the force applied by fixed frame 1 to link 2 through the pinned connection near the center of the link).

Solution A (Analytical)

1. Summation of moments about the pivot pin requires that F_{42} = 60 lb (Note that 40 lb × 1½ in. = 60 lb × 1 in.) ■
2. Dividing F_{12} into horizontal and vertical components, and setting the summation of vertical and horizontal forces acting on link 2 equal to zero yields: V_{12} = 60 lb sin

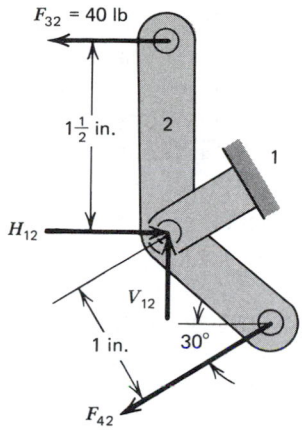

FIGURE 2.7.
Bell-crank forces—analytical solution.

30° = 30 lb; H_{12} = 40 lb + 60 lb cos 30° = 92 lb. The magnitude of F_{12} is $\sqrt{30^2 + 92^2}$ = 97 lb; its direction is upward and to the right at an angle of $\tan^{-1}$ 30/92 = 18° from horizontal. ■

Solution B (Graphical)

The analytical solution merely solved the three equations for equilibrium in a plane for three unknowns. This same solution of simultaneous equations can be accomplished graphically, as in Fig. 2.8. An understanding of the graphical procedure adds to one's insight into nature of the force directions and magnitudes necessary for equilibrium of the link.

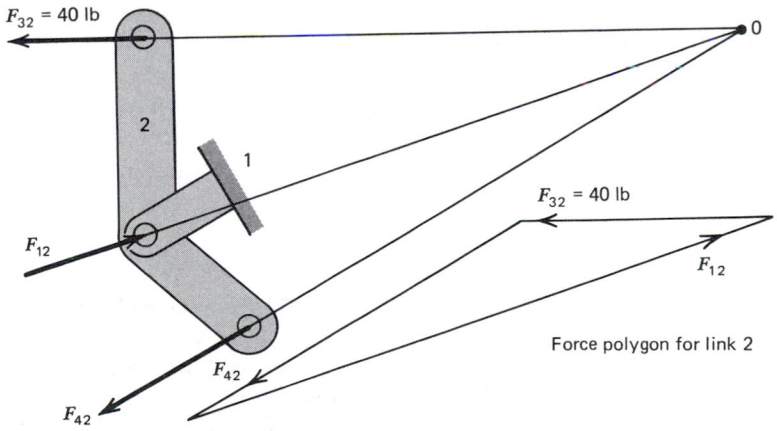

FIGURE 2.8.
Bell-crank forces—graphical solution.

1. For equilibrium, the summation of moments of all forces acting on link 2 must be equal to zero when these moments are taken about *any* point, including point "0," which is the intersection of the two known force lines of action. Since two of the three forces have no moment about point "0," equilibrium requires that the third force also have no moment about "0." This can only be satisfied if the line of action of F_{12} also passes through "0."
2. We know one force completely, and the other two in direction only. A graphical solution for summation of forces equals zero is provided by the force polygon shown in Fig. 2.8. This is constructed by first drawing known force F_{32} in proper direction and with length representing its 40 lb magnitude to any convenient scale. A line with the direction of F_{12} is drawn through either end of the vector representing F_{32} and a line with the direction of F_{42} is drawn through the other end of this vector. Magnitudes of the two unknown forces can now be scaled from the polygon. (Note that the same result is obtained if a line of the direction of F_{42} is drawn through the *tail* of vector F_{32}, with the direction of F_{12} being drawn through the *tip* of F_{32}.) ■

SAMPLE PROBLEM 2.7 Human Finger as Three-Force Member

The principles of Mechanical Engineering Design that are traditionally applied to components of inanimate machines and structures are being increasingly applied in the relatively new field of *Bioengineering*. A case in point is the application of free-body load analysis procedures to the internal load-carrying components of the human finger in studies of arthritic deformity [2,3]. Figure 2.9 illustrates one simplified portion of this study wherein a 10 lb pinch force at the tip of a finger is created by muscle contraction causing tendon force F_t. Since the tendon force has a moment arm about pivot point "0" of only 1/6 that of the pinch force, the tensile force in the tendon must be 60 lb. The force polygon shows that the compressive force between the finger bone (proximal phalanx) and mating bone in the hand (metacarpal) is about 55 lb—a value that may cause crushing of deteriorated arthritic bone tissue.

Other examples of free-body diagram analysis are given in Section 20.3 and in [1], Chapter 1.

2.3 BEAM LOADING

"Beam loading" refers to the lateral loading of members that are relatively long in comparison with their cross-sectional dimensions. Torsional and/or axial loading may or may not be involved as well. By way of review, two cases are shown in Fig. 2.10. Note that each incorporates three basic diagrams: external loads, internal transverse shear forces (V) and internal bending moments (M). All expressions for magnitudes are the result of calculations the reader is advised to verify as a review exercise. (Reactions R_1 and R_2 are calculated first, on the basis of $\Sigma F = 0$ and $\Sigma M = 0$, with distributed load w treated as a concentrated load wb acting in the middle of span b.)

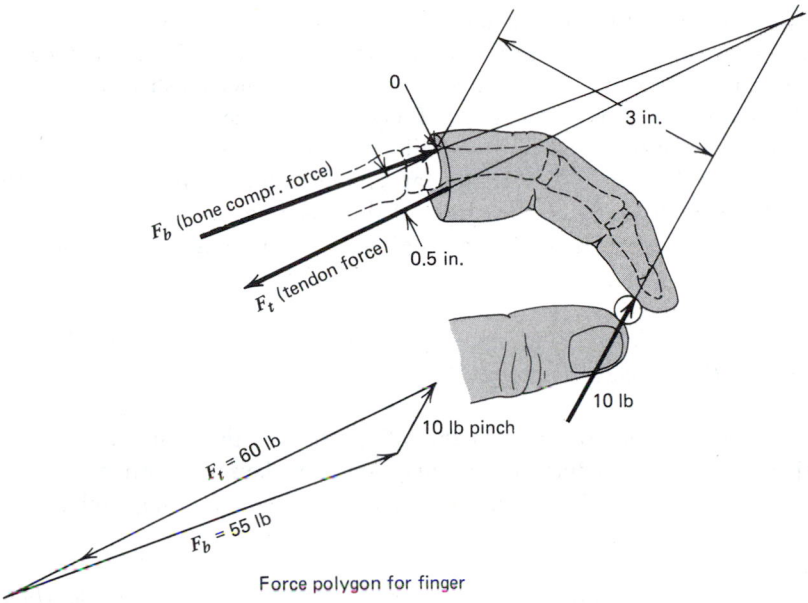

FIGURE 2.9.
Force study of human finger.

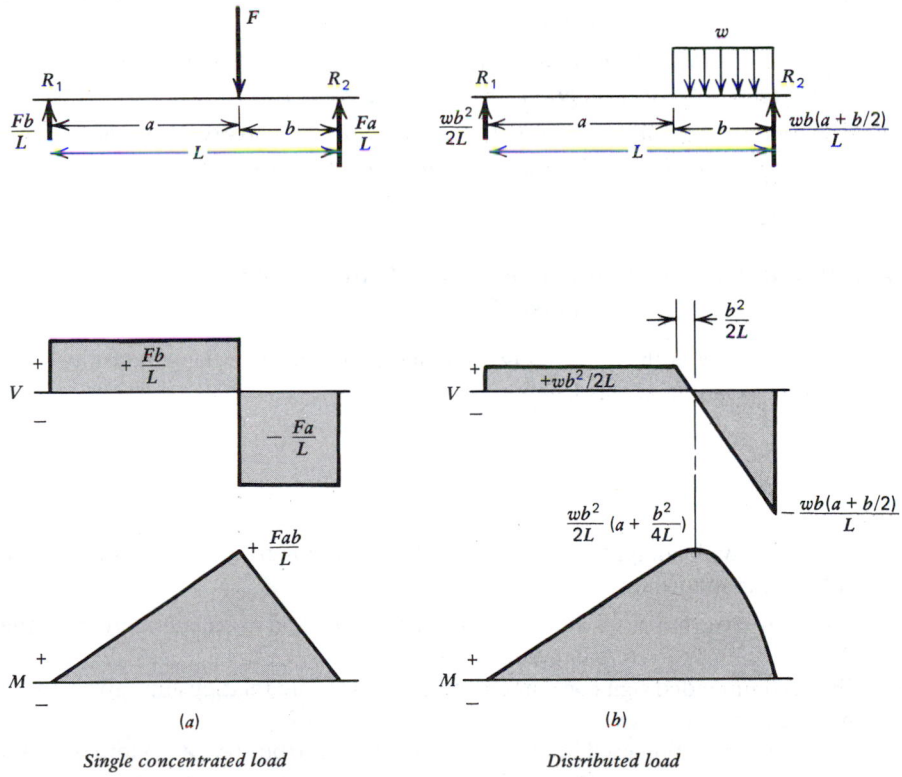

FIGURE 2.10.
Examples of beam load, shear force, and bending moment diagrams.

The sign convention of the shear diagram is arbitrary, but the one used here is recommended: proceed from left to right, following the direction of the applied loads. In this case there are no loads to the left of reaction R_1 and hence no shear forces. At R_1 an upward force of Fb/L is encountered. Proceeding to the right, there are no loads—hence no *change* in the shear force—until the downward load of F is reached. At this point the shear diagram drops an amount F, and so forth. The diagram must come to zero at R_2, as no loads exist to the right of this reaction.

The (arbitrary) sign convention recommended here for bending follows from the relationship that

> *the value of the shear force (V) at any point along the beam is equal to the slope of the bending moment diagram (M) at that point.*

Thus, a constant positive value of shear in the left portion of the beam results in a constant positive slope of the bending moment diagram over the same portion.

The student is reminded of three other important rules or relationships relative to load, shear, and moment diagrams:

1. *The value of local load intensity at any point along the beam is equal to the slope of the shear force diagram (V) at that point.* (For example, the end supports, acting as "points," produce a theoretically infinite upward force intensity. Hence, the slope of the shear diagram at these points is infinite.)
2. *The difference in shear load value between any two points along the beam is equal to the area under the load diagram between these same two points.*
3. *The difference in bending moment value between any two points along the beam is equal to the area under the shear diagram between these two points.*

SAMPLE PROBLEM 2.8 Internal Loads in a Transmission Countershaft

Locate the cross section of the shaft in Fig. 2.4*c* that is subjected to the greatest loading, and determine the loading at this location.

Solution

1. Figure 2.11 shows loading, shear, moment, and torque diagrams for this shaft. Note particularly the following:
   - **a.** The load diagram is in equilibrium—the forces and moments acting in the plane of the paper are balanced.
   - **b.** The recommended sign convention and the four basic relationships given above in italics are illustrated.
   - **c.** The sign convention used in the *torque diagram* is arbitrary. Zero torque exists

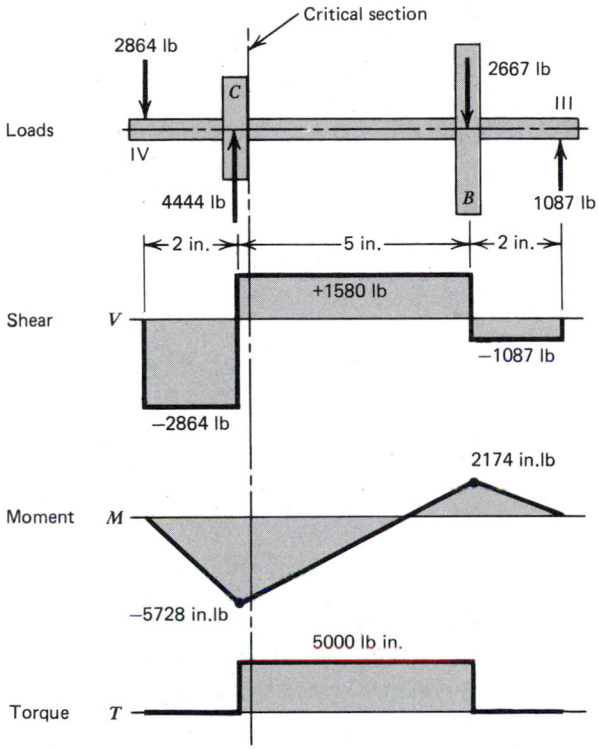

FIGURE 2.11.
Countershaft loading, and determination of critical section.

outboard of the gears, as bearing friction would normally be neglected. Torques of (4444 lb) (2.25 in./2) and (2667 lb)(3.75 in./2) are applied to the shaft at *C* and *B*.

2. The critical location of the shaft is just to the right of gear C. Here we have maximum torque together with essentially maximum bending. (The transverse shear force is less than maximum, but except for highly unusual cases involving extremely short shafts, shear loads are unimportant in comparison with bending loads.) Figure 2.12 shows the portion of the countershaft to the left of the critical section as a free body. Note that this *partial* member constitutes a *free body in equilibrium* under the action of all loads external to it. These include the external loads shown, and also the *internal* loads applied *to* the free body *by* the right-hand portion of the countershaft.

2.4 LOCATING CRITICAL SECTIONS – FORCE-FLOW CONCEPT

The sections chosen for load determination in the previous examples (i.e., in Figs. 2.5, 2.6, and 2.12) were, by simple inspection, clearly those subjected to the most critical

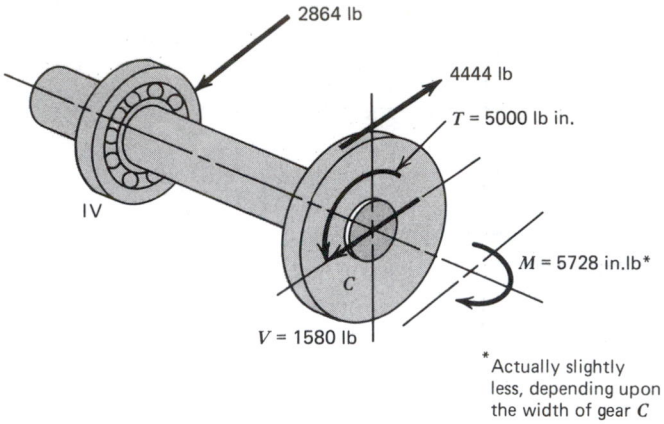

FIGURE 2.12.
Loading at countershaft critical section.

loading. In more complicated cases, however, several sections may be critical, and their locations less obvious. In such instances it is often helpful to employ an orderly procedure of following the "lines of force" (approximate paths taken by the force, determined by simple inspection) through the various parts, and noting along the way any sections suspected of being critical. Such a procedure is illustrated in the following example.

SAMPLE PROBLEM 2.9 Yoke Connection

Using the force-flow concept, locate the critical sections and surfaces in the members shown in Fig. 2.13.

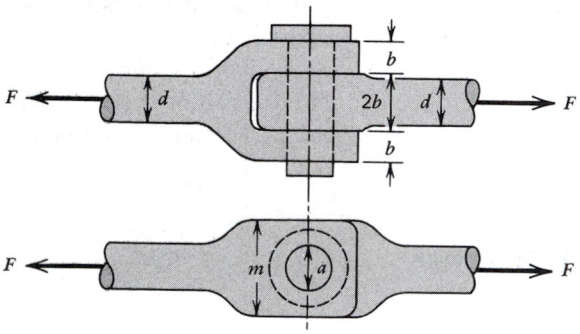

FIGURE 2.13.
Yoke connection.

Solution[2]

A force-flow path through each member is indicated by the dashed lines in Fig. 2.14. Following along this path from left to right, the major critical areas are indicated by the jagged lines identified by the circled numbers.

a. Tensile loading exists at section ① of the fork. If the transition sections have ample material and generous radii, the next critical location is ②, where the force flow encounters a bottleneck because of the reduced area due to the holes. Note that with this symmetrical design, force F is divided into four identical paths, each having an area at location ② of $\frac{1}{2}(m - a)b$.

b. Assuming that the section at ② is adequate to prevent failure, the force flow proceeds on to the next questionable section, which is at ③. Here, the turning of the flow path is associated with shearing stresses tending to "push out" the end segments bounded by jagged lines ③.

c. The next critical area is interface ④ and ④′, where bearing loading exists between the fork-hole and pin surfaces, respectively. In like manner, equal bearing loads are developed at the interface between the pin and the blade-hole surfaces.

d. The forces at ④′ load the pin as a beam, causing direct shear loading of sections ⑤ (note that the pin is in "double shear," as the two surfaces ⑤ are loaded in parallel, each carrying a shear load of $F/2$). Also, the bearing loads produce a maximum bending moment at area ⑥, in the center of the pin.

e. After the forces emerge from the pin and enter the blade, they flow across critical areas ④,③, ②, and ①, which correspond directly to the like-numbered sections of the fork.

f. Although this is not brought out in the simplified force-flow pattern of Fig. 2.14, it

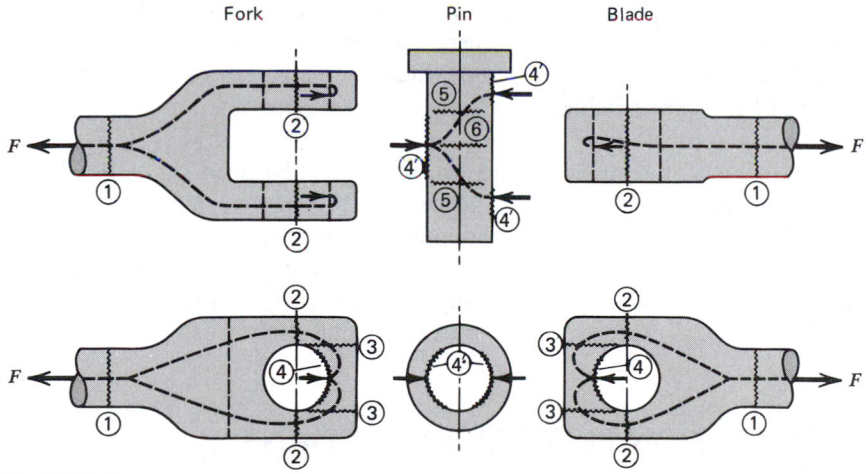

FIGURE 2.14.
Force-flow lines and critical sections in yoke connection.

[2]Adapted from [1].

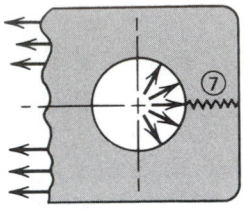

FIGURE 2.15.
Distributed bearing loading may cause hoop tension failure at 7.

should be noted that the bearing loads applied to the surfaces of the holes are not concentrated on the load axis but are *distributed* over these surfaces as shown in Fig. 2.15. This gives rise to *hoop tension* (or circumferential tensile loading) tending to cause tensile failure in the section identified as ⑦.

Although the determination of stresses at the various critical sections is beyond the scope of this chapter, this is a good time to give a word of caution regarding the simplifying assumptions that one might make when these stresses are calculated:

a. The section at ② (in the above sample problem) might be assumed to be in uniform tension. Actually, bending is also present, which *adds* to the tension at the inner, or hole surface, and subtracts at the outer surface. This can be visualized by imagining the distortion of fork and blade members made of rubber and loaded through a metal connecting pin. The *quantitative* evaluation of this effect involves details of the geometry, and is not a simple matter.

b. The distribution of compressive loading on surfaces ④ and ④′ might be assumed uniform, but this could be far from the actual case. The major factors involved are the fit of the pin in the hole, and the rigidity of the members. For example, bending of the pin tends to cause highest bearing loading near the fork-blade interfaces. Moreover, the extent of pin bending depends not only on its own flexibility but also on the tightness of fit. Also, the degree to which the pin is restrained from bending by a close fit has a major influence on the pin bending stresses.

Like many engineering problems, this one illustrates the need (1) to be able to make reasonable simplifying assumptions to get useable answers quickly, (2) to be *aware* that such assumptions were made and interpret the results accordingly, and (3) to make a good "engineering judgment as to whether a simplified solution is adequate in the situation involved, or whether a more sophisticated analysis is justified, involving more advanced analytical procedures and/or experimental studies.

2.5 LOAD DIVISION BETWEEN REDUNDANT SUPPORTS

A *redundant* support is one that could be removed and still leave the supported member in equilibrium. For example, in Fig. 2.16, if the center support (which is redundant) were removed, the beam would be held in equilibrium by the end supports. When redun-

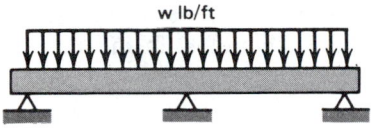

FIGURE 2.16.
Redundant support.

dant supports (reactions) are present, the simple equations of equilibrium no longer suffice to determine the magnitude of load carried by *any* of the supports. This is true because there are more unknowns than equilibrium equations.

A redundant support *adds stiffness* to the structure, and carries a portion of the load proportional to its stiffness. For example, the 100 lb weight in Fig. 2.17*a* is supported from below by a coil spring in compression and above by a coil spring in tension. (Assume that except for the 100 lb load, the springs would be stress-free.) The top spring has a *rate* or *stiffness constant* of 10 lb/in.; the lower spring a constant of 40 lb/in. Under the pull of gravity the weight moves downward, stretching the top spring by an amount δ and compressing the lower spring by the same amount. Thus, the weight is in equilibrium under the action of a gravity force of 100 lb and spring forces of 10δ and 40δ (Fig. 2.17*b*.). From $\Sigma F = 0$, we have $\delta = 2$ in. Hence, the top and bottom spring forces are 20 lb and 80 lb, respectively, illustrating that the *load is divided in proportion to the stiffness of the redundant supports.*

Suppose now that in order to reduce the deflection a third support is added—this in the form of a 10 in. length of steel wire, 0.020 in. in diameter inside the top spring,

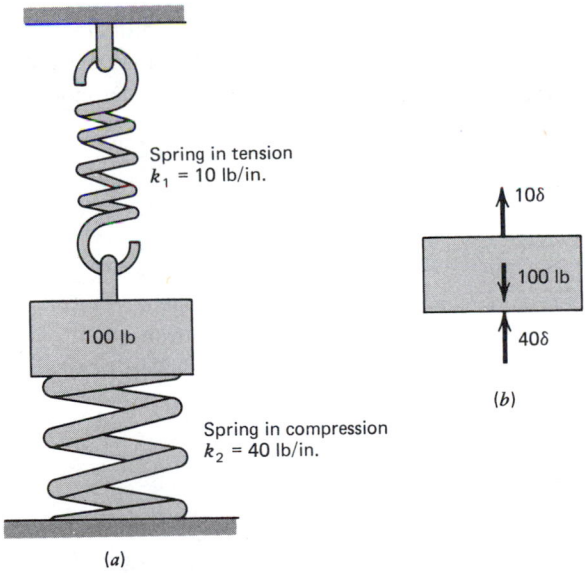

FIGURE 2.17.
Weight supported redundantly from springs above and below.

with the top end of the wire attached to the upper support and the lower end attached to the weight. The spring rate (AE/L) of the wire would be about 942 lb/in., and the *total* stiffness supporting the weight would then be 992 lb/in. Hence the wire would carry 942/992 of the 100 lb load, or 95 lb. Its stress would be

$$\frac{P}{A} = \frac{95 \text{ lb}}{0.000314 \text{ in.}^2}, \text{ or } 303{,}000 \text{ psi}$$

which is a value in excess of its probable tensile strength; hence it would fail (break or yield). Thus, *the strengths of redundant load carrying members should be made approximately proportional to their stiffnesses.*

To illustrate this point further, consider the angle-iron shown in Fig. 2.18. Suppose that when installed as part of a machine or structure, the angle-iron has inadequate rigidity in that the 90° angle deflects more than desired, although this does not cause breakage. To reduce the angular deflection, the small triangular web is welded in place as shown. It becomes a redundant support that limits angular deflection. But it may well add stiffness far out of proportion to its strength. Cracks may appear in or near the welded joint, thereby eliminating the adding stiffness. Furthermore, the cracks so formed may propagate through the main angle-iron. If so, the addition of the triangular "reinforcement" would actually *weaken* the part. Sometimes failures in a complicated structural member (as a casting) can be corrected by *removing* stiff but weak portions (such as thin webs), *provided* the remaining portions of the part are sufficiently strong to carry the increased load imposed upon them due to the increased deflection, and provided, of course, the increased deflection is itself acceptable.

As noted in Section 1.2, properly designed redundant supports can add to the safety of a load-carrying member.

A very useful procedure (Castigliano's method) for calculating redundant load reactions for a completely *elastic* system is given in Section 5.9. At the other extreme, the loading pattern associated with *ductile* failure of a set of redundant supports is discussed in the next article.

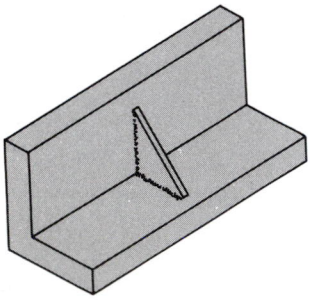

FIGURE 2.18.
Web reinforcement added to an angle-iron.

2.6 FORCE-FLOW CONCEPT APPLIED TO REDUNDANT DUCTILE STRUCTURES

As noted in Section 2.5, loads shared among parallel redundant paths are divided in proportion to path stiffnesses. If the paths are brittle and the load is increased to failure, one path will fracture first, thereby transferring its share of the load to the other paths, and so on, to failure of all paths. For the usual case involving materials of some ductility, one path will *yield* first, thereby reducing its stiffness (stiffness then being proportional to the *tangent* modulus),[3] which allows some of its load to be transferred to other paths. For the ductile case, general yielding of the total structure occurs only after the load has been increased sufficiently to bring all of the parallel paths to their yield strengths.

The force-flow concept, introduced in Section 2.4, is helpful in dealing with ductile redundant structures. This is illustrated in the following example:

SAMPLE PROBLEM 2.10 Riveted Joint

Figure 2.19*a* shows a triple-riveted butt joint, wherein two plates are butted together and loads are transmitted across the joint by top and bottom straps. Each strap has a thickness of 2/3 the plate thickness, and is made of the same material as the plates. Three rows of rivets attach each plate to the straps, as shown. The rivet pattern that is drawn for a width of one pitch is repeated over the full width of the joint.

Determine the critical sections, and discuss the strength of the joint, using the force-flow concept.

Solution[4]

1. Figure 2.19*b* shows a diagrammatic sketch of the force-flow pattern in the plate. A portion of the load is transferred to each of the three rows of rivets. (Since static equilibrium could be satisfied by using any one row, the structure is redundant.) Note that at the outer row the entire force flows across a section containing one rivet hole. At the middle row, a section containing two rivet holes is subjected to all of the force not going to the outer row of rivets. The section of plate at the inner row transmits only the force going to the inner row of rivets. This relationship between force and area at each section of plate is represented diagrammatically in Fig. 2.19*c*. Note the representation of the reduction in area due to rivet holes at the middle and inner rows being twice that at the outer row.
2. Figure 2.19*d* shows how each rivet is associated with three important loadings: bearing with the plate, shear, and bearing with the strap. A diagrammatic force-flow representation of this is given in Fig. 2.19*e*, where the force path encounters five critical sections in series: the reduced tensile area containing the holes in the

---

[3]See Section 3.2.
[4]From [1].

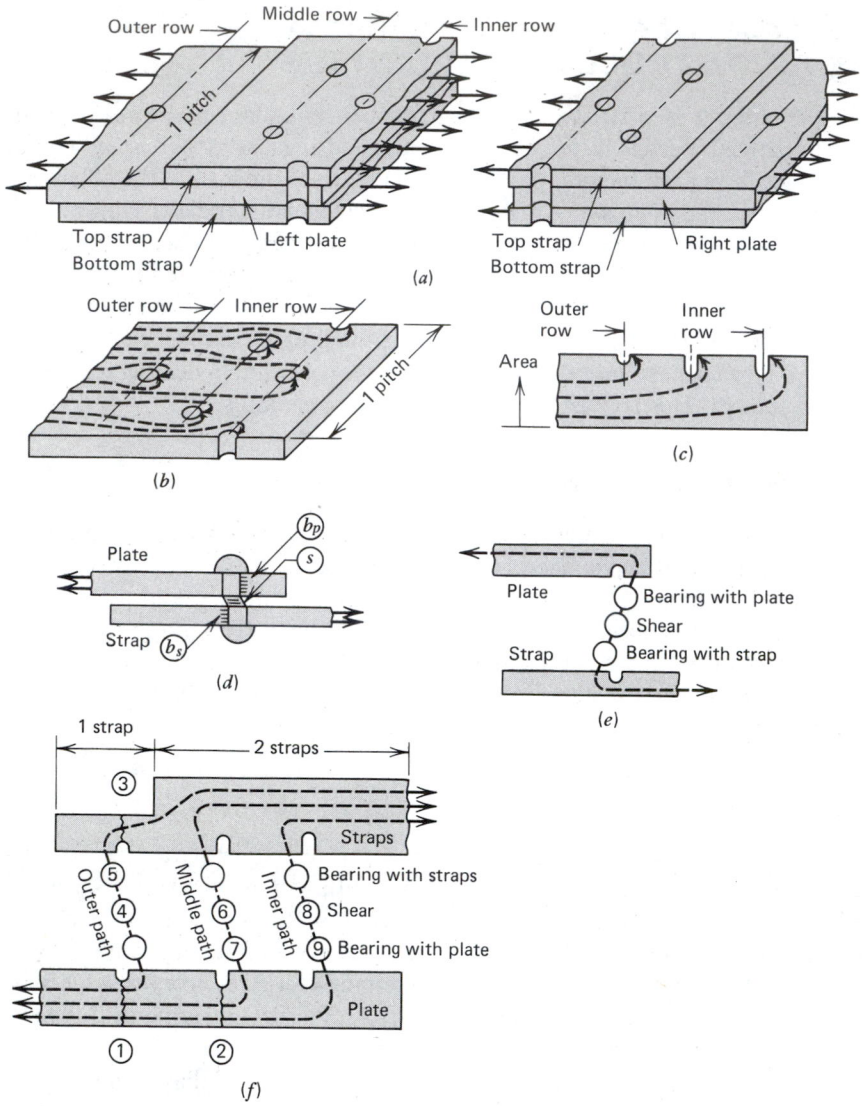

FIGURE 2.19.
Force-flow concept applied to triple-riveted butt joint. (a) Complete joint, broken at center, showing total load carried by straps. (b) Force flow through plate to rivets. (c) Diagram of force flow vs. plate cross-sectional area. (d) Force flow through rivet. (e) Diagrammatic representation of force flow through rivet. (f) Complete diagrammatic representation of force flow.

plate, the sections corresponding to the three loadings involving the rivet, and the reduced tensile area of the strap.

3. Figure 2.19*f* shows a similar representation of the entire joint. Critical sections are identified and numbered ① to ⑨. Basically involved are three parallel force paths,

one going through each row of rivets. Starting at the lower left, all three paths flow across the reduced plate section at the outer row. This is critical area ①. Failure at this point severs all force-flow paths, causing total fracture of the joint. Only two paths flow across the plate at ②, but since the area here is less than at ①, failure is possible. Failure is not possible in the plate at the inner path, as only one force path flows across an area identical with that at ②.

4. Turning to the possibilities of tensile failure in the straps, note that the relative thickness of plate and straps is such that strap tensile failure is possible only at the outer row at critical point ③.

5. With respect to failure possibilities involving the rivets themselves, each rivet is vulnerable in shear and at *one* bearing area (whichever is smaller). In the outer row, the vulnerable bearing area involves the strap, because the strap is thinner than the plate. In the other rows, the vulnerable bearing area involves the plate, as the strap bearing load is shared by two straps, the combined thickness of which exceeds that of the plate. Note also that the middle- and inner-row rivets divide their shear load between *two* areas (i.e., they are loaded in *double shear*), whereas the outer-row rivets have but a single shear area. The possibilities of rivet failure are numbered ④ to ⑨.

6. The distribution of load among the three redundant paths depends on relative stiffnesses; but since riveted joints are invariably made of ductile materials, slight local yielding permits a redistribution of the load. Thus, final failure of the joint will occur only when the external load exceeds the combined load carrying capacity of all three paths. This involves simultaneous failure of all paths and can occur in three possible ways.

   a. tensile failure at ①,
   b. simultaneous failure of the weakest link in each of the three paths, or
   c. simultaneous failure at ② and at the weakest link in the outer path (③, ④, or ⑤).

REFERENCES

1. Juvinall, Robert C., *Engineering Consideration of Stress, Strain and Strength,* McGraw-Hill, New York, 1967.
2. Juvinall, Robert C., "An Engineering View of Musculoskeletal Deformities," Proceedings of the IVth International Congress of Physical Medicine, Paris, 1964.
3. Smith, Edwin M., R. C. Juvinall, L. F. Bender, and J. R. Pearson, "Flexor Forces and Rheumatoid Metacarpophalangeal Deformity," *Journal of the American Medical Association,* **198** (Oct. 10, 1966).

CHAPTER THREE

Materials

3.1 INTRODUCTION

The selection of materials and the processes used in fabrication are integral parts of the design of any machine component. Strength and rigidity are traditionally key factors involved in the selection of a material. Equally important is the relative reliability and durability of the part when made from alternative material possibilities. When the component is expected to operate at extreme temperatures, this must be considered carefully when making the material selection. In recent years, materials choices have been increasingly influenced by recyclability, energy requirements, and environmental pollution. Cost and availability are also vitally important. The cost to be considered is the *total cost of the fabricated part,* including labor and overhead as well as the material itself. The relative cost and availability of various materials varies with time, with the result that the engineer is frequently called upon to evaluate alternative materials in the light of changing market conditions. In summary, the best material for a particular application is the one that provides the best value, defined as the ratio between *overall* performance and *total* cost.

It is assumed that the reader has a previous background in the fundamentals of engineering materials. Also, it is acknowledged that the practicing engineer is faced with a career-long challenge to remain abreast of new materials developments as they apply to the products with which one is concerned. Fitting in between these two phases of engineering materials study, this chapter attempts to summarize some of the relevant basic information, and to emphasize the increasing importance of a rational approach to the use of empirical material properties test data.

The useful life of most machine and structural components ends with *fatigue fail-*

ure or *surface deterioration.* Further information on the resistance of various materials to these modes of failure is included in Chapters 8 and 9, respectively.

Material properties information is given in Appendix C. Appendix C-1 gives physical constants for a variety of engineering materials. Mechanical properties are given in subsequent tables and graphs.

Finally, the reader is reminded that the scope and complexity of the subject is such that consultation with professional metallurgists and materials specialists is often desirable.

3.2 THE STATIC TENSILE TEST – "ENGINEERING" STRESS-STRAIN RELATIONSHIPS

The basic engineering test of material strength and rigidity is the standard tensile test, from which stress-strain curves as shown in Fig. 3.1 are obtained. The stresses and strains plotted are the nominal or so-called *engineering* values, defined as:

$\sigma = P/A_0$, where P is the load and A_0 the original *unloaded* cross-sectional area, and

$\epsilon = \Delta L/L_0$, where ΔL is the change in length due to load and L_0 is the original unloaded length.

Note an important notation convention that will be observed throughout this book: the Greek letter σ denotes normal *stress,* which is a function of the applied loads; S (with appropriate subscripts) designates *strength properties of the material.* For example, Fig. 3.1 shows that when $\sigma = 39$ ksi,[1] the material begins to yield. Hence, $S_y = 39$ ksi. Likewise, the greatest (ultimate) load that the test specimen can withstand corresponds to an engineering stress of 66 ksi. Hence, $S_u = 66$ ksi.

Whereas S (with suitable subscripts) is used for all strength values including those

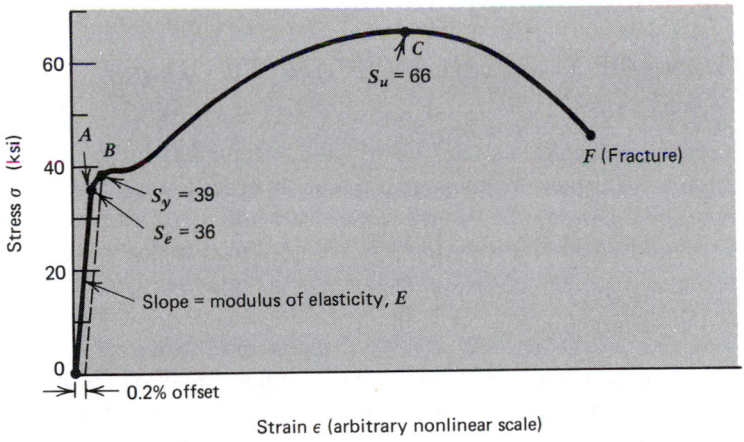

FIGURE 3.1.
Engineering stress-strain curve—hot rolled 1020 steel.

[1]Kilopounds (thousands of pounds) per square inch

for torsion or shear, the letter σ is used for normal stresses only (i.e., stresses caused by tensile, compressive, or bending loads). Shear stresses (caused by torsional or transverse shear loads) are designated by the Greek letter τ.

Returning to Fig. 3.1: several other mechanical properties are indicated on the stress-strain curve. Point A represents the *elastic limit,* S_e. It is the highest stress the material can withstand and still return exactly to its original length when unloaded. When loaded beyond point A, the material exhibits a partially plastic response. For most engineering materials, point A also approximates the *proportional limit,* defined as the stress at which the stress-strain curve first deviates (ever so slightly) from a straight line. Below the proportional limit, Hooke's law applies. The constant of proportionality between stress and strain (which is the slope of the curve between the origin and the proportional limit) is the *modulus of elasticity,* or *Young's modulus,* E. For some materials, a slight deviation from linearity occurs between the origin and a point such as A, at which the deviation begins to become more apparent. Such a material has no true proportional limit. Also, its modulus of elasticity is not definitely defined. The computed value depends on the portion of the curve used for measuring the slope, and on whether the slope is characterized by the tangent or the secant.

Point B on Fig. 3.1 represents the *yield strength,* S_y. It is the value of stress at which significant plastic yielding first occurs. In some ductile materials, notably soft steel, marked yielding occurs suddenly at a clearly defined value of stress. In other materials the onset of appreciable yielding occurs gradually, and the determination of yield strength is somewhat arbitrary. In such cases point B is usually taken as the point where the strain is offset 0.2 percent from the extrapolated elastic curve (see Fig. 3.1). If the load is then removed, the specimen exhibits a 0.2 percent permanent elongation. Yield strength corresponding to a specified (very small) offset is a standard laboratory determination, whereas elastic limit and proportional limit are not.

3.3 IMPLICATIONS OF THE "ENGINEERING" STRESS-STRAIN CURVE

Figures 3.1 and 3.2 represent identical stress-strain relationships, but differ in two respects: (a) Fig. 3.1 uses an arbitrary nonlinear strain scale in order to illustrate more clearly the points previously discussed, whereas the strain scale in Fig. 3.2 is linear; and (b) Fig. 3.2 contains two additional strain scales that are described in point 3 below. Several important concepts are related to these two figures:

1. At the 36-ksi elastic limit of this particular steel, strain (ϵ) has a value of $\sigma/E = 0.0012$. Figure 3.2 shows the strain at ultimate strength and at fracture to be about 250 and 1350 times this amount. Obviously, to the scale plotted, the elastic portion of the curve in Fig. 3.2 is virtually coincident with the vertical axis.
2. Suppose that a tensile member made from this steel has a notch (or hole, groove, slot, etc.) such that the strain at the notch surface is three times the nominal P/AE value. A tensile load causing a nominal stress (P/A) of 30 ksi and nominal

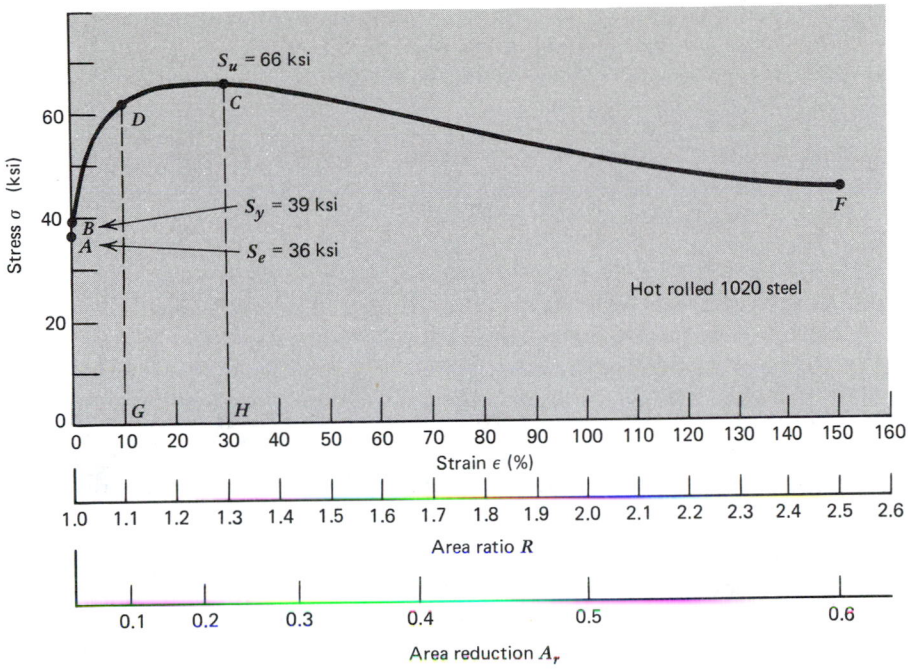

FIGURE 3.2.
Figure 3.1 replotted to linear strain scale.

strain (P/AE) of 0.001 produces a strain three times this large (0.003) at the notch surface. Since even this strain is almost imperceptible in Fig. 3.2, the member would not be altered in any normally detectable way despite the fact that the *calculated elastic* (but totally fictitious) stress at the notch would be 90 ksi, a value well above the ultimate strength.

3. It is difficult to measure accurately the large strains that occur after "necking" of a tensile specimen. This is because the *local* elongation is tremendously greater in the necking zone than elsewhere, and the calculated value of elongation depends on the gage length used. A more accurate determination of elongation at fracture in the *immediate region of failure* can be obtained indirectly by measuring the cross-sectional area at the fracture location. Elongation is then computed by assuming that there has been a negligible change in material volume. For example, let both the original cross-sectional area and extremely short gage length be unity. Suppose the area after fracture is 0.4. This gives a *ratio, R, of initial area to final area of 1/0.4 = 2.5.* Likewise, the *reduction in area, A_r, would be 60 percent of the original area, or 0.6.* If volume remains constant, the gage length must have increased to 2.5, thereby giving a strain (ϵ) due to elongation of 1.5. The three abscissa scales in Fig. 3.2 all represent quantities commonly used in the literature. Relationships between them will now be derived: For constant volume, and using subscripts "0" and "*f*" to denote initial and final values, respectively,

$$A_0L_0 = A_f(L_0 + \Delta L) = A_fL_0(1 + \epsilon), \text{ or}$$

$$A_f = \frac{A_0}{1 + \epsilon} \tag{3.1}$$

$$\text{Area ratio} = R = \frac{A_0}{A_f} = 1 + \epsilon \tag{3.2}$$

$$\Delta A = A_0 - A_f = A_0\left(1 - \frac{1}{1 + \epsilon}\right)$$

$$\text{Area reduction} = A_r = \frac{\Delta A}{A_0} = 1 - \frac{1}{1 + \epsilon} = 1 - \frac{1}{R} \tag{3.3}$$

For practice, it is suggested that the reader verify the agreement of the three abscissa scales in Fig. 3.2 at one or two points.

4. When experimentally determining the σ-ϵ curve for most engineering materials, the load can be removed at any point and then restored without significantly altering the subsequent test points. Thus, if the load is removed at point D of Fig. 3.2, the stress reduces to zero along line DG, which has a slope of $E = 30 \times 10^6$ psi. Reapplication of the load brings the material back essentially to point D, and further load increases produce the same result as if the load removal had not taken place. Suppose that we regard the test specimen at G as a *new* specimen and determine *its* yield strength, ultimate strength, and reduction in area when fractured. The "new" specimen will have a yield strength higher than the original—in fact, its yield strength will be higher than the 62 ksi shown at point D because the area of the new specimen is less than that of the original. At point G the specimen has been permanently stretched to 11/10 of its initial length; hence, its area is only 10/11 of the original. Based on the new area, the yield strength of the "new" specimen is 62 ksi divided by 10/11, or $S_y = 68$ ksi. Similarly, the ultimate strength of the "new" specimen is 66 divided by 10/11, or $S_u = 73$ ksi. The reduction in area at fracture for the "new" specimen is from 10/11 (area at G) to 10/25 (area at F), or 56 percent. This compares with $A_r = 60$ percent based on the original area.

In like manner, consider unloading the specimen at point C (bringing it to point H) and then treating this severely cold-worked material as a "new" specimen. At point H, the specimen length has been increased by a factor of 1.3 and its area reduced by the same factor. Based on this area, the "new" specimen will have values of S_y *and* S_u of 66(1.3), or 86 ksi. The reduction in area at fracture is from an initial area (point H) of 1/1.3 to a final area (point F) of 1/2.5, or 48 percent. Note the fundamental concept that severe cold work, as in loading to point C or beyond, exhausts the material's ductility, causing S_y and S_u to be equal.

An important implication of the above is that the strength and ductility characteristics of metals change substantially during fabrication processes involving cold working.

3.4 THE STATIC TENSILE TEST – "TRUE" STRESS-STRAIN RELATIONSHIPS

A study of Fig. 3.2 revealed that whenever a material is elongated to many times its maximum elastic strain (perhaps 20 or 30 times), the calculated "engineering" stress becomes somewhat fictitious because it is based on an area significantly different than that which actually exists. In such cases this limitation can be avoided by computing the *"true"* stress (in this text designated as σ_T), defined as load divided by the cross-sectional area that exists when the load is acting. Thus: $\sigma = P/A_0$, and $\sigma_T = P/A_f$.

Substituting the equivalent of A_f from Eq. (3.1):

$$\sigma_T = (P/A_0)(1 + \epsilon) = \sigma(1 + \epsilon) = \sigma R \tag{3.4}$$

If one were to replot Fig. 3.2 using true stress, the values at points B, C, and F would be 39(1) = 39, 66(1.3) = 86, and 45(2.5) = 113 ksi, respectively. Note that true stress increases continuously to the point of fracture.

In like manner, "engineering" strain is not a realistic measure where large strains are involved. In such cases it is appropriate to use "true" strain values, ϵ_T. Consider, for example, a very ductile specimen of unit length that is stretched to a length of 5 units, and then stretched further to 5.1 units. The "engineering" strain added by the final 0.1 unit of stretch is 0.1/1 or 0.1. The corresponding "true" strain, however, is only 0.1/5 or 0.02 (change in length divided by the length existing *immediately prior to* the last small strain increment). Mathematically, true strain is defined as

$$\epsilon_T = \sum_{L_0}^{L_f} \frac{\Delta L}{L} = \int_{L_0}^{L_f} \frac{dL}{L} = \ln R = \ln(1 + \epsilon) \tag{3.5}$$

where L_0 and L_f represent the initial and final lengths, respectively, and ln denotes the natural logarithm. For metals, engineering and true strains are essentially the same when they are less than several times the maximum elastic strain.

Figure 3.3 is a true stress-true strain plot of the data represented in Fig. 3.2. Such

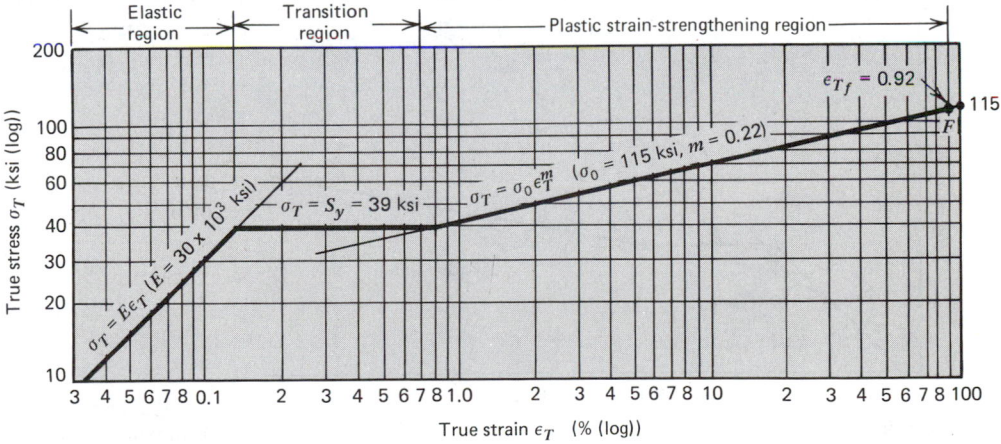

FIGURE 3.3.
True stress-true strain curve—hot rolled 1020 steel (corresponds to Figures 3.1 and 3.2).

plots illustrate general relationships that are helpful in predicting the effect of cold working on the strength properties of many metals. A study of the three regions identified in Fig. 3.3 reveals several important relationships and concepts:

1. *Elastic region*—Strictly speaking, Young's modulus is the ratio of *engineering* stress and strain, but with negligible error it is also the ratio of *true* stress and strain; hence:

$$\sigma_T = E\epsilon_T \tag{3.6}$$

   On the log-log coordinates of Fig. 3.3 this equation plots as a straight line of unit slope, positioned so that the line (extended) passes through the point ($\epsilon_T = 1$, $\sigma_T = E$). Note that E can be thought of as the value of stress required to produce an elastic strain of unity.

2. *Plastic strain-strengthening region*—This region corresponds to the strain-strengthening equation:

$$\sigma_T = \sigma_0 \epsilon_T^m \tag{3.7}$$

   Note that this equation has the same form as Eq. (3.6) except for the exponent m, which is the slope of the line when plotted on log-log coordinates. The strain-strengthening proportionality constant, σ_0,[2] is analogous to the elastic proportionality constant, E, in that σ_0 can be regarded as the value of true stress associated with a true strain of unity.

3. *Transition region*—For an "ideal" material, the value of the elastic limit (approximated as the yield point) corresponds to the intersection of the elastic and plastic lines, as shown in Fig. 3.4. Actual materials may have values of S_e, which are either higher or lower, requiring the addition of an empirical transition curve, as I or II of Fig. 3.4.

Unfortunately, numerical values of the strain hardening characteristics of many engineering materials are not yet available. Much of the available information has been

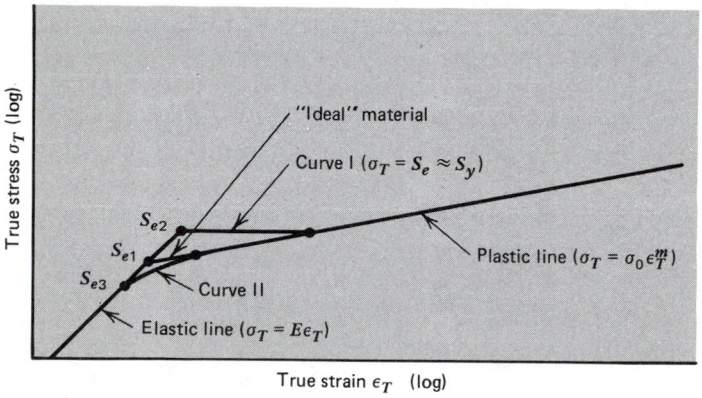

FIGURE 3.4.
True stress-true strain curves showing transition region variations.

[2]Since σ_0 is a material property, the symbol S_0 might be more appropriate; however, σ_0 is used because of its general acceptance in the materials engineering literature.

obtained by Datsko [2]. Values for several materials are given in Appendix C-2. These values are typical, but vary somewhat from heat to heat due to small differences in chemical composition and processing history. This is particularly true of the true strain at fracture (ϵ_{Tf}).

3.5 ENERGY ABSORBING CAPACITY

Some parts must be designed more on the basis of absorbing energy than of withstanding loads. Since energy involves both loads and deflections, stress-strain curves are particularly relevant. Figure 3.5 will be used for illustration. It is essentially the same as Fig. 3.1 except that the strain scale has been further expanded near the origin in the interest of clarity.

The capacity of a material to absorb energy within the elastic range is called *resilience*. Its standard measure is *modulus of resilience*, R_m, defined as the energy absorbed by a unit cube of material when loaded in tension to its elastic limit.[3] This is equal to the triangular area under the elastic portion of the curve (Fig. 3.5), thus:

$$R_m = \tfrac{1}{2}(S_e)(S_e/E) = S_e^2/2E \quad \text{(see footnote 3)} \tag{3.8}$$

The total capacity of a material to absorb energy without fracture is called *toughness*. The *modulus of toughness*, T_m, is the energy absorbed per unit volume of material when loaded in tension to fracture. This is equal to the total shaded area under the curve in Fig. 3.5:

$$T_m = \int_0^{\epsilon_f} \sigma \, d\epsilon \tag{3.9}$$

It is often convenient to perform this integration graphically. A rough approximation sometimes used is:

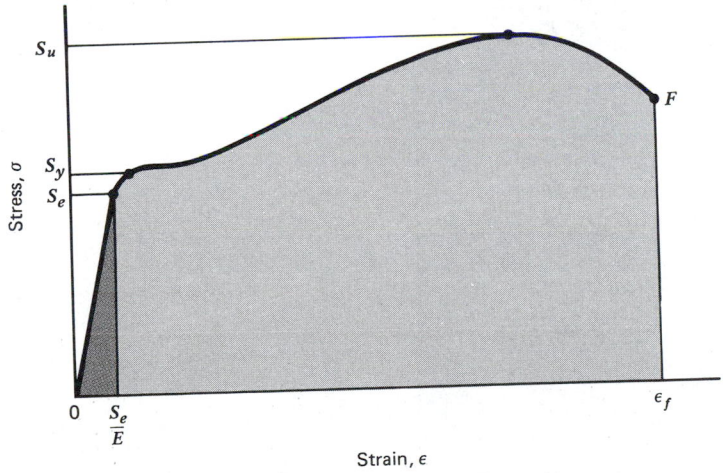

FIGURE 3.5.
Resilience and toughness as represented by the stress-strain curve.

[3]In practice, S_y is usually substituted for S_e.

$$T_m \approx \frac{S_y + S_u}{2} \epsilon_f \tag{3.10}$$

It should be noted that members designed for energy absorption are commonly subjected to impact loading, and that special tests (traditionally, the Charpy and Izod) are used to estimate more accurately the impact energy absorbing capacity of various materials at various temperatures.

3.6 ESTIMATING STRENGTH PROPERTIES FROM PENETRATION HARDNESS TESTS

Penetration hardness tests (usually Brinell or Rockwell) provide a convenient and nondestructive means of estimating the strength properties of metals. Basically, penetration hardness testers measure the resistance to permanent deformation of a material when subjected to a particular combination of triaxial compressive stress and steep stress gradient.

Results of the Brinell hardness test have been found to correlate well with ultimate tensile strength, the relationship being

$$S_u = K_B H_B \tag{3.11}$$

where H_B is the Brinell hardness number, K_B is a constant of proportionality, and S_u is the tensile strength in psi. For most steels, $K_B \approx 500$. Datsko [1,2] has shown a rational basis for K_B being a function of the strain hardening exponent, m. Figure 3.6 gives empirical curves representing this relationship.

Figure 3.7 gives approximate relationships between Brinell, Rockwell, and other hardness numbers.

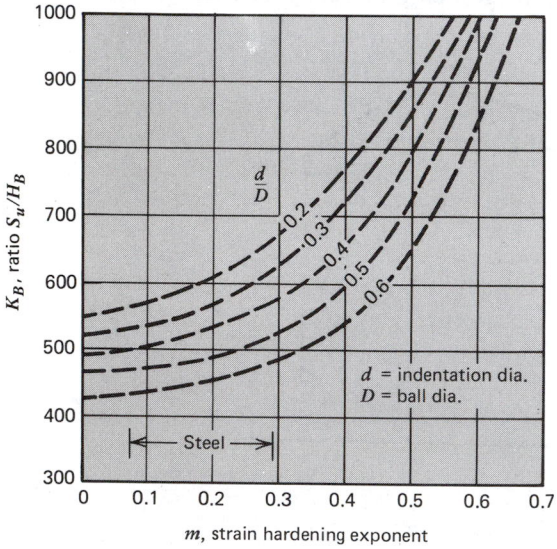

FIGURE 3.6.
Approximate relationships between K_B and m [2].

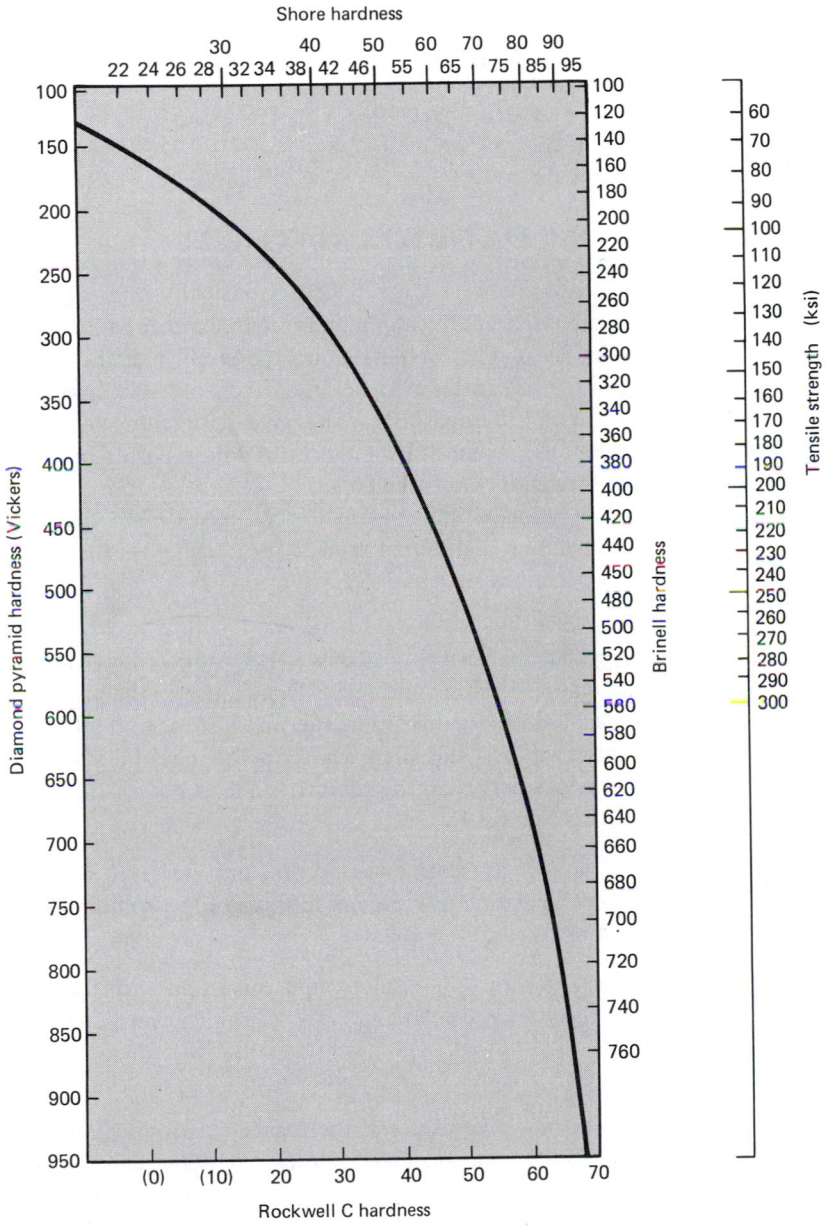

FIGURE 3.7.
Approximate relationships between hardness scales and tensile strength of steel. (Courtesy International Nickel Co., Inc.)

After analyzing extensive data, Datsko [2] concluded that reasonably good estimates of the tensile yield strength of stress-relieved (*not* cold-worked) steels can be made from the equation

$$S_y = 1.05S_u - 30{,}000 \text{ psi} \tag{3.12}$$

or, by substituting Eq. (3.11) with $K_B = 500$,

$$S_y = 525H_B - 30{,}000 \text{ psi} \tag{3.13}$$

3.7 USE OF "HANDBOOK" DATA FOR MATERIAL STRENGTH PROPERTIES

Ideally, an engineer would always base strength calculations on actual test data for the exact material used in the part involved. This would require the use of test specimens that correspond to the material in the fabricated part not only in chemical composition but also in all details of mechanical and thermal history. Because data from such specimens are seldom available, frequent use is made of standard test data reported in handbooks and other sources, such as Appendix C of this book.

There are pitfalls in using "handbook" data, as evidenced by the fact that one frequently finds contradictory information in different references. In using this data, the engineer must be concerned with questions such as:

1. Do the published values represent the results of a single test, or are they average, median, typical, or minimum values from several tests? Depending on the precision with which the variables associated with composition, thermal history, and mechanical history are controlled, there will be a statistical scatter in the material strength. In many situations it is well to consider strength properties in terms of mean values and standard deviations (see Section 6.14).
2. Do the composition, size, previous heat-treatment, and previous mechanical working of the specimens tested correspond *closely enough* to those of the actual part in its final as-fabricated condition?
3. Are the published data consistent within themselves, and consistent with the general pattern of accepted test results for similar materials? In other words, are the data reasonable?

Many tables of materials properties give values of the tensile elastic modulus and Poisson's ratio (ν). From elastic theory, the torsional or shear modulus can then be calculated as:

$$G = \frac{E}{2(1 + \nu)} \tag{3.14}$$

3.8 MACHINABILITY

The cost of producing a machined part is obviously influenced both by the cost of the material and by the ease with which it can be machined. Empirically determined ratings of machinability (defined as relative cutting speed for a fixed tool life under prescribed standard cutting conditions) are published for various materials. Although often useful, these data are sometimes unreliable and even contradictory. In an effort to relate machinability to material parameters on a rational basis, Datsko [1] showed that machinability is a secondary material property that is a function of three primary material physical properties:

$$V_{60} = \frac{1150k}{H_B}(1 - A_r)^{1/2} \tag{3.15}$$

where V_{60} = cutting speed in ft/min for 60-min tool life under standard cutting conditions, k = thermal conductivity, H_B = Brinell hardness number, and A_r = area reduction at fracture. Since the value of k is about the same for all metals, the machinability of metal is essentially a function of *hardness* and *ductility*.

3.9 CAST IRON

Cast iron is a four-element alloy containing iron, carbon (between 2 and 4 percent), silicon, and manganese. Additional alloying elements are sometimes added. The physical properties of an iron casting are strongly influenced by its cooling rate during solidification. This, in turn, depends on the size and shape of the casting and on details of foundry practice. Because of this (and unlike other engineering materials), cast iron is usually specified by its *mechanical properties* rather than by chemical analysis.

The distinctive properties of cast iron result largely from its carbon content: (a) The high carbon content makes molten iron very fluid, so that it can be poured into intricate shapes. (b) The precipitation of carbon during solidification counteracts normal shrinkage to give sound sections. (c) The presence of graphite in the metal provides excellent machinability (even at wear-resisting hardness levels), damps vibration, and aids boundary lubrication at wearing surfaces. When "chilled" (heat removed rapidly from the surface during solidification), virtually all of the carbon near the surface remains combined as iron carbides, giving an extremely hard, wear-resistant surface.

Mechanical properties of several cast irons are given in Appendix C-3.

Gray Iron

The appearance of gray iron comes from the carbon that is precipitated in the form of graphite flakes. Even the softer grades have good wear resistance. Higher hardnesses (giving even better wear resistance) are obtainable by using special foundry techniques, heat-treatment, or additional alloying elements. Because the graphite flakes markedly weaken cast iron in tension, the compressive strength is three to five times as high.

Advantage is often taken of this strength differential, as by incorporating ribs on the compression side of a member loaded in bending.

Typical applications of gray iron include gasoline and diesel engine blocks, machine bases and frames, gears, flywheels, and brake disks and drums.

Ductile (Nodular) Iron

Ductile iron is alloyed with magnesium, which causes the excess carbon to precipitate in the form of small spheres or nodules. These nodules disrupt the structure less than do the graphite flakes of gray iron, thereby giving substantial ductility along with improved tensile strength, stiffness, and impact resistance. Ductile iron is specified by three numbers, as 60-40-18, which denote tensile strength (60 ksi), yield strength (40 ksi), and elongation (18 percent). Typical applications include engine crankshafts, heavy-duty gears, and hardware items such as automobile door hinges.

White Iron

White iron (so-called because of the white appearance of fracture surfaces) is produced in outer portions of gray and ductile iron castings by *chilling* selected surfaces of the mold, thereby denying time for carbon precipitation. The resulting structure is extremely hard, wear-resistant, and brittle. Typical applications are found in ball mills, extrusion dies, cement mixer liners, railroad brake shoes, rolling mill rolls, crushers, and pulverizers.

Malleable Iron

Malleable iron is made from white iron by a two-stage heat-treating process. The resulting structure contains excess graphite in the form of tempered nodules. The necessity of starting with a solid white iron casting limits section thicknesses to about 3 in. Malleable iron castings usually cost slightly less than ductile iron castings. Typical uses include heavy-duty parts having bearing surfaces and used in trucks, railroad equipment, construction machinery, and farm equipment.

3.10 STEEL

Steel is the most extensively used material for machine components. By suitably varying the composition, thermal treatment, and mechanical treatment a tremendous range of mechanical properties can be obtained. Three basic relationships are fundamental to the appropriate selection of steel composition:

1. All steels have essentially the same moduli of elasticity. Thus, if *rigidity* is the critical requirement of the part, *all steels perform equally* and the least costly (including fabricating costs) should be selected.

2. Carbon content, almost alone, determines the maximum hardness that can be developed in steel. Maximum potential hardness increases with carbon content up to about 0.7 percent. This means that relatively small, regularly shaped parts can be heat-treated to give essentially the same hardness and strength with plain carbon steel as with more costly alloy steels.
3. Alloying elements (manganese, molybdenum, chromium, nickel, and others) add *hardenability* to steel. Thus, the potential hardness and strength (which is controlled by carbon content) can be realized with less drastic heat-treatments when these alloying elements are used. This means that with an alloy steel (a) parts with large sections can achieve higher hardnesses in the center or core of the section, and (b) irregularly shaped parts, subject to warpage during a drastic quench, can achieve the desired hardness from a more moderate heat-treatment.

Mechanical properties of a few of the more commonly used steels are given in Appendixes C-4–C-8.

Plain Carbon Steels

These steels contain only carbon as a significant alloying element. Low carbon steels have less than 0.3 percent C, medium carbon steels have 0.3 to 0.5 percent and high carbon steels above 0.5 percent. Appendix C-4b gives typical uses for steels having various levels of carbon content.

Alloy Steels

As indicated above, the basic purpose of adding alloying elements to steel is to increase hardenability. Hardenability is commonly measured by the Jominy end-quench test (ASTM A-255 and SAE J406b), originated by the late Walter Jominy of the Chrysler Corporation. This test involves suspending a 1-in. diameter by 4 in. long bar, heating it above its critical temperature, and then quenching the bottom end with water while the top end air cools. Hardness is then measured at 1/16-in. increments from the quenched end. The distance to which the hardening action extends is a measure of the hardenability imparted by the alloying elements.

Table 3.1 represents the results of a study by Datsko [2] of the relative effectiveness of various alloying elements in imparting hardenability to steel. The elements are listed in order of decreasing effectiveness. The equations give a relative hardenability factor f as a function of the concentration of the element used. For example, manganese is the second most effective element, and the equation is valid for concentrations of up to 1.2 percent. If 1 percent Mn is used, the hardenability factor is 4.46. Similarly, a 2 percent concentration of chromium is only a little more effective, giving a factor of 5.36. Table 3.1 is not complete (a notable omission being vanadium), and alloying elements

TABLE 3.1
Relative Effectiveness of Steel Alloying Elements [2]

Element	Concentration (%)	Relative Hardenability Effectiveness
Boron	$B < 0.002$	$f_B = 17.23B^{0.0268}$
Manganese	$Mn < 1.2$	$f_{Mn} = 3.46Mn + 1$
Manganese	$1.2 < Mn < 2.0$	$f_{Mn} = 5.125Mn - 1$
Molybdenum	$Mo < 1.0$	$f_{Mo} = 3.09Mo + 1$
Chromium	$Cr < 2.0$	$f_{Cr} = 2.18Cr + 1$
Silicon	$Si < 2.0$	$f_{Si} = 0.7Si + 1$
Nickel	$Ni < 2.0$	$f_{Ni} = 0.4Ni + 1$

may also have significant secondary influences. Nevertheless, the equations of Table 3.1 provide a useful guide to the selection of the most economical means for obtaining a required level of hardenability.

Most alloy steels can be classed as either "through-hardening" or "carburizing," the latter being used where a tough core and relatively shallow, hard surface layer are desired. Nitriding and other surface hardening processes are also used.

HSLA Steels

High-strength low-alloy (HSLA) steels were first developed around 1940 as a class of relatively low-cost steels providing much of the advantage of more costly regular alloy steels. In many applications, their greater strength with respect to plain carbon steel permits a weight reduction with little if any increase in the total cost of the part. In recent years, substantial use of HSLA steels has been made in automobiles.

Case Hardening Steels

Case hardening involves a hardening of only the surface material (called the case). It is usually accomplished by carburizing, cyaniding, nitriding, induction hardening, or flame hardening. *Carburizing* involves introducing additional carbon into the surface of an otherwise low carbon steel and then heat-treating to give a high surface hardness. Some specific materials and hardnesses involved are listed in Appendix C-7.

Cyaniding is a similar process that adds nitrogen as well as carbon to the surfaces of low and medium carbon steels.

Nitriding adds nitrogen to an already machined and heat-treated part. The temperature involved in the process is 1000°F (538°C) or less, and no quenching is involved. This feature eliminates possible distortion problems. For maximum case hardness, special "nitralloy" steels (containing aluminum as an alloy) are often used. Medium carbon alloy steels (notably 4340) are also nitrided.

Induction and flame hardening involve heating only the surfaces of parts made of medium carbon and alloy steels, then quenching and tempering.

Stainless Steels

Stainless steels contain, by definition, a minimum of 10.5 percent chromium. Wrought stainless steels are austenitic, ferritic, martensitic, or precipitation-hardening. Mechanical properties of some wrought stainless steels are given in Appendix C-8. Cast stainless steels are usually classed as heat-resistant or corrosion-resistant.

Iron-Based Superalloys

These materials are used primarily for elevated-temperature applications, as in turbines. Properties of typical iron-based superalloys are listed in Appendix C-9. Some authorities consider only the austenitic materials to be true superalloys. In general, they are used at temperatures above 1000°F (538°C), and the martensitic materials are used at lower temperatures. Significant properties of superalloys include high-temperature strength, and high-temperature resistance to creep, oxidation, corrosion, and wear. Typical uses are for parts (including bolts) of gas turbines, jet engines, heat exchangers, and furnaces.

3.11 NON-FERROUS ALLOYS

Aluminum Alloys

Literally hundreds of aluminum alloys are available, in both wrought and cast forms. Properties of a few of the more common ones are listed in Appendixes C-10 and C-11. The chemical composition of aluminum alloys is designated by four digits (wrought forms) or by three digits (cast alloys). Thermal and/or mechanical treatment is indicated by a temper designation that follows the alloy identification number. Temper designations are given in Appendix C-12.

The heat-treatment of aluminum alloys to increase hardness and strength is quite different from the heat-treatment of steel. Aluminum alloys are first held at an elevated temperature long enough to bring the hardening constituents (as Cu, Mg, Mn, Si, Ni) into solution, then quenched, and finally age-hardened. The latter causes some of the hardening elements to precipitate throughout the structure. Some alloys precipitate at room temperature; others require an elevated temperature (artificial aging).

Although aluminum is a readily castable metal serving a host of useful applications, casting problems do exist. Shrinkage during casting is relatively large (3.5 to 8.5 percent by volume), and there is no mechanism analagous to the beneficial carbon precipitation in cast iron to counteract shrinkage. Hot shortness and/or gas absorption can be problems unless appropriate foundry practice details are specified and controlled.

Copper Alloys

Copper alloys include a variety of *brasses* (alloys principally of copper and zinc) and *bronzes* (alloys principally of copper and tin). As a class, copper alloys have good electrical conductivity, thermal conductivity, and resistance to corrosion, but relatively low ratios of strength to weight. They can be hot or cold worked, but strain harden in the process. Ductility can be restored by annealing or by heat associated with welding or brazing. Specific desired properties (as greater strength, resistance to heat-softening, or machinability) can often be markedly improved by adding small amounts of additional alloying agents.

Properties of several common copper alloys are given in Appendix C-13.

Magnesium Alloys

Magnesium alloys are the lightest engineering metals. They are designated by a system established by the American Society for Testing and Materials (ASTM), which covers both chemical composition and tempers. The designation begins with two letters representing alloying elements of the greatest and second greatest concentration. The letter designations are:

A—Aluminum

E—Rare earths

H—Thorium

K—Zirconium

L—Lithium

M—Manganese

Q—Silver

S—Silicon

Z—Zinc

Next are two digits that represent the respective percentages of these two elements, rounded off to whole numbers. Following these digits is a serial letter that indicates some variation in composition or minor alloying constituents or impurities. The temper designations at the end are identical with those used with aluminum (Appendix C-12). For example, alloy AZ31B-H24 contains 3 percent aluminum, 1 percent zinc, and is strain hardened.

Mechanical properties of a few common magnesium alloys are given in Appendix C-14.

Nickel Alloys (Including Nickel-Based Superalloys)

Nickel alloys are used in a variety of structural applications that usually require specific corrosion resistance and/or strength and toughness at temperature extremes as great as 2000°F (1093°C) and −400°F (−240°C).

Typical physical properties are given in Appendix C-15. The nickel and Duranickel alloys contain over 94 percent nickel. Monel represents a series of nickel-copper alloys, based on the mutual solubility of these two elements in all proportions. They are strong and tough at sub-zero temperatures, and especially resistant to stress corrosion cracking (Section 9.5). Hastelloy designates a series of Ni-Mo and Ni-Mo-Cr superalloys. Several Hastelloys resist oxidation and maintain useful strength and creep properties in the range of 2000°F (1093°C). The Inconel, Incoloy, Rene, and Udimet alloys listed in Appendix C-15 are Ni-Cr and Ni-Cr-Fe alloys.

Titanium Alloys

Titanium alloys are nonmagnetic, extremely corrosion-resistant, have low thermal conductivity, and have outstanding strength/weight ratios. On the negative side, they are very expensive and difficult to machine. Mechanical properties of some of the more common alloys are given in Appendix C-16.

Zinc Alloys

Zinc is a relatively inexpensive metal with moderate strength. It has a low melting temperature and so is readily and economically die-cast. Typical zinc die castings include automotive parts, building hardware, office machine components, and toys. Limited use is made of the metal in other forms. Mechanical properties of common zinc die-cast alloys are listed in Appendix C-17. Also included is a relatively new alloy (ZA-12) that can be cast using various methods.

3.12 PLASTICS

The information contained in this section is but a brief overview of an extensive and complex field. In order to deal effectively with the selection and use of plastics in the design of machine components, the reader must study further and/or seek the advice of competent specialists.

Plastics constitute a large and varied group of synthetic organic materials. Although solid in the finished state, plastics are liquid in some stage of manufacture, permitting them to be formed into various shapes. The forming usually involves application of heat and/or pressure. The term "plastic" is really a misnomer, as the material is truly plastic only during a stage of its manufacture.

The basic chemical units of plastic materials are *monomers*. Under appropriate conditions, usually involving heat and/or pressure, *polymerization* takes place, combining monomers into *polymers*. Typical monomers and their corresponding repeating polymer units are shown in Fig. 3.8.

The addition of more and more monomers to form longer and longer polymer chains increases molecular weight and vastly alters physical properties. For example,

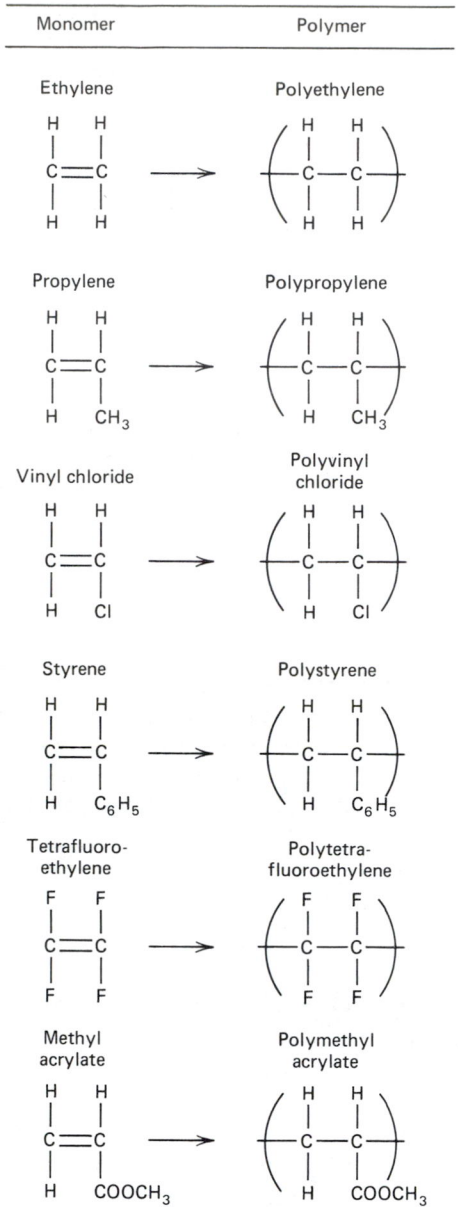

FIGURE 3.8.
Typical monomers and their repeating polymer units.

Fig. 3.9 shows CH_4, which is methane gas. Adding one CH_2 unit gives heavier ethane gas (C_2H_6). Continued addition of CH_2 units gives pentane, a liquid (C_5H_{12}), and paraffin wax ($C_{18}H_{38}$). At approximately $C_{100}H_{202}$, the material is tough enough to be a useful plastic, known as *low-molecular-weight polyethylene.* The toughest polyethylene, called *high-molecular-weight polyethylene,* contains nearly a half-million CH_2 units in a single polymer chain.

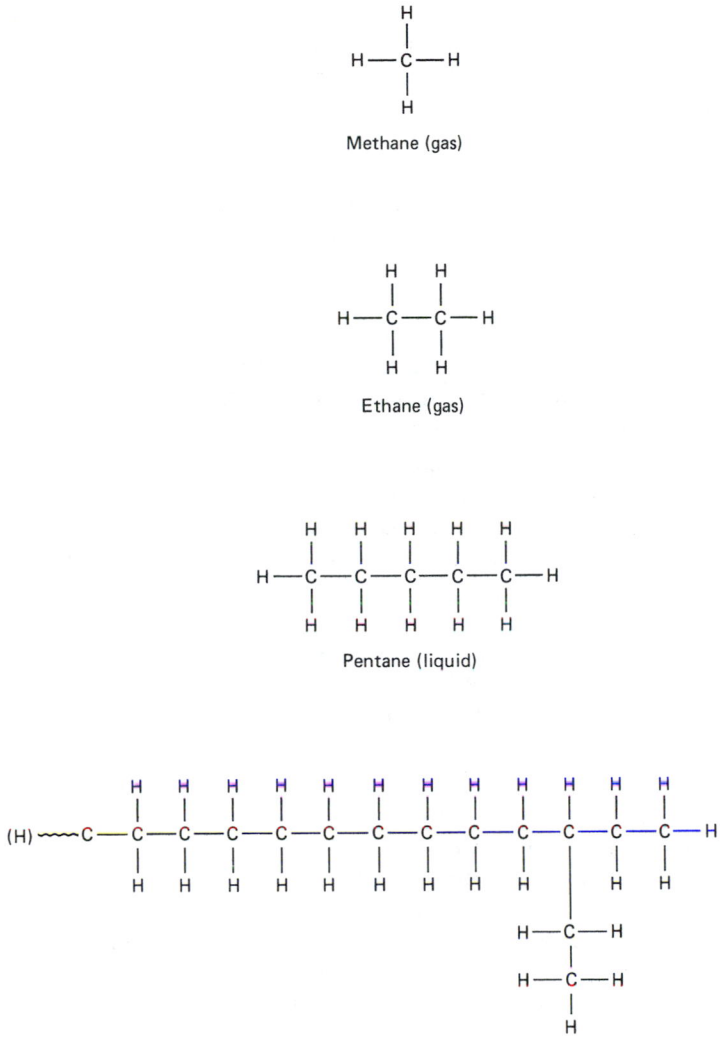

FIGURE 3.9.
Molecular chains.

Polymer chain structures can incorporate side branching, also shown in Fig. 3.9. The degree of branching influences the closeness with which the chains fit together. This, in turn, influences physical properties. Minimal branching promotes tight packing of the polymer chains (hence, high intermolecular attractive forces), giving relatively high density, rigid crystalline structures, and also relatively high mold shrinkage. Extensive branching produces a more flexible, amorphous material with less mold shrinkage and distortion. Physical properties of the finished plastic can also be altered by *copolymerization* (involving two monomers in the building of polymer chains) and by *alloy-*

ing (a strictly mechanical mixing or blending of constituents, not involving chemical bonds).

Plastics have been traditionally designated as *thermoplastic* (softens with heat) and *thermosetting* (does not soften with heat). A preferred designation is *linear* and *cross-linked*. The polymer chains in linear plastics remain linear and separate after molding. The chains in cross-linked plastics are initially linear, but become joined *irreversibly* during molding into an interconnected molecular network.

Cross-linking can be initiated by heat, chemical agents, irradiation, or a combination of these. Some plastics can be either cross-linked or linear. The cross-linked form is more resistant to heat, chemical attack, and creep (better dimensional stability). On the other hand, the linear form is less brittle (more impact resistant), more easily processed, and better adapted to complex shapes.

Glass fiber reinforcement improves most mechanical properties of plastics by a factor of two or more. At substantially increased cost, a further marked improvement is obtainable by carbon-fiber reinforcement. These relatively new materials (with 10 to 40 percent carbon) have tensile strengths as high as 40 ksi. Compared to glass-reinforced resins, they have less mold shrinkage, lower coefficients of expansion, and improved creep resistance, wear resistance, and toughness. The new fiber-reinforced plastics are being increasingly used for machine and structural components requiring light weight and high strength-to-weight ratios.

Properties of a few common plastics are given in Appendix C-18a. A comparison of properties of thermoplastics with and without glass reinforcement is given in Appendix C-18b. Thermosetting plastics benefit similarly from glass reinforcement, the most commercially important being polyester and epoxy resins. In using tables giving properties of plastics, the reader should recall Section 3.7, which gives pitfalls in the use of such "handbook" material properties data. This is particularly true in the case of plastics. Published values reflect values obtained from *standardized* molding conditions that are simple, economical, and readily reproduced. Strength values corresponding to *actual* molding conditions may differ significantly. Furthermore, temperature and rate of loading influence the strength of plastics to a greater extent than metals.

It should be evident that the proper selection of a plastic for a particular design application often requires consultation with a plastics authority.

Appendix C-18c gives a listing of typical applications of the more common plastics. Comments relating to each of these are given below.

Common Plastics [3]

Thermoplastics

ABS (acrylonitrile-butadiene-styrene): Very tough, yet hard and rigid; fair chemical resistance; low water absorption, hence good dimensional stability; high abrasion resistance; easily electroplated.

Acetal: Very strong, stiff engineering plastic with exceptional dimensional stability and resistance to creep and vibration fatigue; low coefficient of friction; high resistance to abrasion and chemicals; retains most properties when immersed in hot water; low tendency to stress-crack.

Acrylic: High optical clarity; excellent resistance to outdoor weathering; hard, glossy surface; excellent electrical properties, fair chemical resistance; available in brilliant, transparent colors.

Cellulosics: Family of tough, hard materials; cellulose acetate, propionate, butyrate, and ethyl cellulose. Property ranges are broad because of compounding; available with various degrees of weather, moisture, and chemical resistance; fair to poor dimensional stability; brilliant colors.

Fluoroplastics: Large family (PTFE, FEP, PFA, CTFE, ECTFE, ETFE, and PVDF) of materials characterized by excellent electrical and chemical resistance, low friction, and outstanding stability at high temperatures; strength is low to moderate; cost is high.

Nylon (polyamide): Family of engineering resins having outstanding toughness and wear resistance; low coefficient of friction, and excellent electrical properties and chemical resistance. Resins are hygroscopic; dimensional stability is poorer than that of most other engineering plastics.

Phenylene oxide: Excellent dimensional stability (very low moisture absorption); superior mechanical and electrical properties over a wide temperature range. Resists most chemicals but is attacked by some hydrocarbons.

Polycarbonate: Highest impact resistance of any rigid, transparent plastic; excellent outdoor stability and resistance to creep under load; fair chemical resistance; some aromatic solvents cause stress cracking.

Polyester: Excellent dimensional stability, electrical properties, toughness, and chemical resistance, except to strong acids or bases; notch-sensitive; not suitable for outdoor use or for service in hot water; also available in thermosetting formulations.

Polyethylene: Wide variety of grades: low, medium, and high density formulations. LD types are flexible and tough. MD and HD types are stronger, harder, and more rigid; all are lightweight, easy to process, low-cost materials; poor dimensional stability and heat resistance; excellent chemical resistance and electrical properties. Also available in ultrahigh-molecular weight grades.

Polyimide: Outstanding resistance to heat (500°F continuous, 900°F intermittent) and to heat aging. High impact strength and wear resistance; low coefficient of thermal expansion; excellent electrical properties; difficult to process by conventional methods; high cost.

Polyphenylene sulfide: Outstanding chemical and heat resistance (450°F continuous); excellent low-temperature strength; inert to most chemicals over a wide temperature range; inherently flame-retardant; requires high processing temperature.

Polypropylene: Outstanding resistance to flex and stress cracking; excellent chemical resistance and electrical properties; good impact strength above 15°F; good thermal stability; light weight, low cost, can be electroplated.

Polystyrene: Low cost, easy to process, rigid, crystal-clear, brittle material; low moisture absorption, low heat resistance, poor outdoor stability; often modified to improve heat or impact resistance.

Polysulfone: Highest heat-deflection temperature of melt-processable thermoplastics; requires high processing temperature; tough (but notch-sensitive), strong, and stiff; excellent electrical properties and dimensional stability, even at high temperature; can be electroplated; high cost.

Polyurethane: Tough, extremely abrasion and impact resistant material; good electrical properties and chemical resistance; can be made into films, solid moldings, or flexible foams; UV exposure produces brittleness, lower properties, and yellowing; also made in thermoset formulations.

Polyvinyl chloride: Many formulations available; rigid grades are hard, tough, and have excellent electrical properties, outdoor stability, and resistance to moisture and chemicals; flexible grades are easier to process but have lower properties; heat resistance is low to moderate for most types of PVC; low cost.

Thermosets

Alkyd: Excellent electrical properties and heat resistance; easier and faster to mold than most thermosets; no volatile byproducts.

Allyl (diallyl phthalate): Outstanding dimensional stability and electrical properties; easy to mold; excellent resistance to moisture and chemicals at high temperatures.

Amino (urea, melamine): Abrasion and chip resistant; good solvent resistance; urea molds faster and costs less than melamine; melamine has harder surface and higher heat and chemical resistance.

Epoxy: Exceptional mechanical strength, electrical properties, and adhesion to most materials; low mold shrinkage; some formulations can be cured without heat or pressure.

Phenolic: Low-cost material with good balance of mechanical, electrical, and thermal properties; limited in color to black and brown.

Polyester: Excellent balance of properties; unlimited colors, transparent or opaque; gives off no volatiles during curing, but mold shrinkage is high; can use low-cost molds without heat or pressure; widely used with glass reinforcement to produce "fiberglass" components; also available in thermoplastic formulations.

Polyurethane: Can be flexible or rigid, depending on formulation; outstanding toughness and resistance to abrasion and impact; particularly suitable for large foamed parts, in either rigid or flexible types; also produced in thermoplastic formulations.

Silicone: Outstanding heat resistance (from −100 to +500°F), electrical properties, and compatibility with body tissue; cures by a variety of mechanisms; high cost; available in many forms; laminating resins, molding resins, coatings, casting or potting resins, and sealants.

REFERENCES

1. Datsko, J., *Material Properties and Manufacturing Processes,* John Wiley & Sons, Inc., New York, 1966.
2. Datsko, J., *Materials in Design and Manufacturing,* Malloy, Inc., Ann Arbor, MI, 1977.
3. *Machine Design 1981 Materials Reference Issue,* Penton/IPC, Cleveland, OH, Vol. 53, No. 6, March 19, 1981.
4. *ASM Metals Reference Book,* American Society for Metals, Metals Park, OH, 1981.

CHAPTER FOUR

Static Body Stresses

4.1 INTRODUCTION

Once the external *loads* applied to a member have been determined (see Chapter 2), the next item of interest is often the resulting *stresses*. This chapter is concerned with *body* stresses, existing within the member as a whole, as distinguished from *surface* or *contact* stresses in localized regions where external loads are applied. This chapter is also concerned with stresses resulting from essentially *static* loading, as opposed to stresses caused by impact or fatigue loading. (Impact, fatigue, and surface stresses are considered in Chapters 7, 8, and 9, respectively.)

As noted in Section 3.2, this book follows the convention of reserving the capital letter S for *material strength* (i.e., S_u for ultimate strength, S_y for yield strength, etc.) and using Greek letters σ and τ for normal and shear stress, respectively.

4.2 AXIAL LOADING

Figure 4.1 illustrates a case of simple *tension*. If external loads P are reversed in direction (i.e., have negative values), then the bar is loaded in simple *compression*. In either case, the loading is *axial*. Small block E represents an arbitrarily located infinitesimally small element of material that is shown by itself in parts b and c. Just as equilibrium of the bar as a whole requires the two external forces P to be equal, equilibrium of the element requires the tensile stresses acting on the opposite pair of elemental faces to be equal. Such elements are commonly shown as in part c, where it is important to remem-

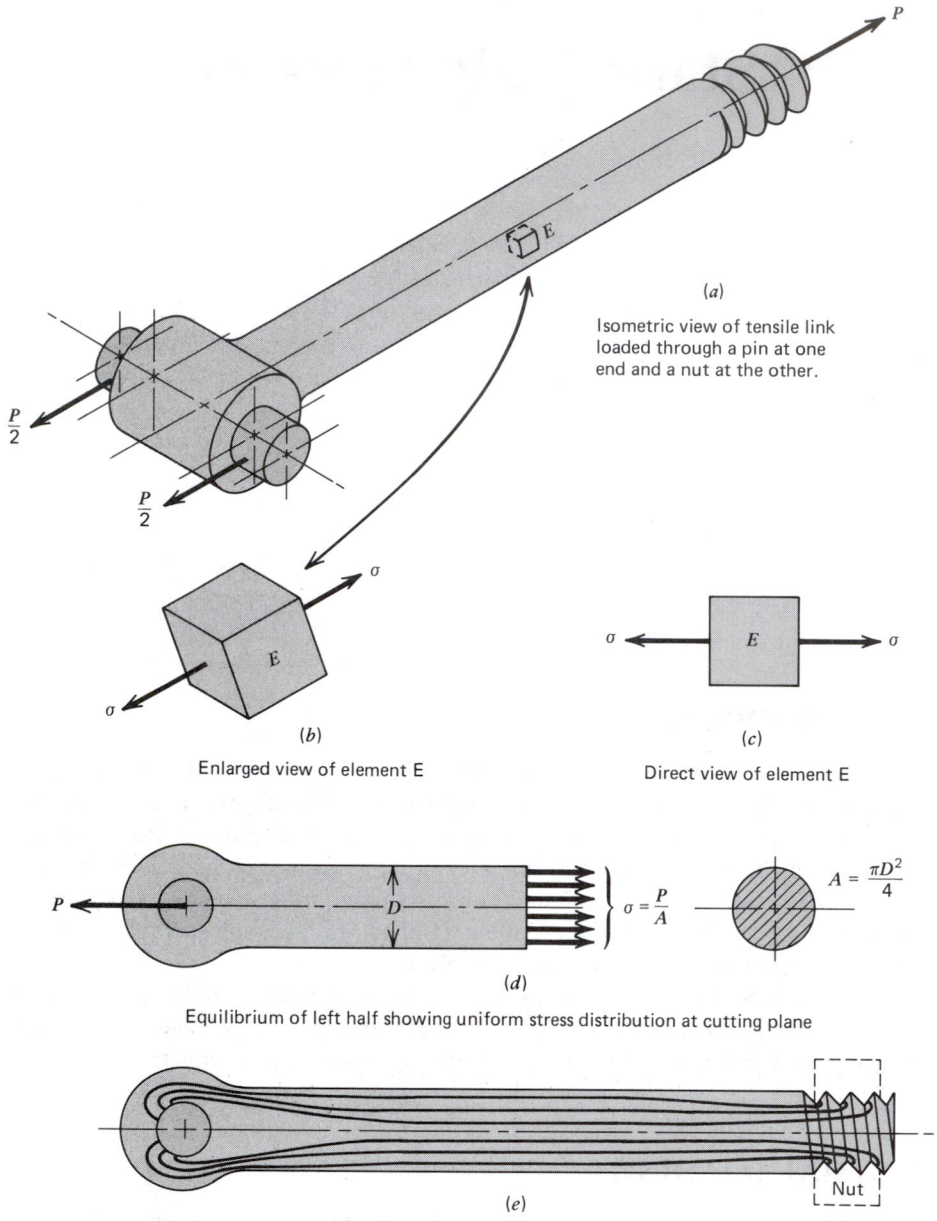

FIGURE 4.1.
Axial loading.

ber that the stresses are acting on faces *perpendicular to the paper*. This is made clear by the isometric view in part b.

Figure 4.1*d* illustrates equilibrium of the left portion of the link under the action of the external force at the left and the tensile stresses acting on the cutting plane. From this equilibrium we have perhaps the simplest formula in all of engineering:

$$\sigma = P/A \tag{4.1}$$

It is important to remember that while this formula is always correct as an expression for the *average* stress in any cross section, disastrous errors can be made by naively assuming that it also gives the correct value of *maximum* stress in the section. Unless several important requirements are fulfilled, the maximum stress will be greater than P/A, perhaps by several hundred percent. The maximum stress is equal to P/A only if the load is *uniformly distributed* over the cross section. This requires that:

1. The section being considered is well removed from the loaded ends. Figure 4.1*e* shows "lines of force flow" to illustrate the general nature of the stress distribution in cross sections at various distances from the ends. A substantially uniform distribution is reached at points about three diameters from the end fittings in most cases.
2. The load is applied *exactly* along the centroidal axis of the bar. If, for example, the loads are applied a little closer to the top, the stresses will be highest at the top of the bar and lowest at the bottom. (Looking at it another way, if the load is eccentric by amount e, a bending moment of intensity Pe is superimposed on the axial load.)
3. The bar is a perfect straight cylinder, with no holes, notches, threads, internal imperfections, or even surface scratches. Any of these give rise to *stress concentration,* which will be dealt with in Section 4.12.
4. The bar is totally free of stress when the external loads are removed. This is frequently not the case. The manufacture of the part and its subsequent mechanical and thermal loading history may have created *residual stresses,* as described in Sections 4.14, 4.15, and 4.16.
5. The bar comes to stable equilibrium when loaded. This requirement is violated if the bar is relatively long and loaded in compression. Then it becomes elastically unstable, and *buckling* occurs. (See Sections 5.10–5.15.)
6. The bar is homogeneous. A common example of *non*-homogeniety is a composite material, as glass or carbon fibers in a plastic matrix. In that case, the matrix and the fibers carry the load *redundantly* (see Section 2.5), and the stiffer material (i.e., having the higher modulus of elasticity) is the more highly stressed.

Figure 4.2 shows an example where unexpected failure can easily result from the naive assumption that the calculation of axial stress involves no more than "P/A." Suppose the load P is 600 N and that six identical welds are used to attach the bracket to a fixed flat surface. The *average* load per weld would be, of course, 100 N. However, the six welds represent redundant force paths of very different stiffnesses. The paths to welds 1 and 2 are *much* stiffer than the others; hence, these two welds may carry nearly all of the load. A much more uniform distribution of load among the six welds could be obtained by adding the two side plates shown dotted in Fig. 4.2*b*, as these would stiffen the force paths to welds 3 to 6.

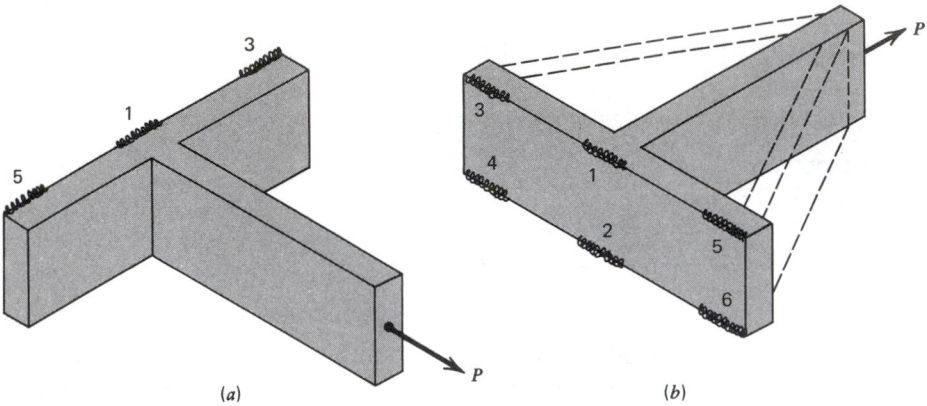

FIGURE 4.2.
Tensile loaded "T" bracket attached by six welds.

At this point one might despair of *ever* using P/A as an acceptable value of maximum stress for relating to the strength properties of the material. Fortunately, such is not the case. The student should acquire increasing insight for making "engineering judgments" relative to the above factors as his or her study and experience progress.

4.3 DIRECT SHEAR LOADING

Direct shear loading involves the application of equal and opposite forces so nearly collinear that the material between them experiences shear stress, with negligible bending. Figure 4.3 shows a bolt serving to restrain relative sliding of two plates subjected to opposing forces P. Neglecting plate interface friction, the bolt cross section marked 1 (of area A) experiences direct shear stress of *average* value

$$\tau = P/A \tag{4.2}$$

If the nut in Fig. 4.3 is tightened to produce an initial bolt tension of P, the direct shear stresses at the root of the bolt threads (area 2), and at the root of the nut threads

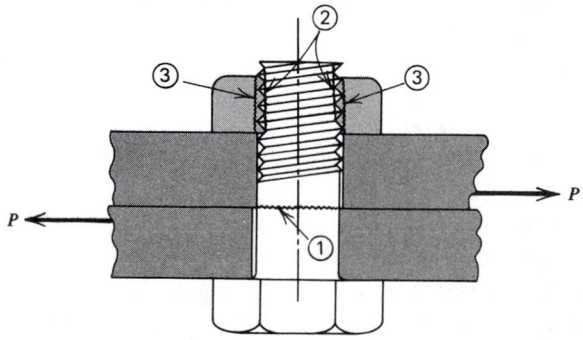

FIGURE 4.3.
Bolted joint, showing three areas of direct shear.

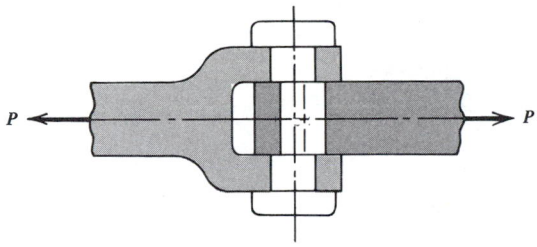

FIGURE 4.4.
Direct shear loading (showing failure in double shear).

(area 3), have *average* values in accordance with Eq. (4.2). The thread root areas involved are cylinders of a height equal to the nut thickness.[1] If the shear stress is excessive, shearing or "stripping" of the threads occurs in the bolt or nut, whichever is weaker.

Similar examples of direct shear occur in rivets, pins, keys, splines, and so on. Also, direct shear loading is commonly used for cutting, as in ordinary household *shears* or scissors, paper cutters, and industrial metal shears.

Figure 4.4 shows a hinge pin loaded in *double* shear, where the load P is carried in shear through two areas in parallel; hence the area A used in Eq. (4.2) is *twice* the cross-sectional area of the pin. Examples of pins loaded in double shear are common: cotter pins used to prevent threaded nuts from rotating (as with automobile wheel bearing retaining nuts), shear pins used to drive boat propellers (the pin fails in double shear when the propeller strikes a major obstruction, thus protecting more expensive and difficult-to-replace members), transverse pins used to hold telescoping tubular members in a fixed position, and many others.

Direct shear loading does not produce *pure* shear (as does torsional loading), and the actual stress distribution is complex. It involves fits between the mating members, and relative stiffnesses. The maximum shear stress will always be somewhat in excess of the P/A value given by Eq. (4.2). In the design of machine and structural members, however, Eq. (4.2) is commonly used in conjunction with appropriately conservative values of working shear stress. Furthermore, to produce total shear fracture of a ductile member, the load must simultaneously overcome the shear strength in every element of material in the shear plane. Thus, for total fracture, Eq. (4.2) would apply, with τ being set equal to the ultimate shear strength, S_{us}.

4.4 TORSIONAL LOADING

Figure 4.5 illustrates torsional loading of a round bar. Note that the direction of the applied torque (T) determines that the left face of element E is subjected to a *downward* shear stress, and the right face to an *upward* stress. Together, these stresses exert a *counterclockwise* couple on the element that must be balanced by a corresponding

[1]Strictly true only for threads with a sharp "V" profile. Shear areas for standard threads are a little less. See Section 10.4.5.

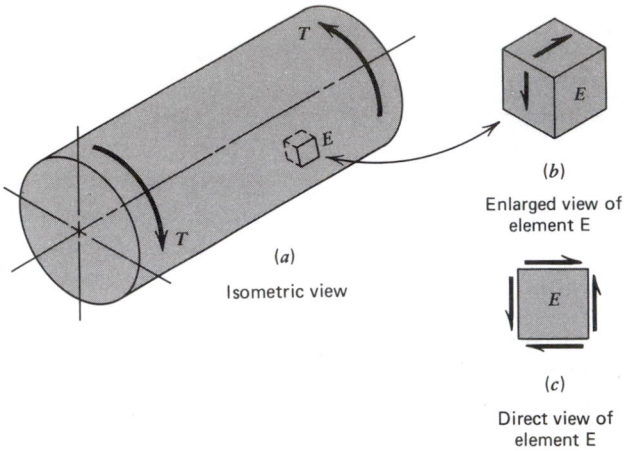

FIGURE 4.5.
Torsional loading of a round bar.

clockwise couple, created by shear stresses acting on the top and bottom faces. The state of stress shown on element E is *pure shear.*

The sign convention for axial loading (positive for tension, negative for compression) distinguishes between two basically different types of loading: compression can cause buckling whereas tension cannot, a chain or cable can withstand tension but not compression, concrete is strong in compression but weak in tension, and so on. The sign convention for *shear* loading serves no similar function—positive and negative shear are basically the same—and the sign convention is purely arbitrary. Any shear sign convention is satisfactory so long as the *same* convention is used throughout any one problem. This book uses the convention of *positive-clockwise;* that is, the shear stresses on the top and bottom faces of element E (in Fig. 4.5) tend to rotate the element *clockwise,* hence are regarded as *positive.* The vertical faces are subjected to *counterclockwise* shear, which is *negative.*

For a round bar in torsion, the stresses vary linearly from zero at the axis to a maximum at the outer surface. Strength of materials texts contain formal proofs that the shear stress intensity at any radius r is

$$\tau = Tr/J \tag{4.3}$$

Of particular interest, of course, is the stress at the surface, where r is equal to the outside radius of the bar. J is the polar moment of inertia of the cross section, which is equal to $\pi d^4/32$ for a solid round bar of diameter d (see Appendix B-1). Simple substitution of this expression in Eq. (4.3) gives the equation for surface torsional stress in a solid round bar of diameter d:

$$\tau = 16T/\pi d^3 \tag{4.4}$$

The corresponding equation for torsional stress in a *hollow* round bar (i.e., round tubing or pipe) follows from substitution of the appropriate equation for polar moment of inertia (see Appendix B-1).

The important assumptions associated with Eqs. (4.3) and (4.4) are:

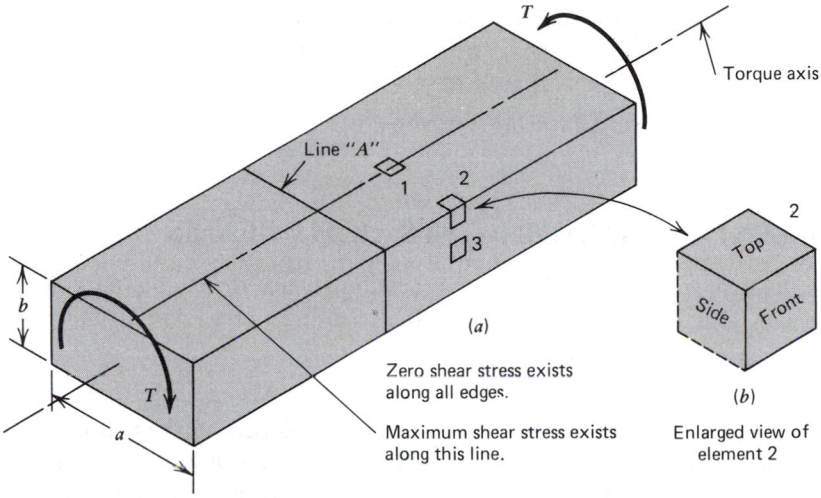

FIGURE 4.6.
Rubber eraser marked to illustrate torsional deformation (hence stresses) in a rectangular bar.

1. The bar must be straight and round (either solid or hollow), and the torque must be applied about the longitudinal axis.
2. The material must be homogeneous, and perfectly elastic within the stress range involved.
3. The cross section considered must be sufficiently remote from points of load application and from stress raisers (i.e., holes, notches, keyways, surface gouges, etc.)

For bars of nonround cross section, the foregoing analysis gives *completely* erroneous results. This can be demonstrated for rectangular bars by marking an ordinary rubber eraser with small square elements 1, 2, and 3 as shown in Fig. 4.6. When the eraser is twisted about its longitudinal axis, Eq. (4.3) implies that the highest shear stress would be at the corners (element 2) because these are farthest from the neutral axis. Likewise, the lowest surface stress should be at element 1 because it is closest to the axis. Observation of the twisted eraser shows exactly the opposite—element 2 (if it could be drawn small enough) does not distort at all, whereas element 1 experiences the *greatest* distortion of any element on the entire surface!

A review of a formal derivation of Eq. (4.3) reminds us of the basic assumption that *transverse planes before twisting remain plane after twisting.* If such a plane is represented by drawing line "A" on the eraser, obvious distortion occurs upon twisting; therefore, the assumption is not valid for a rectangular section.

The equilibrium requirement of corner element 2 makes it clear that this element *must* have zero shear stress: (1) the "free" top and front surfaces do not contact anything that could apply shear stresses; (2) this being so, equilibrium requirements prevent any of the other four surfaces from having shear. Hence, there is zero shear stress along all edges of the eraser.

Torsional stress equations for nonround sections are summarized in references such as [8]. For example, the maximum shear stress for a rectangular section, as shown in Fig. 4.6, is

$$\tau_{\max} = T(3a + 1.8b)/a^2b^2 \tag{4.5}$$

4.5 PURE BENDING LOADING, STRAIGHT BEAMS

Figures 4.7 and 4.8 show beams loaded *only* in bending; hence the term, "pure bending." From studies of the strength of materials, the resulting stresses are given by the equation

$$\sigma = My/I \tag{4.6}$$

where I is the moment of inertia of the cross section with respect to the neutral axis, and y is the distance from the neutral axis. Bending stresses are *normal* stresses, the same as axial stresses. Sometimes the two are distinguished by using appropriate subscripts, as σ_b for bending stresses and σ_a for axial stresses. For the bending shown in Figs. 4.7 and 4.8, tensile stresses exist above the neutral axis of the section (or above the neutral surface of the beam), and compressive stresses below. Maximum values are at the top and bottom surfaces.

Equation (4.6) applies to any cross section (such as the several that are illustrated), with these important limitations:

1. The bar must be initially straight, and loaded in a plane of symmetry.
2. The material must be homogeneous, and all stresses within the elastic range.
3. The section for which stresses are calculated must not be too close to significant stress raisers or to regions where external loads are applied.

Figure 4.7 shows a bending load applied to a beam of cross section having two axes

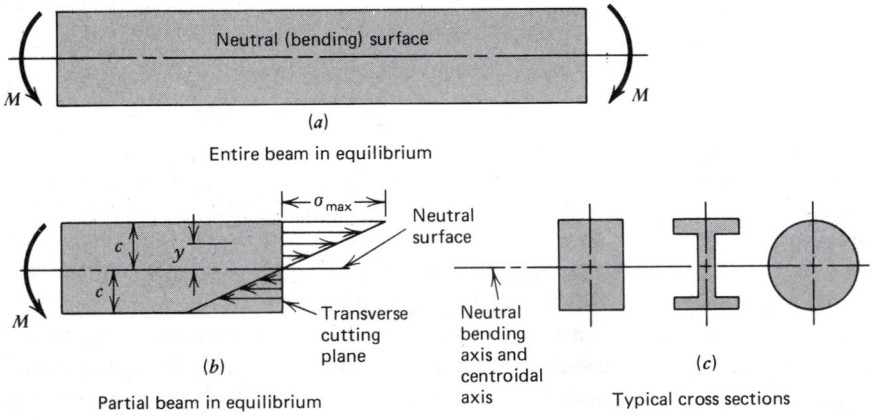

FIGURE 4.7.
Pure bending of sections with two axes of symmetry.

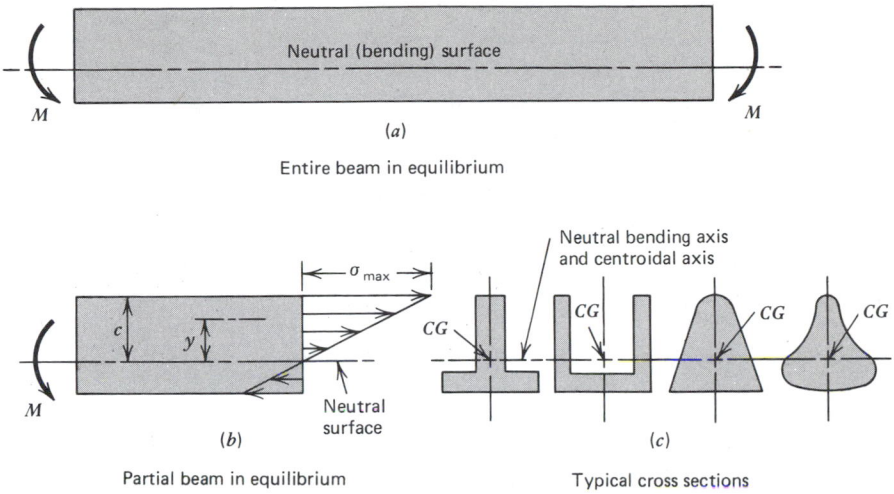

FIGURE 4.8.
Pure bending of sections with one axis of symmetry.

of symmetry. Note that the cutting-plane stresses marked "σ_{max}" are obtained from Eq. (4.6) by substituting c for y, where c is the distance from the neutral axis to the extreme fiber. Often the *section modulus* Z (defined as the ratio I/c) is used, giving the equation for maximum bending stress as

$$\sigma_{max} = M/Z \tag{4.7}$$

For a solid round bar, $I = \pi d^4/64$, $c = d/2$, and $Z = \pi d^3/32$. Hence, for this case

$$\sigma_{max} = 32M/\pi d^3 \tag{4.8}$$

Properties of various cross sections are given in Appendix B-1.

Figure 4.8 shows bending of sections having a single axis of symmetry, and where the bending moment lies in the plane containing the axis of symmetry of each cross section. At this point the reader will find it profitable to spend a few moments verifying that the offset stress distribution pattern shown is necessary to establish equilibrium in Fig. 4.8*b* (i.e., $\Sigma F = \Sigma\sigma\, dA = 0$, and $\Sigma M = M + \Sigma\sigma\, dA\, y = 0$).

4.6 PURE BENDING LOADING, CURVED BEAMS

When initially curved beams are loaded in the plane of curvature, the bending stresses are only approximately in accordance with Eqs. (4.6)–(4.8). Since the shortest (hence stiffest) path along the length of a curved beam is at the inside surface, a consideration of the relative stiffnesses of redundant load paths suggests that the stresses at the inside surface are *greater* than indicated by the straight-beam equations. Figure 4.9 illustrates that this is indeed the case. This figure also shows that equilibrium requirements result in the neutral axis shifting inward (toward the center of curvature) an amount e, and

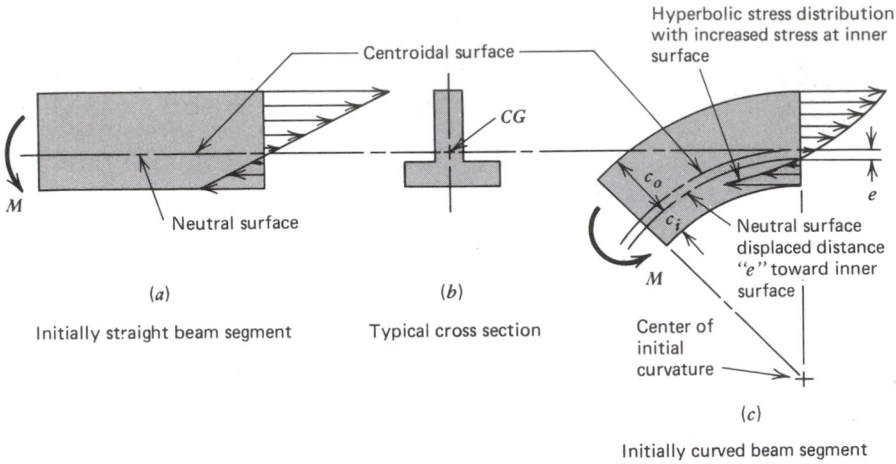

FIGURE 4.9.
Effect of initial curvature, pure bending of sections with one axis of symmetry.

the stress distribution becoming hyperbolic. These deviations from straight-beam behavior are important in severely curved beams, such as those commonly encountered in C-clamps, punch press and drill press frames, hooks, brackets, and chain links.

To understand more clearly the behavior pattern shown in Fig. 4.9*c*, let us develop the basic curved-beam stress equations. With reference to Fig. 4.10, let *abcd* represent an element bounded by plane of symmetry *ab* (which does not change direction when moment M is applied) and plane *cd*. Moment M causes plane *cd* to rotate through angle $d\phi$ to new position $c'd'$. (Note the implied assumption that initially plane sections remain plane after loading.) Rotation of this plane is, of course, about the neutral bending axis, displaced an as-yet-unknown distance e from the centroidal axis.

The strain on the fiber shown at distance y from the neutral axis is

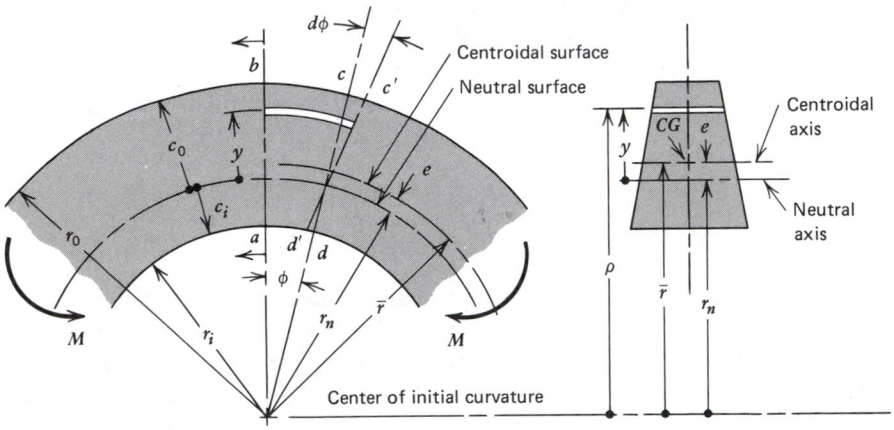

FIGURE 4.10.
Curved beam in bending.

$$\epsilon = \frac{y\,d\phi}{(r_n + y)\phi} \tag{a}$$

For an elastic material, the corresponding stress is

$$\sigma = \frac{Ey\,d\phi}{(r_n + y)\phi} \tag{b}$$

Note that this equation gives a hyperbolic distribution of stress, as illustrated in Fig. 4.9*c*.

Equilibrium of the beam segment on either side of plane *cd* (Fig. 4.10) requires

$$\Sigma F = 0: \quad \int \sigma\,dA = \frac{E\,d\phi}{\phi} \int \frac{y\,dA}{r_n + y} = 0$$

and, since $E \neq 0$,

$$\int \frac{y\,dA}{r_n + y} = 0 \tag{c}$$

$$\Sigma M = 0: \quad \int \sigma y\,dA = \frac{E\,d\phi}{\phi} \int \frac{y^2\,dA}{r_n + y} = M \tag{d}$$

The quantity $y^2/(r_n + y)$ in Eq. (d) can be replaced by $y - r_n y/(r_n + y)$, giving

$$M = \frac{E\,d\phi}{\phi}\left[\int y\,dA - r_n \int \frac{y\,dA}{r_n + y}\right] \tag{e}$$

The second integral in Eq. (e) is equal to zero because of Eq. (c). The first integral is equal to eA. (Note that this integral would be equal to zero if y were measured from the centroidal axis. Since y is measured from an axis displaced distance e from the centroid, the integral has a value of eA.)

Substituting the above expressions into Eq. (e) gives

$$M = \frac{E\,d\phi}{\phi}\,eA \quad \text{or} \quad E = \frac{M\phi}{d\phi\,eA} \tag{f}$$

Substituting Eq. (f) into Eq. (b) gives

$$\sigma = \frac{My}{eA(r_n + y)} \tag{g}$$

Substituting $y = -c_i$ and $y = c_0$ in order to find maximum stress values at the inner and outer surfaces, we have

$$\sigma_i = \frac{-Mc_i}{eA(r_n - c_i)} = \frac{-Mc_i}{eAr_i}$$

$$\sigma_0 = \frac{Mc_0}{eA(r_n + c_0)} = \frac{Mc_0}{eAr_0}$$

The signs of these equations are consistent with the compressive and tensile stresses produced in the inner and outer surfaces of the beam in Fig. 4.10, where the direction of moment M was chosen in the interest of clarifying the analysis. More commonly, a positive bending moment is defined as one tending to *straighten* an initially curved beam. In terms of this convention:

$$\sigma_i = +\frac{Mc_i}{eAr_i} \quad \text{and} \quad \sigma_0 = -\frac{Mc_0}{eAr_0} \tag{4.9}$$

Before using Eq. (4.9), it is necessary to develop an equation for distance e. Beginning with the force equilibrium requirement, Eq. (c), and substituting ρ for $(r_n + y)$:

$$\int \frac{y\,dA}{\rho} = 0$$

But $y = \rho - r_n$; hence:

$$\int \frac{(\rho - r_n)\,dA}{\rho} = 0, \text{ or}$$

$$\int dA - \int \frac{r_n\,dA}{\rho} = 0$$

$\int dA = A$; hence,

$$A = r_n \int dA/\rho \quad \text{or} \quad r_n = \frac{A}{\int dA/\rho} \tag{h}$$

Distance e is equal to $\bar{r} - r_n$; hence:

$$e = \bar{r} - \frac{A}{\int dA/\rho} \tag{4.10}$$

Stress values given by Eq. (4.9) differ from the straight-beam "Mc/I" value by a curvature factor, K. Thus, using subscripts i and o to denote inside and outside fibers, respectively:

$$\sigma_i = K_i Mc/I = K_i M/Z \quad \text{and} \quad \sigma_0 = -K_0 Mc/I = -K_0 M/Z \tag{4.11}$$

Values of K for beams of representative cross sections and various curvatures are plotted in Fig. 4.11. This illustrates a common "rule of thumb": "If $\bar{r}$ is at least 10 times $\bar{c}$, inner fiber stresses are usually not more than 10 percent above the Mc/I value." Values of K_0, K_i, and e are tabulated for several cross sections in [8]. Of course, any section can be handled by using Eqs. (4.9) and (4.10). If necessary, the integral in Eq. (4.10) can be evaluated graphically. Use of these equations is illustrated by the following sample problem.

SAMPLE PROBLEM 4.1

A rectangular beam has an initial curvature, $\bar{r}$, equal to the section depth, h, as shown in Fig. 4.12. How do its extreme-fiber bending stresses compare with those of an otherwise identical straight beam?

Solution

1. For the direction of loading shown in Fig. 4.12, the conventional straight-beam formula gives

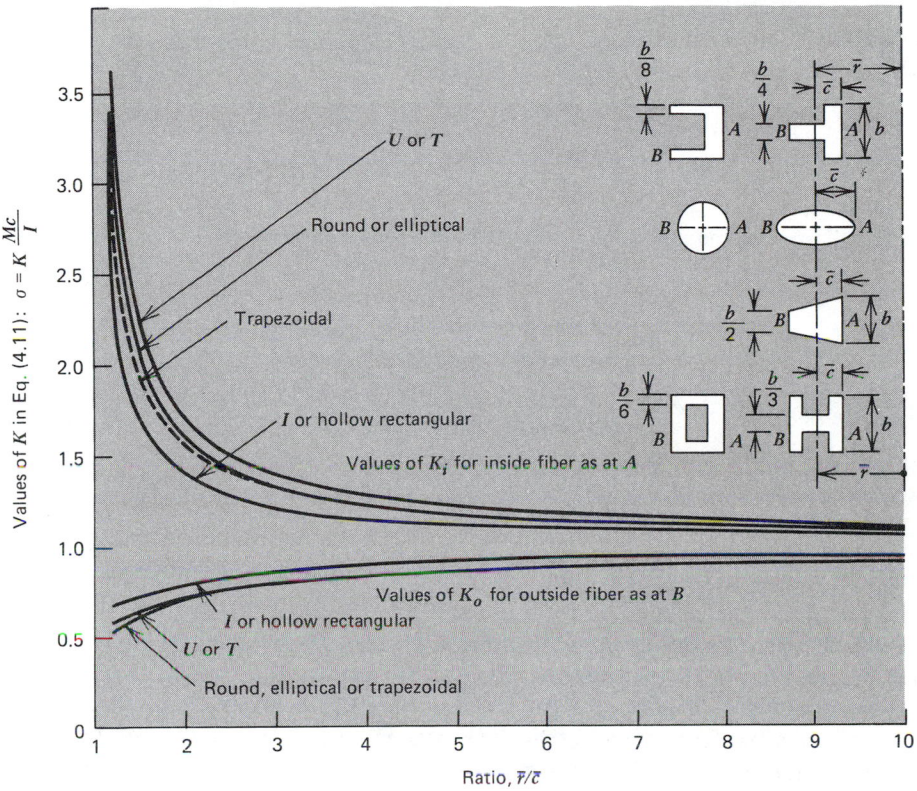

FIGURE 4.11.
Effect of curvature on bending stresses, representative cross sections. (Extracted from [8]. See section 8.1 of [8] for further details and other sections.)

$$\sigma_i = +\frac{Mc}{I} = \frac{6M}{bh^2} \qquad \sigma_0 = -\frac{6M}{bh^2}$$

2. From Eq. (4.10):

$$e = \bar{r} - \frac{A}{\int dA/\rho} = h - \frac{bh}{b\int_{r_i}^{r_0} d\rho/\rho} = h - \frac{h}{\ln(r_0/r_i)} = h\left(1 - \frac{1}{\ln 3}\right)$$

$$e = 0.089761h$$

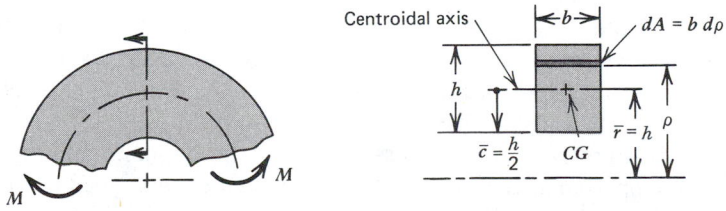

FIGURE 4.12.
Rectangular bar bent to radius of curvature, $\bar{r}$, equal to section depth, h (giving $\bar{r}/\bar{c} = 2$).

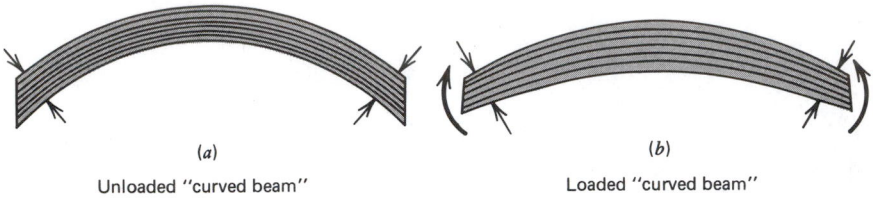

FIGURE 4.13.
Paper pad illustrating radial tension in a curved beam loaded in bending.

3. From Eq. (4.9):

$$\sigma_i = +\frac{M(0.5h - 0.089761h)}{(0.089761h)(bh)(0.5h)} = \frac{9.141M}{bh^2}$$

$$\sigma_0 = -\frac{M(0.5h + 0.089761h)}{(0.089761h)(bh)(1.5h)} = -\frac{4.380M}{bh^2}$$

4. From Eq. (4.11) with $Z = bh^2/6$:

$$K_i = \frac{9.141}{6} = 1.52 \quad \text{and} \quad K_0 = \frac{4.380}{6} = 0.73$$ ■

(These values are consistent with those shown for other sections in Fig. 4.11.)

Note that the stresses dealt with in this article are *circumferential*. Additionally, *radial* stresses are present that are, in some cases, significant. To visualize these, take a paper pad and bend it in an arc, as shown in Fig. 4.13*a*. Apply compressive forces with the thumbs and forefingers so that the sheets will not slide. Next, carefully superimpose (with the thumbs and forefingers) a small bending moment, as in 4.13*b*. Note the separation of the sheets in the center of the "beam," indicating the presence of *radial tension* (radial compression for opposite bending). These radial stresses are small if the center portion of the beam is reasonably heavy. But for an I-beam with a thin web, for example, the radial stresses can be large enough to cause damage—particularly if the beam is made of a brittle material or is subjected to fatigue loading. Further information on curved beam radial stresses is contained in [8] and [9].

4.7 TRANSVERSE SHEAR LOADING IN BEAMS

Although the *average* transverse shear stress in beams such as the shaft in Fig. 2.12 is equal to V/A (i.e., 1580 lb divided by the cross-sectional area in the critical shaft section shown in Fig. 2.12), the *maximum* shear stress is substantially higher. We will now review an analysis of the distribution of this transverse shear stress, with emphasis on an understanding of the basic concepts involved.

Figure 4.14 shows a beam of an arbitrary cross section that is symmetrical about the plane of loading. It is supported at the ends, and carries a concentrated load at the center. We wish to investigate the distribution of transverse shear stress in a plane located distance x from the left support, and at a distance y above the neutral axis. A small square element at this location is shown in the upper-right drawing. The right

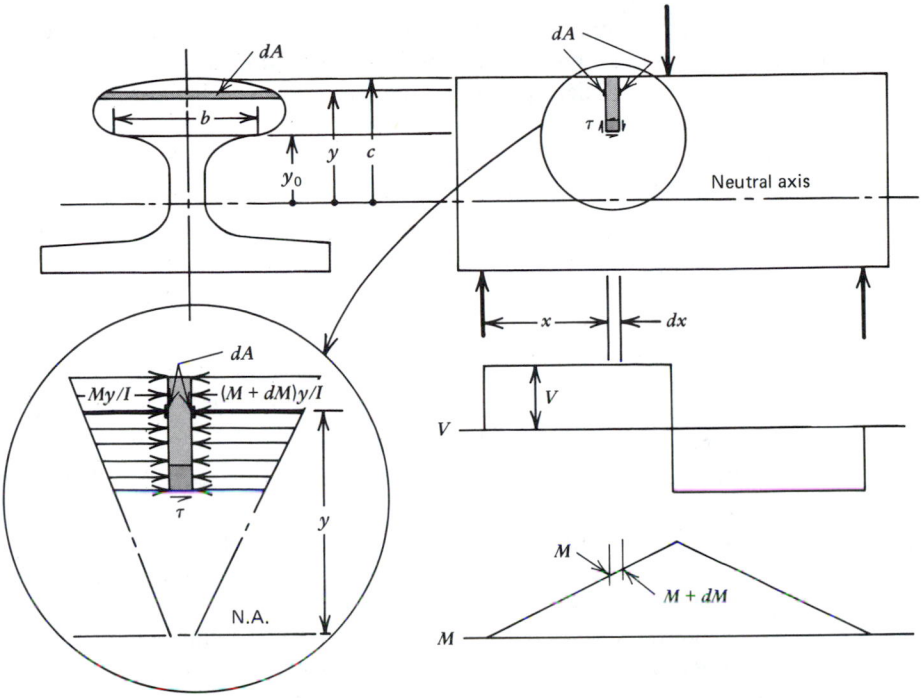

FIGURE 4.14.
Analysis of transverse shear stress distribution.

and left faces of the element are subjected to shear stresses (the magnitude of which is to be determined) with directions established by the fact that the only external force to the left of the element is directed upward, and the resultant of external forces on the right is downward. If only these two vectors acted on the element it would tend to rotate clockwise. This is prevented by the counterclockwise shear stresses shown on the top and bottom surfaces of the element. The reality of these horizontal shear stresses is easy to visualize: if one loads a book or paper tablet with the forces in Fig. 4.14, the pages slide on each other; if the plastic playing cards in a long-unused deck are stuck together, flexing the deck with this three-point beam loading breaks them loose. Coming back to the small element in the figure, the magnitude of all four shear stresses can be determined by evaluating any one of them. We now proceed to evaluate the shear stress on the *bottom* of the element.

Imagine two transverse saw cuts, distance dx apart, starting at the top of the beam and continuing down just to include the sides of the square element. This serves to isolate a segment of the beam, the bottom surface of which is the bottom surface of the element acted upon by shear stress τ. Note that the beam segment involves the full width of the beam. Its bottom surface, acted upon by the unknown shear stress, has a rectangular area of dimensions dx and b. Dimension b will, of course, be different for various values of y_0 (i.e., for various depths of "saw cut").

The enlarged view in Fig. 4.14 shows the forces acting on the beam segment. A key point is that the bending stresses are *slightly greater on the right side* where the bending moment is greater than on the left side by amount dM. The unknown shear stress at the bottom must be sufficiently large to compensate for this inequality. Because the sum of horizontal forces must be zero:

$$\int_{y=y_0}^{y=c} \frac{dM\,y}{I}\,dA = \tau b\,dx$$

But $dM = V\,dx$; hence,

$$\int_{y=y_0}^{y=c} \frac{V\,\cancel{dx}\,y}{I}\,dA = \tau b\,\cancel{dx}$$

Solving for τ:

$$\tau = \frac{V}{Ib}\int_{y=y_0}^{y=c} y\,dA \tag{4.12}$$

Let us now make a few important observations concerning this equation. First, the shear stress is zero at the top (and bottom) surfaces. This is true because the "saw cuts" have no depth, so there is no inequality of bending forces on the two sides to be compensated for by shear stress at the bottom. (Looking at it another way, if the small element in the upper right of Fig. 4.14 is moved to the very top, then the top surface of the element is part of the free surface of the beam. There is nothing in contact with this surface that could impose a shear stress. If there is no shear stress on the top of the element, then the requirements of equilibrium prohibit shear stresses on any of the other three sides.) As the "saw cuts" acquire increasing depth, larger and larger surfaces are exposed to the inequality of bending stress; hence the compensating shear stress must increase correspondingly. Note that at the "saw cut" depth shown in Fig. 4.14, a great increase in shear stress would result from cutting just a little deeper (i.e., slightly reducing y_0) because the area over which the compensating shear stress acts is rapidly decreasing (i.e., b decreases rapidly as y_0 is decreased). Note further that the maximum shear stress is experienced at the neutral axis. This is a most gratifying situation! The maximum shear stress exists precisely where it can best be tolerated—at the neutral axis where the bending stress is zero. At the critical extreme fibers where the bending stress is maximum, the shear stress is zero. (A study of Eq. (4.12) indicates that for unusual sections having a width, b, *at* the neutral axis substantially greater than the width *near* the neutral axis, the maximum shear stress will not be at the neutral axis. However, this is seldom of significance.)

It often helps to establish concepts clearly in mind if one can visualize them on a physical model. Figure 4.15 shows an ordinary rubber eraser ruled with a row of elements that indicates relative shear strains (hence, stresses) when the eraser is loaded as a beam (as shown in part b). If the eraser is loaded carefully, one can see that the top and bottom elements are negligibly distorted (i.e., the initial right angles remain right angles) while the greatest distortion in the right angle corners occurs in the center elements.

Applying Eq. (4.12) to solid round and rectangular sections, we find the parabolic shear stress distributions shown in Fig. 4.16, with maximum values at the neutral axis of

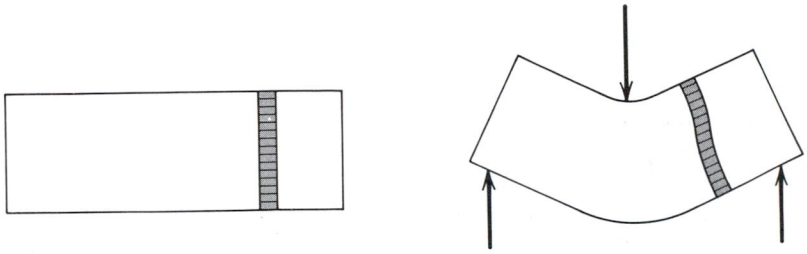

(a)
Marked and unloaded

(b)
Loaded as a beam

FIGURE 4.15.
Transverse shear strain (hence, stress) distribution shown by rubber eraser.

$$\tau_{max} = \tfrac{4}{3}V/A \tag{4.13}$$

for solid round sections and

$$\tau_{max} = \tfrac{3}{2}V/A \tag{4.14}$$

for solid rectangular sections. For a hollow round section, the stress distribution depends on the ratio of inside to outside diameter, but for thin-wall tubing, a good approximation of the maximum shear stress is

$$\tau_{max} = 2V/A \tag{4.15}$$

For a conventional I-beam section, width b is so much less in the web than in the flanges that the shear stresses are much higher in the web. In fact, the shear stresses throughout the web are often approximated by dividing the shear force, V, by the area of the web only, with the web considered as extending the entire depth of the beam.

In the foregoing analysis the tacit assumption was made that the shear stress is uniform across the beam width, b, at any distance, y_0, from the neutral axis (see Fig. 4.14). While not strictly correct, this assumption seldom leads to errors of engineering significance. The variation of shear stress across the width of a beam is treated in [8] and [11]. Another topic left to advanced texts in strength of materials is the loading of beams in other than planes of symmetry.

A final point to be noted is that only in very *short* beams are the transverse shear stresses likely to be of importance *in comparison with the bending stresses*. The principle

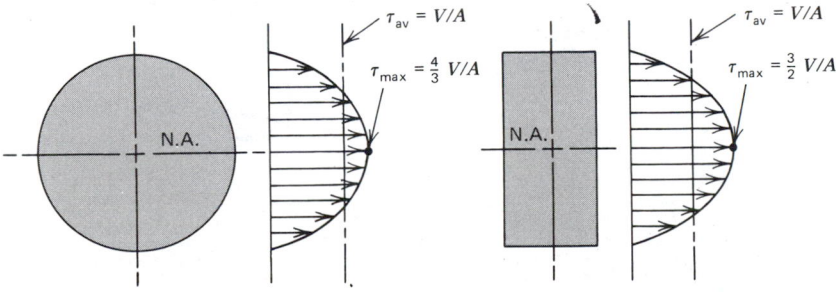

FIGURE 4.16.
Transverse shear stress distribution in solid round and rectangular sections.

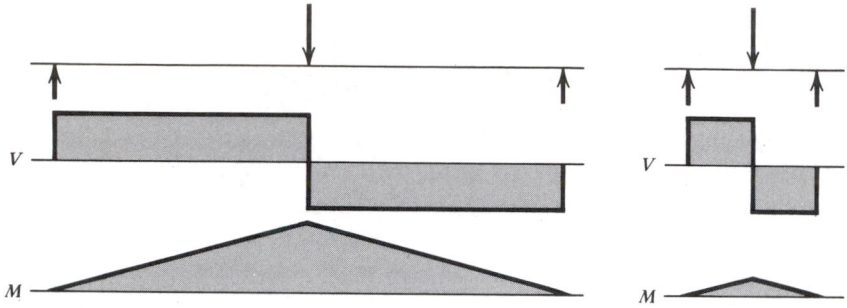

FIGURE 4.17.
Effect of beam length on bending and shear loading.

behind this generalization is illustrated in Fig. 4.17, where the same loads are shown applied to a long and short beam. Both beams have the same shear load and the same *slope* of the bending moment diagram. As the beam length approaches zero, the bending moment (and bending stresses) approaches zero, while the shear load and stresses remain unchanged.

SAMPLE PROBLEM 4.2

Determine the shear stress distribution for the beam and loading shown in Fig. 4.18. Compare this with the maximum bending stress.

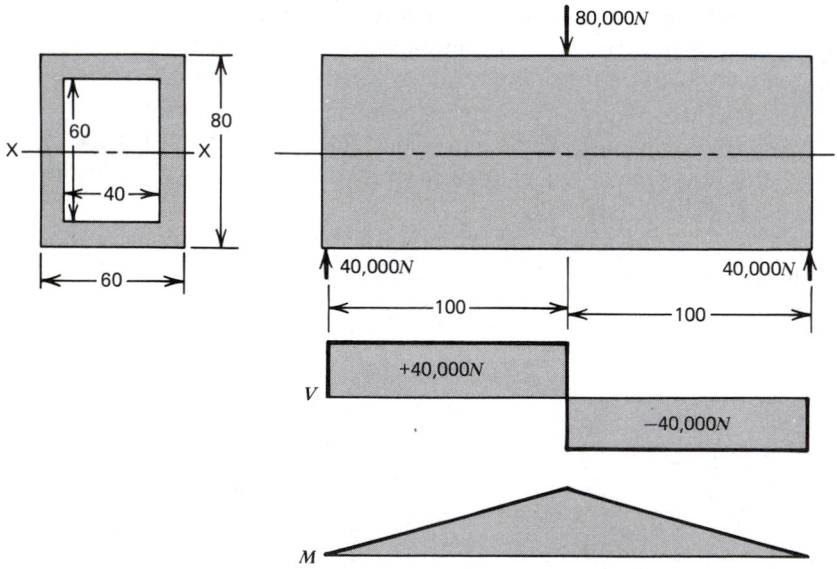

FIGURE 4.18.
Sample Problem 4.2. Beam shear stress distribution. Note: all dimensions are in mm; section properties are $A = 2400\ \text{mm}^2$; and $I_x = 1840 \times 10^6\ \text{mm}^4$.

Solution

1. With reference to Fig. 4.14 and Eq. (4.12), it is known at the outset that $\tau = 0$ at the top and bottom surfaces. This gives a start in plotting the shear stress distribution in Fig. 4.20. As the imaginary parallel "saw cuts" (described in connection with Fig. 4.14) proceed down from the top to increasing depth, the areas exposed to the slightly unbalanced bending stresses increase, thereby causing the compensating shear stress at the bottom of the imaginary segment to increase parabolically. This continues to a "saw cut" depth of 10 mm. Figure 4.19*a* illustrates the imaginary segment just before the "saw cuts" break through the interior surface of the section. The shear stress at this level (which acts on bottom area 60 · dx) is calculated as

$$\tau = \frac{V}{Ib}\int_{y=y_0}^{y=c} y\,dA = \frac{40{,}000}{(1.840 \times 10^6)(60)}\int_{y=30}^{y=40} y(60\,dy)$$

$$\tau = \frac{40{,}000}{(1.840 \times 10^6)60}(60)\left[\frac{y^2}{2}\right]_{y=30}^{y=40} = 7.61\ \text{N/mm}^2, \text{ or } 7.61\ \text{MPa}$$

2. With a slightly deeper "saw cut," the inner surface is broken through, and the area over which the shear stress acts is suddenly reduced to 20 dy, as shown in Fig. 4.19*b*. The unbalanced bending forces acting on the segment sides are virtually unchanged. Thus, the only term that changes in Eq. (4.12) is b, which is reduced by a factor of 3, thereby giving a shear stress three times as high, or 22.83 MPa.

3. As the "saw cut" depth increases until it reaches the neutral axis, the area over which the shear stress acts remains the same, while greater and greater imbalances build up as additional areas dA are exposed. But, as shown in Fig. 4.19*c*, these added areas dA are only ⅓ as large as those in the top portion of the section. Hence, the increased shear stress at the neutral axis is not as great as might at first be

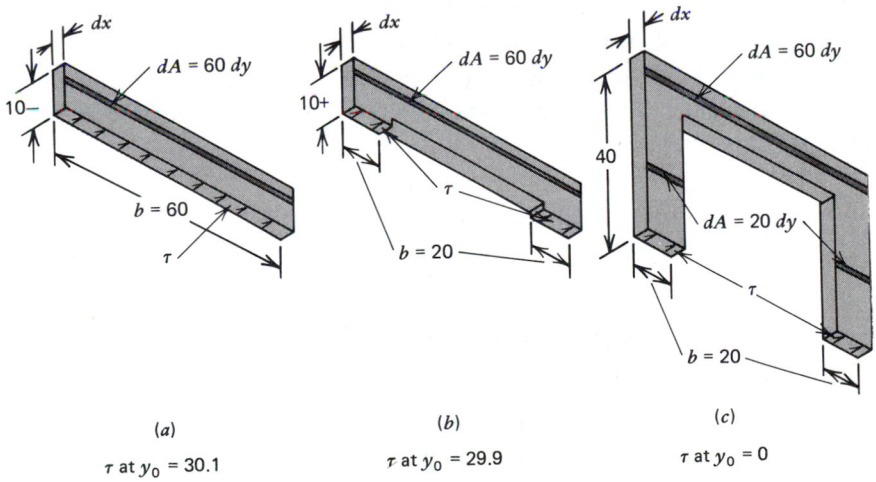

FIGURE 4.19.
Sample Problem 4.2 partial solution —τ at three levels.

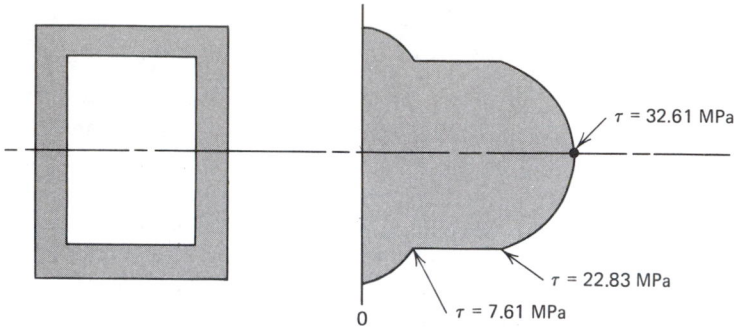

FIGURE 4.20.
Plot of shear stress distribution—Sample Problem 4.2.

expected. When using Eq. (4.12) to find τ at the neutral axis, note that two integrals are involved, one covering the range of y from 0 to 30 mm and the other from 30 to 40 mm (The latter integral, of course, has already been evaluated).

$$\tau = \frac{V}{Ib}\int_{y=y_0}^{y=c} y\,dA = \frac{40{,}000}{(1.840 \times 10^6)(20)}\left[\int_{y=0}^{y=30} y(20\,dy) + \int_{y=30}^{y=40} y(60\,dy)\right]$$

$$\tau = \frac{40{,}000}{(1.840 \times 10^6)(20)}(20)\left[\frac{y^2}{2}\right]_{y=0}^{y=30} + 22.83$$

$$\tau = 32.61\ \text{N/mm}^2, \text{ or } 32.61\ \text{MPa}$$

The above calculations enable the shear stress plot in Fig. 4.20 to be drawn. ■

[Note: Recalling that the shear stress must be zero at the exposed inner surface of the section, it is apparent that the evenly distributed shear stress assumed in Fig. 4.19*a* is incorrect, and that the shear stresses in the outer supported portions of the section at this level will be higher than the calculated value of 7.61 MPa. This is of little importance because, to the degree that shear stresses are of concern, attention will be focused at the level just below, where the calculated value of τ is three times as high, or at the neutral axis where it is a maximum.]

4. By way of comparison, the maximum bending stresses occur in the top and bottom surfaces, at the center of the beam where the bending moment is highest. Here, the bending stress is computed as

$$\sigma = \frac{Mc}{I} = \frac{(40{,}000 \times 100)(40)}{1.84 \times 10^6} = 86.96\ \text{N/mm}^2, \text{ or}$$

$$\sigma = 86.96\ \text{MPa}$$ ■

4.8 INDUCED STRESSES, MOHR-CIRCLE REPRESENTATION

Simple tensile or compressive loading produces induced shear stresses on certain planes; likewise, pure shear loading produces induced tension and compression. In some cases the induced stresses can be more damaging to the material than the direct stresses.

Figure 4.21*a* shows an ordinary rubber eraser marked with two large square ele-

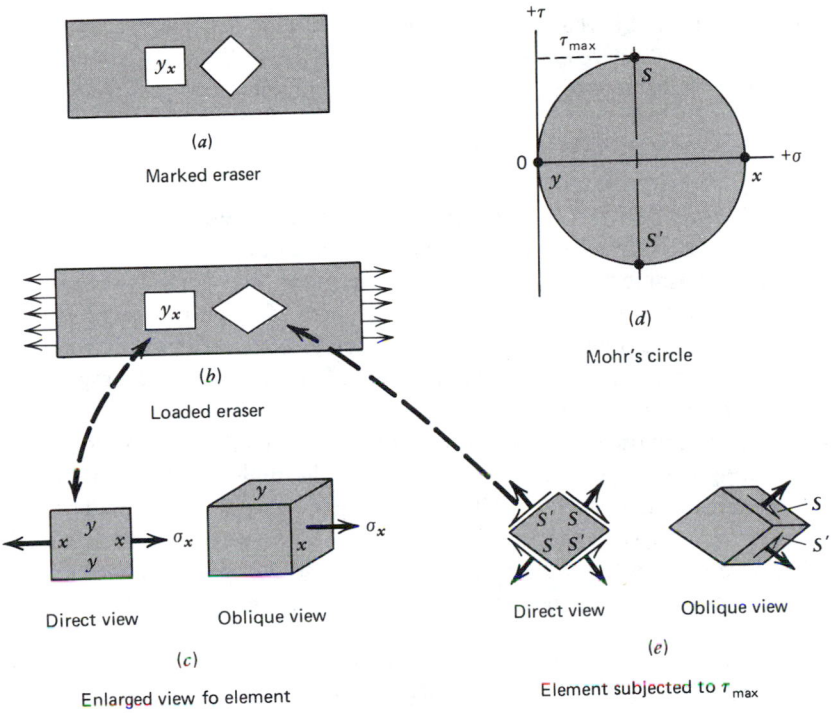

FIGURE 4.21.
Induced shear stress from pure tensile loading.

ments, one oriented in the direction of the sides of the eraser, and the other at 45°. Figure 4.21*b* shows the marked surface subjected to tension (as by flexing the eraser). A *shear* distortion of the 45° square is clearly evident. If the eraser surface is loaded in compression, the shear distortion of the 45° square is reversed.

Figure 4.21*c* shows an enlarged view of a marked element with vertical and horizontal faces marked x and y, and with tensile stress σ_x acting on the x faces. The x and y faces of the element are, of course, perpendicular to the eraser surface, as is made clear by the oblique view, also shown in 4.21*c*.

A Mohr-circle plot of the stresses on the element is shown in 4.21*d*. Points x and y are plotted to represent the normal and shear stresses acting on the x and y faces. The circle is then drawn with the line xy as a diameter.

Proof of the Mohr-circle relationships is left to elementary texts in strength of materials. The emphasis here is on obtaining a clear understanding of the significance and interpretation of the Mohr plot. First, note that as an imaginary cutting plane through the element is rotated (always keeping it perpendicular to the surface), one goes from an x plane (vertical), to a y plane (horizontal), and on to an x plane again in *only 180°*. The normal and shear stresses acting on all of these cutting planes are spread out over the full *360°* of the Mohr circle in 4.21*d*. Thus, angles measured on the circle are *twice* the corresponding angles on the element itself. For example, x and y are 90° apart on the element and 180° apart on the circle.

A second important point is that if one adheres to the shear stress sign convention given in Section 4.4 (i.e., positive-clockwise), rotation of the cutting plane in either direction on the element corresponds to rotating *twice* as far on the circle and *in the same direction.*

Points S and S' on the circle (Fig. 4.21*d*) represent planes of maximum positive and negative shear. On the circle, point S is 90° *counterclockwise* of x. Hence, on the element plane S is *45° counterclockwise* of vertical plane x. The element drawn in Fig. 4.21*e* shows the correctly oriented planes S (which are infinitesimally close together; hence, really represent a single plane). The Mohr circle shows that S planes are acted upon by a positive axial stress and a *positive* shear stress, both of magnitude $\sigma_x/2$. These stresses are shown in Fig. 4.21*e*. The S' plane orientation and stresses are correspondingly determined. (Note that the 45° square in Figs. 4.21*a* and 4.21*b* represents an element subjected to *maximum* shear.)

Figure 4.22*a* shows a marked rubber eraser prior to being loaded in torsion. When the torsional or twisting load is applied, all of the initial right angles in the square lined up with the eraser sides change substantially, indicating shear. In contrast, the right angles in the 45° square do *not* change. When twisting in one direction, the parallel lines in the 45° square get shorter and farther apart. Reverse the twisting and they get longer and closer together. But, in neither case is there an angle change indicating shear.

Figure 4.22*b* shows the direct shear stresses acting on the faces of the element lined up with the eraser. Note that the x faces experience negative (counterclockwise) shear because the direction of load torque is such as to displace the left face downward and the right face upward. Corresponding positive shear is, of course, required on the y faces in order to provide equilibrium. The direct stresses are plotted for x and y faces to establish the Mohr circle in 4.22*c*. Planes subjected to zero shear (also to the extreme values of tension and compression) are called *principal planes.* These are designated as #1 and #2 on the circle. A corresponding *principal element* is shown in 4.22*d*.

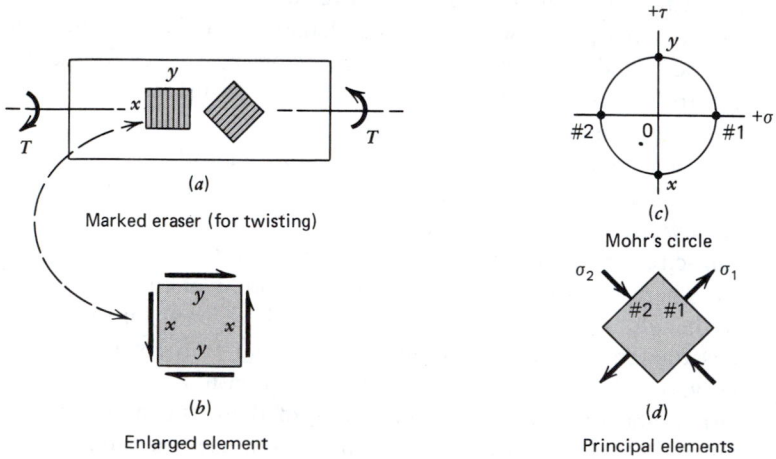

FIGURE 4.22.
Induced axial stress from pure shear loading.

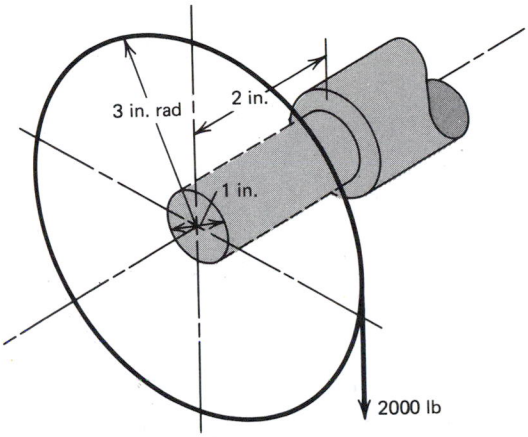

FIGURE 4.23.
Shaft subjected to combined loading. Note that for a solid 1 in. diameter shaft: $A = \pi d^2/4 = 0.785$ in.^{2}; $I = \pi d^4/64 = 0.049$ in.^{4}; and $J = \pi d^4/32 = 0.098$ in.^{4} (see Appendix B-1).

Mohr's circle is named after Otto Mohr, a distinguished German structural engineer and professor who proposed it in 1880 and described it in a published article [4] in 1882. This graphical technique is extremely useful in solving problems and in visualizing the nature of stress states at points of interest.

4.9 COMBINED STRESSES – MOHR-CIRCLE REPRESENTATION

This topic can best be presented through the use of a typical example. Figure 4.23 represents a stationary shaft and pulley subjected to a 2000-lb static load. We wish to determine the location of highest stresses in the 1-in. diameter section, and to analyze the stresses at that point.

1. The shaft is subjected to torsion, bending, and transverse shear. Torsional stresses are a maximum over the entire shaft surface. Bending stresses are a maximum at points A and B, shown in Fig. 4.24. Note that both the bending moment and the distance from the neutral bending axis are a maximum at these two locations. Transverse shear stresses are relatively small compared to bending stresses, and equal to zero at points A and B (see Section 4.7). Thus, they can be neglected. Clearly it is the section containing points A and B that must be investigated.
2. In Fig. 4.25, imagine the shaft to be cut off at the section containing A and B, and consider the member thus obtained as a *free-body in equilibrium.* This is a convenient way of being certain that all loads acting on the cutting plane are identified. In this case there are the three loads, M, T, and V, as shown. Note that the free-

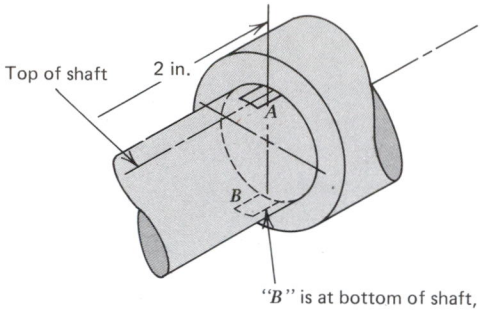

FIGURE 4.24.
Location of highest stresses.

body is indeed in equilibrium, the summation of all forces and moments being zero. Also in Fig. 4.25 are the load, shear, and moment diagrams for the isolated free-body.

3. Compute the direct stresses associated with loads.

Bending stresses (tension at A; compression at B):

$$\sigma_x = \frac{Mc}{I} = \frac{(4000 \text{ in. lb})(\tfrac{1}{2} \text{ in.})}{0.049 \text{ in.}^4} = 40{,}816 \text{ psi} \approx 40.8 \text{ ksi}$$

Torsional stresses (over the entire surface):

$$\tau_{xy} = \frac{Tr}{J} = \frac{(6000 \text{ lb in.})(\tfrac{1}{2} \text{ in.})}{0.098 \text{ in.}^4} = 30{,}612 \text{ psi} \approx 30.6 \text{ ksi}$$

(In support of neglecting the transverse shear stress in step 1, it is of interest to note that its maximum value at the neutral bending axis is $4V/3A = 3.4$ ksi.)

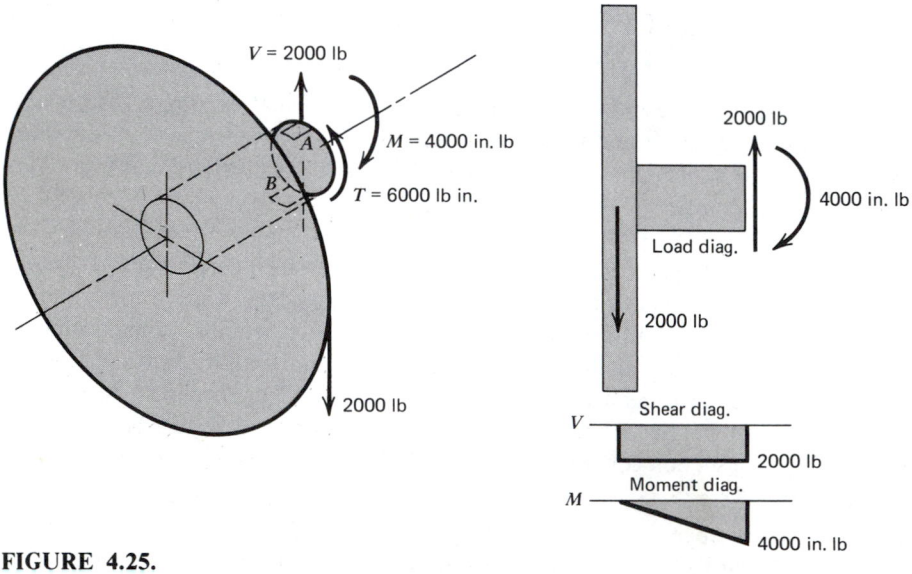

FIGURE 4.25.
Free-body and load diagrams.

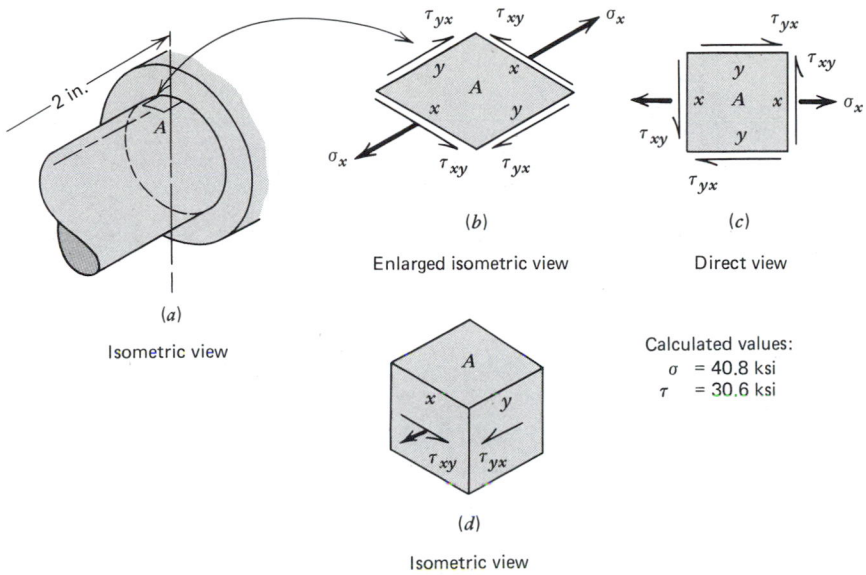

FIGURE 4.26.
Various views of element *A*.

4. Figure 4.26 shows the stresses acting on an element at *A*. (Stresses at *B* are the same except that the bending stress is compressive.) Note that the directions of the two counterclockwise shear stress vectors follow directly from the direction of twisting of the shaft. Then the clockwise direction of the shear vectors on the other pair of faces follows from the requirement of equilibrium. (Note: Subscripts used with the shear stresses in Fig. 4.26 illustrate a common convention, but one that is not of importance in this text: τ_{xy} acts *on* an x face, in the y direction; τ_{yx} acts on a y face, in the x direction. No difficulties would be encountered if both were regarded as τ_{xy}, and the "positive-clockwise" rule followed in order to keep the signs straight.)

   The isometric views are shown for direct comparison with previous figures. The direct view is the conventional way to show a stressed element. The three-dimensional representation shows how the stresses really act on *planes perpendicular to the surface*. The shaft surface itself is unloaded, except for atmospheric pressure, which is negligible.

5. Figure 4.26 shows all of the stresses acting on an element at the most critical stress location. However, the analysis can be carried further. First, recall that the cubical element is infinitesimally small, and its x and y faces represent *only two* of the infinite number of planes perpendicular to the shaft surface passing through *A*. In general, there will be other planes subjected to higher levels of normal stress and shear stress. Mohr's circle provides a convenient means for determining and representing the normal and shear stresses acting on *all* planes through *A*, and perpendicular to the surface. This circle is constructed in Fig. 4.27 by first plotting points representing stresses on the x and y planes, connecting these points with a

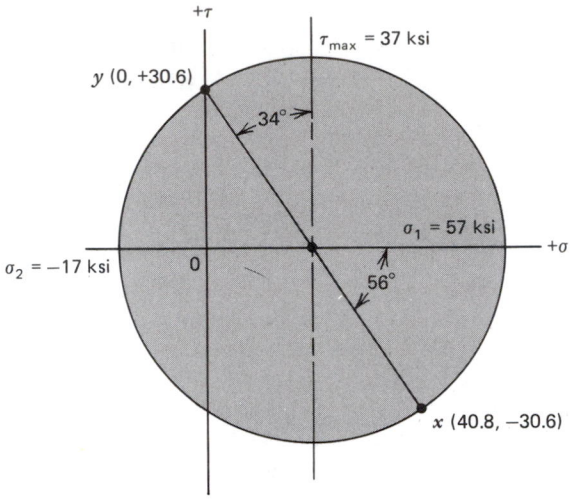

FIGURE 4.27.
Mohr-circle representation of stresses at point *A* of Fig. 4.25.

straight line, and then drawing the circle with the line *xy* as a diameter. The circle provides a convenient graphical solution for the magnitude and orientation of principal stresses σ_1 and σ_2. These stresses are shown on a *principal element* at point *A*, drawn in Fig. 4.28. Note that the #1 principal plane is located by starting with the *x* plane and rotating counterclockwise half of the 56° measured on the circle, and so on.

6. Figure 4.28 shows the magnitude and orientation of the highest normal stresses. It may also be of interest to represent similarly the highest shear stresses. This is done in Fig. 4.29. Observe again the rules of (a) rotating in the *same direction* on the element and the circle, and (b) using angles on the circle that are *twice* those on the element.

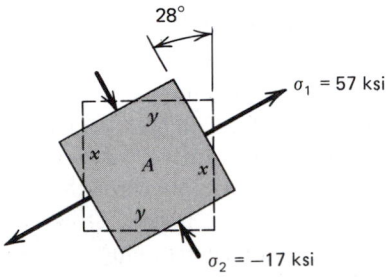

FIGURE 4.28.
Principal element at *A* (direct view) shown in relation to *x* and *y* faces.

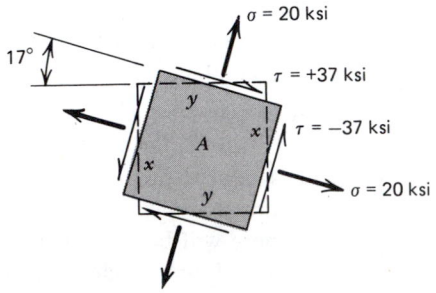

FIGURE 4.29.
Maximum shear element at *A* (direct view) shown in relation to *x* and *y* faces.

4.10 STRESS EQUATIONS RELATED TO MOHR'S CIRCLE

The derivation of the analytical expressions relating normal and shear stresses to the angle of the cutting plane is given in elementary texts on strength of materials and need not be repeated here. The important equations are given below.

If the stresses on an element of given orientation are known (as in Fig. 4.26), the principal stresses, principal directions, and maximum shear stress can be found from a Mohr-circle plot or from the following equations:

$$\sigma_1, \sigma_2 = \frac{\sigma_x + \sigma_y}{2} \pm \sqrt{\tau_{xy}^2 + \left(\frac{\sigma_x - \sigma_y}{2}\right)^2} \tag{4.16}$$

$$2\phi = \tan^{-1} \frac{2\tau_{xy}}{\sigma_x - \sigma_y} \tag{4.17}$$

$$\tau_{\max} = \pm \sqrt{\tau_{xy}^2 + \left(\frac{\sigma_x - \sigma_y}{2}\right)^2} \tag{4.18}$$

where ϕ is the angle between the principal axes and the x and y axes (or the angle between the principal planes and the x and y planes). When ϕ is *positive*, the principal axes (or planes) are *clockwise* from the x and y axes (or planes).

Where the principal stresses are known and it is desired to determine the stresses acting on a plane oriented at any angle ϕ from the principal plane numbered 1, the equations are

$$\sigma_\phi = \frac{\sigma_1 + \sigma_2}{2} + \frac{\sigma_1 - \sigma_2}{2} \cos 2\phi \tag{4.19}$$

$$\tau_\phi = \frac{\sigma_1 - \sigma_2}{2} \sin 2\phi \tag{4.20}$$

Equations (4.16)–(4.18) can readily be developed from the Mohr circle shown in Fig. 4.30, and Eqs. (4.19) and (4.20) from Fig. 4.31. This provides a welcome substitute for rote memory, and one that aids in understanding the physical significance of the equations.

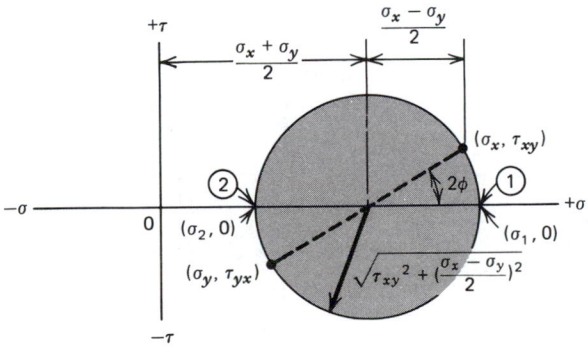

FIGURE 4.30.
Mohr's circle illustrating Eqs. (4.16) to (4.18).

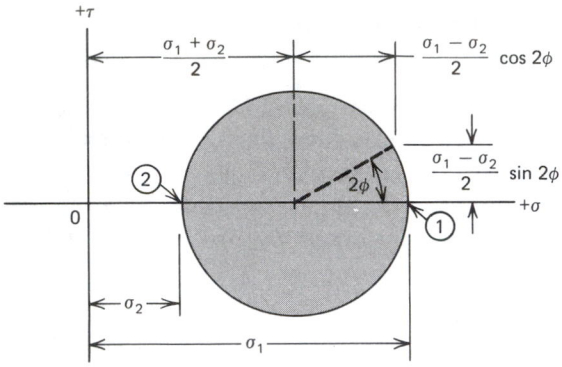

FIGURE 4.31.
Mohr's circle illustrating Eqs. (4.19) and (4.20).

4.11 THREE-DIMENSIONAL STRESSES

Since stresses only exist in real bodies that are three-dimensional, it is best always to think of stresses in three-dimensional terms. Uniaxial stress states (pure tension or compression) involve three principal stresses, but two of them are zero. Likewise, biaxial stresses (as pure shear, or the problem represented in Figs. 4.23 to 4.29) involve one principal stress of zero. Forgetting about a zero principal stress can lead to serious errors, as is illustrated later in this section.

Let us extend the analysis of the state of stress at point A of Fig. 4.24 by treating this as a three-dimensional problem. Figure 4.32 shows five views of stress elements at point A: (a) an oblique view, showing the original x and y planes and the stresses on

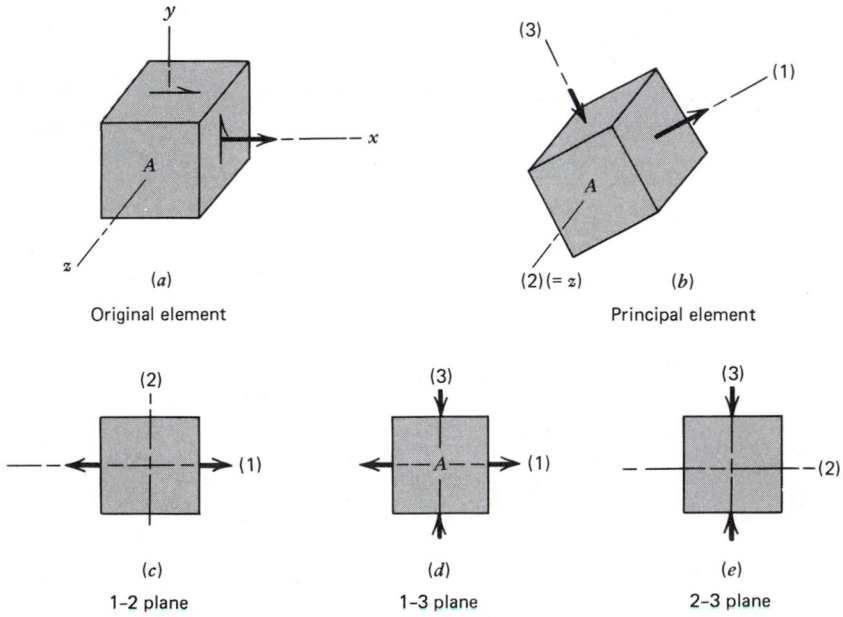

FIGURE 4.32.
Elements representing state of stress at point A.

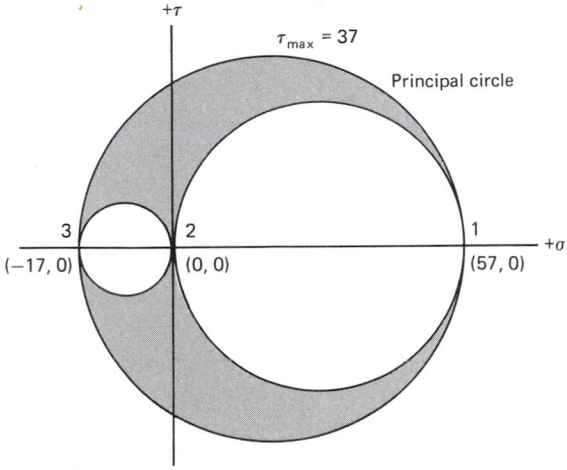

FIGURE 4.33.
Complete Mohr-circle representation of stress state at point *A* of Fig. 4.25.

these planes, (b) a principal element, obtained by rotating 28° about the z axis, (c) to (e) direct, or true views of the 1-2, 1-3, and 2-3 planes of the principal element.

A complete Mohr-circle representation of this state of stress is shown in Fig. 4.33. The large circle between points 1 and 3 represents stresses on all planes through point A, which contain the 2, or z axis. Likewise, the small circle between 2 and 3 gives stresses on all planes containing the 1 axis, and the circle between 1 and 2 represents stresses on planes containing the 3 axis.

Although each of the three circles represents an infinite number of planes through point A, a higher order of infinity remains that does not contain *any* of the principal axes. It can be shown that *all* such planes are subjected to stresses represented by points in the shaded area between the circles. The location of the specific point in this area that corresponds to any given plane is seldom of concern, but the interested reader can find the procedure involved in references such as [1, Section 3.7].

Since the largest of the three Mohr circles always represents the maximum shear stress as well as the two extreme values of normal stress, Mohr called this the *principal circle.*

A common example where the maximum shear stress would be missed if one failed to include the zero principal stress in the Mohr plot is the outer surface of a pressurized cylinder. Here, the axial and tangential stresses constitute tensile principal stresses, while the unloaded outer surface insures that the third principal stress is zero. Figure 4.34 illustrates both the correct value of maximum shear stress, and the incorrect value obtained from a simple two-dimensional analysis. The same situation exists at the inner surface of the cylinder except that the third principal stress (which acts on the surface) is not zero, but a negative value numerically equal to the internal fluid pressure.

For the rare case involving significant shear stresses on *all* faces of the stress element, the reader is referred to detailed works on theoretical stress analysis—for example, [1, 11].

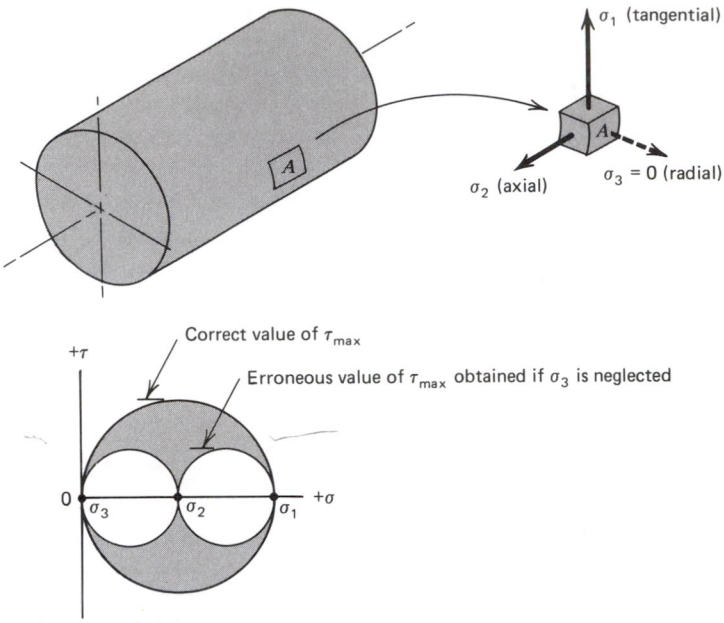

FIGURE 4.34.
Example of biaxial stress where correct determination of τ_{max} requires taking σ_3 into consideration. Internally pressurized cylinder illustrates biaxial stress states where correct determination of τ_{max} requires taking σ_3 into account. Note that (1) for an element on the inside surface, σ_3 is negative, and numerically equal to the internal fluid pressure, and (2) for thin-wall cylinders, $\sigma_2 \approx \sigma_1/2$.

4.12 STRESS CONCENTRATION FACTORS (K_t)

In Section 4.2, Fig. 4.1*e* indicated lines of force flow through a tensile link. It was noted at that time that a uniform distribution of these lines (hence, a uniform distribution of stress) existed only in regions substantially removed from the ends. Near the ends, the force-flow lines indicate a *concentration* of stress near the outer surface. This same stress concentration effect exists for bending and torsional loading as well. We now wish to evaluate the stress concentration associated with various geometric configurations so that the *maximum stresses* existing in a part can be determined.

The first mathematical treatments of stress concentration were published shortly after 1900. In order to handle other than very simple cases, experimental methods for measuring highly localized stresses were developed and used. In recent years, computerized finite element studies have also been employed. The results of many of these studies are available in the form of published graphs, such as those in Figs. 4.35 to 4.41. These give values of the *theoretical stress concentration factor*, K_t (based on a theoretical elastic, homogeneous, isotropic material), for use in the equations:

$$\sigma_{max} = K_t\sigma_{nom} \quad \text{and} \quad \tau_{max} = K_t\tau_{nom} \tag{4.21}$$

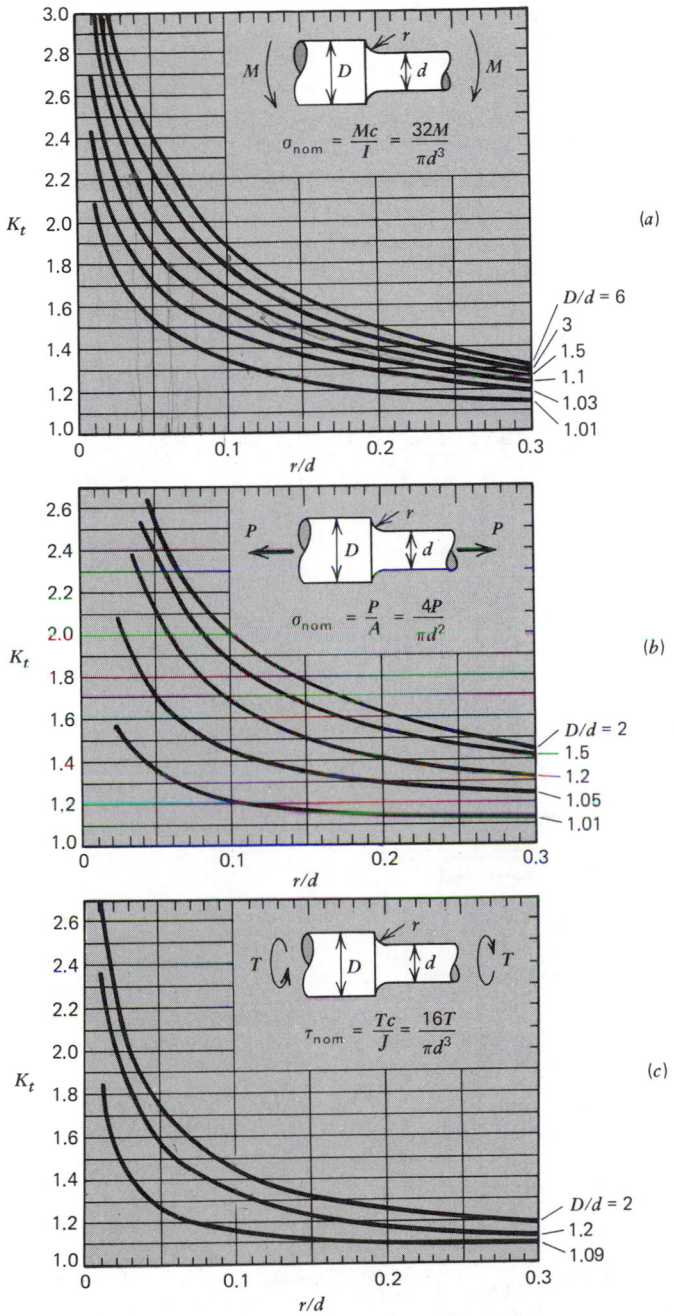

FIGURE 4.35.
Shaft with fillet (a) bending; (b) axial load; (c) torsion [7].

For example, the maximum stress for axial loading (of an ideal material) would be obtained by multiplying P/A by the appropriate value of K_t.

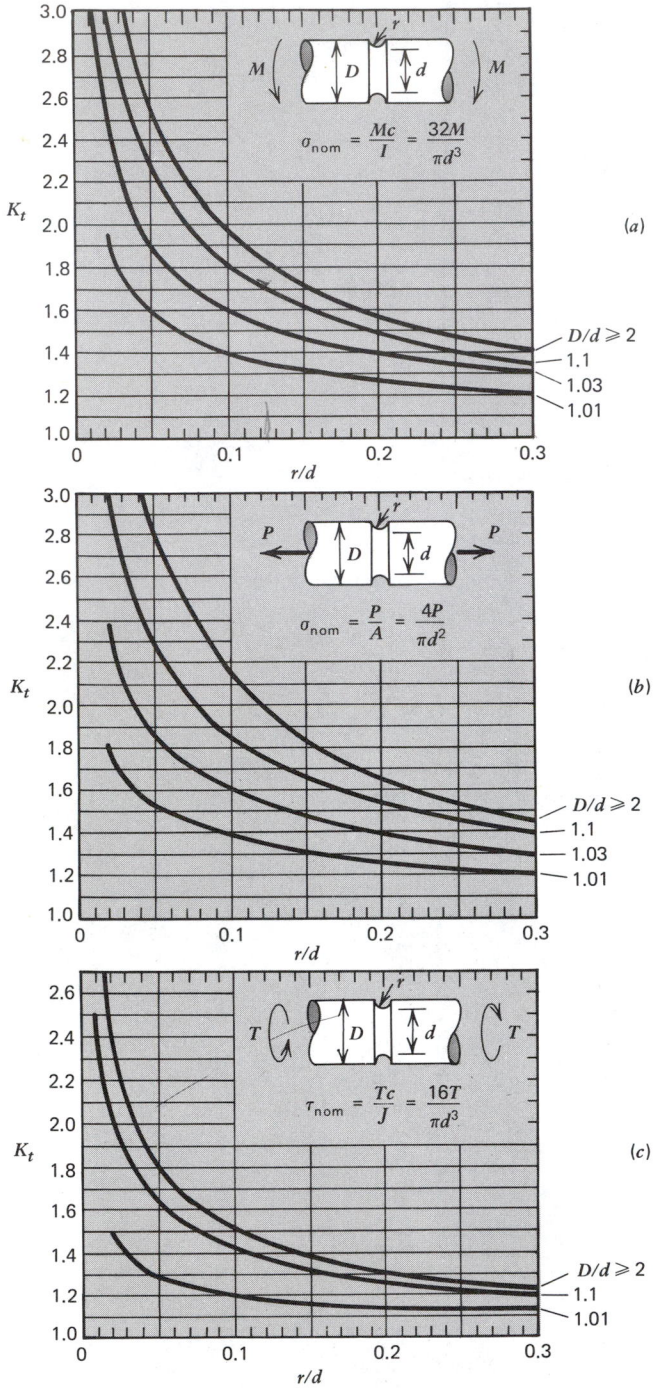

FIGURE 4.36.
Grooved shaft (a) bending; (b) axial load; (c) torsion [7].

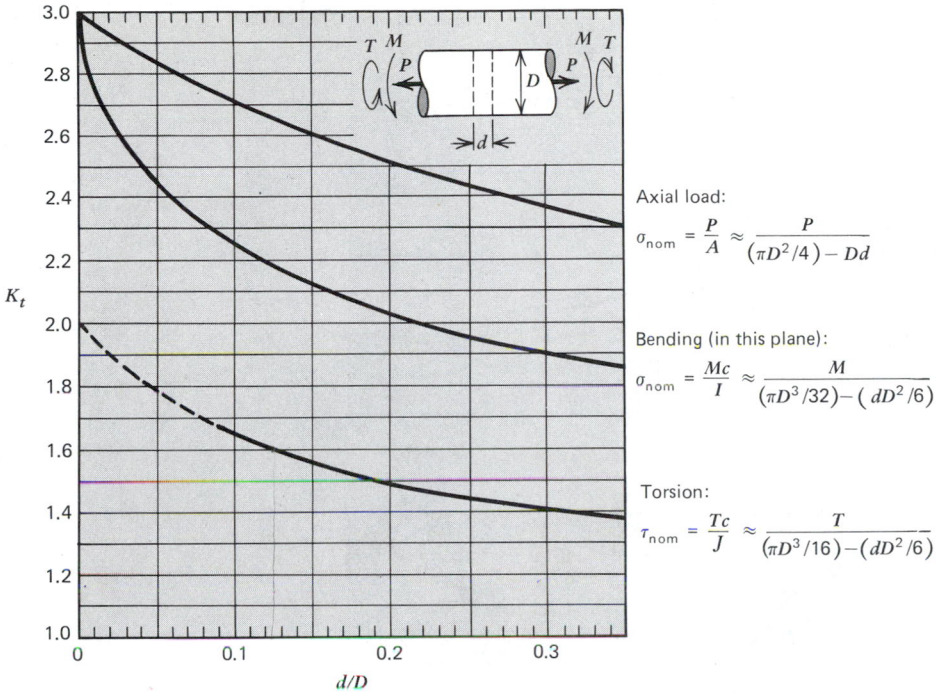

FIGURE 4.37.
Shaft with radial hole [7].

Note that the stress concentration graphs are plotted on the basis of *dimensionless ratios,* indicating that only the *shape* (not the size) of the part is involved. Also note that stress concentration factors are different for axial, bending, and torsional loading. Among the most extensive and authoritative references on stress concentration factors are those of R. E. Peterson [6, 7].

In many situations involving notched parts in tension or bending, the notch not only increases the primary stress but also causes one or both of the other principal stresses to take on nonzero values. This is referred to as the *biaxial or triaxial effect of stress raisers* ("stress raiser" is a general term applied to notches, holes, threads, etc.). Although this is a small secondary effect that will not be pursued further in this book, it is desirable to be able to visualize how these additional stress components can arise. Consider, for example, a soft rubber model of the grooved shaft in tension illustrated in Fig. 4.36*b*. As the tensile load is increased, there will be a tendency for the outer surface to pull into a smooth cylinder. This will involve an *increase* in the diameter and circumference of the section in the plane of the notch. The increased circumference gives rise to a *tangential* stress, which is a maximum at the surface. The increase in diameter is associated with the creation of *radial* stresses. (Remember, though, that this radial stress must be zero at the surface because there are no external radial forces acting there.)

Stress concentration factor graphs, such as those in Figs. 4.35 to 4.41, pertain to

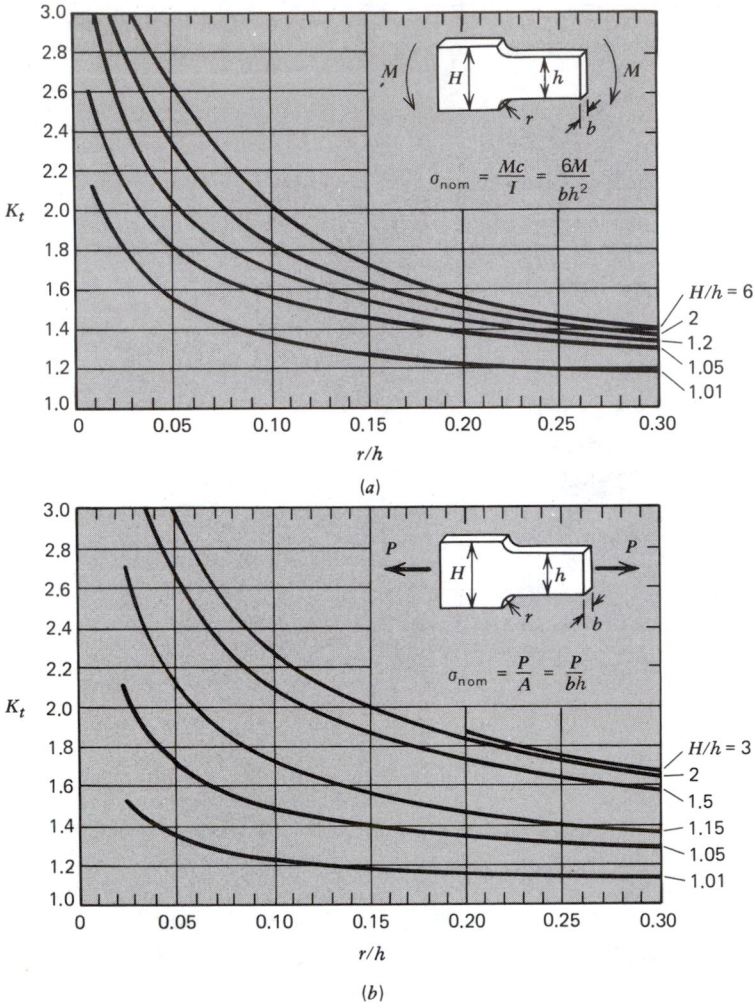

FIGURE 4.38.
Bar with shoulder fillet (a) bending; (b) axial load [7].

the maximum stress, existing at the surface of the stress raiser. The lower values of stress elsewhere in the cross section are seldom of interest but can be determined analytically from theory of elasticity in simple cases, or they can be approximated by finite element techniques or by experimental procedures, such as photoelasticity. The variation in stress over the cross section (i.e., the stress *gradient*) is given for a few cases in [3].

4.13 IMPORTANCE OF STRESS CONCENTRATION

It should be emphasized that the stress concentration factors given in the graphs are *theoretical* (hence, the subscript t) or *geometric* factors based on a theoretical homo-

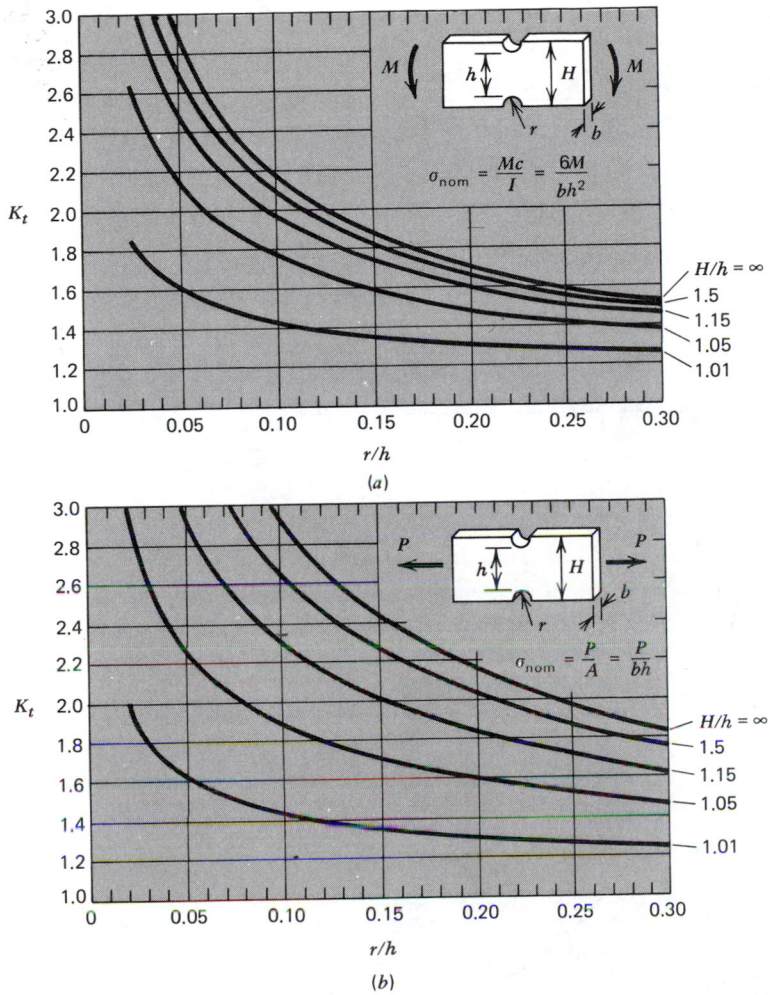

FIGURE 4.39.
Notched flat bar (a) bending; (b) tension [7].

geneous, isotropic, and elastic material. Real materials involve microscopic irregularities causing a certain nonuniformity of microscopic stress distribution, even in notch-free parts. Hence, the introduction of a stress raiser may not do as much *additional* damage as indicated by the theoretical factor. Also, real parts—even if free of stress raisers—involve surface irregularities resulting from processing and use that can be considered as extremely small notches.

The extent to which the engineer must take stress concentration into account depends on (a) the extent to which the real material deviates from the theoretical, and (b) whether the loading is static, or involves impact or fatigue. For materials permeated with internal discontinuities, as gray cast iron, stress raisers usually have little effect, regardless of the nature of loading. This is because surface or geometric irregularities seldom cause more severe stress concentration than that already associated with the

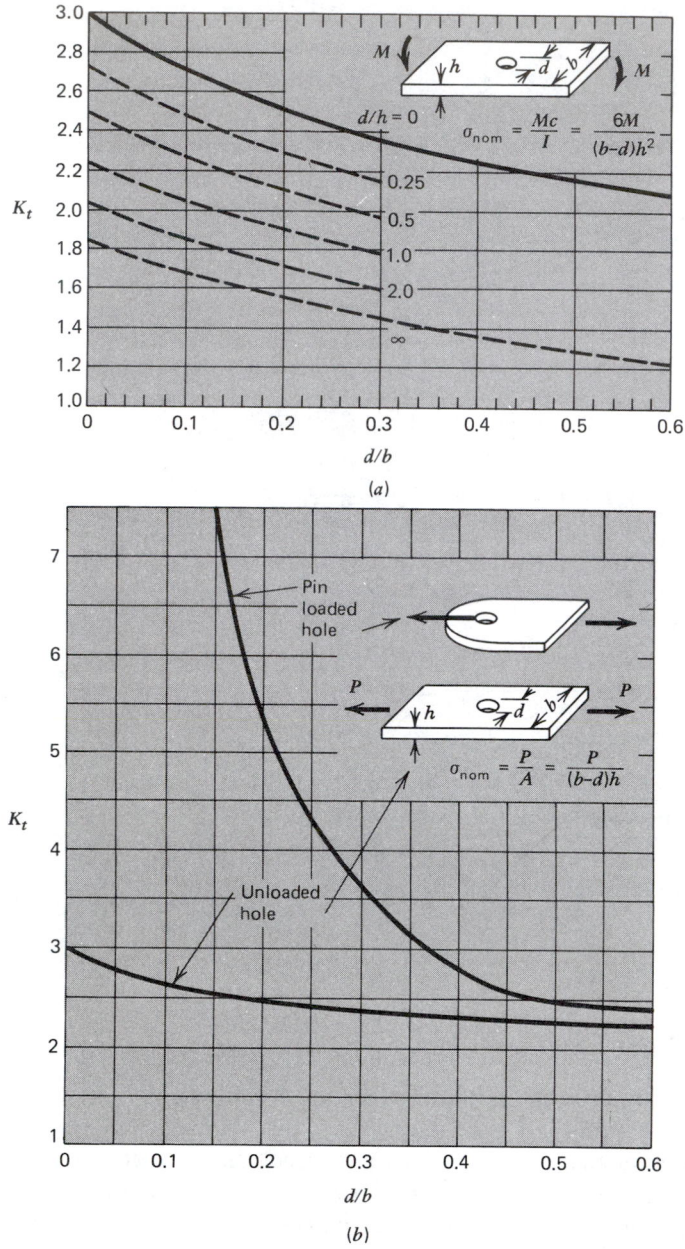

FIGURE 4.40.
Plate with central hole (a) bending [7]; (b) axial load [10].

internal irregularities. For fatigue and impact loading of most engineering materials, stress concentration must be considered, as will be seen in subsequent chapters. For the case of static loading being treated in this chapter, stress concentration is important

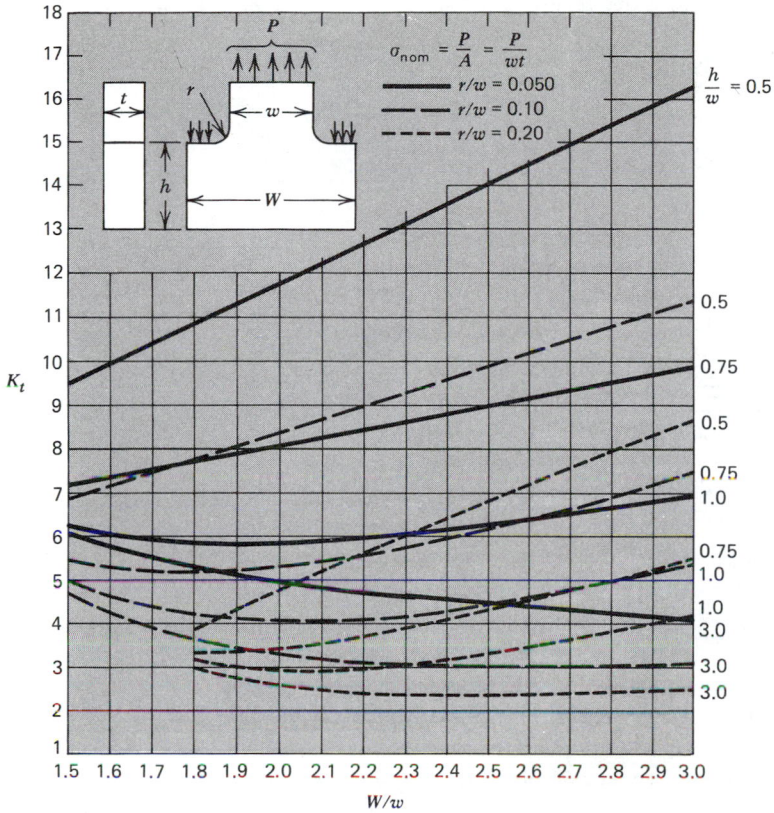

FIGURE 4.41.
T-head, axial load [7].

only with unusual materials that are both brittle and relatively homogeneous[2]; or for normally ductile materials that, under special conditions, *behave* in a brittle manner (see Chapter 6 for further discussion). For the usual engineering materials involving some ductility (and under conditions such that they *behave* in a ductile manner), it is customary to ignore stress concentration for static loads. The basis for this is illustrated below.

Figures 4.42*a* and 4.42*b* show two flat tensile bars each having a minimum cross-sectional area of A, and each made of a ductile material having the "idealized" stress-strain curve shown in 4.42*e*. The load on the unnotched bar (4.42*a*) can be increased to a value equal to the product of area times yield strength before failure (gross yielding) occurs. This is represented in 4.42*c*. Since the grooved bar in 4.42*b* has a stress concentration factor of 2, yielding will *begin* at only half the load, as shown in 4.42*d*. This is repeated as curve "a" of 4.42*f*. As the load is increased, the stress distribution (shown in part f) becomes "b," "c," and finally "d." These curves reflect a continuous

[2]A common example: When tearing open a package wrapped in clear plastic film, a sharp notch in the edge is *most* helpful!

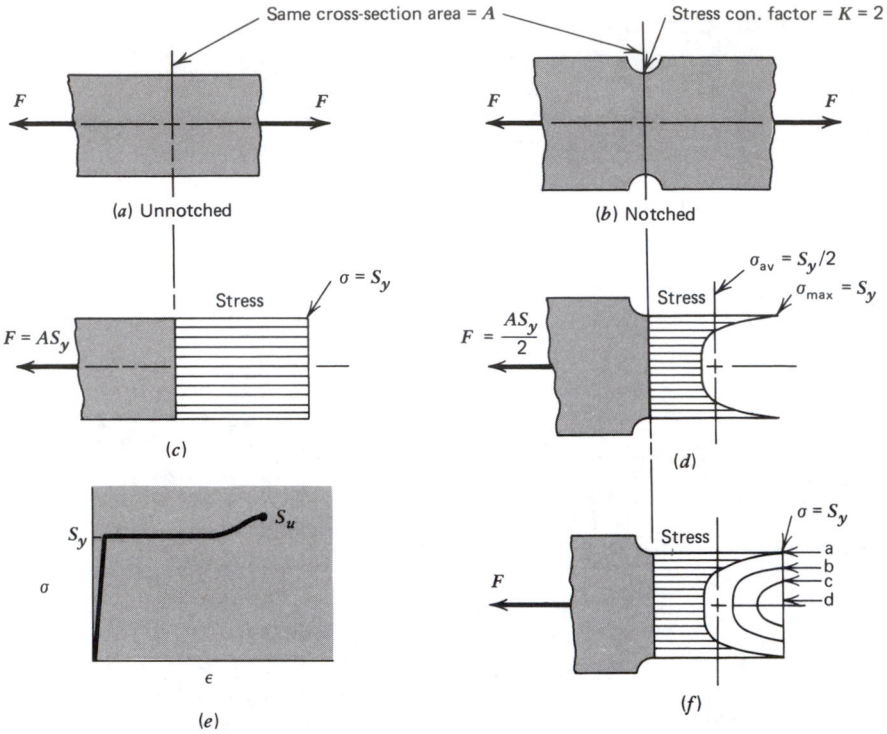

FIGURE 4.42.
Tensile stress distribution of an unnotched and a notched ductile part.

deepening of local yielding, which began at the root of the groove; but gross (or general) yielding involving the entire cross section is not ready to begin until "d" is reached. Note that the load associated with curve "d" is identical to the unnotched load capacity, shown in 4.42*c*. Also note that curve "d" can be achieved without significant stretching of the part. The part as a whole cannot be significantly elongated without yielding the entire cross section, including the portion at the center. Thus, for most practical purposes, the grooved bar will carry the same static load as the ungrooved bar.

4.14 RESIDUAL STRESSES DUE TO YIELDING – AXIAL LOADING

When a part is yielded nonuniformly throughout a cross section, *residual stresses* remain in that cross section after the external load is removed. Consider, for example, the four levels of loading of the notched flat tensile bar shown in Fig. 4.42*f*. This same bar and the four levels of loading are represented in the left column of Fig. 4.43. Note that only *slight* yielding is involved—not major yielding such as often occurs in processing. The middle column in this figure shows the *change* in stress due to removing the load. Except for 4.43*a*, where the load was not quite enough to cause yielding at the notch root, the stress change upon load removal does not exactly cancel the stresses

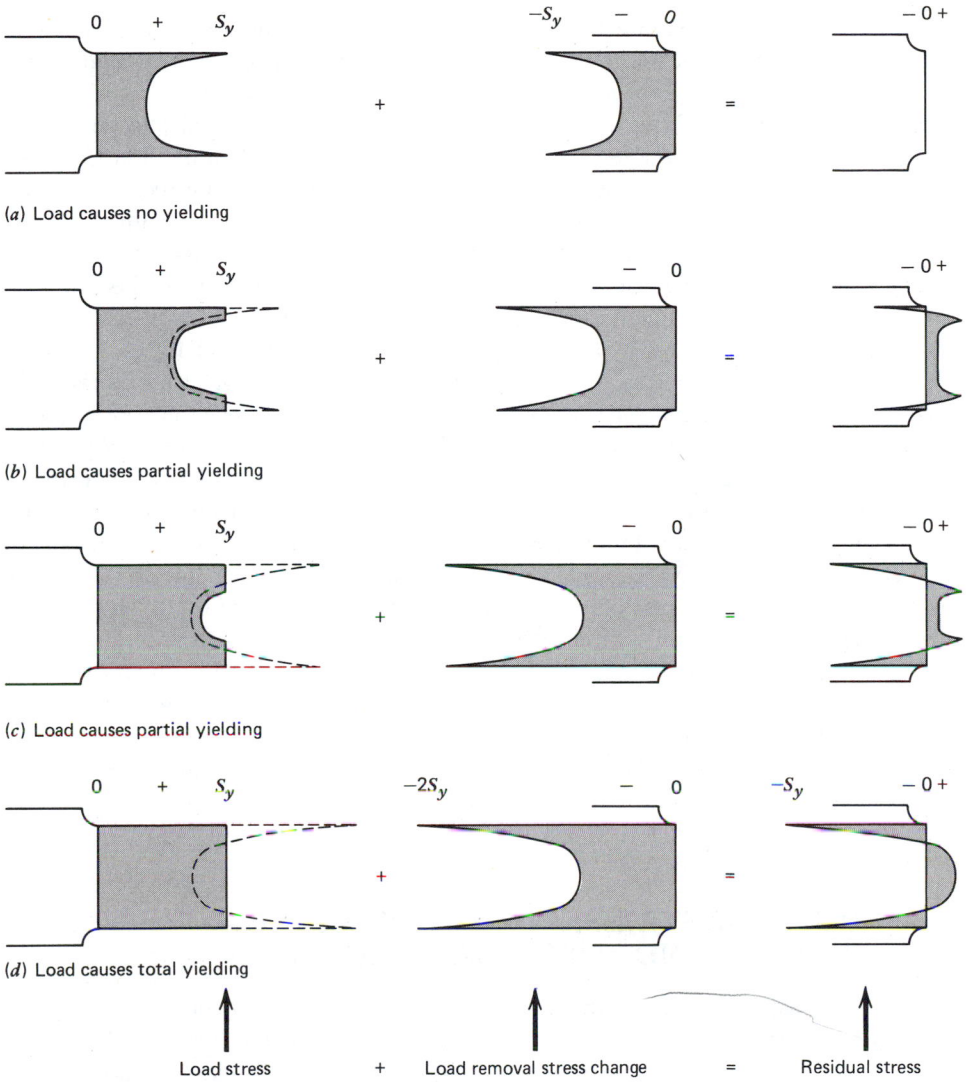

FIGURE 4.43.
Residual stresses due to yielding of a notched tensile bar of $K = 2$ for stress gradients a-d in Fig. 4.42(f).

due to load application. Hence, *residual stresses* remain after the load is removed. These are shown in the right column of Fig. 4.43.

Note that in each case shown in Fig. 4.43, the stress *change* due to load removal is *elastic.*

It is often helpful in visualizing the development of residual stresses such as those shown in Fig. 4.43 to imagine a column of small strain gages mounted from the top to bottom of the notched section. If these gages are attached *while the load is applied to the bar,* they will initially all read zero, although the actual stresses in the cross section are as shown in the left column. As the tensile load is released, the gages will all indicate *compression,* as shown in the middle column of the figure. The *average* compressive stress indicated by the gages when the load is completely removed will, of course, be P/A, but the distribution of this compressive stress will be completely elastic, so long as no yielding occurs *during the load release.* This provision is satisfied in all of the cases shown. Even in 4-43*d*, where the elastic change in stress at the notch root is $2S_y$ (the average change in stress is S_y, and at the notch root it is KS_y), no yielding occurs. Assuming equal yield strengths in tension and compression, the notch root material goes from S_y in tension with the load applied to S_y in compression when the load is released.

The elastic stress gradient curves associated with the various loads can be estimated graphically, as shown by the dotted lines in the left column of Fig. 4.43. (Note that in each case the dotted curve corresponds to the same *average* stress as the solid curve, and that the maximum stress shown in the dotted curve is twice the average stress because $K = 2$.) After the dotted curves are sketched, the center column load release curves are obtained by merely changing the sign. Once this procedure is understood, the center column plots can be dispensed with, and the residual stress curves obtained by merely subtracting the dotted curves from the solid curves in the left column.

Without determining the actual shape of the stress distribution curves (i.e., stress gradients), the residual stress curves obtained in Fig. 4.43 are admittedly approximations. They do, however, reflect the correct surface residual stress and general shape of residual stress distribution curve, and these are usually the matters of primary interest. It must be remembered, too, that the above development of residual stress curves was based on assuming that the material conforms to the idealized stress-strain curve of Fig. 4.42*e*. For that reason also, the residual stress curves in Fig. 4.43 can be no better than good approximations.

4.15 RESIDUAL STRESSES DUE TO YIELDING – BENDING AND TORSIONAL LOADING

Figure 4.44 illustrates residual stresses due to the bending of an unnotched rectangular beam. The figure illustrates the specific case of a 25 × 50 mm beam made of steel having an idealized stress-strain curve with $S_y = 300$ MPa. Unknown moment M_1 produces the stress distribution shown in 4.44*a*, with yielding to a depth of 10 mm. Let us first determine the magnitude of moment M_1.

If the distributed stress pattern is replaced with concentrated forces F_1 and F_2 at the centroids of the rectangular and triangular portions of the pattern, respectively, M_1 is equal to the sum of the couples produced by F_1 and F_2. The magnitude of F_1 is equal to the product of the average stress (300 MPa) times the area over which it acts (10 mm × 25 mm). Similarly, F_2 is equal to an average stress of 150 MPa times an area of 15 mm × 25 mm. The moment arms of the couples are 40 mm and 20 mm, respectively. Hence,

$$M_1 = (300 \text{ MPa} \times 250 \text{ mm}^2)(0.040 \text{ m}) + (150 \text{ MPa} \times 375 \text{ mm}^2)(0.020 \text{ m})$$
$$= 4125 \text{ N}\cdot\text{m}$$

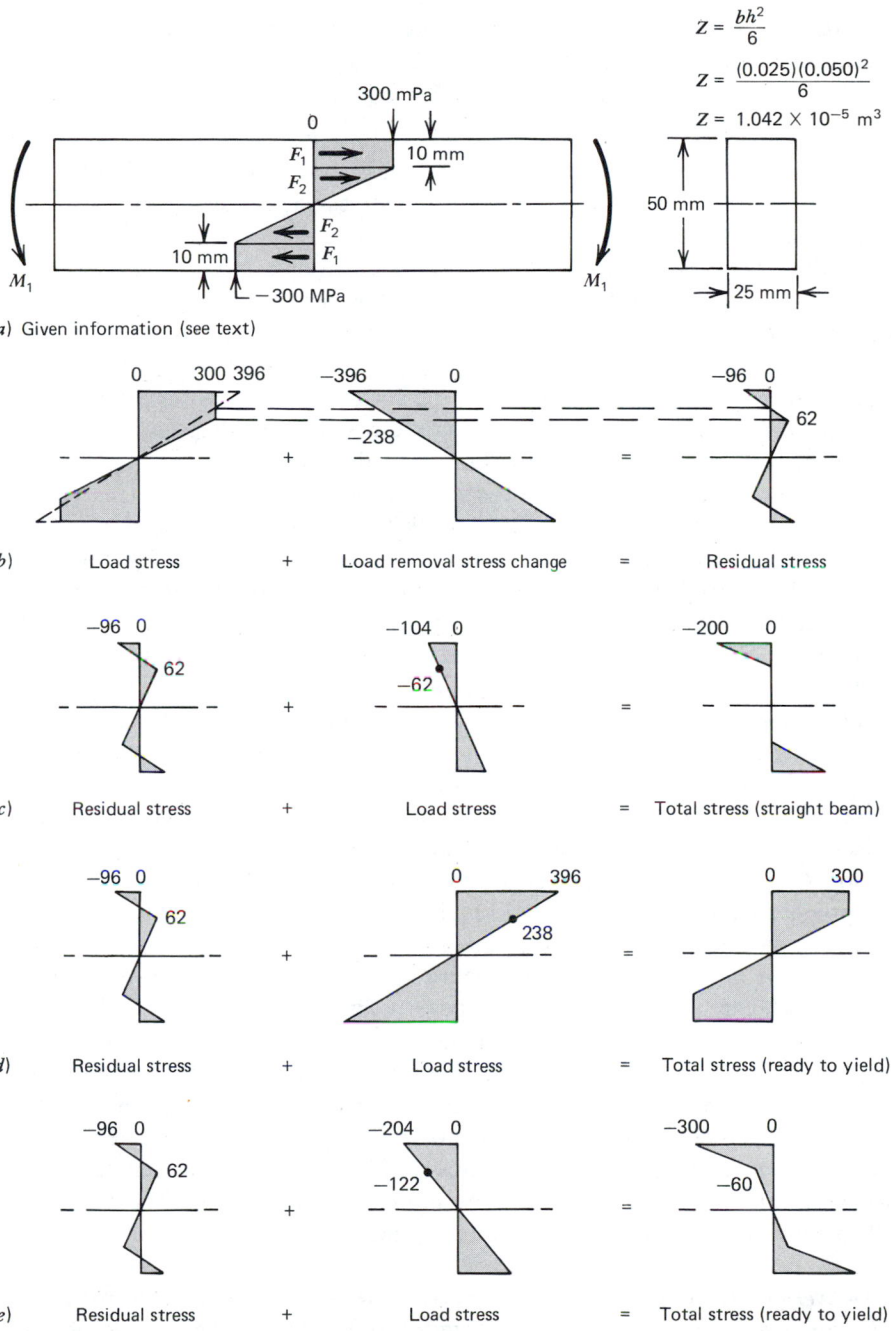

FIGURE 4.44.
Residual stresses in an unnotched rectangular beam.

Next, let us determine the residual stresses remaining after moment M_1 is removed. The elastic stress change due to removal of M_1 is

$$\sigma = M/Z = 4125 \text{ N}\cdot\text{m}/1.042 \times 10^{-5} \text{ m}^3$$
$$= 3.96 \times 10^8 \text{ Pa} = 396 \text{ MPa}$$

The elastic stress distribution due to load removal is shown in the center plot of 4.44*b*. This, added to the load stress, gives the residual stress pattern shown at the right side of that figure.

The dotted line plotted on the load stress diagram of 4.44*b* is the negative of the load removal stress. Since both the solid and dotted patterns on that diagram correspond to the same value of bending moment, one can observe the graphical relationship that the moment of the solid pattern is equal to the moment of the dotted pattern. In retrospect, this fact could have been used to draw the dotted pattern fairly accurately without making any calculations. Notice how points on the load stress diagram serve to locate the points of zero and 62 ksi on the residual stress diagram.

Note that at this point the beam is *slightly* bent. The outer portions that were yielded by the load do not want to come back to their initial positions, whereas the center portions that did not yield do. Thus, a balance of these opposing tendencies is reached, with the residual stress pattern satisfying the equilibrium requirements of $\Sigma F = 0$ and $\Sigma M = 0$. We *know* the beam is slightly bent just by looking at the residual stress pattern. The center portion that was initially straight and stress-free was not yielded; hence, it can again be straight only if it is stress-free.

Figure 4.44*c* shows that the desired center portion stress-free condition requires superimposing a load that develops a compressive stress of 62 MPa 10 mm below the surface. With this load in place, total stresses are as shown at the right of the figure. Since center portion stresses are zero, the beam is indeed straight. Let us compute the magnitude of the moment required to hold the beam straight. It is already known that an elastic surface stress of 396 MPa is associated with a moment of 4125 N·m. By simple proportion, a stress of 104 MPa requires a moment of *1083 N·m.*

Let us now determine the *elastic* bending moment capacity of the beam *after the residual stresses have been established.* Figure 4.44*d* shows that a moment in the same direction as M_1 can be added that superimposes a surface stress of +396 MPa without yielding. From previous calculations, it is known that this stress is associated with a moment of *4125 N·m.* A moment's reflection indicates that this conclusion is obvious: The *release* of original moment M_1 = 4125 N·m caused no yielding; hence, it can be *reapplied* without yielding. Figure 4.44*e* shows that in the direction opposite the original moment M_1, a moment giving a surface stress of 204 MPa is all that can be elastically withstood. Again, by simple proportion, this corresponds to a moment of *2125 N·m.*

The above study illustrates an important principle:

> *An overload causing yielding produces residual stresses that are favorable to future loads in the same direction, and unfavorable to future loads in the opposite direction.*

Furthermore, based on the idealized stress-strain curve, the increase in load capacity in one direction is exactly equal to the decrease in capacity in the opposite direction. These principles can also be illustrated for tensile loading, using Fig. 4.43.

The example of Fig. 4.44 could be carried a step further by considering the external moment required to permanently straighten the beam (so that the center section is again stress-free and therefore straight after the straightening moment is removed), and the new residual stress pattern resulting therefrom. This is done in [2].

Round bars overloaded in torsion can be treated in the same way as the above example for the rectangular bar in bending. The introduction of stress concentration in either the bending or torsional case involves no new concepts beyond those presented in this and the previous article.

4.16 THERMAL STRESSES

Thus far, only stresses caused by the application of external loads have been considered. Stresses can also be caused by nonuniform expansion and contraction of the material due either to temperature changes or to a material phase change. In actual mechanical and structural parts, an accurate quantitative evaluation of these stresses is, in general, beyond the scope of this text. It is important, however, for the student to become familiar with the basic principles involved. From these, important qualitative information can often be gained.

When the temperature of an unrestrained homogeneous, isotropic body is uniformly changed, it expands (or contracts) uniformly in all directions, according to the relationship:

$$\epsilon = \alpha \, \Delta T \tag{4.21}$$

where α is the coefficient of thermal expansion and ΔT is the temperature change. Values of α for several common metals are given in Appendix C-1. This uniform, unrestrained volume change produces no shear strain and no axial or shear stresses.

If restraints are placed on the member during the temperature change, the resulting stresses can be determined by (a) computing the dimensional changes that would take place in the *absence* of constraints, (b) determining the restraining loads necessary to *enforce* the restrained dimensional changes, and (c) computing the stresses associated with these restraining loads. This procedure is illustrated by the following sample problem.

SAMPLE PROBLEM 4.3

A 10-in. length of steel tubing (with properties of $E = 30 \times 10^6$ psi and $\alpha = 7 \times 10^{-6}$ per degree Fahrenheit) having a cross-sectional area of 1 in.^{2} is installed with "fixed" ends so that it is stress-free at 80°F. In operation, the tube is heated throughout to a uniform 480°F. Careful measurements indicate that the "fixed" ends separate by 0.008 in. What loads are exerted on the ends of the tube, and what are the resultant stresses?

Solution

1. For the unrestrained tube

$$\epsilon = \alpha\,\Delta T = (7 \times 10^{-6})(400) = 2.8 \times 10^{-3}$$
$$\Delta L = L\epsilon = 10 \text{ in. } (2.8 \times 10^{-3}) = 0.028 \text{ in.}$$

2. Since the measured expansion was only 0.008 in., the constraints must apply forces sufficient to produce a deflection of 0.020 in. From the relationship

$$\delta = \frac{PL}{AE}$$

which is from elementary elastic theory, and reviewed in Chapter 5,

$$0.020 = \frac{P(10)}{(1)(30 \times 10^6)}, \text{ or } P = 60{,}000 \text{ lb}$$ ■

3. Because the area is unity, $\sigma = 60$ *ksi.* ■

Since these answers are based on elastic relationships, they are valid only if the material has a yield strength of at least 60 ksi *at 480°F.*

If stresses due to temperature change are undesirably large, the best solution is often to reduce the constraint. For example, in the above problem, elimination or drastic reduction of the end fixity would correspondingly eliminate or drastically reduce the 60-ksi computed stress. This is commonly done by using expansion joints, or telescopic joints with appropriate seals.

Thermal stresses also result from the introduction of *temperature gradients* within a member. For example, if a thick metal plate is heated in the center of one face with a torch, the hot surface is restrained from expanding by the cooler surrounding material; consequently, it is in a state of compression. Likewise, the remote cooler metal is forced to expand, causing tensile stresses. A thick plate that is heated on both faces has the outer surface material in biaxial compression and the interior in biaxial tension. The laws of equilibrium require that all forces and moments arising from these internal stresses balance within themselves. If this is not the case for the original geometry of the part, it will distort or warp to a size and shape that *does* result in internal equilibrium. As long as all stresses so introduced are within the elastic limit at the temperatures involved, the part will revert to its original geometry when the initial temperature conditions are restored. If some portion of the part yields, this portion will not tend to revert to the initial geometry, and there will be warpage and internal (residual) stresses when initial temperature conditions are restored. The warpage or distortion of the part is such as to satisfy the requirements of equilibrium. This must be taken into account, for example, in the design of brake drums.

Residual stresses are commonly produced by the thermal gradients associated with heat-treating, flame cutting, welding, and, to a lesser extent, by grinding and some machining operations. For example, when a uniformly heated part is quenched, the surface cools first, and at its lower temperature, has a relatively high yield strength. The subsequent thermal contraction of the core material is resisted by the outer skin, which is thereby placed in residual compression. The core is left in triaxial tension, following the rule "what cools last is in tension." (Note that the surface stresses cannot be triaxial because of the unloaded exposed surface.)

This same principle explains why flame cutting and most welding operations tend to leave surfaces in residual tension: if heating occurs predominantly near the surface, the tendency for surface thermal expansion is resisted by the cooler core. Having a relatively low yield strength at high temperature, the surface yields in compression. Upon cooling, the skin tends to contract, but is again largely restrained by the core. Thus, the surface material is left in biaxial tension.

A related phenomenon producing residual stresses in steel is phase transformation. When steel with sufficient carbon content is quenched from above its critical temperature to form martensite, the new lattice structure is slightly less dense, causing the transformed material to expand slightly. With thorough hardening, the transformation normally occurs last in the interior. This causes undesirable residual tensile stresses in the surface. Special processing can cause the transformation to occur last in the outer skin, giving favorable residual compressive stresses in the surface material.

Residual stresses are added to any subsequent load stresses in order to obtain the total stresses. Furthermore, if a part with residual stresses is subsequently machined, the removal of residually stressed material causes the part to warp or distort. This is true because the removal of this material upsets the internal equilibrium of the part. Subsequent warpage *must* take place to arrive at a new geometry satisfying equilibrium requirements. In fact, a common (destructive) method for determining the residual stress in a particular zone of a part is to very carefully remove material from that zone and then to make a precision measurement of the resulting change in geometry.

Residual stresses are often removed by annealing. This involves uniformly heating the unrestrained part (to a sufficiently high temperature and for a sufficiently long period of time) to cause a virtually complete relief of the internal stresses by localized yielding. The subsequent slow cooling operation introduces no yielding. Hence, the part reaches room temperature in a virtually stress-free state.

For a more detailed discussion of phenomena related to residual stresses, see [2].

4.17 IMPORTANCE OF RESIDUAL STRESSES

In general, residual stresses are important in situations where stress concentration is important. These include brittle materials involving all loading types, and the fatigue and impact loading of ductile as well as brittle materials. For the static loading of ductile materials, harmless local yielding can usually occur to relieve local high stresses resulting from either (or both) stress concentration or superimposed residual stress.

It is easy to overlook residual stresses because they involve nothing that ordinarily brings them to the attention of the senses. When one holds an unloaded machine part, for example, there is normally no way of knowing whether the stresses are all zero or whether large residual stresses are present. Usually there are no readily available means for determining residual stresses. However, a reasonable qualitative estimate can often be made by considering the thermal and mechanical loading history of the part, both during and after manufacture.

Almen and Black[3] cite an interesting example showing that residual stresses remain in a part so long as heat or external loading does not remove them by yielding. The Liberty Bell, cast in 1753, has residual tensile stresses in the outer surface because the casting cooled most rapidly from the *inside* surface (the principle that "what cools last is in residual tension"). After 75 years of satisfactory service, the bell cracked, probably as a result of fatigue from superimposed vibratory stresses caused by ringing the bell. Attempts were made to arrest the crack growth by drilling holes at the ends, but further crack extension subsequently occurred. Almen and Black cite this as proof that residual stresses are still present in the bell.

REFERENCES

1. Durelli, A. J., E. A. Phillips, and C. H. Tsao, *Introduction to the Theoretical and Experimental Analysis of Stress and Strain,* McGraw-Hill, New York, 1958.
2. Juvinall, R. C., *Engineering Considerations of Stress, Strain, and Strength,* McGraw-Hill, New York, 1967.
3. Lipson, C., and R. C. Juvinall, *Handbook of Stress and Strength,* Macmillan, New York, 1963.
4. Mohr, O., *Zivilingenieur,* p. 113, 1882.
5. Neuber, Heinz, *Theory of Notch Stresses,* J. W. Edwards, Inc., Ann Arbor, MI, 1946 (translation of the original German version published in 1937).
6. Peterson, R. E., *Stress Concentration Factors,* John Wiley & Sons, Inc., New York, 1974.
7. Peterson, R. E., *Stress Concentration Design Factors,* John Wiley & Sons, Inc., New York, 1953.
8. Roark, R. J., and W. C. Young, *Formulas for Stress and Strain,* 5th ed., McGraw-Hill, New York, 1975.
9. Seely, F. B., and J. O. Smith, *Advanced Mechanics of Materials,* 2nd ed., John Wiley & Sons, Inc., New York, 1952. (Also 3rd ed. by A. P. Boresi, O. M. Sidebottom, F. B. Seely, and J. O. Smith, John Wiley & Sons, 1978.)
10. Smith, Clarence R., "Tips on Fatigue," Report NAVWEPS 00-25-559, Bureau of Naval Weapons, Washington, D.C., 1963.
11. Timoshenko, S., and J. N. Goodier, *Theory of Elasticity,* 2nd ed., McGraw-Hill, New York, 1951. (Also, Timoshenko, S., *Theory of Elasticity,* Engineering Societies Monograph, McGraw-Hill, New York, 1934.)

[3]Almen, John O., and Paul H. Black, *Residual Stresses and Fatigue in Metals,* McGraw-Hill, New York, 1963.

CHAPTER FIVE

Elastic Strain, Deflection, and Stability

5.1 INTRODUCTION

Elastic strain, deflection, stiffness, and stability are considerations of basic importance to the engineer. Deflection or stiffness, rather than stress, is often the controlling factor in the design of a part. Machine tool frames, for example, must be extremely rigid to maintain manufacturing accuracy. When made massive enough to satisfy rigidity requirements, the stresses may be insignificantly low. Other parts may require great stiffness in order to eliminate vibration problems. Excessive deflection can cause interference between components, and disengagement of gears.

Another important aspect of elastic strains is their involvement in experimental techniques for measuring stresses. Stress is not, in general, a directly measurable quantity; strain is. When the elastic constants of a material are known, experimentally determined strain values can be transposed into corresponding stress values by means of the elastic stress-strain relationships reviewed in Section 5.5.

Figures 5.1*a*–5.1*d* illustrate elastically stable systems. In such cases a small disturbance of the equilibrium illustrated results in corrective elastic restoring forces and/or moments. Such may not be the case in the slender column shown in Fig. 5.1*e*. Here, if the column is slender enough, the elastic modulus low enough, and the load large enough, then the compression member will be *elastically unstable* and the slightest disturbance will cause *buckling* or collapse. This is true even though the P/A stress may be *well below* the elastic limit of the material. The last six sections in this chapter deal with this phenomenon.

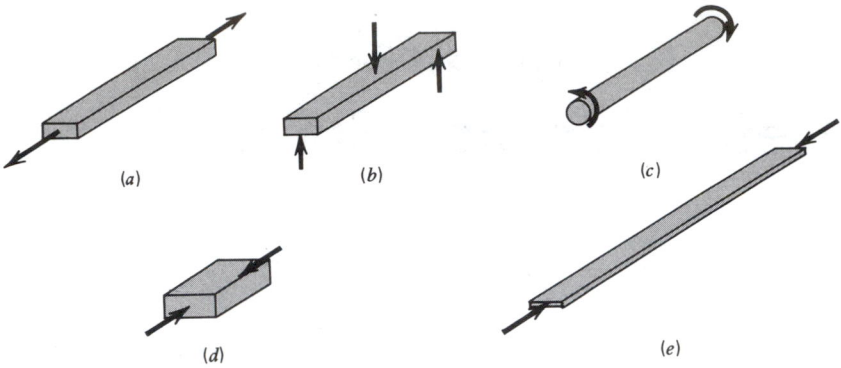

FIGURE 5.1.
Elastically stable (a, b, c, d) and potentially elastically unstable (e) loaded members.

5.2 STRAIN DEFINITION, MEASUREMENT, AND MOHR-CIRCLE REPRESENTATION

Figure 5.2 shows strains experienced by an element subjected to uniaxial tension. Equations defining the three linear components of strain are given on the figure. The loaded element experiences no change in any of the initial right angles. Shear strains γ_{xy}, γ_{xz}, and γ_{yz} are therefore zero, and the element shown is a principal element. Because of this, ϵ_x, ϵ_y, and ϵ_z are also ϵ_1, ϵ_2, and ϵ_3, where subscripts 1, 2, and 3 denote the principal directions. The negative values shown for ϵ_y and ϵ_z result from the material having a positive Poisson's ratio.

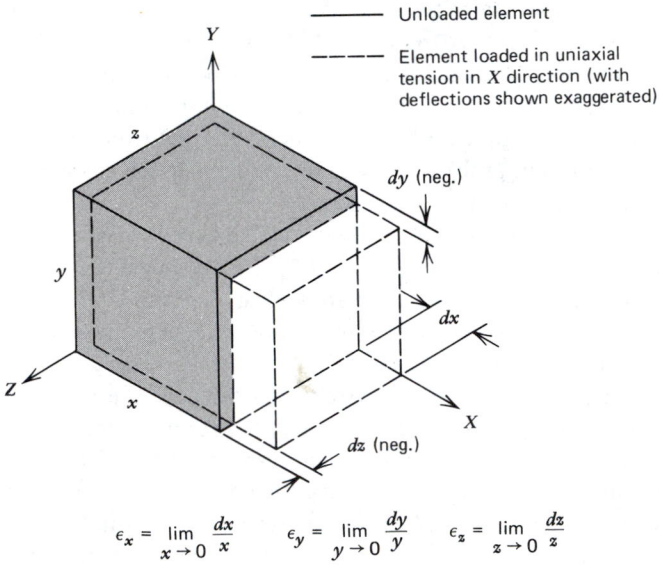

FIGURE 5.2.
Linear strain illustrated for uniaxial tensile loading.

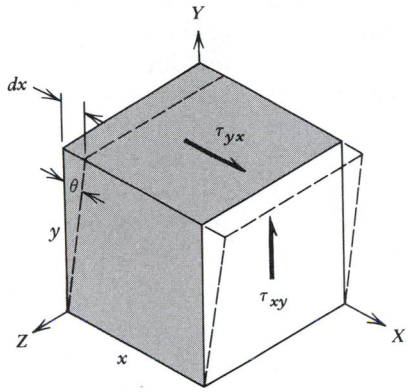

γ_{xy} (shown counterclockwise, hence negative)
γ_{yx} (shown clockwise, hence positive)
$\Big\}$ absolute value $= \lim_{y \to 0} \frac{dx}{y} = \tan\theta \approx \theta$

FIGURE 5.3.
Shear strain illustrated for pure shear loading.

Figure 5.3 shows a similar element subjected to pure shear. The resulting shear strains are defined on the figure. The double subscript notation corresponds to the convention mentioned briefly in Section 4.9 in connection with Fig. 4.26. As was true in the case of shear stresses, there is no need here to be concerned with a distinction between γ_{xy} and γ_{yx}.

Figure 5.4 illustrates how a Mohr strain circle can be plotted from a knowledge of ϵ_x, ϵ_y, and γ_{xy}. The procedure is identical to that used with the Mohr stress circle except that care must be taken to plot as ordinates only *half* the shear strain, γ. Analytical stress equations (4.15) to (4.19) have direct counterparts for strain, with ϵ being substituted for σ, and $\gamma/2$ substituted for τ.

When experimentally determining stresses at critical points on machine and struc-

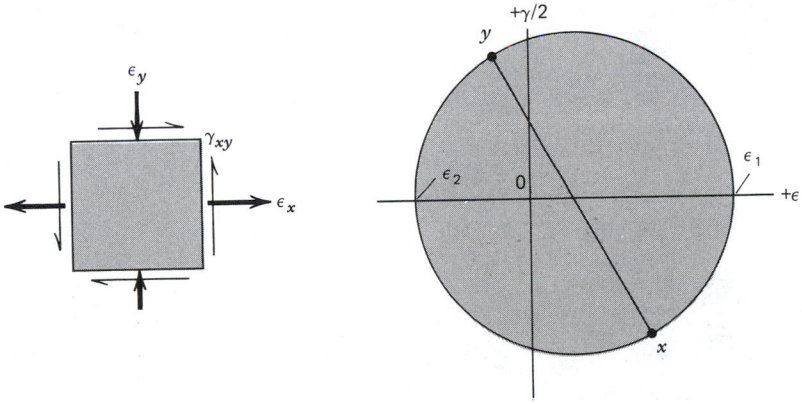

FIGURE 5.4.
Mohr strain circle drawn for known values of ϵ_x, ϵ_y, and γ_{xy}.

tural parts, one begins by measuring strains, and then computes the corresponding stresses for materials of known elastic constants. If the directions of the principal axes are known, principal strains ϵ_1 and ϵ_2 can be measured directly, as by using electric-resistance strain gages of the single-element type (Fig. 5.5*a*) or two-element rosettes (Fig. 5.5*b*). If principal directions are not known, one could theoretically determine ϵ_1 and ϵ_2 by first measuring arbitrarily oriented strains ϵ_x, ϵ_y, and γ_{xy}, and then determining ϵ_1 and ϵ_2 from a Mohr-circle plot as in Fig. 5.4. Unfortunately, direct experimental measurement of shear strain is not normally practical. Hence, use is made of three-

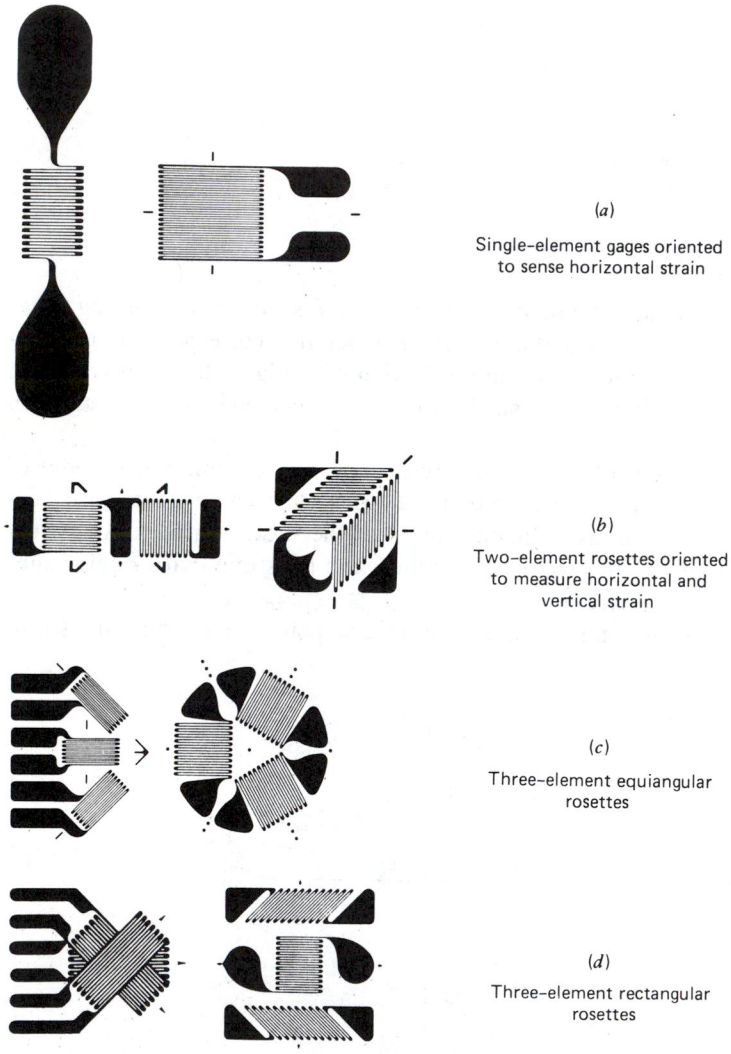

FIGURE 5.5.
Grid configurations of typical metal-foil electric-resistance strain gages.

element strain gage rosettes, illustrated in Figs. 5.5*c* and 5.5*d*. The rationale associated with this is as follows:

First, note that the conventional Mohr-circle construction—either for stress, as shown in Fig. 4.27, or for strain, as shown in Fig. 5.4—provides a convenient graphical solution of three equations in three unknowns. The unknowns are two principal stress (or strain) values, and the angle between the principal planes and the reference planes (usually designated by x and y). A solution requires three known values. These are σ_x, σ_y, and τ_{xy} (or ϵ_x, ϵ_y, and γ_{xy}). Three-element rosettes permit solutions for the same three unknowns, but the "knowns" are all *linear* strains. No simple graphical Mohr-circle type solution is available, so an analytical solution is used. Such a solution is available when given three linear strains in *any* direction [5], but only the two commonly used arrangements will be considered here: equiangular rosettes (Fig. 5.5*c*), and rectangular rosettes (Fig. 5.5*d*).

5.3 ANALYSIS OF STRAIN – EQUIANGULAR ROSETTES

All equiangular rosettes can be represented diagrammatically by Fig. 5.6*a*, which shows individual gages at 0°, 120°, and 240°, in *counterclockwise progression*. Angle α is measured from the 0° gage to the as-yet-unknown principal axes. Figures 5.6*b* and 5.6*c* show other combinations of gage orientations that are equivalent. Using the above notation, the equations for principal strain magnitude and direction are [3]:

$$\epsilon_{1,2} = \frac{\epsilon_0 + \epsilon_{120} + \epsilon_{240}}{3} \pm \sqrt{\frac{(2\epsilon_0 - \epsilon_{120} - \epsilon_{240})^2}{9} + \frac{(\epsilon_{120} - \epsilon_{240})^2}{3}} \tag{5.1}$$

$$\tan 2\alpha = \frac{\sqrt{3}(\epsilon_{120} - \epsilon_{240})}{2\epsilon_0 - \epsilon_{120} - \epsilon_{240}} \tag{5.2}$$

A *positive* value of α means that one measures *clockwise* from the ϵ_0 gage orientation to each of the two principal axis orientations given by Eq. (5.2). To tell which direction

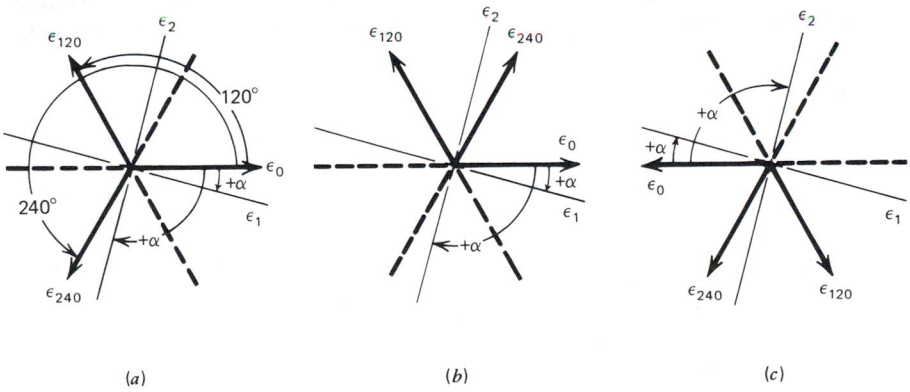

FIGURE 5.6.
Equiangular strain rosette representations.

goes with which principal axis, use the rule that the higher principal strain always lies within 30° of the algebraically highest of ϵ_0, ϵ_{120}, and ϵ_{240}. The use and interpretation of Eqs. (5.1) and (5.2) are illustrated by the following example.

SAMPLE PROBLEM 5.1

The following strains are obtained from an equiangular strain gage rosette:

$$\epsilon_0 = -0.00075 \text{ m/m}$$
$$\epsilon_{120} = +0.0004 \text{ m/m}$$
$$\epsilon_{240} = +0.00185 \text{ m/m}$$

Determine the magnitude and orientation of the principal strains, and check the results by plotting a Mohr circle.

Solution

From direct substitution in Eq. (5.1):

$$\epsilon_{1,2} = 0.0005 \pm 0.0015, \quad \epsilon_1 = 0.0020 \text{ m/m}, \quad \epsilon_2 = -0.0010 \text{ m/m}$$

From Eq. (5.2):

$$\tan 2\alpha = +0.67, \quad 2\alpha = 34°, 214°, \quad \alpha = 17°, 107°$$

The principal strain axes are 17° and 107° clockwise from the 0° gage axis. According to the 30° rule, it is the axis of ϵ_2 that is 17° from ϵ_0 (also, it is intuitively apparent that since ϵ_0 is the only negative gage reading, the principal strain closest to it will be the negative principal strain).

A vector representation of the magnitudes and directions of the gage readings and the principal strains is shown in Fig. 5.7.

The correctness of the solution is verified in Fig. 5.8, where a Mohr circle is drawn based on the computed values of ϵ_1 and ϵ_2. Points are marked on the circle corresponding to the angular orientations of ϵ_0, ϵ_{120}, and ϵ_{240}, as shown in Fig. 5.7 (remembering that

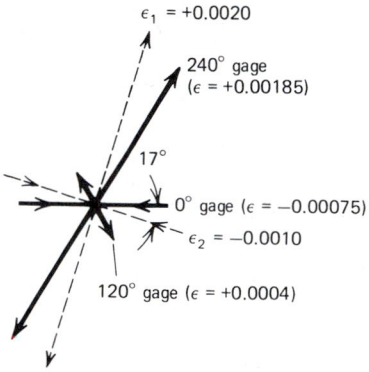

FIGURE 5.7.
Vector representation of Sample Problem 5.1 solution.

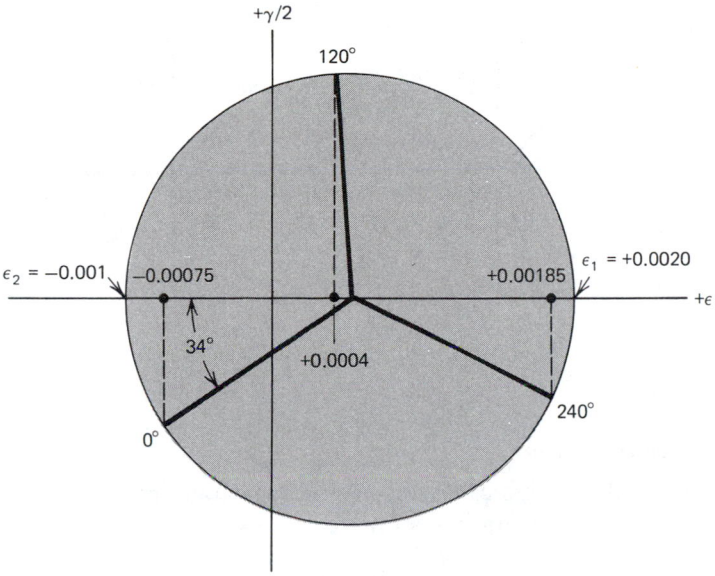

FIGURE 5.8.
Mohr circle check of Sample Problem 5.1 solution.

actual angles are doubled when representing them on the circle). Since the abscissas of these three points correspond to the given gage readings, the solution is correct.

A careful study of Figs. 5.7 and 5.8 is helpful in gaining an intuitive understanding of the significance of the calculations performed with Eqs. (5.1) and (5.2), and of the associated conventions for gage labeling and for discrimination between orientations of ϵ_1 and ϵ_2.

5.4 ANALYSIS OF STRAIN – RECTANGULAR ROSETTES

Figure 5.9*a* shows the basic configuration of the rectangular rosette, with three gage directions progressing *counterclockwise* at 45° increments from an arbitrarily oriented 0° gage. As with the equiangular rosette, there are several possible versions of the basic rectangular orientation. Two others are shown in Figs. 5.9*b* and 5.9*c*. Again, angle α is measured from the 0° gage direction to each of the principal directions. The equations for principal strain magnitude and direction are [5]:

$$\epsilon_{1,2} = \frac{\epsilon_0 + \epsilon_{90}}{2} \pm \sqrt{\frac{(\epsilon_0 - \epsilon_{45})^2 + (\epsilon_{45} - \epsilon_{90})^2}{2}} \tag{5.3}$$

$$\tan 2\alpha = \frac{\epsilon_0 - 2\epsilon_{45} + \epsilon_{90}}{\epsilon_0 - \epsilon_{90}} \tag{5.4}$$

As with the equiangular rosette, a positive value of α means that one measures *clockwise* from the ϵ_0 gage to a principal axis. Discrimination between the two principal axes

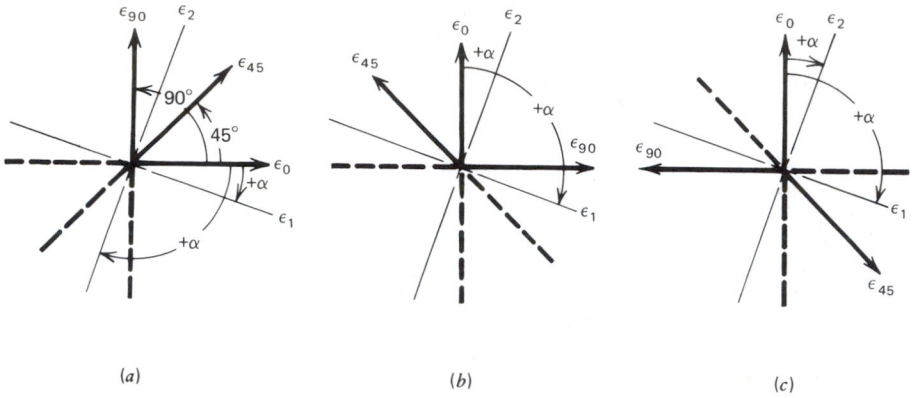

FIGURE 5.9.
Rectangular strain rosette representations.

can be based on the rule that the algebraically greater principal strain makes an angle of less than 45° with the algebraically larger of strains ϵ_0 and ϵ_{90}.

SAMPLE PROBLEM 5.2

Readings obtained from a rectangular rosette are as shown in Fig. 5.10*a* (the readings are in micrometers per meter, which is, of course, the same as microinches per inch). Determine the magnitude and orientation of the principal strains, and check the result by plotting a Mohr circle.

Solution

1. First note that in order to conform with the 45° increment counterclockwise progression, the gages must be labeled as shown in Fig. 5.10*b*.

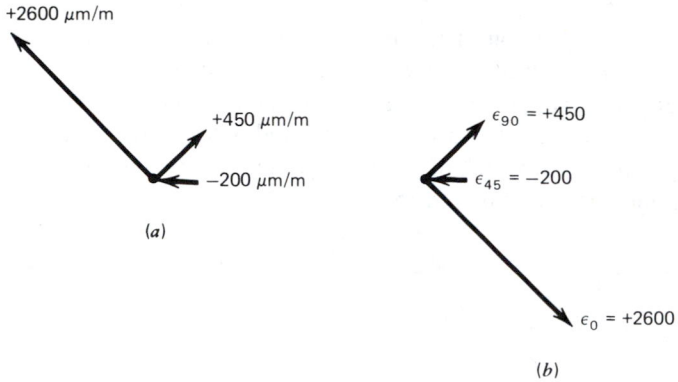

FIGURE 5.10.
Sample Problem 5.2 given data. (a) Gage readings. (b) Equivalent rosettes.

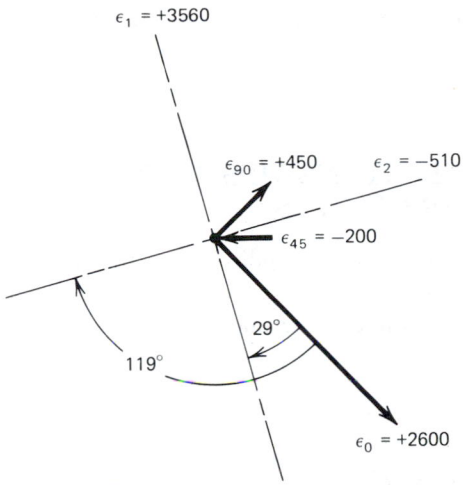

FIGURE 5.11.
Vector representation of Sample Problem 5.2 solution.

2. Substitution in Eqs. (5.3) and (5.4) gives:

$$\epsilon_{1,2} = 1525 \pm 2033, \quad \epsilon_1 = 3558\ \mu\text{m/m}, \quad \epsilon_2 = -508\ \mu\text{m/m}$$

$$\tan 2\alpha = 1.605, \quad 2\alpha = 58^\circ, 238^\circ, \quad \alpha = 29^\circ, 119^\circ$$

These results are represented in Fig. 5.11. ■

3. A Mohr circle is drawn in Fig. 5.12, based on the computed values of ϵ_1 and ϵ_2. Points are marked on the circle corresponding to the angular orientations of ϵ_0, ϵ_{45}, and ϵ_{90} shown in Fig. 5.11. The abscissas of these points are found to check with the given gage readings. ■

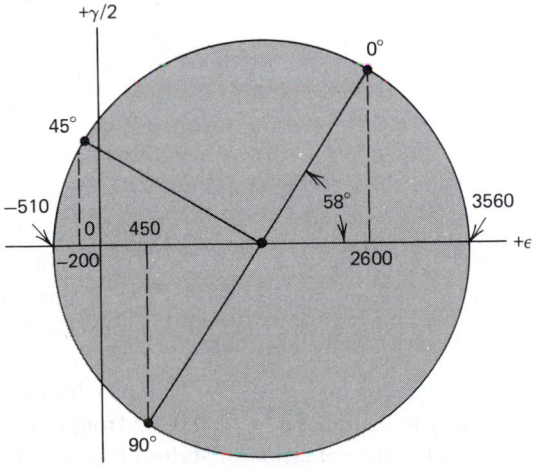

FIGURE 5.12.
Mohr-circle check of Sample Problem 5.2 solution.

Again, a careful study of Figs. 5.11 and 5.12 is advised in order to gain a clear understanding of the physical situation.

5.5 ELASTIC STRESS-STRAIN RELATIONSHIPS AND THREE-DIMENSIONAL MOHR CIRCLES

Elastic, or more precisely, *linearly elastic,* stress-strain relationships for the general case of triaxial stress are given in various references, such as [3]. For the frequently encountered *biaxial stress* case (where ϵ_1 and ϵ_2 are determined experimentally), these reduce to

$$\left.\begin{aligned} \sigma_1 &= \frac{E}{1-\nu^2}(\epsilon_1 + \nu\epsilon_2) \\ \sigma_2 &= \frac{E}{1-\nu^2}(\epsilon_2 + \nu\epsilon_1) \\ \sigma_3 &= 0 \\ \epsilon_3 &= \frac{-\nu}{1-\nu}(\epsilon_1 + \epsilon_2) \end{aligned}\right\} \quad \textbf{(5.5)}$$

Or, if stresses are known and strains are to be computed,

$$\left.\begin{aligned} \epsilon_1 &= \frac{1}{E}(\sigma_1 - \nu\sigma_2) \\ \epsilon_2 &= \frac{1}{E}(\sigma_2 - \nu\sigma_1) \\ \epsilon_3 &= -\frac{\nu}{E}(\sigma_1 + \sigma_2) \end{aligned}\right\} \quad \textbf{(5.6)}$$

For the case of *uniaxial stress* these equations reduce to

$$\left.\begin{aligned} \sigma_1 &= E\epsilon_1 \\ \sigma_2 &= \sigma_3 = 0 \end{aligned}\right\} \quad \textbf{(5.7)}$$

$$\left.\begin{aligned} \epsilon_1 &= \frac{\sigma_1}{E} \\ \epsilon_2 &= \epsilon_3 = -\frac{\nu\sigma_1}{E} \end{aligned}\right\} \quad \textbf{(5.8)}$$

A common serious error is to use $\sigma = E\epsilon$ for all stress states. For example, suppose that the equiangular strain gage rosette in Sample Problem 5.1 was mounted on an aluminum member of elastic constants $E = 71 \times 10^9$ Pa, $\nu = 0.35$. From $\epsilon_1 = 0.0020$ and $\epsilon_2 = -0.0010$, Eq. (5.5) gives $\sigma_1 = 136$ MPa, $\sigma_2 = -24$ MPa. The erroneous computation of principal stresses by merely multiplying each principal strain by E gives $\sigma_1 = 142$ MPa and $\sigma_2 = -71$ MPa.

To complete the determination of states of stress and strain for Sample Problem 5.1 with aluminum material, it is only necessary to compute $\epsilon_3 = -0.0005$ from either Eq. (5.5) or Eq. (5.6). The complete Mohr-circle representation of states of stress and strain is plotted in Fig. 5.13.

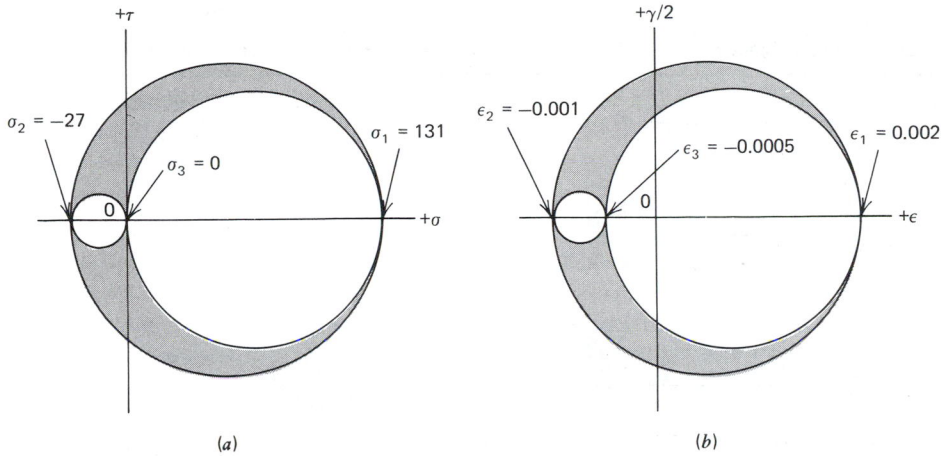

FIGURE 5.13.
State of (a) stress and (b) strain for Sample Problem 5.1 and aluminum material.

5.6 DEFLECTION AND SPRING RATE – SIMPLE CASES

The basic deflection and spring rate formulas are given in Table 5.1, supplemented, in the case of torsion, by Table 5.2. Derivations are not included, as the student has presumably covered these in previous courses. No effort should be made to memorize these equations. Instead, one should be sure to understand the rationale and pattern behind them—then most of the equations will be memorized automatically.

Note that the first three cases involve the deflection at the point of load application, and in the direction of load application. In each of these cases the equation merely states that deflection varies linearly with load (which it must, for the linear-elastic portion of the stress-strain curve) and with length, and inversely with a geometric rigidity property of the cross section and with the appropriate elastic rigidity property of the material. "Spring rate" is also known as *spring constant* or *spring scale*. For linear deflections the spring rate is designated by k (measured in pounds per inch, newtons per meter, etc.). For angular deflections, the symbol K is used (with units of lb·ft per radian, N·m per radian, etc.).

In case 4, note that the length must be squared. This must be so because linear deflection increases both with length and with end slope, the latter being itself a function of length. Likewise, in case 5, length must be *cubed* because the bending moment is an additional factor increasing with length. Incidentally, in case 5 the equation given for deflection results *only* from bending stresses—the contribution to deflection made by transverse shear being considered negligible. Castigliano's method, treated in Section 5.8, enables one to evaluate the shear contribution. For now, let us merely note that the shear contribution to deflection follows exactly the pattern of cases 1, 2, and 3: the shear deflection varies linearly with shear load and length, and inversely with shear modulus of elasticity and with a geometric shear rigidity property of the cross section. The latter is roughly equal to the area for most sections (it is equal to ⅚ times the area for a rectangular section).

TABLE 5.1
Deflection and Stiffness Formulas for Straight Bars (Rods, Beams) of Uniform Section

No.	Case	Deflection	Spring Rate
1.	Tension or Compression L, P Cross-section area = A	$\delta = \dfrac{PL}{AE}$	$k = \dfrac{P}{\delta} = \dfrac{AE}{L}$
2.	Torsion L, T K'[a] = section property. For solid round section, $K' = J = \pi d^4/32$.	$\theta = \dfrac{TL}{K'G}$ For solid round bar and deflection in degrees, $\theta^\circ = \dfrac{584TL}{d^4G}$	$K = \dfrac{T}{\theta} = \dfrac{K'G}{L}$
3.	Bending (angular deflection) L, θ, M I = moment of inertia about neutral bending axis	$\theta = \dfrac{ML}{EI}$	$K = \dfrac{M}{\theta} = \dfrac{EI}{L}$
4.	Bending (linear deflection) L, δ, M I = moment of inertia about neutral bending axis	$\delta = \dfrac{ML^2}{2EI}$	$k = \dfrac{M}{\delta} = \dfrac{2EI}{L^2}$
5.	Cantilever beam loaded at end L, P, δ I = moment of inertia about neutral bending axis	$\delta = \dfrac{PL^3}{3EI}$	$k = \dfrac{P}{\delta} = \dfrac{3EI}{L^3}$

Note: See Appendices A-4 and A-5 for appropriate SI units and prefixes.
[a]See Table 5.2 for values of K' for other sections.

Tables that give deflection equations for beams with a wide variety of loading conditions are contained in various handbooks and references, such as [4].

5.7 BEAM DEFLECTION

Beams are structural members subjected to transverse loads. Examples include machinery shafts, building floor joists, leaf springs, automobile frame members, and

TABLE 5.2
Formulas for Torsional Deflection

$\theta = \dfrac{TL}{K'G}$, where K' is given below.

Cross section	Formula for K'
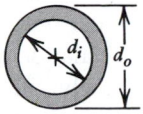	$K' = J = \frac{\pi}{32}(d_o^4 - d_i^4)$
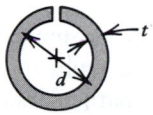	$K' = \frac{\pi d t^3}{3}$
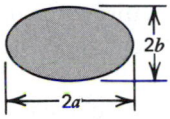	$K' = \frac{\pi a^3 b^3}{a^2 + b^2}$
	$K' = 0.0216a^4$
	$K' = 2.69a^4$
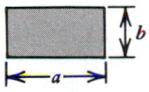	$K' = \frac{ab^3}{16}\left[\frac{16}{3} - 3.36\frac{b}{a}\left(1 - \frac{b^4}{12a^4}\right)\right]$
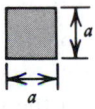	$K' = 0.1406a^4$

numerous other machine and structural components. A beam often requires a larger cross section to limit deflection than it does to limit stress. Hence, many steel beams are made of low-cost alloys because these have the same modulus of elasticity (thus, the same resistance to elastic deflection) as stronger, high-cost steels.

The reader has undoubtedly studied previously one or more of the many methods for computing beam deflection (as area-moment, integration by singularity functions, graphical integration, and numerical integration). Appendix D contains a summary of equations for deflection (as well as shear and moment) pertaining to beams of uniform section with commonly encountered loads. More complete tables are given in many handbooks, such as [4]. When more than one load is applied to a beam (and its response is within the linear elastic range), the deflection at any point on the beam is the sum of the individual deflections produced at that point by each of the loads acting singly. This *method of superposition* (used with information such as that provided in Appendix D) often provides the easiest and quickest solution to beam deflection problems involving several loads.

For various reasons, many beams do not have a uniform cross section. For example, machinery rotating shafts are usually "stepped," as shown in Fig. 5.14, in order best to accommodate the bearings and other parts assembled on the shaft. The calculation of beam deflections in such cases is tedious using conventional methods. Fortunately, these problems can be solved readily on a computer. Table 5.3 gives a BASIC program for determining the deflection of stepped shafts with a microcomputer.

Since, because of computers, it is now unusual for an engineer to solve stepped-shaft deflection problems by "longhand" methods, the following sample problem is included primarily to aid the student in understanding the fundamental mathematical manipulations involved. Such an understanding is vital to the intelligent engineering use of the computer. Moreover, it is always well for the engineer to have the satisfaction of knowing how to solve the problem without a computer, if this should sometimes be necessary for simple cases.

The solution to the following sample problem is based directly on the fundamental equations:

$$\left.\begin{aligned}
&\text{Load intensity at any point} && x = \frac{d^4\delta}{dx^4} EI = w \\
&\text{Shear force at any point} && x = \frac{d^3\delta}{dx^3} EI = V \\
&\text{Bending moment at any point} && x = \frac{d^2\delta}{dx^2} EI = M \\
&\text{Slope at any point} && x = \frac{d\delta}{dx} = \Theta \\
&\text{Deflection at any point} && x = \delta
\end{aligned}\right\} \quad (5.9)$$

The concepts involved in integrating or differentiating to obtain any one of the above quantities from any of the others are fundamental, and the engineering student is urged to study the following sample problem carefully. An understanding of the solution to this problem will carry over to other fields that involve essentially the same differential equations. A prominent example is in dynamics, where successive time derivatives of displacement give velocity, acceleration, and jerk.

SAMPLE PROBLEM 5.3

The stepped shaft at the top of Fig. 5.14 is made of steel (E = 207 GPa), and loaded (as by gears or pulleys mounted on the shaft) with the 1.5 kN and 2.5 kN forces shown.

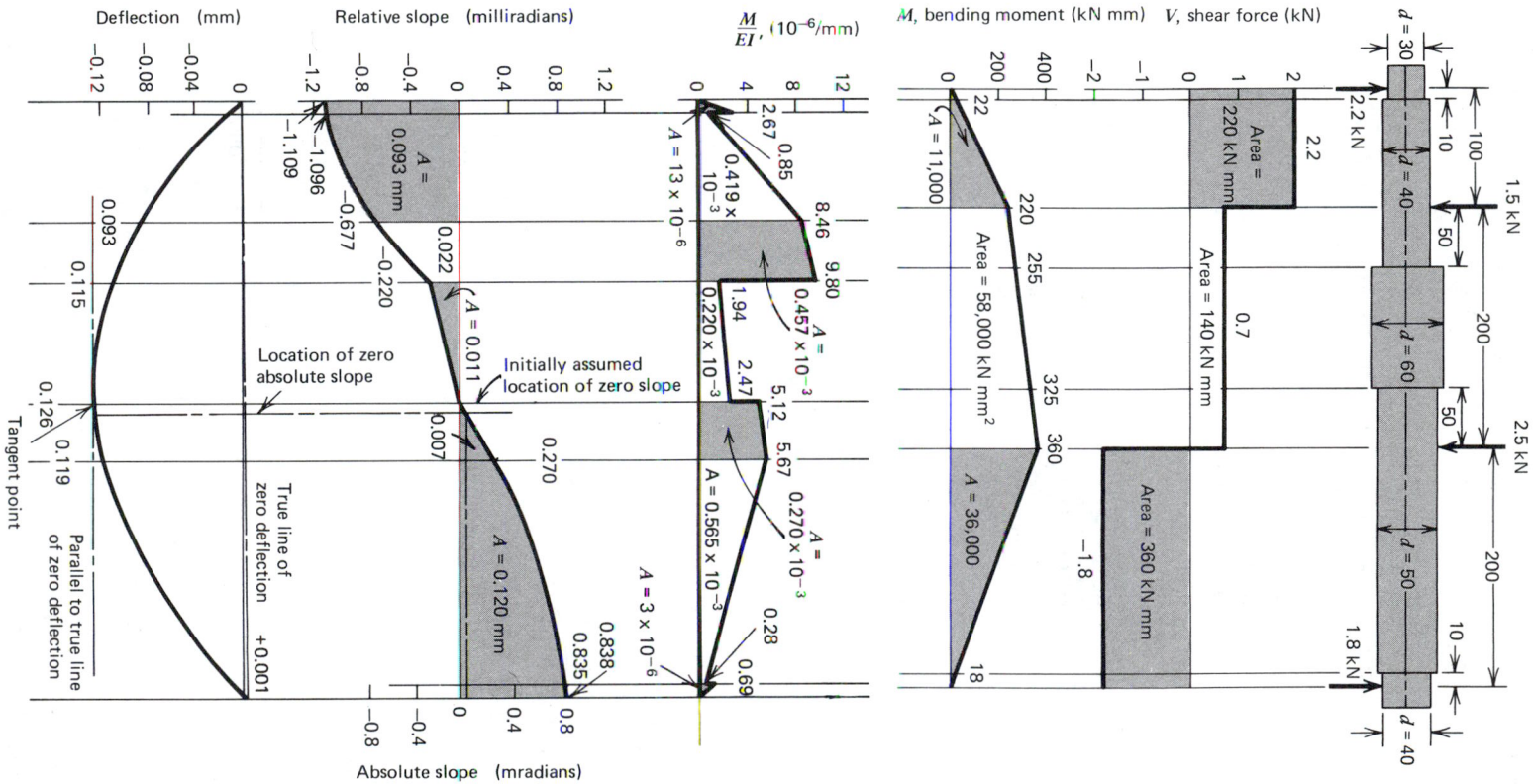

FIGURE 5.14.
Deflection determination for an end-supported stepped steel shaft with two concentrated loads. (Sample Problem 5.3)

TABLE 5.3
Applesoft BASIC Program for Determining Elastic Deflections of Stepped Shafts

The program is applicable to all cases of stepped cylindrical shafts subjected to concentrated forces and moments. It is based on the numerical double integration of the M/EI curve by Simpson's rule.

[Program written by M. Y. Zarrugh, and based on C. R. Mischke, "An Exact Method for Determining the Bending Deflection and Slope in Stepped Shafts," *Advances in Reliability and Stress Analysis,* John J. Burnes (ed.), ASME, 1978, pp. 101–115.]

```
300  HOME : REM  ENTER DATA
310  PRINT "DEFLECTION ANALYSIS IN STEPPED SHAFTS"
320  PRINT "       WRITTEN BY M. Y. ZARRUGH"
330  PRINT "        UNIVERSITY OF MICHIGAN"
340  INPUT "HOW MANY NODES? ";NP
360  DIM X(NP),Y(NP),S(NP),I1(NP),I2(NP),M(2 * NP),ME(2 * NP)
380  FOR I = 1 TO NP: PRINT "DISTANCE X FOR NODE NO. ";I;: INPUT X(I): NEXT I
390  INPUT "HOW MANY DIA CHANGES? ";ND: DIM JD(ND),DO(ND)
400  FOR I = 1 TO ND - 1: PRINT "NODE # AND DIA. FOR DIA CHANGE NO. "I;: INPUT JD(I),DO(I): NEXT I: PRINT
420  INPUT "NODE NO. FOR LEFT BEARING? ";IA: INPUT "NODE NO. FOR RIGHT BEARING? ";IB: PRINT
440  FOR I = 1 TO NF : PRINT "NODE # AND VALUE OF FORCE NO. "; I;: INPUT JF(I),F(I): NEXT I: PRINT
460  INPUT "HOW MANY APPLIED MOMENTS? ";NM: IF NM = O GOTO 470
465  DIM AM(NM),JM(NM): FOR I = 1 TO NM: PRINT "NODE # AND VALUE OF MOMENT NO. ";I;: INPUT JM(I),AM(I): NEXT I
467  GOTO 480
470  DIM AM(1),JM(1)
480  INPUT "YOUNG'S MODULUS? ";E
500  REM  COMPUTE REACTIONS
510 SF = O:SM = O: FOR I = 1 TO NF:SF = SF + F(I):SM = SM + F(I) * (X(JF(I)) - X(IA)): NEXT I: IF NM = O GOTO 530
520  FOR I = 1 TO NM:SM = SM + AM(I): NEXT I
530 RB =  - SM / (X(IB) - X(IA)):RA =  - SF - RB
540  REM  COMPUTE BENDING MOMENTS
550  REM  ADD RA AND RB TO ARRAY F
560  FOR I = 1 TO NF: IF IA < JF(I) THEN  GOTO 580
570  NEXT I: GOTO 600
580  FOR J = NF TO I STEP  - 1:F(J + 1) = F(J):JF(J + 1) = JF(J): NEXT J
600 F(I) = RA:JF(I) = IA: FOR I = 1 TO NF + 1: IF JF(I) > IB THEN  GOTO 610
605  NEXT I: GOTO 620
610  FOR J = NF + 1 TO I STEP  - 1:F(J + 1) = F(J):JF(J + 1) = JF(J): NEXT J
620 F(I) = RB:JF(I) = IB
630  REM ---EVALUATE M VALUES
640 VV = O:FP = 1:MP = 1:M(1) = O: FOR I = 1 TO NP:J = (I - 1) * 2 + 1: IF I = 1 THEN  GOTO 670
660 DX = X(I) - X(I - 1):M(J) = M(J - 1) + VV * DX:M(J + 1) = M(J)
670  IF MP > NM THEN  GOTO 710
680  IF I <  > JM(MP) THEN  GOTO 710
700 M(J + 1) = M(J) - AM(MP):MP = MP + 1
710  IF FP > NF + 2 THEN  GOTO 740
720  IF I <  > JF(FP) THEN  GOTO 740
730 VV = VV + F(FP):FP = FP + 1
740  NEXT I: PRINT "BENDING MOMENTS ARE READY"
750  REM  COMPUTE M/EI
760 ME(1) = O.O:PI = 3.141593:DP = 1: FOR I = 1 TO NP:J = (I - 1) * 2 + 1: IF I = 1 THEN  GOTO 790
770 ME(J) = M(J) / (E * II):ME(J + 1) = M(J + 1) / (E * II): IF I <  > JD(DP) THEN  GOTO 800
790 DD = DO(DP):II = PI * DD ^ 4 / 64:ME(J + 1) = M(J + 1) / (E * II):DP = DP + 1
800  NEXT I: PRINT "M/EI VALUES ARE READY"
810  REM  COMPUTE I1=INTGRL(M/EI) VS. X
820 I1(1) = O:I2(1) = O
830  FOR I = 2 TO NP:J = (I - 1) * 2 + 1:DX = X(I) - X(I - 1):MM = (ME(J - 1) + ME(J)) / 2.O
840 I1(I) = I1(I - 1) + MM * DX:IM = I1(I - 1) + (ME(J - 1) + MM) * DX / 4
850 DI = (I1(I) + 4 * IM + I1(I - 1)) / 6.O:I2(I) = I2(I - 1) + DI * DX
856  NEXT I
860  REM  COMPUTE C1 AND C2
870 DX = X(IA) - X(IB):C1 =  - (I2(IA) - I2(IB)) / DX:C2 = (X(IB) * I2(IA) - X(IA) * I2(IB)) / DX
900  REM  COMPUTE DEFLECTIONS AND SLOPES
910  FOR I = 1 TO NP:S(I) = I1(I) + C1:Y(I) = I2(I) + C1 * X(I) + C2: NEXT I
920  PRINT : PRINT "DEFLECTION ANALYSIS OF STEPPED SHAFTS": PRINT :
925  PRINT "NODE";"  ";"DIST";"   ";"SLOPE";"     ";"DEFLECT"
930  FOR I = 1 TO NP: PRINT " ";I;"    ";X(I);"  ";S(I);"  ";Y(I): NEXT I
940  END
```

It is simply supported by bearings at each end. Determine the deflection at all points along the shaft.

Solution

1. The bearing supporting forces (2.2 kN and 1.8 kN) and the shear and moment diagrams are determined as in Fig. 2.11.

TABLE 5.3 (*continued*)

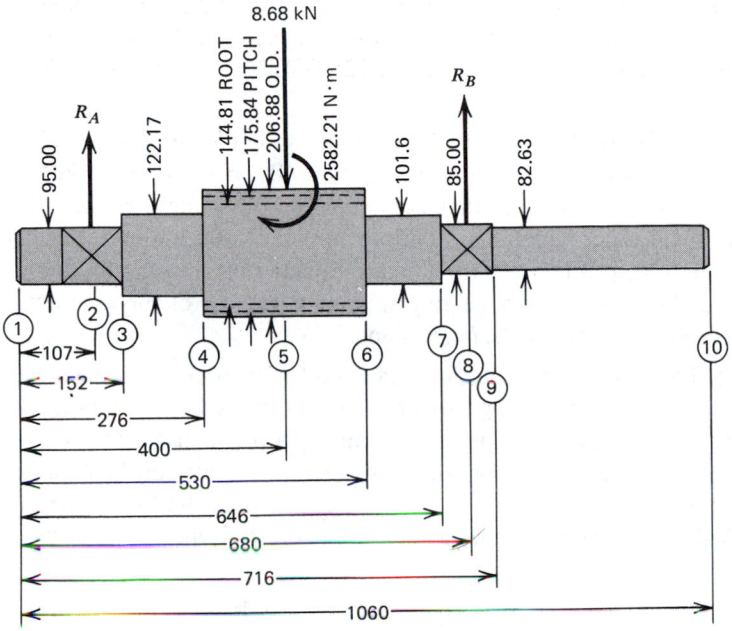

A shaft with integral worm, dimensions in millimeters.

```
HOW MANY NODES? 10
DISTANCE X FOR NODE NO. 1?0
DISTANCE X FOR NODE NO. 2?107
DISTANCE X FOR NODE NO. 3?152
DISTANCE X FOR NODE NO. 4?276
DISTANCE X FOR NODE NO. 5?400
DISTANCE X FOR NODE NO. 6?530
DISTANCE X FOR NODE NO. 7?646
DISTANCE X FOR NODE NO. 8?680
DISTANCE X FOR NODE NO. 9?716
DISTANCE X FOR NODE NO. 10?1060
HOW MANY DIA CHANGES? 7
NODE # AND DIA. FOR DIA CHANGE NO. 1?1
??95
NODE # AND DIA. FOR DIA CHANGE NO. 2?3
??122.17
NODE # AND DIA. FOR DIA CHANGE NO. 3?4
??144.81
NODE # AND DIA. FOR DIA CHANGE NO. 4?6
??101.6
NODE # AND DIA. FOR DIA CHANGE NO. 5?7
??85
NODE # AND DIA. FOR DIA CHANGE NO. 6?9
??82.63

NODE NO. FOR LEFT BEARING? 2
NODE NO. FOR RIGHT BEARING? 8
```

```
HOW MANY APPLIED FORCES? 1
NODE # AND VALUE OF FORCE NO. 1?5
??-8.68

HOW MANY APPLIED MOMENTS? 1
NODE # AND VALUE OF MOMENT NO. 1?5
??-2582.21
YOUNG'S MODULUS? 207
BENDING MOMENTS ARE READY
M/EI VALUES ARE READY

DEFLECTION ANALYSIS OF STEPPED SHAFTS

NODE  DIST    SLOPE          DEFLECT
 1     0   -3.55164105E-05  3.80025592E-03
 2    107  -3.55164105E-05  0
 3    152  -3.58405307E-05  -1.60310027E-03
 4    276  -3.73934634E-05  -6.12501141E-03
 5    400  -3.90918827E-05  -.0108576819
 6    530  1.68612857E-05  -.0119361558
 7    646  1.05028693E-04  -3.79191686E-03
 8    680  1.14776103E-04  -2.72848411E-12
 9    716  1.14776103E-04  4.13193969E-03
 10   1060  1.14776102E-04  .0436149188
```

2. Note carefully the key factors involved in the graphical integration of any curve (as the shear diagram) to obtain the next lower curve (as the moment diagram).
   a. The *difference in ordinate* between any two points along the lower curve is equal to the *area under the curve* above between these two points.
   b. The absolute value of the ordinates of the lower curve is determined from known end conditions. For example, the moment at the bearing supports is known to be zero.

c. The *slope* at any point on the lower curve is equal to the *ordinate* of the curve above.

3. To account for the differences in diameter along the shaft, each segment of the moment curve is divided by the product EI for that segment. (Values of I for the 30, 40, 50, and 60 mm segments are 39,761, 125,664, 306,796, and 636,173 mm^4, respectively.)
4. When integrating the M/EI curve to obtain slope, we encounter the problem that the location of zero slope is not known at this time. By visualizing or sketching a rough deflection curve, it is clear that zero slope occurs somewhere in the mid-portion of the shaft. Accordingly, an "initially assumed location of zero slope" is chosen as shown. (Note: Final accuracy is not affected by this assumption—zero slope could even be assumed at a bearing support.) Because of the assumption, it is necessary to call the ordinate "*relative* slope."
5. Integration of the slope curve to obtain deflection begins with the known location of zero deflection at the left bearing support. *If* the estimated location of zero slope was correct, the calculated deflection at the right support will also come out to zero. In this case, the assumed zero-slope location was only slightly off. To correct for this, connect the two known points of zero deflection with the "true line of zero deflection." Values of true deflection at any point *must be measured perpendicularly from this line.*
6. Finally, the correct location of zero slope is determined by drawing a line tangent to the deflection curve, parallel to the "true line of zero deflection." This enables the "absolute slope" scale to be added.
7. Note that in this case negligible error would have been introduced by eliminating the steps at the shaft ends; that is, the problem could have been simplified by carrying out the 40 mm and 50 mm diameters to the ends.

Shaft design and analysis is discussed further in Chapter 17.

5.8 ELASTIC DEFLECTIONS BY CASTIGLIANO'S METHOD

Situations frequently arise where it is necessary to compute elastic deflections not covered by the simple cases given in Table 5.1, and involving more than the bending loading treated in the previous article. Castigliano's method is selected here for dealing with these situations. It is selected because of its versatility in handling a wide range of deflection problem types, and because it is also useful in determining redundant reactions (treated in the next section). The theorem underlying the method was published as part of a thesis by Alberto Castigliano while a student at the Turin (Italy) Polytechnical Institute in the mid-nineteenth century.

Figure 5.15 shows an elastic load-deflection curve for the completely general case. The load can be any force or moment with the displacement being the corresponding linear or angular displacement. The single-crosshatched area (U) under the curve is

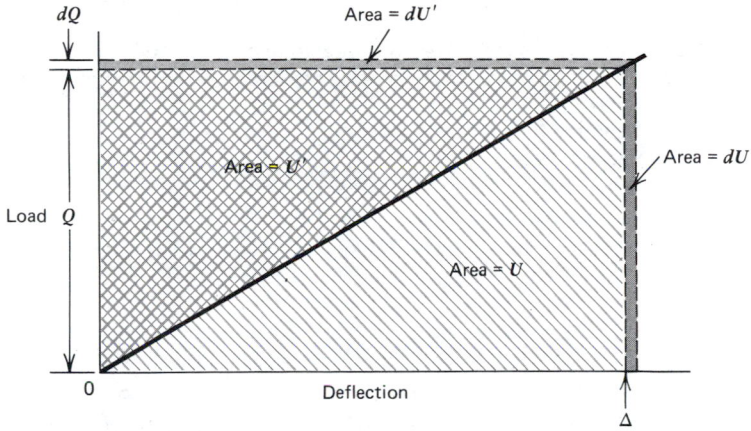

FIGURE 5.15.
General load-deflection curve for elastic range.

equal to the stored elastic energy. The double-crosshatched area (U') is known as the complementary energy. By simple geometry,

$$U = U' = Q\Delta/2$$

that is, stored elastic energy is equal to deflection times average force. The additional energy associated with incremental load dQ is

$$dU' = dU = \Delta\, dQ$$

Solving for the deflection, Δ, we have

$$\Delta = dU/dQ$$

In the general case, Q may be only one of many loads acting on the member. In order to find the deflection in the direction of Q and at the point where Q is applied, the derivative must be taken while all other loads are held constant. Thus, the general equation for deflection, which constitutes Castigliano's theorem, is

$$\Delta = \partial U/\partial Q \tag{5.9}$$

The great importance of this simple equation justifies restating it in words:

When a body is elastically deflected by any combination of loads, the deflection at any point and in any direction is equal to the partial derivative of strain energy (computed with all loads acting) with respect to a load at that point and acting in that direction.

The theorem can be used to find a deflection at any point and in any direction, even if there is no load at that point and in that direction. It is only necessary to apply an *imaginary* load (force or moment), commonly designated as Q, at the point and in the direction desired and to solve for deflection as a function of Q. After this expression is obtained, Q is equated to zero to obtain the final answer.

TABLE 5.4
Summary of Energy and Deflection Equations for Use With Castigliano's Method

Load Type (1)	Factors Involved (2)	Energy Equation for Special Case Where All Three Factors are Constant With x (3)	General Energy Equation (4)	General Deflection Equation (5)
Axial	P, E, A	$U = \dfrac{P^2L}{2EA}$	$U = \displaystyle\int_0^L \frac{P^2}{2EA}\,dx$	$\Delta = \displaystyle\int_0^L \frac{P(\partial P/\partial Q)}{EA}\,dx$
Bending	M, E, I	$U = \dfrac{M^2L}{2EI}$	$U = \displaystyle\int_0^L \frac{M^2}{2EI}\,dx$	$\Delta = \displaystyle\int_0^L \frac{M(\partial M/\partial Q)}{EI}\,dx$
Torsion	T, G, K'	$U = \dfrac{T^2L}{2GK'}$	$U = \displaystyle\int_0^L \frac{T^2}{2GK'}\,dx$	$\Delta = \displaystyle\int_0^L \frac{T(\partial T/\partial Q)}{GK'}\,dx$
Transverse shear (rectangular section)	V, G, A	$U = \dfrac{3V^2L}{5GA}$	$U = \displaystyle\int_0^L \frac{3V^2}{5GA}\,dx$	$\Delta = \displaystyle\int_0^L \frac{6V(\partial V^a/\partial Q)}{5GA}\,dx$

[a]Change constant from 6/5 to 1 for quick *estimates* involving nonrectangular sections.

In all but the simplest cases, the expression for U in Eq. (5.9) will involve more than one term, each of which expresses the energy associated with one component of loading. The equations for elastic strain energy associated with the various types of loading are summarized in the third and fourth columns of Table 5.4. To review where these come from, consider the case of axial loading, where, as was noted in Fig. 5.15,

$$U = Q\Delta/2$$

In terms of the particular case involving an axial load, designated as P, and an axial deflection, designated δ,

$$U = P\delta/2 \tag{a}$$

But, from case 1 in Table 5.1,

$$\delta = PL/AE \tag{b}$$

Substitution of Eq. (b) into Eq. (a) gives

$$U = P^2L/2EA \tag{c}$$

For the case of a bar length L that has a varying cross section and possibly a varying modulus of elasticity,

$$U = \int_0^L \frac{P^2}{2EA}\,dx \tag{d}$$

The other equations in Table 5.4 are similarly derived (see [3]).

The most straightforward application of Castigliano's method involves (1) obtaining the proper expression for all components of energy, using the equations for U in Table 5.4, and then (2) taking the appropriate partial derivative to obtain deflection. It is recommended that the student work one or two problems in this way to be sure that the basic procedure is clearly understood. A shorter and more expedient solution can often be obtained by applying the technique of differentiating under the integral sign, represented by the deflection equations in the final column of Table 5.4.

The first two of the examples below illustrate a comparison of the above two procedures. Several examples are included to illustrate the wide variety of types of problems to which Castigliano's theorem can be applied. The student is urged to study each of them carefully. Note that the key part of the solution is always setting up the proper expressions for load. After that, the mathematical manipulation is routine.

SAMPLE PROBLEM 5.4 Simply Supported Beam With Concentrated Center Load

Determine the deflection at the center of a centrally loaded beam, shown in Fig. 5.16.

Solution 1

First, compute energy; then take the partial derivative to get deflection. The beam experiences two kinds of loading—bending and shear—with magnitudes at any point x (see Fig. 5.16) of

$$M = \frac{P}{2}x \quad \text{and} \quad V = \frac{P}{2}$$

Correspondingly, the equation for energy has two terms:

$$U = 2\int_0^{L/2} \frac{M^2}{2EI}\,dx + \int_0^L \frac{3V^2}{5GA}\,dx$$

$$U = 2\int_0^{L/2} \frac{P^2x^2}{8EI}\,dx + \int_0^L \frac{3(P/2)^2}{5GA}\,dx$$

$$U = \frac{P^2}{4EI}\int_0^{L/2} x^2\,dx + \frac{3P^2}{20GA}\int_0^L dx$$

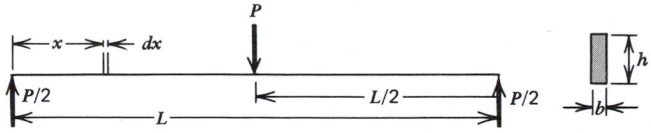

FIGURE 5.16.
Sample Problem 5.4—simply loaded, simply supported beam.

$$U = \frac{P^2L^3}{96EI} + \frac{3P^2L}{20GA}$$

Taking the partial derivative to obtain deflection:

$$\delta = \frac{\partial U}{\partial P} = \frac{PL^3}{48EI} + \frac{3PL}{10GA}$$

Before continuing, three points in the above solution merit attention:

1. The equation for bending load, $M = Px/2$, is valid only between $x = 0$ and $x = L/2$. The simplest way to handle this situation was to recognize that the beam is symmetrical about the center, with energy in the right and left halves being equal. Hence, we accounted for bending energy by integrating between 0 and $L/2$ and doubling.
2. Since V, G, and A are all constant with x, we could have obtained the transverse shear energy term more quickly by using the equation in column 3 of Table 5.4. (This is illustrated in Solution 2, below.)
3. The final answer, consisting of two terms, may surprise those who recognize the first term *only* as being the traditional deflection equation for this case. Obviously, when one uses only the first term, the deflection due to transverse shear is neglected. As will be seen in Sample Problem 5.5, this is justified in almost all cases. However, it is important for the engineer to be *aware* of neglecting the shear term. And whenever there is any doubt about the shear term being negligible, Castigliano's method enables one to evaluate quickly its magnitude.

Solution 2

Compute deflection by differentiating the bending term under the integral sign, and using column 3 of Table 5.4 for the shear term.

As in the first solution, write the equations for the two kinds of loading present (at any location x along the beam):

$$M = \frac{P}{2}x \quad \text{and} \quad V = \frac{P}{2}$$

Again, handling the bending term by integrating over half the length and then doubling, we determine deflection directly:

$$\delta = 2\int_0^{L/2} \frac{M(\partial M/\partial P)}{EI}\,dx + \frac{\partial}{\partial P}(U \text{ for trans. shear})$$

$$\delta = \frac{2}{EI}\int_0^{L/2} \left(\frac{Px}{2}\right)\left(\frac{x}{2}\right) dx + \frac{\partial}{\partial P}\left(\frac{3(P/2)^2L}{5GA}\right)$$

$$\delta = \frac{2}{EI}\int_0^{L/2} \frac{Px^2}{4}\,dx + \frac{3PL}{10GA}$$

$$\delta = \frac{P}{2EI}\left[\frac{(L/2)^3}{3} - 0\right] + \frac{3PL}{10GA}$$

$$\delta = \frac{PL^3}{48EI} + \frac{3PL}{10GA}$$

■

SAMPLE PROBLEM 5.5 Comparison of Bending and Shear Terms in Sample Problem 5.4

Using the formula derived in Sample Problem 5.4, evaluate the magnitude of the two deflection terms for the case of $P = 5000$ N, $L = 400$ mm, $b = 25$ mm, $h = 50$ mm, and the beam is made of steel, having $E = 207$ GPa and $G = 80$ GPa.

$$\delta = \frac{PL^3}{48EI} + \frac{3PL}{10GA}$$

$$\delta = \frac{5000(0.400)^3}{48\left(207 \times 10^9\right)\left(\frac{25[50]^3}{12} \times 10^{-12}\right)} + \frac{3(5000)(0.400)}{10(80 \times 10^9)([0.025][0.050])}$$

$$\delta = 1.237 \times 10^{-4} + 6.000 \times 10^{-6} = 1.297 \times 10^{-4} \text{ m}$$ ■

This calculation illustrates the general rule that

for rectangular-section beams of length at least eight times depth, transverse shear deflection is less than 5 percent of bending deflection.

In fact, it is unusual for transverse shear to make a significant contribution to *any* engineering deflection problem.

SAMPLE PROBLEM 5.6 Use of Dummy Load

Determine the vertical deflection at the free end of the 90° bent cantilever shown in Fig. 5.17.

Solution

Since there is no given load acting *at the point* and *in the direction* of the desired deflection, such a load *must* be added (Q in Fig. 5.17).

If transverse shear is neglected, four components of energy are present:

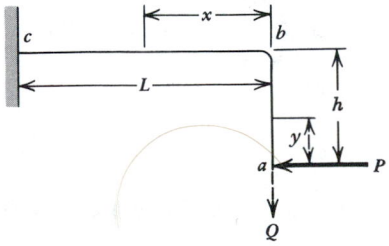

FIGURE 5.17.
Sample Problem 5.6—end-loaded 90 deg. bent cantilever.

1. Bending in portion ab, where $M_{ab} = Py$ (note that y was arbitrarily defined so as to give a simple expression for M_{ab}).
2. Bending in portion bc, where $M_{bc} = Qx + Ph$ (again, x was arbitrarily defined for convenience).
3. Tension in ab, of magnitude Q.[1]
4. Compression in bc, of magnitude P.[1]

Deflection for each of the four terms is determined as

$$\delta = \int_0^h \frac{M_{ab}(\partial M_{ab}/\partial Q)}{EI}\,dy + \int_0^L \frac{M_{bc}(\partial M_{bc}/\partial Q)}{EI}\,dx + \int_0^h \frac{Q(\partial Q/\partial Q)}{EA}\,dx + \int_0^L \frac{P(\partial P/\partial Q)}{EA}\,dx$$

$$\delta = \int_0^h \frac{(Py)(0)}{EI}\,dy + \int_0^L \frac{(Qx + Ph)x}{EI}\,dx + \frac{Qh}{EA} + \int_0^L \frac{P(0)}{EA}\,dx$$

Now that the partial derivatives have been taken, we can set

$$Q = 0$$

to simplify the remaining mathematics:

$$\delta = 0 + \int_0^L \frac{Phx}{EI}\,dx + 0 + 0, \qquad \delta = \frac{PhL^2}{2EI}$$ ■

The solution to this problem by (1) evaluating energy, U, and then (2) taking the partial derivative, is of comparable length and constitutes a good exercise for those not previously acquainted with Castigliano's method.

SAMPLE PROBLEM 5.7 Tangential Deflection of a Split Ring

Figure 5.18*a* shows a piston ring being expanded by a tool to facilitate installation. The ring is sufficiently "thin" to justify use of the straight-beam bending formula. Derive an expression relating the separating force F and the separation δ. Include all terms.

Solution

Figure 5.18*b* shows a typical segment of the ring bounded by angle θ (defined in the figure) as a free body. The deflection has three components: those due to bending, axial load (compression in the bottom half, tension in the top), and transverse shear. Corresponding equations are:

$$M = FR(1 - \cos\theta)$$
$$P = F\cos\theta$$
$$V = F\sin\theta$$

The above equations are valid for all values of θ. Hence, we can integrate from 0 to 2π:

[1]It is noted in Sample Problem 5.8 that axial loading terms are almost always negligible if bending and/or torsion terms are present. The axial loading terms are included here to illustrate general procedure.

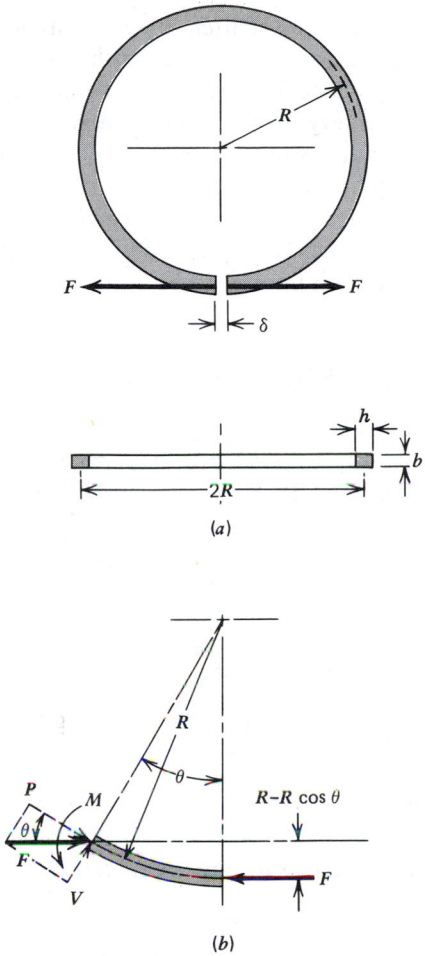

FIGURE 5.18.
Sample Problem 5.7—piston ring loaded by an expander.

$$\delta = \frac{1}{EI}\int_0^{2\pi} M\frac{\partial M}{\partial F} R\, d\theta + \frac{1}{EA}\int_0^{2\pi} P\frac{\partial P}{\partial F} R\, d\theta + \frac{6}{5GA}\int_0^{2\pi} V\frac{\partial V}{\partial F} R\, d\theta$$

$$\delta = \frac{1}{EI}\int_0^{2\pi} FR(1 - \cos\theta)R(1 - \cos\theta)R\, d\theta + \frac{1}{EA}\int_0^{2\pi} F(\cos^2\theta)R\, d\theta$$

$$+ \frac{6}{5GA}\int_0^{2\pi} F(\sin^2\theta)R\, d\theta$$

$$\delta = \frac{FR^3}{EI}\int_0^{2\pi} (1 - 2\cos\theta + \cos^2\theta)\, d\theta + \frac{FR}{EA}\int_0^{2\pi} \cos^2\theta\, d\theta + \frac{6FR}{5GA}\int_0^{2\pi} \sin^2\theta\, d\theta$$

$$\delta = \frac{FR^3}{EI}\left(2\pi - 0 + \pi\right) + \frac{RF\pi}{EA} + \frac{6FR\pi}{5GA}$$

$$= \frac{3\pi FR^3}{EI} + \frac{\pi FR}{EA} + \frac{6\pi FR}{5GA}$$

■

In the above solution the values of the definite integrals were written directly, without the bother (and possibility for error) of integrating the expressions and then substituting upper and lower limits. Evaluation of definite integrals such as these can be accomplished expediently by taking advantage of the elementary graphical interpretations in Fig. 5.19. Such figures can be reproduced readily from the simplest concepts of integral calculus, thereby avoiding dependence upon memory or integral tables.

Note that the ring is symmetrical about the vertical, but not the horizontal axis. Hence, we could have integrated between 0 and π and then doubled each integral, but we could not have integrated between 0 and $\pi/2$ and then multiplied by 4.

SAMPLE PROBLEM 5.8 Comparison of Terms in Solution to Sample Problem 5.7

Determine the numerical values for deflection in Example 4 if $R = 2$ in., $b = 0.2$ in., $h = 0.3$ in., and the ring is made of cast iron, having $E = 18 \times 10^6$ psi and $G = 7 \times 10^6$ psi.

Solution

For the dimensions given, $A = 0.06$ in.^{2} and $I = 0.00045$ in.^{4}. Substitution of these values in the deflection equation gives

$$\delta = (930 + 0.58 + 1.79)F \times 10^{-5} \text{ in.}$$ ■

This problem illustrates the general rule that

if bending and/or torsion terms are present, axial and transverse shear terms are almost always negligible.

Note the hedging in saying "*almost* always." A nice thing about Castigliano's method is that if there is any question about encountering an unusual situation where axial and transverse shear terms might be important, they can readily be evaluated.

5.9 REDUNDANT REACTIONS BY CASTIGLIANO'S METHOD

As noted in Chapter 2, a redundant reaction is a supporting force or moment that is not necessary for equilibrium. Thus, as the magnitude of a redundant reaction is varied, deflections change but equilibrium remains. Castigliano's theorem tells us that the deflection associated with any reaction (or applied load) that can be varied without upsetting equilibrium is equal to the partial derivative of total elastic energy with respect to that reaction (or load). In the previous article, the loads were all known and a desired deflection solved for. In this article, the deflection is known and we will solve for the corresponding load. Again, the procedure is best understood by careful study of typical sample problems.

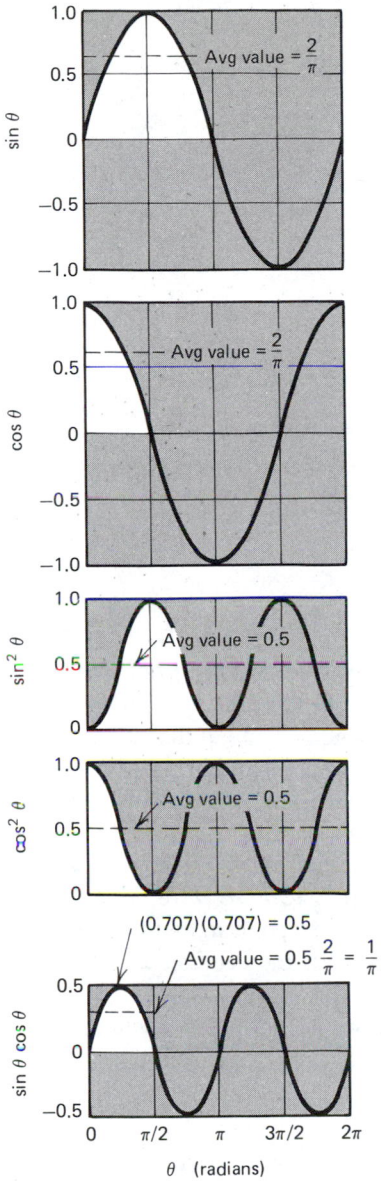

$$\int_0^{\pi/2} \sin\theta\, d\theta = 1; \quad \int_0^{\pi} \sin\theta \quad d\theta = 2; \quad \int_0^{2\pi} \sin\theta\, d\theta = 0.$$

(From this, average value $= \dfrac{2}{\pi}$ can be determined)

$$\int_0^{\pi/2} \cos\theta\, d\theta = 1; \quad \int_0^{\pi} \cos\theta\, d\theta = 0; \quad \int_0^{2\pi} \cos\theta\, d\theta = 0$$

$$\int_0^{\pi/2} \sin^2\theta\, d\theta = \frac{\pi}{2}(0.5) = \frac{\pi}{4}; \quad \int_0^{\pi} \sin^2\theta\, d\theta = \pi\,(0.5) = \frac{\pi}{2}$$

$$\int_0^{2\pi} \sin^2\theta\, d\theta = \pi$$

$$\int_0^{\pi/2} \cos^2\theta\, d\theta = \frac{\pi}{2}(0.5) = \frac{\pi}{4}; \quad \int_0^{\pi} \cos^2\theta\, d\theta = \pi(0.5) = \frac{\pi}{2}$$

$$\int_0^{2\pi} \cos^2\theta\, d\theta = \pi$$

$$\int_0^{\pi/2} \sin\theta\cos\theta\, d\theta = \frac{\pi 1}{2\pi} = \frac{1}{2}; \quad \int_0^{\pi} \sin\theta\cos\theta\, d\theta = 0$$

$$\int_0^{2\pi} \sin\theta\cos\theta\, d\theta = 0$$

FIGURE 5.19.
Graphical evaluation of definite integrals, such as those encountered in Sample Problem 5.7.

SAMPLE PROBLEM 5.9 Guy Wire Tension to Prevent Pole Deflection

Figure 5.20 represents a pole supporting an eccentric mass. The pole is "fixed" at the bottom, and supported horizontally by a guy wire at point *a*. What tension F in the guy wire is required to make the pole deflection equal to zero where the wire is attached?

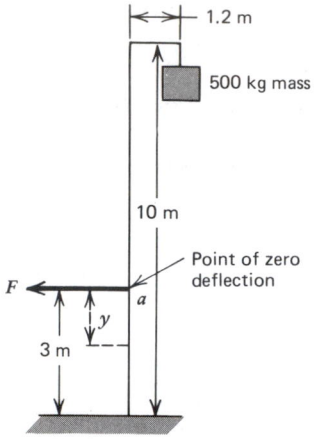

FIGURE 5.20.
Sample Problem 5.9—eccentrically loaded, redundantly supported column.

Solution

Assuming the pole is on the earth, where the gravitational field is 9.8 m/s^2, the gravitational force on the mass is 4900 N.

Tension F is a redundant reaction (i.e., the pole would be in equilibrium without the guy wire), and the deflection at the point where F is applied and in the direction of F is to be zero. Hence, the partial derivative of the total elastic energy in the system with respect to F must be zero.

Before writing the complete expression for energy, let us omit those terms that will later drop out because their partial derivatives with respect to F are zero. This will be the case with all terms representing energy in the pole above point a, and the term for compression of the pole below a. The only remaining term is the bending energy below a.

Defining y for convenience as shown in Fig.15.20, we have, for this bending term,

$$M = (4900 \text{ N})(1.2 \text{ m}) - Fy \quad \text{or} \quad M = 5880 - Fy$$

Bending energy below point a is

$$U = \int_0^3 \frac{M^2}{2EI}\,dy = \frac{1}{2EI}\int_0^3 (5880^2 - 11{,}760Fy + F^2y^2)\,dy$$

$$U = \frac{1}{2EI}(5880^2 \times 3 - 52{,}920F + 9F^2)$$

The horizontal deflection at point a is

$$\delta = 0 = \frac{\partial U}{\partial F} = \frac{1}{2EI}(-52{,}920 + 18F)$$

Because E and I can only have finite values, the term in parentheses must be equal to zero, giving $F = 2940$ N. ■

Note the seemingly remarkable result that the answer is completely independent of E and I. The stiffness of the pole is not a factor *so long as the deflection does not cause the moment arm of the gravity force to significantly differ from 1.2 m.* This is a condition that applies to all Castigliano problems. The load equations are written in terms of a given geometry. To the degree that subsequent deflections alter the geometry, errors are introduced into the load equations that cause corresponding errors in the computed deflection. Although this should be kept in mind, it is seldom a cause for concern in engineering problems.

The next sample problem might appear to belong more properly in the previous article. Careful study indicates, however, that the desired deflection cannot be computed without *first* evaluating a redundant torsional reaction.

SAMPLE PROBLEM 5.10 Deflection of Redundantly Supported Bracket

Determine the deflection at the load P for the three-sided bracket shown in Fig. 5.21. The bracket has the same material and cross section throughout.

Solution

Figure 5.21*b* shows an exploded view of each portion of the bracket as free bodies in equilibrium. All loads can be expressed in terms of P and dimensions a and b *except* for redundant moment M_0. We *must* determine an expression for M_0 before Castigliano's theorem can be applied to determine the deflection.

To visualize more clearly the redundant moment M_0, it may be helpful to consider the following. Imagine that instead of being welded to the fixed vertical support, the bracket slips into bushings or bearings mounted in this support. Then, when load P is applied, deflections in the bracket will cause a slight rotation of the ends that extend inside the bushings. Now, imagine that with load P applied, pipe wrenches are placed adjacent to the fixed support on each end of the bracket, and just enough torque is applied to rotate the ends back to their original positions. The wrenches are then providing the redundant reaction torque, M_0. With the bracket welded to the vertical support before the load is applied, this redundant torque is applied to the bracket by the support, through the weld.

Castigliano's method will now be used to determine M_0 as being that torque necessary to enforce zero angular deflection at the points of bracket attachment.

When writing expressions for the various load terms, advantage can be taken of symmetry; energies in the right and left halves of the bracket are identical. Dimensions x and y in Fig. 5.21*a* may be defined as desired, and this is done so as to simplify the load equations. Assuming transverse shear to be negligible, three sources of energy are present. Considering the right side only, these are:

Bending between 1 and 2, where $M_{1,2} = -M_0 + Px/2$

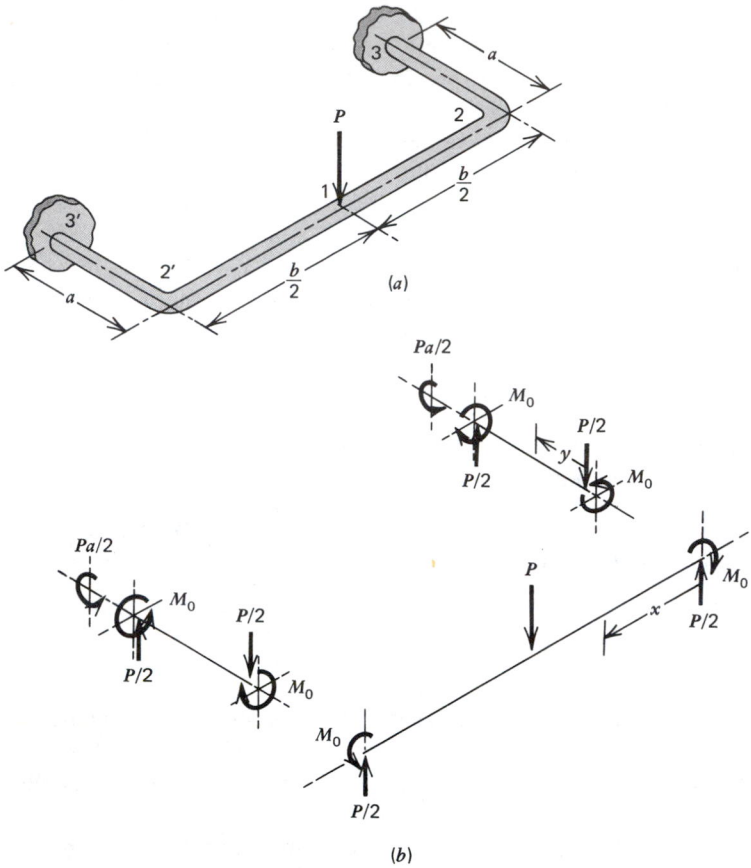

FIGURE 5.21.
Sample Problem 5.10—centrally loaded rectangular bracket.

Bending between 2 and 3, where $M_{2,3} = Py/2$

Torsion between 2 and 3, where $T_{2,3} = M_0$

Writing directly the equation for a zero angular deflection at the point where M_0 is applied:

$$\theta = 0 = 2\int_0^{b/2} \frac{M_{1,2}(\partial M_{1,2}/\partial M_0)}{EI}\,dx + 2\int_0^{a} \frac{M_{2,3}(\partial M_{2,3}/\partial M_0)}{EI}\,dy + 2\frac{T_{2,3}a}{GK'}$$

Canceling the 2's and substituting values,

$$0 = \frac{1}{EI}\int_0^{b/2}\left(-M_0 + \frac{Px}{2}\right)(-1)\,dx + \frac{1}{EI}\int_0^{a}\frac{Py}{2}(0)\,dy + \frac{M_0 a}{GK'}$$

Integrating and substituting limits,

$$0 = \frac{1}{EI}\left(\frac{M_0 b}{2} - \frac{Pb^2}{16}\right) + 0 + \frac{M_0 a}{GK'}$$

Solving for M_0,

$$M_0\left(\frac{b}{2EI}+\frac{a}{GK'}\right)=\frac{Pb^2}{16EI}$$

which reduces to

$$M_0=\frac{Pb^2GK'}{8(bGK'+2aEI)}$$

In proceeding to determine the deflection, the algebra is made a bit more manageable by letting $M_0 = PZ$, where Z is a constant defined by

$$Z=\frac{b^2GK'}{8(bGK'+2aEI)} \tag{e}$$

The same three energy terms are involved as before. This time the partial derivative is taken with respect to P, and the result is the desired deflection:

$$\delta=\frac{2}{EI}\int_0^{b/2}M_{1,2}\frac{\partial M_{1,2}}{\partial P}\,dx+\frac{2}{EI}\int_0^{a}M_{2,3}\frac{\partial M_{2,3}}{\partial P}\,dy+\frac{2}{GK}\int_0^{a}T_{2,3}\frac{\partial T_{2,3}}{\partial P}\,dy$$

$$\delta=\frac{2}{EI}\int_0^{b/2}\left(-PZ+\frac{Px}{2}\right)\left(-Z+\frac{x}{2}\right)dx+\frac{2}{EI}\int_0^{a}\frac{Py}{2}\frac{y}{2}+\frac{2}{GK}\int_0^{a}(PZ)(Z)\,dy$$

Evaluation of the three definite integrals yields

$$\delta=\frac{P}{EI}\left(Z^2b-\frac{Zb^2}{4}+\frac{b^2}{48}\right)+\frac{P}{EI}\frac{a^3}{6}+\frac{2PZ^2a}{GK'}$$

Substitution of Eq. (e) into the above gives, after routine but admittedly tedious algebra,

$$\delta=\frac{P}{48EI}(b^2+8a^3)-\frac{Pb^4GK'}{64EI(bGK'+2aEI)}$$ ■

5.10 EULER COLUMN BUCKLING—ELASTIC INSTABILITY

We normally think of deflections within the elastic range as varying linearly with load. Several notable exceptions occur, all of which involve subjecting relatively long, thin portions of material to compressive stress. Perhaps the most common of these are long, slender columns loaded in compression. Examples include columns in buildings, structural compression links (as in bridges), piston connecting rods, coil springs in compression, and jack screws. These correspond to the general case treated by Leonhard Euler pronounced *oil'er*) in 1744 when he published the first known treatise on elastic stability.

Euler's analysis pertains to Fig. 5.22, where a long, slender column—such as an ordinary yard stick—is loaded in compression. Euler assumed the ideal case of a perfectly straight column, with the load precisely axial, the material perfectly homogeneous, and stresses within the linear elastic range. If such a column is loaded below a certain value, P_{cr}, any slight lateral displacement given to the column (as shown exaggerated in Fig. 5.22) results in an internal elastic restoring moment more than adequate

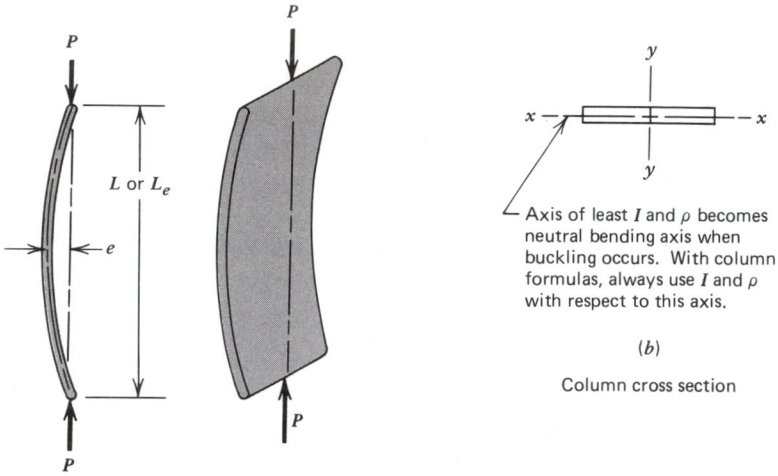

FIGURE 5.22.
Initially straight column in Euler buckling.

to restore straightness to the column when the lateral displacing force is removed. The column is then *elastically stable.* When P_{cr} is exceeded, the slightest lateral displacement results in an eccentric bending moment, Pe, greater than the internal elastic restoring moment and the column collapses. Thus, loads in excess of P_{cr} render the column *elastically unstable.*

Euler's classical equation for P_{cr} is derived in almost all texts in strength of materials. We give it here without repeating the derivation:

$$P_{cr} = \frac{\pi^2 EI}{L_e^2} \tag{5.10}$$

where

E = modulus of elasticity

I = moment of inertia of the section with respect to the buckling-bending axis. This is the *smallest* value of I about any axis, as illustrated in Fig. 5.22.

L_e = equivalent length of column. This is the same as the actual length, L, for the hinged end connections shown in Fig. 5.22. Values of L_e for columns having other end conditions are given in the next section.

Substituting into Eq. (5.10) the relationship $I = A\rho^2$ (i.e., moment of inertia = area times radius of gyration squared[2]) gives

[2]Various symbols are used for radius of gyration, the most common one being r. The symbol ρ is used here (and in some other references) to avoid confusion with the actual radius of a round-section column.

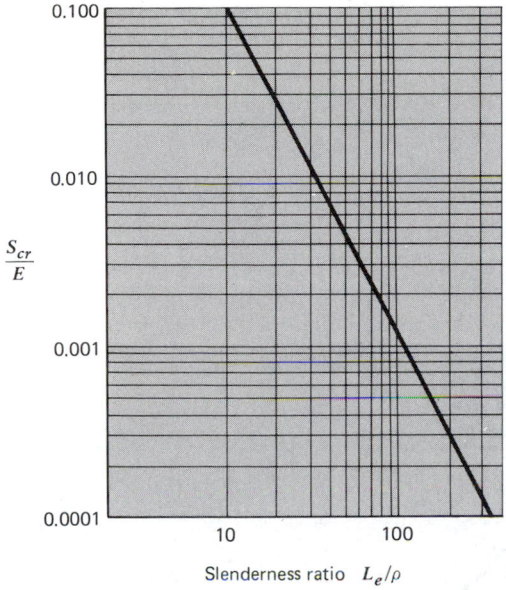

FIGURE 5.23.
Log-log plot of Euler Eq. (5.11). (Dimensionless, hence applies to all materials within their elastic range.)

$$S_{cr} = \frac{P_{cr}}{A} = \frac{\pi^2 E}{(L_e/\rho)^2} \quad \text{or} \quad \frac{S_{cr}}{E} = \frac{\pi^2}{(L_e/\rho)^2} \tag{5.11}$$

where the ratio, L_e/ρ, is known as the *slenderness ratio* of the column. Note that this equation gives the value of the P/A stress at which the column becomes elastically unstable. It has nothing to do with the yield strength or ultimate strength of the material.

Equation (5.11) is plotted in Fig. 5.23 using log-log coordinates. Note that this single straight line represents a general relationship that applies to all (elastic) materials. Also, being dimensionless, it is equally useable for either SI or English units. The plot shows that the critical buckling P/A load, as a percentage of the modulus of elasticity, depends only on the slenderness ratio.

Euler curves corresponding to the elastic modulus of steel and of aluminum are plotted on linear coordinates in Fig. 5.24. Also shown are direct compressive yield curves for S_y = 496 MPa (72 ksi), and S_y = 689 MPa (100 ksi). A steel compression member having a yield strength of 689 MPa would, according to Euler theory, fail if its combination of loading and geometry plotted above curve *ACE*. Likewise, an aluminum member having a yield strength of 496 MPa would theoretically fail under conditions plotting above curve *BDF*. We will see in Section 5.11 that in real life failures would be expected to occur at lower values of P/A, particularly in the regions of transition points *C* and *D*.

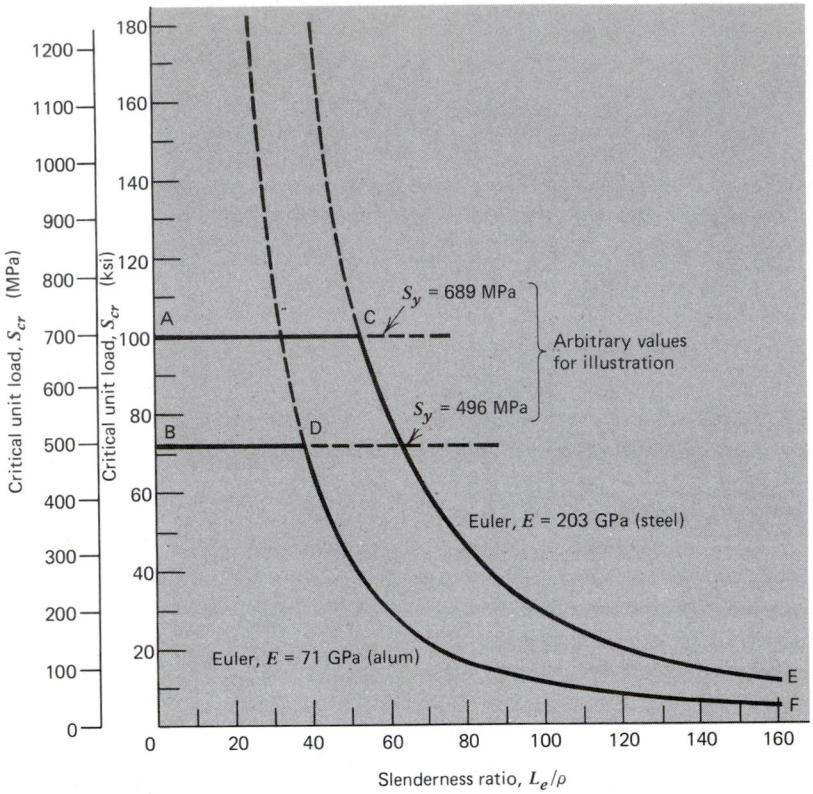

FIGURE 5.24.
Euler column buckling curves illustrated for two values of E and S_y.

5.11 EFFECTIVE COLUMN LENGTH FOR VARIOUS END CONDITIONS

Euler's analysis indicates that the theoretical shape of the deflection curve in Fig. 5.22 is one half sine wave. In order to use a single Euler equation—as (5.10) or (5.11)—for all end conditions, it is customary to work with *equivalent* column length, defined as the length of an equivalent column, pinned at both ends (or the length corresponding to a half sine wave, or the length between points of zero bending moment).

Figure 5.25 shows the most common types of column end conditions. The theoretical values of equivalent length correspond to *absolute rigidity* of all fixed ends (i.e., zero rotation due to the bending reaction). In actual practice this can only be approached; hence, columns having one or both ends fixed always have equivalent lengths longer than the theoretical. The "minimum AISC recommendations" tabulated in Fig. 5.25 apply to end constructions where "ideal conditions are approximated." When fixed end attachments are less rigid, judgment must be used. In cases of questionable rigidity, it is sometimes prudent to make the conservative assumption that

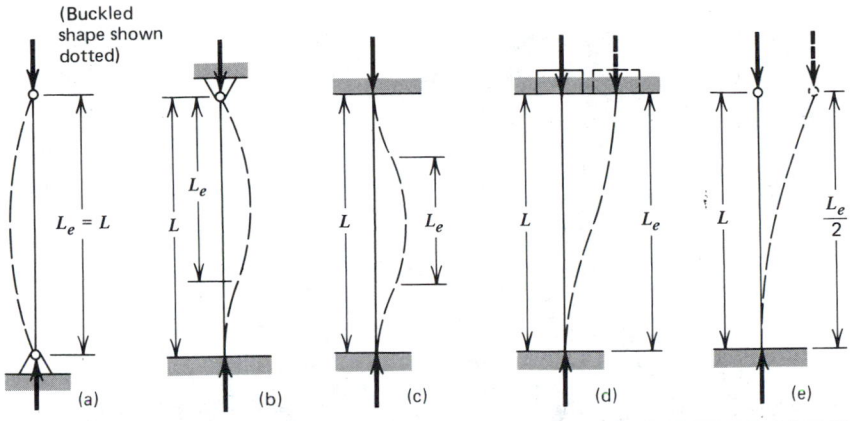

Theoretical $L_e = L$	$L_e = 0.707L$	$L_e = 0.5L$	$L_e = L$	$L_e = 2L$
Minimum AISC Recommend. $L_e = L$	$L_e = 0.80L$	$L_e = 0.65L$	$L_e = 1.2L$	$L_e = 2.1L$

(From Manual of Steel Construction, 7th Ed., 1970, pg. 5–138, American Institute of Steel Construction, Inc., New York.)

FIGURE 5.25.
Equivalent column lengths for various end conditions.

bending rigidity of the "fixed" support is negligible, and therefore equivalent to a pinned end.

5.12 COLUMN DESIGN EQUATIONS – J. B. JOHNSON PARABOLA

Because of inevitable deviations from the ideal situation represented by the curves *ACE* and *BDF* in Fig. 5.24, column failures occur at smaller loads than predicted by theory, particularly in the vicinity of points *C* and *D*. Many empirical modifications have been proposed to deal with this. Some of these are embodied in codes that pertain to the design of specific equipment involving columns. Perhaps the most widely used modification is the parabola proposed by J. B. Johnson around the turn of the century. This is shown for two cases in Fig. 5.26. The equation of the parabola is

$$S_{cr} = \frac{P_{cr}}{A} = S_y - \frac{S_y^2}{4\pi^2 E}\left(\frac{L_e}{\rho}\right)^2 \tag{5.12}$$

Although there is much scatter in the test data, the Johnson parabola has been found to agree reasonably well with experimental results.

As illustrated in Fig. 5.26, the parabola is always tangent to the Euler curve at the point

$$S_{cr} = \frac{S_y}{2}, \qquad \frac{L_e}{\rho} = \left(\frac{2\pi^2 E}{S_y}\right)^{1/2} \tag{5.13}$$

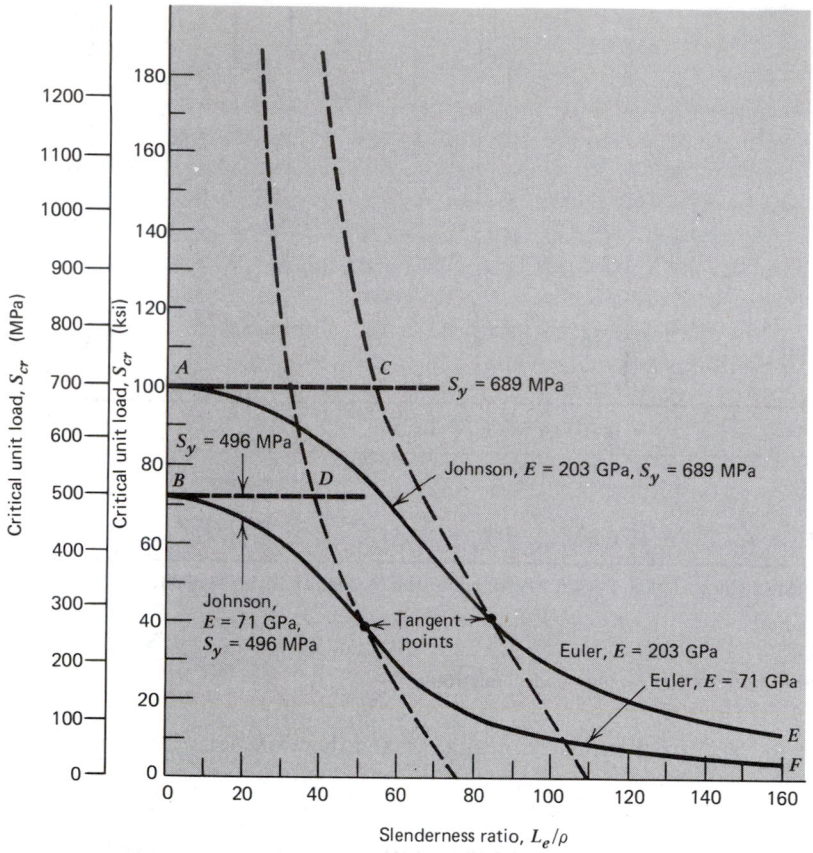

FIGURE 5.26.
Euler and Johnson column curves illustrated for two values of E and S_y (Used in Sample Problems 5.11 and 5.12).

This tangency point often serves to distinguish between "intermediate" columns (parabola range) and "long" columns (Euler range). "Short" columns are commonly regarded as those having L_e/ρ less than 10, in which case the critical unit load can be taken as S_y.

Equation (5.12) was written to correspond to the general equation of a parabola:

$$y = a - bx^2$$

Sometimes constants other than those of Eq. (5.12) are used in order to obtain a better fit with specific experimental data.

SAMPLE PROBLEM 5.11 Determine the Required Diameter of a Steel Piston Rod

An industrial machine requires a solid round piston connecting rod 1 m long (between pinned ends) that is subjected to a maximum compressive force of 80,000 N. Using a safety factor of 2.5, what diameter is required if steel is used, having properties of $S_y = 689$ MPa, $E = 203$ GPa?

Solution

The material corresponds to line AE of Fig. 5.26; rod construction corresponds to $L_e = L = 1$ m. Tentatively assuming that the Euler relationship applies:

$$\frac{P}{A} = \frac{\pi^2 E}{(L_e/\rho)^2}$$

where the *design overload, P*, is 80,000 N × 2.5 or 200,000 N. For the solid round section specified here,

$$A = \pi D^2/4, \qquad \rho = D/4.$$

Hence,

$$\frac{4P}{\pi D^2} = \frac{\pi^2 E D^2}{16 L_e^2}, \qquad 64PL_e^2 = \pi^3 E D^4$$

$$D = \left(\frac{64PL_e^2}{\pi^3 E}\right)^{1/4} = \left(\frac{64(200{,}000)(1)^2}{\pi^2(203 \times 10^9)}\right)^{1/4} = 0.0378 \text{ m}$$

This diameter gives a slenderness ratio of

$$\frac{L_e}{\rho} = \frac{1}{0.0378/4} = 106$$

Figure 5.26 shows this to be well beyond the tangent point and into the range where the Euler relationship can indeed be applied. Hence, the final answer (slightly rounded off) is 38 mm. ■

SAMPLE PROBLEM 5.12 Required Diameter of an Aluminum Piston Rod

Repeat Sample Problem 5.11, except reduce the length to 200 mm and use aluminum with properties of $S_y = 496$ MPa, $E = 71$ GPa.

Solution

Again, tentatively assuming the column to be in the Euler range:

$$D = \left(\frac{64PL_e^2}{\pi^2 E}\right)^{1/4} = \left(\frac{64(200{,}000)(0.2)^2}{\pi^3(71 \times 10^9)}\right)^{1/4} = 0.0220 \text{ m}$$

$$\frac{L_e}{\rho} = \frac{0.20}{0.0220/4} = 36.4$$

Figure 5.26 shows that this is too "short" a column for the Euler relationship to apply. The required diameter will be greater than 22 mm, and the Johnson equation is applicable:

$$\frac{P}{A} = S_y - \frac{S_y^2}{4\pi^2 E}\left(\frac{L}{\rho}\right)^2, \qquad A = \frac{\pi D^2}{4}, \qquad \rho = \frac{D}{4}$$

$$\frac{200{,}000(4)}{\pi D^2} = (496 \times 10^6) - \frac{(496 \times 10^6)^2}{4\pi^2(71 \times 10^9)}\frac{(0.2)^2(16)}{D^2}$$

$$\frac{254{,}648}{D^2} = (496 \times 10^6) - \frac{56{,}172}{D^2}$$

$$D = \left(\frac{254{,}648 + 56{,}172}{496 \times 10^6}\right)^{1/2} = 0.025 \text{ m}$$

$$\frac{L_e}{\rho} = \frac{0.2(4)}{0.025} = 32$$

This result is about as expected. The final answer is a diameter of 25 mm. ■

5.13 ECCENTRIC COLUMN LOADING – THE SECANT FORMULA

If the line of action of the resultant column load, P, does not pass through the centroidal axis of the cross section, the column is loaded eccentrically. The distance between the load axis and the column axis is the eccentricity, e. When the eccentric moment Pe is taken into account, the following analytical equation, known as the *secant formula,* can be derived:[3]

$$S_{cr} = \frac{P_{cr}}{A} = \frac{S_y}{1 + (ec/\rho^2)\sec\left[(L_e/\rho)\sqrt{P_{cr}/4AE}\right]} \tag{5.14}$$

where c is the distance from the neutral bending plane to the extreme fiber, and where ec/ρ^2 is known as the *eccentricity ratio.*

It is important to note that the above equation pertains to buckling *in the plane of bending moment Pe.* Hence, the radius of gyration, ρ, must be taken with respect to the corresponding bending axis. If this is not also the least radius of gyration, then one must also check for buckling about the axis of least ρ, using the procedures for concentric column loading described in the preceding articles.

To illustrate this point, suppose the column shown in Fig. 5.22 is loaded with a force whose line of action is displaced a small distance along the X axis. Although this eccentricity increases any tendency to buckle about the Y axis, it has no effect on buckling about the X axis. If the column section shown were more nearly square, it is easy to visualize buckling about the Y axis for eccentricities (along the X axis) greater than some critical value, and buckling about the X axis for smaller eccentricities.

The secant formula is inconvenient to use for design purposes because of the involved way the various column dimensions appear in the equation. Curves, such as

[3]See almost any basic texts in strength of materials.

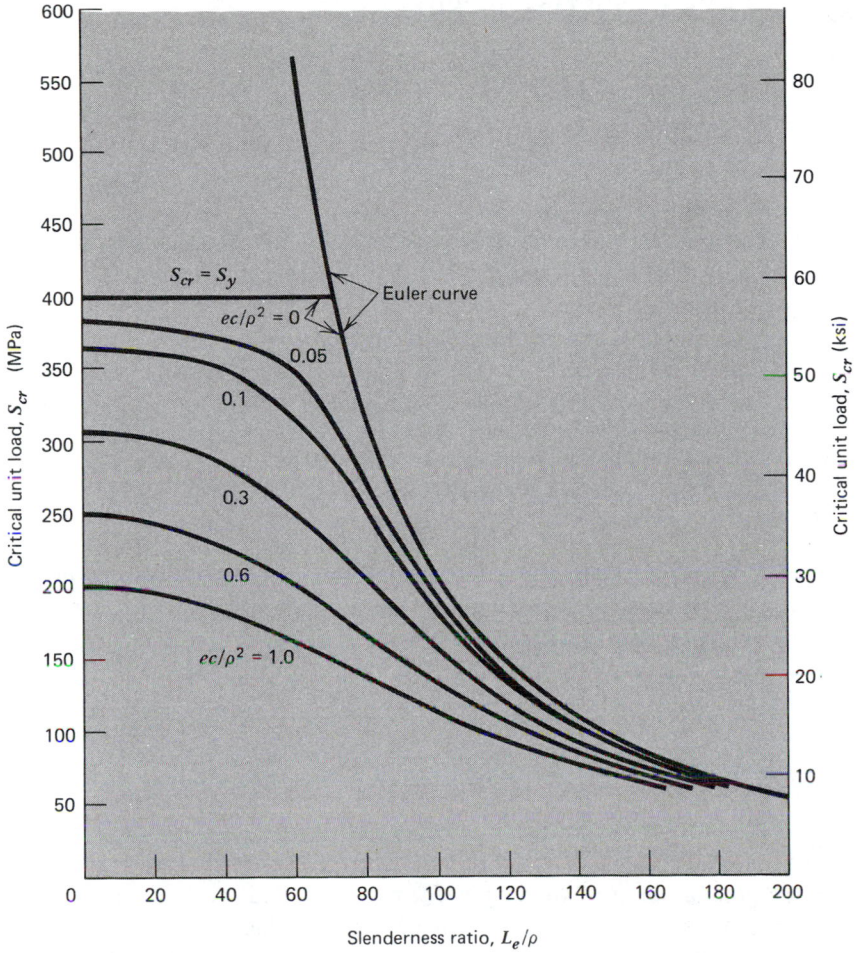

FIGURE 5.27.
Comparison of Secant and Euler formulas for E = 207 GPa, S_y = 400 MPa.

those in Fig. 5.27, can be prepared for eccentrically loaded column design and analysis involving a material with specific values of E and S_y.

The secant formula can also be used with centrally loaded columns if one assumes that some estimated small eccentricity would inevitably be present in any realistic situation. An assumed eccentricity equal to $L_e/400$ is sometimes suggested [8]. For "centrally" loaded structural columns, an assumed eccentricity ratio (ec/ρ^2) of 0.025 is often used, as the result of an extensive study in 1933 by a Committee of the American Society of Civil Engineers.[4]

[4]Report of a Special Committee on Steel Column Research, Trans. Amer. Soc. Civil Engrs., Vol. 98, 1933.

5.14 EQUIVALENT COLUMN STRESSES

As noted above, column formulas (as the Euler and the Johnson) provide equations for S_{cr} with which an applied P/A load can be compared. S_{cr} can be thought of as being related to S_y by the equation

$$S_{cr} = \frac{S_y}{\alpha} \tag{a}$$

Where α is a factor by which the compressive strength is reduced due to buckling tendencies. For extremely short columns (as $L_e/\rho < 10$), α is essentially unity. For longer columns, α takes on increasing values.

In the Euler range, it follows from Eq. (5.11) that

$$\alpha = \frac{S_y}{S_{cr}} = \frac{S_y(L_e/\rho)^2}{\pi^2 E} \tag{5.15}$$

Similarly, in the Johnson range, and using Eq. (5.12):

$$\alpha = \frac{S_y}{S_{cr}} = \frac{4\pi^2 E}{4\pi^2 E - S_y(L_e/\rho)^2} \tag{5.16}$$

It is sometimes convenient to use α as a stress multiplier. Then, one compares $\alpha P/A$ directly with S_y. This concept is particularly useful when working with combined stresses. For example, if a direct compressive stress is involved in the calculation of σ_x or σ_y in Eqs. (4.15)–(4.17), use $\alpha P/A$ to make allowance for buckling tendency.

5.15 OTHER TYPES OF BUCKLING

Columns designed for structures requiring very high strength-weight ratios frequently use non-ferrous materials not having a sharply defined yield point. For these materials in particular, the gradual onset of yielding as S_{cr} is approached involves a progressive reduction of slope of the stress-strain curve, with a resulting reduction in effective elastic modulus, E. Methods have been developed based on the "tangent modulus" concept for dealing more effectively with this situation.

The buckling stability of a long column of circular cross section can be greatly increased with no increase in weight by redistributing the same material into a tubular cross section. There is a limit to how far one can go in this direction because very-thin-wall tubing tends to buckle locally—wrinkling in accordion fashion—while the axis of the column itself may remain straight. This is illustrated in Fig. 5.28*a*, and can readily be demonstrated using an ordinary sheet of paper and a little transparent tape. The proportions are not at all critical, but try rolling an 8½ in. × 11 in. sheet into a tube 8½ in. long with either single thickness plus small overlap (about 3¼ in. diameter) or double thickness (about 1¾ in. diameter). If the paper is of reasonable quality, the resulting "column" will easily support the weight of this book. Pushing down on the book, using care to keep the load concentric, will cause a wrinkling, or "accordion-type" failure. The best failure patterns are usually obtained by pushing the book down quickly.

If a thin-wall tube is subjected to external pressure (as in boiler tubes, oil well

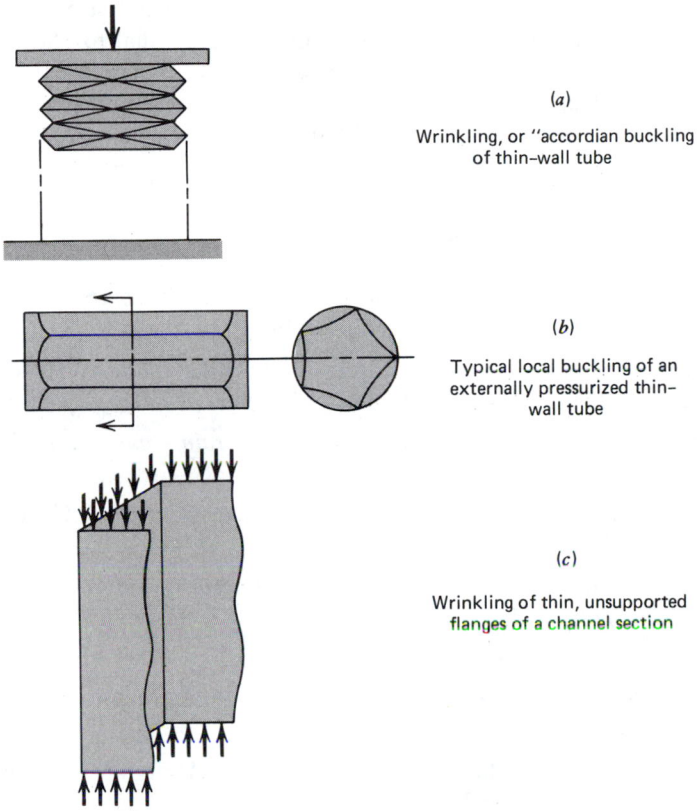

FIGURE 5.28.
Examples of local buckling.

casings, vacuum tanks, suction pipes, etc.), then the circumferential compressive stresses can cause local buckling in the form of flutes or longitudinal corrugations, as shown in Fig. 5.28*b*. When attempting to bend a thin-wall tube into a circular arc, common experience tells us that local buckling tends to occur on the compression side. The desired bending can sometimes be accomplished by providing lateral support to the compression surface, either externally by a special bending fixture, or internally by filling with sand or other suitable material.

A thin plate bent in the form of an angle or channel can fail in local buckling, or wrinkling, as shown in Fig. 5.28*c*. Similar failure can occur in a thin "I" beam flange, subjected to compression due to bending. Also, a thin "I" beam web can "wrinkle" due to compression associated with a high shear stress. Similar wrinkling due to shear-induced compression occurs in stressed-skin construction, as used in aircraft panels. (This may be permitted, to a degree, without harm.)

Appropriate formulas for dealing with a wide variety of local buckling situations have been summarized by Roark and Young [4]. Analytical treatments are given in advanced texts, as [5].

It is interesting to note the many columns appearing in nature.[5] Plant stems are generally tubular, and well into the Euler range with values of L_e/ρ equal to 150 or more. Wall thicknesses are adequate to provide greater reserve for local buckling than for general Euler buckling. The long bones of vertebrates provide interesting studies in column design. An example is the eccentrically loaded human femor (thigh bone).

REFERENCES

1. Durelli, A. J., E. A. Phillips, and C. H. Tsao, *Introduction to the Theoretical and Experimental Analysis of Stress and Strain,* McGraw-Hill, New York, 1958.
2. Hetenyi, M. (ed), *Handbook of Experimental Stress Analysis,* Chapter 9, "Strain Rosettes," by J. H. Meier, John Wiley & Sons, Inc., New York, 1950.
3. Juvinall, R. C., *Engineering Considerations of Stress, Strain, and Strength,* McGraw-Hill, New York, 1967.
4. Roark, R. J. and W. C. Young, *Formulas for Stress and Strain,* 5th ed., McGraw-Hill, New York, 1975.
5. Seely, F. B. and J. O. Smith, *Advanced Mechanics of Materials,* 2nd ed., John Wiley & Sons, Inc., New York, 1952.
6. Shanley, F. R., *Strength of Materials,* McGraw-Hill, New York, 1957.
7. Timoshenko, S. and J. N. Goodier, *Theory of Elasticity,* 2nd ed., McGraw-Hill, New York, 1951.
8. Timoshenko, S. and G. H. McCullough, *Elements of Strength of Materials,* D. Van Nostrand Company, Inc., New York, 1935.

[5]See *Mechanical Design in Organisms* by Wainwright, Biggs, Currey, and Gosline, John Wiley & Sons, Inc., New York, 1975.

CHAPTER SIX

Failure Theories, Safety Factors, and Reliability

6.1 INTRODUCTION

Previous chapters have dealt with the determination of loads (Chapter 2), with the stresses and deflections caused by those loads (Chapters 4 and 5), and with the ability of representative materials to resist *standard* test loads (Chapter 3). The present chapter is concerned with (a) predicting the capability of materials to withstand the infinite combination of *nonstandard* loads to which they are subjected as fabricated machine and structural components, and (b) the selection of appropriate *safety factors* to provide the required safety and reliability. As with the earlier chapters, the concern in this chapter is primarily with static loads.

Reliability is a matter of extreme importance in the engineering of a product, and this point is becoming increasingly recognized. On the other hand, it is important that components not be *overdesigned* to the point of making them unnecessarily costly, heavy, bulky, or wasteful of resources.

The fundamental concept of designing every component of a machine to satisfy a given life and reliability requirement without overdesigning a single part was perhaps never so well phrased as by Oliver Wendell Holmes in 1858 when he wrote "The One-Hoss Shay." This classical bit of literary (and perhaps technical) heritage is all the more remarkable because Holmes was not an engineer, but a physician and professor of anatomy at the Harvard Medical School. He was the son of a Congregational minister, and the father of Oliver Wendell Holmes, Jr., the noted Supreme Court Justice.

In his remarkable poem, reproduced below, Holmes tells of a colonial deacon possessed of such uncanny engineering genius that he was able to design and build a little one-horse carriage so that every single component had a useful life of exactly 100 years (to the minute!), at the end of which time they all failed, causing the little wagon to collapse in a heap of rubble! (Imagine, if we could only design automobiles this way—so that they ran trouble-free for, say, 200,000 miles, at the expiration of which the owner had better be driving into a convenient junk yard!)

The One-Hoss Shay
(The Deacon's Masterpiece)

Oliver Wendell Holmes

Have you heard of the wonderful one-hoss shay,
That was built in such a logical way
It ran a hundred years to a day,
And then, of a sudden, it—ah, but stay,
I'll tell you what happened without delay,
Scaring the parson into fits,
Frightening people out of their wits,—
Have you ever heard of that, I say?

Seventeen hundred and fifty-five,
Georgius Secundus *was then alive,—*
Snuffy old drone from the German hive.
That was the year when Lisbon-town
Saw the earth open and gulp her down,
And Braddock's army was done so brown,
Left without a scalp to its crown.
It was on the terrible Earthquake-day
That the Deacon finished the one-hoss shay.

Now in building of chaises, I tell you what,
There is always ***somewhere*** *a weaker spot,—*
In hub, tire, felloe,[1] in spring or thill,[2]
In panel, or crossbar, or floor, or sill,
In screw, bolt, thoroughbrace,[3]—lurking still,
Find it somewhere you must and will,—
Above or below, or within or without,—
And that's the reason, beyond a doubt,
A chaise breaks down, but doesn't wear out.

But the Deacon swore (as Deacons do),
With an "I dew vum," or an "I tell yeou,"
He would build one shay to beat the taown
'N' the keounty 'n' all the kentry raoun';
It should be so built that it ***couldn'*** *break daown:*
—"Fur," said the Deacon, "'t's mighty plain
Thut the weakes' place mus' stan' the strain;
'N' the way t' fix it, uz I maintain,
Is only jest
T' make that place uz strong uz the rest."

So the Deacon inquired of the village folk
Where he could find the strongest oak,
That couldn't be split nor bent nor broke,—
That was for spokes and floor and sills;
He sent for lancewood to make the thills;
The crossbars were ash, from the straightest trees,
The panels of white-wood, that cuts like cheese,
But lasts like iron for things like these;
The hubs of logs from the "Settler's ellum,"—
Last of its timber,—they couldn't sell 'em,
Never an axe had seen their chips,
And the wedges flew from between their lips,
Their blunt ends frizzled like celery tips;
Step and prop-iron, bolt and screw,
Spring, tire, axle, and linchpin[4] too,
Steel of the finest, bright and blue;
Thoroughbrace bison-skin, thick and wide;
Boot, top, dasher, from tough old hide
Found in the pit when the tanner died.
That was the way he "put her through."—
"There!" said the Deacon, "naow she'll dew!"

Do! I tell you, I rather guess
She was a wonder, and nothing less!

Colts grew horses, beards turned gray,
Deacon and Deaconess dropped away,
Children and grandchildren—where were they?
But there stood the stout old one-hoss shay
As fresh as on Lisbon-earthquake-day!

EIGHTEEN HUNDRED;—it came and found
The Deacon's masterpiece strong and sound.
Eighteen hundred increased by ten;—
"Hahnsum kerridge" they called it then.
Eighteen hundred and twenty came;—
Running as usual; much the same.
Thirty and forty at last arrive,
and then come fifty, and FIFTY-FIVE.

Little of all we value here
Wakes on the morn of its hundredth year
Without both feeling and looking queer.
In fact, there's nothing that keeps its youth,
So far as I know, but a tree and truth.
(This is a moral that runs at large;
Take it.—You're welcome.—No extra charge.)

FIRST OF NOVEMBER—the-Earthquake-day,—
There are traces of age in the one-hoss shay.
A general flavor of mild decay,
But nothing local, as one may say.
There couldn't be,—for the Deacon's art
Had made it so like in every part
That there wasn't a chance for one to start.
For the wheels were just as strong as the thills,
And the floor was just as strong as the sills,
And the panels just as strong as the floor,
And the whipple-tree neither less nor more.
And the back-cross bar as strong as the fore,
And spring and axle and hub ***encore****.*
And yet, as a whole, it is past a doubt
In another hour it will be ***worn out!***

First of November, 'Fifty-five!
This morning the parson takes a drive.
Now, small boys, get out of the way!
Here comes the wonderful one-hoss shay,
Drawn by a rat-tailed, ewe-necked bay.
"Huddup!" said the parson. Off went they.

The parson was working his Sunday text,
Had got to ***fifthly****, and stopped perplexed*
At what the—Moses—was coming next.
All at once the horse stood still,
Close by the meet'n' house on the hill.
—First a shiver, and then a thrill.
Then something decidedly like a spill,—
And the parson was sitting up on a rock,
At half-past nine by the meet'n' house clock,—
Just the hour of the Earthquake shock!
—What do you think the parson found,
When he got up and stared around?
The poor old chaise in a heap or mound,
As if it had been to the mill and ground!
You see, of course, if you're not a dunce,
How it went to pieces all at once,—
All at once, and nothing first,—
Just as bubbles do when they burst.

End of the wonderful one-hoss shay,
Logic is logic. That's all I say.

[1]rim.
[2]shaft on each side of horse.
[3]leather strap between front and rear spring, supporting body (one on each side).
[4]pin through axle to keep the wheel on.

6.2 TYPES OF FAILURE

Failure of a loaded member can be regarded as any behavior that renders it unsuitable for its intended function. At this point we are concerned only with *static* loading, saving for later chapters impact, fatigue, and surface wear (all of which, incidentally, the deacon must have been intimately concerned with in the design of his masterpiece). Static loading can result in objectionable deflection and elastic instability (Chapter 5) as well as *plastic distortion* and *fracture,* with which the present chapter is concerned.

Distortion, or plastic strain, is associated with shear stresses, and involves slip (or sliding) along natural slip planes. Failure is defined as having occurred when the plastic deformation reaches an arbitrary limit, as the 0.2 percent offset in a standard tensile test. Often, appreciably more yielding can occur without harm, as (a) in localized areas of stress concentration, and (b) in some members subjected to bending or torsion where yielding is restricted to the outer surface. The definition of failure by distortion is arbitrary, and not always easy to make (i.e., how much distortion is too much?). Fracture, on the other hand, is clearly defined as the separation or fragmentation of a member into two or more pieces. It normally constitutes a "pulling apart," associated with a tensile stress.

In general, materials prone to distortion failure are classed as *ductile,* and those prone to fracture without significant prior distortion as *brittle.* Unfortunately, there is an intermediate "gray area" wherein a given material can fail in either a ductile or brittle manner depending on circumstances. It is well known that materials that are normally ductile can fracture in a brittle manner at sufficiently low temperatures. Other factors promoting brittle fracture are sharp notches and impact loading. An important concept in this connection is that of *transition temperature*—that is, a fairly narrow temperature range above which the material and associated geometry and loading conditions produce ductile failure, with brittle fracture occurring at lower temperatures.

For generations, a matter of concern to engineers and metallurgists has been the brittle fracture of structural steels that behave in a completely ductile manner during the ordinary laboratory strength tests. Figure 6.1 shows a rather spectacular example of a World War II tanker broken in two by a brittle fracture despite the normal ductility associated with the grade of steel used. The mechanisms of brittle fracture are the concern of a relatively new discipline, *fracture mechanics.*

To treat adequately the subject of fracture mechanics would require a far more lengthy presentation than could possibly be justified in this text. Yet the subject is so important that any engineer concerned with the design and development of mechanical and structural components should have at least a knowledge of its basic principles. It is with this purpose in mind that the next two articles are presented. Hopefully, the interested student will find an opportunity to continue his or her study in this area (see [1,9]).

6.3 FRACTURE MECHANICS – BASIC CONCEPTS

Very briefly, the fracture mechanics approach begins with the assumption that all real materials contain cracks of some size—even if only submicroscopic. If brittle fracture occurs, it is because the conditions of loading and environment (primarily temperature) are such as to cause an almost instantaneous propagation to failure of one or more of the original cracks. If fatigue loading is involved, the initial cracks may grow very slowly until one of them reaches a *critical size* (for the loading, geometry, and environment involved), at which time total fracture occurs.

FIGURE 6.1.
"S.S. Schenectady," T-2 tanker, broken in two in fitting-out dock, Portland, Oregon, Jan. 16, 1943. (Courtesy Ship Structures Committee, U.S. Government.)

Theoretically, the stress concentration factor at the base of a crack approaches infinity because the radius at the crack-root approaches zero (as with r/d approaching zero in Fig. 4.35). This means that if the material has any ductility, yielding will occur within some small volume of material at the crack-root, resulting in a redistribution of stress. Thus, the effective stress concentration factor is considerably less than infinity, and furthermore it varies with the intensity of the applied nominal stress. In the fracture mechanics approach, one does not attempt to evaluate an effective stress concentration *per se*; rather, a *stress intensity factor, K*, is evaluated. This can be thought of as a measure of the effective local stress at the crack-root. Once evaluated, K is then compared with a *limiting value of K* that is necessary for crack propagation in that material. This limiting value is a characteristic of the material, called *fracture toughness,* or *critical stress intensity factor*, K_c, which is determined from standard tests.

Most currently available values of K and K_c are for *tensile loading,* which is called *Mode I.* Accordingly, these values are designated as K_I and K_{Ic}. Modes II and III involve shear loading. The treatment here will be concerned only with Mode I.

Most available values of K_{Ic} (as those in Table 6.1) are for relatively *thick* members, such that the material at the crack-root is approximately in a state of *plane strain* (low-stressed material surrounding the crack resists the "Poisson's ratio" contraction of material at the crack-root, thereby enforcing $\epsilon_3 \approx 0$ in the thickness direction). Crack-root material in sufficiently *thin* members is free to contract in the thickness direction, giving $\sigma_3 \approx 0$ or a condition of *plane stress.* The plane *strain* tensile loading, with σ_3 being tensile, offers less opportunity for redistributing high crack-root stresses by shear-yielding. (This is evident by considering three-dimensional Mohr stress circles for $\sigma_3 =$

TABLE 6.1
Strength Properties of 1-in.-Thick Plates

Material	Temperature	S_u (ksi)	S_y (ksi)	K_{Ic} (ksi $\sqrt{\text{in.}}$)
7075-T651 Aluminum	room	78	70	27
Ti-6Al-4V (annealed)	room	130	120	65
D6AC Steel	room	220	190	70
D6AC Steel	−40°F	227	197	45
4340 Steel	room	260	217	52

Source: Mr. A. Gomza, Grumman Aerospace Corp.

0 and for σ_3 = a positive value.) Because of this, values of K_{Ic} for plane *strain* are substantially lower than those for plane *stress*. Thus, the more readily available plane strain values of K_{Ic} are often used for conservative calculations when the value of K_{Ic} for the actual thickness is not known.

6.4 FRACTURE MECHANICS – APPLICATIONS

Figure 6.2*a* shows a "thin" plate (for example, the "skin" of an airplane) with a crack of length $2c$ extending through the full thickness. If the crack length is a small fraction of the plate width, and if the P/A stress figured on the basis of the net area, $t(2w - 2c)$, is less than the yield strength, then the stress intensity factor at the edges of the crack is approximately

$$K_I = (1.8\sqrt{c})\sigma_g \tag{6.1}$$

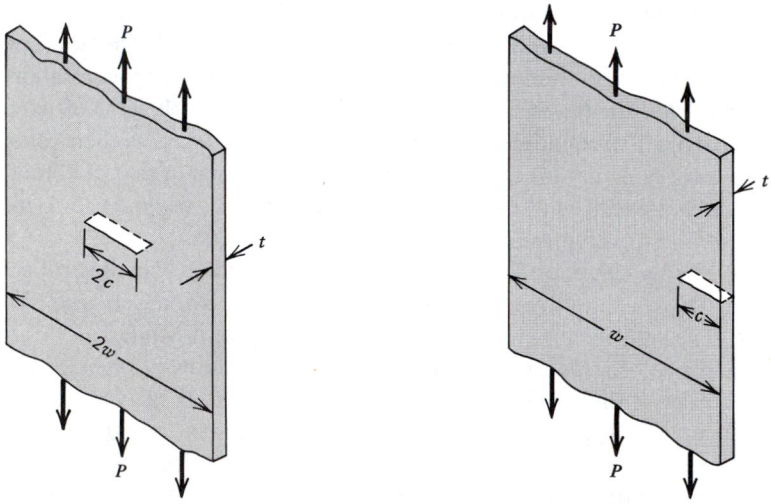

(*a*) Center crack (*b*) Edge crack

FIGURE 6.2.
Through-the-thickness cracks in thin plates.

where σ_g is the *gross*-section tensile stress, $P/2wt$. Rapid fracture occurs when K_I becomes equal to K_{Ic}, the fracture toughness value for the material. In this case of a *thin* plate, the plane *stress* value of K_{Ic} would be preferred. Thus, failure occurs when three basic variables reach the following approximate relationship:

$$K_{Ic} = (1.8\sqrt{c})\sigma_g \tag{6.2}$$

For a crack occurring at the edge of a plate, as in Fig. 6.2*b*, the above equations apply, with only a small increase in the value of the constant. Thus, the failure criterion for Fig. 6.2*b* becomes, approximately,

$$K_{Ic} = (2.0\sqrt{c})\sigma_g \tag{6.3}$$

SAMPLE PROBLEM 6.1 Determine the Critical Load for a "Thin" Plate With a Central Crack

A plate of width $2w = 6$ in., and thickness $t = 0.06$ in., is made of 7075-T651 aluminum ($S_u = 78$ ksi, $S_y = 70$ ksi), and a *plane stress* $K_{Ic} = 60$ ksi $\sqrt{\text{in.}}$ It is used in an aircraft application where periodic inspection for cracks will be made. Estimate the highest load, P (see Fig. 6.2*a*), that can be applied without causing sudden fracture when a central crack grows to a length, $2c$, of 1 in.

Solution

From Eq. (6.2),

$$\sigma_g = \frac{K_{Ic}}{1.8\sqrt{c}} = \frac{60}{1.8\sqrt{0.5}} = 47.14 \text{ ksi}$$

$$P = \sigma_g(2wt) = 47{,}140 \text{ psi } (6 \text{ in.} \times 0.06 \text{ in.}) = 16{,}970 \text{ lb}$$ ■

Cracks in thick plates generally begin at the surface, taking a somewhat elliptical form, as shown in Fig. 6.3. If $2w/t > 6$, $a/2c =$ about 0.25, $w/c > 3$, $a/t < 0.5$, and $\sigma_g/\sigma_y < 0.8$, the stress intensity factor at the edges of the crack is approximately

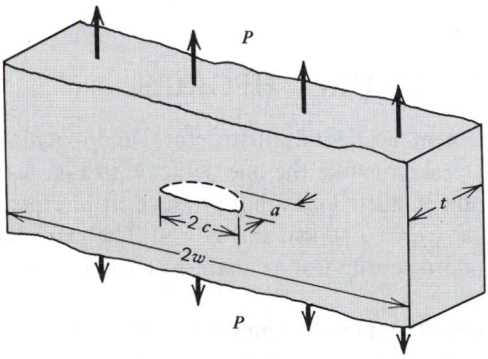

$$K = \frac{\sigma_g \sqrt{a}}{\sqrt{0.39 - 0.053(\sigma_g/S_y)^2}} \tag{6.4}$$

Fracture would be predicted for values of K exceeding K_{IC}.

Table 6.1 gives typical mechanical properties of 1- in.-thick plates made of various structural materials commonly used in aircraft applications. Note particularly (1) the relatively high fracture toughness of the titanium alloy in comparison to its ultimate strength, (2) the room temperature comparison of K_{IC} for the two steels of nearly equivalent ultimate strengths, and (3) the reduction in K_{IC} with temperature for the high-toughness D6AC steel.

SAMPLE PROBLEM 6.2 Determine the Critical Crack Depth for a Thick Plate

A Ti-6A1-4V (annealed) titanium plate is loaded as in Fig. 6.3 to a gross-area stress of $0.73S_y$. For dimensions $t = 1$ in., $2w = 6$ in., and $a/2c = 0.25$, estimate the critical crack depth, a_{cr}, at which rapid fracture will occur.

Solution

From Eq. (6.4), and setting $K = K_{Ic}$,

$$a_{cr} = \left(\frac{K_{Ic}\sqrt{0.39 - 0.053(\sigma_g/S_y)^2}}{\sigma_g} \right)^2$$

$$= \left(\frac{65\sqrt{0.39 - 0.053(0.73)^2}}{(0.73)(120)} \right)^2 = 0.20 \text{ in.}$$

■

An important design requirement of internally pressurized members is that a crack be able to propagate through the full wall thickness (thereby causing a leak that can be readily detected) without becoming unstable and leading to total fracture.

6.5 THE "THEORY" OF STATIC FAILURE THEORIES

Engineers involved in the design and development of all kinds of structural and machine components are repeatedly confronted with problems like the one depicted in Fig. 6.4: A proposed application involves a combination of static loads that produce, at a critical location, stresses of (for example) $\sigma_1 = 80$ ksi, $\sigma_2 = -40$ ksi, and $\sigma_3 = 0$. The material being considered was found to fail on a standard tensile test at a stress of 100 ksi. Will this material fail in the proposed application?

Since it is impractical to test every material and every combination of stresses, σ_1, σ_2, and σ_3, a failure theory is needed for making predictions on the basis of a material's performance on the tensile test, of how strong it will be under any other conditions of static loading. The "theory" behind the various classical failure theories is that *whatever*

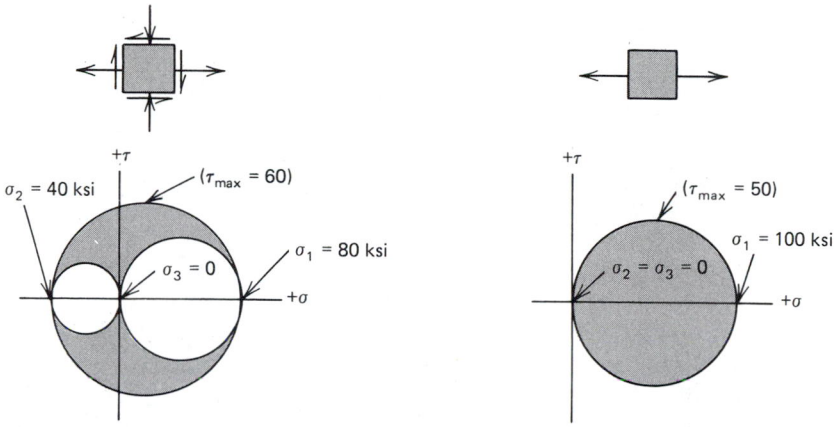

(*a*) Proposed application involving $\sigma_1 = 80$, $\sigma_2 = -40$, $\sigma_3 = 0$

(*b*) Standard tensile test of proposed material. Tensile strength, S = 100 ksi.

FIGURE 6.4.
Typical situation requiring a failure theory.

is responsible for failure in the standard tensile test will also be responsible for failure under all other conditions of static loading. For example, suppose the theory is that failure occurred during the tensile test represented by Fig. 6.4*b* simply because the material is unable to withstand a tensile stress above 100 ksi. The theory then predicts that under *any* conditions of loading, the material will fail if, and only if, σ_1 exceeds 100 ksi. Since the proposed application in Fig. 6.4*a* involves a maximum tensile stress of only 80 ksi, no failure is predicted.

On the other hand, suppose that it is postulated that failure during the tensile test occurred because the material is limited by its inherent capacity to resist *shear* stress, and that, based on the tensile test, the shear stress capacity is 50 ksi. On this basis, failure *would* be predicted in Fig. 6.4*a*.

The reader probably recognized the above examples as illustrating the *maximum-normal-stress* and *maximum-shear-stress* theories, respectively. Other theories have been advanced that would interpret the information in Fig. 6.4*b* as establishing limiting values of other allegedly critical quantities, such as normal strain, shear strain, total energy absorbed, and distortion energy absorbed.

Sometimes one of the above theories is modified empirically in order to obtain better agreement with experimental data. It should be emphasized that in any case the failure theories presented in this chapter apply only to situations where the same type of failure (i.e., ductile or brittle) occurs in the application as in the standard test.

6.6 MAXIMUM-NORMAL-STRESS THEORY

This theory, generally credited to the English scientist and educator W. J. M. Rankine (1802–1872) is perhaps the simplest of all failure theories. It contends merely that failure will always occur whenever the greatest tensile stress tends to exceed the uniaxial

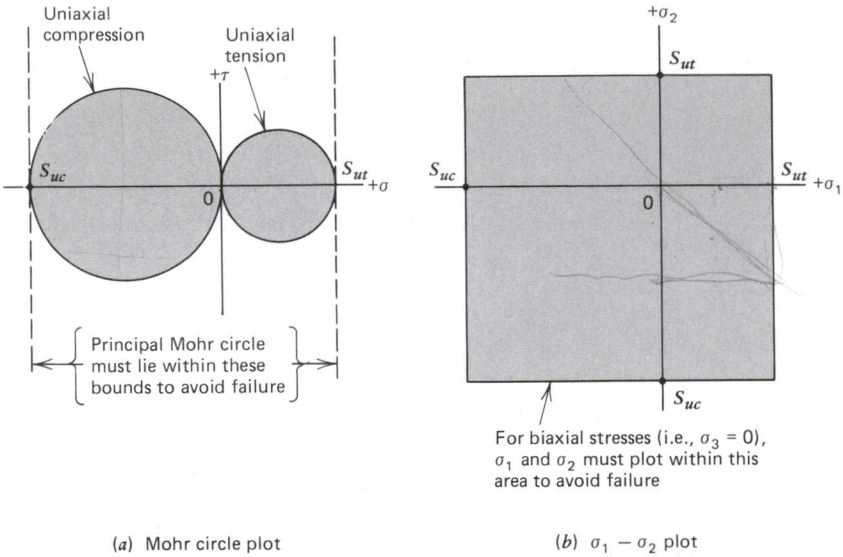

(*a*) Mohr circle plot (*b*) $\sigma_1 - \sigma_2$ plot

FIGURE 6.5.
Two graphical representations of the maximum-normal-stress theory.

tensile strength, or whenever the largest compressive stress tends to exceed the uniaxial compressive strength. With respect to the Mohr-circle plot in Fig. 6.5*a*, failure is predicted for any state of stress for which the principal Mohr circle extends beyond either of the dotted vertical boundaries. On the σ_1-σ_2 plot for biaxial stresses (i.e., $\sigma_3 = 0$) shown in Fig. 6.5*b* failure is predicted for all combinations of σ_1 and σ_2 falling *outside* the shaded area.

This theory has been found to correlate reasonably well with test data for brittle fractures. As might be expected, it is not suited for predicting ductile failures. For this reason, the test points in Fig. 6.5 have been marked "S_{ut}" and "S_{uc}," ultimate strengths in tension and compression, respectively, of an assumed brittle material.

6.7 MAXIMUM-SHEAR-STRESS THEORY

This is thought to be the oldest failure theory, being originally proposed by the great French scientist C. A. Coulomb (1736–1806), who made major contributions to the field of mechanics as well as to electricity. (The reader is undoubtedly familiar with Coulomb's law of electromagnetic force and the Coulomb as the standard unit of quantity of electrical charge.) Tresca wrote an important paper relating to the maximum-shear-stress theory in 1864, and J. J. Guest of England conducted tests around 1900 that led to wide usage of the theory. For these reasons the maximum-shear-stress theory is sometimes called the Tresca theory or Guest's law.

This theory is represented graphically in Fig. 6.6. Note carefully in Fig. 6.6*b* that in the first and third quadrants the zero principal stress is involved in the principal Mohr circle, whereas it is not in the second and fourth quadrants. The single test point is

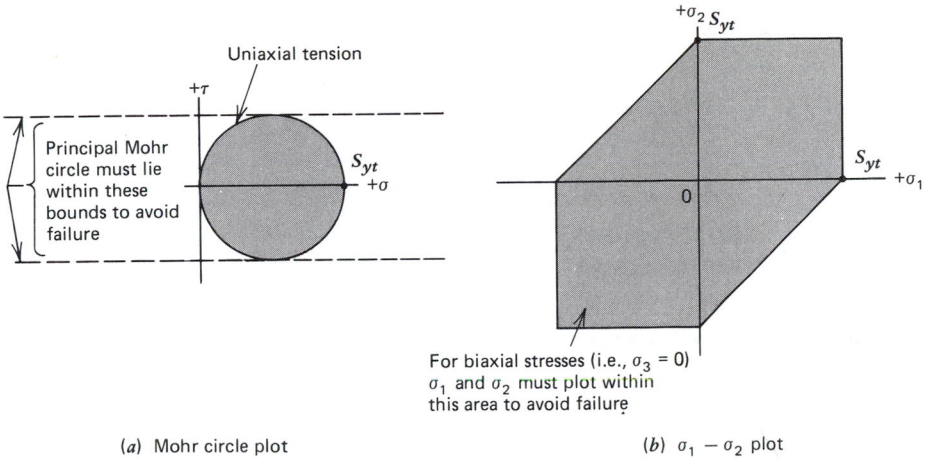

(*a*) Mohr circle plot (*b*) $\sigma_1 - \sigma_2$ plot

FIGURE 6.6.
Two graphical representations of the maximum-shear-stress theory.

marked "S_{yt}," yield strength in tension, of an assumed ductile material. A data point in compression or torsion could serve as well, but the tensile test is the most common and the most accurate; hence, it is normally used. Of course, if the material truly behaves in accordance with the maximum-shear-stress theory, all test data would agree on the level of shear stress associated with failure.

This theory correlates reasonably well with the yielding of ductile materials. However, the maximum-distortion-energy theory, discussed in the next section, is recommended because it correlates better with actual test data for the yielding of ductile materials.

6.8 MAXIMUM-DISTORTION ENERGY THEORY (MAXIMUM-OCTAHEDRAL-SHEAR-STRESS THEORY)

A remarkable thing about this theory is that the equations involved can be derived from at least five different hypotheses (see [5], pp. 117–122). The two most important pertain to the names given to the theory in the above article heading. Credit for the theory is often given to M. T. Hueber (Poland), R. von Mises (Germany and the U.S.A.), and H. Hencky (Germany and the U.S.A.), who contributed to it in 1904, 1913, and 1925, respectively. More recently, Timoshenko[1] has brought to light the fact that it was proposed in 1856 by James Clerk Maxwell of England who, like Coulomb, is better remembered for his contributions to electrical engineering than for his important contributions to the field of mechanics.

Briefly, the contention of the maximum-distortion-energy theory is that any elastically stressed material undergoes a (slight) change in shape and/or volume. The energy

[1]Timoshenko, Stephen P., *History of Strength of Materials*, McGraw-Hill, New York, 1953.

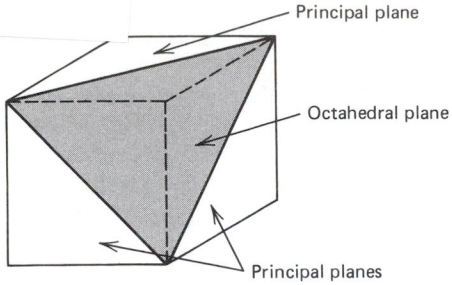

FIGURE 6.7.
An octahedral plane, shown with respect to principal planes.

required to produce this change is stored in the material as elastic energy. It was early recognized that engineering materials could withstand enormous hydrostatic pressures (i.e., $\sigma_1 = \sigma_2 = \sigma_3 =$ large compression) without damage. It was therefore postulated that a given material has a definite limited capacity to absorb energy of *distortion* (i.e., energy tending to change shape but not size), and that attempts to subject the material to greater amounts of *distortion* energy result in yielding.

When using this theory it is convenient to work with an *equivalent stress*, σ_e, defined as the value of uniaxial tensile stress that would produce the same level of distortion energy (hence, according to the theory, the same likelihood of failure) as the actual stresses involved. In terms of the existing principal stresses, the equation for equivalent stress is

$$\sigma_e = \frac{\sqrt{2}}{2}[(\sigma_2 - \sigma_1)^2 + (\sigma_3 - \sigma_1)^2 + (\sigma_3 - \sigma_2)^2]^{1/2} \tag{6.5}$$

For the usual case of biaxial stress, this reduces to

$$\sigma_e = (\sigma_1^2 + \sigma_2^2 - \sigma_1\sigma_2)^{1/2} \tag{6.6}$$

If the direct stresses σ_x, σ_y, and τ_{xy} are more readily obtainable, a convenient form of the equivalent stress equation is

$$\sigma_e = (\sigma_x^2 + \sigma_y^2 - \sigma_x\sigma_y + 3\tau_{xy}^2)^{1/2} \tag{6.7}$$

If only σ_x and τ_{xy} are present, the equation reduces to the

$$\sigma_e = (\sigma_x^2 + 3\tau_{xy}^2)^{1/2} \tag{6.8}$$

Once the equivalent stress is obtained, this is compared with the yield strength from the standard tensile test. If σ_e exceeds S_{yt}, yielding is predicted.

These same equations can readily be derived on the basis of shear stress on an octahedral plane. Figure 6.7 illustrates the relationship of an octahedral plane to the faces of a principal element. There are eight octahedral planes, all of which have the same intensity of normal and shear stress. σ_e can be defined as that intensity of uniaxial tensile stress that produces the same level of shear stress on the octahedral planes (hence, according to the theory, the same likelihood of failure) as do the actual stresses involved.

Figure 6.8 shows that a σ_1-σ_2 plot for the maximum-distortion-energy theory is an

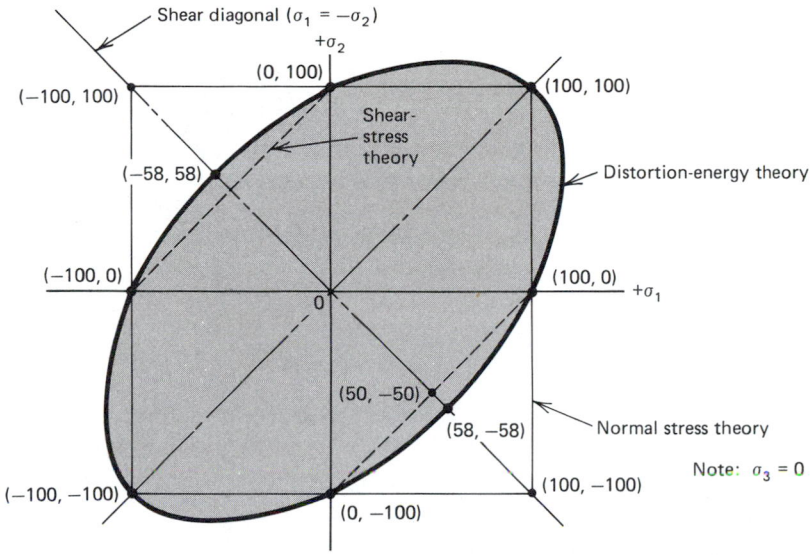

FIGURE 6.8.
$\sigma_1 - \sigma_2$ plot of distortion-energy and other theories, for a ductile material of $S_{yt} = S_{yc} = 100$ ksi. (The D.E. theory predicts failure for all points outside the ellipse.) Note that the point (58, −58) is actually ($\sqrt{3}/3$, $-\sqrt{3}/3$). Distortion-energy theory predicts shear yield strength is $\sqrt{3}/3$, or 0.577 times tensile yield strength, whereas shear-stress theory predicts 0.500 times tensile yield, and normal stress theory predicts 1.0 times tensile yield strength.

ellipse. This is shown in comparison with corresponding plots for the maximum-shear-stress and maximum-normal-stress theories for the case of a ductile material having $S_{yt} = S_{yc} = 100$ ksi. The distortion-energy and the shear-stress theories agree pretty well, with the distortion-energy theory giving the material credit for 0 to 15 percent more strength, depending on the ratio of σ_1 to σ_2. Also shown on this figure is the *shear diagonal,* or locus of all points corresponding to pure shear ($\sigma_1 = -\sigma_2$; $\sigma_3 = 0$). It is interesting to note the wide variation in shear strengths predicted by the various theories. As previously noted, actual tests of ductile materials usually agree quite well with the distortion-energy theory, which predicts (Eq. (6.8), or Fig. 6.8) that $S_{sy} = 0.58S_y$.

The complete derivation of the equations for equivalent stress, from both the distortion-energy and octahedral-shear-stress points of view, is contained in several references, such as [2].

6.9 MODIFIED MOHR THEORY

Over the years, various empirical modifications to the basic failure theories have been proposed, one of which is the *Mohr theory* (also known as the Coulomb–Mohr theory), represented in Fig. 6.9. This theory was suggested for brittle materials, for which the compressive strength far exceeds the tensile strength. (Although the theory is generally

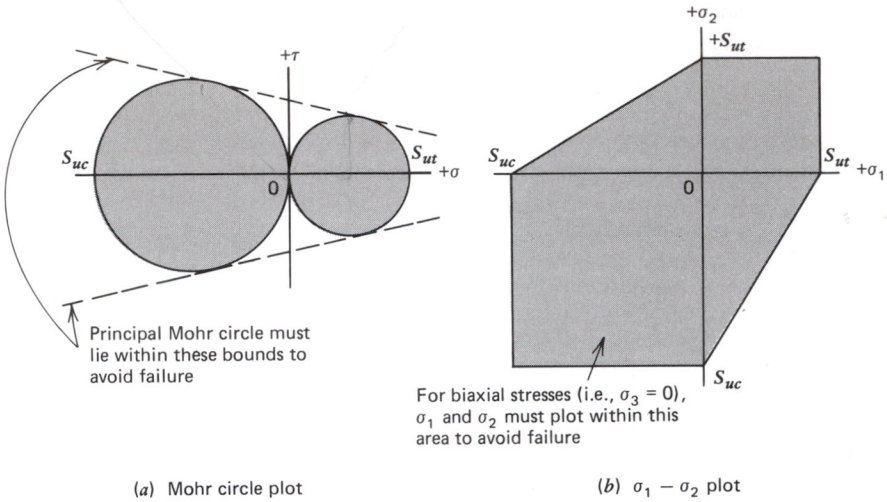

(*a*) Mohr circle plot (*b*) $\sigma_1 - \sigma_2$ plot

FIGURE 6.9
Two graphical representations of the Mohr (or Coulomb-Mohr) theory.

thought of as an empirical modification of the maximum-shear-stress theory, based on using experimental values for both tensile and compressive strengths, it can be derived analytically on the basis of including the effect of *internal friction.* See [5], pp. 122–127.)

A modification of the Mohr theory, illustrated in Fig. 6.10, is recommended for predicting the fracture of brittle materials. It correlates better with most experimental data than do the Mohr or maximum-normal-stress theories, which are also used.

It is well to remember that at best, a failure theory is a substitute for good test data pertaining to the actual material and combination of stresses involved. Also, any

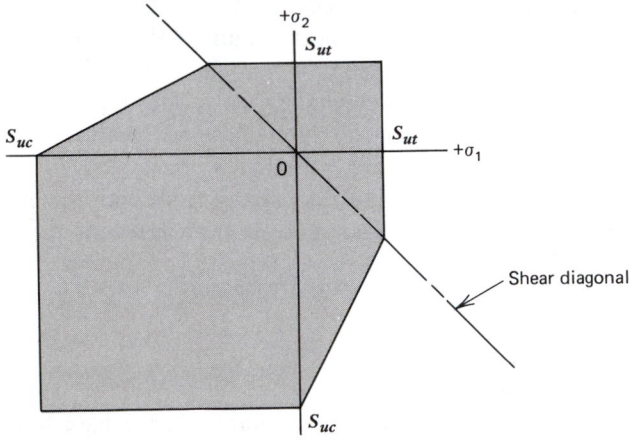

FIGURE 6.10.
Graphical representation of the modified Mohr theory for biaxial stresses ($\sigma_3 = 0$).

additional good test data can be used to improve a theoretical failure theory curve for a given material. For example, suppose that the material involved in Fig. 6.8 was known to have an experimentally determined torsional yield strength of 60 ksi (but before accepting this value, one should be well aware of the inherent difficulty of making an accurate experimental determination of S_{ys}). One might then conclude that the material did indeed appear to behave in general accordance with the distortion-energy theory, but not exactly. By empirically modifying the ellipse just enough to pass through the experimental point, a presumably better failure theory curve would result for making predictions in the second and fourth quadrants.

6.10 SELECTION AND USE OF FAILURE THEORIES

In situations where one can reasonably expect an overloaded part in service to fail in the same manner as the standard tensile test bar made of the same material, it is recommended that (a) the maximum-distortion-energy theory be used to predict ductile yielding, and (b) the modified Mohr theory be used to predict brittle fracture.

SAMPLE PROBLEM 6.3 Estimate the Safety Factor of a Steel Part

Strain gage tests have established that the critical location on the surface of a steel part is subjected to principal stresses of $\sigma_1 = 35$ ksi, and $\sigma_2 = -25$ ksi. (Being an exposed, unloaded surface, $\sigma_3 = 0$.) The steel has a yield strength of 100 ksi. Estimate the safety factor with respect to initial yielding, using the preferred theory. As a matter of interest, compare this with results given by other failure theories.

Solution

Figure 6.11 depicts a graphical solution. Starting from the "nominal load point," the stresses can be proportionally increased until σ_1 reaches values of 58 ksi, 66 ksi, and 100 ksi, according to the shear-stress, distortion-energy, and normal-stress theories, respectively. Corresponding safety-factor estimates are $58/35 = 1.7$, $66/35 = 1.9$, and $100/35 = 2.9$. (Final answers were given with only two significant figures to emphasize that neither the inherent validity of the theories nor the accuracy of graphical construction justifies an implication of highly precise answers.) It is concluded that:

a. The "best" prediction of safety factor is 1.9, based on the distortion-energy theory.

b. The shear-stress theory is in reasonably good agreement (it is often used by engineers to obtain quick estimates).

c. The normal-stress theory has no validity in this case. (To mistakenly use it would give an answer indicating a degree of safety that does not exist.)

The graphical solution was illustrated because it is quick, it is at least as accurate as

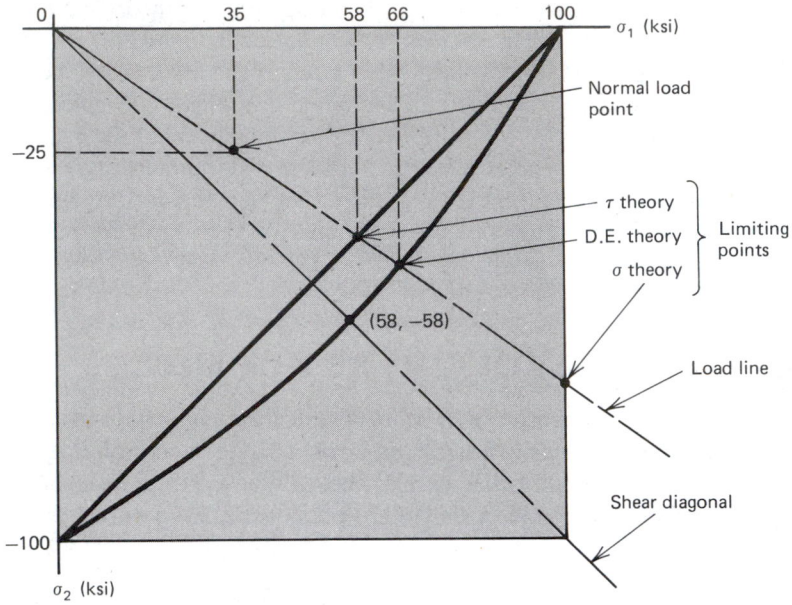

FIGURE 6.11.
Graphical solution to Sample Problem 6.3.

the theories themselves, and it gives one a good intuitive feel for what is going on. Analytical solutions are, of course, equally valid, and are given below:

a. For the distortion-energy theory, Eq. (6.6) gives

$$\sigma_e = (\sigma_1^2 + \sigma_2^2 - \sigma_1\sigma_2)^{1/2}$$
$$\sigma_e = [(35)^2 + (-25)^2 - (35)(-25)]^{1/2} = 52.2$$

Thus, the given stresses are *equivalent* to a simple tensile stress of 52.2 ksi. The tensile test established that the material can withstand a tensile stress of 100 ksi. The safety factor is therefore 100/52.2, or 1.9.

b. Using the shear-stress theory, the principal stresses define a principal Mohr circle having a diameter of 60, and a radius of 30. Thus, the maximum shear stress in the part is 30 ksi. The standard tensile test gave a principal Mohr circle having a *radius* of 50. Thus the material is capable of withstanding a shear stress of 50 ksi. The safety factor is therefore 50/30, or 1.7.

c. Using the normal-stress theory, the application involves a maximum normal stress of 35 ksi, whereas the standard tensile test established that the material is capable of withstanding a normal stress of 100 ksi. The safety factor (by wishful thinking!) is therefore 100/35, or 2.9.

The above discussion of failure theories applies to *isotropic* materials. For cases involving anisotropic materials subjected to various combinations of stress, the reader is referred to special references, as [6].

6.11 SAFETY FACTORS – CONCEPT AND DEFINITION

The original meaning of safety factor was a number by which the ultimate tensile strength of a material was divided in order to obtain a value of "working stress" or "design stress." These design stresses, in turn, were often used in highly simplified calculations that made no allowance for such factors as stress concentration, impact, fatigue, differences between properties of the material in the standard test specimen and in the manufactured part, and so on. As a result, one can still find in handbooks safety factor recommendations as high as 20–30. Modern engineering design involves a rational accounting for all factors possible, leaving relatively few items of uncertainty to be covered by a safety factor that is commonly in the range of 1.5–3.

Modern engineering practice also bases the safety factor on the *significant strength* of the material—not necessarily the static tensile strength. For example, if failure involves static yielding, the safety factor relates (through the use of an appropriate failure theory) the static stress due to the anticipated load (called the *significant stress*) to the static yield strength of the material (called the *significant strength)*, exactly as illustrated in Sample Problem 6.3. If the significant stresses involve fatigue, then the safety factor is based on the fatigue strength; if brittle fracture is the expected mode of failure, the factor is based on tensile strength, and so forth. Thus, safety factor, N, can be defined as,

$$N = \frac{\text{significant strength of the material}}{\text{corresponding significant stress, due to normal loads}^2} \tag{6.9}$$

Safety factor can also be defined in terms of loads:

$$N = \frac{\text{design overload}}{\text{normal load}^2} \tag{6.10}$$

where the *design overload* is defined as being just sufficient to cause failure.

In most cases, the two definitions of safety factor are equivalent. For example, if a material has a significant strength of 200 MPa and the significant stress is 100 MPa, the safety factor is 2. Looking at it the other way, the design overload needed to bring the stress up to the limiting value of 200 MPa is twice the normal load, thus giving a safety factor of 2.

Although the distinction between Eqs. (6.9) and (6.10) may seem trivial, it is recommended that the design overload concept and Eq. (6.10) be used. These are always valid, whereas there are cases where Eq. (6.9) cannot be properly applied. For example, consider the design of a slender column for a safety factor of 2. Figure 6.12 shows how load and stress increase to buckling failure, and where operation with a safety factor of 2 would be, calculated both ways. The discrepancy occurs due to nonlinearity of the load-stress curve. It is clear which interpretation is most conducive to the engineer's peace of mind.

Another example in support of the design overload concept involves the general case of fatigue loading (treated in Chapter 8), which consists of a combination of mean

[2]This is one of those seemingly inevitable cases where a word, or letter, has more than one meaning. Here, a "normal load" is distinguished from an "abnormal load" or an "overload." From the context, it should be clear that the meaning is not "normal load" as distinguished from a "shear load."

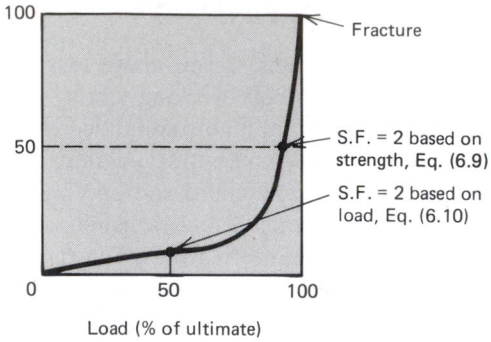

FIGURE 6.12.
Two concepts of safety factor for a buckling column.

(or static) and alternating loads. The kind of overload most likely to occur may involve increasing either or both of these load components. The design overload concept permits the safety factor to be computed with respect to whatever kind of overload is of interest.

A word of caution: It follows from the above that there are cases where the term "safety factor" is ambiguous. It is therefore necessary to be sure that it is clearly defined in all cases where ambiguity could occur.

6.12 SAFETY FACTORS – SELECTION OF A NUMERICAL VALUE

After going as far as is practical in determining the significant strength of the actual fabricated part and the details of the loading to which it will be subjected, there always remains some margin of uncertainty that must be covered by a safety factor. The part *must* be designed to withstand a "design overload" somewhat larger than the normally expected load.

In the last analysis, selection of the safety factor comes down to engineering judgment based on experience. Sometimes these selections are formalized into design codes covering specific situations—for example, the ASME Pressure Vessel Codes, the various building codes, and stipulated safety factor values in legal contracts covering the design and development of special machines. Also, safety factors are often embodied into computer programs for the design of specific components. In these cases, the responsibility for making the engineering judgment falls upon the engineer responsible for the code or computer program. But only partly so, because the engineer *using* the code or program must be satisfied that this detail of the code or program is indeed appropriate for the particular application.

The selection of an appropriate value of safety factor is based primarily on the following five factors:

1. *Degree of loading uncertainty.* In some cases, loads can be determined with virtual certainty. The centrifugal forces in the rotor of an a.c. motor cannot exceed those

calculated for synchronous speed. The loads acting on an engine valve spring are definitely established by the "valve open" and "valve closed" positions (however, in a later chapter we will mention "spring surge," which could introduce a degree of uncertainty). But what loads should be used for the design of automotive suspension components, where loads can vary tremendously depending on the severity of use and abuse? And what about a comparable situation in a completely new kind of machine for which there is no previous experience to serve as a guide? The greater the uncertainty, the more conservative the engineer must be in selecting an appropriate design overload, or safety factor.

2. *Degree of material strength uncertainty.* Ideally, the engineer would have available extensive data pertaining to the strength of the material *as fabricated* into the actual (or very similar) parts, and tested at temperatures and in environments similar to those actually encountered. But this is seldom the case. More often, the available material strength data pertain to samples smaller than the actual part, without experiencing any cold working that may be involved in part fabrication, and tested at room temperature in ordinary air. Moreover, there is bound to be some variation in strength from one test specimen to another. Sometimes the engineer must work with material test data for which such information as specimen size and degree of data scatter (and the relationship between the reported single value and the total range of scatter) are unknown. Furthermore, the material properties in some cases may change significantly over the service life of the part. Considering all such factors, the greater the uncertainty, the larger the safety factor that must be used.

3. *Uncertainties in relating applied loads to material strength.* At this point the reader is already familiar with a number of possible uncertainties, such as (a) validity of the assumptions involved in the standard equations for calculating nominal stresses, (b) accuracy in determining the effective stress concentration factors involved, (c) accuracy in estimating residual stresses, if any, introduced in fabricating the part, and (d) suitability of any failure theories and other relationships used to estimate "significant strength" from available laboratory strength test data.

4. *Consequences of failure—human safety and economics.* If the consequences of failure are catastrophic, relatively large safety factors must, of course, be used. Also, if the failure of some relatively inexpensive part could cause extensive shut-down of a major assembly line, simple economics dictates increasing the cost of this part severalfold (if necessary) in order to virtually eliminate the possibility of its failure.

An important item involved is the *nature* of a failure. If failure involves ductile yielding, the consequences are likely to be less severe than for brittle fracture. Accordingly, safety factors recommended in handbooks are invariably larger for brittle materials.

5. *Cost of providing a large safety factor.* This cost always involves a monetary consideration, and may also involve important consumption of resources. In some cases, a safety factor larger than needed may have serious consequences. A dramatic example is a hypothetical aircraft with excessive safety factors making it too heavy to fly! With respect to the design of an automobile, it would be possible to increase safety factors on structural components to the point where a "maniac" driver could hardly cause a failure even when trying. But to do so would penalize "sane" drivers by requiring them to pay

for stronger components than they could use. More likely, of course, it would motivate them to buy competitive cars! Consider this situation: Should an automotive engineer increase the cost per car by $10 in order to avoid 100 failures in a production run of a million cars, where the failures would not involve safety, but would entail a $100 repair (i.e., should $10,000,000 be spent in order to save $10,000 plus some customer inconvenience?)?

A key point in safety-factor selection is *balance*. All parts of a machine or system should have *consistent* safety factors. Components involving possible human injury or major cost should have the greatest safety factors; components that are comparable in these respects should generally have about the same safety factor, and so on. In fact, balance is perhaps the key to proper safety factor selection—balance based on good engineering judgment—which is in turn based on all available information and experience. (Now, marvel again at the balance achieved by the deacon in his design of the amazing "One-Hoss Shay"!)

Having read through this much philosophy of safety-factor selection, the reader is entitled to have some suggestions of "ball park" values of safety factor that have been found useful, at least as a guide. For this purpose, the following recommendations of Prof. Joseph Vidosic [8] are suggested. These safety factors are based on yield strength.

1. N = 1.25–1.5 for exceptionally reliable materials used under controllable conditions and subjected to loads and stresses that can be determined with certainty—used almost invariably where low weight is a particularly important consideration.
2. N = 1.5–2 for well-known materials, under reasonably constant environmental conditions, subjected to loads and stresses that can be determined readily.
3. N = 2–2.5 for average materials operated in ordinary environments and subjected to loads and stresses that can be determined.
4. N = 2.5–3 for less tried or for brittle materials under average conditions of environment, load, and stress.
5. N = 3–4 for untried materials used under average conditions of environment, load, and stress.
6. N = 3–4 should also be used with better known materials that are to be used in uncertain environments or subjected to uncertain stresses.
7. Repeated loads: the factors established in items 1–6 are acceptable but must be applied to the *endurance limit* rather than the yield strength of the material.
8. Impact forces: the factors given in items 3–6 are acceptable, but an *impact factor* should be included.
9. Brittle materials: where the ultimate strength is used as the theoretical maximum, the factors presented in items 1–6 should be approximately doubled.
10. Where higher factors might appear desirable, a more thorough analysis of the problem should be undertaken before deciding on their use.

6.13 RELIABILITY

A concept closely related to safety factor is *reliability*. If 100 "identical" parts are put into service and two fail, then the parts proved to be 98 percent reliable (which might or might not be good enough). Although the reliability concept finds considerably more application with parts subjected to wear and fatigue loading, we introduce it here in the simpler context of static loading. The utility of the reliability approach is dependent upon having adequate information on the statistical distribution of (a) *loading* applied to parts in service, from which can be calculated the significant stress, and (b) the significant *strength* of production runs of manufactured parts.

Figure 6.13 shows hypothetical distribution curves for significant stress and for corresponding significant strength. The mean value of strength is 70 and the mean value of stress 40. That means that if an "average" part from the production run were put into service under "average" conditions of loading, there would be a *margin of safety* of 30. However, the unshaded area of the curve indicates that there is *some* possibility of a weak part (as strength equals 50) being installed in a particularly severe application (as stress equals 60), in which case failure would occur. Thus, even though the margin of safety is, on the average, 30, there would be a few cases where the margin of safety would be negative, and failure expected. Figure 6.14 shows a corresponding plot of the distribution of margin of safety. In most instances, interest would be focused on the size of the unshaded area at the left, indicating failures.

In order to obtain quantitative estimates of the percentage of anticipated failures from a study like the above, one must look into the nature of the distribution curves for significant stress and strength. We will consider here only the case involving *normal* or *Gaussian* distributions. Although this is only one of the "mathematical models" sometimes found best suited to actual cases, it is probably the most common.

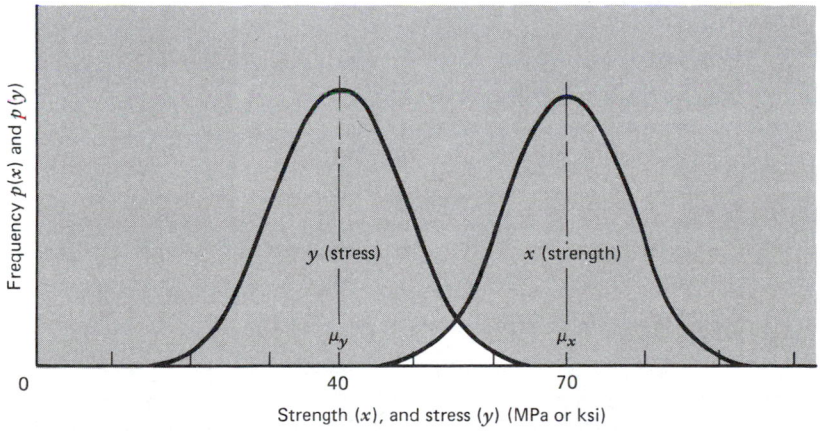

FIGURE 6.13.
Distribution curves for significant strength (x) and significant stress (y).

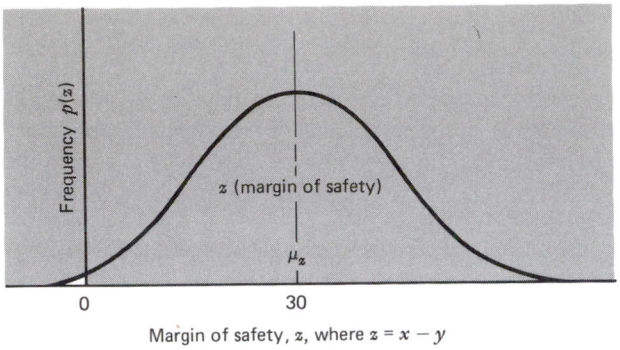

FIGURE 6.14.
Distribution curve for margin of safety (z).

6.14 NORMAL DISTRIBUTIONS

The normal distribution function is most commonly credited to Gauss, but was also discovered independently by two other eighteenth-century mathematicians: Laplace and DeMoivre. Several normal distribution curves are plotted in Fig. 6.15. They have the equation

$$p(x) = \frac{1}{\sqrt{2\pi}\,\sigma} \exp\left(-\frac{(x-\mu)^2}{2\sigma^2}\right), \quad -\infty < x < \infty \tag{6.11}$$

where μ is the mean value of the quantity, and σ the standard deviation[3] (more about that a little later).

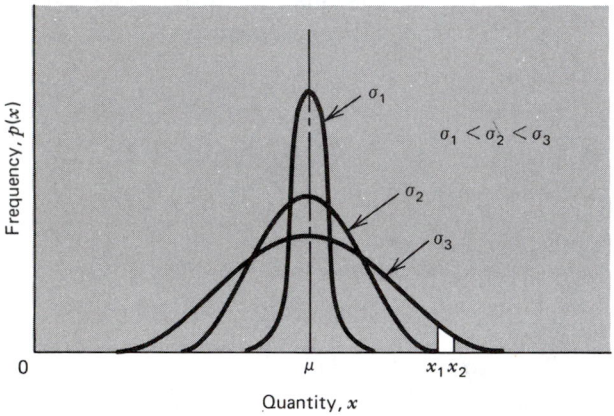

FIGURE 6.15.
Normal distribution curves having a common μ and various σ.

[3]We apologize for using these letters again, but our ancestors utterly disregarded the needs of future generations of engineers and scientists when they put only 26 and 24 letters in the Roman and Greek alphabets, respectively. But perhaps an engineer who cannot tell a normal stress from a standard deviation has more serious problems anyway!

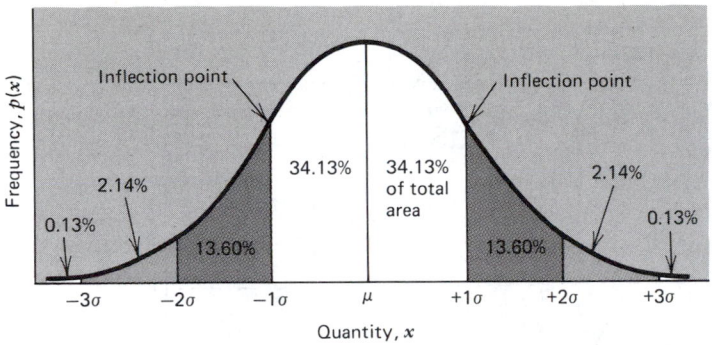

FIGURE 6.16.
Properties of all normal distribution curves.

One limitation of this mathematical model for many applications is the fact that the curve extends assymptotically all the way to plus and minus infinity. Since the probability that any individual value of x will fall between plus and minus infinity is one, the area under each of the curves in Fig. 6.15 is unity. Likewise, the probability that the value will be between any specific values x_1 and x_2 is equal to the area under the curve between those values of x, as shown in Fig. 6.15.

Varying μ with constant σ merely shifts the curve to the right or left. Varying σ with constant μ changes the shape of the curve, as shown in Fig. 6.15. σ can be thought of as the standard index of *dispersion* or scatter of the particular quantity. Mathematically, μ and σ are defined as:

$$\mu = \text{mean} = \frac{1}{n}\sum_{i=1}^{n} x_i \tag{6.12}$$

$$\sigma = \text{standard deviation} = \sigma = \sqrt{\frac{1}{n-1}\sum_{i=1}^{n}(X_i - \mu)^2} \tag{6.13}$$

Figure 6.16 illustrates a particularly useful property of *all* normal distribution curves: 68 percent of the population represented fall within the band $\mu \pm 1\,\sigma$, 95 percent fall within the band $\mu \pm 2\sigma$, and so on. Percentages of the population corresponding to any other portions of the distribution can be determined from Fig. 6.17. (Note: It is suggested that the reader not familiar with normal distributions verify the numerical values given in Fig. 6.16 by using Fig. 6.17.)

6.15 INTERFERENCE THEORY OF RELIABILITY PREDICTION

This theory has already been illustrated by the "interference" or shaded area of overlap in Fig. 6.13, and by Fig. 6.14. It can be shown that

$$\mu_z = \mu_x - \mu_y \tag{6.14}$$

and

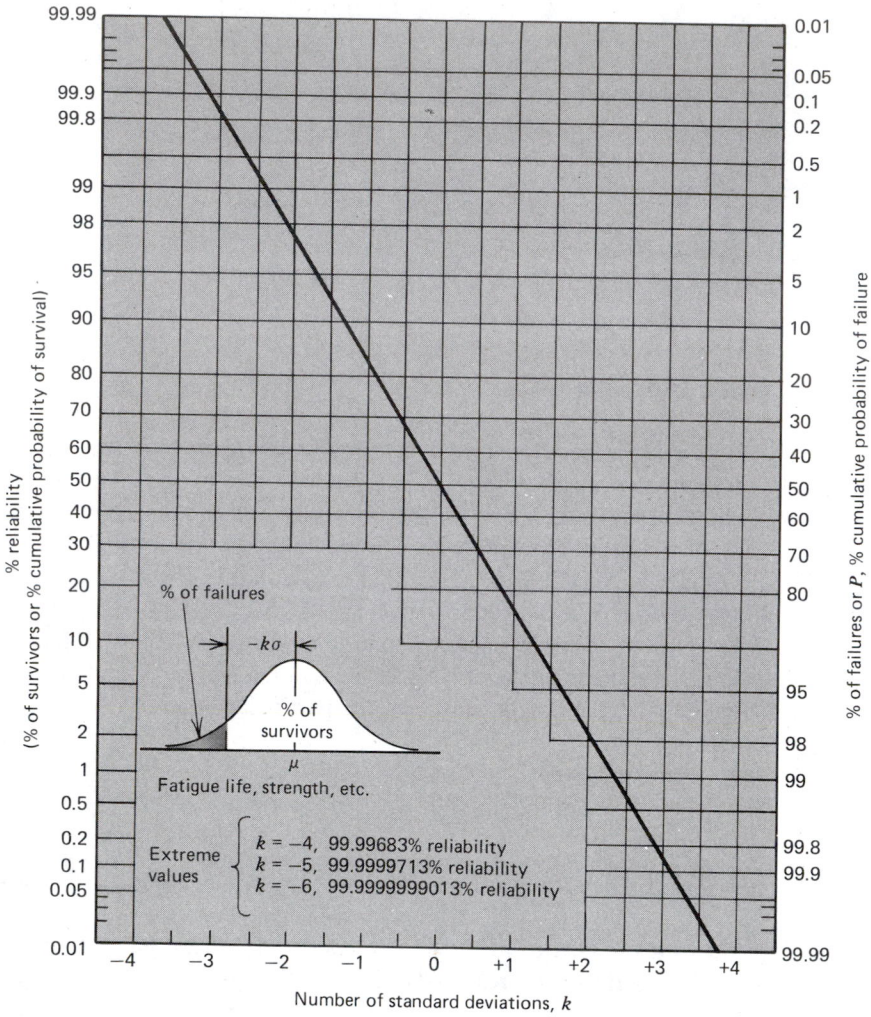

FIGURE 6.17.
Generalized normal distribution curve plotted on special probability coordinates.

$$\sigma_z = \sqrt{\sigma_x^2 + \sigma_y^2} \tag{6.15}$$

By definition, the probability of failure is $p(z < 0)$.

If the strength, x, and stress, y, are normally distributed, then it can be shown that the margin of safety, z, also has a normal distribution.

The following example illustrates a typical application of the interference theory.

SAMPLE PROBLEM 6.4 Application of Interference Theory of Reliability

Bolts installed on a production line are tightened with automatic wrenches. They are to be tightened sufficiently to yield the full cross section in order to produce the highest possible initial tension. The limiting condition is twisting off the bolts during assembly. The bolts have a mean twisting-off strength of 20 N·m with a standard deviation of 1 N·m. The automatic wrenches have a standard deviation of 1.5 N·m. What mean value of wrench torque setting would result in an estimated 1 bolt in 500 twisting off during assembly?

Solution

a. $\sigma_x = 1$ N·m, $\sigma_y = 1.5$ N·m. From Eq. (6.15), $\sigma_z = 1.80$ N·m.

b. Figure 6.17 shows that a failure percentage of 0.2 corresponds to 2.9 standard deviations below the mean. (2.9)(1.80 N·m) = 5.22 N·m. Hence, $\mu_z = 5.22$ N·m.

c. Since $\mu_x = 20$ N·m, we obtain from Eq. (6.14) $\mu_y = 14.78$ N·m. This is the required value of wrench setting. ■

REFERENCES

1. Hertzberg, Richard W., *Deformation and Fracture Mechanics of Engineering Materials,* John Wiley & Sons, Inc., New York, 1976.
2. Juvinall, Robert C., *Engineering Considerations of Stress, Strain, and Strength,* McGraw-Hill, New York, 1967.
3. Lipson, Charles, and R. C. Juvinall, *Handbook of Stress and Strength,* Macmillan, New York, 1963.
4. Lipson, Charles, and J. Sheth, *Statistical Design and Analysis of Engineering Experiments,* McGraw-Hill, New York, 1973.
5. Marin, Joseph, *Mechanical Behavior of Engineering Materials,* Prentice-Hall, Inc., Englewood Cliffs, N.J., 1962.
6. Marin, Joseph, "Theories of Strength for Combined Stresses and Nonisotropic Materials," *J. Aeron. Sci.,* **24,** No. 4. (April, 1957), pp. 265–269.
7. Tipper, C. F., *The Brittle Fracture Story,* Cambridge University Press, New York, 1962.
8. Vidosek, Joseph P., *Machine Design Projects,* Ronald Press, New York, 1957.
9. Wilhem, D. P., "Fracture Mechanics Guidelines for Aircraft Structural Applications," U.S. Air Force Technical Report AFFDL-TR-69-111, Feb., 1970.

CHAPTER SEVEN

Impact

7.1 INTRODUCTION

The previous chapters have dealt almost exclusively with *static* loading. We turn now to the more commonly encountered case of *dynamic* loading. Dynamic loading includes both *impact,* the subject of this chapter, and *fatigue,* which will be introduced in Chapter 8.

Impact loading is also called *shock, sudden,* or *impulsive* loading. The reader has inevitably experienced and observed many examples of impact loading—driving a nail or stake with a hammer, breaking up concrete with an air hammer, automobile collisions (even minor ones, involving bumper impact during careless parking), dropping of cartons by freight handlers, razing of buildings with an impact ball, automobile wheels dropping into pot-holes, and so on.

Impact loads may be divided into three categories in order of increasing severity: (a) rapidly moving loads of essentially constant magnitude, as produced by a vehicle crossing a bridge, (b) suddenly applied loads, as those resulting from an explosion, or from combustion in an engine cylinder, and (c) direct-impact loads, as produced by a pile driver, drop forge, or vehicle crash. These are illustrated schematically in Fig. 7.1. In 7.1*a*, mass m is held so that it just touches the top of spring k and is suddenly released. Dashpot c (also called a damper or shock absorber) adds a frictional supporting force that prevents the full gravitational force mg from being applied to the spring immediately. In 7.1*b* there is no dashpot, so the release of mass m results in an instantaneous application of the full force mg. In 7.1*c*, not only is the force applied instantaneously, but the mass acquires kinetic energy before it strikes the spring.

The significant thing about the dashpot action in Fig. 7.1*a* is that it results in a

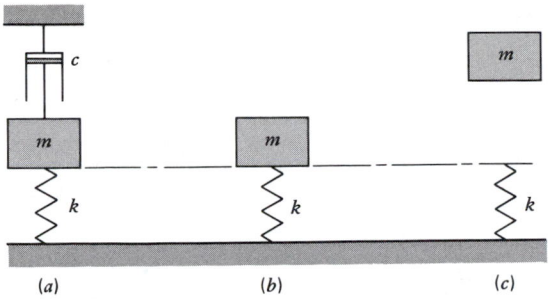

FIGURE 7.1.
Three levels of impact loading produced upon instantaneous release of mass *m*.

gradual application of the load *mg*. If the load is applied slowly enough, it can be considered as static. The usual way of distinguishing between impact and static loading in this situation is to compare the time required for load application with the natural period of vibration of the undamped mass on the spring.

(For the reader not yet acquainted with elementary vibration theory, imagine that the mass in Fig. 7.1*b* is attached to the spring, that it is pushed down, and suddenly released. The mass will then vibrate up and down, with a fixed interval between consecutive times that it is in the "full up" or "full down" position. This time interval is the *natural period of vibration* of the mass on the spring. The relationship between this period (τ, sec), the mass (m, kg or lb sec^2/in.), and the spring constant (k, N/m or lb/in.) is

$$\tau = 2\pi \sqrt{\frac{m}{k}} \tag{a}$$

Thus, the *greater* the mass and the *softer* the spring the *longer* the period of vibration (or, the lower the natural frequency of vibration).)

If the time required to apply the load (i.e., to increase it from zero to its full value) is greater than three times the natural period, dynamic effects are negligible, and static loading may be assumed. If the time of loading is less than half the natural period, impact is definitely involved. Of course, there is a "gray area" in between.

Impact loads can be compressive, tensile, bending, torsional, or a combination of these. The sudden application of a clutch and the striking of an obstruction by the bit of an electric drill are examples of torsional impact.

An important difference between static and impact loading is that statically loaded parts must be designed to *carry loads,* whereas parts subjected to impact must be designed to *absorb energy*.

Material strength properties usually vary with speed of load application. In general, this works out favorably because both the yield and ultimate strengths tend to increase with speed of loading. (Remember though, that rapid loading tends to promote brittle fracture, as noted in Section 6.2.) Figure 7.2 shows the effect of strain rate on tensile properties of mild steel.

One of the problems in applying a theoretical analysis of impact to actual engi-

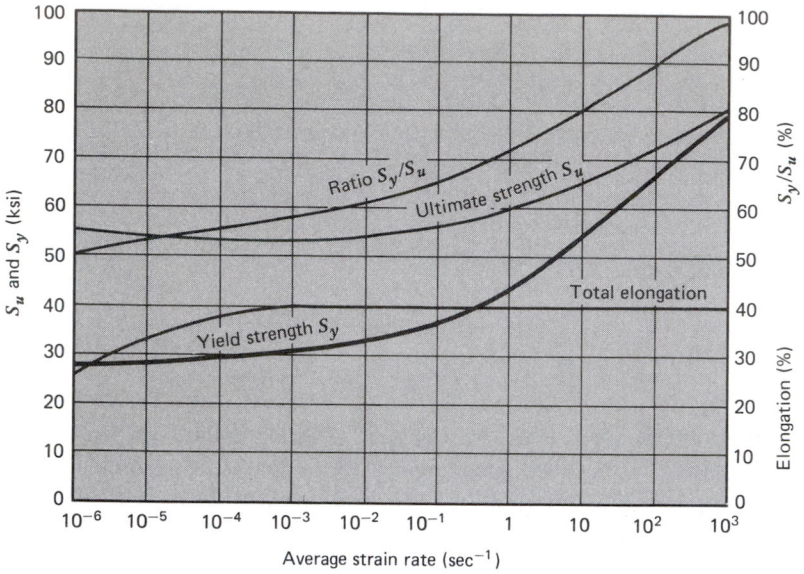

FIGURE 7.2.
Effect of strain rate on tensile properties of mild steel at room temperature [2].

neering problems is that often the time rates of load application and of strain development can only be approximated. This sometimes leads to the use of empirically determined stress impact factors, together with the static strength properties of the material. This practice works out well when good empirical data are available that apply closely to the part being designed. An example is the use of a stress impact factor of 4 in designing automotive suspension parts. Even in cases where the use of these empirical factors is justified, it is important for the engineer to have a good understanding of the basic fundamentals of impact loading.

7.2 STRESS AND DEFLECTION DUE TO LINEAR AND BENDING IMPACT

Figure 7.3 shows an idealized version of a freely falling mass (of weight W) impacting a structure. (The structure is represented by a spring, which is appropriate because *all* structures have *some* elasticity.) To derive from Fig. 7.3 the simplified equations for stress and deflection, the same assumptions are made as when deriving the equation for the natural frequency of a simple spring-mass system: (1) the mass of the structure (spring) is negligible, (2) deflections within the mass itself are negligible, and (3) damping is negligible. These assumptions have some important implications:

1. The first assumption implies that the dynamic deflection curve (i.e., the instantaneous deflections resulting from impact) is identical to that caused by the static application of the same load, multiplied by an *impact factor*. In reality, the dynamic

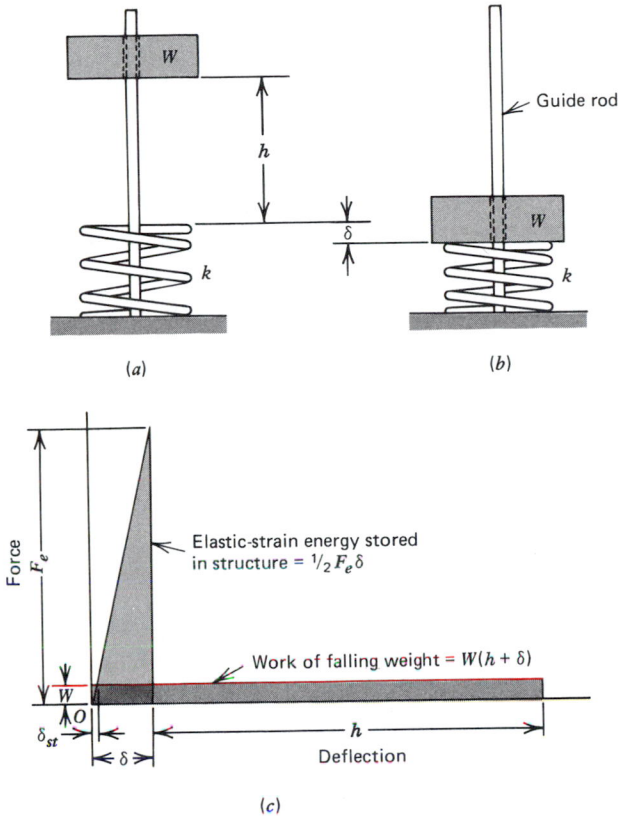

FIGURE 7.3.
Impact load applied to elastic structure by falling weight. (a) Initial position; (b) position at instant of maximum deflection; (c) force-deflection-energy relationships.

deflection curve inevitably involves points of higher *local* strain (hence, higher local stress) than does the static curve.

2. *Some* deflection must inevitably occur within the impacting mass itself. To the extent that it does, a portion of the energy is absorbed within the mass, thereby causing the stresses and deflections in the structure to be a little *lower* than the calculated values.
3. Any actual case involves some (though perhaps very little) friction damping in the form of windage, rubbing of the mass on the guide rod and end of the spring (in Fig. 7.3), and internal friction within the body of the deflecting structure. This damping can cause the actual stresses and deflections to be significantly less than those calculated from the idealized case.

Keeping the above limitations in mind, the following analysis of the idealized case provides an understanding of basic impact phenomena, together with equations that are very helpful as a *guide* in dealing with linear impact.

In Fig. 7.3, the falling mass is such that (in the gravitational field involved) it has a weight, W (newtons or pounds). The structure is assumed to respond to the impact elastically, with a spring constant of k (newtons per meter or pounds per inch). The maximum value of the deflection due to impact is δ (meters or inches). F_e is defined as an *equivalent static force* that would produce the same deflection, δ; that is, $F_e = k\delta$. The static deflection that exists after the energy is damped out and the weight comes to rest on the structure is designated by δ_{st}, where $\delta_{st} = W/k$.

Equating the potential energy given up by the falling mass with the elastic energy absorbed by the spring (structure),

$$W(h + \delta) = \tfrac{1}{2}F_e\,\delta \tag{b}$$

Note that the factor of ½ appears because the spring takes on the load *gradually*.

By definition,

$$F_e = (\delta/\delta_{st})\,W \quad \text{or} \quad \delta/\delta_{st} = F_e/W \tag{c}$$

Substituting (c) into (b),

$$W(h + \delta) = \frac{1}{2}\frac{\delta^2}{\delta_{st}}W \tag{d}$$

Equation (d) is a quadratic equation in δ, which is solved routinely to give

$$\delta = \delta_{st}\left(1 + \sqrt{1 + \frac{2h}{\delta_{st}}}\right) \tag{7.1}$$

Substitution of (c) in (7.1) gives

$$F_e = W\left(1 + \sqrt{1 + \frac{2h}{\delta_{st}}}\right) \tag{7.2}$$

Since the structure (spring) is assumed to respond elastically to the impact, the stress produced is proportional to the load. The term in parentheses in Eqs. (7.1) and (7.2) is called the *impact factor*. It is the factor by which the load, stress, and deflection caused by the dynamically applied weight, W, exceed those caused by a slow, static application of the same weight.

In some cases it is more convenient to express Eqs. (7.1) and (7.2) in terms of velocity at impact v (meters per second or inches per second) instead of height of fall h. For free fall, the relationship between these quantities is

$$v^2 = 2gh \quad \text{or} \quad h = \frac{v^2}{2g} \tag{e}$$

where g is the acceleration of gravity measured in meters per second per second or inches per second per second.

Substitution of (e) in (7.1) and (7.2) gives

$$\delta = \delta_{st}\left(1 + \sqrt{1 + \frac{v^2}{g\delta_{st}}}\right) \tag{7.1a}$$

and

$$F_e = W\left(1 + \sqrt{1 + \frac{v^2}{g\delta_{st}}}\right) \tag{7.2a}$$

Reducing distance h to zero gives the special case of a *suddenly applied load,* for which the impact factor—in Eqs. (7.1) and (7.2)—is equal to 2. This may have been one basis for designers in the past sometimes doubling safety factors when impact was expected.

In many problems involving impact, the deflection is almost insignificant in comparison to h (see Fig. 7.3). For this case, Eqs. (7.1) and (7.2) can be simplified to

$$\delta = \delta_{st}\sqrt{\frac{2h}{\delta_{st}}} = \sqrt{2h\delta_{st}} \tag{7.3}$$

$$F_e = W\sqrt{\frac{2h}{\delta_{st}}} = \sqrt{2Whk} \tag{7.4}$$

Similarly, Eqs. (7.1a) and (7.2a) simplify to

$$\delta = \delta_{st}\sqrt{\frac{v^2}{g\delta_{st}}} = \sqrt{\frac{\delta_{st}\,v^2}{g}} \tag{7.3a}$$

$$F_e = W\sqrt{\frac{v^2}{g\delta_{st}}} = \sqrt{\frac{v^2kW}{g}} \tag{7.4a}$$

In the above four equations, gravity was considered *only* as the means for developing the velocity of the weight at the point of impact (the further action of gravity after impact being neglected). Hence, Eqs. (7.3a) and (7.4a) apply also to the case of a *horizontally* moving weight striking a structure, where the impact velocity v is developed by means other than gravity. In this case, δ_{st} is the static deflection that *would* exist *if* the entire system were rotated 90° to allow the weight to act vertically upon the structure. Thus, regardless of the actual orientation,

$$\delta_{st} = W/k \tag{f}$$

It is useful to express the equations for deflection and equivalent static force as functions of the impact kinetic energy U, where, from elementary physics,

$$U = \tfrac{1}{2}mv^2 = Wv^2/2g \tag{g}$$

Substitution of (f) and (g) into Eqs. (7.3a) and (7.4a) gives

$$\delta = \sqrt{\frac{2U}{k}} \tag{7.3b}$$

$$F_e = \sqrt{2Uk} \tag{7.4b}$$

An important special case of linear impact is that of a straight rod or bar impacted in compression or in tension. The tensile case is illustrated schematically in Fig. 7.4*a*. The tensile rod sometimes takes the form of a bolt. *If* the impact load is applied concentrically, and *if* stress concentration can be neglected (mighty big "ifs," usually!), then one can substitute into Eq. (7.4b) the elementary expressions

$$\sigma = F_e/A \tag{h}$$

and

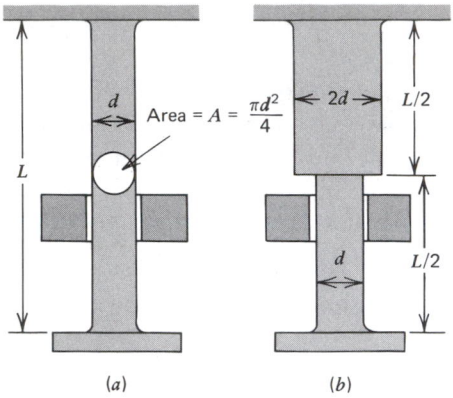

FIGURE 7.4.
Tensile impact.

$$k = AE/L \tag{i}$$

where A and L are the rod cross-sectional area and length, respectively. The resulting equation is

$$\sigma = \sqrt{\frac{2UE}{AL}} = \sqrt{\frac{2UE}{V}} \tag{7.5}$$

where V is the volume of material in the rod.

Note the important implication of Eq. (7.5)—the stress developed in the rod is a function of its *volume* irrespective of whether this volume involves a long rod of small area or a short rod of large area.

Solving Eq. (7.5) for U gives

$$U = \frac{\sigma^2 V}{2E} \tag{7.5a}$$

This shows the impact energy capacity of a straight rod to be a remarkably simple function of its volume, its modulus of elasticity, and the *square* of the allowable stress.

Despite the importance of this basic relationship, it should be emphasized that Eqs. (7.5) and (7.5a) may, in practice, give results that are considerably optimistic—that is, give a calculated stress *lower* than the actual peak stress, and, correspondingly, a calculated energy capacity *greater* than that which actually exists. The main reasons for this are: (1) the stresses are not uniform throughout the member, due to stress concentration and nonuniformity of loading on the impacted surface, and (2) the impacted member has mass. The inertia resulting from the rod mass causes the impacted end of the rod to have a greater *local* deflection (hence, stress) than it would if inertial effects did not prevent the instantaneous distribution of deflection throughout the length of the rod. The effect of stress raisers is considered in Section 7.4. The quantitative effect of the mass of the struck member is left for more advanced works; see [1] and [5].

SAMPLE PROBLEM 7.1 Axial Impact – Importance of Section Uniformity

Figure 7.4 shows two round rods subjected to tensile impact. How do their elastic energy-absorbing capacities compare? (Neglect stress concentration, and use S_y as an approximation of the elastic limit.)

Solution

The elastic capacity for Fig. 7.4*a* is determined directly from Eq. (7.5a), where $\sigma = S_y$:

$$U_a = \frac{S_y^2 V}{2E}$$

In 7.4*b*, the energy absorbed by the upper and lower halves must be determined separately. The smaller lower half is critical; it can be brought to a stress of S_y, and its volume is $V/2$ (where V = volume of the full-length rod in 7.4*a*). Thus, energy capacity of the lower half is

$$U_{bl} = \frac{S_y^2 V/2}{2E} = \frac{1}{2} U_a$$

The same force is transmitted through the full length of the rod. The upper half has four times the area of the lower half; hence, it has four times the volume and only ¼ the stress. Thus, the energy capacity of the upper half is

$$U_{bu} = \frac{(S_y/4)^2(2V)}{2E} = \frac{1}{8} U_a$$

The total energy capacity is the sum of U_{bl} and U_{bu}, which is *only five-eighths the energy capacity of 7.4a.* Since the rod in 7.4*b* has 2½ times the volume and weight of the straight rod, it follows that the *energy capacity per pound* is *four times as great* with the uniform-section rod. Moreover, the stress concentration in the middle of the stepped bar would further reduce its capacity and would tend to promote brittle fracture. This point is treated further in the next section. ■

SAMPLE PROBLEM 7.2 Relative Energy Absorption Capacity of Various Materials

Figure 7.5 shows a falling weight that impacts on a block of material serving as a bumper. Estimate the relative elastic-energy-absorption capacities of the following bumper materials:

Material	Density (kN/m³)	Elastic modulus (E)	Elastic Limit (S_e, MPa)
Soft steel	77	207 GPa	207
Hard steel	77	207 GPa	828
Rubber	9.2	1.034 MPa	2.07

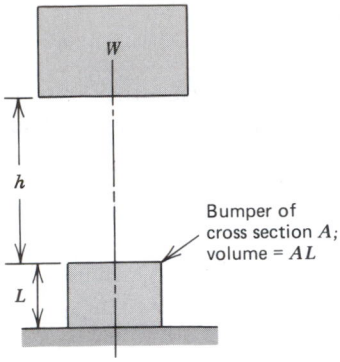

FIGURE 7.5.
Impact loading of compression bumper.

Solution

From Fig. 7.3, the elastic-strain-energy absorbed is ½ $F_e\delta$, or the area under the force-deflection curve. At the elastic limit, $F_e = S_e A$, and $\delta = F_e L/AE$. Substitution of these values gives

$$U = \frac{1}{2} F_e \delta = \frac{S_e^2 AL}{2E} = \frac{S_e^2 V}{2E} \tag{h}$$

which, not surprisingly, corresponds exactly with Eq. (7.5a).

Substitution of the given material properties in Eq. (h) indicates that on the basis of unit volume, the relative elastic-energy-absorption capacities of the soft steel, hard steel, and rubber are 1:16:20. On a unit mass or weight basis the relative capacities are 1:16:168. ■

The *elastic*-energy-absorption capacity per unit volume of a material is equal to the area under the elastic portion of the stress-strain diagram and is called the *modulus of resilience* of the material. The *total* energy-absorption capacity in tension per unit volume of the material is equal to the total area under the stress-strain curve (extending out to fracture) and is sometimes called the *modulus of toughness* of the material. In the above problem the two steels differed markedly in their moduli of resilience, but their relative toughnesses would likely be comparable.

SAMPLE PROBLEM 7.3 Bending Impact – Effect of Compound Springs

Figure 7.6 shows a wood beam supported on two springs and loaded in bending impact. Estimate the maximum stress and deflection in the beam, based on the assumption that the masses of the beam and spring can be neglected.

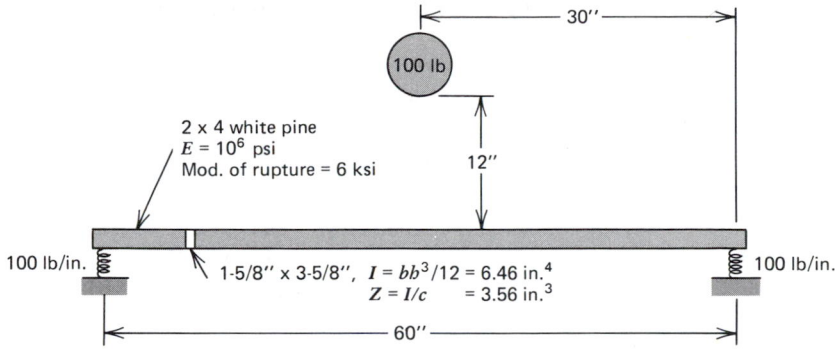

FIGURE 7.6.
Bending impact, with compound spring.

Solution

The static deflection for the beam only, supporting springs only, and total system are

$$\delta_{st}(\text{beam}) = \frac{PL^3}{48EI} = \frac{100(60)^3}{48(10^6)(6.46)} = 0.070 \text{ in.}$$

$$\delta_{st}(\text{springs}) = \frac{P}{2k} = \frac{100}{2(100)} = 0.50 \text{ in.}$$

$$\delta_{st}(\text{total}) = 0.070 + 0.50 = 0.57 \text{ in.}$$

From Eq. (7.1) or (7.2) the impact factor is

$$1 + \sqrt{1 + \frac{2h}{\delta_{st}}} = 1 + \sqrt{1 + \frac{24}{0.57}} = 7.6$$

Hence, the total impact deflection is $0.57 \times 7.6 = 4.3$ in., but the deflection of the beam itself is only $0.07 \times 7.6 = 0.53$ in. The extreme-fiber beam stress is estimated from $F_e = 100 \times 7.6 = 760$ lb:

$$\sigma = \frac{M}{Z} = \frac{F_e L}{4Z} = \frac{760(60)}{4(3.56)} = 3200 \text{ psi}$$ ■

The estimated stress is well within the given *modulus of rupture* of 6000 psi. (The modulus of rupture is the computed value of M/Z at failure in a standard static test.)

It is interesting to note that if the supporting springs are removed, the total static deflection is reduced to 0.07 in., and the impact factor increased to 19.6. This would give a computed maximum beam stress of 8250 psi, which is greater than the modulus of rupture. If the inertial effect of the beam mass does not cause the actual stress to be very much higher than 8250 psi, it is possible that the "dynamic-strengthening effect" shown in Fig. 7.2 would be sufficient to prevent failure. Because this effect is usually appreciable for woods, the results of standard beam impact tests are often included in references giving properties of woods.

7.3 STRESS AND DEFLECTION DUE TO TORSIONAL IMPACT

The analysis of the preceding article could be repeated for the case of torsional systems, and a corresponding set of equations developed. Instead, advantage will be taken of the direct analogy between linear and torsional systems to write the final equations directly. The analogous quantities involved are:

Linear	Torsional
δ, deflection (m or in.)	θ, deflection (rad)
F_e, equivalent static force (N or lb)	T_e, equivalent static torque (N · m or lb-in.)
m, mass (kg or lb sec^2/in.)	I, moment of inertia (N · sec^2 · m or lb sec^2 in.)
k, spring rate (N/m or lb/in.)	K, spring rate (N · m/rad or lb-in./rad)
v, impact velocity (m/sec or in./sec)	ω, impact velocity (rad/sec)
U, kinetic energy (N · m or in.-lb)	U, kinetic energy (N · m or in.-lb)

The two following equations have the letter "t" added to the equation number to designate torsion.

$$\theta = \sqrt{\frac{2U}{K}} \tag{7.3bt}$$

$$T_e = \sqrt{2UK} \tag{7.4bt}$$

For the important special case of torsional impact of a solid round bar of diameter d:

1. From Table 5.1,

$$K = \frac{T}{\theta} = \frac{K'G}{L} = \frac{\pi d^4 G}{32L} \tag{i}$$

2. From Eq. (4.4) with T replaced by T_e,

$$\tau = \frac{16T_e}{\pi d^3} \tag{j}$$

3. Volume, $V = \pi d^2 L/4$ **(k)**

Substitution of (i), (j), and (k) into Eq. (7.4bt) gives

$$\tau = 2\sqrt{\frac{UG}{V}} \tag{7.6}$$

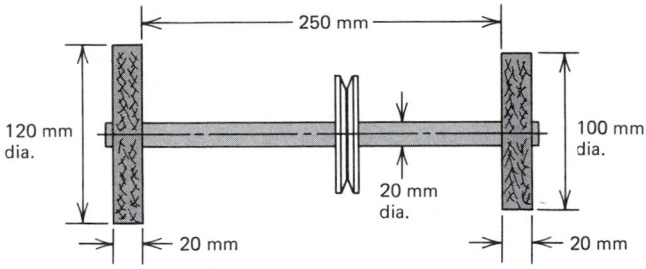

FIGURE 7.7.
Torsional impact of grinder shaft.

SAMPLE PROBLEM 7.4 Torsional Impact

Figure 7.7 shows the shaft assembly of a grinder, with an abrasive wheel at each end and a belt-driven sheave at the center. When turning 2400 rpm the smaller abrasive wheel is accidentally jammed, causing it to stop "instantly." Estimate the resulting maximum torsional stress and deflection of the shaft. Consider the abrasive wheels as solid disks of density $\rho = 2000\ \text{kg/m}^3$. The shaft is steel ($G = 79$ GPa), and its weight may be neglected.

Solution

It is the energy in the 120-mm wheel that must be absorbed by the shaft. From the torsional equivalent of (g), this is

$$U = \tfrac{1}{2} I \omega^2$$

where

$$I = \tfrac{1}{2} m r^2$$

and

$$m = \pi r^2 t\, \rho$$

Combining the above:

$$U = \tfrac{1}{4} \pi r^4 t\, \rho\, \omega^2$$

Substituting numerical values (with units of meters, kilograms, and seconds):

$$U = \frac{1}{4}\pi(0.060)^4(0.020)(2000)\left(\frac{2400 \times 2\pi}{60}\right)^2$$

$$U = 25.72\,\frac{\text{kg}\cdot\text{m}^2}{\text{s}^2} = 25.72\ \text{N}\cdot\text{m}$$

From (7.6)

$$\tau = 2\sqrt{\frac{UG}{V}}$$

$$= 2\sqrt{\frac{(25.72)(79 \times 10^9)}{\pi(0.010)^2(0.250)}} = 321.7 \times 10^6 \text{ Pa}$$

or,

$$\tau = 322 \text{ MPa}, \qquad \blacksquare$$

$$\theta = \frac{TL}{JG}$$

where $T = \tau J/r$ (i.e., $\tau = Tr/J$); hence

$$\theta = \frac{\tau L}{rG} = \frac{(321.7 \times 10^6)(0.250)}{(0.010)(79 \times 10^9)} = 0.10 \text{ rad} = 5.7^\circ \qquad \blacksquare$$

The above calculations assume, of course, that the stresses are within the elastic range. Note that no provision was made for stress concentration or for any superimposed bending load that would also be present as a result of the accident. Torque applied to the sheave by the belt was also neglected, but this would likely be negligible because of belt slippage. Also, it is only because of assumed belt slippage that the inertia of the driving motor is not a factor.

7.4 EFFECT OF STRESS RAISERS ON IMPACT STRENGTH

Figure 7.8 shows the same tensile impact bar as Fig. 7.4*a* except that recognition is given to the fact that stress concentration exists at the ends of the bar. As with static loading it is *possible* that local yielding would redistribute the stresses so as to virtually nullify the effect of the stress raiser. But under impact loading the *time* available for plastic action is likely to be so short that brittle fracture (with an effective stress concentration factor almost as high as the theoretical value, K_t, obtained from a chart sim-

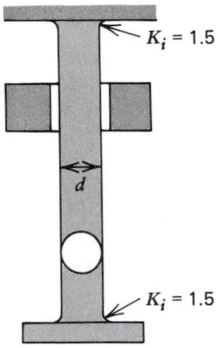

FIGURE 7.8.
Plain impact bar.

ilar to Fig. 4.35*b*) will sometimes occur even in a material that exhibits ductile behavior in the tensile test. In terms of the discussion in Section 6.2, adding a stress raiser and applying an impact load are both factors tending to raise the *transition temperature*—that is, cause brittle fracture without dropping to as low a temperature.

Because of the difficulty in predicting impact notch effects from theoretical considerations, standard notched impact tests are used, such as the *Charpy* and *Izod*. These, too, have their limitations, as notched impact strength varies markedly with size, shape, and nature of impact. Because of this, special laboratory tests that more closely simulate actual conditions are sometimes used.

SAMPLE PROBLEM 7.5 Notched Tensile Impact

Suppose that from special tests it has been determined that the effective stress concentration factor for impact loading, K_i, at the ends of the rod in Fig. 7.8 is 1.5, as shown. How much does the stress raiser decrease the energy absorbing capacity of the rod, as estimated from Eq. (7.5a)?

Solution

First, two observations: (a) if the rod is sufficiently long, the volume of material in the region of the end fillets is a very small fraction of the total, and (b) the material at the critical fillet location cannot be stressed in excess of the material strength, S. That means that *nearly all* of the material can be considered as stressed to a uniform level that cannot exceed S/K_i, or, in this case, $S/1.5$. Thus, a good approximation is that after considering the stress raiser, the same volume of material is involved, but at a stress level reduced by a factor of 1.5. Since the stress is squared in Eq. (7.5a), taking the notch into consideration *reduces the energy capacity by a factor of* 1.5^2, *or 2.25.* ■

SAMPLE PROBLEM 7.6 Notched Tensile Impact

Figure 7.9 shows the same impact bar as Fig. 7.8 except that a sharp groove, with $K_i = 3$, has been added. Compare the impact energy capacities of the bars in Figs. 7.8 and 7.9.

Solution

In Fig. 7.9, the impact capacity is limited to the value that brings the stress at the groove to the material strength, S. Since the effective stress concentration factor is 3, the nominal stress level in the section of the groove is $S/3$. Because of the 4:1 area ratio, the nominal stress in the bulk of material (*not* in the groove plane) is only $S/12$. For a long bar, the percentage of volume near the groove is very small. Thus, with reference to Eq.

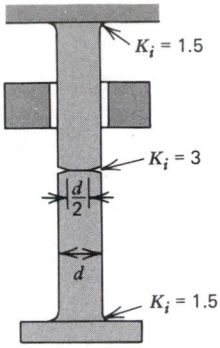

FIGURE 7.9.
Grooved impact bar.

(7.5a), the only substantial difference made by introducing the groove is to reduce the value of σ from $S/1.5$ to $S/12$. Since σ is squared in the equation, the groove reduces the energy capacity by a factor of 64; that is, the grooved bar has *less than 2 percent* of the energy absorbing capacity of the ungrooved bar! ■

From the above discussion, it follows that the effective design of an efficient energy absorbing member involves two key steps:

1. Minimize stress concentration as much as possible. (Always try to reduce the stress at the point where it is highest.)
2. Having done that, remove all possible "excess material" so that the stress everywhere is as close as possible to the stress at the most critical point. By removing this excess material, the *load* that the member can carry is not reduced, and the *deflection* is increased. Since energy absorbed is the integral of force times deflection, energy absorbing capacity is thereby increased. (Recall the dramatic example of this principle in Sample Problem 7.1.)

SAMPLE PROBLEM 7.7 Modifying a Bolt Design for Greater Impact Strength

Figure 7.10*a* shows a bolt that is subjected to tensile impact loading. Suggest a modified design that would have greater energy absorbing capacity. How much increase in capacity would the modified design provide?

Solution

1. Following the two-step procedure suggested above, first try to reduce stress concentration where it is most critical. The highest stress is at the thread ($K_i = 3.5$, acting at an area of only 600 mm^2). Assume that by modifying the thread slightly to provide a smooth, generous fillet at the root, K_i can be reduced to 3.0, as shown in

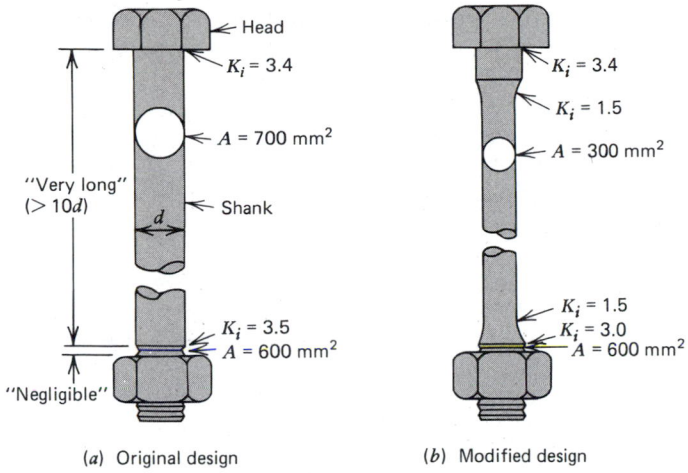

FIGURE 7.10.
Bolt subjected to tensile impact.

Fig. 7.10*b*. The other point of stress concentration is in the fillet under the bolt head. This fillet must be small in order to provide adequate flat area for contact. Actually, there is no incentive to reduce stress concentration at that point because the stress there will be less than at the thread root, even with the modified thread design.[1]

$$\left(\sigma = \frac{P}{A} K_i; \quad \frac{P}{700} \times 3.4 < \frac{P}{600} \times 3.0\right)$$

2. Leaving a short length of full diameter shank under the bolt head to serve as a pilot to center the bolt in the bolt hole, the diameter in the rest of the shank can be reduced to make the shank stress nearly equal to the stress at the thread root. Figure 7.10*b* shows a reduced shank diameter that is flared out to the full diameter with a large radius, to give minimal stress concentration. Based on a conservative stress concentration estimate of 1.5, the shank area can be reduced to half the effective stress area at the thread:

$$A = 600 \times \frac{1.5}{3.0} = 300 \text{ mm}^2$$

3. Assume that the bolt is sufficiently long that the volume of uniformly stressed material in the central portion of the shank is the only volume that need be considered, and that the volumes in the two cases are proportional to the areas, 700 and 300 mm². From Eq. (7.5a), $U = \sigma^2 V/2E$. Since E is a constant, the ratio of energy capacities for Figs. 7.10*b* and 7.10*a* is

$$\frac{U_b}{U_a} = \frac{\sigma_b^2 V_b}{\sigma_a^2 V_a} \qquad \textbf{(m)}$$

In Fig. 7.10*a*, the stress in the large volume of material in the shank is less than

[1]This is usually, but not *always*, the case.

the material strength S because of *both* stress concentration *and* the difference in area between the thread and shank. Thus, if $\sigma = S$ at the thread root, the shank stress is

$$\sigma_a = \frac{S}{3.5}\left(\frac{600}{700}\right) = 0.245S$$

Let the shank volume in Fig. 7.10*a* be designated as V. In Fig. 7.10*b*, the stress at the thread can again be S. Corresponding shank stress is

$$\sigma_b = \frac{S}{3.0}\left(\frac{600}{300}\right) = .667S$$

The shank volume in Fig. 7.10*b* is $V(300/700)$, or $0.429V$. Substituting these values in Eq. (m):

$$\frac{U_b}{U_a} = \frac{(0.667S)^2(0.429V)}{(0.245S)^2(V)} = 3.18$$ ■

Thus, the redesign has over three times the capacity of the original, as well as being lighter.

Two other designs of bolts with increased energy absorbing capacity are illustrated in Fig. 7.11. In 7.11*a*, the full-diameter piloting surface has been moved to the center of the shank in order to provide alignment of the two clamped members. Figure 7.11*b* shows a more costly method of removing excess shank material, but it preserves nearly all of the original torsional and bending strength of the bolt. The torsional strength is often important, as it influences how much the nut can be tightened without "twisting off" the bolt.

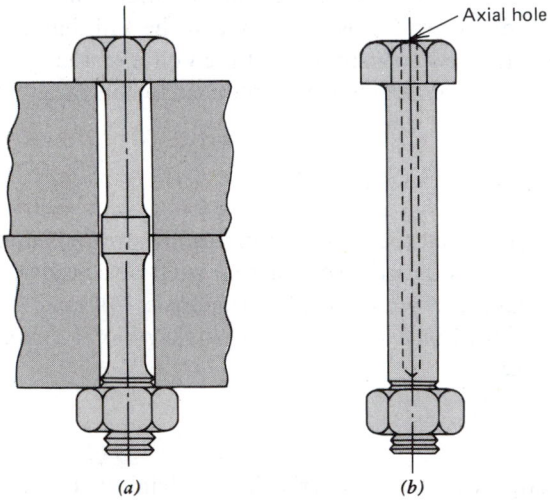

FIGURE 7.11.
Bolts designed for energy absorption.

REFERENCES

1. Juvinall, R. C., *Engineering Considerations of Stress, Strain, and Strength,* McGraw-Hill, New York, 1967.
2. Manjoine, M. J., "Influence of Rate of Strain and Temperature on Yield Stresses of Mild Steel," *J. Appl. Mech.,* **66,** (1944), pp. A-211–A-218.
3. Marin, Joseph, *Mechanical Behavior of Engineering Materials,* Prentice-Hall, Inc., Englewood Cliffs, N.J., 1962.
4. Rinehart, John S., and John Pearson, *Behavior of Metals under Impulsive Loads,* The American Society for Metals, Cleveland, 1954.
5. Timoshenko, S., and J. N. Goodier, *Theory of Elasticity,* 2nd ed., McGraw-Hill, New York, 1951.
6. Vigness, Irwin, and W. P. Welch, "Shock and Impact Considerations in Design," in Oscar J. Horger (ed.), *ASME Handbook: Metals Engineering—Design,* 2nd ed., McGraw-Hill, New York, 1965.

CHAPTER EIGHT

Fatigue

8.1 INTRODUCTION

Until about the middle of the nineteenth century engineers treated fluctuating or repeated loading the same as static loading, except for the use of larger safety factors. Use of the term *fatigue* in these situations appears to have been introduced by Poncelet of France in a book published in 1839. Modern authorities suggest that the term *progressive fracture* would perhaps have been more appropriate.

"Fatigue" fractures begin with a minute (usually microscopic) crack at a critical area of high local stress. This is almost always at a geometric stress raiser. Additionally, minute material flaws or pre-existing cracks are commonly involved (recall from Section 6.3 that the fracture mechanics approach assumes pre-existing cracks in *all* materials). An inspection of the surfaces after final fracture (as in Fig. 8.1) often reveals where the crack has gradually enlarged from one "beach mark" to the next until the section is sufficiently weakened that final fracture occurs on one final load application. This can be because the stress exceeds the ultimate strength, with fracture occurring as in a static tensile test. Usually, however, the final fracture is largely "brittle," and takes place in accordance with the fracture mechanics concepts treated in Sections 6.3 and 6.4. (Recall that brittle fracture is promoted by high stress concentration and rapid load application, both of which are normally present when final fatigue fracture occurs.)

In Fig. 8.1, the curvature of the "beach marks" serves to indicate the failure origin. The beach-marked area is known as the *fatigue zone*. It has a smooth, velvety texture developed by the repeated pressing together and separating of the mating crack surfaces. This contrasts with the relatively rough final fracture. A distinguishing characteristic of fatigue fracture of a ductile material is that little if any macroscopic distortion

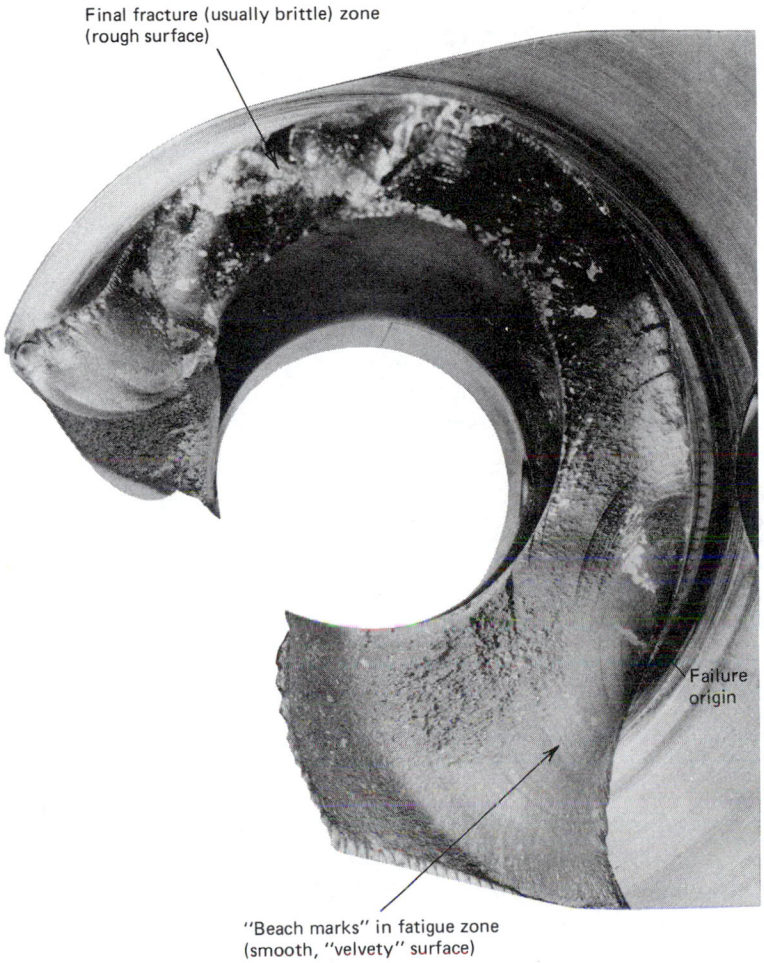

FIGURE 8.1.
Fatigue failure originating in fillet of aircraft crankshaft (SAE 4340 Steel, 320 Bhn).

occurs during the entire process, whereas, failure due to static overload produces gross distortion.

8.2 BASIC CONCEPTS

Extensive research over the last century has resulted in a partial understanding of the basic mechanisms associated with fatigue failures. Reference 3 contains a summary of much of the current knowledge as it applies to engineering applications. The following are a few fundamental and elementary concepts that are helpful in understanding the patterns of observed fatigue behavior:

1. Fatigue failure results from *repeated plastic deformation,* such as the breaking of a wire by bending it back and forth repeatedly. Without repeated plastic yielding, fatigue failures cannot occur.
2. Whereas a wire can be broken after a very few cycles of gross plastic yielding, fatigue failures typically occur after thousands or even millions of cycles of minute yielding that often exists only on a *microscopic level.* This can occur at stress levels far below the conventionally determined yield point or elastic limit.
3. Because highly localized plastic yielding can be the beginning of a fatigue failure, it behooves the engineer to focus attention on all potentially vulnerable locations such as holes, sharp corners, threads, keyways, surface scratches, and corrosion. Such a location is shown at the root of a notch in Fig. 8.2. *Strengthening these vulnerable locations is often as effective as making the entire part from a stronger material.*
4. If the local yielding is sufficiently minute, the material *may* strain-strengthen, causing the yielding to cease. The part will then have actually benefited from this mild overload. But if the local yielding is any more than this, the repeated load cycling will cause a loss of local ductility (in accordance with concepts discussed in Section 3.3) until the cyclic strain imposed at the vulnerable spot in question can no longer be withstood without fracture.
5. The initial fatigue crack usually results in an increase in local stress concentration. As the crack progresses, the material at the crack-root at any particular time is subjected to the destructive localized reversed yielding. As the crack deepens, thereby reducing the section and causing increased stresses, crack propagation rate increases until the remaining section is no longer able to support a single load application and final fracture occurs, usually in accordance with the principles of fracture mechanics. (There are cases where a fatigue crack advances into a region of lower stress and/or greater material strength, and the crack ceases to propagate, but these cases are unusual.)

Present engineering practice relies heavily on the wealth of empirical data that has accumulated from fatigue tests of numerous materials, in various forms, and subjected

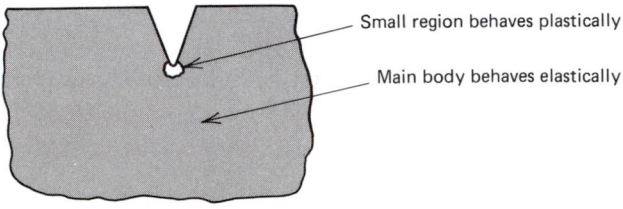

FIGURE 8.2.
Enlarged view of a notched region.

to various kinds of loading. The remainder of this chapter is based largely on these data. The next article describes the standardized R. R. Moore fatigue test, which is used to determine the fatigue strength characteristics of materials under a standardized and highly restricted set of conditions. After reviewing the pattern of results obtained from this test, succeeding sections deal with the influence of deviating from the standard test in various ways, so as to work toward the completely general case of fatigue in an orderly fashion.

The generalizations or patterns of fatigue behavior developed in the remainder of this chapter enable the engineer to estimate the fatigue behavior for combinations of materials, geometry, and loading for which test data are not available. This is an extremely important practice in modern engineering. The preliminary design of critical parts normally involves this procedure. Then prototypes of the preliminary design are built and fatigue-tested. The results provide a basis for refining the preliminary design to arrive at a final design, suitable for production.

8.3 STANDARD ROTATING-BENDING FATIGUE STRENGTHS (S_n')

Figure 8.3 represents a standard R. R. Moore rotating-beam fatigue testing machine. The reader should verify that the loading imposed by the four symmetrically located bearings causes the center portion of the specimen to be loaded in *pure bending* (i.e., zero transverse shear), and that the stress at any point goes through a cycle of tension-to-compression-to-tension with each shaft rotation. The highest stress level is at the center, where the diameter is a standard 0.300 in. The large radius of curvature prevents stress concentration. Various weights are chosen to give the desired stress levels. The motor speed is usually 1750 rpm. When the specimen fails, the weight drops, opening contact points *C*, which stops the motor. The number of cycles to failure is indicated by the revolution counter.

A series of tests made with various weights and using test specimens carefully made to be as nearly identical as possible gives results that are plotted as *S-N curves*. As illustrated in Fig. 8.4, *S-N* curves are plotted either on semilog or log-log coordinates. Note that the intensity of reversed stress causing failure after a given number of

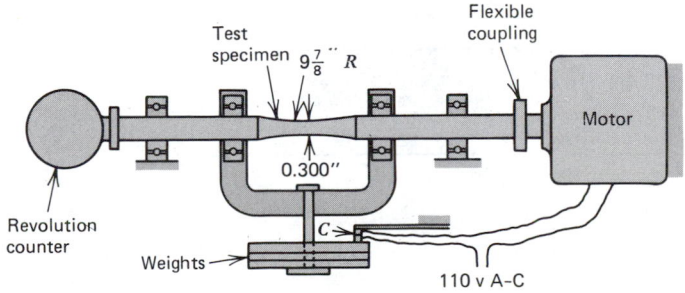

FIGURE 8.3.
R. R. Moore rotating-beam fatigue-testing machine.

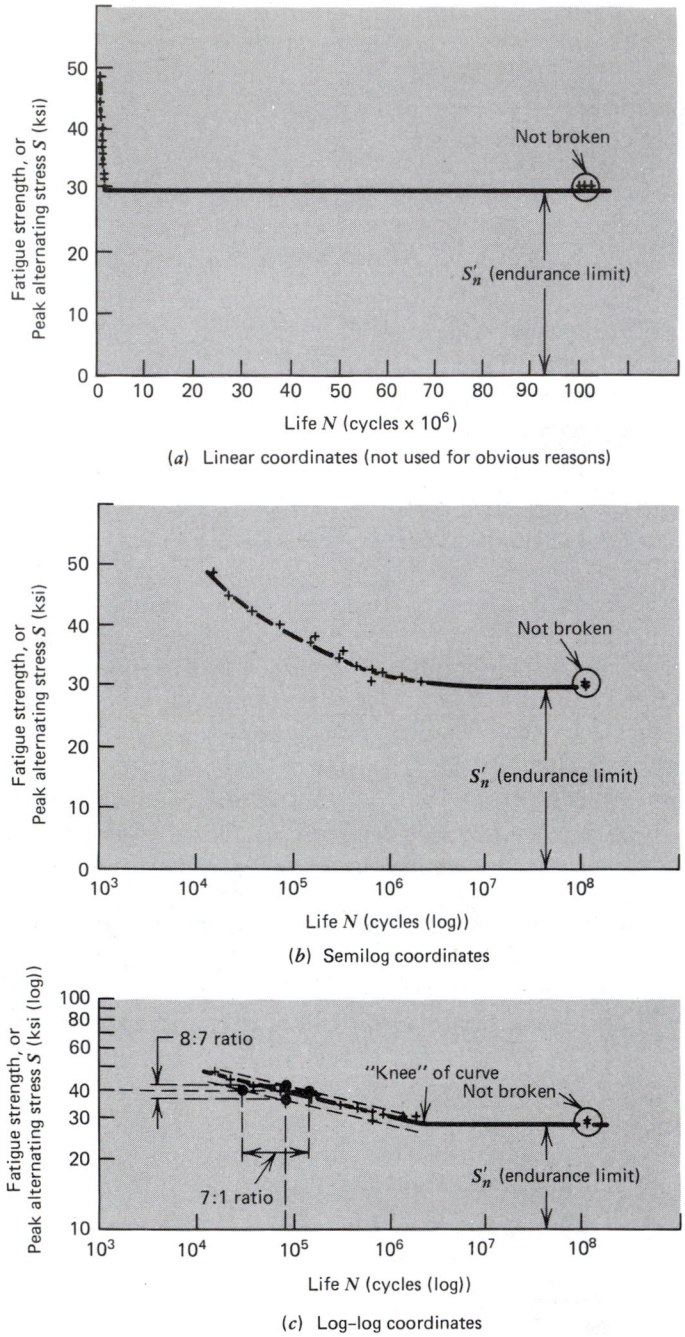

FIGURE 8.4.
Three *S-N* plots of representative fatigue data for 120 Bhn steel.

cycles is called the *fatigue strength* corresponding to that number of loading cycles. Numerous tests have established that *ferrous materials* have an *endurance limit,*

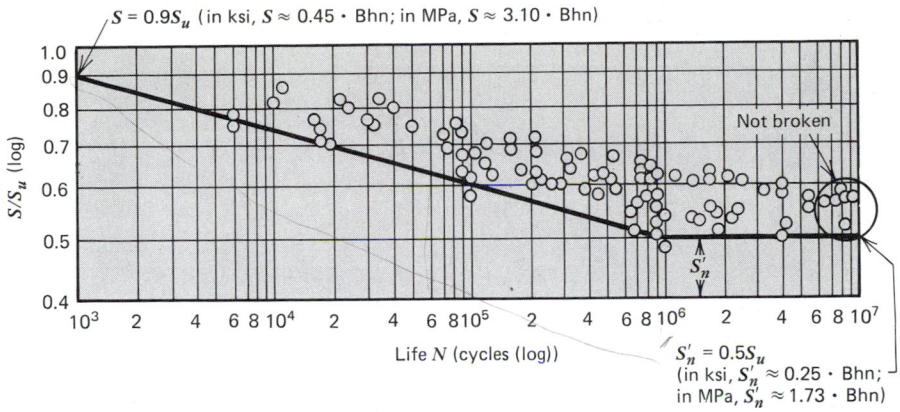

FIGURE 8.5.
Generalized *S-N* curve for wrought steel with superimposed data points [7].

defined as the highest level of alternating stress that can be withstood indefinitely without failure. The usual symbol for endurance limit is S_n. It is designated as S'_n in Fig. 8.4, where the prime indicates the special case of the standard test illustrated in Fig. 8.3. Log-log coordinates are particularly convenient for plotting ferrous *S-N* curves because of the straight-line relationship shown.

Figure 8.4*c* illustrates the "knee" of *S-N* curves for materials that have a clearly defined endurance limit. This "knee" normally occurs between 10^6 and 10^7 cycles. It is customary to make the conservative assumption that ferrous materials must not be stressed above the endurance limit if a life of 10^6 or more cycles is required. This assumption is illustrated in the generalized *S-N* curve for steel shown in Fig. 8.5.

Because fatigue failures originate at *local* points of relative weakness, the results of fatigue tests have considerably more scatter than do those of static tests. For this reason the statistical approach to defining strength (see Sections 6.13–6.15) takes on greater importance. Standard deviations of endurance limit values are commonly in the range of 4 to 9 percent of the nominal value. Ideally, the standard deviation is determined experimentally from tests corresponding to the specific application involved. Often, 8 percent of the nominal endurance limit is used as a conservative estimate of the standard deviation when more specific information is not available.

The data scatter illustrated in Fig. 8.4 is typical for carefully controlled tests. The scatter band marked on Fig. 8.4*c* illustrates an interesting point: The scatter in *fatigue strength* corresponding to a given life is small; the scatter in *fatigue life* corresponding to a given stress level is large. Even in carefully controlled tests, these life values can vary over a range of five or ten to one.

A multitude of standard fatigue tests (Fig. 8.3) have been conducted over the last few decades, with results tending to conform to certain generalized patterns. The most commonly used of these is shown in Fig. 8.5. With a knowledge of only the ultimate tensile strength, a good approximation of the complete *S-N* curve for steel can quickly be made. Furthermore, tensile strength can be estimated from a nondestructive hardness test. For steels, the tensile strength in psi is about 500 times the Brinell hardness; hence, a conservative estimate of endurance limit is about 250 H_B. *The latter relationship can be counted on only up to Brinell hardness values of about 400*. The endurance limit *may*

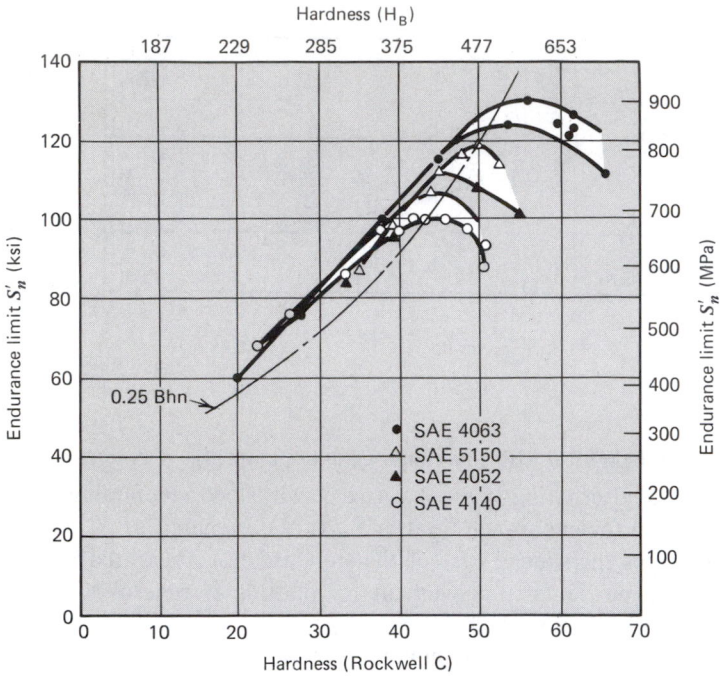

FIGURE 8.6.
Endurance limit vs. hardness for four alloy steels. (From M. F. Garwood, H. H. Zurburg, and M. A. Erickson, *Interpretation of Tests and Correlation with Service,* American Society for Metals, 1951, p. 13.)

or may not continue to increase for higher hardnesses, depending on the composition of the steel. This is illustrated in Fig. 8.6.

Although the 10^3 cycle fatigue strength in Fig. 8.5 is computed as being about 90 percent of the ultimate strength, the *actual* stress is not that high. The reason is that fatigue strength values corresponding to the test points in Fig. 8.4 are *computed* according to the elastic formula, $\sigma = Mc/I$. Loads large enough to cause failure in 1000 cycles usually cause significant yielding, resulting in *actual* stresses that are *lower* than calculated values. This point is illustrated in Fig. 8.7.

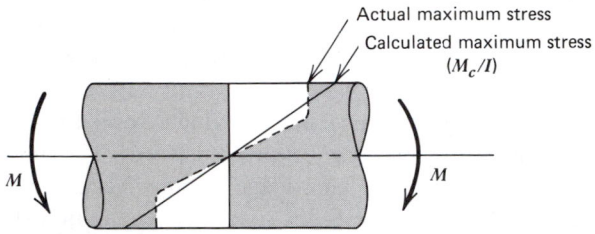

FIGURE 8.7.
Representation of maximum bending stress at low fatigue life (as 1000~) (Note: *Calculated* maximum stress is used in *S-N* plots.)

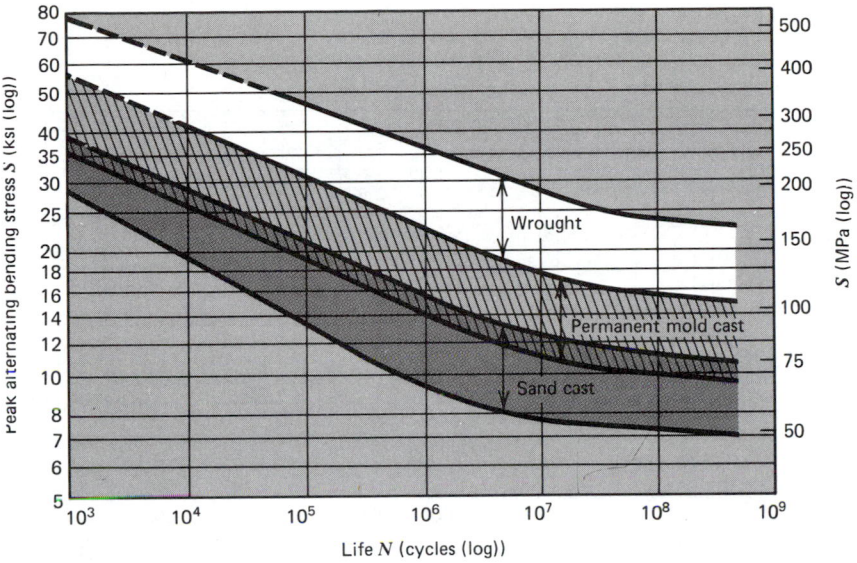

FIGURE 8.8.
***S-N* bands for representative aluminum alloys, excluding wrought alloys with $S_u <$ 38 ksi.**

The fatigue strength characteristics of cast iron are similiar to those of steel with the exception that the endurance limit corresponds to about 0.4 (rather than 0.5) times the ultimate strength.

Representative *S-N* curves for various aluminum alloys are shown in Fig. 8.8. Note the absence of a sharply defined "knee" and true endurance limit. This is typical of nonferrous metals. In the absence of an endurance limit, the fatigue strength at 10^8 or 5×10^8 cycles is often used. (To give a "feel" for the time required to accumulate this many cycles, an automobile would typically travel nearly 400,000 miles before any one of its cylinders fired 5×10^8 times.) For typical wrought aluminum alloys, the 5×10^8 cycle fatigue strength is related to the static tensile strength as indicated in Fig. 8.9.

Typical *S-N* curves for magnesium alloys are shown in Fig. 8.10. The 10^8 cycle fatigue strength is about 0.35 times tensile strength for most wrought and cast alloys.

For copper alloys (including brasses, bronzes, cupro-nickels, etc.), the ratio of 10^8 cycle fatigue strength to static tensile strength ranges between 0.25 and 0.5 in most cases. For nickel alloys, the ratio of these strengths is usually between 0.35 and 0.5.

Titanium and its alloys behave like steel in that they tend to exhibit a true endurance limit in the range of 10^6 to 10^7 cycles, with the endurance limit being between 0.45 and 0.65 times the tensile strength.

8.4 FATIGUE STRENGTHS FOR REVERSED BENDING AND REVERSED AXIAL LOADING

If a test specimen, similar to the one used in the R. R. Moore testing machine, is not rotated but mounted horizontally with one end fixed and the other pushed alternately

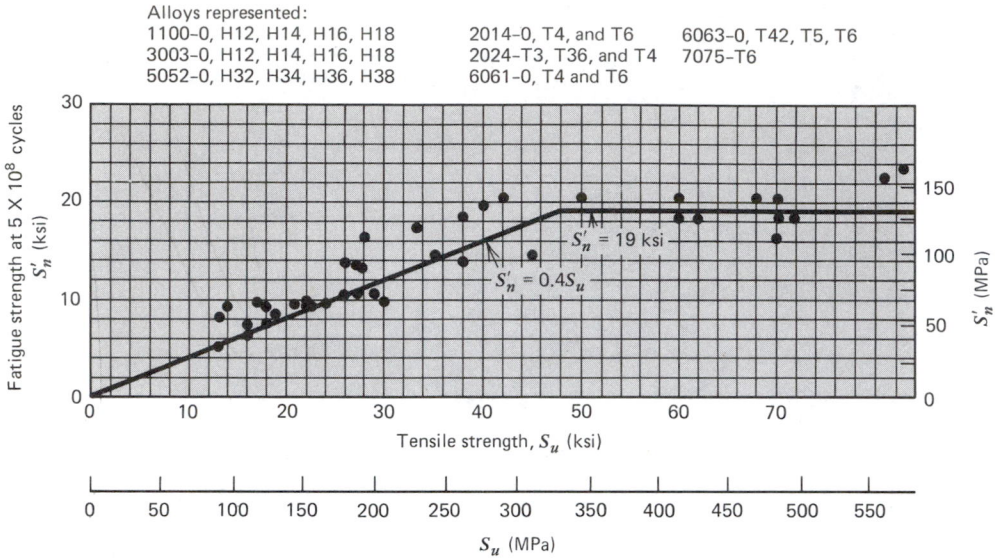

FIGURE 8.9.
Fatigue strength at 5 × 10⁸ cycles, common wrought-aluminum alloys.

up and down, *reversed* bending stresses are produced. These differ from stresses caused by *rotating* bending only in that the maximum stresses are limited to the top and bottom, whereas rotating bending produces maximum stresses all around the circumference. In rotating bending, a fatigue failure will originate from the weakest point on the surface; in reversed bending there is an excellent statistical probability that the weakest point will not be at exactly the top or bottom. This means that the fatigue strength in

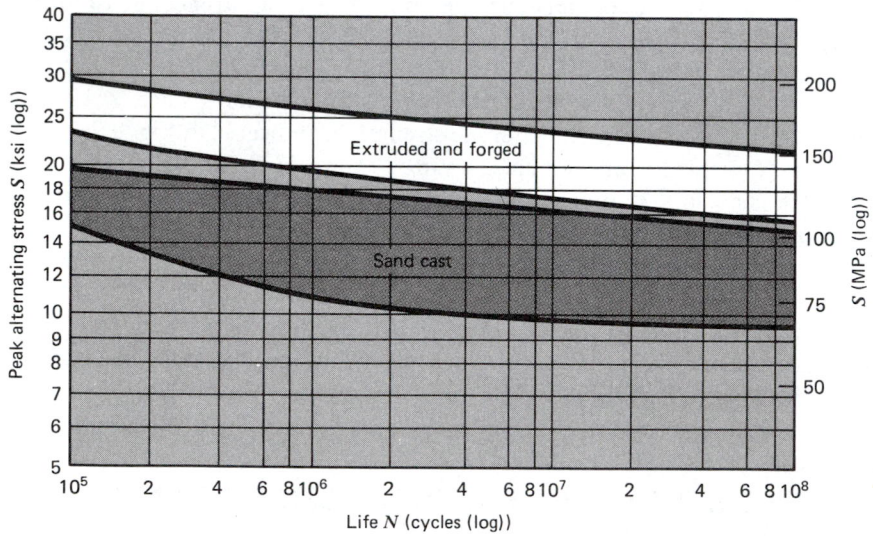

FIGURE 8.10.
General range of *S-N* curves for magnesium alloys.

reversed bending is usually slightly greater than in rotating bending. The difference is small, and usually neglected. Thus, when dealing with problems involving reversed bending, a small error on the conservative side is deliberately introduced.

Similar reasoning indicates that a reversed *axial* loading—which subjects the *entire cross section* to the maximum stress—should give *lower* fatigue strengths than rotating bending. This is indeed the case, and this difference should be taken into account. Axial or "push-pull" tests give endurance limits about 10 percent lower than rotating bending. Furthermore, if the supposedly axial load is just a *little* off-center (as with nonprecision parts having as-cast or as-forged surfaces), slight bending is introduced which causes the stresses on one side to be a little higher than P/A. Ideally, one would determine the load eccentricity and calculate the peak alternating stress as $P/A + Mc/I$, but the magnitude of the unwanted eccentricity is often not known. In such cases it is customary to take this into account by using only the P/A stress, and reducing the rotating bending endurance limit by a little *more* than 10 percent (perhaps by 20 to 30 percent).

Since this reduction of 10 percent or more in the rotating bending endurance limit is associated with *differences in stress gradient,* we will take this into account by multiplying the basic endurance limit, S'_n, by a *gradient factor* or *gradient constant,* C_G, where $C_G = 0.9$ for pure axial loading of precision parts and C_G ranges from perhaps 0.7 to 0.9 for axial loading of nonprecision parts.

Stress gradient is also responsible for the 10^3 cycle fatigue strength being lower for axial loading than for bending loads. Recall, from Fig. 8.7, that the $0.9S_u$ strength for rotating bending was in most cases an artificial calculated value that neglected the effect of yielding at the surface. Yielding cannot reduce the surface stress in the case of axial loading. Accordingly, tests indicate that the 10^3 cycle strength for this loading is only about $0.75S_u$.

The above points are illustrated in Fig. 8.11. The top two curves show comparative estimated S-N curves for bending and axial loading.

8.5 FATIGUE STRENGTH FOR REVERSED TORSIONAL LOADING

Since fatigue failures are associated with highly localized yielding, and since yielding of ductile materials has been found to correlate well with the distortion-energy theory, it is perhaps not surprising that this theory has been found useful to predict the endurance limit of ductile materials under various combinations of reversed biaxial loading, including torsion. This is illustrated in Fig. 8.12. *Thus, for ductile metals, the endurance limit (or long-life endurance strength) in reversed torsion is about 58 percent of the endurance limit (or long-life endurance strength) in reversed bending.* This is taken into account by multiplying the basic endurance limit, S'_n, by a *load factor,* C_L, of 0.58.

Since torsional stresses involve stress gradients similar to bending, it is not surprising that, like in bending, the 10^3 cycle fatigue strength is generally about 0.9 times the *appropriate* ultimate strength. Thus, for reversed torsion the 10^3 cycle strength is approximately 0.9 times the ultimate *shear* strength. Experimental values for ultimate

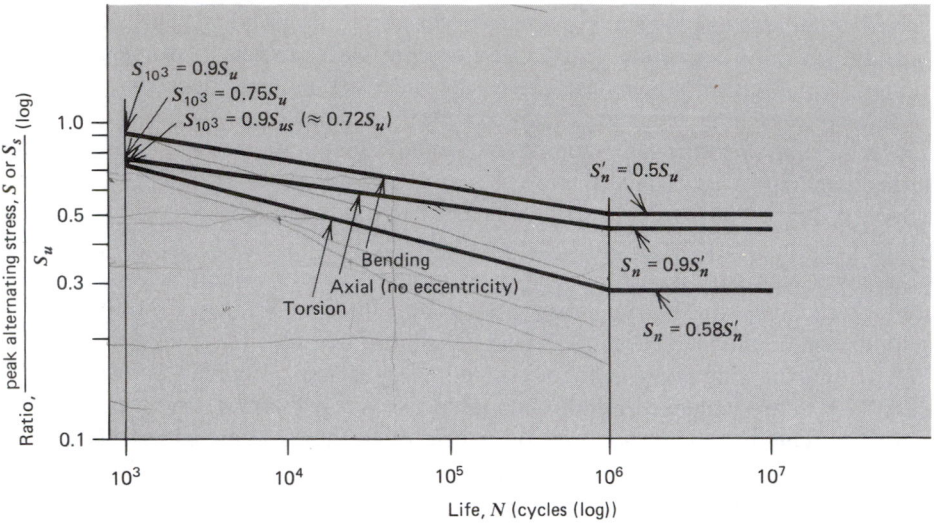

FIGURE 8.11.
Generalized S-N curves for polished 0.3-in.-diameter steel specimens (based on calculated elastic stresses, ignoring possible yielding).

torsional shear strength should be used if available. If not, they may be *roughly* approximated as:

$$S_{us} = 0.8S_u \quad \text{(for steel)}$$
$$S_{us} = 0.7S_u \quad \text{(for other ductile metals)}$$

The bottom curve of Fig. 8.11 shows an estimated torsional S-N curve for steel based on the above relationships.

There are fewer data available to support a generalized procedure for estimating torsional S-N curves for *brittle* materials, and this makes it all the more desirable to obtain actual experimental fatigue data for the specific material and loading condition involved in the problem at hand. In the absence of such data, torsional S-N curves for brittle materials are sometimes *estimated* on the basis of (a) assuming an endurance limit at 10^6 cycles of 0.8 times the standard reversed bending endurance limit (this correlates somewhat with using the Mohr theory of failure to relate bending and torsion in the same way that the distortion-energy theory is used for ductile materials), and (b) assuming a 10^3 cycle strength of $0.9S_{us}$, the same as for ductile materials.

8.6 FATIGUE STRENGTH FOR REVERSED BIAXIAL LOADING

Figure 8.12 illustrates the good general agreement of the distortion-energy theory with the endurance limit (or long-life fatigue strength) of ductile materials subjected to all combinations of reversed biaxial loading. For shorter-life fatigue strengths of ductile materials, and for brittle materials, we are not in a very good position to make fatigue

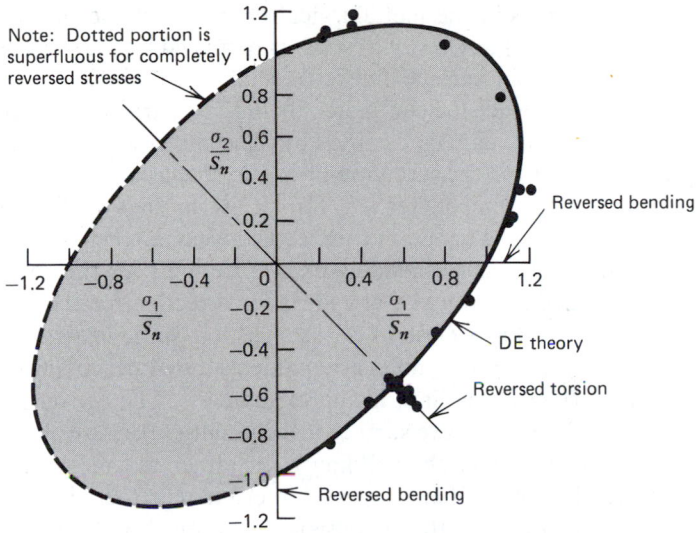

FIGURE 8.12.
$\sigma_1 - \sigma_2$ plot for completely reversed loading, ductile materials. [Data from Walter Sawert, Germany, 1943, for annealed mild steel; and H. J. Gough, "Engineering Steels under Combined Cyclic and Static Stresses," *J. Appl. Mech.*, 72 (March 1950), pp. 113–125.]

strength predictions without the benefit of directly applicable experimental data. With this reservation in mind, the following procedure is tentatively recommended:

1. For *ductile* materials, use the *distortion-energy* theory (usually Eq. (6.8)) to convert from the actual load stresses to an equivalent stress that is regarded as a reversed *bending* stress. Then proceed to relate this stress to the fatigue properties of the material (i.e., the *S-N* curve) in reversed bending.
2. For *brittle* materials, use the *Mohr* theory to obtain an equivalent reversed stress that is regarded as a reversed *bending* stress, and relate this to the bending fatigue properties (i.e., *S-N* curve) of the material. (A convenient graphical procedure for determining the equivalent bending stress is to draw a σ_1-σ_2 plot like Fig. 6.9*b* for the material, and then plot the point corresponding to the actual reversed stresses. Next, draw a line through this point and parallel to the failure line. The intersection of this line with the σ_1 axis gives the desired equivalent bending stress.)

8.7 INFLUENCE OF SURFACE AND SIZE ON FATIGUE STRENGTH

Up to this point, all discussions of fatigue strength have assumed the surface to have a special "mirror polish" finish. This involves a costly laboratory procedure, but serves to minimize (a) surface scratches and other geometric irregularities acting as points of stress concentration, (b) any differences in the metallurgical character of the surface layer of material and the interior, and (c) any residual stresses produced by the surface finishing procedure. Normal commercial surface finishes usually result in localized points of greater fatigue vulnerability; hence the commercially finished parts have lower fatigue strengths. The amount of "surface damage" caused by the commercial processes depends not only on the process, but also on the susceptibility of the material to damage. Figure 8.13 gives estimated values of surface factor, C_S, for various finishes applied to steels of various hardnesses. In all cases the endurance limit for the laboratory polished surface is multiplied by C_S to obtain the corresponding endurance limit for the commercial finish. It is standard practice *not* to make any surface correction for the 10^3 cycle strength—the reason being that this is close to the strength for static loads, and that the static strength of ductile parts is not significantly influenced by surface finish.

The surface factor for ordinary gray cast iron is approximately 1. The reason for this is that even the mirror polished samples have surface discontinuities because of the graphite flakes in the cast iron matrix, and that adding even rather severe surface scratches does not make the situation much, if any, worse. Unfortunately, there is little published information on surface factors for other materials. For critical parts, actual fatigue tests involving the material and surface in question must be conducted.

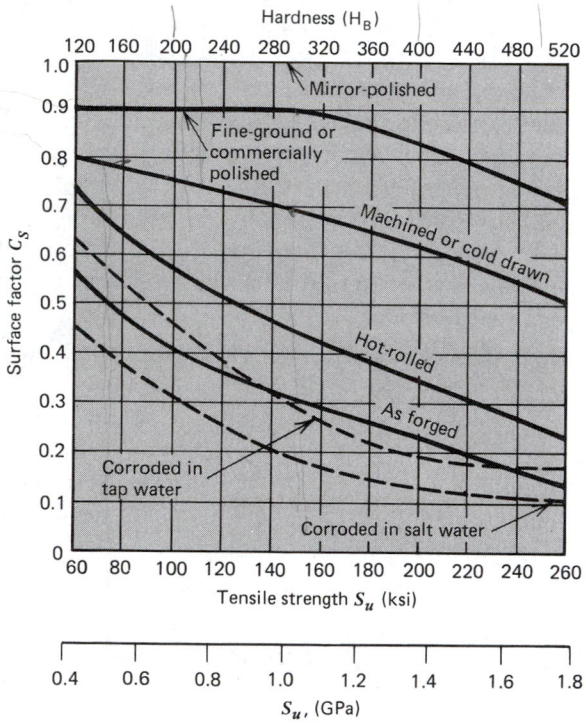

FIGURE 8.13.
Reduction in endurance limit due to surface finish—steel parts.

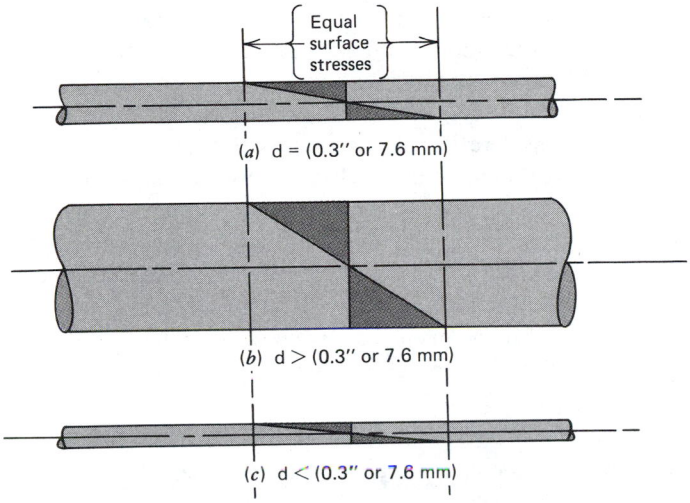

FIGURE 8.14.
Stress gradients vs. diameter for bending and torsion.

In Section 8.4 it was pointed out that the endurance limit for reversed axial load is about 10 percent lower than for reversed bending because of *stress gradient*. For the 0.3-in.-diameter bending specimen, the rapid drop in stress level below the surface is somehow beneficial. The 0.3-in.-diameter *axial* specimen does not enjoy this benefit. A comparison of stress gradients in Figs. 8.14*a* and 8.14*b* shows that large specimens in bending or torsion do not have the same favorable gradients as the standard 0.3-in. specimen. Experiments show that if the diameter is increased to much more than 0.4 in., most of the beneficial gradient effect is lost. *Hence, parts of diameter more than 0.4 in. (or 10 mm) that are subjected to reversed bending or torsion should carry a gradient factor, C_G of 0.9, the same as parts subjected to axial loading.* Figure 8.14*c* shows that very small parts have an even more favorable gradient than the standard R. R. Moore specimen. Thus, one might expect the endurance limit for such parts to be *greater* than for 0.3-in.-diameter parts. *Sometimes* this has been found to be the case—but unless specific data are available to substantiate this increase, it is best to use a gradient factor of unity for these small parts.

Consider the question of what gradient factor to use with the bending of a rectangular section, say 6 mm × 12 mm. If the bending is about the neutral axis that places the tension and compression surfaces 6 mm apart, use $C_G = 1$; if the tension and compression surfaces are 12 mm apart, use $C_G = 0.9$. Thus, the gradient factor is determined on the basis of an equivalent round section having the same stress gradient as the actual part.

Recall that a gradient factor of 0.9 (or lower) was specified (Section 8.4) for *all* axially loaded parts because the stress gradient is unfavorable, regardless of size.

Parts with sections larger than about 50 mm equivalent diameter are usually found to have lower endurance limits than those computed using the gradient factors recommended above. This is due in part to metallurgical factors such as hardenability, as the interior of large-section parts is usually metallurgically different than the surface metal.

The extent to which the endurance limit of very large parts is reduced varies substantially, and generalizations are hardly warranted. If the part in question is critical, there is no substitute for pertinent test data. As a *very rough guide,* however, an indication of values sometimes used is given in Table 8.1.

The foregoing recommended consideration of specimen size was focused on the influence of size on stress gradient. It should be noted that a more extensive treatment of this subject would consider other facets. For example, the larger the specimen, the greater the statistical probability that a flaw of given severity (from which fatigue failure could originate) will exist somewhere near the surface (with bending or torsional loads) or somewhere within the entire body of material (with axial loads). Also, the influence of processing on metallurgical factors is often more favorable for smaller parts, even in the size range below 50 mm equivalent diameter.

8.8 SUMMARY OF ESTIMATED FATIGUE STRENGTHS FOR COMPLETELY REVERSED LOADING

The foregoing articles have emphasized the desirability of obtaining actual fatigue test data that pertain as closely as possible to the application involved. Generalized empirical factors were given for use when such data are not available. These factors can be

TABLE 8.1
Generalized Fatigue Strength Factors for Ductile Materials

	Bending	Axial	Torsion
a. Endurance Limit[a]			
$S_n = S'_n C_L C_G C_S$, where S'_n is the R. R. Moore endurance limit, and			
C_L (load factor)	1	1	0.58
C_G (gradient factor):			
dia < (0.4 in. or 10 mm)	1	.7 to .9	1
(0.4 in. or 10 mm) < dia < (2 in. or 50 mm)[b]	0.9	.7 to .9	0.9
C_S (surface factor)	——See Fig. 8.13.——		
b. 10^3 cycle strength[c]	$0.9S_u$	$0.75S_u$	$0.9S_{us}$[d]

Note: 1. For all biaxial loading except torsion, convert actual stresses to an *equivalent bending stress* using the distortion-energy theory. Then relate this equivalent stress to the *bending fatigue strength.*
2. For other than 50 percent reliability, multiply S_n by C_R. (10^3 cycle strength is usually not multiplied by C_R.)

[a]For materials not having an endurance limit, apply the factors to the 10^8 or 5×10^8 cycle strength.
[b]For (2 in. or 50 mm) < dia < (4 in. or 100 mm) reduce these factors by about 0.1. For (4 in. or 100 mm) < dia < (6 in. or 150 mm) reduce these factors by about 0.2.
[c]No corrections for gradient or surface are normally made, but the experimental value of S_u or S_{us} should pertain to sizes reasonably close to those involved.
[d]$S_{us} \approx 0.8S_u$ for steel; $S_{us} \approx 0.7S_u$ for other ductile metals.

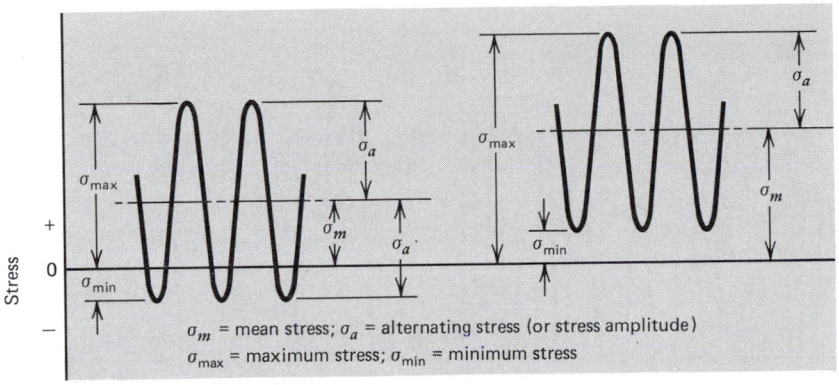

FIGURE 8.15.
Fluctuating stress notation illustrated with two examples.

applied with greatest confidence to steel parts because most of the data on which they are based came from testing steel specimens.

Three of these factors are involved in the estimate for endurance limit:

$$S_n = S'_n C_L C_G C_S \tag{8.1}$$

Table 8.1 gives a summary of all factors used for estimating the fatigue strength of ductile materials (when subjected to completely reversed loading). It serves as a convenient reference for solving problems.

8.9 INFLUENCE OF MEAN STRESS ON FATIGUE STRENGTH

Machine and structural parts seldom encounter completely reversed stresses; rather, they typically encounter a *fluctuating* stress that is a combination of static plus completely reversed stress. A fluctuating stress is usually characterized by its *mean* and *alternating* components. However, the terms *maximum* stress and *minimum* stress are also used. All four of these quantities are defined in Fig. 8.15. Note that if any two of them are known, the others are readily computed. This text uses primarily mean and alternating stress components, as in Fig. 8.17. The same information can be portrayed graphically with *any* combination of two of the stress components shown in Fig. 8.15. For example, σ_m-σ_{max} coordinates are often found in the literature. For convenience, some graphs use all four quantities, as in Figs. 8.18–8.20.

The existence of a static tensile stress reduces the amplitude of reversed stress that can be superimposed. Figure 8.16 illustrates this concept. Fluctuation (a) is a completely reversed stress corresponding to the endurance limit—the mean stress is zero and the alternating stress S_n. Fluctuation (b) involves a tensile mean stress. In order to have an equal (in this case, "infinite") fatigue life, the alternating stress must be less than S_n. In going from (b) to (c), (d), (e), and (f), the mean stress continually increases; hence, the alternating stress must correspondingly decrease. Note that in each case the stress fluctuation is shown as starting from zero, and that the stresses are *computed*

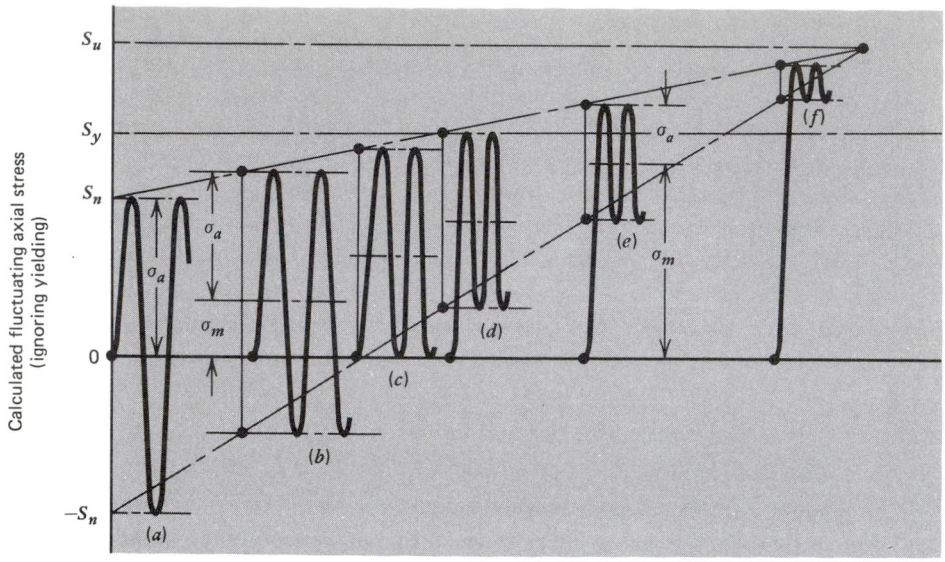

FIGURE 8.16.
Various fluctuating uniaxial stresses, all of which correspond to equal fatigue life.

P/A values. Microscopic yielding occurs even at (a), as has been previously noted. Upon reaching (d), macroscopic yielding begins. Although load fluctuations (e) and (f) give "infinite" life, the part is yielded on the first load application.

Figure 8.17 gives a convenient graphical representation of various combinations of mean and alternating stress in relation to criteria both for yielding and for various fatigue lives. It is often called a *constant life fatigue diagram* because it has lines corresponding to a constant 10^6 cycle (or "infinite") life, constant 10^5 cycle life, and so forth.

To begin the construction of this diagram, put on it first the information that is already known. The horizontal axis ($\sigma_a = 0$) corresponds to static loading. Yield and ultimate strengths are plotted at points A and B. For ductile materials, the compressive yield strength is minus S_y, and this is plotted at point A'. If the mean stress is zero and the alternating stress is equal to S_y (point A''), the stress fluctuates between $+S_y$ and $-S_y$. All points along the line AA'' correspond to fluctuations having a tensile peak of S_y; all points on $A'A''$ correspond to compressive peaks equal to $-S_y$. All combinations of σ_m and σ_a causing no (macroscopic) yielding are contained within triangle $AA'A''$.

All *S-N* curves considered in this chapter correspond to $\sigma_m = 0$. Hence, we can read from these curves points like C, D, E, and F for any fatigue life of interest. Connecting these points with B gives estimated lines of constant life. This empirical procedure for obtaining constant life lines is credited to Goodman; hence the lines are commonly called *Goodman lines.*

Laboratory tests have consistently indicated that compressive mean stresses do *not* reduce the amplitude of allowable alternating stress; if anything, they slightly *increase* it. Figure 8.17 is thus conservative in showing the constant life lines as horizontal to the

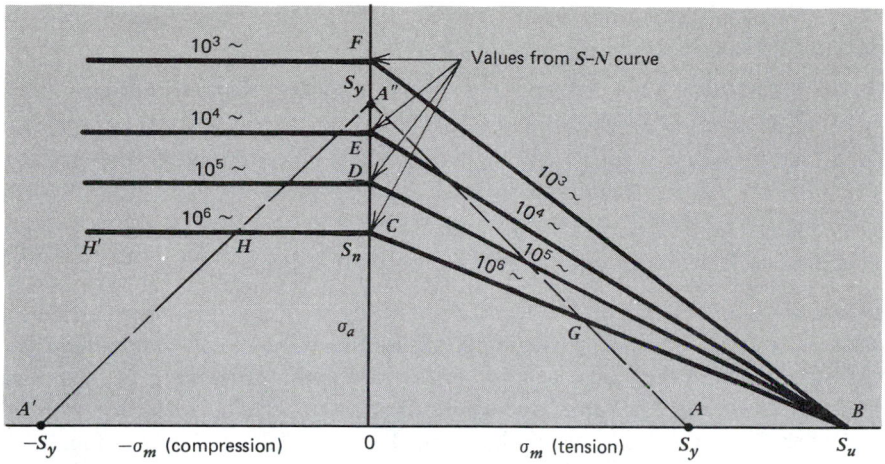

FIGURE 8.17.
Constant life fatigue diagram—ductile materials.
Bending loads: **Construct diagram as shown; take points *C, D,* and so on from *S-N* curve for reversed bending.**
Axial loads: **Construct diagram as shown; take points *C,* and so on from *S-N* curve for reversed axial loads.**
Torsional loads: **Omit left half of diagram (*any* torsional mean stress is considered positive); take points *C,* and so on from *S-N* curve for reversed torsion; use S_{ys} and S_{us} instead of S_y and S_u. (For steel, $S_{us} \approx 0.8S_u$, $S_{ys} \approx 0.58S_y$.)**
General biaxial loads: **Construct the diagram as for *bending* loads, and use it with *equivalent* load stresses, computed as follows. (Note that these equations apply to the generally encountered case where σ_a and σ_m exist in one direction only. Corresponding equations for the more elaborate general case are also tentatively applicable.)**

1. **Equivalent *alternating* bending stress, σ_{ea}, is calculated from the *distortion energy theory* as being *equivalent* to the combination of existing *alternating* stresses:**

$$\sigma_{ea} = \sqrt{\sigma_a^2 + 3\,\tau_a^2} \tag{a}$$

2. **Equivalent *mean* bending stress, σ_{em}, is taken as the *maximum principal stress* resulting from the superposition of all existing static (mean) stresses. Use Mohr circle, or:**

$$\sigma_{em} = \frac{\sigma_m}{2} + \sqrt{\tau_m^2 + \left(\frac{\sigma_m}{2}\right)^2} \tag{b}$$

(Various other suggested equations for σ_{em} are found in the literature.)

left of points *C, D,* and so on. (The lines apparently extend indefinitely as far as *fatigue* is concerned, the limitation being only static compression failure.)

Detailed modifications of the diagram for various types of loading are given in Fig. 8.17. Let us observe the significance of various areas on the diagram:

1. If a life of at least 10^6 cycles is required *and no* yielding is permitted (even at

extreme fibers in bending or torsion, where a little yielding might be difficult to detect), we must stay inside area *A'HCGA*.

2. If *no* yielding but less than 10^6 cycles of life are required, then we can also work within some or all of area *HCGA''H*.
3. If 10^6 cycles of life are required but yielding is acceptable, area *AGB* (and area to the left of *A'H*) may be used, in addition to area *A'HCGA*.
4. Area above *A''GB* (and above *A''HH'*) corresponds to yielding on the first load application *and* fatigue fracture prior to 10^6 cycles of loading.

The procedure for general biaxial loads given in Fig. 8.17 should be recognized as a substantial simplification of a very complex situation. It applies best to situations involving long life, where the loads are all in phase, where the principal axes for mean and alternating stresses are the same, and where these axes are fixed with time. For an illustration where these conditions would be fulfilled, consider the example in Fig. 4.25 with the shaft stationary, and with the 2000 lb static load changed to a load that fluctuates between 1500 and 2500 lb. The *static* stresses on element *A* would be unchanged, but *alternating* stresses would be added. The alternating bending and the alternating torsion would obviously be in phase, the principal planes for mean and alternating stresses would be the same, and these planes would remain the same as the load fluctuated.

Figures 8.18–8.20 give constant life fatigue strengths for certain steel and aluminum materials. They differ from Fig. 8.17 in the following respects:

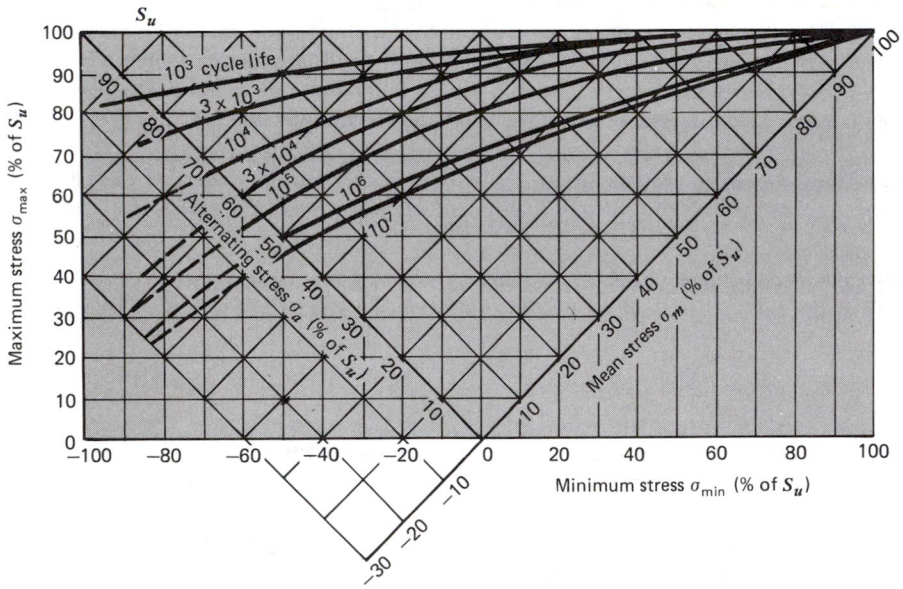

FIGURE 8.18.
Fatigue-strength diagram for alloy steel, S_u = 125 to 180 ksi, axial loading. Average of test data for polished specimens of AISI 4340 steel (also applicable to other alloy steels, such as AISI 2330, 4130, 8630). (Courtesy Grumman Aerospace Corp.)

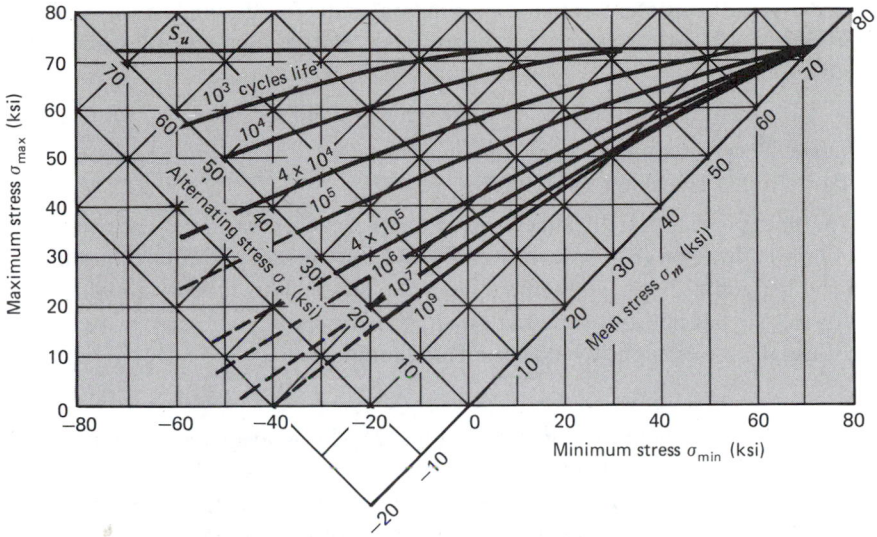

FIGURE 8.19.
Fatigue-strength diagram for 2024-T3, 2024-T4, and 2014-T6 aluminum alloys, axial loading. Average of test data for polished specimens (unclad) from rolled and drawn sheet and bar. Static properties for 2024; S_u = 72 ksi, S_y = 52 ksi; for 2014: S_u = 72 ksi, S_y = 63 ksi. (Courtesy Grumman Aerospace Corp.)

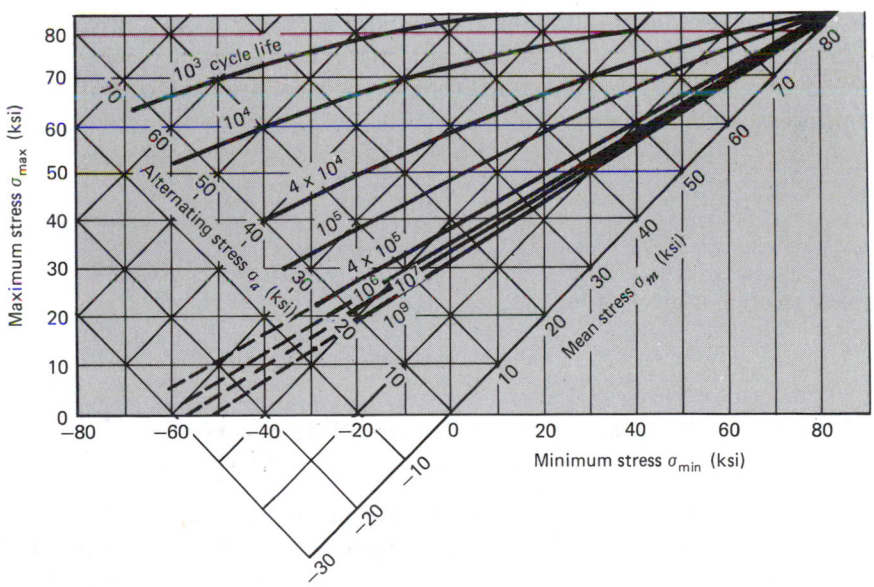

FIGURE 8.20.
Fatigue-strength diagram for 7075-T6 aluminum alloy, axial loading. Average of test data for polished specimens (unclad) from rolled and drawn sheet and bar. Static properties: S_u = 82 ksi, S_y = 75 ksi. (Courtesy Grumman Aerospace Corp.)

1. Figures 8.18–8.20 represent actual experimental data for the materials involved, whereas Fig. 8.17 shows conservative empirical relationships that are generally applicable.
2. Figures 8.18–8.20 are "turned 45°," with scales added to show σ_{max} and σ_{min} as well as σ_m and σ_a.
3. Yield data are not shown on these figures.
4. The experimental constant life lines shown have some curvature, indicating that Fig. 8.17 errs a little on the conservative side in both the straight "Goodman lines" and in the horizontal lines for compressive mean stress. This conservatism usually exists for ductile but not for brittle materials. Experimental points for brittle materials are usually on or slightly below the Goodman line.

When experimental data like those given in Figs. 18–20 are available, these are to be preferred over the estimated constant life fatigue curves constructed in Fig. 8.17.

The reader will find that Fig. 8.17 and Table 8.1 provide helpful summaries of information pertaining to the solution of a large variety of fatigue problems.

SAMPLE PROBLEM 8.1 Estimation of S-N and Constant Life Curves from Tensile Test Data

Using the empirical relationships given in this section, estimate the *S-N* curve and a family of constant life fatigue curves pertaining to the axial loading of precision steel parts having $S_u = 150$ ksi, $S_y = 120$ ksi, and commercially polished surfaces. All cross-section dimensions are to be under 2 in.

Solution

The complete solution is given in Fig. 8.21.

SAMPLE PROBLEM 8.2 Determine the Required Size of a Tensile Link Subjected to Fluctuating Loading

A round tensile link with negligible stress concentration is subjected to a load fluctuating between 1000 and 5000 lb. It is to be a precision member (so that use of $C_G = 0.9$ is justified) with commercially polished surfaces. The material is to be steel, with $S_u = 150$ ksi, $S_y = 120$ ksi. A safety factor of 2 is to be used, applied to all loads.

a. What diameter is required if infinite life is needed?
b. What diameter is required if only 10^3 cycles of life are needed?

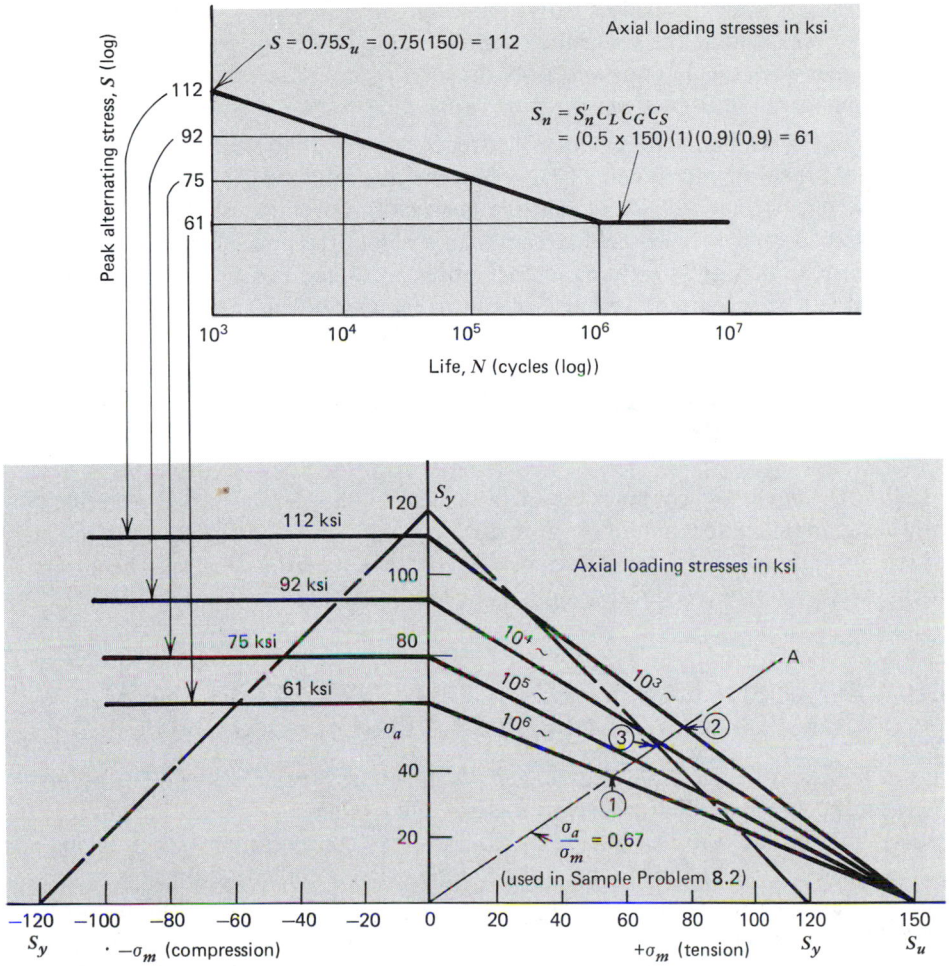

FIGURE 8.21.
Sample Problem 8.1—Estimate S-N and σ_m, $-$ σ_a curves for steel, S_u = 150 ksi, axial loading, commercially polished surfaces.

Solution

1. The fatigue strength properties of the material conform to those represented in Fig. 8.21 *provided* the diameter comes out to be under 2 in.

2. At the *design overload:* $\sigma_m = 6000/A$, $\sigma_a = 4000/A$. Thus, regardless of the area, $\sigma_a/\sigma_m = 0.67$. This is represented by line OA on Fig. 8.21. Note the interpretation of this line: If area A is infinite, both σ_m and σ_a are zero, and the stresses are represented by the origin, point O. Moving out along line OA corresponds to progressively decreasing values of A. For part a of the problem we need to determine the area corresponding to the intersection of OA with the infinite life line (same as 10^6 cycles, in this case), which is labeled (1). At this point, σ_m = 57 ksi; from σ_m =

$6000/A$, A is determined as 0.107 in.^{2} From $A = \pi d^2/4$, $d = 0.369$ *in.* This is indeed well within the size range for the value of $C_G = 0.9$, which had to be assumed when the diagram was constructed. In many cases, the final answer might be rounded off to $d = \frac{3}{8}$ *in.* ■

3. For part *b*, with only 10^3 cycles of life required, we can move out along line *OA* of Fig. 8.21, seemingly to point ②, where the line intersects the 10^3 cycle life line. However, if point ③ is crossed, the peak design overload of 10,000 lb imposes stresses in excess of the yield strength. In a notch-free tensile bar the stresses are uniform so that gross yielding of the entire link would occur. Assuming that this could not be permitted, we must choose a diameter based on point ③, not point ②. Here, $\sigma_m = 71$ ksi, from which $A = 0.086$ in.^{2}, and $d = 0.328$ *in.*, perhaps rounded off to $d = \frac{5}{16}$ *in.* This diameter corresponds to an estimated life greater than required, but to make it any smaller would cause general yielding on the first overload application. ■

Probably the most common use of fatigue strength relationships is in connection with designing parts for infinite (or 5×10^8 cycle) life, or in analyzing parts intended for infinite fatigue life. In these cases no *S-N* curve is required. It is only necessary to calculate the estimated endurance limit and plot the infinite life Goodman line.

8.10 INFLUENCE OF STRESS CONCENTRATION WITH COMPLETELY REVERSED FATIGUE LOADING

Figure 8.22 shows typical *S-N* curves for (a) unnotched specimens and (b) otherwise identical specimens except for a stress raiser. Unlike other *S-N* curves we will use, the stresses plotted are *nominal* stresses; that is, stress concentration is not taken into account. The specimen dimensions in the section where fatigue fractures occur are the same for both (a) and (b). Hence, any given *load* causes the same *calculated stress* in both cases. As shown in the figure, the ratio of the unnotched to notched endurance limit is the *fatigue stress concentration factor,* designated as K_f. Theoretically, one might expect K_f to be equal to the theoretical or geometric factor K_t, discussed in Section 4.12. Fortunately, tests show that K_f is often less than K_t. This is apparently due to internal irregularities in the structure of the material. An "ideal" material would have internal stresses in exact accordance with elastic theory; actual materials have internal irregularities causing local points to have higher stresses. Thus, even unnotched samples suffer from these internal "notches." Adding an external geometric notch (groove, thread, hole, etc.) to this material may not do as much *additional* damage as it would if the material itself were "perfect." An extreme case in point is ordinary (not "high-strength") gray cast iron. The internal stress raisers caused by graphite flakes in the matrix are such that the addition of a geometric stress raiser has little or no effect. This means that if the material in Fig. 8.22 were one of the lower grades of gray cast iron, the two *S-N* curves would nearly coincide. A material with a uniform fine-grain matrix is *highly sensitive* to notches (i.e., $K_f \approx K_t$); cast iron is insensitive to notches (i.e., $K_f \approx 1$).

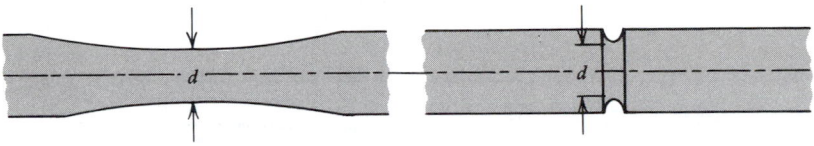

(*a*) Unnotched specimen ("u") (*b*) Notched specimen ("n")

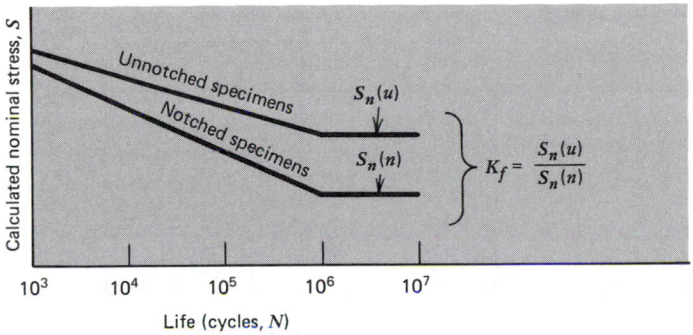

(*c*) Illustration of fatigue stress concentration factor, K_f

FIGURE 8.22.
Reversed-load fatigue tests, notched vs. unnotched specimens.

The above situation is commonly dealt with by using a *notch sensitivity factor, q,* defined by the equation

$$K_f = 1 + (K_t - 1)q \tag{8.2}$$

where q ranges between zero (giving $K_f = 1$) and unity (giving $K_f = K_t$). Thus, to determine fatigue stress concentration factors from corresponding theoretical (or geometric) factors, it is necessary to know the notch sensitivity of the material.

The situation is a little more complicated than it appears above because notch sensitivity depends not only on the material, but also on the relative radius of the geometric notch and the dimensions of characteristic internal imperfections. Notch radii so small that they approach the imperfection size give zero notch sensitivity. This is indeed fortunate; otherwise even minute scratches (which give extremely high values of K_t) on what would generally be called a smooth, polished surface would disastrously weaken the fatigue strength. Figure 8.23 is a plot of notch sensitivity vs. notch radius for some common materials. Note that in all cases the notch sensitivity approaches zero as the notch radius approaches zero. Also, note that the data for steels illustrate the fundamental tendency for the harder and stronger materials to be more notch sensitive. This means that changing from a soft to a harder and stronger steel normally results in increased part fatigue strength, but not in as *great* an increase as might be expected because of the increased notch sensitivity. Finally, Fig. 8.23 shows that a given steel is a little more notch sensitive for torsional loading than for bending and axial loading. For example, a 0.04-in.-radius notch in a 160-Bhn steel part has a notch sensitivity of about 0.71 if the loading is bending or axial, and about 0.76 if the load is torsional.

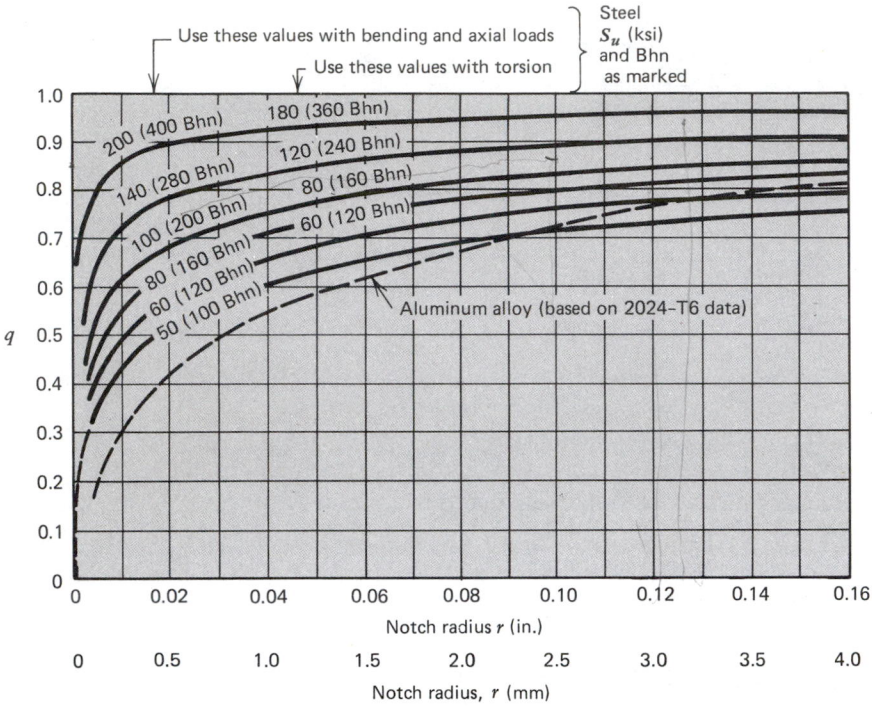

FIGURE 8.23.
Notch sensitivity curves (after [9]). (Note: (1) r is the radius at the point of potential fatigue crack origin. (2) For $r > 0.16$ in., extrapolate or use $q \approx 1$.)

Figure 8.22 shows that the influence of the notch at 10^3 cycles is considerably less than at 10^6 cycles. Some references advise neglecting the influence of the stress raiser at 10^3 cycles. Although certain data support this recommendation, a closer study indicates that this is valid only for a relatively soft metal (steel, aluminum, magnesium, and probably others); but that for relatively hard and strong alloys of these same metals, the notch influence at 10^3 cycles can be nearly as great as at 10^6 cycles (see [6], Fig. 13.26).

There is a fundamental difficulty in analyzing notch effects at the low-cycle end of curves like Fig. 8.22*c*. That is because the *calculated nominal stress* used in the plot does not relate very closely with the *actual loading conditions* imposed upon the local region at the root of the notch where a fatigue crack starts. Figure 8.2 shows an enlarged view of the notch region of a notched specimen, as in Fig. 8.22*b*. Under reversed loading sufficient to cause fatigue failure after, say, 10^3 cycles, plastic yielding will occur throughout some small region at the base of the notch. This region contributes little to the rigidity of the total part; thus, the *strains* within this zone are determined almost entirely by the stable elastic resistance of the great bulk of material outside this zone. This means that during a fatigue test with constant maximum load, the maximum *strain* within the "vulnerable" zone will stay constant from cycle to cycle. The *actual stress* within the zone may vary widely with time, depending on the strain hardening or strain softening characteristics of the material. Hence, a meaningful study of low-cycle fatigue must deal with actual local strain rather than calculated nominal local stress. This "strain cycling" approach is beyond the scope of this book. (See references such as [3].) For our purposes, it is recommended that the full fatigue stress concentration factor K_f be used in all cases. For relatively short-life situations, this may be overly conservative (i.e., the effect of the stress concentration may be substantially less than K_f).

Another question should be considered here: Is it better to treat K_f as a stress concentration factor or as a strength reduction factor? Authorities differ on this point, but in this book K_f will be regarded as a *stress concentration factor*. Looking at Fig. 8.22, it would be easy to regard K_f as a strength reduction factor and calculate a "notched endurance limit" as equal to $S'_n C_L C_G C_S / K_f$. This would not be incorrect, but it has the disadvantage of implying that the *material itself* is weakened by the notch. It is not, of course—the notch merely caused higher local stresses. Also, by using K_f as a stress multiplier (instead of a strength reducer), *S-N* curves and constant life fatigue strength curves are independent of notch geometry, and the same curves can be used repeatedly for members with various stress raisers. Finally, when considering residual stresses caused by peak loads (as in Fig. 4.43), it is necessary that K_f be regarded as a stress concentration factor.

8.11 INFLUENCE OF STRESS CONCENTRATION WITH MEAN PLUS ALTERNATING LOADS

It was shown in Sections 4.14 and 4.15 that peak loads, causing calculated elastic stresses to exceed the yield strength, cause yielding and resultant residual stresses. Furthermore, the residual stresses always serve to lower the *actual* stresses when the same peak load is applied again. To illustrate the effect of residual stresses on fatigue life where mean, as well as alternating, stresses are involved, consider the examples developed in Fig. 4.43.

Suppose this notched tensile bar is made of steel, with $S_u = 450$ MPa, $S_y = 300$ MPa, and that its size and surface are such that the estimated constant life fatigue curves are as shown at the bottom of Fig. 8.24. The top of Fig. 8.24 shows a fluctuation of the notch stress calculated without taking yielding into account. The first three cycles correspond to the loading and unloading involved in Fig. 4.43*a*. The next two (dotted) cycles represent progressively increasing the load to that involved in Fig. 4.43*b*. Note that these dotted cycles show an elastically computed stress at the notch root of about $\frac{7}{6}S_y$. Correspondingly, the three cycles drawn with solid lines at (b) in Fig. 8.24 show calculated stresses fluctuating between zero (when the load is removed) and 400 MPa, which is $\frac{4}{3}S_y$. This procedure continues along the top of Fig. 8.24 until the condition shown in Fig. 4.43*d* is reached. Here, the calculated stress is zero when the load is off, and $2S_y$ when it is on.

Just below this in Fig. 8.24 is a corresponding plot of *actual* stresses at the notch root. It is based on the assumption that the ductile material used can be approximated (within the limited strain range involved) by a "flat-top" idealized stress-strain curve,

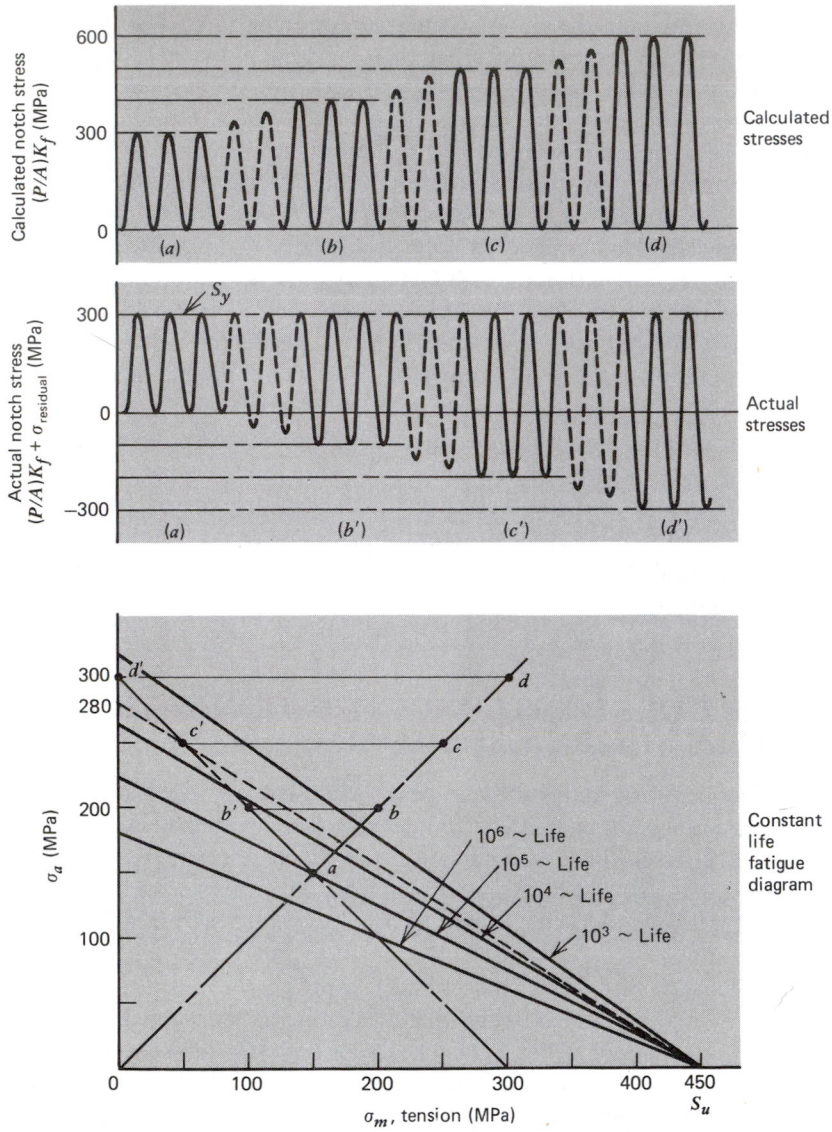

FIGURE 8.24.
Fatigue life estimation for repeated application of stresses shown in Fig. 4.43; with steel, S_u = 450 mPa, S_y = 300 mPa.

like the one drawn in Fig. 4.42*e*. No yielding occurs during the first three cycles at (a), but yielding *does* occur at the notch root when the calculated stress exceeds 300 MPa during the first dotted cycle. A little more yielding takes place during each cycle that the load goes higher than it had been previously. When the load is "on" during one of the cycles at (b), the stress distribution corresponds to the solid line in the left drawing of Fig. 4.43. When the load is "off," the stresses are not zero but correspond to the residual stress pattern in the right-hand drawing of Fig. 4.43. At the notch, these stresses range from S_y when the load is on to the residual stress of $-S_y/3$ when the load is off. This procedure continues along the "actual notch stress" plot of Fig. 8.24 until the condition shown in Fig. 4.43*d* is reached. Here, the actual notch root stress is S_y when the load is applied, and a residual stress of $-S_y$ when the load is off.

At the bottom of Fig. 8.24, stresses resulting from "on and off" application of the Fig. 4.43 loads are represented in comparison to the fatigue strength characteristics of the material. Points *a, b, c,* and *d* correspond to the *calculated* notch-root stresses (which, because of the yielding and residual stress, have little significance). Points *a, b′, c′,* and *d′* correspond to *actual* stresses (based on an ideal stress-strain curve) and are fairly realistic. Note that in every case, yielding has reduced the *mean* stress; it does *not* change the *alternating* stress.

Based on the plot in Fig. 8.24, the estimated fatigue lives corresponding to repeated application of the various levels of tensile loading would be 10^5 cycles for loading (a), perhaps 1½ or 2 times 10^4 cycles for loading (b), about 6×10^3 cycles for (c), and about $2\frac{1}{2} \times 10^3$ cycles for (d). These estimates represent rough visual interpolations between adjacent constant life lines. The dotted line through point *c′* illustrates a better procedure. This line is a "Goodman line" corresponding to some unknown life. All points on the line correspond to the *same* life; in particular, point *c′* corresponds to the same life as a completely reversed stress of 280 MPa. Now, we can go to the *S-N* curve (not shown) and take from it the life corresponding to 280 MPa. To keep the matter of life predictions in proper perspective, remember that such predictions are inherently very rough, except on a statistical basis—as was illustrated by the scatter band of the *S-N* curve in Fig. 8.4*c*. Also, do not forget the previously mentioned limitations of the procedure for making predictions in the low-cycle range.

In the above example the stress concentration factor of 2, originally used in Fig. 4.43, was taken as a *fatigue* stress concentration factor in Fig. 8.24. Assuming that the material has a notch sensitivity, *q*, of something less than unity, the theoretical stress concentration factor, K_t, would be greater than 2.

Life predictions like the above can be made very conveniently from fatigue strength diagrams in the form of Figs. 8.18–8.20. On these diagrams, points *a, b, c,* and *d* would lie on the vertical axis (i.e., $\sigma_{\min} = 0$), and points *b′, c′,* and *d′* would lie on the horizontal line, $\sigma_{\max} = S_y$.

At first, one may be inclined to become a little alarmed with the appearance of points like *b, c,* and *d* on the σ_m-σ_a plot. Even point *c*—let alone point *d*—shows a peak stress in excess of the ultimate strength! One must remember that these are fictitious calculated stresses, and that the extent of yielding they represent is usually very small. With the tensile bar in this example, there is no way that *very much* yielding can occur at the notch root without yielding the entire cross section—and this is only on the verge of happening at point *d*.

In summary, the procedure recommended here for fatigue life prediction of notched parts subjected to combinations of mean and alternating stress is:

All stresses (both mean and alternating) are multiplied by the fatigue stress concentration factor, K_f, and correction is made for yielding and resultant *residual stresses if the calculated peak stress exceeds the material yield strength.*

This procedure is sometimes called the *residual stress method* because of the recognition it gives to the development of residual stresses. An alternative procedure sometimes used is to apply the stress concentration factor to the alternating stress *only,* and *not* take residual stresses into account. One can see that in *some* cases this reduction in mean stress due to not multiplying it by K_f might be about the same as the reduction in mean stress achieved with the residual stress method by taking yielding and residual stress into account. Because of not multiplying the mean stress by a stress concentration factor, this alternative procedure is sometimes called the *nominal mean stress method.*

SAMPLE PROBLEM 8.3 Determine the Required Diameter of a Shaft Subjected to Mean and Alternating Torsion

A shaft must transmit a torque of 1000 N·m, with superimposed torsional vibration causing an alternating torque of 250 N·m. A safety factor of 2 is to be applied to both loads. A heat-treated alloy steel is to be used, having $S_u = 1.2$ GPa, and $S_y = 1.0$ GPa (unfortunately, test data are not available for S_{us} or S_{ys}). It is required that the shaft have a shoulder, with $D/d = 1.2$ and $r/d = 0.05$ (as shown in Fig. 4.35). A good quality commercial ground finish is to be specified. What diameter is required for infinite fatigue life?

Solution

1. Before even beginning to solve a problem like this, an engineer should carefully review the design with regard to the critical fillet. Is it really necessary that the radius be so small? If so, is the quality control in the production and inspection departments such that the part will not be made with merely a "sharp corner"? And what about the control of surface finish? As far as shaft fatigue strength is concerned, a high quality finish *in the fillet* is very important. Will the production and inspection departments be aware of this? The other 99.9 percent of the shaft surface is of little consequence unless a high quality finish is needed for other reasons (as to provide a good bearing surface or close-tolerance fit). If the quality finish is not needed on these other portions of the shaft, cost might be saved by changing to an ordinary machined surface. For this example, let us assume that the shaft *should* be and *will* be made as specified.
2. Construct the fatigue strength diagram shown in Fig. 8.25. (Since infinite life is required, there is no need for an *S-N* curve.) In computing an estimated value for S_n it was necessary to *assume* that the diameter will come out to be between 10 and 50 mm. If it does not, the solution will have to be repeated with a more appropriate value of C_G.

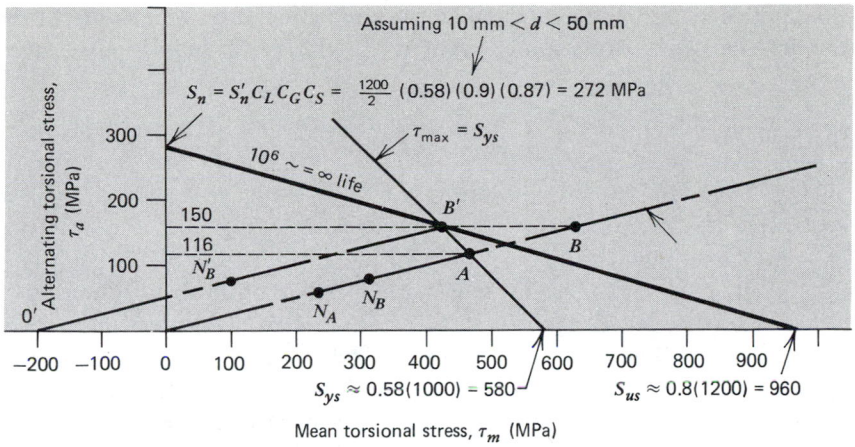

FIGURE 8.25.
Fatigue strength diagram for Sample Problem 8.3.

3. The *calculated* notch-root stresses (i.e., not yet taking any possible yielding into account) are:

$$\tau_m = (16T_m/\pi d^3)K_f$$
$$\tau_a = (16T_a/\pi d^3)K_f$$

In order to find K_f from Eq. (8.2), K_t and q must first be determined. K_t is readily found from Fig. 4.35*c* as 1.57, but the determination of q from Fig. 8.23 again requires an assumption of the final diameter. This presents little difficulty, however, as the curve for torsional loading of steel of this strength (S_u = 1.2 GPa = 174 ksi, or very close to the top curve on the figure) gives $q \approx 0.95$ for $r \geq 1.5$ mm, which in this case corresponds to $d \geq 30$ mm. With the given loading, intuition (or subsequent calculation) tells us that the shaft will have to be at least this large. Substitution of these values, together with the given values for design overload (nominal load times safety factor), gives:

$$K_f = 1 + (K_t - 1)q = 1 + (1.57 - 1)0.95 = 1.54$$

$$\left.\begin{aligned} \tau_m &= [(16 \times 2000\ \text{N}\cdot\text{m})/\pi d^3]\ 1.54 = 15{,}685/d^3 \\ \tau_a &= [(16 \times 500\ \text{N}\cdot\text{m})/\pi d^3]\ 1.54 = 3{,}922/d^3 \end{aligned}\right\} \frac{\tau_a}{\tau_m} = 0.25$$

4. Starting at the origin of Fig. 8.25 (which corresponds to making the diameter infinite) and moving to the right along the line of slope = 0.25, we tentatively stop at point *A*. If no yielding is to be permitted, the stresses can go no higher than this. At *A*, τ_a = 116 MPa or 0.116 GPa. Thus, $3{,}922/d^3 = 0.116$ or $d = 32.2$ *mm*.

5. In most cases, perhaps a little yielding in the localized zone of the fillet under "design overload" conditions could be permitted. If so, the diameter can be further reduced until the *calculated* stresses reach point *B* on Fig. 8.25, because yielding and residual stresses bring the *actual* stresses back to point *B'*, which is right on

the infinite life line. Yielding did not affect the alternating stress magnitude, so the equation for alternating stress can be equated to 150 MPa, giving $d = 29.7$ *mm.*

Before accepting either the $d = 32.3$ mm or the $d = 29.7$ mm answer, it is important to go back and see if the assumed values for C_G and q are consistent with the diameter finally chosen. In this case they are. ■

Before leaving this example, it is interesting to note on Fig. 8.25 the operating stresses for *normal* operation (i.e., $T_m = 1000$ N·m, $T_a = 250$ N·m). If point A is selected as the required overload point, then normal operation involves operating at point N_A (mid-point between O and A). If B' is the selected overload point, normal operation would be at N_B, the mid-point between O and B. But if the machine is operated at the design overload and *subsequently* operated normally, a residual stress, represented by O', is involved. With this residual stress present, stresses are at O' when the load is off, at N'_B when the load is normal, and at B' with the design overload.

SAMPLE PROBLEM 8.4 Estimate the Safety Factor of a Disk Sander Shaft

Figure 8.26 pertains to the shaft of a disk sander that is made of steel having $S_u = 900$ MPa, and $S_y = 750$ MPa. The most severe loading occurs when an object is held near the periphery of the disk (100 mm radius) with sufficient force to develop a friction torque of 12 N·m (which approaches the stall torque of the motor). Assume a coefficient of friction of 0.6 between the object and the disk. What is the safety factor with respect to eventual fatigue failure of the shaft?

Solution

1. Looking first at design details relating to shaft fatigue, the relatively very large 5 mm radius is excellent for minimizing stress concentration at this necessary step in the shaft. It would be desirable to reduce the 50-mm disk overhang, but we will assume that for this particular application the overhang is necessary.
2. The 12-N·m torque specification requires that the tangential force, F_t, be 120 N. With a coefficient of friction of 0.6, this requires a normal force, F_n, of 200 N.
3. These two force components produce the following loading at the shaft fillet:

Torque: $T = 12$ N·m $= 12{,}000$ N·mm

Axial load: $P = 200$ N

Bending: In the horizontal plane, $M_h = 120$ N $\times$ 50 mm
in the vertical plane, $M_v = 200$ N $\times$ 100 mm
The resultant is $M = \sqrt{M_h^2 + M_v^2} = 20{,}900$ N·mm

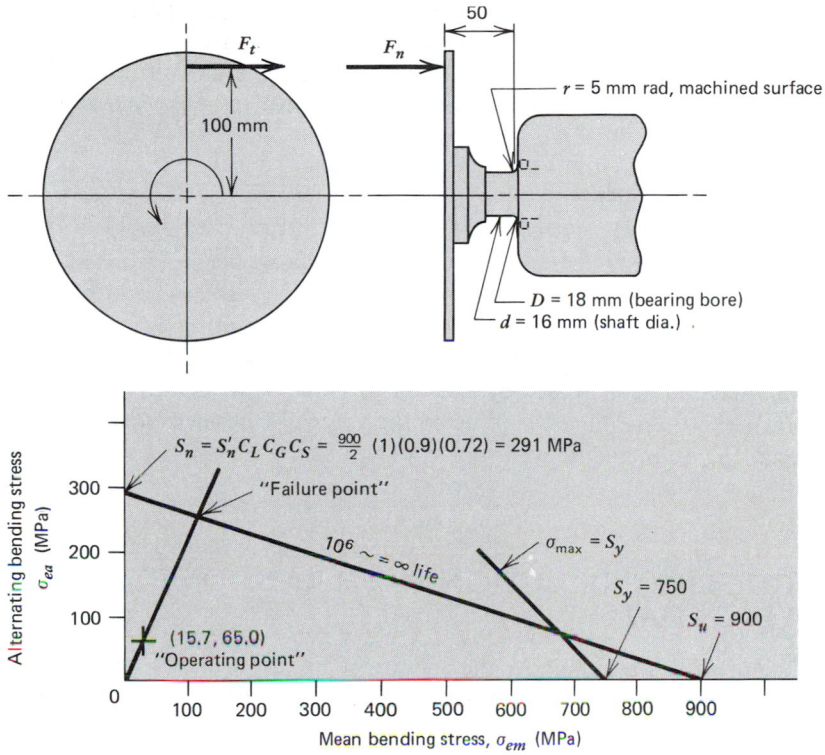

FIGURE 8.26.
Sample Problem 8.4—Disk sander.

4. From Fig. 4.35, geometric stress concentration factors for torsion, axial, and bending loads are about

$$K_{t(t)} = 1.10, \quad K_{t(a)} = 1.28, \quad K_{t(b)} = 1.28$$

From Fig. 8.23, estimated notch sensitivities, q, are 0.93 for torsion, and 0.91 for bending and axial loads. From Eq. (8.2), values of K_f are estimated as 1.09, 1.25, and 1.25 for torsional, axial, and bending loads, respectively.

5. The three stress components at the fillet are:

$$\tau = \frac{16T}{\pi d^3} K_{f(t)} = \frac{16(12{,}000)}{\pi(16)^3}(1.09) = 16.3 \text{ MPa}$$

$$\sigma_{(a)} = \frac{P}{A} K_{f(a)} = \frac{-200(4)}{\pi(16)^2}(1.25) = -1.24 \text{ MPa}$$

$$\sigma_{(b)} = \frac{32M}{\pi d^3} K_{f(b)} = \frac{32(20{,}900)}{\pi(16)^3}(1.25) = 65.0 \text{ MPa}$$

6. Applying the procedure specified for "general biaxial loads" in Fig. 8.17, an estimated infinite life Goodman line for *bending* loads is constructed in Fig. 8.26. Next, an "operating point" that corresponds to the *equivalent* mean and alternating bending stresses is placed on the diagram. Of the three stress components determined above, torsional and axial stresses are constant for steady-state operating conditions; the bending stress is completely reversed (the bending stress at any point on the fillet goes from tension-to-compression-to-tension during each shaft revolution). Using the recommended procedure to determine the equivalent mean and alternating stresses:

$$\sigma_{em} = \frac{\sigma_m}{2} + \sqrt{\tau^2 + \left(\frac{\sigma_m}{2}\right)^2} = \frac{-1.24}{2} + \sqrt{(16.3)^2 + \left(\frac{-1.24}{2}\right)^2} = 15.7 \text{ MPa}$$

$$\sigma_{ea} = \sqrt{\sigma_a^2 + 3\tau_a^2} = \sqrt{(65.0)^2 + 0} = 65.0 \text{ MPa}$$

7. By drawing a line through the origin and the "operating point" we see that all stresses would have to be increased by a factor of about 4 to reach the estimated "failure point" where conditions would be on the verge of causing eventual fatigue failure. Hence, the estimated safety factor is 4. ■

8.12 FATIGUE LIFE PREDICTION WITH RANDOMLY VARYING LOADS

Predicting the life of parts stressed above the endurance limit is at best a rough procedure. This point is illustrated by the typical scatter band of 7:1 life ratio shown in Fig. 8.4*c*. For the large percentage of mechanical and structural parts subjected to randomly varying stress cycle intensity (for example, automotive suspension and aircraft structural components), the prediction of fatigue life is further complicated. The procedure given here for dealing with this situation was proposed by Palmgren of Sweden in 1924 and, independently, by Miner of the United States in 1945. The procedure is often called the *linear cumulative damage rule,* with the names of Miner and/or Palmgren attached.

Palmgren and Miner very logically proposed the simple concept that if a part is cyclically loaded at a stress level causing failure in 10^5 cycles, then each cycle of that loading consumes one part in 10^5 of the life of the part. If other stress cycles are interposed corresponding to a life of 10^4 cycles, each of these consumes one part in 10^4 of the life, and so on. When, on this basis, 100 percent of the life has been consumed, fatigue failure is predicted.

The Palmgren or Miner rule is expressed by the following equation in which n_1, n_2, . . . , n_k represent the number of cycles at specific overstress levels, and N_1, N_2, . . . , N_k represent the life (in cycles) at these overstress levels, as taken from the appropriate *S-N* curve. Fatigue failure is predicted when

$$\frac{n_1}{N_1} + \frac{n_2}{N_2} + \frac{n_k}{N_k} = 1 \quad \text{or} \quad \sum_{j=1}^{j=k} \frac{n_j}{N_j} = 1 \tag{8.3}$$

Use of the linear cumulative damage rule is illustrated in the following example.

SAMPLE PROBLEM 8.5 Fatigue Life Prediction with Randomly Varying Completely Reversed Stresses

Stresses (including stress concentration factor, K_f) at the critical notch of a part fluctuate randomly as indicated in Fig. 8.27*a*. The stresses could be bending, torsional, or axial—or even equivalent bending stresses resulting from general biaxial loading. The plot shown represents what is believed to be a typical 20 seconds of operation. The material is steel, and the appropriate *S-N* curve is given in 8.27*b*. This curve is corrected for load, gradient, and surface. Estimate the fatigue life of the part.

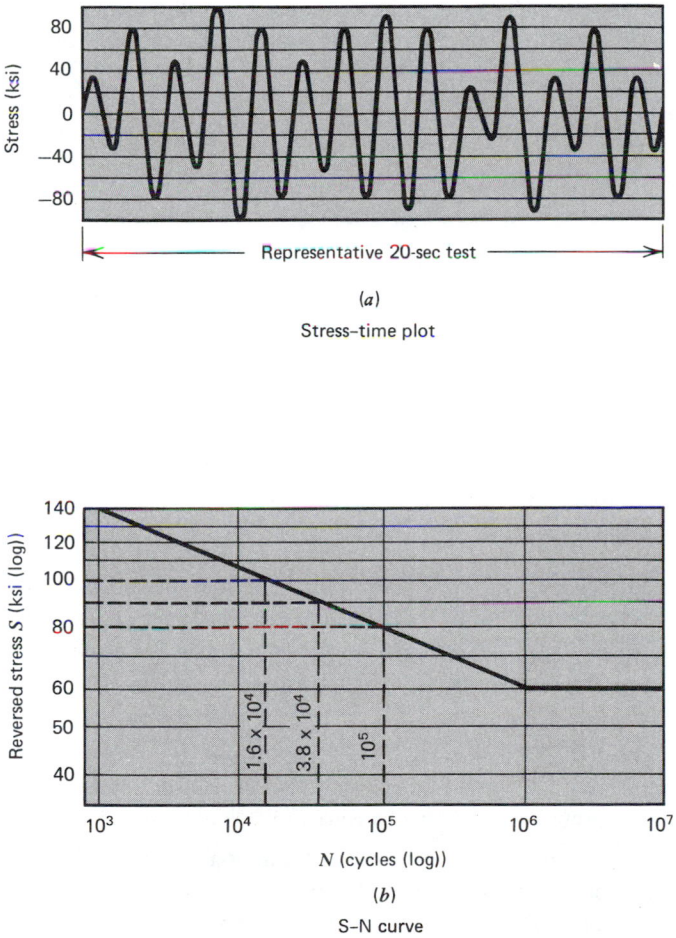

FIGURE 8.27.
Sample Problem 8.5—Fatigue life prediction, reversed stresses.

Solution

In Fig. 8.27*a* there are eight stress cycles above the endurance limit of 60 ksi: five at 80 ksi, two at 90 ksi, and one at 100 ksi. The *S-N* curve shows that each 80 ksi cycle uses one part in 10^5 of the life, each at 90 ksi uses one part in 3.8×10^4, and the one at 100 ksi uses one part in 1.6×10^4. Adding these fractions of life used gives

$$\frac{n_1}{N_1} + \frac{n_2}{N_2} + \frac{n_3}{N_3} = \frac{5}{10^5} + \frac{2}{3.8 \times 10^4} + \frac{1}{1.6 \times 10^4} = 0.0001651$$

For the fraction of life consumed to be unity, the 20-second test time must be multiplied by $1/.0001651 = 6059$. This corresponds to 2019 minutes, or *about 30–35 hours.* ■

The linear cumulative damage rule can easily be extended to problems involving mean as well as alternating stresses. The next example illustrates this for the case of fluctuating bending stresses. The procedure would be the same for a fluctuating *equivalent* bending stress, computed in accordance with the instructions for "general biaxial loads" on Fig. 8.17, and illustrated in Sample Problem 8.4.

SAMPLE PROBLEM 8.6 Fatigue Life Prediction – Randomly Varying Fluctuating Bending Stresses

Figure 8.28*a* represents the stress fluctuation at the critical notch location of a part during what is believed to be a typical 6 seconds of operation. The bending stresses plotted include the effect of stress concentration. The part is made of an aluminum alloy having $S_u = 480$ MPa, and $S_y = 410$ MPa. The *S-N* curve for bending is given in 8.28*c*. This curve is corrected for stress gradient and surface. Estimate the life of the part.

Solution

1. The 6-second test period includes, in order, two cycles of fluctuation *"a,"* three cycles of fluctuation *"b,"* two cycles of *"c,"* one cycle of *"d,"* and one of *"b."* Each of these fluctuations corresponds to a combination of mean and alternating stress plotted as a point in Fig. 8.28*b*. For example, *"a"* consists of $\sigma_m = 50$ MPa, $\sigma_a = 100$ MPa.
2. Points *"a"* through *"d"* on Fig. 8.28*b* are connected by straight lines to the point $\sigma_m = S_u$ on the horizontal axis. This gives a family of four "Goodman lines," each corresponding to some constant (but as yet unknown) life.
3. The four Goodman lines intercept the vertical axis at points a' through d'. According to the Goodman concept, points a through d correspond to exactly the same fatigue lives as points a' through d'. These lives are determined from the *S-N* curve in Fig. 8.28*c*. Note that the life corresponding to conditions a and a' can be considered infinite.

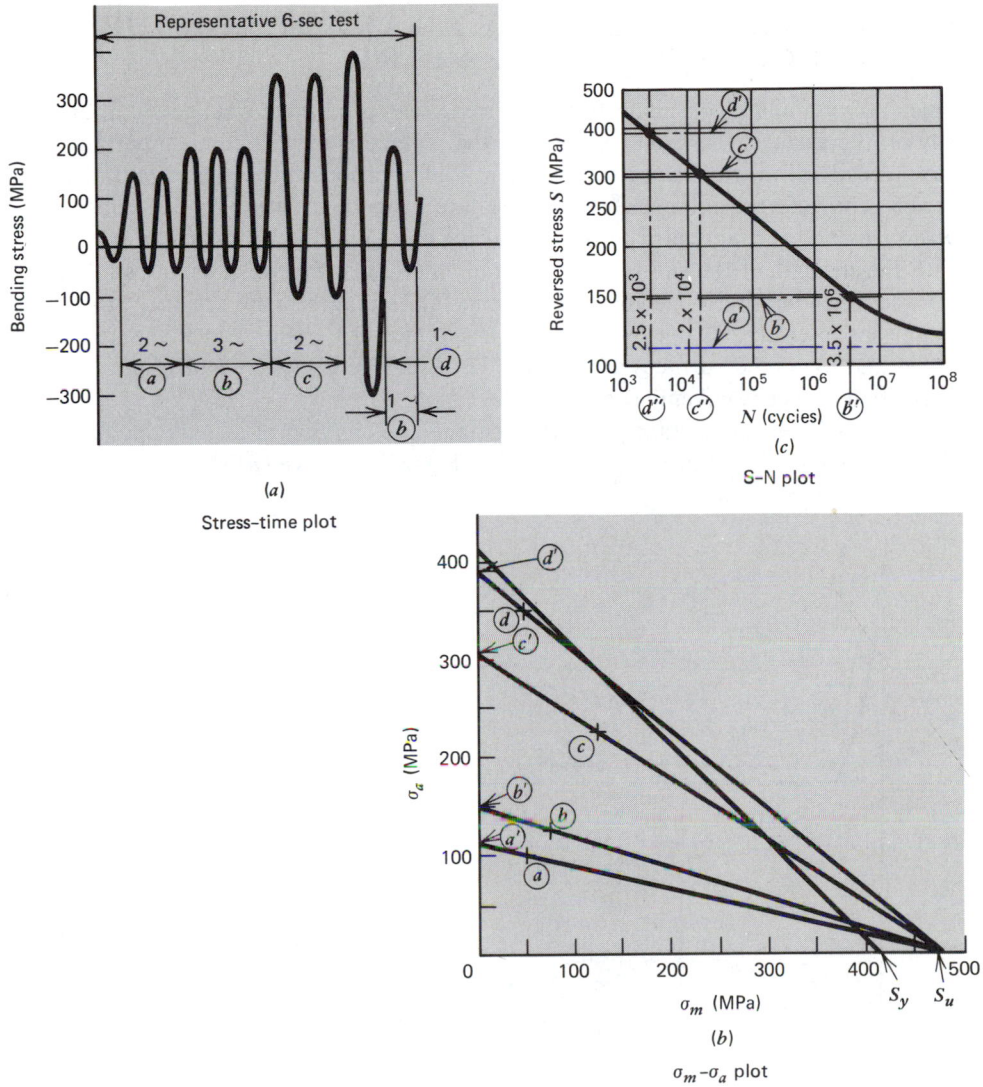

FIGURE 8.28.
Fatigue life prediction, randomly varying stresses (Sample Problem 8.6).

4. Adding up the portions of life consumed by overload cycles b, c, and d gives

$$\frac{n_b}{N_b} + \frac{n_c}{N_c} + \frac{n_d}{N_d} = \frac{4}{3.5 \times 10^6} + \frac{2}{2 \times 10^4} + \frac{1}{2.5 \times 10^3} = 0.0005011$$

This means that the estimated life corresponds to 1/0.0005011 or 1996 periods of 6-second duration. This is equivalent to 199.6 minutes, or about 3⅓ hours. ■

8.13 INFLUENCE OF SURFACE TREATMENTS ON THE FATIGUE STRENGTH OF A PART

Since fatigue failures originate from localized areas of relative weakness that are usually on the surface, the local condition of the surface is of particular importance. We have already dealt with the surface constant C_S for various categories of *finishing* operations. This and the following two articles are concerned with various surface *treatments,* giving specific regard to their influence on (1) surface strength, in comparison with the strength of the sub-surface material, and (2) surface residual stress. All three surface considerations—geometry (smoothness), strength, and residual stress—are somewhat interrelated. For example, the low values of C_S shown in Fig. 8.13 for hot-rolled and as forged surfaces are due partly to surface geometry and partly to decarburization (hence, weakening) of the surface layer.

The influence of surface strengthening and the creation of a favorable (compressive) residual surface stress are illustrated in Fig. 8.29. Curve (*a*) shows the stress gra-

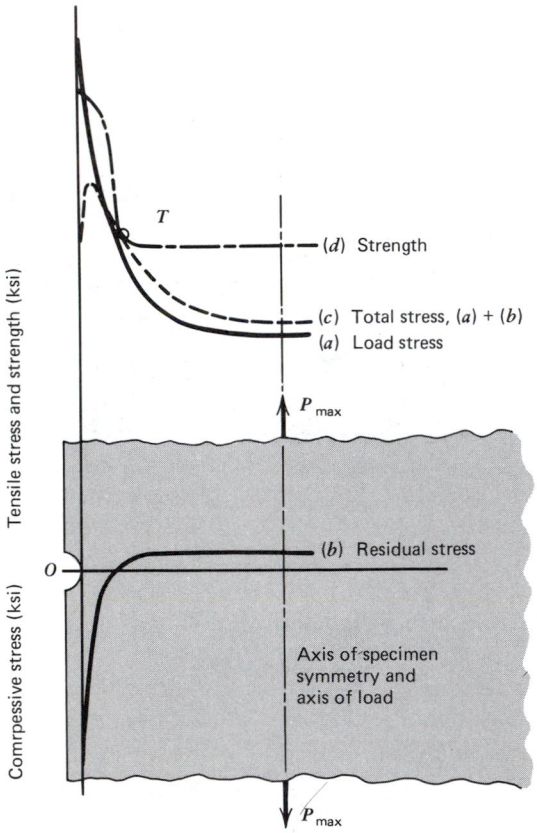

FIGURE 8.29.
Stress and strength gradients, surface-strengthened notched part subjected to axial load.

dient in the vicinity of a notch, due to a tensile load. Curve (*b*) shows a desirable residual stress gradient, giving compression in the groove surface. Curve (*c*) shows total stress, the sum of (*a*) and (*b*). Curve (*d*) shows a desirable *strength* gradient, resulting from a treatment causing surface strengthening. Note that (1) the surface strengthening and compressive residual stresses have substantially increased the load that can be carried, and (2) the point of potential failure origin has been moved under the surface to point *T*, where curves (*c*) and (*d*) are tangent. This means that the surface of the groove could deteriorate somewhat in service (as by corrosion, surface scratches, etc.) without reducing the load carrying capacity of the part. A further benefit to fatigue strength that is not evident in Fig. 8.29 is that compressive residual stresses shift the "operating point" on the mean stress-alternating stress diagram (like points *a–d* in Fig. 8.28*b*) to the left, which increases fatigue life.

The concept of comparing strength gradients with total stress gradients provides ready explanation of the fact that parts having steep load stress gradients and poor surfaces (low values of C_S) benefit most from surface treatment. Unnotched parts with axial loading benefit very little unless they have poor initial surface finishes. Severely notched parts loaded in bending or torsion benefit most. Since practically all parts have critical areas with stress concentration, surface strengthening treatments are usually quite effective. For example, Fig. 8.29 illustrates about a 60 percent increase in allowable load stress due to the combination of surface strengthening plus residual stress. For applications involving bending or torsional loads, it is not uncommon for the fatigue load capacity to be more than doubled.

The next two sections discuss processes for surface strengthening. It is also important to be aware of processes causing surface *weakening*. Grinding, for example, if not performed carefully and with slow or moderate feed rates can cause harmful surface residual tensile stresses, and even minute surface cracks. Chrome and nickel plating, even though good for the surface in providing corrosion protection, can substantially reduce the endurance limit of steel parts by causing hydrogen gas to be adsorbed. This is known as *hydrogen embrittlement*. This damage can be minimized by taking special care, as in using low plating current densities and baking the parts (usually in the range of 600° to 900°F) after plating. Properly performed, electroplating steel parts with soft metals like copper, cadmium, zinc, lead, and tin causes little if any fatigue weakening. Relatively little information is available on the effect of electroplating and anodizing non-ferrous metals. Both beneficial and harmful effects have been reported in specific instances. Welding and flame-cutting operations tend to produce harmful surface residual tensile stresses, unless special precautions are taken (as subsequent thermal stress-relieving).

The following listing of a few basic principles may help to place the subject of fatigue-strengthening surface treatments in proper perspective.

The engineer concerned with the design and development of machine and structural components subjected to dynamic loads should:

1. Seek to identify all local areas of stress concentration where fatigue failures could conceivably start.

2. Review possibilities for modifying the design to reduce stress concentration and/or to move the stress raiser to an area of lower nominal stress.
3. Pay particular attention to the surface finish (C_S) *in these areas.*
4. Consider what can be done in the manufacture of the part to strengthen the surface layer and provide a compressive residual stress at potentially critical stress raisers.

8.14 MECHANICAL SURFACE TREATMENTS (SHOT PEENING, ETC.)

Mechanical surface treatments cold work the surface material so as to cause compressive residual stresses and, depending on the properties of the material, often produce strain strengthening. Also, the surface geometry (smoothness) is altered—usually for the better unless the surface was initially polished or fine-ground.

The most common and versatile of the cold-working treatments is *shot peening.* It is widely used with springs, gears, shafts, connecting rods, and many other machine and structural components. Shot peening involves bombarding the surface with high-velocity iron or steel shot discharged from a rotating wheel or pneumatic nozzle. The resulting lightly hammered or *peened* effect tends to reduce the thickness and therefore increase the area of the exposed skin. Since the area increase is resisted by sub-surface material, the skin is placed in residual compression. The thickness of the compressive layer is usually less than a millimeter. The highest compressive stresses occur slightly below the surface and are commonly of the order of half the yield strength. Sometimes greater compressive residual stresses are obtained by loading the part in tension while it is being peened. This is called *strain peening.*

With steel parts, shot peening is more effective with harder steels because the yield strength is a greater percentage of the ultimate strength. This means that the resulting residual stresses are less easily "wiped out" by subsequent load stresses that cause the total (load plus residual) stress to exceed the yield strength. With reference to the relationship illustrated in Fig. 8.6, endurance strength increases with hardness up to substantially higher values with shot peening. Machine parts made of very high-strength steels (tensile strengths above about 1400 MPa or 200 ksi) are particularly benefited.

A related mechanical surface treatment is *cold rolling.* This usually involves rotating the part while suitable contoured rollers are pressed against the surface to be strengthened, as a shaft fillet or groove. This can create compressive residual stresses to a depth of a cm or more. Cold rolling has been applied to parts of all sizes, including large railway crankpins and axles of diameters up to 400 mm. Cold rolling is particularly effective for fatigue-strengthening shafts used with pressed-fit hubs (this helps to compensate for the high stress concentration in the shaft at the edge of the hub).

The fatigue-strengthening advantages of cold rolling are sometimes obtained as a by-product of a roll-forming operation. Under enough pressure, and with a suitable material, screw threads, shaft splines, and even fine gear teeth can be formed by cold rolling. The properties of the material then reflect the severe cold working. Also, residual compressive stresses are usually created.

Coining is another cold-forming operation that increases fatigue strength. An example is the pressing of a cone or oversize ball into the surface at the end of a hole, leaving a residual compressive stress at the vulnerable intersection of the hole and surface. Another example is cold-pressing round grooves into a shaft on both sides of a transverse hole.

In the absence of specific data, *it is usually conservative to account for the influence of shot peening or other cold-working treatments by using a surface factor C_S of unity, regardless of the prior surface finish.*

8.15 THERMAL AND CHEMICAL SURFACE-HARDENING TREATMENTS (INDUCTION HARDENING, CARBURIZING, ETC.)

The purpose of these treatments is usually to provide surfaces with increased resistance to wear; however, they also serve to increase fatigue strength, and for this reason are considered here.

The strictly thermal processes of flame and induction hardening of steel parts containing sufficient carbon produce surface residual compressive stresses (due to a phase transformation tending to slightly increase the *volume* of the surface layer) as well as surface hardening. As expected, maximum benefits are obtained with notched parts having steep applied stress gradients. In such cases, fatigue strengths can often be more than doubled.

Carburizing and nitriding are examples of chemical-thermal processes that add carbon or nitrogen to the surface layer, together with appropriate heat-treatment. The resulting hardened skin (or "case"), together with surface residual compressive stresses, can be very effective in increasing fatigue strength. In fact, nitriding has been found capable of rendering parts nearly immune to weakening by ordinary stress raisers. This point is illustrated by the following example given by Floe (Part 2, Section 8.6 of [5]).

	Endurance Limit (ksi)	
Geometry	**Nitrided**	**Not Nitrided**
Without notch	90	45
Half-circle notch	87	25
V notch	80	24

REFERENCES

1. American Society for Testing and Materials, *Achievement of High Fatigue Resistance in Metals and Alloys* (Symposium), American Society for Testing Materials, Philadelphia, 1970.
2. Boyer, H. E. (ed.), *Metals Handbook No. 10: Failure Analysis and Prevention,* 8th ed., American Society for Metals, Metals Park, Ohio, 1975.

3. Fuchs, H. O., and R. I. Stephens, *Metal Fatigue in Engineering,* John Wiley & Sons, Inc., New York, 1980.

4. Graham, J. A. (ed.), *Fatigue Design Handbook,* Society of Automotive Engineers, Inc., New York, 1968.

5. Horger, O. J. (ed.), *ASME Handbook: Metals Engineering—Design,* 2d ed., McGraw-Hill, New York, 1965.

6. Juvinall, R. C., *Engineering Considerations of Stress, Strain, and Strength,* McGraw-Hill, New York, 1967.

7. Lipson, C., and R. C. Juvinall, *Handbook of Stress and Strength,* Macmillan, New York, 1963.

8. Madayag, A. F., *Metal Fatigue: Theory and Design,* John Wiley & Sons, Inc., New York, 1969.

9. Sines, G., and J. L. Waisman (eds.), *Metal Fatigue,* McGraw-Hill, New York, 1959.

CHAPTER NINE

Surface Damage

9.1 INTRODUCTION

Previous chapters have dealt with damage occurring within the body of a part (yielding, fracture, excessive deflection, buckling). In addition, various kinds of damage can occur to the *surface* of a part, which render it unfit for use. To begin the list, the surface may corrode, either in a normal atmosphere or in other, usually more corrosive, environments such as salt water. Surface corrosion may combine with static or fatigue stresses to produce a *more* destructive action than would be expected by considering the actions of corrosion and stress separately. High relative velocities between solid parts and liquid particles can cause *cavitation* of the liquid, which may be destructive to the surface of the part. When two solid members are pressed together, *contact stresses* are produced, and these require special consideration. When the members are in *sliding* contact, several types of deterioration can occur that come under the general heading of *wear*. The severity of wear can be reduced by using a *lubricant* (as an oil, grease, or solid film) between the rubbing surfaces.

Experience indicates that more machine parts "fail" due to surface deterioration than by breakage. In an automobile, for example, consider the range of surface damage represented by corroded exhaust systems, rusted body panels, and worn piston rings, suspension joints, and other rubbing parts (not to mention upholstery and floor covering).

The economic cost of surface damage underscores its importance. In 1978 the National Bureau of Standards reported to Congress that the estimated total annual costs of corrosion and wear in the U.S. were $70 billion and $20 billion, respectively.[1]

[1]Bennett, L. H., *Economics Effects of Metallic Corrosion in the U.S.*, a Report to Congress, National Bureau of Standards, March 1978.

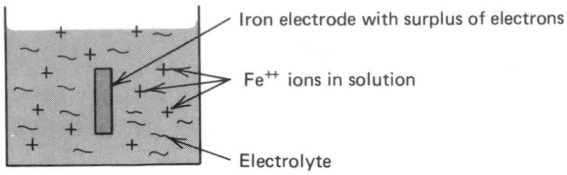

FIGURE 9.1.
Iron and electrolyte in equilibrium (no current flowing).

9.2 CORROSION – FUNDAMENTALS

Corrosion is the degradation of a material (normally a metal) by chemical or electro-chemical reaction with its environment. Most corrosion results from electro-chemical, or *galvanic* action. This is a complex phenomenon, giving rise to the specialized discipline of *corrosion engineering.*

Figure 9.1 shows an *electrode* of chemically pure iron in a homogeneous *electrolyte* (ion carrier, as fresh water, salt water, moist atmosphere, mud, etc.). Positively charged Fe^{++} ions go into solution, leaving an excess of electrons (i.e., a negative charge) on the iron electrode. When an equilibrium *electrode potential* is reached, no further electro-chemical action takes place.

In Fig. 9.2, the iron electrode is incompletely plated, leaving exposed iron, as shown. If the plating is tin, for example, the plating tends to lose positively charged ions, with a corresponding negative charge remaining in the tin coating. At the same time, the exposed iron surface tends to lose positive iron ions, leaving a negative charge in the body of the iron. Reference to Table 9.1 shows that iron (no. 17 in the table) is more active than tin (no. 14) in the galvanic series. This means that iron has the greater tendency to ionize, and develops the greater negative charge (electrode potential) in the body of the metal. Hence, an electrical current will flow within the metals, with electrons going from the iron (which becomes the *anode*) to the tin coating (now the *cathode*). A flow of ions through the electrolyte completes the circuit. This process involves a continuous discharge of iron ions; hence the iron *corrodes.* The current is in the direction to *prevent* a discharge of tin ions; thus the tin cathode does *not* corrode. This phenomenon is commonly observed in rusted "tin" cans, where breaks in the tin coating result in corrosion of the exposed steel. The iron ions discharged into the electrolyte

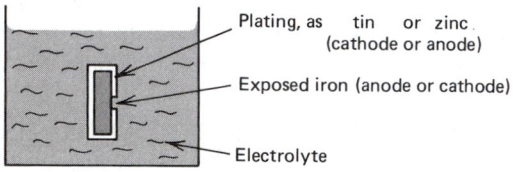

FIGURE 9.2.
Imperfectly plated iron in electrolyte. (Current flows continuously—direction depends on plating material.)

TABLE 9.1 Galvanic Series—Corrosion Compatibility Chart*

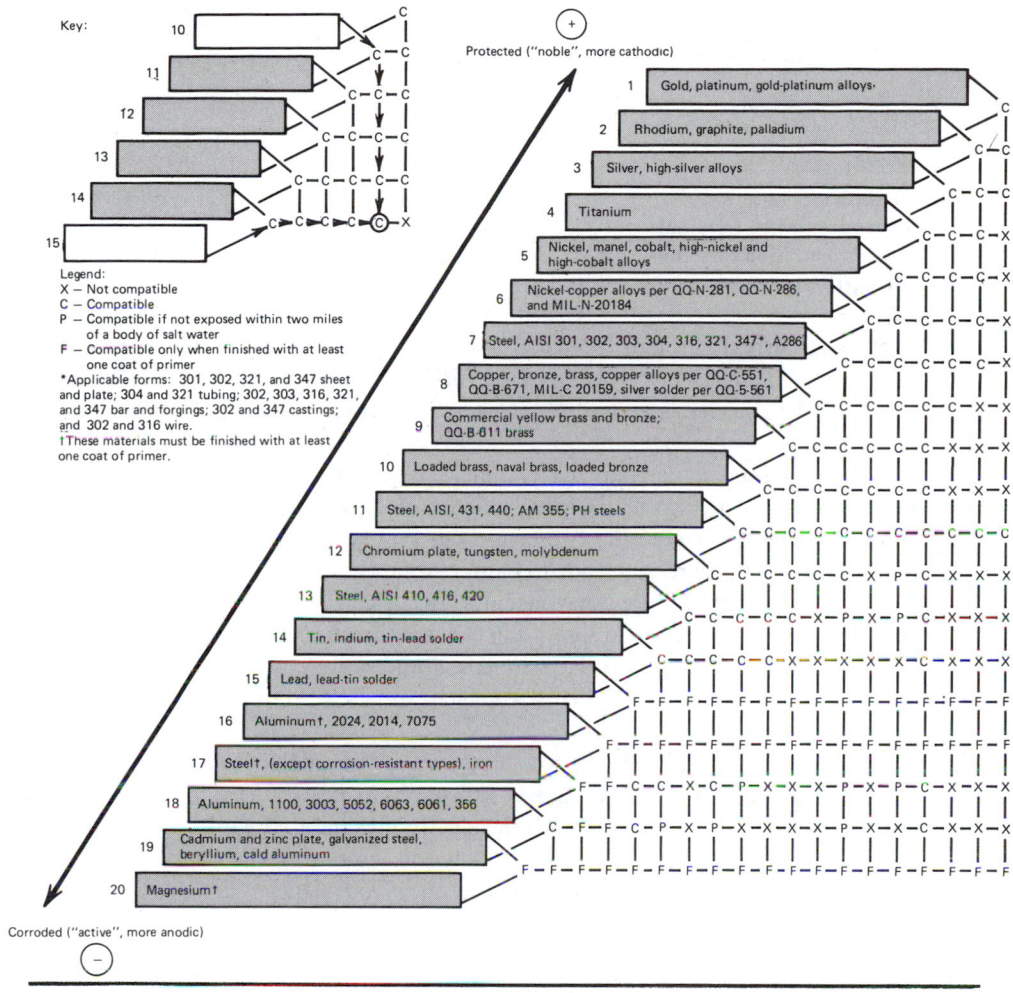

*From C.F. Littlefield and E. C. Groshart, Galvanic Corrosion, *Machine Design*, Vol. 35, p. 243 (May 9, 1963).

commonly combine with hydroxyl and oxygen ions and precipitate out as ferric hydroxide and ferric oxide, or *rust*.

Although the relative order of metals in the galvanic series is generally similar for most commonly encountered electrolytes, exceptions do occur. An interesting case in point is the previously mentioned "tin" can. The various acids, alkalies, and organic substances in canned foods provide electrolytes in which the inner steel surface of the can is *cathodic* with respect to tin, and is therefore protected. Furthermore, tin salts, which may be present in extremely low concentration due to corrosion of the tin, are nontoxic. Hence, tin plate is considered ideal for handling beverages and foods.

Suppose, in Fig. 9.2, that the plating is not tin but *zinc* (no. 19). Reference to Table 9.1 indicates that in ordinary environments the zinc will go in solution (corrode), and

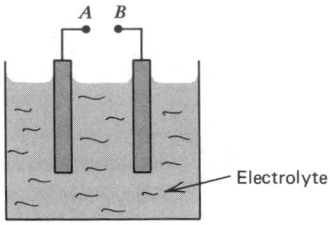

FIGURE 9.3.
Two electrodes, with terminals for external connection to a conductor or battery.

the direction of current through the electrolyte will oppose the discharge of positively charged iron ions. Thus, the iron becomes the *cathode,* and does not corrode. Zinc coating represents the common practice of *galvanizing* ferrous materials to protect them against corrosion.

Most corrosion phenomena involve two metal electrodes in contact, as the iron and zinc, or iron and tin, in Fig. 9.2. Closely related electrochemical phenomena can be illustrated with Fig. 9.3, where the two electrodes (*A* and *B*) are not in direct contact. Suppose these electrodes are zinc (no. 19) and carbon (graphite) (no. 2). Reference to Table 9.1 indicates that the greater concentration of electrons would be on the zinc. (Zinc has the greater tendency to lose positive ions, thereby leaving it with a negative charge with respect to the carbon). If terminals *A* and *B* are connected by a wire, electrons will flow through the wire. This is what happens in an ordinary carbon-zinc dry cell.

Again, referring to Fig. 9.3, suppose electrode *A* is iron (no. 17) and *B* is copper (no. 8), and that the two terminals are connected by a battery that *forces* a flow of positive charges from the copper, through the electrolyte, and to the iron. This is accomplished by copper ions (positive) going into solution, corroding the copper away, and then migrating to the iron electrode where they are deposited. This process is *electroplating.*

Figures 9.4 and 9.5 illustrate practical applications of the above principles in suppressing corrosion. In Fig. 9.4, the natural flow of galvanic current is such that the equipment to be protected is the *cathode,* and the zinc and magnesium plates are *sacrificial anodes.* (When these become depleted, they can be easily replaced.) In Fig. 9.5, an external direct voltage source forces a flow of electrons to the equipment to be protected, making it the cathode.

9.3 CORROSION – ELECTRODE AND ELECTROLYTE HETEROGENEITY

In the previous section it was assumed that the electrodes and the electrolyte were homogeneous. Many actual corrosion situations deviate substantially from this "ideal" condition, and these deviations markedly affect the nature and extent of the resulting

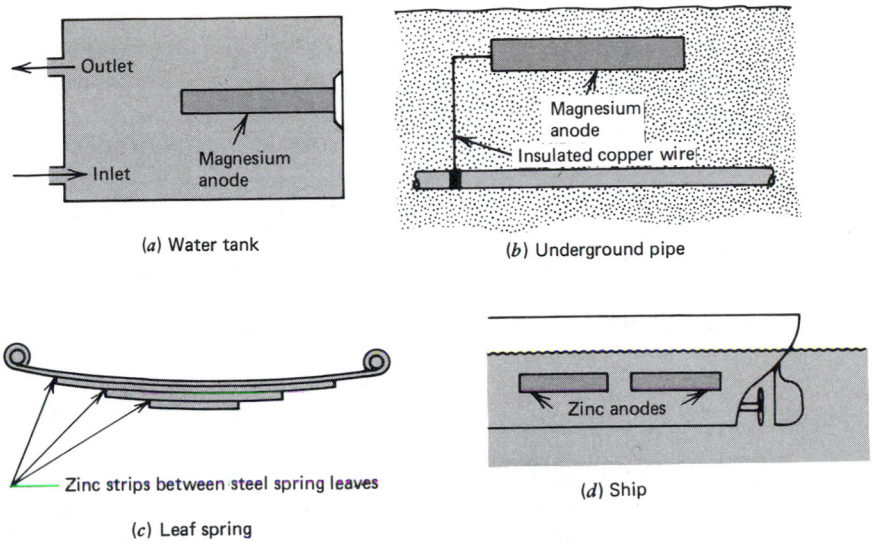

FIGURE 9.4.
Cathodic protection of steel using sacrificial anodes.

corrosion. For example, aluminum and titanium exposed to ordinary atmosphere form protective oxide films (Al_2O_3, and TiO_2) on their surfaces that electrically insulate the body of the material so that the corrosion current density is almost nil. This explains why aluminum can be used for boats, despite its position in the galvanic series (no. 18 in Table 9.1). Iron, chromium, nickel, titanium, and many of their important alloys exhibit the phenomenon of *passivation,* which means that insulating oxide films are maintained *in certain environments.* When the metal is in the *active* condition (without the oxide film), corrosion current densities are frequently 10^4 to 10^6 times those for the *passive* state.

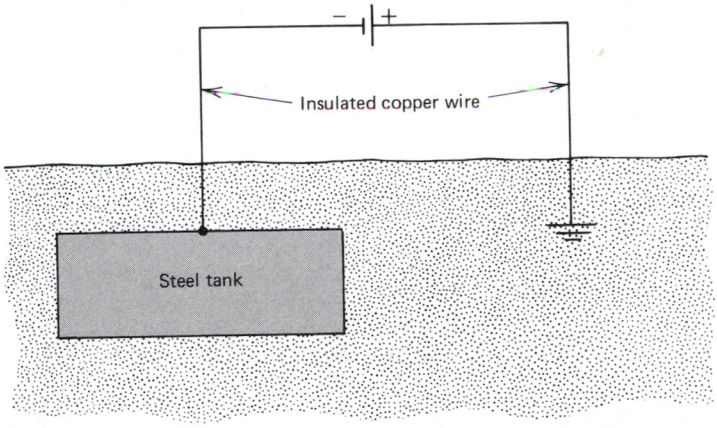

FIGURE 9.5.
Cathodic protection of underground tank using small d.c. impressed voltage.

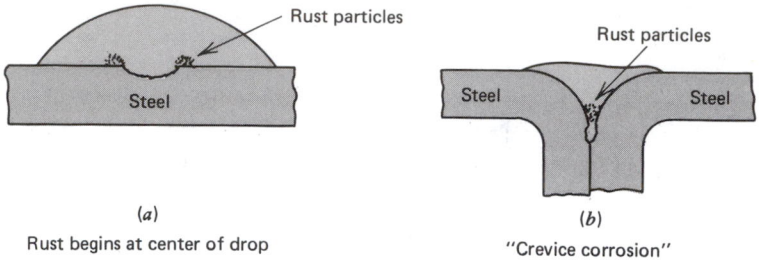

FIGURE 9.6.
Corrosion due to oxygen-starvation of electrolyte at center of droplet or stagnant puddle.

Another important form of electrode heterogeneity is the microstructure of the metal. For example, when pearlite is etched with a mild acid, the microstructure can be seen because the ferrite and carbide constituents become anodes and cathodes in a multitude of minute galvanic cells. Whichever is the anode corrodes, thus enabling it to be visually distinguished. (Ferrite and carbide have very close electrode potentials, and their relative position in the galvanic series depends on the electrolyte used.)

Another important cause of local galvanic action (and corrosion) is heterogeneity of the electrolyte. A common (and enormously costly!) example is the variation in composition of the salt-mud-water deposit under car bodies in the northern U.S., which sets up vigorous localized galvanic action. Another example is the corrosion of buried pipe, which passes through strata of soil of varying salt content.

Figure 9.6 illustrates two cases where a drop of electrolyte (usually water) creates electrolyte heterogeneity by sealing off atmospheric oxygen from the center of the drop, compared to the oxygen-rich electrolyte near the edge. Corrosion occurs at the oxygen-starved interior. Figure 9.6*b* illustrates what is commonly called *crevice corrosion.* A related observation that the reader may have made is that smooth surfaces tend to corrode less rapidly than rough surfaces.

9.4 DESIGN FOR CORROSION CONTROL

First, consider material selection. Materials toward the inactive, or "noble," end of the galvanic series for the electrolyte involved develop lower electrical potentials tending to send their ions into solution, and thus corrode more slowly. This is not the whole story, though, because, as noted above, the *rate* with which metal ions go into solution can be dramatically reduced by oxide films. Thus, aluminum corrodes extremely slowly in sea water, despite its position in the galvanic series. The same thing is true of stainless steels when, in most environments, these films change the metal's action from "active" to "passive."

Differences in microstructure due to heat-treating can influence the intensity of galvanic action within the multitude of surface galvanic cells. Certain low-alloy steels (called "weathering" steels) have been developed that effectively resist atmospheric corrosion through the formation of protective rust films. These steels are used primarily for architectural construction. After initial weathering, they have a uniform reddish-brown (rust) color, and galvanic action nearly stops. Because weathering steels need not be painted, their use often results in a substantial cost saving over the life of the structure.

Chemical treatment of metal surfaces may result in surface films that offer some degree of insulation of the base metal from the environment. Common examples are phosphate coatings on steel ("Parkerizing," or "Bonderizing") and oxide coatings on steel (brown, black, or blue in color). The effectiveness of these films is usually minimal, unless combined with rubbing with oils or waxes (as with gun barrels). Phosphate coatings on steel help provide good adherence of paint and decrease the tendency for corrosion to undercut the paint film at scratches or other discontinuities. *Anodizing* of aluminum alloys produces a stable aluminum oxide film that has excellent corrosion resistance, and that can be dyed in a variety of colors. (Common experience indicates that the protective oxide film on unanodized aluminum rubs off to produce a black stain.)

Metal surfaces can be sealed from possible contact with an electrolyte by nonporous coatings, such as procelain enamel applied to bathroom fixtures, or rubber coatings vulcanized to steel plates.

Ordinary paints provide a good barrier for the diffusion of water and oxygen to a metal surface, but are permeable to some extent. Because of this, and because paint films can become scratched and otherwise damaged, it is important to use an effective primer. The pigments in the prime coat applied to bare metal should be effective corrosion inhibitors. Any water reaching the metal surface then dissolves a small amount of primer pigment, making the water less corrosive. (The action of inhibitors is discussed later in this section.)

Where appearance and weight considerations are not stringent, it is sometimes more economical to make a part heavier than necessary to allow for anticipated future corrosion than it is to provide effective corrosion protection. Also, the impact of corrosion damage can often be reduced by designing equipment so that parts vulnerable to corrosion damage can be easily replaced.

The above discussion assumed that only a single metal was involved. Where minimum corrosion of two or more metals connected by an electrolyte is desired, the metals should be as close together as possible in the galvanic series. For example, cadmium is very close to aluminum in most environments; hence, cadmium-plated screws can be used in direct contact with aluminum parts. But if a copper washer is introduced, rapid corrosion of the aluminum normally occurs. In fact, aluminum has been found to corrode rapidly in rain water that previously contacted copper flashings or gutters. Other common examples of galvanic corrosion include the corroding of aluminum car door sills in contact with a steel body structure in a salt environment (from road salt in northern areas, or "salt air" near the coasts), steel pipe connected to aluminum plumbing, steel screws in brass marine hardware, and lead-tin solder on copper wire.

Table 9.1 shows the compatibility with respect to galvanic corrosion of commonly used metals and alloys when placed together in most atmospheric environments. Anodized aluminum is compatible with all other metals. The chart assumes that the two metals have comparable exposed areas. If one metal has only a small fraction of the

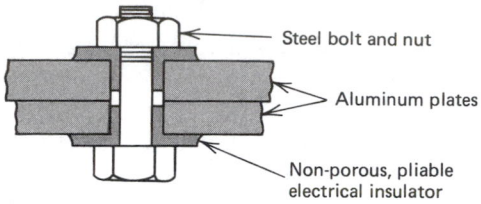

FIGURE 9.7.
Insulator breaking galvanic circuit between steel bolt and aluminum plates.

exposed surface of a second metal with which it is normally compatible, it may suffer galvanic attack. Also, where large temperature differences are involved, thermoelectric action may be a significant factor.

When it is necessary to use dissimilar metals together, the corroding galvanic circuit can often be broken by using an electrical insulator, as shown in Fig. 9.7.

Sometimes galvanic cells are deliberately designed into a component or system by providing a *sacrificial anode* to protect a metal cathode. This was illustrated by the zinc coating (galvanizing) of steel in connection with Fig. 9.2, and other examples in Fig. 9.4. In the United States alone, thousands of tons of magnesium are used annually as sacrificial electrodes.

Figure 9.5 showed an alternative method of cathodically protecting underground tanks or pipe. The cost of the external electrical energy consumed is an obvious factor limiting the application of this method.

An important factor in the design of metal parts to resist corrosion is the *area effect*. To minimize corrosion of the anode, its exposed surface area should be large in comparison to the cathode surface area. This results in a *low current density* at the anode; hence a low corrosion rate. The area effect is clearly illustrated by the examples depicted in Fig. 9.2. Where the plating was zinc (as in galvanized steel), the large zinc surface results in low current density and a low rate of corrosion. Where the plating was tin, the small area of exposed ferrous material results in rapid corrosion, as observed in discarded and rusted "tin" cans.

Design for corrosion control also requires careful attention to factors relating to the *electrolyte*. The electrolyte may be a liquid in which metal electrodes are totally immersed, or its function may be provided by exposure to spray or mist, by alternate wetting and drying, as by rain, by contact with moist earth, or merely from humidity in the atmosphere. It is important to eliminate or minimize crevices where corrosion can occur, as shown in Fig. 9.6*b*. Parts should be designed so that moisture can drain completely, without leaving residual liquid to encourage corrosion, as shown in Fig. 9.6*a*. This means that smooth butt-welded joints tend to corrode more slowly than bolted or riveted lap joints. Surfaces that are smooth tend to corrode less than rough surfaces, as they have less tendency to retain liquid electrolyte. Unavoidable crevices should be sealed, ideally with vulcanized rubber, or the equivalent.

Surfaces subject to retaining deposits of mud and salt (as auto underbody surfaces) should be designed to facilitate cleaning, as heterogeneous mixtures of soil and salt in contact with a metal can cause destructive galvanic action.

Electrolytes tend to cause the least galvanic action when stagnant. Agitation and temperature gradients tend to remove ion concentration near the electrode surfaces, thus increasing galvanic currents. Hence, temperature gradients and fluid velocities should be minimized.

Atmospheric corrosion tends to be greatest at high temperatures and high humidity. For example, corrosion rates of structural steel in tropical climates are reportedly at least twice that encountered in temperate climates.

In recirculating cooling systems, *inhibitors* (chemicals added to the liquid coolant in small concentration) are used to make the liquid a less effective electrolyte. These inhibitors act to encourage *passivation* (recall Section 9.3) of the metals subject to corrosion, and otherwise impede the movement of ions in the region of the electrodes. The inhibitor must be appropriate to the liquid and the metals involved.

For further study, Refs. 11 and 3 are particularly suggested. Refs. 1, 2, 7, and 12 also contain helpful information with regard to corrosion.

9.5 CORROSION PLUS STATIC STRESS

When static *tensile* stresses exist in a metal surface subjected to certain corrosive environments, the combined action can cause brittle cracking and fracture that would not be predicted on the basis of considering these two factors separately. Such cracks are called *stress-corrosion* cracks, and have been known to engineers since at least 1895, when it was noted that surface cracks developed on iron tires of wagon wheels after periods of exposure to humid atmosphere. These tires were subjected to residual tensile stresses due to their being forced onto the wheels with an interference fit.

Although the vulnerability of engineering metals to stress-corrosion cracking varies greatly, nearly all are susceptible to some degree and in some environments. Stress-corrosion cracking is a complex phenomenon not yet fully understood. Environments causing severe galvanic corrosion of a metal are not necessarily the same as those associated with serious stress-corrosion cracking of that metal. The relative resistance of materials to ordinary galvanic corrosion in a specific environment is not usually the same as the relative resistance of these materials to stress-corrosion cracking. For example, stress-corrosion cracking has been observed in certain stainless steels that are completely corrosion resistant when unstressed.

The stress associated with stress-corrosion cracking is *always* tensile, and is the sum of the residual and operating stresses existing at the *local site* of crack initiation and propagation. (Residual stresses are caused by the processing and assembly of parts, as opposed to operating stresses caused by the applied loads; see Sections 4.14 and 4.15.) This total stress necessary for stress-corrosion cracking is most often of the order of 50 to 75 percent of the tensile yield strength. Residual stresses alone can easily be this high.

Stress-corrosion cracks may occur after a period of time varying from a few minutes to several years, depending upon the corrosive environment and the surface tensile stress. Attempts to determine stress levels below which stress-corrosion cracks would never occur (analogous to the "endurance limit" for fatigue loading of ferrous mate-

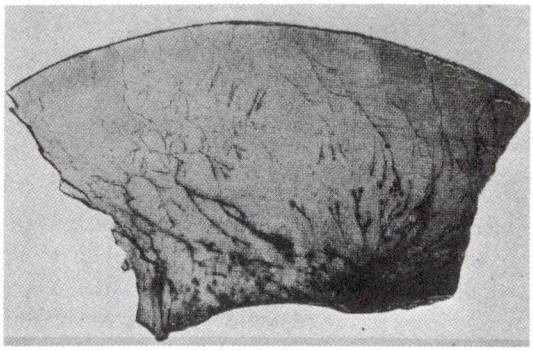

FIGURE 9.8.
Stress-corrosion cracking of a stainless-steel conveyor blade [4a].

rials) have not yet been successful. However, an increased understanding of stress-corrosion cracking failures is being gained through studies involving the application of fracture mechanics (see Sections 6.3 and 6.4).

When stress-corrosion cracking occurs, there are usually multiple cracks originating at the surface, with fracture resulting from a single crack propagating normal to the resultant tensile stress. The appearance of a typical failure is shown in Fig. 9.8.

An early reported example of stress-corrosion cracking was in riveted steam boilers, where alkaline boiler water in crevices around the rivets contacted boiler plate surfaces that were in tension (because of internal boiler pressure), with these stresses being accentuated by stress concentration. Explosions resulted in some cases. Also, mild steel boiler tubes operating with water containing sodium hydroxide have failed because of stress-corrosion cracking. Other examples of this type of failure include stainless steel aircraft parts exposed to sea coast salt environment (together with elevated operating temperatures in some cases); thin hubs or rims pressed onto an inner member with a heavy interference fit; bridge cables, where the steel is inappropriate for the chemicals in the local atmosphere; steel brackets that support heavy static weights; and "season cracking" of brass cartridge cases (deep-drawn cylindrical cases having residual tensile stresses in the sharp bottom corner, with cracking occurring after storage, with no applied loads).

The methods for reducing stress-corrosion cracking are primarily:

1. changing to a more stress-corrosion resistant material for the environment involved;
2. reducing corrosive action, as by providing cathodic protection, using protective coatings on the vulnerable surfaces, or making the environment less corrosive, as by adding inhibiters;
3. reducing tensile stress, as by reducing interference fits, using heavier sections, annealing (caution: in some cases this may render the material *more* susceptible to stress-corrosion cracking), and by shot-peening or hammer-peening vulnerable surfaces.

The latter alternative of peening is of particular practical importance, as this serves to overcome previous surface residual tension and impart helpful residual compressive stresses (see Section 8.14).

A summary of the stress-corrosion performance of several metals and alloys is given in [4a].

9.6 CORROSION PLUS CYCLIC STRESS

The combined action of corrosion and fatigue loading usually causes earlier failure than would be expected from a consideration of these two effects separately. This phenomenon is called *corrosion fatigue*. It occurs with most metals, but most markedly with those having low corrosion resistance. Corrosion fatigue is a complex action, not yet fully understood. A simplified explanation which has been proposed begins with initial corrosion pits that serve as points of stress concentration. Protective films, formed as a result of the corrosion, are usually weak and brittle. Hence, they are ruptured by the imposed cyclic strain. This rupture exposes unprotected metal, which quickly corrodes, forming another film, which is also ruptured by cyclic strain, and so on. Thus, the initial corrosion pit becomes a fatigue crack that propagates more rapidly than can be explained by considering corrosion and cyclic loading separately. As would be expected, corrosion fatigue failures show discoloration of the crack propagation surfaces, whereas ordinary fatigue crack surfaces are corrosion-free (like the surfaces characterized as "smooth, velvety" in Fig. 8.1).

The fatigue strength of corroding parts depends on the *elapsed time* involved, as well as the cyclic stress and the corroding environment. The fatigue strength for a given number of stress cycles is obviously greater if these cycles are imposed quickly, without allowing much time for corrosion to occur. Test results tend to support the following generalizations:

1. Corrosion fatigue strengths do *not* correlate with tensile strengths. This is likely because of, in part, the greater sensitivity to corrosion pits ("notches") of the stronger metals.
2. Medium-alloy steels have only slightly higher corrosion fatigue strengths than carbon steels, and in neither case is corrosion fatigue strength improved by heat-treatment.
3. Corrosion-resistant steels, such as those containing chromium, have higher corrosion fatigue strengths than other steels. *Good corrosion resistance is more important than high tensile strength.*
4. Residual tensile stresses are harmful; residual compressive stresses, such as those caused by shot-peening, are beneficial.

Remedial measures for corrosion fatigue are similar to those for stress-corrosion cracking: (1) use a more corrosion-resistant material, (2) reduce corrosive action by protective coatings, inhibitors, or cathodic protection, and (3) minimize tensile stresses, and introduce residual compressive stresses.

Much of the experimental data on corrosion fatigue strength of various metals that are still referred to was reported by D. J. McAdam, Jr.,[2] the investigator who originally coined the term "corrosion fatigue." Some of McAdam's results are summarized in [5] and [12].

Note that Fig. 8.13 gives surface factors for test specimens with *previously* corroded surfaces. These factors serve only as a rough guide, and do not apply at all to situations where the part is subjected to corrosion *while the cyclic stress is applied.*

9.7 CAVITATION DAMAGE

Cavitation is the formation of gas bubbles or "cavities" in a liquid that is moving with respect to a nearby solid surface. Bubbles are formed when the liquid pressure drops below its vapor pressure. When these bubbles subsequently collapse at or near the solid surface, pressure waves impinge upon the surface causing local stresses that can be high enough to cause plastic deformation of many metals. Often, metal damage is evident only after repeated bombardment of these pressure waves, much as in the case of damage due to ordinary metal fatigue.

Cavitation commonly occurs on ship propellers, centrifugal pump and turbine blades, and other surfaces where high local liquid velocities and large liquid static pressure gradients are encountered. The resulting damage to metal surfaces is essentially mechanical. In corrosive environments, however, cavitation can repeatedly damage or remove protective oxide films, thereby increasing galvanic action.

A surface area damaged by cavitation appears roughened, with closely spaced pits. In severe cases, enough material is removed to give the surface a spongy texture.

If it is not feasible to eliminate or reduce cavitation by modifying the liquid composition, velocity, flow pattern, or static pressure, then the most effective means of dealing with cavitation damage is usually to increase surface hardness. Stainless steel is often the most effective material available at reasonable cost. The following materials are listed in decreasing order of resistance to cavitation damage: stellite,[3] 18-8 cast stainless steel, cast magnesium bronze, cast steel, bronze, cast iron, and aluminum. This list is extracted from [3] and [7], which also list other materials.

9.8 TYPES OF WEAR

The previous sections in this chapter dealt with surface damage resulting from contact with fluids. The remaining sections discuss contact with another solid. In many cases, the resulting surface damage is classed as "wear."

The most common types of wear are *adhesive* wear and *abrasive* wear. These are treated in the next three articles. A third type is *corrosion film* wear, wherein the cor-

---

[2]D. J. McAdam, Jr., Corrosion Fatigue of Metals as Affected by Chemical Composition, Heat Treatment, and Cold Working, *Trans* ASTM, Vol. 11 (1927).

[3]Trademark of Union Carbide Corporation.

roded surface film is alternately removed by sliding and then reformed. A typical example is the wear that may occur on cylinder walls and piston rings of diesel engines burning high sulphur fuels. An important type of surface deterioration sometimes incorrectly classed as "wear" is *surface fatigue,* discussed in Section 9.14.

All forms of wear are strongly influenced by the presence of a *lubricant.* The information in Section 13.1 on types of lubricants and in Section 13.14 on boundary lubrication is relevant here. The wear rate for an unlubricated bearing can be 10^5 times that for boundary lubrication.

In typical well-designed machine components, the initial wear rate on rubbing surfaces during "run in" may be relatively high. As the more pronounced surface peaks are worn off causing the actual contact area to increase, the wear rate decreases to a small constant value. After a period of time the wear rate may again increase because of such factors as lubricant contamination or increased surface temperature.

Reference 8 contains 33 articles dealing with the various aspects of wear, and is an excellent reference for further study.

9.9 ADHESIVE WEAR

On a microscopic scale, sliding metal surfaces are never smooth. Although surface roughness may be only a few microinches (or a few hundredths of a micrometer), inevitable peaks (often called "asperities") and valleys occur, as shown in Fig. 9.9. Since the contact pressure and frictional heat of sliding are concentrated at the small local areas of contact indicated by the arrows, local temperatures and pressures are extremely high, and conditions are favorable for welding at these points. (*Instantaneous local* temperatures may reach the melting point of the metal, but with temperature gradients so steep that the part remains cool to the touch.) If melting and welding of the surface asperities (at the arrows in Fig. 9.9) do occur, either the weld or one of the two metals near the weld must fail in shear to permit the relative motion of the surfaces to continue. New welds (adhesions) and corresponding fractures continue to occur, resulting in what is appropriately called *adhesive wear.* Since adhesive wear is basically a *welding phenomenon,* metals that weld together easily are most susceptible. Loose particles of metal and metal oxide resulting from adhesive wear cause further surface wear because of abrasion.

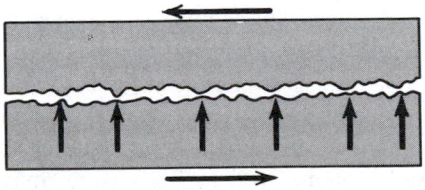

FIGURE 9.9.
Greatly enlarged view of two nominally "smooth" rubbing surfaces.

FIGURE 9.10.
Results of scoring and seizure in a differential. The broken pinion gear resulted from seizure to its shaft. (From Lipson, C., *Basic Course in Failure Analysis,* Penton Publishing Co., Cleveland, 1970.)

If the surface asperity welding and tearing cause a transfer of metal from one surface to the other, the resulting wear or surface damage is called *scoring.* If the local welding of asperities becomes so extensive that the surfaces will no longer slide on each other, the resulting failure is called *seizure.* Figure 9.10 shows severe scoring and seizure of differential pinion gears on their shafts. (This unit is similar to automotive differentials that allow the two driving wheels to turn at different speeds for going around corners.) Perhaps the best known examples of seizure occur in engines that continue to operate (but not for long!) after losing their liquid coolant or oil supply. Pistons may seize to the cylinder walls and/or the crankshaft may seize to its bearings.

Severe adhesive wear is also called *galling.* Mild adhesive wear between piston rings and cylinder walls is often called *scuffing.*

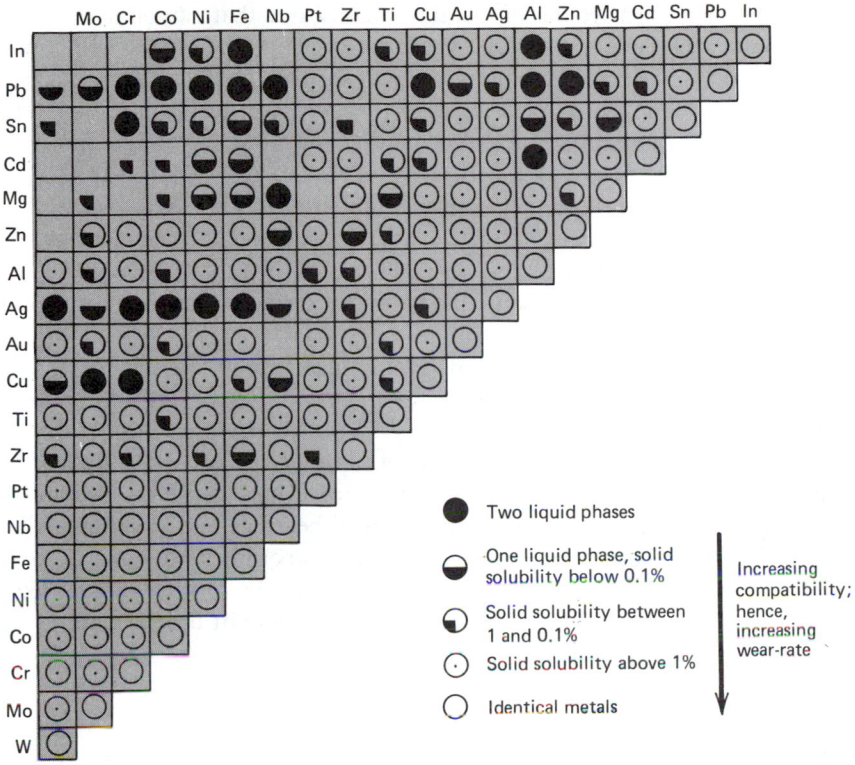

FIGURE 9.11.
Compatibility of various metal combinations. (From Rabinowicz, E., "Wear Coefficients—Metals," Section IV of [8].)

When like metals are rubbed together with suitable pressure and velocity, conditions are ideal for asperity welding because both surfaces have the same melting temperature. Also, the cohesive bonds thus formed are normally stronger than the adhesive bonds between dissimilar welded asperities. For these reasons, like or metallurgically similar metals should not normally be used together under conditions likely to cause wear problems. Metallurgically similar metals are referred to as "compatible." Compatible metals are defined as having complete liquid miscibility and at least 1 percent solid solubility of one metal in the other at room temperature. Figure 9.11 shows the degree of compatibility of various combinations of metals.

In general, the higher the surface hardnesses (more precisely, the higher the ratio of surface hardness to elastic modulus), the greater the resistance to adhesive wear.

9.10 ABRASIVE WEAR

The term "wear" most often refers to *abrasive* wear, which is due to the rubbing of abrasive particles on a surface. These particles are typically small, hard, and have sharp edges—like grains of sand, or particles of metal or metal oxide that rub off a wearing metal surface. Common examples include the wearing down of wood or metal with sand or emery paper or with a grinding wheel, the wearing of shoe leather by scuffing against cement sidewalks, the wearing of a plow blade or earth auger during use, and the removal of metal from the journal surface of a rotating shaft due to foreign abrasive particles in the lubricant.

Usually, the harder the surface the more resistant it is to abrasive wear. Hard metal surfaces are produced by heat-treatment, flame or induction hardening, carburizing, nitriding, electroplating, flame plating, or other means. Not all of these methods are applicable to severe applications because the hardened surfaces must sometimes be at least 3 mm thick to give adequate service life.

In machinery design, it is extremely important to use appropriate oil filters, air cleaners, dust covers, shaft seals, and so on to keep foreign particles away from rubbing metal surfaces.

Sometimes, one of a pair of rubbing members is made relatively soft and is designed to be easily and economically replaced. For example, hard surfaces on rotating shafts are protected by the use of softer, easily replaced bearings and bushings. It is sometimes desirable for the bearing to be sufficiently soft to permit hard abrasive particles to completely imbed so that they do not protrude above the surface and act like abrasive particles on sandpaper. This is one reason why soft babbitt bearings are used with automotive engine crankshafts.

9.11 FRETTING

Fretting, also known as *fretting corrosion,* is classed as a form of adhesive wear, but usually embraces elements of abrasive wear and corrosion film wear as well. Fretting occurs when surfaces, pressed together, experience slight relative motion. Examples include press fits (as bearings pressed onto shafts), and bolted and riveted connections, where fluctuating loads produce slight relative movement. Other examples are leaf spring interfaces and stacks of sheet metal transported long distances by rail or truck. The relative motion is typically of the order of 0.01 to 0.25 mm. The resulting damage can be mere discoloration of mating surfaces (as with the transported metal sheets), the formation of surface pits (most common), or the wearing away of a millimeter of material (extreme case). The roughness and pitting caused by fretting make the surface more vulnerable to fatigue failure. *Reducing fatigue strength is a major consequence of fretting.*

A widely accepted theory is that the oscillatory motion breaks down natural protective surface films, resulting in exposure of bare metal surface "peaks" that weld together and are then torn away by the relative motion. With most engineering materials, the surface debris thus formed oxidizes to form powdery abrasive particles that build up and cause continuing wear. With ferrous surfaces the oxide powder is sometimes called "cocoa" because of its brown color. Magnesium and aluminum oxide particles appear black.

Resistance to frettage varies widely among different materials. Cobalt-base hard-facing alloys are among the best. In general, steel-on-steel and cast iron-on-cast iron are good, but interfaces involving stainless steel or titanium are poor. Brass-on-steel tends to be better than steel-on-steel. Combinations of cast iron with aluminum, magnesium, chrome plate, tin plate, or plastics are poor. The creation of *surface residual compressive stresses* by heat-treatment or cold-working has proven particularly effective in retarding the propagation of fatigue cracks initiated by fretting. The increased surface hardness resulting from these treatments is probably beneficial also. Low viscosity, high tenacity lubricants tend to reduce the intensity of fretting, their main effect apparently being to keep oxygen away from the active interface.

Sometimes fretting can be arrested by increasing surface interface pressure so that relative motion ceases. However, if the relative motion continues, frettage damage will usually increase with the higher pressure.

Further details concerning fretting corrosion are given in Refs. 3, 4, 5, 7, 8, and 11.

9.12 ANALYTICAL APPROACH TO WEAR

Although the design of machine components from the standpoint of resisting wear remains largely empirical, analytical approaches are now available. The generally recognized "wear equation," which emerged in the 1940s, can be written as

$$\text{wear rate} = \frac{\delta}{t} = \left(\frac{K}{H}\right)pv \tag{9.1}$$

where

δ = wear depth, mm (or in.)

t = time, s (sec)

K = wear coefficient (dimensionless)

H = surface hardness, MPa (ksi) (see footnote 4)

p = surface interface pressure, MPa (ksi)

v = sliding velocity, mm/s (in./sec)

For two rubbing surfaces a and b, this equation implies that the rate of wear of surface a is proportional to the wear coefficient (for material a when in contact with material b), inversely proportional to the surface hardness of a, and, assuming a constant coefficient of friction, directly proportional to the rate of friction work.

For a given total compressive force between the surfaces, the volume of material worn away is independent of the area of contact. Thus, another and more commonly used form of the wear equation is

[4]Brinell, Vickers, and Knopp hardness values are in kg/mm^2. To convert to MPa or ksi, multiply by 9.81 or 1.424, respectively.

$$W = \left(\frac{K}{H}\right)FS \tag{9.1a}$$

where

W = volume of material worn away, mm^3 (in.^{3})

F = compressive force between the surfaces, N (kilopounds)

S = total rubbing distance, mm (in.)

The best way to obtain values of K for a particular design application is from experimental data involving the same material combination operating under essentially the same conditions (for example, obtaining wear constants for the design of a "new model" from wear data obtained from a similar "old model"). Also, the literature contains values of K for many material combinations that have been obtained under laboratory conditions. When using the latter, it is important that there be a correspondence between the laboratory test and the expected application with respect to approximate interface temperature as well as to materials and lubrication.

Figure 9.12 illustrates ranges of values typically obtained with various combinations of material compatibility (see Fig. 9.11) and lubrication, and for three wear modes. The values of K pertain to the softer of the two rubbing metals.

Test data for wear coefficients show considerable scatter, typically over a range of plus or minus a factor of 4. This is perhaps to be expected because adhesive wear tends to be proportional to the fourth or fifth power of the friction coefficient, which itself has considerable scatter.

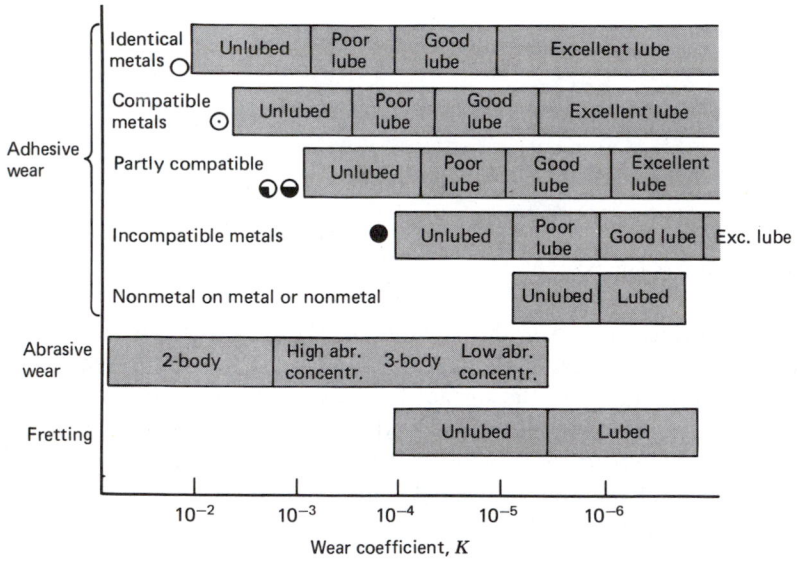

FIGURE 9.12.
Estimated wear coefficients for various sliding situations. (From Rabinowicz, E., "Wear Coefficients—Metals," Section IV of [8].)

The following sample problem illustrates the computation of K from experimental data.

SAMPLE PROBLEM 9.1

A pin-on-disk friction testing apparatus involves the unlubricated rounded end of a copper pin of 80 Vickers hardness being pressed with a force of 20 N against the surface of a rotating steel disk of 210 Brinell hardness. The rubbing contact is at a radius of 16 mm; the disk rotates 80 rpm. After 2 hours the pin and disk are weighed. It is determined that adhesive wear has resulted in weight losses equivalent to wear volumes of 2.7 mm^3 and 0.65 mm^3 for the copper and steel, respectively. Compute the wear coefficients.

Solution

1. $S = 2\pi(16)\,\dfrac{\text{mm}}{\text{rev}} \cdot 80\,\dfrac{\text{rev}}{\text{sec}} \cdot 60\,\dfrac{\text{min}}{\text{hr}} \cdot 2\text{ hr} = 9.65 \times 10^5\text{ mm}$
2. $H = 9.81(80) = 785$ MPa (copper)
   $H = 9.81(210) = 2060$ MPa (steel)
3. From Eq. (9.1a), $K = WH/FS$,

$$K = \frac{2.7(785)}{20(9.65 \times 10^5)} = 1.10 \times 10^{-4} \text{ (for copper)}$$

$$K = \frac{0.65(2060)}{20(9.65 \times 10^5)} = 6.94 \times 10^{-5} \text{ (for steel)}$$

9.13 CURVED SURFACE CONTACT STRESSES

The theoretical contact between curved surfaces is generally a point or a line (as a ball or cylinder and plane, a pair of mating gear teeth, etc.). When curved *elastic* bodies are pressed together, *finite* contact areas are developed because of deflections. These contact areas are so small, however, that corresponding compressive stresses tend to be extremely high. In the case of machine components like ball bearings, roller bearings, gears, and cam and followers, these *contact stresses* at any specific point on the surface are *cyclically applied* (as with each revolution of a bearing or gear), hence *fatigue failures* tend to be produced. These failures involve minute cracks that propagate to permit small bits of material to separate from the surface. This surface damage, sometimes referred to as "wear," is preferably called *surface fatigue*. These failures are discussed more fully in the following section. The present section lays a necessary foundation by considering in more detail the *stresses* caused by pressure plus possible sliding between curved elastic bodies.

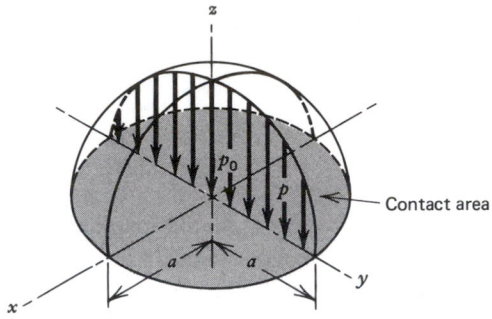

(a)
Two spheres

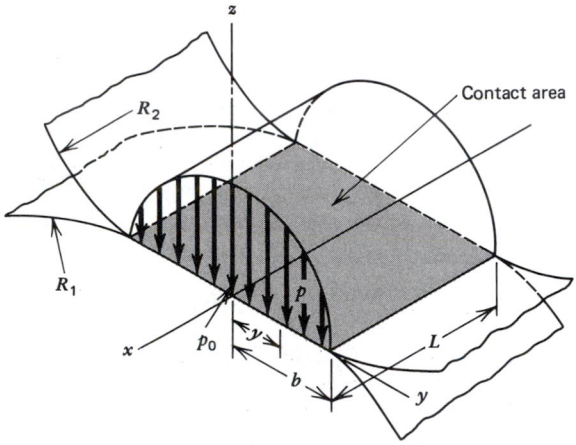

(b)
Two parallel cylinders

FIGURE 9.13.
Contact pressure distribution.

Figure 9.13 illustrates the contact area and corresponding stress distribution between two spheres and two cylinders, loaded with force, F. The maximum contact pressure p_0, exists on the load axis. The area of contact is defined by dimension a for the spheres, and b and L for the cylinders. The equations for p_0, a, and b can be simplified by introducing a quantity, Δ, which is a function of Young's modulus (E) and Poisson's ratio (ν) for the contacting bodies, 1 and 2.

$$\Delta = \frac{1 - \nu_1^2}{E_1} + \frac{1 - \nu_2^2}{E_2} \tag{9.2}$$

For two spheres:

$$p_0 = 0.578 \sqrt[3]{\frac{F(1/R_1 + 1/R_2)^2}{\Delta^2}} \tag{9.3}$$

$$a = 0.908 \sqrt[3]{\frac{F\Delta}{1/R_1 + 1/R_2}} \tag{9.4}$$

For a sphere and flat plate, R_2 is infinite; for a sphere and spherical socket, R_2 is negative.

For two parallel cylinders:

$$p_0 = 0.564 \sqrt{\frac{F(1/R_1 + 1/R_2)}{L\Delta}} \tag{9.5}$$

$$b = 1.13 \sqrt{\frac{F\Delta}{L(1/R_1 + 1/R_2)}} \tag{9.6}$$

For a cylinder and flat plate, R_2 is infinite; for a cylinder and cylindrical groove, R_2 is negative.

For other cases of two curved surfaces pressed together (as a wheel rolling on a crowned rail) see Refs. 5, 6, 9, and 10.

Contact pressure, p_0, is, of course, also the value of the *surface compressive stress,* σ_z, at the load axis. The original analysis of elastic contact stresses was published in 1881 by Heinrich Hertz of Germany, at the age of 24. In his honor, the stresses at the mating surfaces of curved bodies in compression are called *Hertz contact stresses.*

Figure 9.14 shows how the direct compressive stress, σ_z, diminishes below the surface. It also shows corresponding values of σ_x and σ_y. These compressive stresses result from Poisson's ratio—the material along the load axis that is compressed in the z direction tends to expand in the x and y directions, but the surrounding material does not want to move away to accommodate this expansion; hence the compressive stresses in the x and y directions. Because of symmetry of loading, it can be shown that the x, y, and z stresses plotted in Fig. 9.14 are *principal* stresses. Figure 9.15 shows a Mohr-circle plot for stresses on the load axis at depth b below the surface for two parallel cylinders. Note that the value of the maximum shear stress, τ_{max}, in Fig. 9.15 agrees with the value plotted in Fig. 9.14*b*. Other values of τ_{max} plotted in Figs. 9.14*a* and 9.14*b* can be verified in the same manner.

All stresses previously considered in this section exist along the load axis. Figure 9.16 shows an important shear stress existing below the surface and *displaced* from the load axis. Note that if the cylinders are rotating together in the direction indicated, any point below the surface experiences stresses as shown first at A and then B. This is a *completely reversed* shear stress, believed to be very significant in connection with subsurface fatigue crack initiation. This stress is greatest at points below the surface a distance of about $0.5b$ (distance b is defined in Fig. 9.13*b*). As a point at this depth rolls through the contact zone, the maximum values of this shear stress are reached at a distance of about b on either side of the load axis.

Most rolling members also tend to *slide,* even if only slightly (mating gear teeth, a cam and follower, and to some extent the rolling members in ball and roller bearings). The resulting friction forces cause tangential normal and shear stresses that are super-

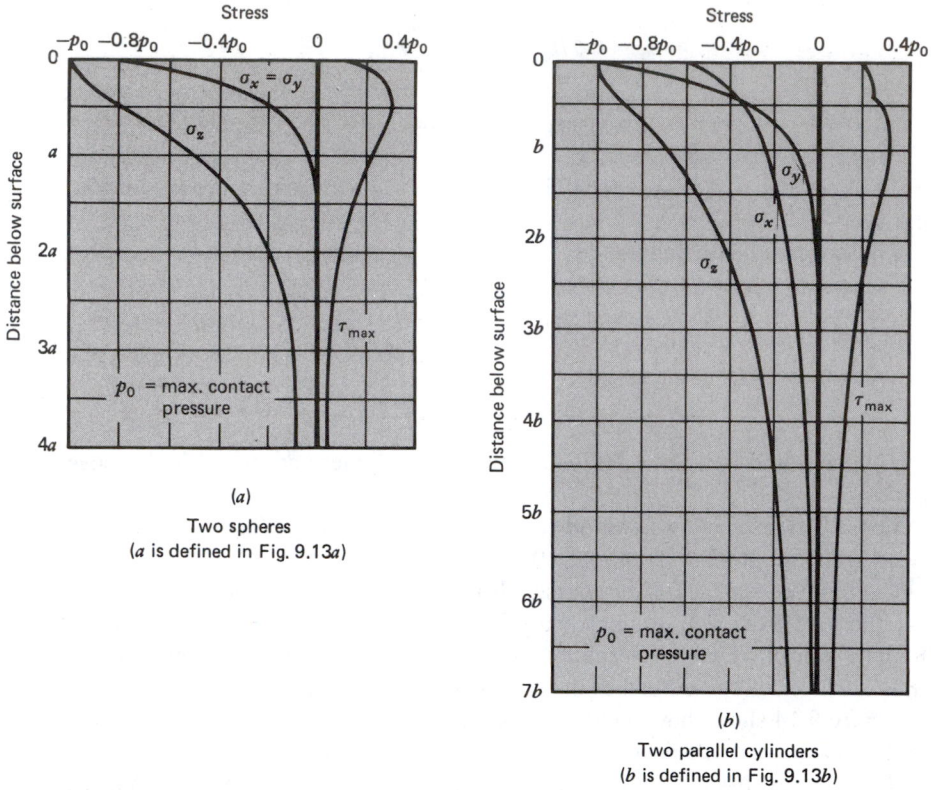

FIGURE 9.14.
Elastic stresses below the surface, along the load axis (for $\nu = 0.3$).

imposed on the stresses caused by the normal loading. These tangential stresses are illustrated in Fig. 9.17. As any given point on the surface rolls through the contact zone the tangential shear stresses vary from zero-to-maximum-to-zero, while the normal stresses vary from zero-to-tension-to-compression-to-zero. The presence of a surface tensile stress is undoubtedly important in the propagation of surface fatigue cracks.

Let us conclude this section by summarizing the more important aspects of stresses associated with contacting curved surfaces. First, the maximum contact pressures, and the "flattened" areas of contact, are given by the classical equations of Heinrich Hertz. Below the surface and on the load axis is an important shear stress associated with the "Poisson's ratio" expansion of the compressed material (τ_{max} in Figs. 9.14 and 9.15). Below the surface, and to each side of the load axis, is shear stress τ_{xy} (Fig. 9.16). This is particularly important in rolling members because it *reverses direction* as any given point below the surface rolls through the contact zone. Some sliding often accompanies rolling, and causes both a tangential surface shear stress and reversed surface tangential stress (Fig. 9.17). Two other very important factors affecting stresses in the contact region are (1) highly localized heating and thermal expansion caused by sliding friction, and (2) the hydrodynamic pressure distribution within the oil film that normally exists

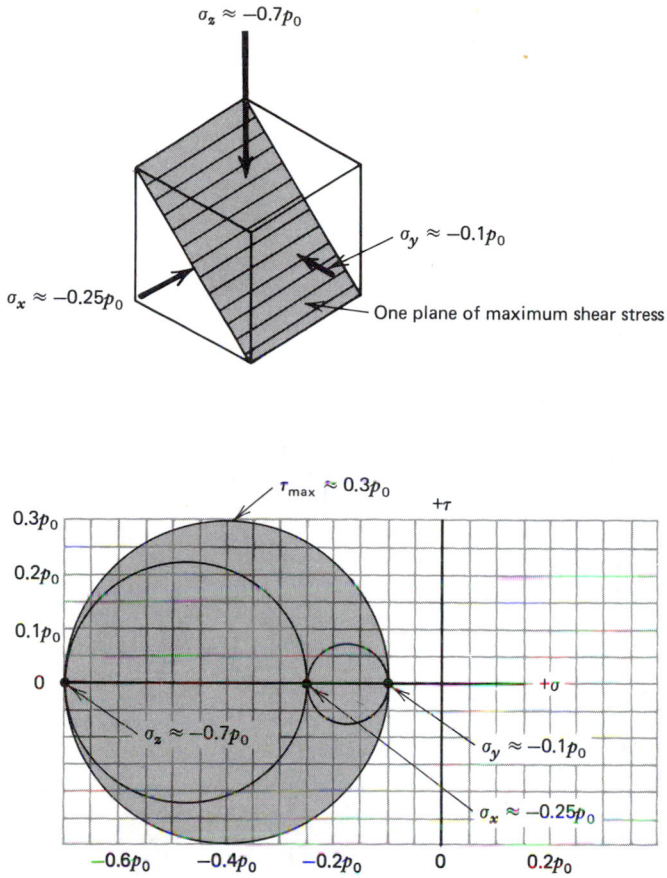

FIGURE 9.15.
Principal element and Mohr-circle representation, two cylinders, on the load axis, distance *b* below surface.

in the contact region. Because of these many factors, the maximum Hertz contact pressure (p_0 in Fig. 9.13) is not *by itself* a very good index of contact loading severity.

9.14 SURFACE FATIGUE FAILURES

Surface fatigue failures result from the repeated application of loads that produce stresses in and under the contacting surfaces, as described in the previous article. Cracks resulting from these stresses propagate until small bits of surface material become separated, producing *pitting* or *spalling*. *Pitting* originates with surface cracks, and each pit involves a relatively small surface area. *Spalling* originates with subsurface cracks, and the spalls are thin "flakes" of surface material. These failure types occur commonly in gear teeth, ball and roller bearings, cams and followers, and metal wheels rolling on rails. Typical examples are illustrated in Fig. 9.18.

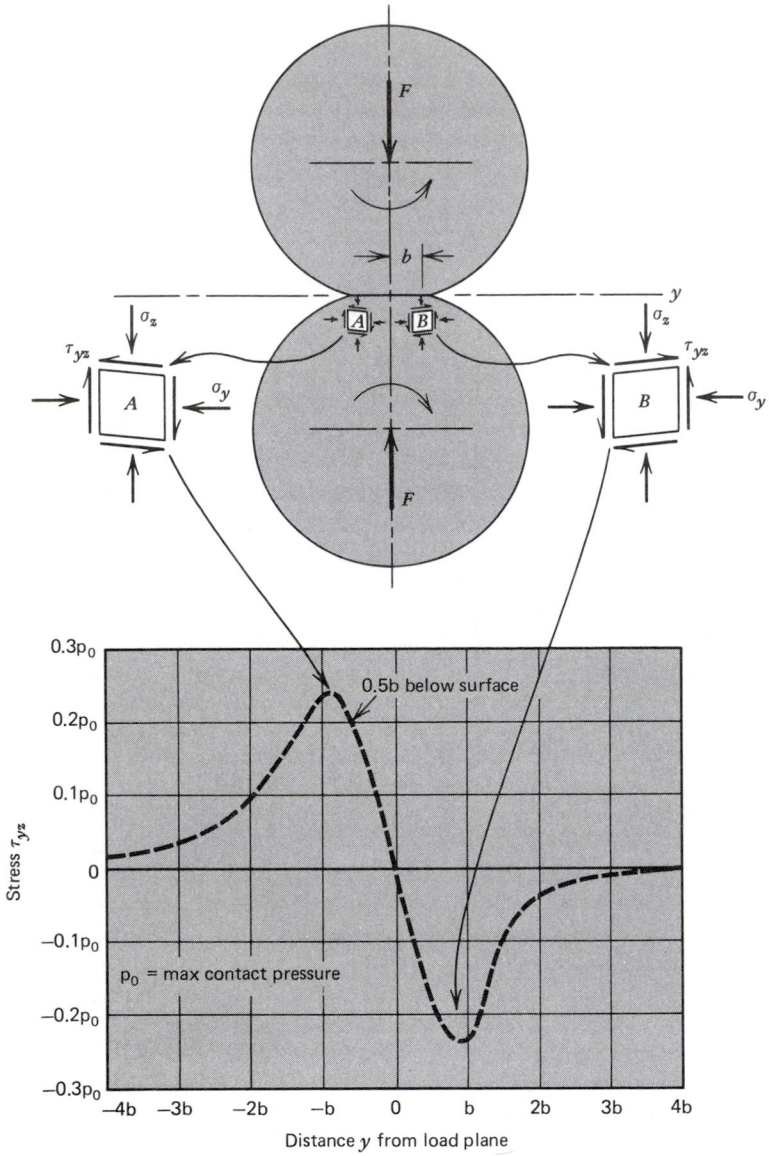

FIGURE 9.16.
Sub-surface shear stress that reverses when rolling through the contact zone. Values plotted are for a depth of 0.5*b* below the surface, and $\nu = 0.3$. Normally loaded parallel cylinders. (Note: τ_{yz} has its maximum value at a depth of 0.5*b* below the surface.) [From J. O. Smith and Chang Keng Liu, "Stresses Due to Tangential and Normal Loads on an Elastic Solid with Application to Some Contact Stress Problems," *J. Appl. Mech.* (March 1953).]

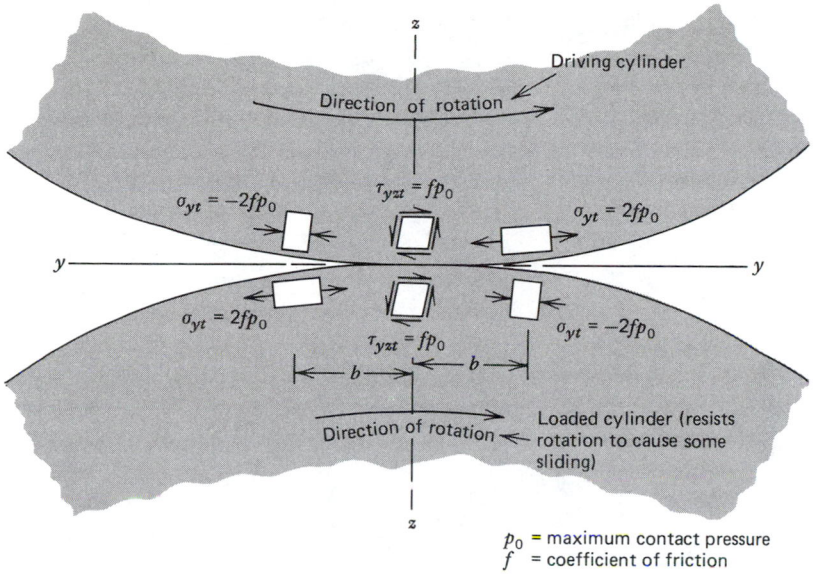

FIGURE 9.17.
Tangential normal and shear stresses caused by sliding friction between two parallel cylinders. Maximum values are at the surface, in the locations shown. (Note: *t* is a subscript denoting that the stress is due to tangential (frictional) loading.)

Figure 9.19 shows typical *S-N* curves based on computed elastic Hertz stress (p_0 in Fig. 9.13). Note that the degree of sliding generally increases from the parallel rollers (which do not transmit a torque) represented by the top line, to the spur gear teeth corresponding to the bottom line.

The tendency for surfaces to fail in fatigue can obviously be reduced by decreasing loads and decreasing sliding. Better lubrication helps in at least three ways: (1) lower friction reduces the surface tangential shear stress and also the tensile stress shown in Fig. 9.17; (2) lower friction plus improved heat transfer reduce thermal stresses; and (3) presence of a good lubricating film usually results in a more favorable distribution of pressure over the contact area.

In general, increased surface hardness increases resistance to surface fatigue. However, the associated increased strength reduces the ability of minute surface imperfections to adjust by wear or surface flow, and thereby reduces localized contact pressures. This is part of the rationale behind the common practice of making one of a pair of mating gears very hard, with the other somewhat softer to allow for "run-in" of the surfaces.

Precise accuracy of surface geometry and extreme surface smoothness are highly beneficial. Exceptions occur when significant sliding is present. Then, surface porosity, or a pattern of minute depressions on one of the mating surfaces, may help by providing tiny reservoirs for holding lubricant.

Compressive residual stresses in the contacting surfaces increase resistance to sur-

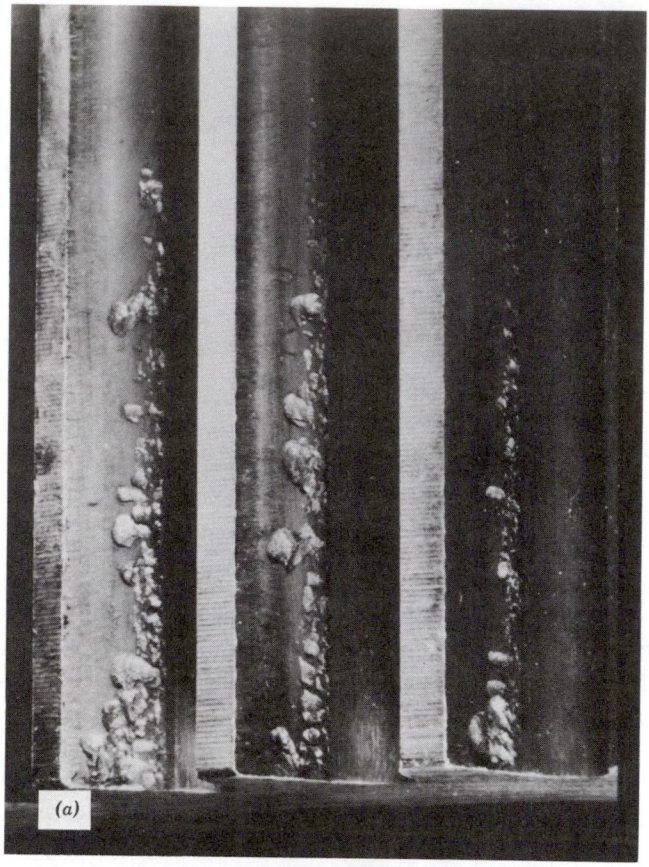

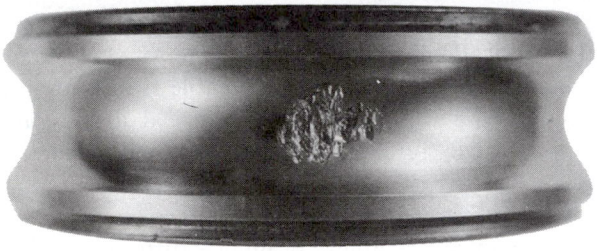

(b)

FIGURE 9.18.
Surface fatigue failures. *a*—(Courtesy American Gear Manufacturers Association); *b*—(Courtesy New Departure-Hyatt Division, General Motors Corp.)

face fatigue failure. This is to be expected, and follows the general pattern of such stresses discouraging fatigue failure and surface damage.

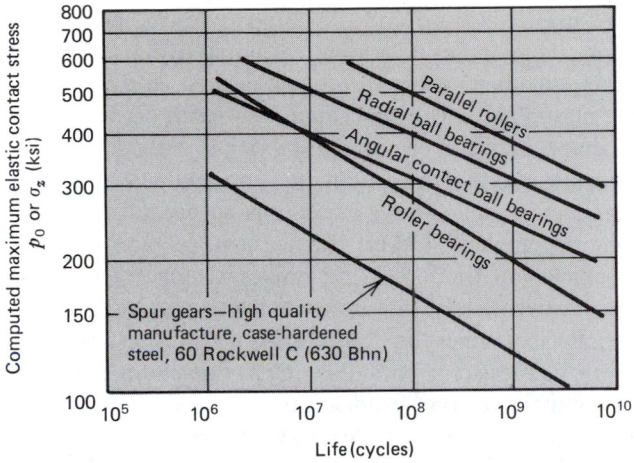

FIGURE 9.19.
Average *S-N* curves for contact stresses—rollers, bearings, and spur gears, 10 percent failure probability [6].

9.15 CLOSURE

As previously noted, corrosion and wear represent an estimated annual cost (1978) in the U.S. of some 90 billion dollars. Furthermore, more machine parts *wear* out than break. The reduction of this enormous economic and ecologic burden presents one of the greatest challenges to modern engineering. The solution would seem to involve (1) designing to reduce surface damage as much as feasible, and (2) providing for easy replacement of machine components most vulnerable to surface deterioration. Almost everyone is familiar with instances where an entire machine (like a washing machine or refrigerator) was scrapped because it was too costly to replace one or two worn-out parts.

Let us review briefly three key aspects of machine component surfaces:

1. *Smoothness* is important to fatigue strength (recall surface factor, C_s, in Fig. 8.13), to wear resistance, and, to some extent, to corrosion resistance.
2. *Hardness* is important to fatigue strength (as in steel, where S'_n in ksi $\approx 250 \cdot H_B$), and very important to resistance to wear and cavitation damage.
3. *Surface residual stress* is very important, with compressive residual stresses increasing fatigue strength and increasing resistance to stress corrosion cracking, corrosion fatigue, surface fatigue (from contact stresses), and damage from fretting corrosion.

An important concept in the modern design of many machine components is that of *choosing different materials for the interior and for the surface.* If the material most suitable for the bulk of the part does not meet the surface requirements, a second material can often be applied to the surface. For example, steel parts can be coated (by electroplating, mechanical plating, hot dipping, cladding, flame spraying, etc.) with zinc, cadmium, chromium, nickel, or other metals to provide desired corrosion resistance. Soft metal parts, or even plastic parts, can be coated with hard, bright surface metal for abrasion resistance and appearance. For extreme hardness, carbides of tungsten and other metals can be commercially applied by flame spraying and other processes. For applications requiring low friction and wear, coatings incorporating fluoroplastics (as Teflon) are commonly applied. Other plastic coatings are used for applications requiring a *high* coefficient of friction (as for brakes, clutches, and belts). Parts having surfaces subjected to extreme heat can be coated with special high-temperature alloys or ceramic materials. Sometimes the desired coating can be incorporated into a paint-type material, as corrosion-resistant paints having a pigment of powdered zinc, or wear-resistant coatings consisting of tiny beads of alumina and ceramic in an epoxy resin. Thus, it is becoming increasingly feasible to avoid serious compromises associated with making parts from a single material.

Ecological and health considerations must be increasingly considered when choosing a coating material and coating process. For example, cadmium in the human body can pose serious danger. Cadmium plating of steel parts has long been extensively used to provide corrosion resistance. (Over 1500 tons were used for this purpose in the U.S. in 1978, according to the U.S. Bureau of Mines.) A byproduct of cadmium electroplating is large quantities of spent fluids of high cadmium content. Disposal of this waste without polluting water or soil is a problem. The development of safe (and economical) processes for cadmium coating thus becomes an important scientific and engineering challenge.

REFERENCES

1. Cocks, F. H. (ed.), *Manual of Industrial Corrosion Standards and Control,* American Society for Testing and Materials, Philadelphia, 1973.
2. Colangelo, V. J., and F. A. Heiser, *Analysis of Metallurgical Failures,* John Wiley & Sons, Inc., New York, 1974.
3. Fontana, M. G., and N. D. Greene, *Corrosion Engineering,* 2nd ed., McGraw-Hill, New York, 1978.
4. Horger, O. J. (ed.), *ASME Handbook: Metals Engineering—Design,* 2nd ed., McGraw-Hill, New York, 1965. (a) Part 3, Sec. 1.1, "Mechanical Factors Influencing Corrosion," by H. R. Copson. (b) Part 3, Sec. 1.2, "Fretting Corrosion and Fatigue," by G. Sachs and O. J. Horger.
5. Juvinall, R. C., *Engineering Considerations of Stress, Strain, and Strength,* McGraw-Hill, New York, 1967.
6. Lipson, C., and R. C. Juvinall, *Handbook of Stress and Strength,* Macmillan, New York, 1963.
7. Lipson, C., *Wear Considerations in Design,* Prentice-Hall, Inc., Englewood Cliffs, N.J., 1967.

8. Peterson, M. B., and W. O. Winer (ed.), *Wear Control Handbook,* The American Society of Mechanical Engineers, New York, 1980.

9. Roark, R. J., and W. C. Young, *Formulas for Stress and Strain,* 5th ed., McGraw-Hill, New York, 1975.

10. Timoshenko, S., and J. N. Goodier, *Theory of Elasticity,* 3rd ed., McGraw-Hill, New York, 1970.

11. Uhlig, H. H., *Corrosion and Corrosion Control,* 2nd ed., John Wiley & Sons, Inc., New York, 1971.

12. Van Vlack, L. H., *Elements of Materials Science and Engineering,* 3rd ed., Addison-Wesley, Reading, Mass., 1975.

PART TWO

Applications

CHAPTER TEN

Threaded Fasteners and Power Screws

10.1 INTRODUCTION

A lay person might consider threaded fasteners (screws, nuts, and bolts) to be the most mundane and uninteresting of all machine elements. Looking deeper, the engineer finds that these seemingly simple components exist in amazing variety, and with design details representing remarkable ingenuity. The economic implications of fastener design (both threaded and non-threaded, as rivets) are tremendous. For example, the airframe of a large jet aircraft has approximately 2.4×10^6 fasteners costing about $750,000. The safety implications of the fasteners used in many machines—particularly vehicles carrying people—are obvious. Corrosion considerations are often critical, as differences in materials used in fasteners and clamped members give rise to potential galvanic cells. Many fasteners must be designed for easy, low-cost (often automated) assembly. Ease of disassembly is often important, too, where servicing and parts replacement must be considered. On the other hand, *difficulty* of disassembly is sometimes important for resisting vandalism. As a further requirement, ease of disassembly for scrapping and recycling of components and materials is becoming increasingly important. (It has been facetiously suggested that automobile fasteners should withstand safely all operating and safety-related crash loads, but be so designed that upon dropping the car from a height of several feet, all fasteners would fail, permitting the pile of components to be easily sorted for recycling!)

In summary, the problem of devising bolts (and other fasteners) that are lighter in weight, cheaper to manufacture *and use,* less susceptible to corrosion, and more resis-

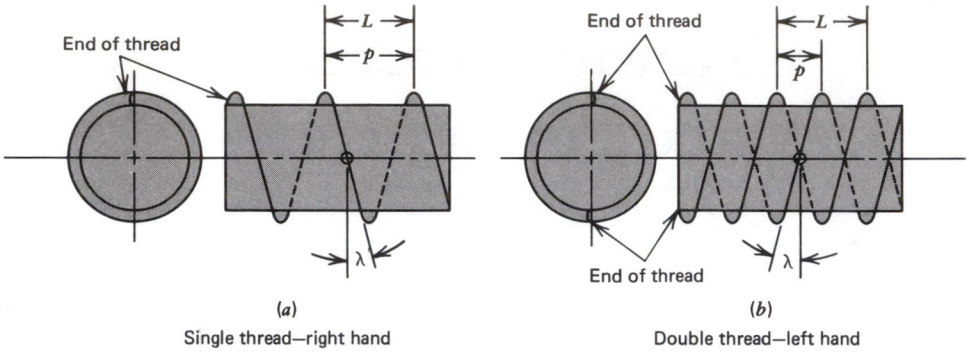

FIGURE 10.1
Helical threads of pitch *p*, lead *L* and lead angle λ.

tant to loosening under vibration presents a never-ending challenge to the engineer working in this field. In addition, nearly *all* engineers are concerned with the selection and use of fasteners, and thus need to be somewhat knowledgeable of the choices available, and the factors governing their selection and use.

Power screws of various descriptions are also commonly encountered machine components. Being threaded members, their engineering and design has much in common with bolts.

10.2 THREAD FORMS, TERMINOLOGY, AND STANDARDS

Figure 10.1 illustrates the basic arrangement of a helical thread wound around a cylinder, as used on screw-type fasteners, power screws, and worms (the kind that are used in worm and worm gear sets—see Chapter 16). Pitch, lead, lead angle, and hand-of-thread are defined by the illustrations. Virtually all bolts and screws have a single thread, but worms and power screws sometimes have double, triple, and even quadruple threads. Unless otherwise noted, all threads are assumed to be right-hand.

Figure 10.2 shows the standard geometry of screw threads used on fasteners. This is basically the same for both *Unified* (inch series) and *ISO* (International Standards Organization, metric) threads. (Detail variations in the root region in order to reduce stress concentration are discussed in Section 10.12.) Standard sizes for the two systems are given in Tables 10.1 and 10.2. Table 10.1 shows both the *fine thread* (UNF, standing for Unified National Fine) and *coarse thread* (UNC, Unified National Coarse) series. The *stress area* tabulated is based on the average of the pitch and root diameters. This is the area used for *"P/A"* stress calculations. It approximates the smallest possible fracture area, considering the presence of the helical thread. The official American National Standard from which the information in Table 10.1 was taken is ANSI (American National Standards Institute) B1.1 (1974), which is published by the American Society of Mechanical Engineers and sponsored jointly by the ASME and the Society of Automotive Engineers. This standard also defines an *extra fine thread* series and

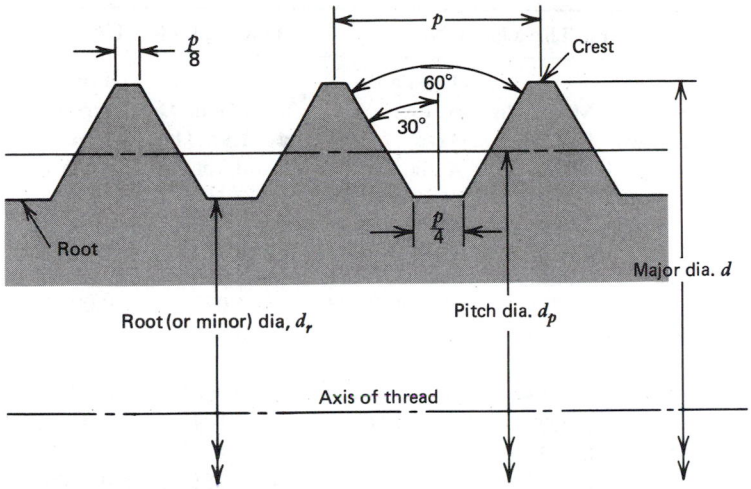

FIGURE 10.2.
Unified and ISO thread geometry. (Basic profile of external thread is shown.)

eight *constant pitch thread* series (each of these covers a range of sizes, with 4, 6, 8, 12, 16, 20, 28, and 32 threads per inch). The great majority of screws and bolts with inch-series threads, however, conform to the standard coarse and fine series listed in Table 10.1.

One of the earliest thread forms to be used was the V thread, essentially like the modern profile shown in Fig. 10.2 except that the 60-deg.-angled sides extended to sharp points at the crest and root. The sharp crests were vulnerable to damage, and the sharp roots caused severe stress concentration. The previous American standard (American National thread) and the former British standard (Whitworth thread) modified the sharp crest and root in slightly different ways. Both countries then agreed on the Unified standard, illustrated in Fig. 10.2. More recently, all major nations have agreed on the ISO (metric) threads.

Different applications require screw threads with different degrees of precision and different amounts of clearance between the mating threaded members. Accordingly, screw threads are manufactured to different classes of fit. For Unified threads, three classes are standard, with Class 1 representing the loosest fit and greatest tolerances, and Class 3 the tightest fit and closest tolerances. Obviously, Class 3 threaded members are also the most expensive. The tolerance zones for nut and screw are illustrated in Fig. 10.3. Detailed information regarding dimensions, fits, and tolerances for the various inch-series threads may be found in ANSI B1.1.

Figure 10.4 illustrates most of the standard thread forms used for power screws. Acme threads are the oldest, and are still in common use. The Acme stub is sometimes used because it is easier to heat-treat. The square thread gives slightly greater efficiency but is seldom used because of difficulties in manufacturing the 0 deg. thread angle. Furthermore, it lacks the ability of the Acme thread to be used with a split nut (split on an axial plane), the two halves of which can be moved together to compensate for

TABLE 10.1
Basic Dimensions of Unified Screw Threads

		Coarse Threads—UNC			Fine Threads—UNF		
Size	**Major Dia. d (in.)**	**Threads per In.**	**Minor Dia. Ext. Thd. d_r (in.)**	**Tensile Stress Area A_t (in.^{2})**	**Threads per In.**	**Minor Dia. Ext. Thd. d_r (in.)**	**Tensile Stress Area A_t (in^2)**
0(.060)	0.0600	—	—	—	80	0.0447	0.00180
1(.073)	0.0730	64	0.0538	0.00263	72	0.0560	0.00278
2(.086)	0.0860	56	0.0641	0.00370	64	0.0668	0.00394
3(.099)	0.0990	48	0.0734	0.00487	56	0.0771	0.00523
4(.112)	0.1120	40	0.0813	0.00604	48	0.0864	0.00661
5(.125)	0.1250	40	0.0943	0.00796	44	0.0971	0.00830
6(.138)	0.1380	32	0.0997	0.00909	40	0.1073	0.01015
8(.164)	0.1640	32	0.1257	0.0140	36	0.1299	0.01474
10(.190)	0.1900	24	0.1389	0.0175	32	0.1517	0.0200
12(.216)	0.2160	24	0.1649	0.0242	28	0.1722	0.0258
¼	0.2500	20	0.1887	0.0318	28	0.2062	0.0364
5/16	0.3125	18	0.2443	0.0524	24	0.2614	0.0580
⅜	0.3750	16	0.2983	0.0775	24	0.3239	0.0878
7/16	0.4375	14	0.3499	0.1063	20	0.3762	0.1187
½	0.5000	13	0.4056	0.1419	20	0.4387	0.1599
9/16	0.5625	12	0.4603	0.182	18	0.4943	0.203
⅝	0.6250	11	0.5135	0.226	18	0.5568	0.256
¾	0.7500	10	0.6273	0.334	16	0.6733	0.373
⅞	0.8750	9	0.7387	0.462	14	0.7874	0.509
1	1.0000	8	0.8466	0.606	12	0.8978	0.663
1⅛	1.1250	7	0.9497	0.763	12	1.0228	0.856
1¼	1.2500	7	1.0747	0.969	12	1.1478	1.073
1⅜	1.3750	6	1.1705	1.155	12	1.2728	1.315
1½	1.5000	6	1.2955	1.405	12	1.3978	1.581
1¾	1.7500	5	1.5046	1.90			
2	2.0000	4½	1.7274	2.50			
2¼	2.2500	4½	1.9774	3.25			
2½	2.5000	4	2.1933	4.00			
2¾	2.7500	4	2.4433	4.93			
3	3.0000	4	2.6933	5.97			
3¼	3.2500	4	2.9433	7.10			
3½	3.5000	4	3.1933	8.33			
3¾	3.7500	4	3.4433	9.66			
4	4.0000	4	3.6933	11.08			

Note: See ANSI standard B1.1-1974 for full details. Unified threads are specified as "½ in.-13UNC," "1 in.-12UNF."

TABLE 10.2
Basic Dimensions of ISO Metric Screw Threads

Nominal Diameter d (mm)	Coarse Threads: Pitch p (mm)	Coarse Threads: Stress Area A_t (mm^2)	Fine Threads: Pitch p (mm)	Fine Threads: Stress Area A_t (mm^2)
3	0.5	5.03		
3.5	0.6	6.78		
4	0.7	8.78		
5	0.8	14.2		
6	1	20.1		
7	1	28.9		
8	1.25	36.6	1	39.2
10	1.5	58.0	1.25	61.2
12	1.75	84.3	1.25	92.1
14	2	115	1.5	125
16	2	157	1.5	167
18	2.5	192	1.5	216
20	2.5	245	1.5	272
22	2.5	303	1.5	333
24	3	353	2	384
27	3	459	2	496
30	3.5	561	2	621
33	3.5	694	2	761
36	4	817	3	865
39	4	976	3	1030

Note: Metric threads are identified by diameter and pitch as "M8 × 1.25."

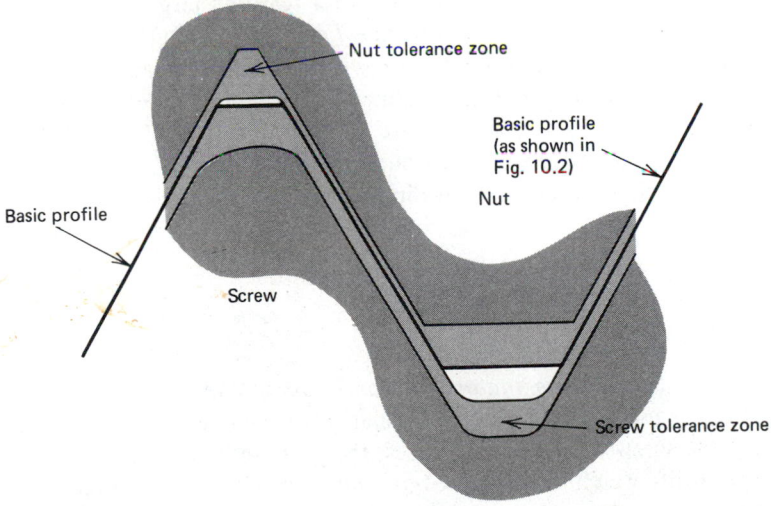

FIGURE 10.3.
Tolerance zones for various classes of unified threads. (Note: Each class—1, 2, and 3—uses a portion of the zones shown.)

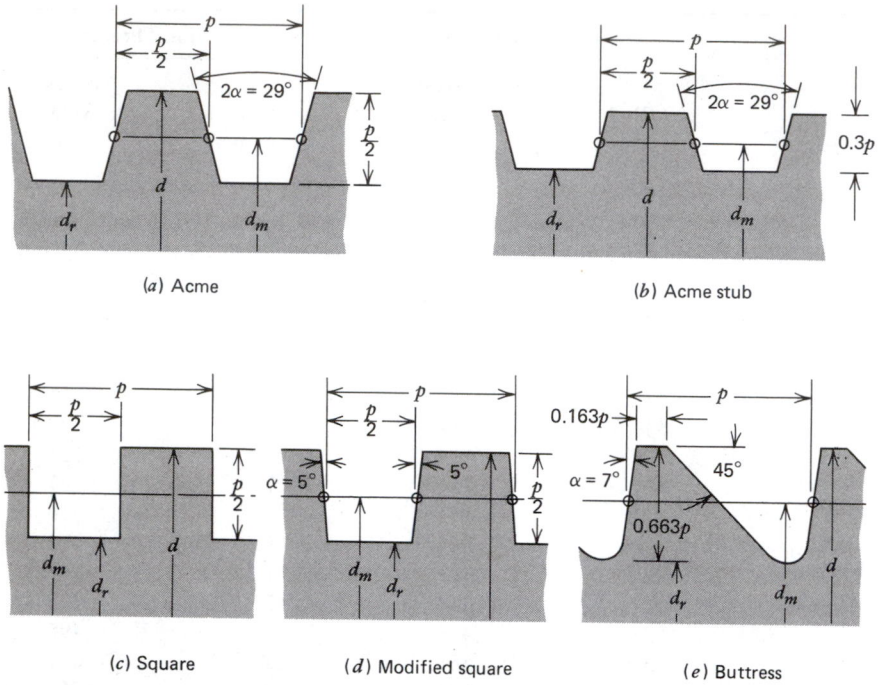

FIGURE 10.4.
Power screw thread forms. (Note: All threads shown are external (i.e., on screw, not on nut.) d_m is mean diameter of thread contact $\approx (d + d_r)/2$).

thread wear. The 5 deg. thread angle of the modified square thread partially overcomes these objections. The buttress thread is sometimes used for resisting large axial forces in one direction (the load is carried on the face with the 7 deg. thread angle). Standard sizes are given in Table 10.3. In the case of power screws with multiple threads, it should be noted that the number of threads per inch is defined as the reciprocal of the pitch, *not* the reciprocal of the lead.

The threads discussed in this article are all wound around a *cylinder,* as shown in Fig. 10.1. Other threads, such as those used on pipe and on wood screws, are wound around a cone.

10.3 POWER SCREWS

Power screws, sometimes called *linear actuators,* or *translation screws,* are used to convert rotary motion of either the nut or the screw to relatively slow linear motion of the mating member along the screw axis. In many cases, the purpose is to obtain a great mechanical advantage to lift weights (as screw-type jacks) or exert large forces (as presses, and tensile testing machines, home garbage compactors, C-clamps, etc.). In other cases the purpose is to obtain precise positioning of the axial movement (as a micrometer screw, or the lead screw of a lathe).

TABLE 10.3
Standard Sizes of Power Screw Threads

Major Dia. d (in.)	Threads per Inch: Acme and Acme Stub[a]	Square and Modified Square	Buttress[b]
¼	16	10	
5/16	14		
⅜	12		
⅜	10	8	
7/16	12		
7/16	10		
½	10	6½	16
⅝	8	5½	16
¾	6	5	16
⅞	6	4½	12
1	5	4	12
1⅛	5		
1¼	5	3½	10
1⅜	4		10
1½	4	3	10
1¾	4	2½	8
2	4	2¼	8
2¼	3	2¼	8
2½	3	2	8
2¾	3	2	6
3	2	1¾	6
3½	2	1⅝	6
4	2	1½	6
4½	2		5
5	2		5

[a]See ANSI standard B1.5-1977 for full details.
[b]See ANSI standard B1.9-1973 for full details.

Figure 10.5 shows a simplified drawing of three different screw jacks supporting a weight. Note in each case that only the shaded member connected to the handle rotates, and that a ball thrust bearing transfers the axial force from the rotating to a nonrotating member. All three jacks are basically the same, but let us choose Fig. 10.5*c* for determining the torque, Fa, that must be applied to the nut in order to lift a given weight.

By turning the nut in Fig. 10.5*c*, each portion of the nut thread is forced to climb an inclined plane. Let us represent this plane by unwinding (or developing) a portion of one turn of the screw thread, as shown in the lower-left portion of Fig. 10.6. If a full turn were developed, a triangle would be formed, illustrating the relationship

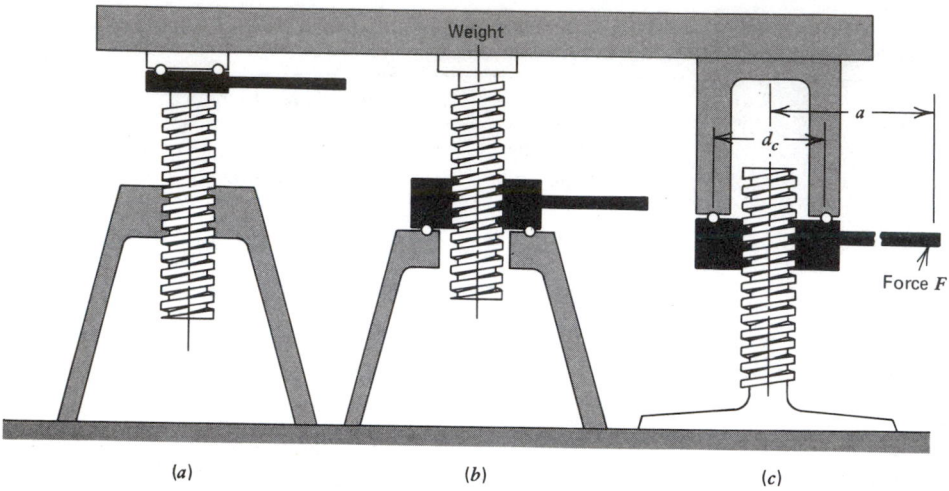

FIGURE 10.5.
Weight supported by three screw jacks. (In each case, only the shaded member rotates.)

$$\tan \lambda = \frac{L}{\pi d_m} \tag{10.1}$$

An infinitesimally small segment of the nut is represented in Fig. 10.6 by the small block acted upon by load w (a portion of the total axial load W), normal force n (shown in true view at the lower right), friction force fn, and tangential force q. Note that force q times $d_m/2$ represents the torque applied to the nut segment.

Summing tangential forces acting on the block (i.e., horizontal forces in the lower left view):

$$\Sigma F_t = 0: \quad q - n(f \cos \lambda + \cos \alpha_n \sin \lambda) = 0 \tag{a}$$

Summing axial forces (vertical forces in the lower left view):

$$\Sigma F_a = 0: \quad w + n(f \sin \lambda - \cos \alpha_n \cos \lambda) = 0$$

or

$$n = \frac{w}{\cos \alpha_n \cos \lambda - f \sin \lambda} \tag{b}$$

Combining (a) and (b):

$$q = w\left(\frac{f \cos \lambda + \cos \alpha_n \sin \lambda}{\cos \alpha_n \cos \lambda - f \sin \lambda}\right) \tag{c}$$

As noted above, the torque corresponding to force q is $q(d_m/2)$. Since the small block represents a typical segment of the nut thread, integration over the entire thread surface in contact results in the same equation except that w and n are replaced by W and N, where the latter represent the *total* vertical and normal loads, respectively, acting on the thread. Thus, the equation for torque required to lift load W is

$$T = Q\left(\frac{d_m}{2}\right) = \frac{Wd_m}{2}\left(\frac{f \cos \lambda + \cos \alpha_n \sin \lambda}{\cos \alpha_n \cos \lambda - f \sin \lambda}\right) \tag{10.2}$$

Note that torque, T, is also equal to Fa in Fig. 10.5c.

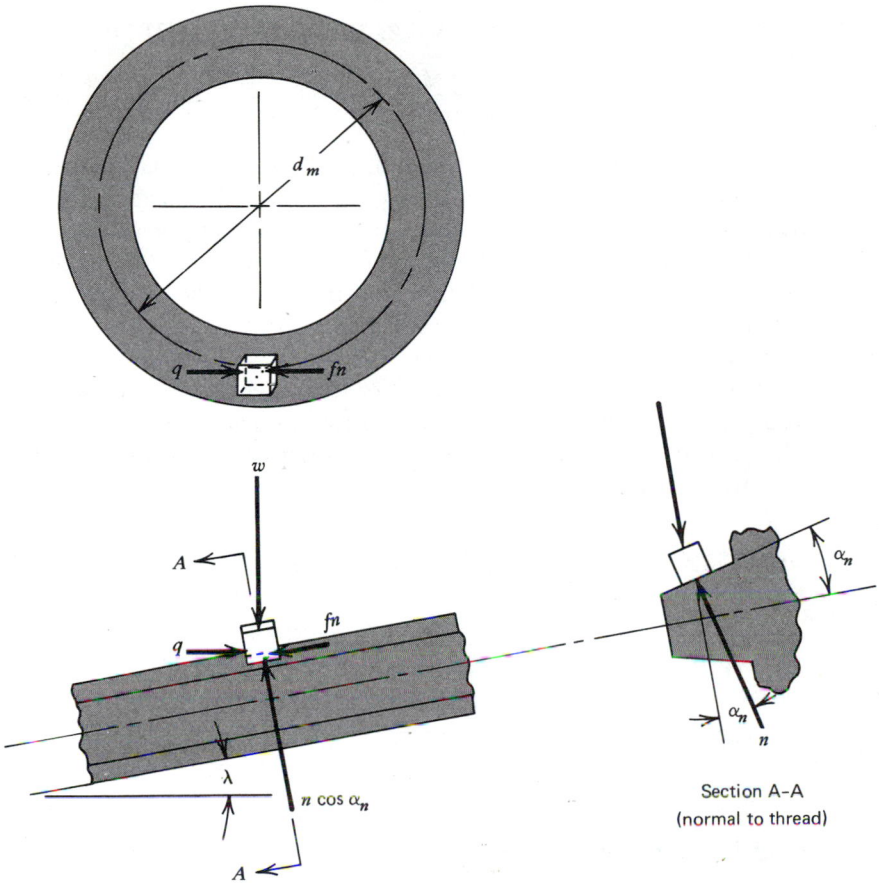

FIGURE 10.6.
Screw thread forces.

Since the lead rather than the lead angle is usually a known standard value, a more convenient form of the torque equation is obtained by dividing the numerator and denominator by cos λ and then substituting $L/\pi d_m$ for tan λ. This gives

$$T = \frac{Wd_m}{2}\left(\frac{f\pi d_m + L\cos\alpha_n}{\pi d_m \cos\alpha_n - fL}\right) \tag{10.3}$$

Most power screw applications require a bearing surface or *thrust collar* between stationary and rotating members. In Fig. 10.5 this function is served by the ball thrust bearing of diameter d_c. In many cases, a simple thrust washer is used. If the coefficient of friction of the collar washer or bearing is f_c, then the added torque required to overcome collar friction is $Wf_cd_c/2$,[1] and the total torque required to lift the load is

$$T = \frac{Wd_m}{2}\left(\frac{f\pi d_m + L\cos\alpha_n}{\pi d_m \cos\alpha_n - fL}\right) + \frac{Wf_cd_c}{2} \tag{10.4}$$

[1]In the unusual case where a more detailed analysis of collar friction might be justified, the procedures used with disk clutches, discussed in Chapter 18, are recommended.

For the special case of the square thread, $\cos \alpha_n = 1$, and Eq. (10.4) simplifies to:

$$T = \frac{Wd_m}{2}\left(\frac{f\pi d_m + L}{\pi d_m - fL}\right) + \frac{Wf_c d_c}{2} \tag{10.4a}$$

For the Acme thread, $\cos \alpha_n$ is so nearly equal to unity that Eq. (10.4a) can usually be used without significant error, particularly when one considers the inherent variability in the coefficient of friction. Several remaining equations in this article will be developed for both the general case and the special case of the square thread. Acme thread problems can usually be handled with sufficient accuracy using the simpler square thread equations. In all cases, however, an awareness of the more complete general equations provides additional insight.

The above analysis pertained to *raising* a load, or to turning the rotating member "against the load." The analysis for *lowering* a load (or turning a rotating member "with the load") is exactly the same except that the directions of q and fn (Fig. 10.6) are reversed. This gives equations for load lowering of

$$T = \frac{Wd_m}{2}\left(\frac{f\pi d_m - L\cos \alpha_n}{\pi d_m \cos \alpha_n + fL}\right) + \frac{Wf_c d_c}{2} \tag{10.5}$$

for the general case, and

$$T = \frac{Wd_m}{2}\left(\frac{f\pi d_m - L}{\pi d_m + fL}\right) + \frac{Wf_c d_c}{2} \tag{10.5a}$$

for the special case of the square thread.

10.3.1 Values of Friction Coefficients

When a ball or roller thrust bearing is used, f_c is usually low enough that collar friction can be neglected, thus eliminating the second term from the above equations.

When a plain thrust collar is used, values of f and f_c vary generally between about 0.08 and 0.20 under conditions of ordinary service and lubrication, and for the common materials of steel against cast iron or bronze. This range includes both starting and running friction, with starting friction being about ⅓ higher than running friction. Special surface treatments and coatings can reduce these values by at least half (see Section 9.15).

10.3.2 Values of Thread Angle in the Normal Plane

Figure 10.7 shows the thread angle measured in the normal plane (α_n, as used in the above equations) and in the axial plane (α, as usually specified, and as shown in Fig. 10.4). It follows from Fig. 10.7 that

$$\tan \alpha_n = \tan \alpha \cos \lambda \tag{10.6}$$

For small helix angles, $\cos \lambda$ is often taken as unity.

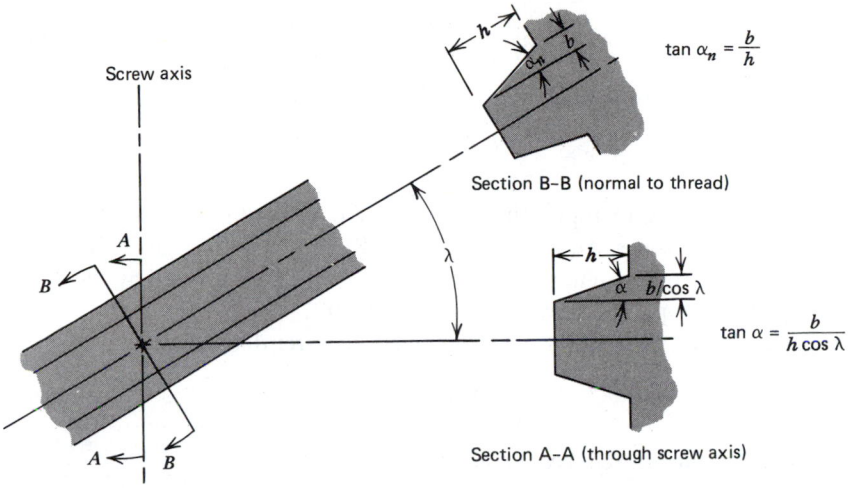

FIGURE 10.7.
Comparison of thread angles measured in axial and normal planes (α and α_n).

10.3.3 Overhauling and Self-Locking

A *self-locking screw* is one that requires a positive torque to lower the load; an *overhauling screw* is one that has low enough friction to enable the load to lower itself; that is, a negative external lowering torque must be maintained to keep the load from lowering. If collar friction can be neglected, Eq. (10.5) shows that a screw is self-locking if

$$f \geqq \frac{L \cos \alpha_n}{\pi d_m} \tag{10.7}$$

For a square thread, this simplifies to

$$f \geqq \frac{L}{\pi d_m}, \quad \text{or } f \geqq \tan \lambda \tag{10.7a}$$

A word of caution is appropriate at this point: even though a screw is self-locking under static conditions, it may overhaul when exposed to *vibration.* This is particularly important in the case of screw fasteners that tend to loosen because of vibration. (Section 10.8 treats special locking devices to prevent screws from overhauling or loosening under vibrating conditions.)

10.3.4 Efficiency

The work output of a power screw (as in the jacks, shown in Fig. 10.5) for one revolution of the rotating member is the product of force times distance, or WL. Corresponding work input is $2\pi T$. The ratio $WL/2\pi T$ is equal to efficiency. Substituting the expression for T in Eq. (10.4), with collar friction neglected, gives

$$\text{efficiency, } e = \frac{L}{\pi d_m}\left(\frac{\pi d_m \cos \alpha_n - fL}{\pi f d_m + L \cos \alpha_n}\right) \tag{10.8}$$

or, for the case of the square thread,

$$e = \frac{L}{\pi d_m}\left(\frac{\pi d_m - fL}{\pi f d_m + L}\right) \tag{10.8a}$$

It is always reassuring to approach an analysis from two different points of view and arrive at the same answer. Accordingly, Eq. (10.8) can be derived by defining efficiency as the ratio of torque required to raise the load with $f = 0$ divided by the actual torque required. This simple exercise is left to the reader.

Substitution of Eq. (10.1) into Eq. (10.8) gives, after routine simplification,

$$e = \frac{\cos \alpha_n - f \tan \lambda}{\cos \alpha_n + f \cot \lambda} \tag{10.9}$$

which, for the square thread, reduces further to

$$e = \frac{1 - f \tan \lambda}{1 + f \cot \lambda} \tag{10.9a}$$

Figure 10.8 gives a plot of efficiency as a function of coefficient of friction and lead angle. It is worth taking time to note that the plots come out as expected, in that:

1. The higher the coefficient of friction, the lower the efficiency.
2. Efficiency approaches zero as lead angle approaches zero because this represents doing a lot of friction work when moving the "block" in Fig. 10.6 around and around the thread without lifting the load very much.
3. Efficiency again approaches zero as lead angle approaches 90 deg., and efficiency also decreases slightly as the thread angle is increased from zero (square thread) to 14½ deg. (Acme thread). Furthermore, efficiency would also approach zero if the thread angle were to approach 90 deg. This is because the greater the angle that the thread surface makes with a plane perpendicular to the screw axis, the greater must be the normal force in order to provide a given load force (or to support a given weight). Increasing the normal force increases the friction force correspondingly because the relationship between them is simply the coefficient of friction. To illustrate this, imagine pushing down on a flat washer and rotating it back and forth, as in Fig. 10.9*a*. Now change the flat washer to a conical washer, as in 10.9*b*, and push it down with the same force, against a mating conical surface. It will be harder to turn back and forth because the "wedging action" increases the normal force, and therefore the friction force. Finally, imagine the cone angle steepened to the extent shown in 10.9*c*. Now the wedging action may be so great that it becomes almost impossible to turn the washer.

10.3.5 Rolling Contact

Figure 10.10 shows a *ball-bearing screw,* where sliding friction between screw and nut threads has been replaced with approximate rolling contact between the balls and the

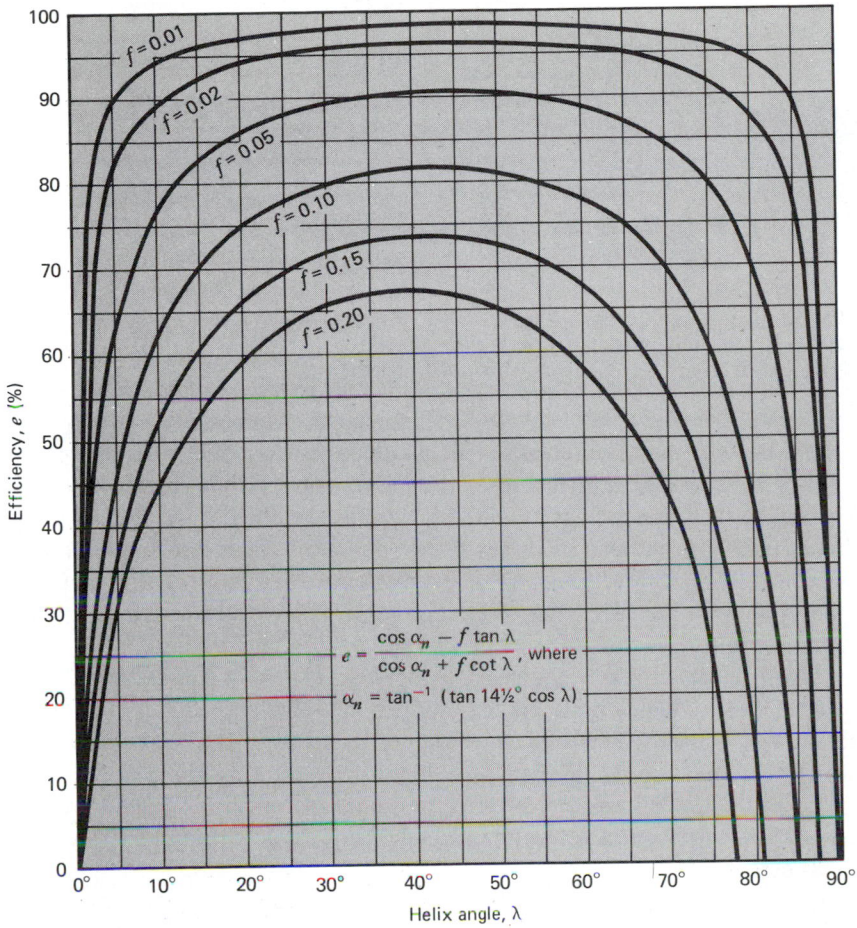

FIGURE 10.8.
Efficiency of Acme screw threads when collar friction is negligible. (Note: Values for square threads are higher by less than 1°.)

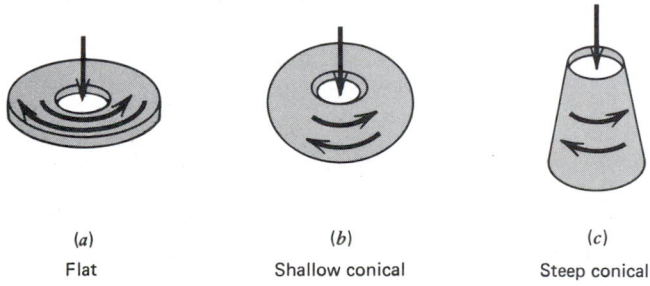

FIGURE 10.9.
Influence of "wedging action" on friction, illustrated by three load-carrying washers.

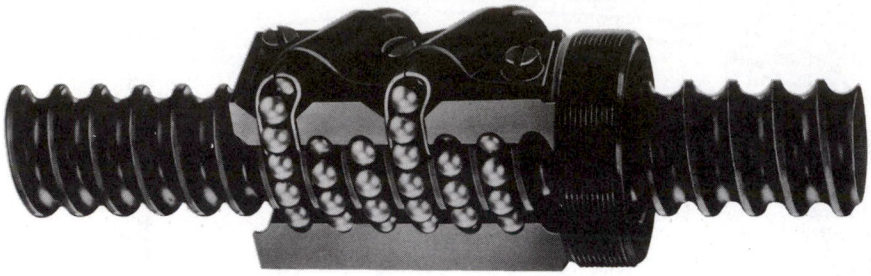

FIGURE 10.10.
Ball-bearing screw assembly with portion of nut cut away to show construction. (Courtesy Saginaw Steering Gear Division, General Motors Corp.)

grooves in the screw and nut. This gives drastically decreased friction, with efficiencies commonly 90 percent or higher. Because of the low friction, ball-bearing screws are usually overhauling. This means that a brake must be used to hold a load in place. On the other hand, it also means that the screw is *reversible* in that linear motion can be converted to relatively rapid rotary motion in applications where this is desirable. Operation is smooth, without the "slip-stick" action commonly observed in regular power screws (because of differences between static and sliding friction).

The load carrying capacity of ball-bearing screws is normally higher than that of regular power screws of the same diameter. The smaller size and lighter weight are often an advantage. On the other hand, buckling problems (of a long screw loaded in compression) and critical speed problems (of high-speed rotating screws) are more severe in some cases. Cleanliness and a thin film of lubricant are important to ball-bearing screws. Where dry operation is unavoidable, load capacity may be reduced by as much as 90 percent.

Ball-bearing screws are commonly used in aircraft landing gear retractors, large jet aircraft engine thrust reverser actuators, automatic door closers, antenna drives, hospital bed adjustors, machine tool controls, and numerous other applications.

Power screws are also available with patented arrangements for positioning *rollers* in the nut that contact the screw thread surface with a line contact rather than a point contact like the ball-bearing screw. These roller-equipped nuts provide low friction together with very high load carrying capacity. Hence, they find use in such applications as offshore oil-rig elevators, heavy construction jacks, and mining machinery.

SAMPLE PROBLEM 10.1 Acme Power Screw

A screw jack with a 1 in., double-thread Acme screw is used to raise a load of 1000 lb. A plain thrust collar of 1½ in. mean diameter is used. Coefficients of running friction are estimated as 0.12 and 0.09 for f and f_c, respectively.

a. Determine the screw pitch, lead, thread depth, mean pitch diameter, and helix angle.

b. Estimate the starting torque for raising and for lowering the load.

c. Estimate the efficiency of the jack when raising the load.

Solution

a. From Table 10.3, there are 5 threads per inch, hence $p = 0.2$ in. ■
Because of the double thread, $L = 2p$, or $L = 0.4$ in. ■
From Fig. 10.4*a*, thread depth $= p/2 = 0.1$ in. ■
From Fig. 10.4*a*, $d_m = d - p/2 = 1$ in. $- 0.1$ in. $= 0.9$ in. ■
From Eq. (10.1), $\lambda = \tan^{-1} L/\pi d_m = \tan^{-1} 0.4/\pi(0.9) = 8.05°$. ■

b. For starting, increase the given coefficients of friction by about ⅓, giving $f = 0.16$ and $f_c = 0.12$. Equation (10.4a) could be used with sufficient accuracy, but we will illustrate the complete solution, using Eq. (10.4):
First, find α_n from Eq. (10.6):

$$\alpha_n = \tan^{-1}(\tan\alpha\cos\lambda)$$
$$= \tan^{-1}(\tan 14.5° \cos 8.05°) = 14.36°$$

Then, substituting in Eq. (10.4),

$$T = \frac{Wd_m}{2}\left(\frac{f\pi d_m + L\cos\alpha_n}{\pi d_m\cos\alpha_n - fL}\right) + \frac{Wf_c d_c}{2}$$
$$= \frac{1000(0.9)}{2}\left(\frac{0.16\pi(0.9) + 0.4\cos 14.36°}{\pi(0.9)\cos 14.36° - 0.16(0.4)}\right) + \frac{1000(0.12)(1.5)}{2}$$
$$= 141.3 + 90: \; T = 231.3 \text{ lb in.}$$ ■

(With reference to Fig. 10.5*c*, this would correspond to a force of 19.3 lb on the end of a 12-in. handle. If Eq. (10.4a) is used, the answer is only slightly less: 228.8 lb in.). For lowering the load, use Eq. (10.5):

$$T = \frac{Wd_m}{2}\left(\frac{f\pi d_m - L\cos\alpha_n}{\pi d_m\cos\alpha_n + fL}\right) + \frac{Wf_c d_c}{2}$$
$$= \frac{1000(0.9)}{2}\left(\frac{0.16\pi(0.9) - 0.4\cos 14.36°}{\pi(0.9)\cos 14.36° + 0.16(0.4)}\right) + 90$$
$$= 10.4 + 90: \quad T = 100.4 \text{ lb in.}$$ ■

(Eq. (10.5a) gives a torque of 98.2 lb in.)

c. Repeating the substitution in Eq. (10.4) except changing the coefficient of friction to the running values of 0.12 and 0.09 gives a torque required to raise the load, once motion is started, of $121.5 + 67.5 = 189$ lb in. Substituting once more in Eq. (10.4) except changing both friction coefficients to zero gives a raising torque requirement of $63.7 + 0 = 63.7$ lb in. Efficiency is the ratio of friction-free torque to actual torque, or

$$e = \frac{63.7}{189} = 33.7 \text{ percent}$$ ■

If a ball thrust bearing were used so that collar friction could be neglected, the efficiency would increase to $63.7/121.5 = 52$ percent. This would correspond to the efficiency of the screw itself, and agrees with the plotted value in Fig. 10.8.

10.4 STATIC SCREW STRESSES

This section applies equally to power screws, covered in the previous section, and to threaded fasteners, discussed in the following sections. Let us consider separately the various stresses to which these members are subjected.

10.4.1 Torsion

Power screws in operation, and threaded fasteners during tightening, are subjected to torsional stresses,

$$\tau = \frac{Tc}{J} = \frac{16T}{\pi d^3} \tag{4.3, 4.4}$$

where d is the thread root diameter, d_r, obtained from Fig. 10.4 (for power screws) or Tables 10.1 and 10.2 (for threaded fasteners). If the screw or bolt is hollow, then $d_r^3 - d_i^3$ should be substituted for d^3 in the stress equation, where d_i represents the inside diameter.

Where collar friction is negligible, the torque transmitted through a power screw is the full applied torque. With threaded fasteners, the equivalent of substantial collar friction is normally present, in which case it is customary to assume that the torque transmitted through the threaded section is approximately half the wrench torque.

10.4.2 Axial Load

Power screws are subjected to direct *"P/A"* tensile and compressive stresses; threaded fasteners are normally subjected only to tension. The effective area for threaded fasteners is the tensile stress area, A_t (see Tables 10.1 and 10.2). For power screws, a similar "tensile stress area" could be computed, but this is not ordinarily done because axial stresses are seldom critical. A simple and conservative approximation of power screw axial stresses can be based on the minor or root diameter, d_r.

The distribution of the axial stress near the ends of the loaded portion of a screw is far from uniform. This is not of concern now because axial stresses are quite low in power screws and because threaded fasteners should always have enough ductility to permit local yielding at thread roots without damage. When considering *fatigue* loading of bolts (Section 10.11), this stress concentration is *very* important.

10.4.3 Combined Torsion and Axial Load

The combination of the above stresses can be treated in the normal fashion presented in Sections 4.9 and 4.10, with the distortion energy theory used as a criterion for yielding. In the case of threaded fasteners, it is normal for some yielding to occur at the thread roots during initial tightening.

10.4.4 Thread Bearing (Compressive) Stress, and Its Distribution Among the Threads in Contact

Figure 10.11 illustrates the "force-flow" through a bolt and nut used to clamp two members together. Compression between the screw and nut threads exists at threads numbered 1, 2, and 3. This type of direct compression is often called *bearing,* and the area used for the P/A stress calculation is the *projected* area that, for each thread, is $\pi(d^2 - d_i^2)/4$. The number of threads in contact is seen from the figure to be t/p. Thus,

$$\sigma = \frac{4P}{\pi(d^2 - d_i^2)} \cdot \frac{p}{t} \tag{10.10}$$

Diameter d_i is the minor diameter of the internal thread. For threaded fasteners this can be approximated by d_r, which is tabulated in Table 10.1. Exact values are given in ANSI B1.1 - 1974 and various handbooks, but these are not usually necessary because thread bearing stresses are seldom critical.

Equation (10.10) gives an *average* value of bearing stress. The stress is *not* uni-

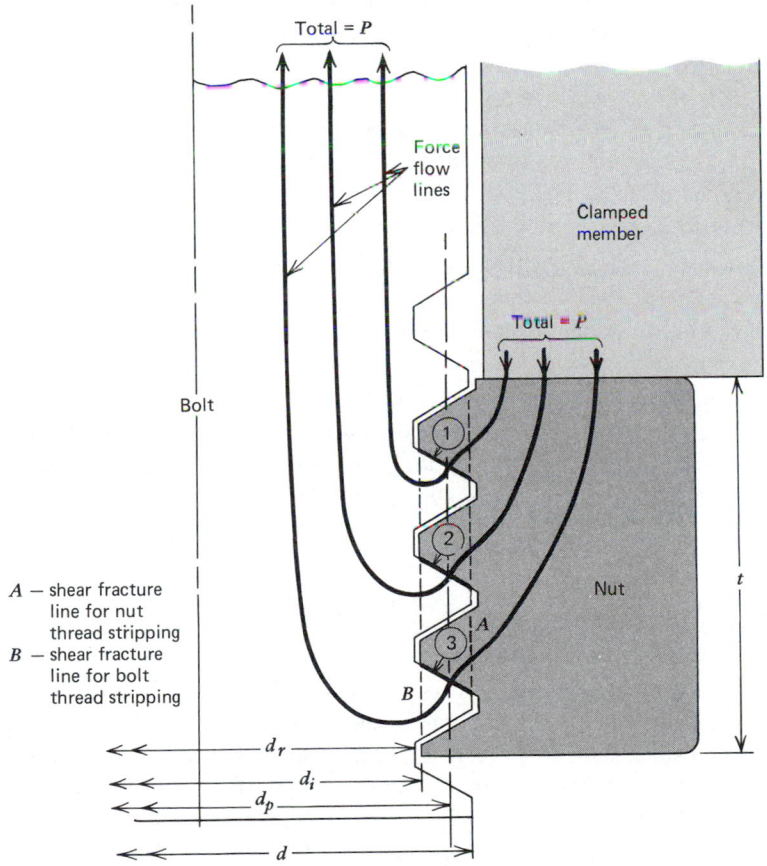

FIGURE 10.11.
Force flow for bolt in tension.

formly distributed because of factors such as bending of the threads as cantilever beams, and manufacturing variations from the theoretical geometry. Furthermore, a study of Fig. 10.11 reveals two important factors causing *thread 1* to carry more than its share of the load:

1. The load is shared among the three threads as *redundant load carrying members.* The shortest (and stiffest) path is through thread 1. Thus, it carries the greatest portion of the load (see Section 2.6).
2. The applied load causes the threaded portion of the bolt to be in *tension,* whereas the mating portion of the nut is in *compression.* The resulting deflections *slightly* increase bolt pitch and decrease nut pitch. This tends to relieve the pressure on threads 2 and 3.

The problem of overcoming the above tendency, and obtaining a more nearly equal distribution of loads among the threads in contact, is an important one, as will be seen when considering the fatigue loading of bolts (Section 10.11). This problem continues to challenge the ingenuity of engineers concerned with fastener design and development. Three ideas that have been used are:

1. Making the nut from a softer material than the bolt, so that the highly loaded first thread will deflect (either elastically or plastically), thereby transferring more of the load to the other threads. This may require increasing the number of threads in contact in order to maintain adequate strength.
2. Manufacturing the nut threads with a *very slightly* greater pitch than the bolt threads so that the two pitches are theoretically equal *after* the load is applied. The thread clearances and precision of manufacture must, of course, be such that the nut and bolt can be readily assembled.
3. Modifying the nut design as shown in Fig. 10.12. Here, the nut loading puts the region of the top threads in *tension,* thus causing elastic changes in pitch that approximately match the changes in bolt pitch. Such special nuts are expensive, and have been used only in critical applications involving fatigue loading.

10.4.5 Thread Shear ("Stripping") Stress, and Nut Thickness Requirement

With reference to Fig. 10.11, if the nut material is weaker than the bolt material in shear (and this is *usually* the case), a sufficient overload would "strip" the threads from the nut along cylindrical surface A. If the bolt material is weaker in shear, the failure surface would be B. From the thread geometry shown in Fig. 10.2, the shear area is equal to $\pi d(0.75t)$, where d is the diameter of the shear fracture surface involved.

Let us now determine the nut thickness (or depth of engagement of a screw in a tapped hole) needed to provide a *balance* between bolt tensile strength and thread stripping strength if both members (bolt, and nut or tapped hole) are made of the same material. The bolt tensile force required to yield the entire threaded cross section is

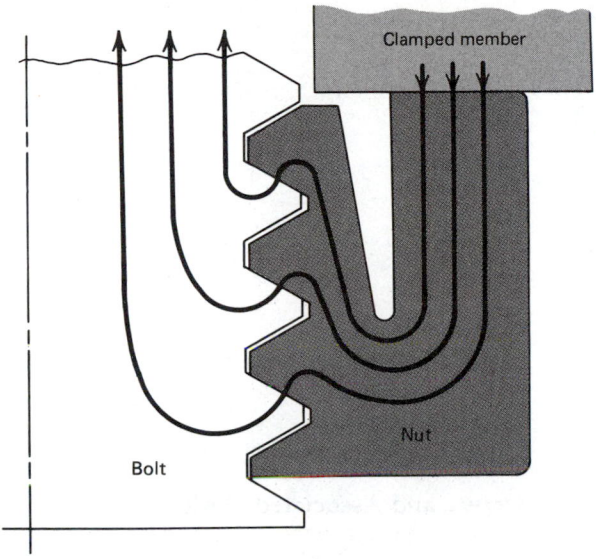

FIGURE 10.12.
Special nut provides more nearly equal distribution of load among threads in contact.

$$F = A_t S_y \approx \frac{\pi}{4}(0.9d)^2 S_y$$

where d is the major diameter of the thread. With reference to Fig. 10.11, the bolt tensile load required to yield the entire thread-stripping failure surface of the nut is

$$F = \pi d(0.75t)S_{sy} \approx \pi d(0.75t)(0.58S_y)$$

Equating the above two expressions for F indicates that bolt tensile and thread-stripping strengths are balanced when the nut thickness is approximately

$$t = 0.47d \qquad \textbf{(d)}$$

Since nuts are usually softer than bolts in order to allow *slight* yielding of the top thread(s) and thus distribute the load more uniformly among the threads in contact (see point number 1 at the end of the previous section), the standard nut thickness is approximately $\frac{7}{8}d$.

10.4.6 Transverse Shear Loading, and Providing Transverse Alignment

Bolts are sometimes subjected to transverse shear loading, as illustrated in Figs. 4.3 and 4.4 and discussed in Section 4.3. Often these shear loads are transmitted by friction, where the friction load carrying capacity is equal to the bolt tension times the coefficient of friction at the clamped interface. For the case of double shear illustrated in Fig. 4.4, the friction load capacity would be twice this amount.

Sometimes bolts are required to provide precise alignment of mating members, and are made with a pilot surface as shown in Fig. 10.13.

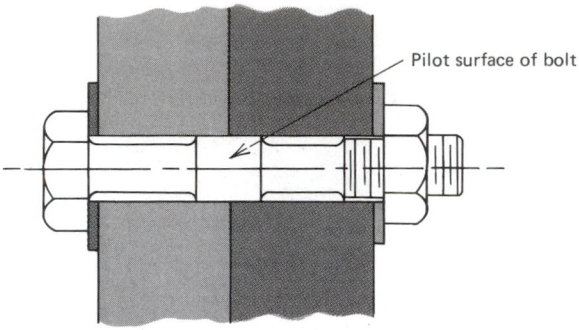

FIGURE 10.13.
Bolt with pilot surface.

10.4.7 Column Loading of Power Screws, and Associated Design Details

Long power screws loaded in compression must be designed with *buckling* in mind. Sections 5.10–5.14 apply. Before getting involved with this, however, it is important first to make sure that it is *necessary* to subject the screws to compression. Often, a simple redesign permits the screws to be in tension. For example, Fig. 10.14*a* shows a press (it could represent a household trash compactor) with the screws in compression. Figure 10.14*b* shows an alternative design with the screws in tension. The latter is obviously to be preferred.

The two presses in Fig. 10.14 provide excellent examples for using the "force-flow concept," described in Section 2.4. In the press shown in 10.14*a*, forces flow from the moving press head through the screws to the ball thrust bearings, then through the top, sides, and bottom of the frame, and finally into the bottom surface of the material being compressed. Because all sections of the frame carry load, they are drawn heavy. In contrast, the press shown in 10.14*b* has a short load path from the press head through the screws to the ball thrust bearings, and then through the bottom of the frame to the material being compressed. Because they carry no significant loads, the sides and top of the frame can be very much lighter. *It is almost always desirable to keep load paths as short and compact as possible.*

Another advantage of the press in 10.14*b* is that the axial free play of the screws is easily controlled because all four thrust bearings involve surfaces of the bottom frame member. In contrast, the screw axial free play in 10.14*a* involves tolerances and deflections of the screw length between thrust bearings and of the frame height between thrust bearings.

A construction detail of the presses is worth noting: In order to permit assembly of the press in 10.14*b*, the thrust collars bearing against the plain thrust washers at the top surface of the bottom frame member must be removable from the screws. Note that the only load carried by these thrust collars is the gravity load of the press head when there is no material being compressed.

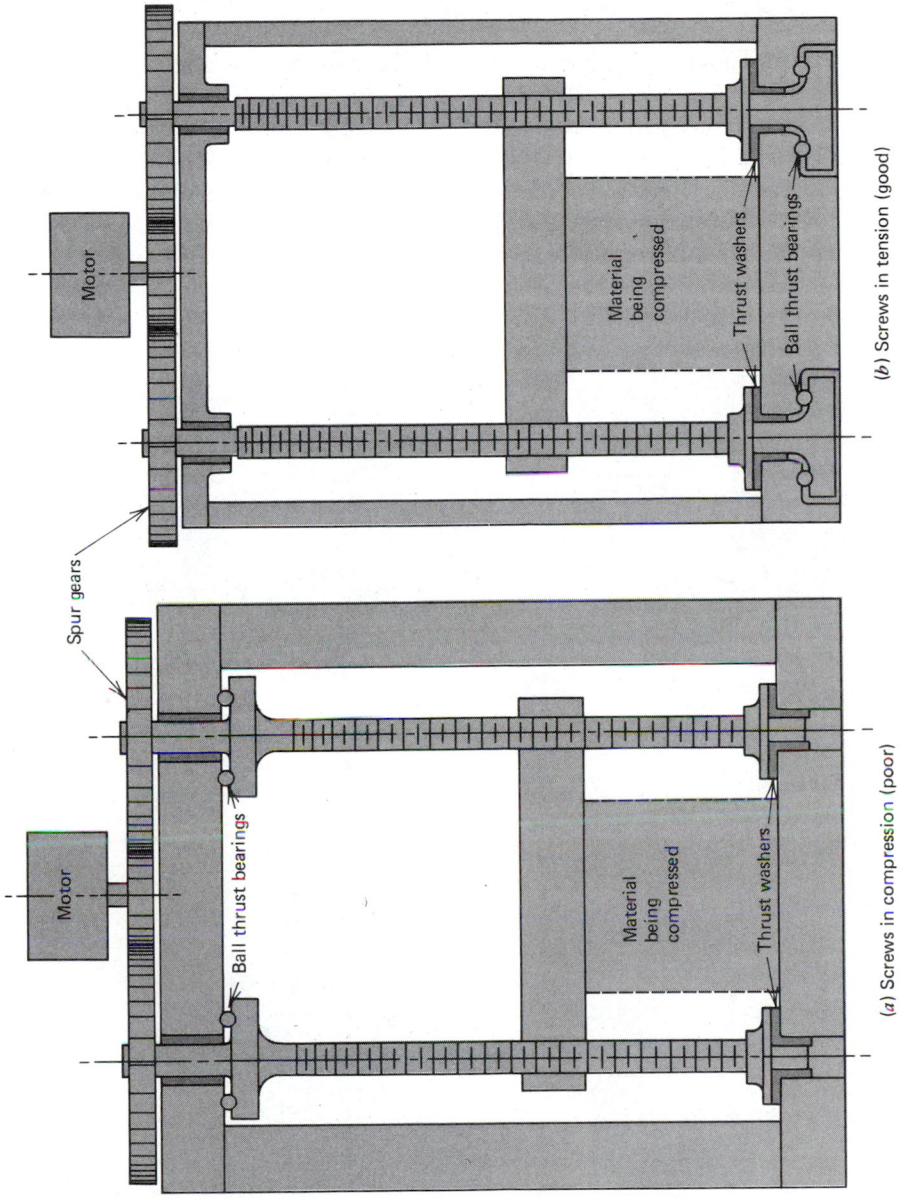

FIGURE 10.14.
Alternative screw press arrangements.

10.5 THREADED FASTENER TYPES

Figure 10.15 shows four basic types of threaded fasteners. Screws and bolts are by far the most common types, and the difference between them is only one of intended use: bolts are intended for use with nuts; screws are intended for screwing into tapped holes. Sometimes screws are supplied with a captive washer (usually a lock washer) under the head, in which case they are called *sems*. Sems save assembly time, and eliminate the possibility that a screw will be installed without its specified washer.

A stud is threaded on both ends, and is usually screwed permanently into a tapped hole. Threads on the two ends may or may not be identical. Threaded rod is the least

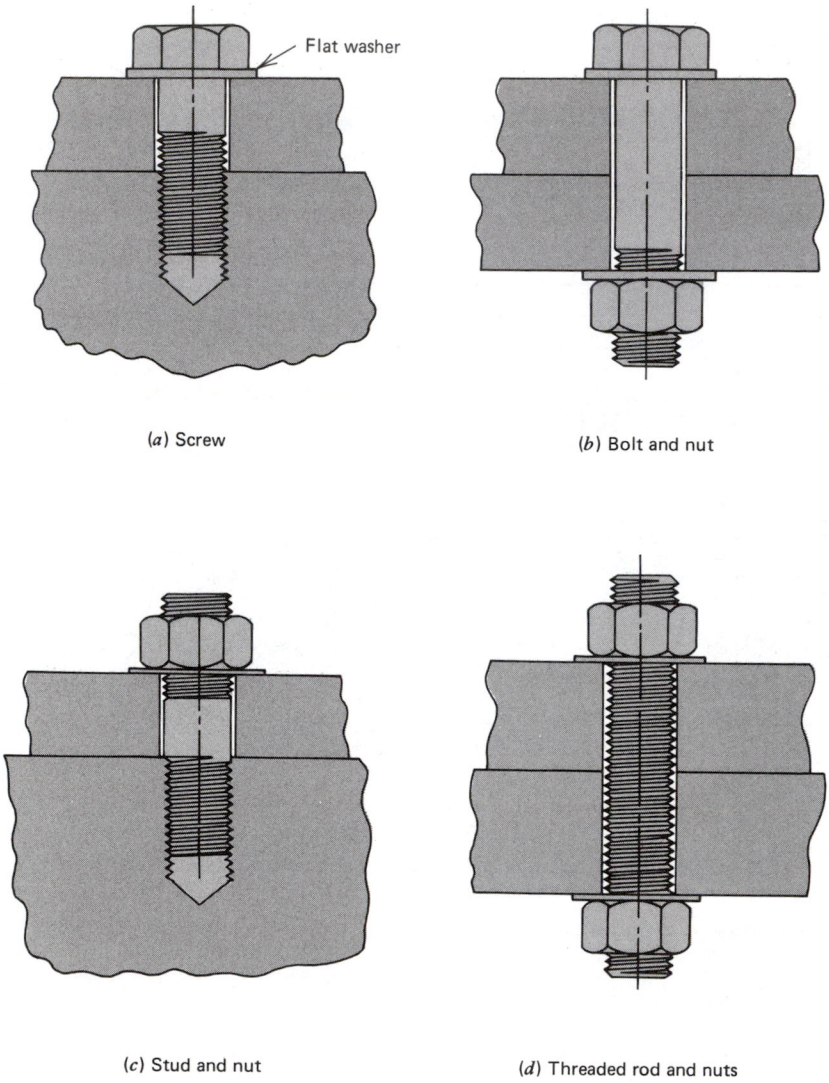

FIGURE 10.15.
Basic threaded fastener types.

common type. It is usually used when a very long threaded member is desired. Threaded rod can often be purchased from a hardware store in lengths of a few feet, and then cut off as required.

Figure 10.16 shows most of the common fastener head types. An interesting exception to the rule that a bolt can also serve as a screw by using it with a tapped hole

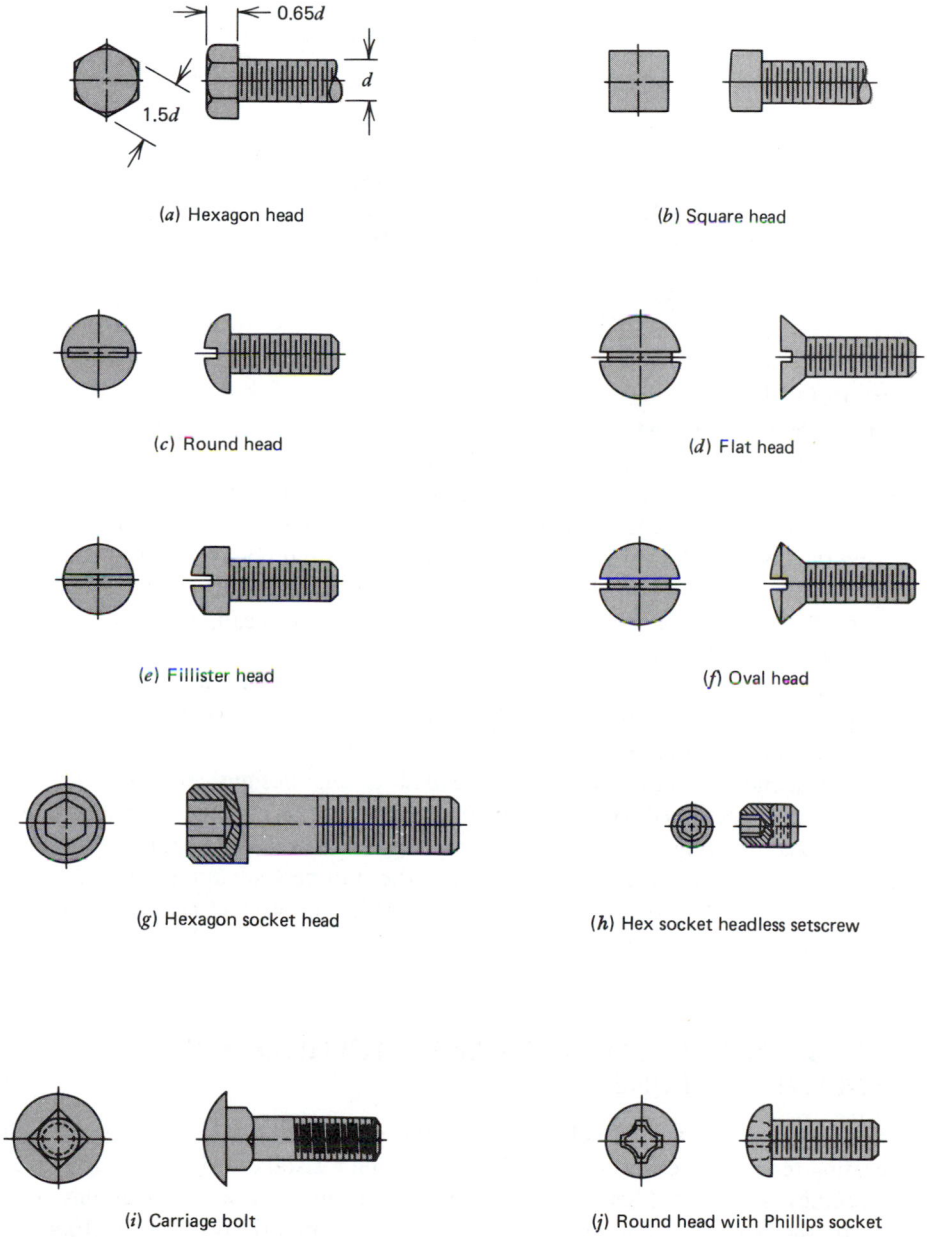

FIGURE 10.16.
Some common screw (and bolt) head types.

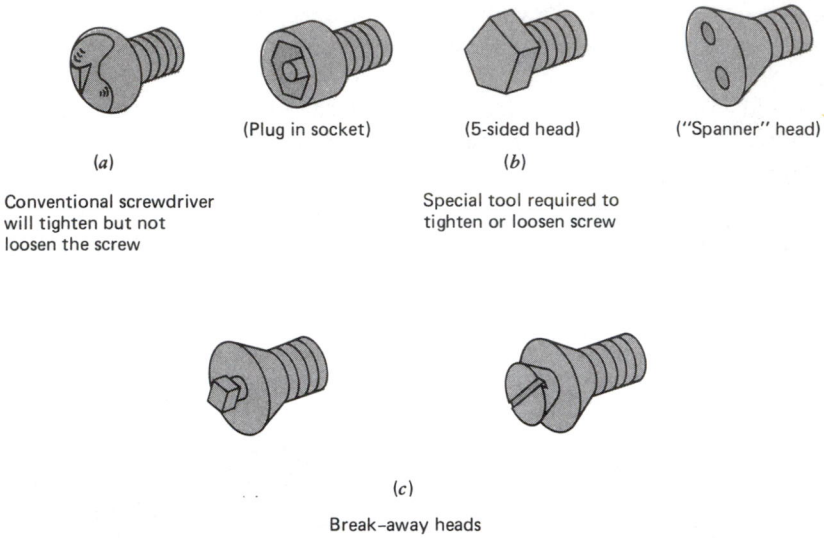

FIGURE 10.17.
"Tamper-resistant" screw heads.

(rather than a nut) is the carriage bolt. These are used with soft materials (particularly wood) so that the square corners under the head can be forced into the material, thereby preventing the bolt from turning. Hexagonal-head screws and bolts are most commonly used for connecting machine components. Sometimes they cannot be used because of insufficient clearance to put a socket or end wrench on the head. In such cases the hexagonal-socket head is most commonly used.

In recent years there has been an increasing need for screws that are resistant to tampering by unauthorized personnel. Figure 10.17 illustrates several proposed solutions that have been marketed.

An almost endless number of special threaded fastener designs continue to appear. Some are specially designed for a specific application. Others embody proprietary features that appeal to a segment of the fastener market. Not only is ingenuity required to devise better threaded fasteners, but also to *use* them to best advantage in the design of a product. Chow [2] gives several examples of reducing product cost through judicious selection and application of fasteners.

10.6 FASTENER MATERIALS AND METHODS OF MANUFACTURE

Materials for screws, nuts, and bolts are normally selected on the basis of strength (at the operating temperatures involved), weight, corrosion resistance, magnetic properties, life expectancy, and cost. Most fasteners are made from steel of specifications standardized by the Society of Automotive Engineers and summarized in Tables 10.4 and 10.5 (this is not true for many aerospace and other highly critical fasteners). Heading

TABLE 10.4
Specifications for Steel Used in Inch Series Screws and Bolts

SAE Grade	Diameter d (in.)	Proof Load (Strength)[a] S_p (ksi)	Yield Strength[b] S_y (ksi)	Tensile Strength S_u (ksi)	Elongation, Minimum (%)	Reduction of Area, Minimum (%)	Core Hardness, Rockwell Min	Core Hardness, Rockwell Max	Grade Identification Marking on Bolt Head
1	¼ thru 1½	33	36	60	18	35	B70	B100	None
2	¼ thru ¾	55	57	74	18	35	B80	B100	None
2	Over ¾ to 1½	33	36	60	18	35	B70	B100	None
5	¼ thru 1	85	92	120	14	35	C25	C34	
5	Over 1 to 1½	74	81	105	14	35	C19	C30	
5.2	¼ thru 1	85	92	120	14	35	C26	C36	
7	¼ thru 1½	105	115	133	12	35	C28	C34	
8	¼ thru 1½	120	130	150	12	35	C33	C39	

[a]Proof load (strength) corresponds to permanent deformation not over 0.0001 in. measured on actual bolts.
[b]Yield strength corresponds to 0.2 percent offset measured on machine test specimens.
Source: Society of Automotive Engineers standard J429k (1979).

TABLE 10.5
Specifications for Steel Used in Millimeter Series Screws and Bolts

SAE Class	Diameter d (mm)	Proof Load (Strength)[a] S_p (MPa)	Yield Strength[b] S_y (MPa)	Tensile Strength S_u (MPa)	Elongation, Minimum (%)	Reduction of Area, Minimum (%)	Core Hardness, Rockwell	
							Min	Max
4.6	5 thru 36	225	240	400	22	35	B67	B87
4.8	1.6 thru 16	310	—	420	—	—	B71	B87
5.8	5 thru 24	380	—	520	—	—	B82	B95
8.8	17 thru 36	600	660	830	12	35	C23	C34
9.8	1.6 thru 16	650	—	900	—	—	C27	C36
10.9	6 thru 36	830	940	1040	9	35	C33	C39
12.9	1.6 thru 36	970	1100	1220	8	35	C38	C44

[a]Proof load (strength) corresponds to permanent deformation not over 0.0025 mm measured on actual bolts.
[b]Yield strength corresponds to 0.2 percent offset measured on machine test specimens.
Source: Society of Automotive Engineers standard J1199 (1979).

is done cold for diameters up to about ¾ in., and heated for larger sizes. Threads are usually formed by rolling between dies that force the material to cold form into the threaded contour of the die grooves. The threads thus formed are stronger than cut or ground threads in fatigue and impact because of cold working, favorable (compressive) residual stresses at the thread roots, and a more favorable grain structure. Because of these advantages, high-strength screws and bolts should have rolled threads. Furthermore, greatest fatigue strength is obtained if the threads are rolled *after* heat-treatment so that the resulting work-hardening and favorable residual stresses are not lost. (It is, of course, less costly to roll-form the threads before hardening.)

Fasteners are also made of aluminum (the most commonly used alloys being 2024-T4, 2111-T3, and 6061-T6), brass, copper, nickel, Monel, Inconel, stainless steel, titanium, beryllium, and various plastics. For any application, the fastener material must be considered in connection with potential corrosion problems associated with the anticipated environment and the other metals involved (Section 9.2). Also, appropriate coatings should be considered with respect to corrosion protection, thread friction, and thread wear considerations (Section 9.15).

10.7 BOLT TIGHTENING AND INITIAL TENSION

For most applications, screws and nut-bolt assemblies should ideally be tightened to produce an initial tensile force, F_i, nearly equal to the full "proof load," which is defined as the maximum tensile force that does not produce a normally measurable permanent set. (This is a little less than the tensile force producing a 0.2 percent offset elongation associated with standard tests to determine S_y.) On this basis, initial tensions are commonly specified in accordance with the equation

$$F_i = K_i A_t S_p \tag{10.11}$$

where A_t is the tensile stress area of the thread (Tables 10.1 and 10.2), S_p is the "proof strength" of the material (Tables 10.4 and 10.5), and K_i is a constant, usually specified in the range of 0.75 to 1.0. For ordinary applications involving static loading, let $K_i \approx 0.9$, or

$$F_i = 0.9 A_t S_p \tag{10.11a}$$

Briefly, the rationale behind so high an initial tension is:

1. For loads tending to separate rigid members (as in Fig. 10.30), the bolt load cannot be increased very much unless the members do actually separate, and the higher the initial bolt tension the less likely the members are to separate; and
2. For loads tending to shear the bolt (as in Fig. 10.31), the higher the initial tension the greater the friction forces resisting the relative motion in shear.

Further implications of the initial bolt tension for the case of fatigue loading will be discussed in Section 10.11.

Another point to be considered is that the tightening of a screw or nut imparts a *torsional* stress to the screw or bolt, along with the initial tensile stress. During initial

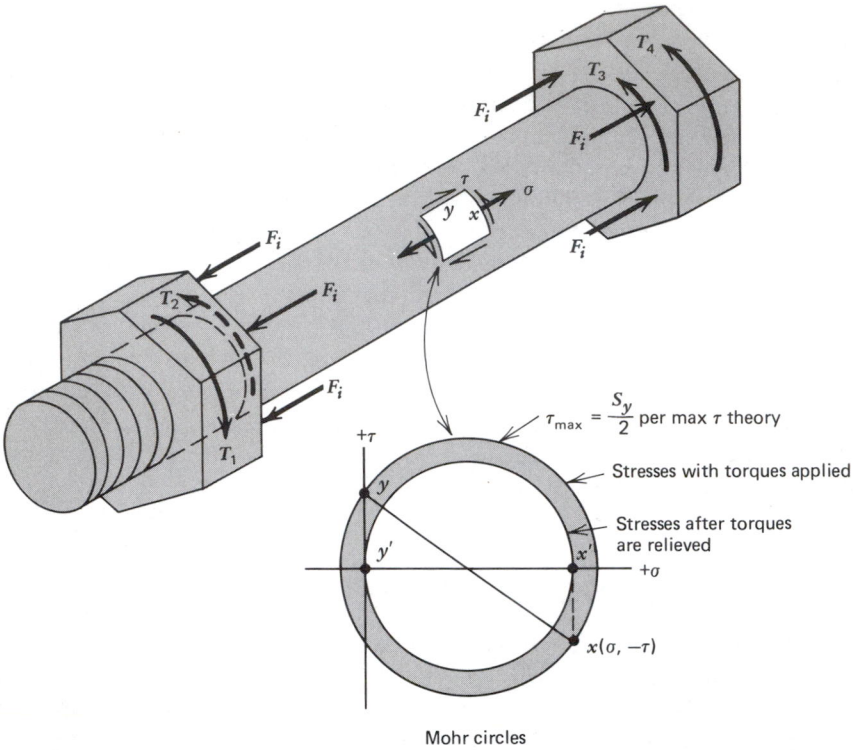

FIGURE 10.18.
Bolt loads and stresses due to initial tightening of nut. $\Sigma M = 0$ for the bolt and nut assembly shown; that is, $T_1 = T_2 + T_3 + T_4$, where: T_1 = nut wrench torque, T_2 = nut face friction torque = $fF_i\, r_n$ (where r_n is the effective radius of nut face friction forces), T_3 = bolt head friction torque $\leqq fF_i\, r_h$ (where r_h is effective radius of bolt head friction forces), T_4 = wrench torque required to keep bolt head from turning. Note that $T_4 = 0$ if $fF_i\, r_h > T_1 - T_2$.

use, the screw or bolt usually "unwinds" very slightly, relieving most or all of the torsion.

To illustrate this point, Fig. 10.18 shows the loads applied to a bolt and nut assembly during tightening. Remember that the frictional forces and associated torques vary considerably with materials, finishes, cleanliness, lubrication, and so on. Also, this figure assumes the ideal case of the bolt axis being *exactly* perpendicular to all clamped surfaces so that no bending is imparted to the bolt. An important point results from the difference between the static and dynamic coefficients of friction: suppose that wrench torque T_1 is progressively increased to the full specified value with *continuous* rotation of the nut. The resulting clamping force, F_i, will be higher than if the nut rotation is allowed to stop momentarily at, say 80 percent of full torque. The greater value of static friction may then be such that subsequent application of full torque will not cause further nut rotation, with the result that F_i will be less than desired.

Figure 10.18 also illustrates bolt stress, both initially (tightening torque still

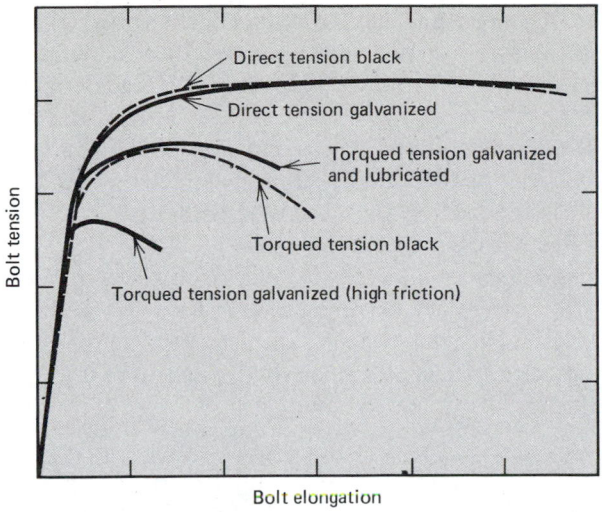

FIGURE 10.19.
Bolt tension versus elongation, resulting from tightening by torquing versus direct tensioning, and for black oxide versus galvanized surfaces [5]. (Note: Direct tension is produced by hydraulic loading; hence no torsional stresses are produced.)

applied) and finally (with "unwinding" having relieved all torsion). For clarity, the stress element shown is in the shank. The yield-limited stresses shown would more appropriately be associated with a cross section in the plane of a thread. An important point that does *not* show up in Fig. 10.18 is that the torsional stress is also dependent on the thread friction between the bolt and nut. For example, if this friction is very high, there will be a substantial holding torque T_4 needed to prevent bolt rotation, and the torsional stresses in the bolt can be so high that yielding is reached at relatively low values of F_i.

Figure 10.19 illustrates the implications of the above in terms of (a) the initial tension that can be achieved with a given bolt and (b) the amount of elongation that can be achieved before over-tightening produces fracture of the bolt.

An accurate determination of the bolt tensile load produced during tightening is not easy to make. One accurate way is to use a special bolt with an axially drilled hole in which an electric-resistance strain gage is mounted. Another method employs ultrasonic equipment to measure the length of the bolt before and after tightening. (Both the actual lengthening of the bolt and the introduction of tensile stresses increase the time required for an ultrasonic pulse to travel from one end of the bolt to the other, and return.) A method long used for low production rate assembly of critical parts is to measure with a micrometer the length of the bolt before assembly, and to then tighten the nut until the bolt elongates the desired amount. This can only be done, of course, if both ends of the bolt are accessible. A modern method suited for automated assembly operations involves continuous monitoring of the wrench torque and nut rotation. When a computer determines that the relationship of these quantities indicates the onset of yielding, the wrench is disengaged. A more mundane, but often effective, method is to "seat" the surfaces by tightening the screw or nut very tight and then loosening it, tighten the screw or the nut "finger tight," and turn it (with a wrench) a predetermined additional angle.

The most common method of tightening a bolt a measured amount is probably to use a torque wrench. The accuracy of this method can be seriously limited by variations in friction. Normal torque-wrench usage controls initial tension within perhaps ± 30 percent; with special care, ± 15 percent is considered reasonable.

An equation relating tightening torque to initial tension can be obtained from Eq. (10.4) by recognizing that load W of a screw jack corresponds to F_i for a bolt, and that collar friction in the jack corresponds to friction on the flat surface of the nut or under the screw head. Using a rough average coefficient of friction (both f and f_c) of 0.15, Eq. (10.4) gives, for standard screw threads,

$$T = 0.2F_i d \tag{10.12}$$

where d is the nominal major diameter of the thread. Remember, this is only an *approximate* relationship, dependent upon "average" conditions of thread friction.

A very common way to tighten a screw or nut is, of course, to merely choose a conventional wrench and tighten it until it "feels" right. Although this method would never be *specified* for a critical fastener, an engineer should have *some* idea of how tight an "ordinary mechanic" might be expected to tighten bolts of various sizes using ordinary wrenches. A study made near the turn of the century indicated this to be

$$F_i \text{ (lb)} \approx 16{,}000d \text{ (in.)} \tag{e}$$

Despite its limitations, Eq. (e) does point up the important fact that whereas the tensile load carrying capacity of bolts increases with the *square* of the diameter (and torsional capacity with the cube of diameter), the natural tendency is to tighten to initial tension values that increase only as the *first power* of the diameter. This means that small bolts and screws tend to get twisted off, and that very large ones tend to be undertightened.

A word of caution is in order regarding the loss of bolt initial tension during operation. When "rigid" parts are bolted together, the elastic deflection of the parts may be only a hundredth of a millimeter or less. Should the loading cause any creep, or subsequent flattening of minute high spots on the surfaces, much of the bolt initial tension will be lost. Loss of initial tension can also be caused by wear or corrosion of the mating surfaces, or by the "squeezing out" of surface films. Tests have shown that a typical joint loses about 5 percent of its initial tension within a few minutes, and that various relaxation effects result in the loss of about another 5 percent within a few weeks. Long-term stress relaxation is a problem with high-temperature joints, as in jet engines and nuclear reactors. These applications require bolts made of special high-temperature alloys. Because of these losses in initial tension (and because of possible thread loosening, which is discussed in the following article), it is sometimes necessary to retighten bolts periodically.

10.8 THREAD LOOSENING AND THREAD LOCKING

An inherent advantage of threaded fasteners (in comparison with riveting, welding, or cementing) for many applications is that they *can* be easily and nondestructively disassembled. A *disadvantage* that has plagued engineers since early days is that they sometimes loosen and disassemble themselves! The concept explaining *why* threads loosen is a very simple one; it is given in the next two paragraphs.

In Fig. 10.6 the tightening of a nut (which is the same as the lifting of a load with a power screw) is represented by a small block on an inclined plane. If friction is sufficient to prevent the block from sliding back down the plane, the thread is said to be self-locking. *All* threaded fasteners are designed to have a small enough helix angle (λ) and a high enough coefficient of friction (f) to be self-locking *under static conditions.* However, if *any* relative motion occurs between the bolt and nut threads (inclined plane and block), the nut tends to loosen (block slides down inclined plane). This relative motion is most often caused by vibration, but can be due to other causes such as differential thermal expansion, or slight dilation and contraction of the nut with changes in axial bolt load. This dilation and contraction represents very small changes in nut thread diameter, and is due to the "wedging action" of the loaded tapered threads. For example, this would *not* occur with a square thread (see Fig. 10.4*c*).

To understand why *any* relative motion tends to cause thread loosening, place a small solid object (as a pencil or pen) on a slightly inclined plane (as a book) and lightly tap the edge of the plane in *any* direction. With the tapping just sufficient to cause the slightest relative motion, the small object inexorably slides down the incline. An analogy familiar to car drivers in northern climates is that of a car on an ice-covered crowned road. If traction is lost (by braking or by applying too much torque to the driving wheels), the car slides laterally to the side of the road. Note that in this case the friction between rubber and ice is ample to prevent lateral sliding as long as there is no relative motion, but *not* when there is sliding in any direction.

The following are among the factors influencing thread loosening:

1. The greater the helix angle (i.e., the greater the slope of the inclined plane), the greater the loosening tendency. Thus, coarse threads tend to loosen more easily than fine threads.
2. The greater the initial tightening, the greater is the frictional force that must be overcome to initiate loosening.
3. Soft or rough clamping surfaces tend to promote slight plastic flow that decreases the initial tightening tension, and thus promotes loosening.
4. Surface treatments and conditions that tend to increase the friction coefficient provide increased resistance to loosening.

The problem of thread loosening has resulted in numerous and ingenious special designs and design modifications, and it continues to challenge the engineer to find increasingly positive and inexpensive solutions. A brief summary of the common solutions used to date is given below.

Figure 10.20 shows the familiar helical and toothed lock washers. They work on

(*a*) Helical (split) type

(*b*) Twisted-tooth type (Teeth may be external, as in this illustration, or internal.)

FIGURE 10.20.
Common types of lock washers.

the principle of flattening the hard, sharp-cornered projections that tend to "bite" into the mating metal surfaces and resist loosening by a "wedging" action.

Slotted and castle nuts (Fig. 10.21) are used with a cotter pin or wire that fits in diametrically opposite slots and passes through a drilled hole in the bolt. This provides a positive lock but may require slight over or under tightening in order to align a pair of slots with the bolt hole.

Locknuts generally are of two types—*free-spinning* and *prevailing-torque*. Figure 10.22 shows three varieties of free-spinning locknuts. These spin on freely until seated, and then produce a deflection upon tightening that causes a gripping of the mating bolt threads. The lockwashers shown in Fig. 10.20 are also "free-spinning" thread locking devices.

Figure 10.23 illustrates representative prevailing-torque locknuts. These develop a friction torque as soon as the threads are fully engaged. Hence, they require wrenching to the final seated position, and tend strongly to stay in position even when not tightened. Prevailing-torque nuts or fasteners usually provide the best solution in critical applications, particularly when extreme vibration is involved.

Sometimes two standard nuts (or one standard nut and a thin "jam nut") are used

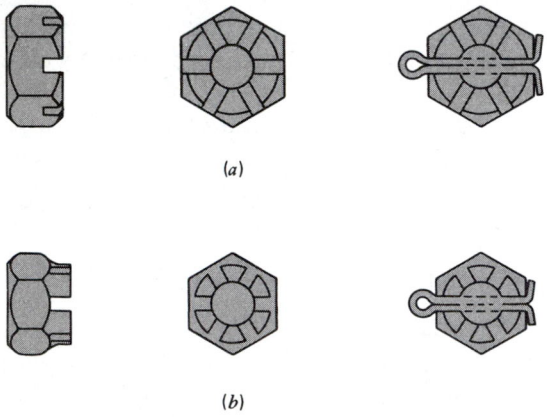

(*a*)

(*b*)

FIGURE 10.21.
(a) Slotted and (b) Castle nuts. Also shown with drilled bolt and cotter pin.

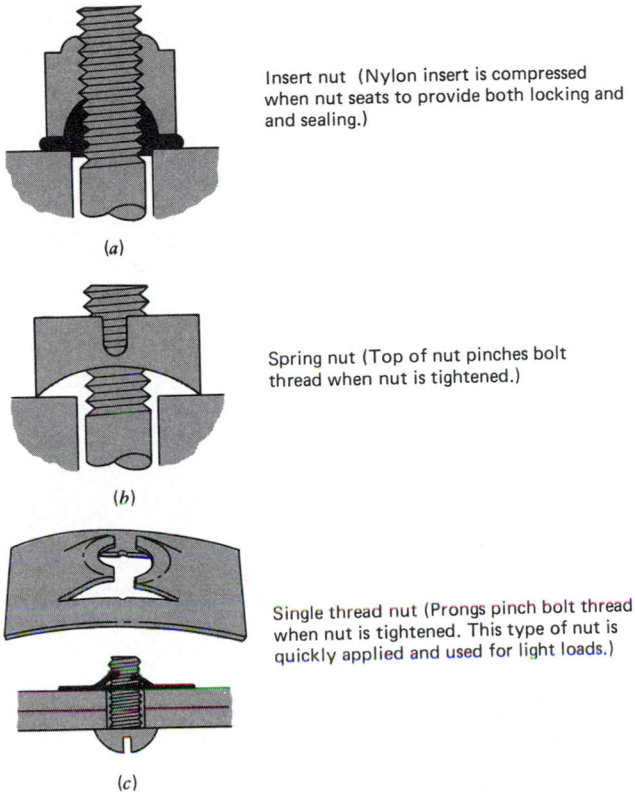

FIGURE 10.22.
Examples of free-spinning lock nuts.

together to obtain a locking action. The extra standard nut or the "jam nut" is installed first and tightened gently. Then the standard nut is installed and fully tightened against the extra nut.

Another means of providing thread locking is to coat the mating threads with a special adhesive or lacquer that may also act to seal the threads against fluid leakage. These adhesives are available in various grades, depending on whether or not ease of subsequent disassembly is important. Sometimes the adhesive coating is contained in a small capsule that is attached to the threads during manufacture. The capsule breaks during assembly, and the adhesive flows over the threaded surfaces.

Note that essentially all of the thread-locking methods discussed above can be applied to screws as well as to nut and bolt combinations.

10.9 BOLT TENSION WITH EXTERNAL JOINT-SEPARATING FORCE

Bolts are typically used to hold parts together in opposition to forces tending to pull, or slide, them apart. Common examples are connecting rod bolts, cylinder head bolts, and

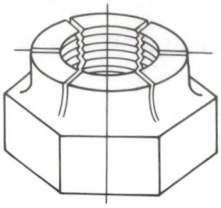

Spring-top nut
(Upper part of nut is tapered. Segments press against bolt threads.)

(*a*)

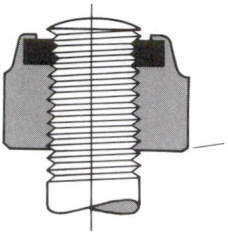

Nylon-insert nuts
(Collar or plug of nylon exerts friction grip on bolt threads.)

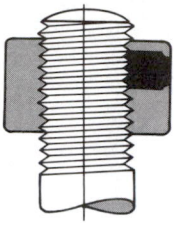

(*b*)

Distorted-thread nut
(Portion of nut thread is distorted to provide friction grip on bolt threads. In this example, the top of the nut has been swedged inward at three points.)

(*c*)

FIGURE 10.23.
Examples of prevailing-torque lock nuts. (*c* **Courtesy Russell, Bindsall & Ward Corp.**)

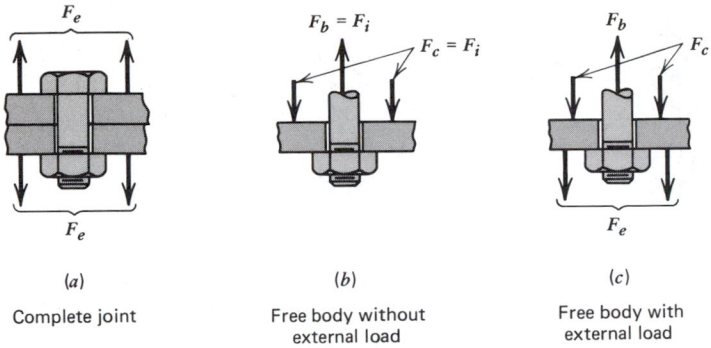

FIGURE 10.24.
Free-body study of bolt tensile loading.

so on. Figure 10.24*a* shows the general case of two parts connected with a bolt, and subjected to an external force F_e tending to separate them. Figure 10.24*b* shows a portion of this assembly as a free body. In this figure the nut has been tightened, but the external force has not yet been applied. The bolt axial load F_b and the clamping force between the two plates F_c are both equal to the initial tightening force F_i. Figure 10.24*c* shows the same members as a free body after external force F_e has been applied. Equilibrium considerations require one or both of the following: (1) an increase in F_b and (2) a decrease in F_c. The relative magnitudes of the changes in F_b and F_c depend on the relative elasticities involved.

To assist the reader in visualizing the significance of the *relative* elasticities in bolted joints, Figs. 10.25 and 10.26 illustrate two extreme cases. Figure 10.25*a* shows a cover plate bolted on the end of a pressure vessel (or cylinder head bolted to the cylinder block of a piston-type compressor). The key feature is the thick rubber gasket that is so soft that the other parts can be considered infinitely rigid *in comparison*. When

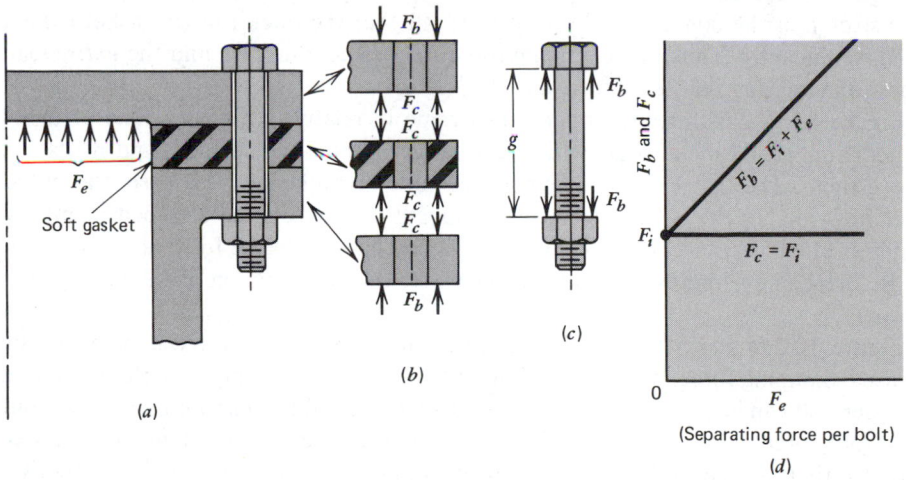

FIGURE 10.25.
F_b and F_c vs F_e for soft clamped members—rigid bolt.

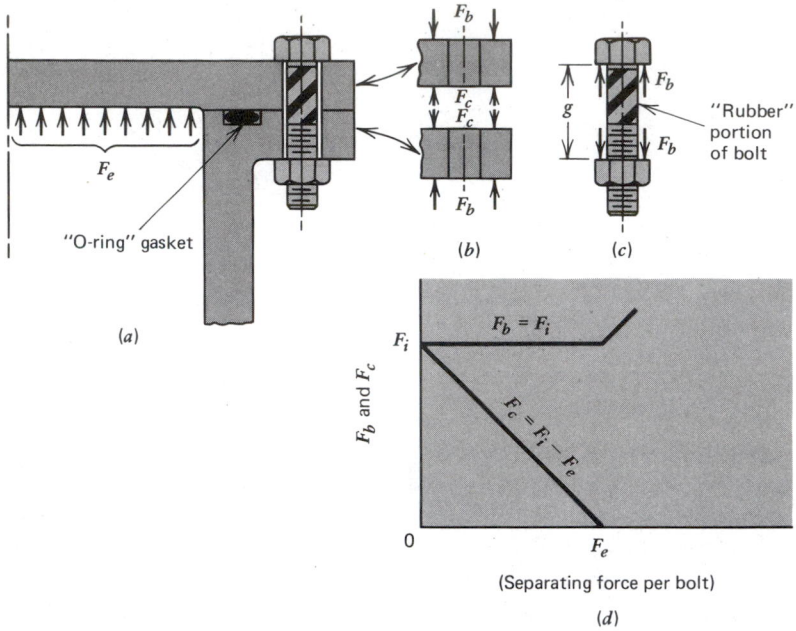

FIGURE 10.26.
***F_b* and *F_c* vs *F_e* for rigid clamped members—soft bolt.**

the nut is tightened to produce initial force F_i, the rubber gasket compresses appreciably; the bolt elongates negligibly. Figures 10.25*b* and 10.25*c* show details of the bolt and the clamped surfaces. Note the distance defined as the *grip, g*. On initial tightening, $F_b = F_c = F_i$.

Figure 10.25*d* shows the change in F_b and F_c as separating load F_e is applied. The elastic stretch of the bolt caused by F_e is so small that the thick rubber gasket cannot expand significantly. Thus, the clamping force F_c does not diminish and the *entire* load F_e goes to increasing bolt tension (review Fig. 10.24).

Figure 10.26 illustrates the opposite extreme in relative stiffnesses. The clamped members are "rigid" metal parts with precision ground mating surfaces and no gasket between them. The bolt has a center portion made of rubber. In this case, the initial tightening stretches the bolt; it does not significantly compress the clamped members. (Sealing of the fluid is accomplished by a "confined gasket" in the form of a rubber O-ring. Before being compressed by the cover plate, the cross section of the O-ring was circular.)

Figure 10.26*d* shows that in this case the entire separating force is balanced by decreased clamping force, with *no* increase in bolt tension. The only way the tension in the rubber bolt can be increased is to increase its length, and this cannot happen without an external force great enough to physically separate the mating clamped surfaces. (Note also that as long as the mating surfaces remain in contact, the sealing effectiveness of the O-ring is undiminished.)

The extreme cases illustrated in Figs. 10.25 and 10.26 can, at most, be only approximated. Let us now investigate the realistic case where both the bolt and the clamped members have applicable stiffness. Joint tightening involves *both* an elongation of the bolt *and* compression of the clamped members. (In ungasketed joints, these deflections can easily be under a thousandth of an inch. For purposes of visualization, it may be helpful to think in terms of the greater deflections that would be associated with using both the rubber gasket of Fig. 10.25 *and* the rubber bolt of Fig. 10.26.) When external force F_e is applied, the bolt and the clamped members elongate *the same amount,* δ (i.e., dimension g increases by amount δ for both).

From Fig. 10.24, the separating force must be equal to the sum of the increased bolt force plus the decreased clamping force, or

$$F_e = \Delta F_b + \Delta F_c \tag{e}$$

By definition,

$$\Delta F_b = k_b\delta \quad \text{and} \quad \Delta F_c = k_c\delta \tag{f}$$

where k_b and k_c are the *spring constants* for the bolt and clamped members, respectively.

Substituting (f) into (e),

$$F_e = (k_b + k_c)\delta, \quad \text{or } \delta = \frac{F_e}{k_b + k_c} \tag{g}$$

Combining (f) and (g),

$$\Delta F_b = \left(\frac{k_b}{k_b + k_c}\right) F_e \quad \text{and} \quad \Delta F_c = \left(\frac{k_c}{k_b + k_c}\right) F_e \tag{h}$$

Thus, the equations for F_b and F_c are

$$F_b = F_i + \left(\frac{k_b}{k_b + k_c}\right) F_e \quad \text{and} \quad F_c = F_i - \left(\frac{k_c}{k_b + k_c}\right) F_e \tag{10.13}$$

Equations (10.13) are represented graphically in Fig. 10.27. Also represented in this figure are two additional points worthy of note:

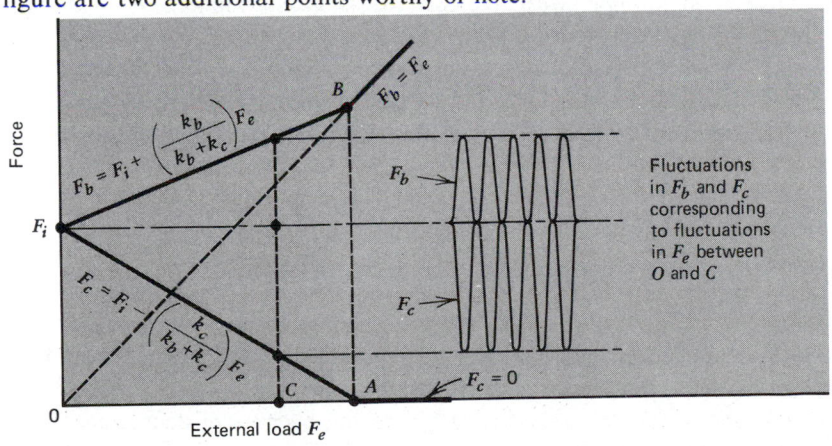

FIGURE 10.27.
Force relationships for bolted connections.

1. When the external load is sufficient to bring the clamping force to zero (point A) the bolt force and the external force must be equal (recall Fig. 10.24). Hence, the figure shows "$F_c = 0$" and "$F_b = F_e$" for values of F_e in excess of A. (This is not a normally usable range of F_e, as physical separation of the clamped surfaces would seldom be acceptable.)
2. When the external load is alternately applied and removed, fluctuations of F_b and F_c are conveniently determined from the figure, as shown. (More about this in Section 10.11 on bolt fatigue.)

Use of Eq. (10.13) requires a determination of k_b and k_c, or at least a reasonable estimation of their relative values. From the basic equations for axial deflection ($\delta = PL/AE$) and for spring rate ($k = P/\delta$):

$$k_b = \frac{A_b E_b}{g} \quad \text{and} \quad k_c = \frac{A_c E_c}{g} \tag{10.14}$$

where the grip, g, represents the approximate effective length for both cases. Two difficulties that commonly arise in estimating k_c are:

1. The clamped members may consist of a stack of different materials, representing "springs" in series. For this case, use the formula for the spring rate of springs in series:

$$1/k = 1/k_1 + 1/k_2 + 1/k_3 + \cdots \tag{10.15}$$

2. The effective cross-sectional area of the clamped members is seldom easy to determine. This is particularly true if the members have irregular shape, or if they extend a substantial distance from the bolt axis. An empirical procedure sometimes used to estimate A_c is illustrated in Fig. 10.28.

An effective experimental procedure for determining the ratio of k_b and k_c for a given joint is to use a bolt equipped with an electric-resistance strain gage or to monitor bolt length ultrasonically. This permits a direct measurement of F_b both before and after F_e is applied. Some handbooks contain rough estimates of the ratio k_c/k_b for various general types of gasketed and ungasketed joints. For a "typical" ungasketed joint, k_c is sometimes taken as three times k_b, but with careful joint design $k_c = 6k_b$ is readily attainable.

10.10 BOLT (OR SCREW) SELECTION FOR STATIC LOADING

The primary loading applied to bolts is tensile, shear, or a combination of the two. Also, some bending is usually present because of the clamped surfaces not being exactly parallel and perpendicular to the bolt axis (Fig. 10.29*a*) and because of deflections of the loaded members (Fig. 10.29*b*).

Before considering more sophisticated examples, it is important to recognize that

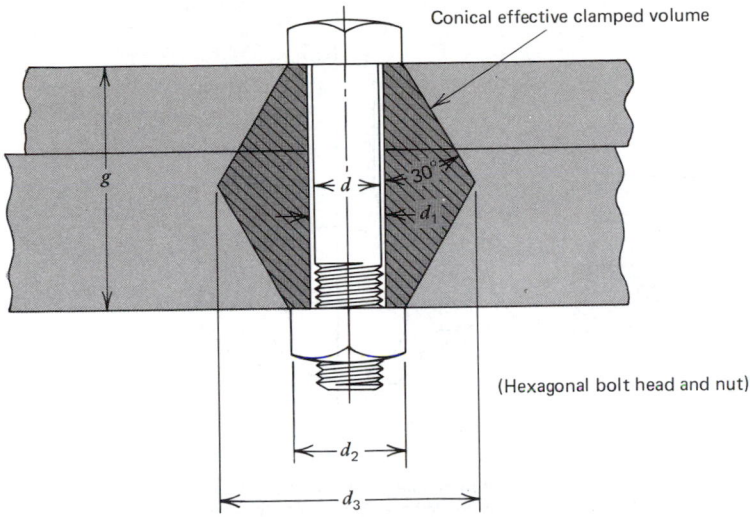

FIGURE 10.28.
One method for estimating effective clamped member area (for calculating k_c). Effective area, A_c, is approximately equal to the average area of the hatched section. Thus;

$$A_c = \frac{\pi}{4}\left[\left(\frac{d_3 + d_2}{2}\right)^2 - d_1^2\right]$$

where

$d_1 \approx d$ **(for small clearances)**

$d_2 = 1.5\,d$ **(for standard hex head bolts—see Fig. 10.16)**

$d_3 = d_2 + g \tan 30° = 1.5\,d + g \tan 30°$

Substitution of these values leads to

$$A_c = \frac{\pi}{16}(5d^2 + 6\,dg \tan 30° + g^2 \tan^2 30°) \approx d^2 + 0.68\,dg + 0.065\,g^2$$

Note: If clamped members are soft, the use of hardened steel flat washers increases the effective value of A_c.

screws and bolts are sometimes selected rather arbitrarily. Such is the case with non-critical applications where loads are small—for example, the attachment of license plates to automobiles. Almost any size would do, including sizes considerably smaller than those used. Selection is a matter of judgment, based on such factors as appearance, ease of handling and assembly, and cost. Even in bolt applications involving known significant loads, larger bolts than necessary are sometimes used because a smaller size "doesn't look right," and the cost penalty of using the larger bolts is minimal.

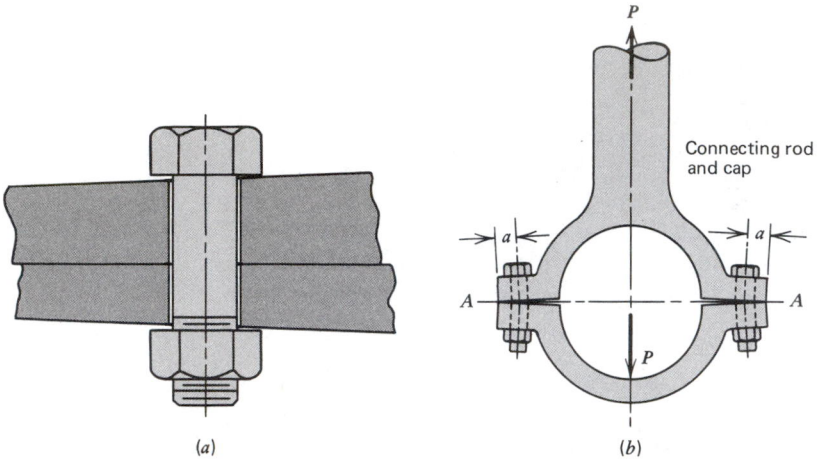

Bolt bending caused by nonparallelism of mating surfaces. (Bolt will bend when nut is tightened.)

Bolt bending caused by deflection of loaded members. (Note tendency to pivot about A; hence, bending is reduced if dimension a is increased.)

FIGURE 10.29.
Examples of nonintended bolt bending.

SAMPLE PROBLEM 10.2 Select Screws for Pillow Block Attachment – Tensile Loading

Figure 10.30 shows a ball bearing encased in a "pillow block" and supporting one end of a rotating shaft. The shaft applies a static load of 9 kN to the pillow block, as shown. Select appropriate metric (ISO) screws for the pillow block attachment, and specify an appropriate tightening torque.

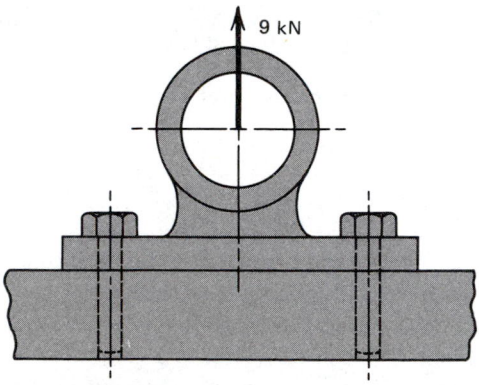

FIGURE 10.30.
Pillow block attached by two machine screws.
(See Sample Problem 10-2.)

Solution

1. Any class of steel could be used, but there appears no reason to specify a costly high-strength steel. Class 5.8, with a proof strength of 380 MPa (Table 10.5), is relatively inexpensive and is arbitrarily chosen.
2. The nominal load for each of the two bolts is 4.5 kN. Reference to Section 6.12 indicates that if screw failure would not endanger human life, cause other damage, or entail costly shut-down, a safety factor of 2.5 would be reasonable. Since in this case the cost of using a larger safety factor is trivial, and since failure might prove rather costly, let us use "engineering judgment" and increase the safety factor to 4. Then, the "design overload" for each bolt is 4.5 kN × 4, or 18 kN.
3. For static loading of a ductile material, stress concentration can be neglected and the simple "$\sigma = P/A$" equation used, with σ being equal to the proof strength when P is equal to the design overload:

$$380 \text{ MPa} = \frac{18{,}000 \text{ N}}{A_t}, \quad \text{or } A_t = 47.4 \text{ mm}^2$$

4. Reference to Table 10.2 indicates an appropriate standard size of class 5.8 screw to be M10 × 1.5 (for which $A_t = 58.0 \text{ mm}^2$). ■
5. Initial tightening tension might reasonably be specified (Eq. 10.11a) as

$$F_i = 0.9A_tS_p = 0.9(58.0 \text{ mm}^2)(380 \text{ MPa}) = 19{,}836 \text{ N}$$

6. This corresponds to an estimated tightening torque (Eq. 10.12) of

$$T = 0.2F_id = 0.2(19.8 \text{ kN})(10 \text{ mm}) = 39.6 \text{ Nm}$$ ■

In step 3, above, the solution for required area was independent of the stiffness ratio, k_c/k_b, and also independent of initial tension, F_i. Regardless of these quantities, static failure of the bolts will occur only when the overload is sufficient to yield the entire bolt cross section, with the pillow block pulled away from the mating fixed surface (i.e., $F_c = 0$). The optimum initial tension would be the highest value that does not yield the bolts enough to damage them after taking them out and re-installing them many times. This provides maximum protection against surface separation ($F_c = 0$) and maximum protection against thread loosening (by providing maximum thread friction). (In the next article it will be seen that both the stiffness ratio and initial tension are important factors in determining bolt size when *fatigue* loading is involved.)

Note that the above solution was based on *slight yielding* of the entire bolt cross section when the design overload is reached. A small additional overload would distort the bolts so that they would not be reusable. However, a *substantially* higher overload would be needed to bring the material to its ultimate strength and fracture the bolts (ratio $S_u/S_p = 520/380 = 1.37$ in this case). In some situations, the design overload might be based on using the ultimate strength of the bolt material rather than its proof or yield strength.

The *shear strengths* of steel bolts of various grades was studied by Fisher and Struik [5], who concluded that a reasonable approximation is

$$S_{us} \approx 0.62S_u \text{ (see footnote 2)} \tag{10.16}$$

[2]For direct (not torsional) shear loading.

SAMPLE PROBLEM 10.3 Determine Shear Load Capacity of a Bolted Joint

Figure 10.31 shows a ½ in.-13UNC grade 5 steel bolt loaded in double shear (i.e., the bolt has two shear planes, as shown). The clamped plates are made of steel, and have clean and dry surfaces. The bolt is to be tightened with a torque wrench to its full proof load; that is, $F_i = S_p A_t$. What force, F, is the joint capable of withstanding? (Note: This double shear bolt loading is the same as that subjected to the pin in Figs. 2.14 and 2.15. It is assumed that the bolt and plates have adequate strength to prevent the other failure modes discussed in connection with Figs. 2.14 and 2.15.)

Solution

1. Specified initial tension is $F_i = S_p A_t = 85{,}000 \text{ psi} \times 0.1419 \text{ in.}^2 = 12{,}060$ lb. But with a roughly estimated ± 30 percent torque-wrench variation and 10 percent initial tension loss during the first few weeks of service (see Section 10.7), a conservative assumption of working value of F_i is about 7600 lb.
2. Reference 5 gives a summary (p. 78) of friction coefficients obtained with bolted plates. The coefficient for semi-polished steel is approximately 0.3, and for sand or grit-blasted steel approximately 0.5. Various paints, platings, and other surface treatments can alter the coefficient markedly, downward in most cases. For this case, a friction coefficient of 0.4 is assumed. This gives a force required to slip each of the two interfaces of 7600 lb $\times$ 0.4 = 3040 lb. Thus, the value of F required to overcome friction is estimated to be in the region of 6000 lb. ■
3. Although it is often desirable to limit applied load F to the value that can be transmitted by friction, we should know the larger value of force that can be transmitted through the bolt itself. For the two shear planes involved, this force is equal to $2S_{ys}A$, where A is the area of the bolt *at the shear planes*—in this case, $\pi(0.5)^2/4 = 0.196 \text{ in.}^2$ Taking advantage of the fact that the distortion-energy theory gives a good estimate of shear yield strength for ductile metals, $S_{ys} = 0.58S_y = 0.58(92 \text{ ksi}) = 53$ ksi. Thus, for yielding of the two shear planes, $F = 2(.196 \text{ in.}^2)(53{,}000 \text{ psi}) = 21{,}000$ lb. ■
4. The estimated 21,000-lb load would bring the shear stress to the yield strength over the entire cross section of the shear planes, and the very small amount of yielding involved would probably result in losing most or all of the clamping and friction forces. A further increase in load would cause total shear failure, as indicated in Fig. 10.31*b*. This total failure load is calculated as in step 3, above, except for replacing S_{ys} with S_{us}. From Eq. (10.16), $S_{us} \approx 74$ ksi; the corresponding estimated load is $F = 29{,}000$ lb. ■

Note that in Fig. 10.31 the threaded portion of the bolt does *not* extend to the shear plane. This is important for a bolt loaded in shear. Extending the thread to the shear plane is conservatively considered to reduce the shear area to a circle equal to the thread root diameter; in this case, $A = \pi(0.4056)^2/4 = 0.129 \text{ in.}^2$, which is a reduction of 34 percent.

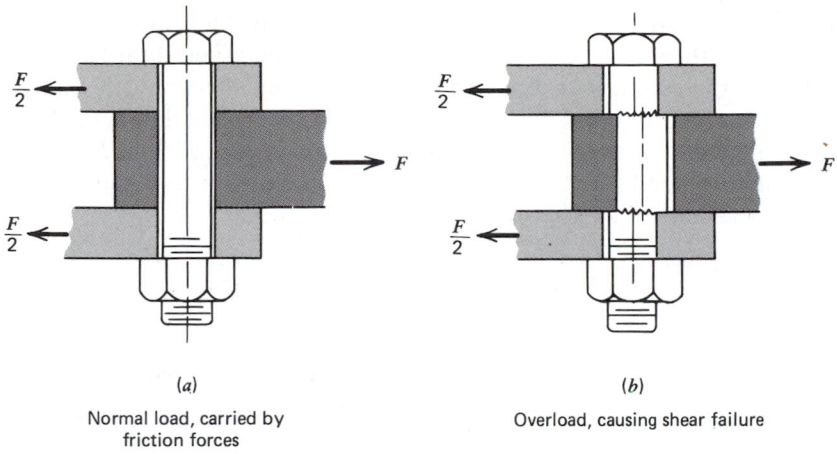

FIGURE 10.31.
Bolt loaded in double shear. (See Sample Problem 10-3.)

SAMPLE PROBLEM 10.4 Select Bolts for Bracket Attachment, Assuming Shear Carried by Friction

Figure 10.32 shows a vertically loaded bracket attached to a fixed member by three identical bolts. Although the 24-kN load is normally applied in the center, the bolts are to be selected on the basis that the load eccentricity shown could occur. Because of safety considerations, SAE class 9.8 steel bolts and a minimum safety factor of 6 (based on proof strength) are to be used. Determine an appropriate bolt size.

Solution

1. Let us make the common assumption that deflections of the clamped members are negligible, and that the load tends to rotate the bracket about point A. (Actually,

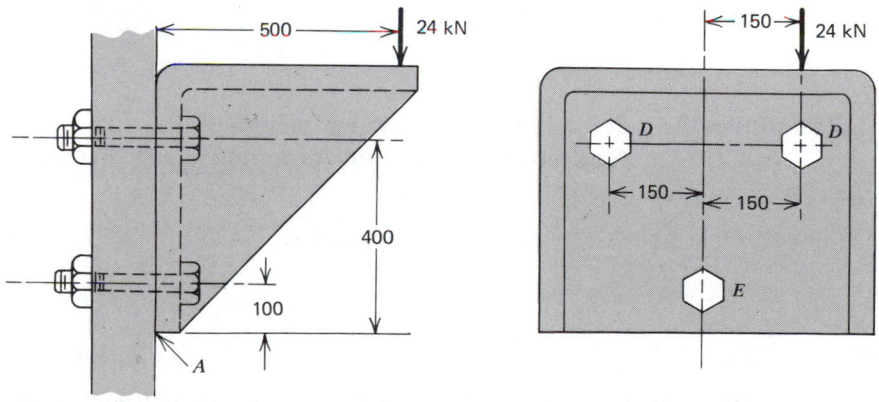

FIGURE 10.32.
Vertically loaded bracket supported by three bolts. (See Sample Problem 10-4.)

the high compressive stresses at A will cause deflections that slightly raise the point of rotation.) With the assumptions of rigid clamped members, and shear loads carried by friction, the eccentricity of the applied load has no effect on bolt loading. With the bracket tending to rotate about point A, the strain (and hence the load) imposed upon the two bolts D is four times that of bolt E. Let F_D and F_E denote the tensile loads carried by bolts D and E. Summation of moments about point A for the *design overload* of 24 kN(6) = 144 kN gives

$$\begin{aligned} 500(144) &= 100F_E + 400F_D + 400F_D \\ &= 25F_D + 400F_D + 400F_D = 825F_D, \text{ or} \\ F_D &= 87.27 \text{ kN} \end{aligned}$$

2. Class 9.8 steel has a proof strength of 650 MPa. Hence the required tensile stress area is

$$A_t = \frac{87{,}270 \text{ N}}{650 \text{ MPa}} = 134 \text{ mm}^2$$

Reference to Table 10.2 indicates the required thread size to be M16 × 2. (Because of appearance, and to provide additional safety, a larger bolt size might be selected.) ■

As in Sample Problem 10.2, the bolt size required is independent of k_b, k_c, and F_i, *except* for the fact that F_i must be large enough to justify the assumption that shear forces are transmitted by friction. With an assumed coefficient of friction of 0.4, and an initial tension (after considering tightening variations and initial relaxation) of at least $0.55S_pA_t$, compare the available shear-friction force (using 16-mm bolts) with the applied shear overload:

$$\begin{aligned} \text{Available friction force} &= (3 \text{ bolts})(0.55S_pA_t)f \\ &= 3(0.55)(650 \text{ MPa})(157.27 \text{ mm}^2)(0.4) \\ &= 67{,}500 \text{ N} \end{aligned}$$

which represents a margin of safety with respect to the 24-kN applied overload, plus the rotational tendency caused by the overload eccentricity. The latter effect is dealt with in the following example.

SAMPLE PROBLEM 10.5 Select Bolts for Bracket Attachment, Neglecting Friction and Assuming Shear Forces Are Carried by the Bolts

Repeat Sample Problem 10.4, except neglect the friction forces.

Solution

1. Neglecting friction has no effect on bolt stresses in the *threaded region,* where attention was focused in Sample Problem 10.4. For this problem, attention is

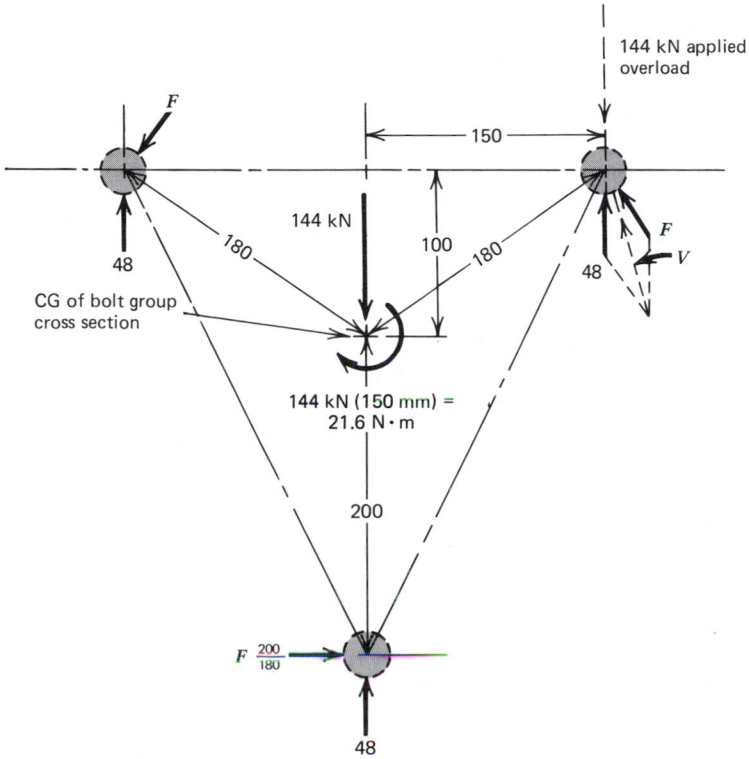

$\Sigma M_{cg} = 0$: 144 kN (150 mm) = F(180 mm) + F(180 mm) + $F(\frac{200}{180})$(200 mm)
$\therefore F = 37.1$ kN

FIGURE 10.33.
Shear-plane force and moment equilibrium for the bracket in Fig. 10.32. (See Sample Problem 10-5.)

shifted to the *bolt shear plane* (at interface between bracket and fixed plate). This plane experiences the tensile force of 87.27 kN calculated in Sample Problem 10.4 in addition to the shear force calculated below in step 2.

2. The applied eccentric shear force of 144 kN tends to displace the bracket downward and also rotate it clockwise about the center of gravity of the bolt-group cross section. For three bolts of equal size, the center of gravity corresponds to the centroid of the triangular pattern, as shown in Fig. 10.33. This figure shows the original applied load (dotted vector) replaced by an equal load applied at the centroid (solid vector) plus a torque that is equal to the product of the force and the distance it was moved. Each bolt is assumed to carry ⅓ of the vertical shear load, plus a tangential force (with respect to rotation about the center of gravity) that is proportional to its distance from the center of gravity. Calculations on the figure show this tangential force to be 37.1 kN for each of the top bolts. The vector sum of the two shear forces is obviously greatest for the upper-right bolt. Routine calculation determines this to be $V = 81.5$ kN.

3. The critical upper-right bolt is thus subjected to a tensile stress, $\sigma = 87{,}270/A$, and a shear stress, $\tau = 81{,}500/A$. Substitution in the distortion energy equation gives an equivalent tensile stress of

$$\sigma_e = \sqrt{\sigma^2 + 3\tau^2} = \frac{1}{A}\sqrt{(87{,}270)^2 + 3(81{,}500)^2} = \frac{166{,}000}{A}$$

4. Equating this to the proof stress gives

$$\frac{166{,}000}{A} = S_p = 650 \text{ MPa} \quad \therefore A = 255 \text{ mm}^2$$

5. Finally,

$$A = \frac{\pi d^2}{4}, \quad \text{or } d = \sqrt{\frac{4A}{\pi}} = \sqrt{\frac{4(255)}{\pi}} = 18.03 \text{ mm}$$

Thus, a *shank* diameter of 18 mm is required. ■

In comparing this solution with that of Sample Problem 10.4, note that *for this particular case,* shear plus tension in the bolt shear plane proved to be more critical than tension alone in the threads.

10.11 BOLT (OR SCREW) SELECTION FOR FATIGUE LOADING – FUNDAMENTALS

Bolt fatigue involves fluctuating *tension,* commonly accompanied by a small amount of alternating bending (as in Fig. 10.29*b*). (Alternating shear loads are usually reacted by clamping friction forces or by separate members such as dowel pins.) This section involves the application of the principles and concepts of Chapter 8 to bolts. Because of initial tightening tension, bolts inherently have high mean stresses. Also, stress concentration is always present at the thread roots. These two points are treated in Sections 8.9 and 8.11.

Table 10.6 gives approximate values of fatigue stress concentration factor K_f for standard screws and bolts. Note that (1) rolled threads have lower values of K_f because

TABLE 10.6
Fatigue Stress Concentration Factors, K_f, for Steel Threaded Members (Approximate Values for Unified and ISO Threads)

Hardness	SAE Grade (Unified Thds.)	SAE Class (ISO Thds.)	K_f[a] Rolled Thds.	K_f[a] Cut Thds.
Below 200 Bhn (annealed)	2 and below	5.8 and below	2.2	2.8
Above 200 Bhn (hardened)	4 and above	8.8 and above	3.0	3.8

[a]With good commercial surfaces, use $C_S = 1$ (rather than a value from Fig. 8.13) when using these values of K_f.

of work-hardening and residual stresses, and (2) hardened threads have higher values of K_f because of their greater notch sensitivity. For thread finishes of good commercial quality, these values may be used with a surface factor, C_S, of unity.

Section 10.4 and Figs. 10.11 and 10.12 explained the tendency for most of the bolt load to be carried by the threads nearest to the loaded face of the nut, and noted that the degree to which the stresses were concentrated in this region was influenced by the nut design. This is one reason why actual values of K_f may differ from those given in Table 10.6.

Figure 10.34 represents an analysis of the fatigue strength of an M10 × 1.5 class 9.8 bolt installed in a joint with $k_c = 2k_b$, and subjected to an external load which fluctuates between 0 and 9 kN. The "case 1" curves in Fig. 10.34*a* correspond to an initial tightening force of 10 kN. (Note that these curves are similar to those in Fig. 10.27.) Each time the external load is applied, the bolt force increases and the clamping force decreases, with the sum of the two effects being equal to the external load of 9 kN. When the external load is removed, F_b and F_c revert to their initial value of 10 kN.

The "case 2" curves correspond to a bolt initial tension equal to the full yield strength ($F_i = S_yA_t = 660 \times 58.27 = 38{,}458$ N ≈ 38.5 kN). It must be understood that this represents an extreme case that would never be specified. (The highest specified initial tension would be the full proof strength—about 9 percent lower in this case.) Following the procedures given in Chapter 8 (recall particularly Section 8.11 and Fig. 8.25, and the treatment of yielding and residual stresses in Section 4.14 and Figs. 4.41 and 4.42), the assumed idealized stress-strain curve in Fig. 10.34*b* is used. With the bolt tightened to its full yield strength, *all* material in the cross section is stressed to S_y. This means that any small additional bolt elongation results in *no* increase in load; that is, the effective modulus of elasticity for an increased load is not 207 GPa, but *zero,* and the associated bolt stiffness is also zero. Hence, the first time joint separating force F_e is applied, *all* of it is reacted by decreased clamping force; *none* goes to increasing bolt load. This is shown in the top curves of Fig. 10.34*a*. When F_e is released, the bolt relaxes (shortens) slightly, and this relaxation is *elastic.* Hence, corresponding changes in F_b and F_c are controlled by the elastic values of k_b and k_c. Elastic bolt relaxation is reversible, which means that the load can be re-applied without yielding. As the external load continues to cycle, the bolt and clamping forces fluctuate, as shown by the case 2 curves.

Let us now determine the stress fluctuation at the critical thread root location for cases 1 and 2. For case 1, initial tightening produces a computed thread root stress of

$$\sigma = \frac{F_i}{A_t}K_f = \frac{10{,}000 \text{ N}}{58.27 \text{ mm}^2}(3) = 515 \text{ MPa}$$

On initial application of the external load, F_b is increased to 13 kN, with an associated elastically computed thread stress of 670 MPa. Slight yielding reduces this stress to 660 MPa. The elastic reduction in stress when F_b is reduced to 10 kN is 154 MPa. Thus, the thread root stress fluctuation corresponds to the case 1 line of Fig. 10.34*b*. This is *exactly* the same fluctuation as shown for case 2. In both cases the maximum stress corresponds to S_y and the minimum stress results from an elastic relaxation during load removal. Hence, the two cases are represented by the same point on a mean stress-alternating stress diagram (point 2 on Fig. 10.34*c*). The only difference between them

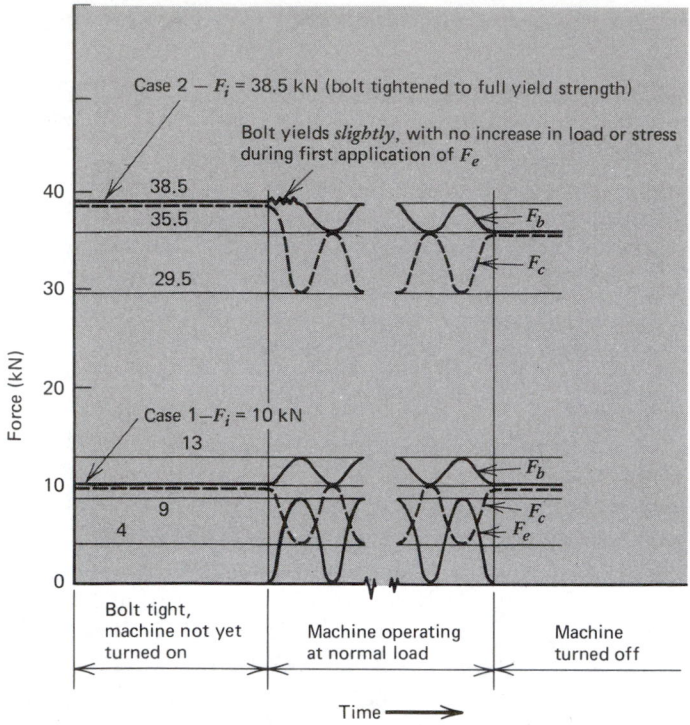

(*a*)

Fluctuations in F_b and F_c caused by fluctuations in F_e

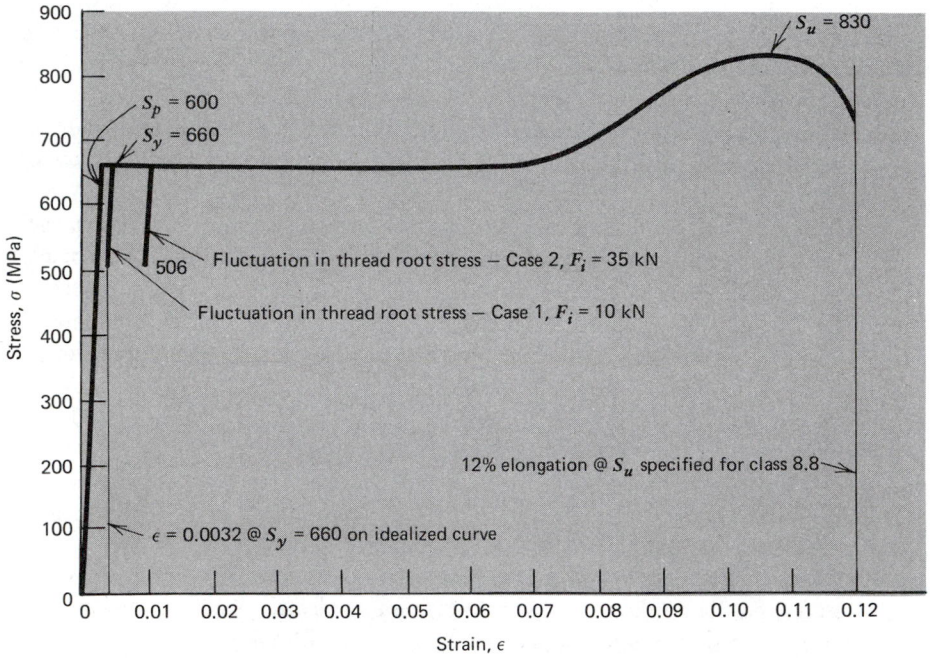

(*b*)

Idealized (*not* actual) stress—strain curve for class 8.8 bolt steel

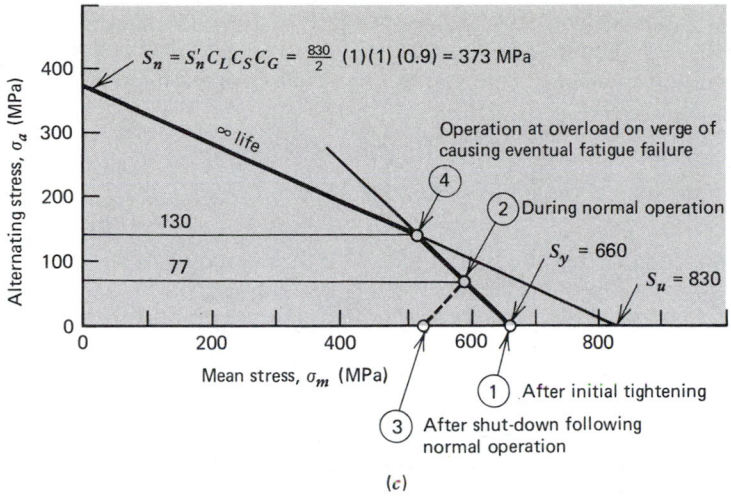

Mean stress–alternating stress diagram for plotting thread root stresses

FIGURE 10.34.
Bolt fatigue loading example. M10 × 1.5 class 9.8 bolt tightened to two different values of F_1. Elastic stiffnesses are such that $k_c = 2k_b$, (rolled threads, $K_f = 3$). Note: Because not all class 9.8 properties are available, class 8.8 properties are used for illustration.

is the greater yielding that occurs during tightening and initial load application in case 2.

Point 2 on Fig. 10.34*c* applies to *any* value of initial tension between 10 and 38.5 kN. If the initial tension exceeds 12.8 kN, the thread root stress reaches point 1 upon initial tightening. Point 3 shows the thread root stress after external loading is discontinued (i.e., machine shut down). The difference between points 1 and 3 is caused by bolt yielding during the initial application of F_e. When the machine operation resumes, the thread root "operating point" moves from 3 back to 2.

The safety factor with respect to eventual fatigue failure is 130/77, or 1.7, because an overload to failure would have to increase the alternating stress to 130 MPa (point 4) in order to reach the infinite life Goodman line. Point 4 corresponds to an external load fluctuation between zero and 1.7 times 9 kN, or 15.3 kN. Note that with case 1 initial tightening, this overload would not be possible because the clamping force would drop to zero and the joint members would separate ($F_c = 10 - (2/3)(15.3) < 0$). With case 2 initial tightening, the joint retains a large clamping force with this overload ($F_c = 38.5 - 15.3 = 23.2$ kN).

It is important to remember that actual bolt fatigue performance will deviate somewhat from that indicated by the foregoing analysis because of deviations of the actual stress-strain curve from the assumed idealized curve.

Although bolt tightening to full yield strength should not be specified (primarily because of the possibility of going too far beyond that point when actual tightening occurs), it is desirable to specify tightening to the full *proof* strength (i.e., $F_i = S_p A_t$), where sufficiently close control over the tightening operation can be maintained. The advantages of tightening this tightly are:

1. Reduced dynamic load on the bolt due to increased *effective* clamped member area. (The greater the initial tightening, the more intimately in contact the clamped surfaces remain during load cycling, particularly when considering the effect of load eccentricity, as illustrated in Fig. 10.29*b*.)
2. Maximum protection against overloads causing the joint to separate.
3. Maximum protection against thread loosening (see Section 10.8).

It is important to recognize that the small amount of thread root yielding involved in tightening bolts to the full proof load is not harmful to any bolt material of acceptable ductility. Note, for example, that all of the steels listed in Tables 10.4 and 10.5 have an area reduction of 35 percent.

SAMPLE PROBLEM 10.6 Importance of Initial Tension on Bolt Fatigue Load Capacity

Figure 10.35*a* represents a model of two steel machine parts clamped together with a single ½ in.-13 UNC grade 5 bolt with cut threads, and subjected to a separating force that fluctuates between zero and F_{max}. What is the greatest value of F_{max} that would give infinite bolt fatigue life (a) if the bolt has *no* initial tension, and (b) if the bolt is initially tightened to its full proof load?

Solution

1. A σ_m-σ_a diagram is routinely constructed in Fig. 10.35*b*. For part (a) the only stresses are due to the fluctuating load, with

$$\sigma_m = \sigma_a = \frac{F_{max}}{2A_t} K_f = \frac{F_{max}}{2(0.1419)}(3.8) = 13.39F_{max}$$

(Units in this case are pounds and inches.)

2. For borderline infinite fatigue life, Fig. 10.35*b* shows

$$\sigma_m = \sigma_a = 37{,}000 \text{ psi}$$

Therefore, $13.39F_{max} = 37{,}000$, or (after rounding off)

$$F_{max} = 2760 \text{ lb}$$

■

3. For part (b), the initial tension is

$$F_i = S_p A_t = (85{,}000)(0.1419) = 12{,}060 \text{ lb}$$

4. If the steel plates have smooth, flat surfaces, and if there is no gasket between them, k_b and k_c are simply proportional to A_b and A_c (see Eq. (10.14)). Assuming that the bolt threads extend only slightly above the nut, and the bolt shank is ½ in. diameter over its full length,

$$A_b = \frac{\pi}{4} d^2 = \frac{\pi}{4}(\tfrac{1}{2} \text{ in.})^2 = 0.196 \text{ in.}^2$$

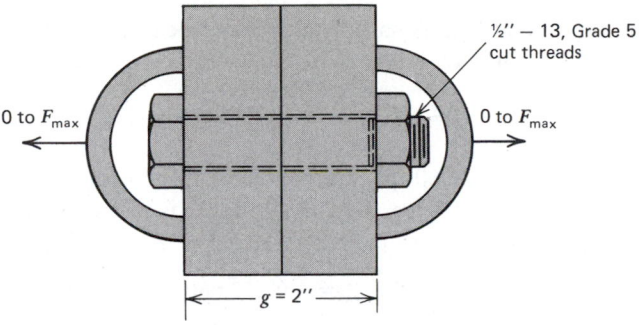

(a)

Simplified model of machine members bolted together

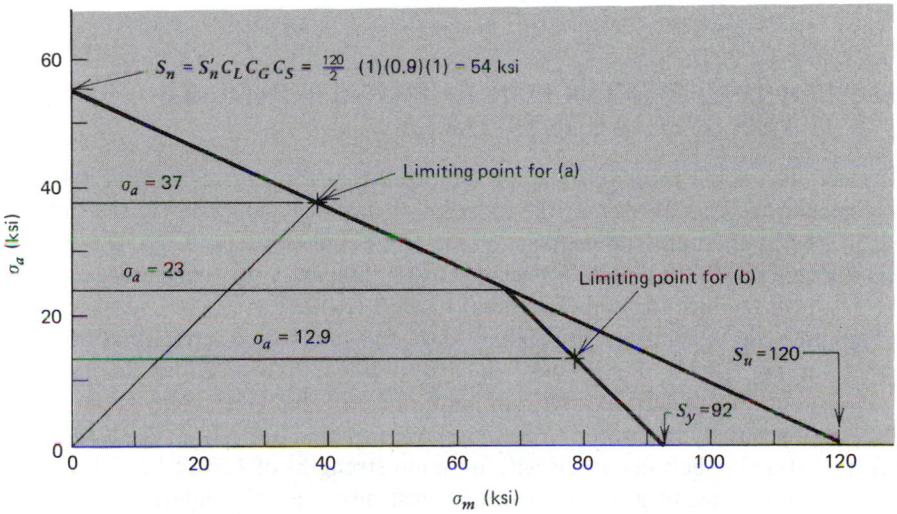

(b)

Fatigue diagram for thread root

FIGURE 10.35.
Bolted plates with fluctuating separating force. (See Sample Problem 10-6.)

Using Fig. 10.28 to estimate A_c,

$$A_c = \frac{\pi}{16}(5d^2 + 6dg + g^2)$$

$$= \frac{\pi}{16}[5(½)^2 + 6(½)(2) + (2)^2] = 2.21 \text{ in.}^2$$

Thus,

$$\frac{k_b}{k_b + k_c} = \frac{A_b}{A_b + A_c} = \frac{0.196}{0.196 + 2.21} = 0.08$$

which means that only 8 percent of the external load fluctuation is felt by the bolt (92 percent goes to decreasing clamping pressure).

5. The alternating bolt load is half of the peak-to-peak load fluctuation, or $0.04F_{max}$. Thus, the alternating bolt stress is

$$\sigma_a = \frac{F_a}{A_t} K_f = \frac{0.04F_{max}}{0.1419}(3.8) = 1.07F_{max}$$

6. With $F_i = S_p A_t = 12{,}060$ lb, external loads up to a little over 12,060 lb will not cause joint separation. Hence, $F_{max} = 12{,}060$ lb is just satisfactory *if* the alternating bolt stress does not cause fatigue failure. For $F_{max} = 12{,}060$ lb:

$$\sigma_a = 1.07F_{max} = 1.07(12{,}060) = 12{,}900 \text{ psi}$$

Figure 10.35*b* shows that this point is well below the infinite life Goodman line (σ_a could go as high as 23 ksi). Hence, the answer to (b) is (rounded off): $F_{max} =$ 12,000 lb, or *4½ times* the value for part (a). ■

10.12 BOLT SELECTION FOR FATIGUE LOADING – USING SPECIAL TEST DATA

The procedures of the preceding section indicated that for any reasonable bolt tightening (causing yielding at least at the thread root, but not going beyond the full proof strength), the alternating load capacity of the bolt is independent of initial tightening. Unless higher values of initial bolt tension cause significantly greater clamped member stiffnesses, tests confirm that this is generally true [3, 4, 11].

The intersection points corresponding to the fatigue-limited alternating bolt stress (as point 4 in Fig. 10.34*c*) for various bolt steels indicate that the allowable alternating stress does not increase with ultimate strength; rather, it remains about constant. This, too, has been generally confirmed by testing. For example, Table 10.7, discussed below, makes no distinction between steels with ultimate strengths of 120 and 260 ksi.

Despite the points of general verification just discussed, the determination of bolt allowable alternating load by the procedures used in the previous section is clearly very rough. The fatigue-limiting intersection point (point 4 in Fig. 10.34*c*) is highly influenced by small variations in the value taken for S_y. (Note that if S_p were used instead, a much higher fatigue strength would be indicated.) Assuming a sharp-cornered idealized stress-strain curve is of questionable validity in this case. Furthermore, the Goodman line is assumed valid for a stress variation far removed from the only fatigue test point (zero mean stress) used as its basis. Similarly, the use of average values of K_f (Table 10.6) cannot always reflect accurately the influence of thread profile, thread surface finish and treatment, and design of the mating threaded member.

In order to provide a much more accurate basis for designing critical fatigue-loaded connections and for selecting appropriate high-strength bolts, extensive fatigue tests of bolted connections have been made. Some of the results are summarized in Table 10.7, which gives fatigue-limited alternating *nominal* stresses for several bolts when properly installed with the corresponding specified nuts.

In view of the fact that the allowable alternating load capacity of a 120-ksi ultimate strength bolt tightened to half its proof load is about the same as that of a 260-ksi bolt tightened to its full proof load, one might be tempted to conclude that the two

TABLE 10.7
Fatigue Strength of Tightened Bolts,[a] S_a

Material	Thread Rolling	Finish	Thread ISO	S_a ksi	S_a MPa
Steel, S_u = 120–260 ksi	Before H.T.	Phos. and oil	Std.	10	69
"	After H.T.	"	"	21	145
"	"	Cad plate	"	19	131
"	"	Phos. and oil	Special[b]	26	179
"	"	Cad plate	"	23	158
Titanium, S_u = 160 ksi			Std.	10	69
"			Special[b]	14	96

[a]Alternating nominal stress defined as alternating bolt force/A_t, 50 percent probability of failure, bolt sizes to 1 in. or 25 mm. [3,4,11].
[b]SPS Technologies, Inc. "Asymetric" thread (incorporates large root radius). (Fillet under bolt head must be rolled to make this point as strong in fatigue as the thread.)

were equivalent for fatigue applications. This is decidedly not the case. Actually, the key to modern successful fasteners designed for critical fatigue application is *maximizing initial tension* (i.e., using very high strength bolts tightened to very nearly their full proof strength). As previously mentioned, increased initial tension (1) usually increases clamped member stiffness (which reduces bolt fluctuating stress), (2) gives greater insurance against joint separation, and (3) gives increased resistance to thread loosening. Furthermore, by obtaining the increased initial tension with bolts that are relatively small but highly stressed, bolt stiffness is reduced (further reducing bolt fluctuating stress). Also, smaller bolts may permit the size and weight of associated parts to be reduced.

SAMPLE PROBLEM 10.7 Selection of Pressure Vessel Flange Bolts

A flanged joint like Fig. 10.26 involves a cylinder internal diameter of 250 mm, a bolt circle diameter of 350 mm, and an internal gage pressure that fluctuates rapidly between zero and 2.5 MPa. Twelve conventional class 8.8 steel bolts with threads rolled before heat-treatment are to be used. The cylinder is made of cast iron (E = 100 GPa) and the cover plate of aluminum (E = 70 GPa). Construction details are such that the effective clamped area, A_c, can conservatively be assumed equal to $5A_b$. Also, the clamped thicknesses of the cast iron and aluminum members are the same. For infinite fatigue life with a safety factor of 2, determine an appropriate bolt size. Assume that after a period of operation, the initial tension may be as low as $0.55S_pA_t$.

Solution

1. The total value of F_e at the design overload (normal load times safety factor) is

$$\frac{\pi}{4} d^2 p_{\max} = \frac{\pi}{4} (250 \text{ mm})^2 (5.0 \text{ MPa}) = 245.4 \text{ kN}$$

which, divided among 12 bolts, gives 20.5 kN per bolt.

2. Stiffness k_c is the resultant of two "springs" in series (let the cast iron be "spring" 1 and the aluminum "spring" 2), for which Eq. (10.15) applies:

$$\frac{1}{k_c} = \frac{1}{k_1} + \frac{1}{k_2}$$

where

$$\frac{1}{k_1} = \frac{L_1}{A_1 E_1} = \frac{g/2}{5A_b(100)}$$

Similarly,

$$\frac{1}{k_2} = \frac{g/2}{5A_b(70)}$$

Substituting,

$$\frac{1}{k_c} = \frac{g}{1000A_b} + \frac{g}{700A_b} = 0.00243g/A_b, \quad \text{or}$$

$$k_c = 412A_b/g$$

Likewise,

$$k_b = \frac{A_b E_b}{L_b} = \frac{A_b(200)}{g}$$

which leads to

$$k_c = 2.06k_b$$

and

$$\frac{k_b}{k_b + k_c} = 0.33$$

3. Let us use the fatigue strength data in Table 10.7, which lists, for this bolt material, a fatigue-limiting value of alternating *nominal* stress of 69 MPa. The actual value of alternating nominal stress is

$$\sigma_a = \frac{F_a}{A_t} = \frac{(0.33/2)(20{,}500)}{A_t} = \frac{3383}{A_t}$$

Hence, the required value of A_t is $3383/69 = 49 \text{ mm}^2$.

4. From Table 10.2, select the next larger standard size: M10 × 1.5. with $A_t = 58.27 \text{ mm}^2$. ■

5. The minimum initial clamping force is given as $0.55S_pA_t = (0.55)(0.600 \text{ GPa})(58.27\text{mm}^2) = 19.2$ kN. Since 33 percent of the applied 20.5-kN load contributed to bolt tension, the remaining 67 percent (i.e., 13.7 kN) will decrease clamping force, thereby leaving a minimum clamping force of 5.5 kN.

The ratio of bolt spacing to bolt diameter in the above problem came out to $350\pi/(12 \times 10)$ or 9.16. Empirical guidelines sometimes used are that this ratio should be (a) less than 10 in order to maintain good flange pressure between bolts, and (b) greater than 5 in order to provide convenient clearance for standard wrenches.

The above problem was based on the use of an O-ring gasket (Fig. 10.26). With an *ordinary* gasket (as in Fig. 10.25, except not nearly so thick):

1. flange rigidity and bolt closeness are more critical, as sealing effectiveness is controlled by the lowest clamping pressure, mid-way between bolts;
2. k_c is *much* lower, thus imposing much higher alternating stresses on the bolts; and
3. F_c must be maintained high enough for sealing when F_{max} is applied, whereas the captive O-ring requires only that $F_c > 0$.
4. On the other hand, the captive gasket is more costly because of the need to machine an accurate groove, and the pliable material of the O-ring may be less resistant to heat, age, and chemical attack.

10.13 INCREASING BOLTED JOINT FATIGUE STRENGTH

It may be helpful at this point to bring together a list of the various ways by which one can improve bolt (and screw) fatigue strength:

1. Modify stiffnesses to decrease the portion of the external load that increases bolt tension.
   - **a.** Increase k_c by using higher modulus materials, flat and smooth mating surfaces (without gaskets in between), greater area and thickness of plates in compression. (Note in Fig. 10.28 how increasing grip can increase A_c.)
   - **b.** Decrease k_b by securing the desired clamping force with smaller bolts of higher strength, and by fully utilizing the material strength by more precise control of initial tensioning. Another effective means of reducing k_b is to reduce the shank area, as in Figs. 7.10*b*, 7.11, and 10.13. Ideally, the shank should be reduced with a very large radius fillet so that stresses in this fillet are very nearly as high as stresses in the critical thread root. Reducing bolt modulus of elasticity helps too, of course, if this can be done without losing material fatigue strength.
2. Modify the nut (or other female threaded member) to equalize the load carried by the several threads in contact (Fig. 10.12), and be sure that the number of threads in contact is adequate.
3. Reduce the thread root stress concentration by using a larger root radius.
   - **a.** A U.S. government standard developed for military aircraft, MIL-B-7838, calls for modifying the basic profile of Fig. 10.2 by using a $0.144p$ thread root

fillet radius. This form is used in some industrial fasteners and in most aerospace tension bolts up to 180 ksi tensile strength.

b. Standard MIL-S-8879 specifies a fillet radius of $0.180p$, and is used with aerospace bolts of 180 ksi tensile strength and higher. (Obviously, these special bolt materials go beyond the SAE grade listing of Table 10.4. Ultra-high strength steel bolts have tensile strengths of 220 and 260 ksi.)

c. Exotic aerospace bolts of columbium, tantalum, beryllium, and other highly notch sensitive materials sometimes use fillet radii of $0.224p$, and even $0.268p$.

A limiting factor to the use of very large radius fillets is the associated reduction in thread depth. However, when difficult bolt fatigue problems are encountered, it is well to consider going to a fillet radius of at least $0.144p$.

4. Use a material of highest practical proof strength in order to obtain maximum initial tension.
5. Use tightening procedures which insure values of F_i as close as possible to A_tS_p.
6. Be sure that the threads are rolled rather than cut, and that threads are rolled *after* heat-treatment.[3] The higher the strength, the more important it is to roll *after* hardening. This has been experimentally verified for tensile strengths as high as 300 ksi.
7. After reducing stress concentration and strengthening the thread as much as possible, be sure that the fillet radius under the bolt head is sufficient to avoid failures at that point. *Cold roll* this fillet if necessary (see Sect. 8.14).
8. Minimize bolt bending (Fig. 10.29).
9. Guard against partial loss of initial tension in service because of thread loosening, or materials taking a permanent set. Retighten bolts as necessary. Also take steps to insure proper tightening after bolts are removed for servicing, and to replace bolts before they yield objectionably because of repeated retightening.

REFERENCES

1. ANSI (American National Standards Institute) standards B1.1-1974, B1.5-1977, B1.9-1973, American Society of Mechanical Engineers, New York.
2. Chow, W. W., *Cost Reduction in Product Design,* Van Nostrand Reinhold, New York, 1978.
3. Finkelston, R. J., and F. R. Kull, "Preloading for Optimum Bolt Efficiency," *Assembly Engineering* (Aug. 1974).
4. Finkelston, R. J., and P. W. Wallace, "Advances in High-Performance Mechanical Fastening," Paper 800451, Society of Automotive Engineers, New York, N.Y., Feb. 25, 1980.
5. Fisher, J. W., and J. H. A. Struik, *Guide to Design Criteria for Bolted and Riveted Joints,* John Wiley & Sons, Inc., New York, 1974.

[3]Ordinary heat-treated bolts are rolled *before* heat-treatment to reduce cost.

6. Juvinall, R. C., *Engineering Considerations of Stress, Strain, and Strength,* McGraw-Hill, New York, 1967.

7. *Machine Design,* 1980 Fastening and Joining Reference Issue, Penton/IPC, Inc., Cleveland, November 13, 1980.

8. Osgood, C. C., *Fatigue Design,* Wiley-Interscience, New York, 1970.

9. Parmley, R. O. (ed.), *Standard Handbook of Fastening and Joining,* McGraw-Hill, New York, 1977.

10. SAE (Society of Automotive Engineers) standards J 429 and J 1199, Society of Automotive Engineers, New York, 1979.

11. Walker, R. A., and R. J. Finkelston, "Effect of Basic Thread Parameters on Fatigue Life," Paper 700851, Society of Automotive Engineers, New York, N.Y., October 5, 1970.

CHAPTER ELEVEN

Rivets, Welding, and Bonding

11.1 INTRODUCTION

As with screws and bolts (Chapter 10), rivets exist in great variety, and often exhibit remarkable ingenuity. The brief treatment given here attempts to help the reader gain some familiarity with the choices available, and gain a better "feel" for applying the basic principles of load and stress analysis to them.

One might question the appropriateness of including material on welding and bonding in this book, as these are not really machine components. But since the engineer is frequently faced with the need to choose between various threaded and nonthreaded fasteners (which *are* machine components) and the alternatives of welding or bonding, it seems appropriate to treat them here, at least briefly.

Had space permitted, another provision for assembly which could have been included is that of designing the members so that they can be joined with a *snap fit*. The snap-fit assembly can be designed to be permanent, or readily disassembled. This potentially very economical and satisfactory joining method is well presented in [3]. Other fastening and joining methods are also discussed in [3] and [5]. The engineer concerned with the design of mechanical equipment should be familiar with all of these.

11.2 RIVETS

Conventional structural rivets, illustrated in Fig. 11.1, are widely used in aircraft, transportation equipment, and other products requiring high joint strength. They are also used in the construction of buildings, boilers, bridges, and ships, but in recent decades these applications have made increasing use of welding. Because of vital safety considerations, the design of riveted connections for these latter applications is governed by construction codes formulated by such technical societies as the *American Institute of Steel Construction (AISC)* and the *American Society of Mechanical Engineers (ASME)*. Figure 2.19 illustrates a typical riveted joint. An analysis of the loading taken by its redundant force paths is discussed in Section 2.6. The analysis of tensile and transverse shear stresses in rivets is comparable to that given for bolts in Section 10.4. Also, much of Sample Problems 10.3, 10.4, and 10.5 (Section 10.10) is applicable to rivets. Significant initial tension is attainable in rivets by installing them at a red heat. The tension develops upon cooling and thermal contraction.

Whereas the development of modern welding equipment has reduced the importance of rivets for heavy structural applications, the development of modern riveting machines has greatly expanded their use in fastening smaller components in a multitude of industrial products associated with the automotive, appliance, electronic, furniture, business machine, and other fields. Rivets have frequently replaced threaded fasteners in these applications because of lower installed cost. Rivets are much cheaper than bolts, and modern high-speed riveting machines—some of which fasten over 1000 assemblies per hour—give low assembly cost. Rivets can also serve as pivot shafts (as in folding lawn furniture), electrical contacts, stops, and inserts.

In comparison with threaded fasteners, rivets are not susceptible to unintended loosening, but in some cases they impede desired disassembly and servicing. Many readers will have had the experience of wanting to disassemble an electrical appliance to make a simple repair (perhaps a broken electrical wire just inside the case of an alarm clock, or the end of an iron cord) only to find that rivets have been used. The appliance is then usually discarded and a new one purchased. Thus, despite the initial manufacturing economy, these applications of rivets sometimes make questionable economic and ecologic sense. On the other hand, making an electrical device so that it cannot be disassembled and tampered with by the layman may constitute a safety feature. (This point illustrates how seemingly simple engineering decisions often require an in-depth consideration of many factors, and that the best overall solution may require imaginative, new design approaches to optimize economic, safety, and ecologic considerations!)

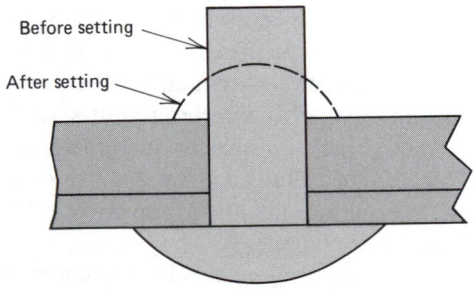

FIGURE 11.1.
Conventional solid rivet before and after setting.

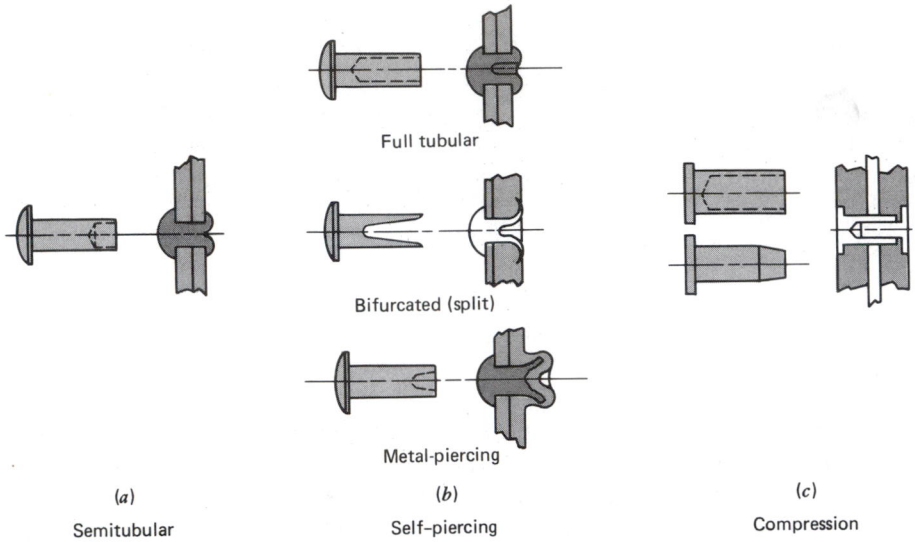

FIGURE 11.2.
Basic types of tubular rivets [7]. (a) Semitubular. (b) Self-piercing. (c) Compression.

Rivets can be made from most any ductile material, with carbon steel, aluminum, and brass being most commonly used. Various platings, paints, and oxide coatings may be applied. In general, a rivet cannot provide as strong an attachment as a bolt or screw of the same diameter. As with bolts, care must be used in selecting materials to be used together because of possible galvanic action.

Industrial rivets are of two basic types: *tubular,* and *blind.* Each type comes in a multitude of varieties.

Several tubular rivets are shown in Fig. 11.2. The semitubular version is the most common. The depth of the hole at the end of the shank does not exceed 112 percent of the shank diameter. Self-piercing rivets make their own holes as they are installed by a riveting machine. Full tubular rivets are generally used with leather, plastics, wood, fabric, or other soft materials. Bifurcated, or split, rivets can be used to join light-gage metals. Metal-piercing rivets can fasten metals like steel and aluminum with hardnesses up to about R_B 50 (approximately 93 Bhn). Compression rivets are made in two parts, as shown. The diameters are selected to provide an appropriate interference fit at each interface. Compression rivets can be used with wood, brittle plastics, or other materials with little danger of splitting during setting (installation). One common application is in the cutlery field, where the flush surfaces and interference fits provide no crevices for food and dirt particles to accumulate, as in riveted knife handles.

Figure 11.3 shows situations where blind rivets (requiring access to only one side of the joint) are appropriate. Figure 11.4 illustrates several varieties of blind rivets.

More detailed information on rivets is available in references such as [4,5,7].

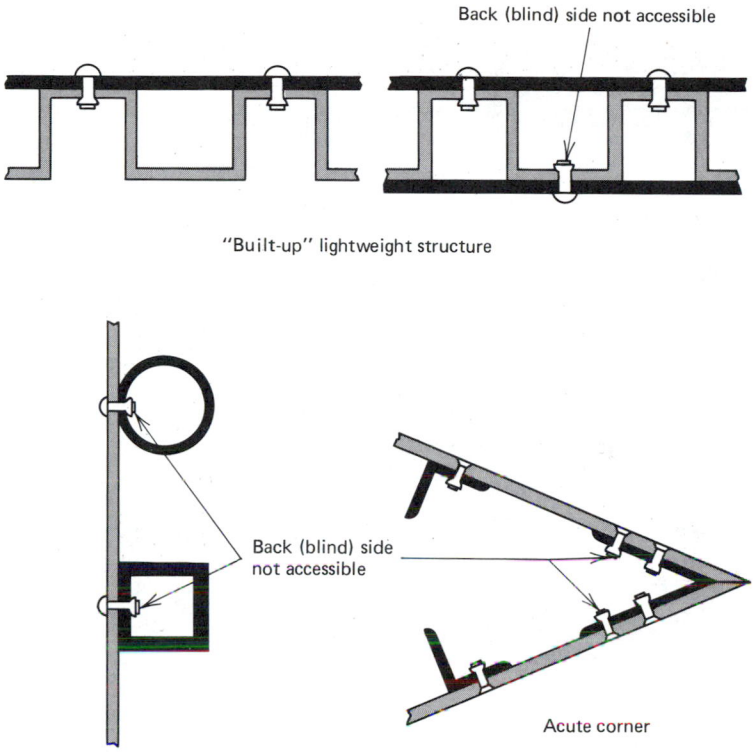

FIGURE 11.3.
Typical applications where blind rivets are used [7].

11.3 WELDING PROCESSES

Modern developments in welding techniques and equipment have provided the engineer with a range of attractive alternatives for fastening (as an alternative to bolts or rivets), and for fabricating parts. Machine members can often be fabricated by welding at lower cost than by casting or forging. Typical examples are shown in Fig. 11.5.

Most industrial welding is done by *fusion,* with the workpieces melting at their common surfaces. Heat is applied by an electric arc passing between an electrode and the work, by high-amperage electric current passing through the mating workpieces, or by gas flame. Electric arc welding takes several forms, depending on (a) how the filler material (consumable welding rod) is applied, and (b) the way the molten weld metal is shielded from the atmosphere:

1. *Shielded metal arc welding (SMAW),* also called stick welding, is the common manual (nonautomated) process used for repair work and the welding of large structures. The welder feeds a consumable, coated electrode (welding rod) in the work

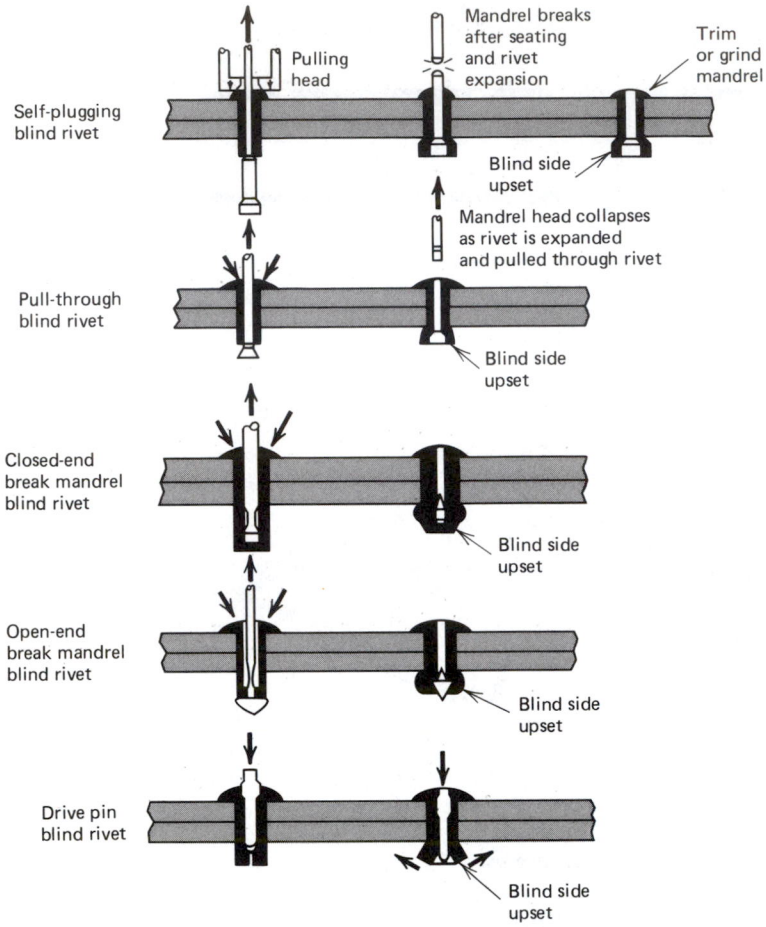

FIGURE 11.4.
Various types of mechanically expanding rivets with installation procedure shown [7].

area. A flux coating on the electrode vaporizes to form a shielding gas around the molten weld metal. SMAW is usually used with steels.

2. *Gas metal arc welding (GMAW),* also called *metal-inert gas welding (MIG),* is a commonly automated process producing high-quality welds at high welding speeds with most metals. The consumable electrode is not coated, but it projects from a nozzle that supplies a shielding gas (argon for aluminum and other non-ferrous metals; low-cost carbon dioxide for steels).
3. *Gas tungsten arc welding (GTAW),* also called *tungsten-inert gas welding (TIG),* employs a nonconsumable tungsten electrode, with a filler wire fed in separately. A nozzle surrounding the tungsten electrode supplies helium or argon gas for shielding. This process is slower than GMAW, but can be used with thinner metals,

either ferrous or non-ferrous. The process gives high-quality welds of non-ferrous and dissimilar metals, and can be fully automated.

4. *Flux core arc welding (FCAW),* is much like SMAW except that flux is in the hollow core of the consumable electrode rather than on its outer surface. Also, a shielding gas (usually carbon dioxide) is sometimes supplied from a nozzle surrounding the electrode. The process produces fast, clean welds on ferrous metals.
5. *Submerged arc welding (SAW)* is done on flat workpieces, with a mound or line of granular flux being deposited on the workpieces in advance of the moving consumable electrode. The flux melts to produce a protective molten slag, under which the fresh weld is "submerged." The process is most commonly used with thick sections requiring deep weld penetration.

Resistance welding involves the passage of electrical current (generating heat at the rate of "I^2R") through the work pieces while they are clamped firmly together. No flux or shielding is used, but the process may be carried out in a vacuum or inert-gas. Filler material is not normally used. Resistance welding is especially suited for mass production of continuous welds (as pipe seams) or spot welds of most steels and aluminum alloys. Material thicknesses range from about 0.004 to 0.75 in.

Gas welding—usually performed manually with an oxyacetylene torch—is relatively slow, involves more generalized heating of the work, and is most often used for making repairs. A filler wire is usually, but not necessarily, used.

Laser-beam welding, plasma-arc welding, electron-beam welding, and electroslag welding are also fusion welding processes, and are used on a small scale for very special applications.

Nonfusion, or *solid state welding* processes use a combination of heat and pressure to join parts, but the temperature involved (except at local asperities) is usually less than the melting point of the materials. Two examples are *inertia welding* (kinetic energy, stored in a flywheel, is converted to friction heat when a workpiece connected to the flywheel is forced against a stationary piece) and *ultrasonic welding* (workpieces clamped between a fixed anvil and vibrating "horn" bond as they are rubbed together at ultrasonic frequency). Note the parallel between these processes and *seizure,* discussed under *adhesive wear* in Section 9.9.

The suitability of several common metals and alloys to arc and gas welding is summarized in Table 11.1.

Thermoplastics can be welded much like metals. Heat can be applied by a *hot gas* (usually inert), contact with *hot metal plates,* or electrically by *dielectric heating* (thin films, mostly used for packaging, break down under high voltage and high frequency to produce heat needed for fusion) and *induction heating* (electromagnetic induction currents are generated in metal inserts or imbedded metal powders). Inertia welding (called *spin welding* when used with plastic materials) and *ultrasonic welding* techniques are applicable to the thermoplastics. Rubbing the mating plastic pieces together at sonic frequencies (usually 120 Hz) is also effective, with bonding usually accomplished after 2 or 3 seconds. This is called *vibration welding. Solvent bonding* of thermoplastic parts is accomplished when the members are softened by coating with a sol-

TABLE 11.1
Weldability of Common Metals

Metal	Arc	Gas	Metal	Arc	Gas
Carbon steel					
Low and medium carbon	G[a]	G	Magnesium alloy	X[b]	G
High carbon	G	F[c]			
Tool steel	F	F	Copper and copper alloys		
			Deoxidized copper	F	G
Cast steel, plain carbon	G	G	Pitch, electrolytic, and lake	G	F
Gray and alloy cast iron	F	G	Commercial bronze, red brass, low brass	F	G
Malleable iron	F	F	Spring, admiralty, yellow, and commercial brass	F	G
Low-alloy, high-strength steels			Muntz metal, naval brass, magnesium bronze	F	G
No-Cr-Mo and Ni-Mo	F	F			
Most others	G	G	Phosphor bronze, bearing bronze, and bell metal	G	G
Stainless steel			Aluminum bronze	G	F
Chromium	G	F	Beryllium copper	G	—
Chromium-nickel	G	G			
			Nickel and nickel alloys	G	G
Aluminum and Al alloys					
Commercially pure	G	G	Lead	X	G
Al-Mn alloy	G	G			
Al-Mg-Mn and Al-Si-Mg alloy	G	F			
Al-Cu-Mg-Mn alloy	F	X			

[a]G—good; commonly used.
[b]X—not used.
[c]F—fair; occasionally used under favorable conditions.

vent, then clamped together for 10–30 seconds during which time the plastic molecules intermingle. The parts bond together when the solvent evaporates.

11.4 WELDED JOINTS SUBJECTED TO STATIC AXIAL AND DIRECT SHEAR LOADING

Weldments, such as those shown in Fig. 11.5, are usually fabricated from carbon steel pieces held in position by suitable clamps or fixtures while the welding is performed. The strength of the welded joints is dependent on many factors that must be properly controlled in order to obtain welds of high quality. The heat of welding may cause metallurgical changes in the parent metal in the weld vicinity. Also, residual stresses (and associated warping of the parts) may be introduced because of thermal gradients causing differential expansion and contraction, the influence of clamping forces, and the changes in yield strength with temperature. Residual stress and warping problems are most pronounced when welding pieces of varying thickness and irregular shape. Measures that can be taken to control these problems include heating the parts to a uniform temperature before welding, following detailed "good-welding practice" for the appli-

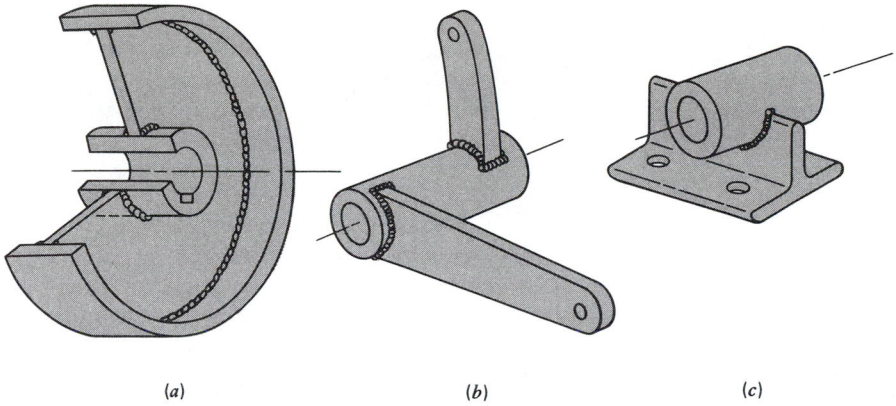

FIGURE 11.5.
Machine parts fabricated by fusion welding.

cation involved, giving the weldment a low-temperature stress-relieving anneal after welding, and shotpeening the weld area after cooling.

For welded construction used in buildings, bridges, and pressure vessels, the law requires that appropriate codes (such as [1]) be adhered to. In critical applications where uncertainties associated with the welding process are involved, laboratory testing of prototype welded joints is often advisable.

The basic concept of fusion welding is to fuse the materials together into a single—hopefully homogeneous—member. Properties of the welding rod (filler material) must, of course, be properly matched with those of the parent material. Ideally, the stress and strength analysis would then proceed as if the part were made from a single piece of stock.

Welding electrode material strength and ductility specifications have been standardized by the American Welding Society (AWS) and the American Society for Testing Materials (ASTM). For example, welding electrodes in the E60 and E70 series are designated E60XX and E70XX, where the 60 and 70 indicate tensile strengths of at least 60 or 70 ksi, respectively. The last two digits (XX) indicate welding process details. Specified yield strengths for E60 and E70 series rods are about 12 ksi below the tensile strengths, and minimum elongation is between 17 and 25 percent. These properties all apply to the *as-welded* material.

Figure 11.6 shows four varieties of full-penetration *butt welds*. If these are of good quality, the weld is fully as strong as the plate, and its *efficiency* (welded member strength/solid member strength) is 100 percent (remember, this is for *static*—not fatigue—loading).

Fillet welds, illustrated in Fig. 11.7, are commonly classified according to the direction of loading: (a) *parallel load* (Fig. 11.7*c*), or (b) *transverse load* (Figs. 11.7*d* and 11.7*e*). For parallel loading, both plates exert a shear load on the weld. With transverse loading, one plate exerts a shear load and the other a tensile (or compressive) load on the weld. The *size* of the weld is defined as the *leg length, h* (Fig. 11.7*a*). Usually, but not necessarily, the two legs are of the same length. Conventional engineering practice

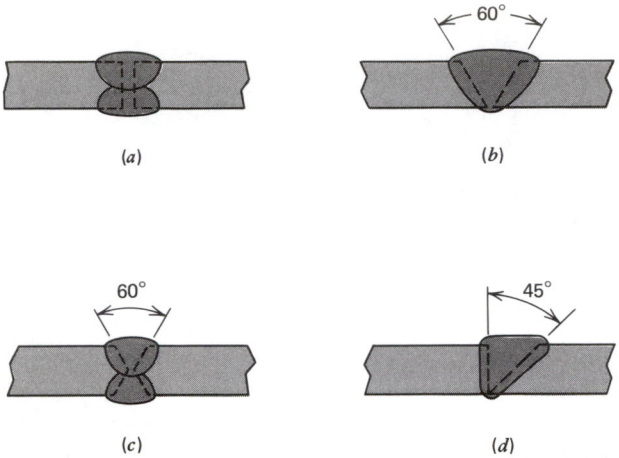

FIGURE 11.6.
Representative butt or groove welds.

considers the significant weld stress to be the *shear stress in the throat section* for either parallel or transverse load. Throat length, t (Fig. 11.7a), is defined as the shortest distance from the intersection of the plates to (a) the straight line connecting the ends of the two legs (Fig. 11.7a'), or (b) to the weld bead surface (Fig. 11.7a''), whichever is less. For the usual case of a convex weld with equal legs, $t = 0.707h$. The throat area used for stress calculations is then the product tL, where L is the length of weld. Figure 11.7b shows that for welds with convex weld beads achieving significant penetration into the parent materials, the actual minimum weld section area can be significantly greater than tL. Although welding codes and conservative standard practice do not give the weld credit for this extra throat area, there is no reason not to consider it in the design of members not covered by codes *provided* the extra calculated load capacity can be verified by test, and the production welding process is sufficiently well controlled to insure that all welds actually achieve the increased section area.

Weld size, h, should be reasonably proportioned in comparison to the thickness of plates being welded. For practical reasons, h is usually a minimum of 3 mm for plates less than 6 mm thick—on up to a minimum of 15 mm for plates over 150 mm thick.

SAMPLE PROBLEM 11.1 Estimate the Static Strength of a Parallel-Loaded Fillet Weld

The plates in Fig. 11.7c are 12 mm thick, and made of steel having $S_y = 350$ MPa. They are welded together by convex fillet welds along sides AB and CD, each of which is 50 mm long. Yield strength of the weld metal is 350 MPa. With a safety factor of 3 (based on yield strength), what static load, F, can be carried, using a 6-mm weld leg?

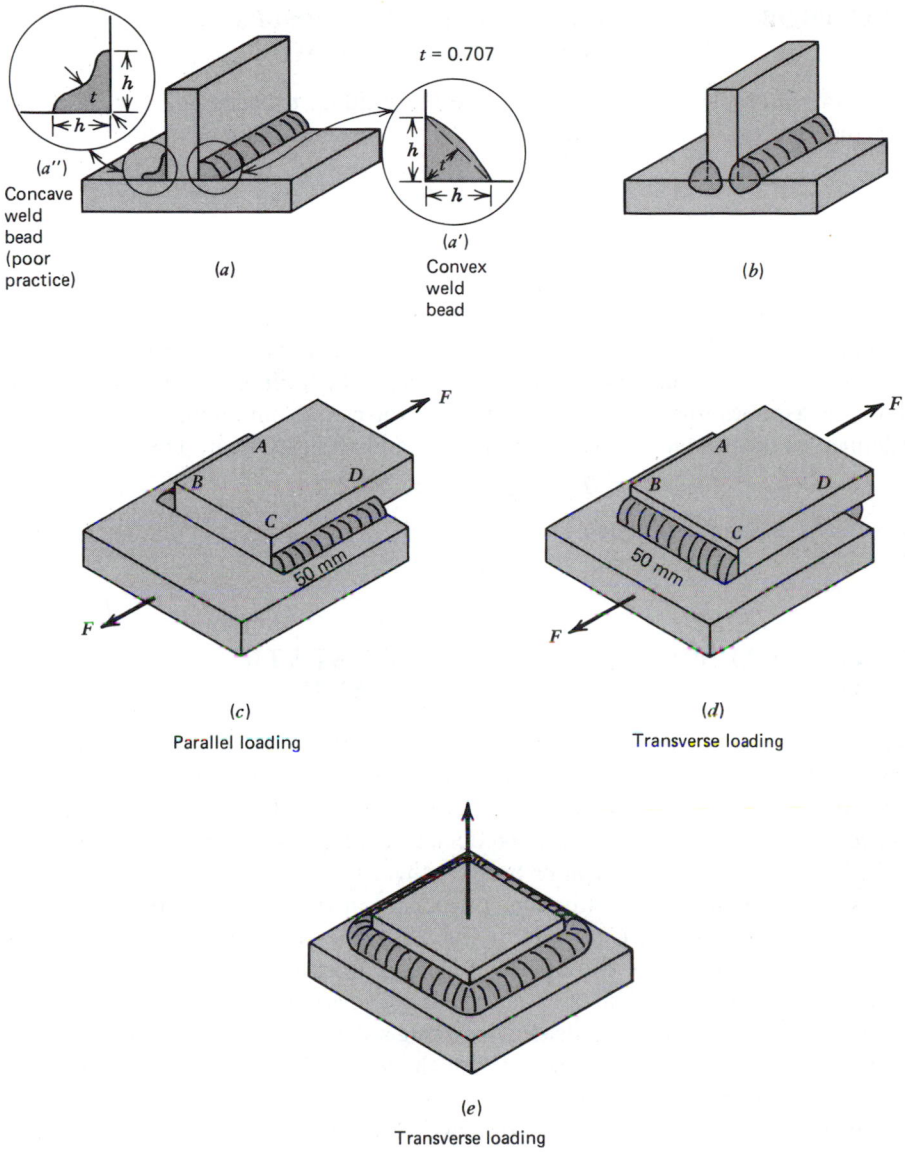

FIGURE 11.7.
Fillet welds.

Solution

1. Weld throat area is a total of (0.707)(6)(100) = 424 mm^2 for the two welds.
2. This weld throat area is stressed in shear. Using the distortion-energy theory, $S_{sy} = 0.58S_y = 203$ MPa.
3. $F = S_{sy}A/\text{S.F.} = (203)(424)/3 = 28{,}700$ N, or *28.7 kN*. ■

SAMPLE PROBLEM 11.2 Estimate the Static Strength of a Transverse-Loaded Fillet Weld

Repeat Sample Problem 11.1 except change the welding to sides *AD* and *BC* (thus making the joint the same as in Fig. 11.7*d*).

Solution

1. The welds are loaded transversely, with the horizontal weldplate interface being loaded in shear and the vertical interface loaded in tension. As the load flows through the weld metal (recall concepts involved in Section 2.6), the loading involves varying proportions of tension and shear. The traditional simplified approximation assumes that the critical stress involves the minimum throat section (defined as the product tL) carrying the entire load, F, in shear. Then,

$$F = S_{sy}A/\text{S.F.} = 28.7\ kN$$

■

exactly as in Sample Problem 11.1. It should be emphasized, though, that the solution in this case is less rigorous than in Problem 11.1.

11.5 WELDED JOINTS SUBJECTED TO STATIC TORSIONAL AND BENDING LOADING

Figures 11.8 and 11.9 illustrate two cases often referred to as representing eccentric loading: (a) In Fig. 11.8, the load is *in the plane of the weld group,* thereby subjecting the welded connection to *torsion* as well as direct loading. (b) In Fig. 11.9, the load is *out of the plane of the weld group,* thus producing *bending* in addition to direct loading. Note that in addition to the torsion or bending there is a direct loading component on the welds corresponding to the situations considered in the previous article: weld segments *BC* in Fig. 11.8, and *BC* and *AD* in Fig. 11.9, are subjected to parallel loading; all other weld segments in the two figures are loaded transversely.

A rigorous analysis of the stresses on various portions of each weld segment in Figs. 11.8 and 11.9 would be very complicated, involving a detailed study of both the rigidity of the parts being joined and the geometry of the weld. The various incremental segments of the weld serve as a multiplicity of *redundant* attachments, each carrying a portion of the load dependent upon its *stiffness* (recall Sections 2.5, 2.6, and 4.2; and particularly Fig. 4.2). The following procedures are based on simplifying assumptions commonly used to obtain results sufficiently accurate for engineering use in *most* (not all) cases.

Stresses on the welds in Figs. 11.8 and 11.9 consist of the vector sum of:

1. direct shear stresses, calculated as in the previous section (these stresses are ordinarily assumed to be uniformly distributed over the length of all welds), and
2. superimposed torsional and/or bending stresses, calculated from the conventional formulas, $\tau = Tr/J$ and $\sigma = Mc/I$ (assuming that the parts being joined are completely rigid).

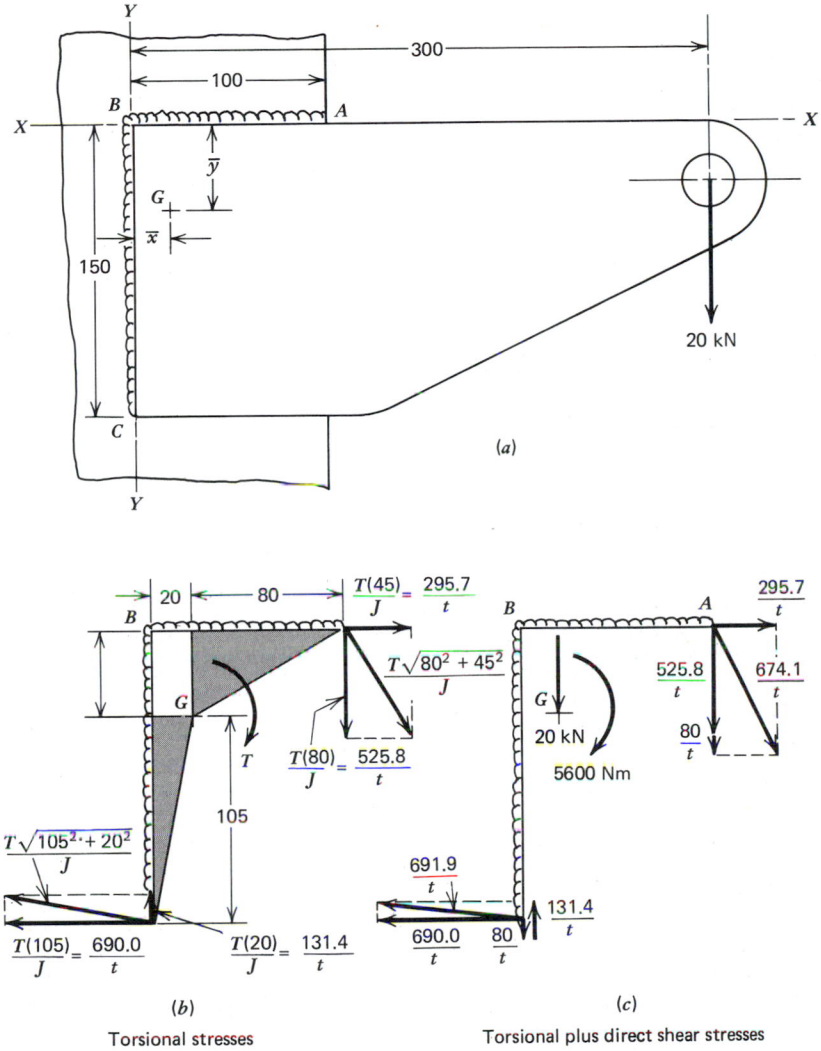

FIGURE 11.8.
In-plane eccentric loading (Sample Problem 11.3).

Values of I and J for the common weld patterns made up of linear segments can be computed with reference to Fig. 11.10. The procedure is illustrated in Sample Problems 11.3 and 11.4.

SAMPLE PROBLEM 11.3 Determine Weld Size with In-Plane Eccentric Loading

Determine the required weld size for Fig. 11.8*a* using E60 welding rod (S_y = 345 MPa) and safety factor of 2.5 based on yielding.

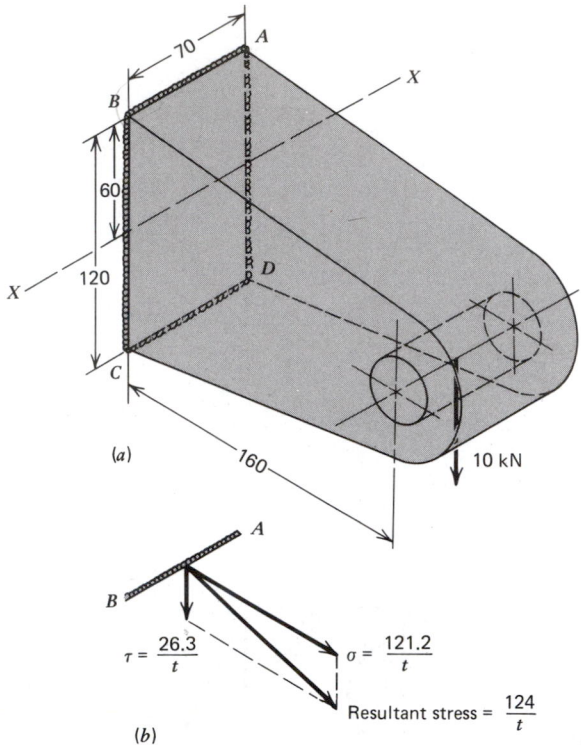

FIGURE 11.9.
Out-of-plane eccentric loading (Sample Problem 11.4).

Solution

1. The applied load of 20 kN is equivalent to the same load acting through center of gravity, G, of the weld pattern, plus a clockwise torque of $(20{,}000)(300 - \bar{x})$ N·mm. We proceed first to locate the center of gravity, G, defined by dimensions $\bar{x}$ and $\bar{y}$.

2. Let A_i denote the weld-segment area, and x_i and y_i the coordinates to the weld-segment center of gravity of any straight-line segment of the weld group. Then

$$\bar{x} = \frac{\Sigma A_i x_i}{\Sigma A} = \frac{(100t)(50) + (150t)(0)}{100t + 150t} = 20 \text{ mm}$$

$$\bar{y} = \frac{\Sigma A_i y_i}{\Sigma A} = \frac{(100t)(0) + (150t)(75)}{100t + 150t} = 45 \text{ mm}$$

3. The polar moment of inertia of the weld pattern about G is the sum of the contributions made by each of the weld segments. From the relationships developed in Fig. 11.10, $J = \Sigma(I_X + I_Y)$ for each weld segment.

   For side weld only,

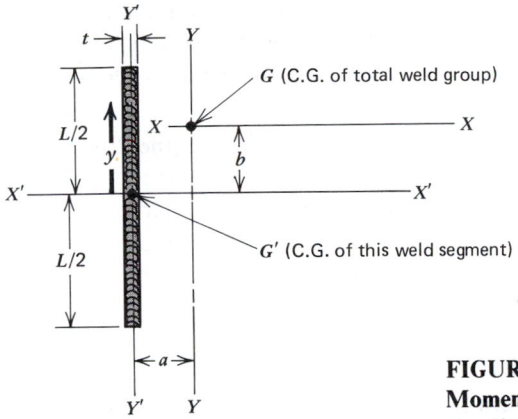

FIGURE 11.10.
Moments of inertia of linear weld segments. (For simplicity it is *assumed* that the effective weld width in the plane of the paper is the same as throat length, *t*, shown in Fig. 11.7*a*'.)
Rectangular moment of inertia about axes of symmetry of the weld segment, X' and Y'
(*t* is assumed to be very small in comparison with other dimensions)

$$I_{X'} = \int y^2\,dA = 2\int_0^{L/2} y^2\,(t\,dy) = \frac{L^3t}{12}$$
$$I_{Y'} = 0$$

Rectangular moment of inertia about axes X and Y thru c.g. of total weld group

$$I_X = I_{X'} + Ab^2 = \frac{L^3t}{12} + Ltb^2$$
$$I_Y = I_{Y'} + Aa^2 = Lta^2$$

Polar moment of inertia about an axis perpendicular to c.g. of weld group

$$J = I_X + I_Y = \frac{L^3t}{12} + Lt(a^2 + b^2)$$

$$J_s = \frac{150^3t}{12} + 150t[20^2 + (75 - 45)^2] = 476{,}250t$$

For top weld only,

$$J_t = \frac{100^3t}{12} + 100t[45^2 + (50 - 20)^2] = 375{,}833t$$
$$J = J_s + J_t = 852{,}083t$$

4. Figure 11.8*b* shows the torsional stresses at weld ends *A* and *C*, which are the two locations where the combined torsional and direct shear stresses are greatest. The resultant torsional stress at each of these points is Tr/J, where $T = (20{,}000)(280)$

N·mm and $J = 852{,}083t$ mm^4. The value of r for each point is equal to the square root of the sum of the squares of the legs of right triangles, but there is no need to compute r as we desire only the horizontal and vertical components of the torsional stress, as computed on the figure.

5. Figure 11.8c shows the vector addition of the torsional stress and the direct shear stress at points A and C. The direct shear stress is, of course, assumed to be merely $V/A = 20{,}000/250t = 80/t$ MPa. In this case, point C has the larger resultant stress of $692/t$ MPa. Equating this to the estimated shear yield strength (using the distortion-energy theory) and applying the safety factor,

$$\frac{692}{t} = \frac{345(0.58)}{2.5}, \quad t = 8.65 \text{ mm}$$

6. From the geometric relationship shown in Fig. 11.7a',

$$h = \frac{t}{0.707} = \frac{8.65}{0.707} = 12.23 \text{ mm}$$

7. Weld size would normally be specified as an integral number of mm. The choice between a 12-mm weld (giving a calculated safety factor of 2.45) and a 13-mm weld (calculated safety factor = 2.66) would depend upon circumstances and engineering judgment. In this case, let us arbitrarily state the final answer as $h = 13$ mm. ■

8. Although a 13-mm weld is required only at point C, the same size weld would normally be specified throughout. The weld might be specified as intermittent, however, with skips in regions where the vector sum of the stress components is relatively small.

Be sure that you understand the approximation in the above procedure, which is conventionally used for simplicity: throat dimension t is assumed to be in a 45 deg. orientation when calculating transverse shear and axial stresses; but the *same* dimension is assumed to be in the plane of the weld pattern when computing torsional stresses. After these stresses are vectorially added and a value for t determined, the final answer for weld dimension h again assumes dimension t to be in the 45 deg. plane. Although of course not rigorously correct, this convenient procedure is considered to be justified when used by the engineer who understands what he or she is doing, and who interprets the results accordingly. Note that the same simplifying approximation appears when handling bending loads, as illustrated by the next sample problem.

SAMPLE PROBLEM 11.4 Determine Weld Size for Out-of-Plane Eccentric Loading

Determine the weld size required for Fig. 11.9a using E60 welding rod ($S_y = 345$ MPa) and a safety factor of 3.

Solution

1. The loading involves direct shear plus bending, with the bending moment being (10,000)(160) N·mm.
2. The rectangular moment of inertia about neutral bending axis, X, consists of contributions made by the two vertical and two horizontal welds; that is, $I_X = 2I_v + 2I_h$. With reference to relationships summarized in Fig. 11.10,

$$I_v = \frac{L^3 t}{12} = \frac{(120)^3 t}{12} = 144{,}000t$$

$$I_h = Lta^2 = 70t(60)^2 = 252{,}000t$$

$$I_X = 2I_v + 2I_h = 792{,}000t$$

3. The (tensile) bending stress on weld AB is

$$\sigma = \frac{Mc}{I_X} = \frac{1{,}600{,}000(60)}{792{,}000t} = \frac{121.2}{t} \text{ MPa}$$

and the transverse shear stress on all welds is

$$\tau = \frac{V}{A} = \frac{10{,}000}{(120 + 120 + 70 + 70)t} = \frac{26.3}{t} \text{ MPa}$$

4. Figure 11.9*b* shows the resultant of the bending and transverse shear stresses to be $124/t$ MPa. The conventional procedure is to regard this resultant as a shear stress acting in the plane of the weld throat and equate it to the allowable shear stress. Using the distortion-energy theory to obtain an estimated shear yield strength (often the less accurate but more conservative maximum shear stress theory is used instead), and applying the safety factor of 3:

$$\frac{124}{t} = \frac{345(0.58)}{3} \text{ MPa}, \quad \text{or } t = 1.86 \text{ mm}$$

5. Although the plane of the above maximum shear stress does not correspond to the 45 deg. minimum throat plane illustrated in Fig. 11.7*a*′, it is customary to use the (smaller, thus more conservative) 45 deg. throat plane area to calculate the required weld leg size, h:

$$h = \frac{t}{0.707} = \frac{1.86}{0.707} = 2.63 \text{ mm}$$

6. It is not normally practical to apply a weld of leg length less than ⅛ in., or 3 mm.

Thus our answer would be: *Use h = 3 mm.* (In practice, the use of a larger weld than theoretically required is often accompanied by welding *intermittently* along the length. Also, in this case the entire weld CD could be omitted. Bending stresses at the bottom are compressive, and these can be carried directly without a weld. Omitting the bottom weld would impose a little higher direct shear stress on the other three welds, but this would have little effect.) ■

11.6 FATIGUE CONSIDERATIONS IN WELDED JOINTS

When welded joints are subjected to fatigue loading, small voids and inclusions (which have little effect on static strength) constitute points of local stress concentration and reduce fatigue strength. Also, the weld material extending beyond the plane of the plate surfaces in butt welds (Fig. 11.6), which is called "reinforcement," causes obvious stress concentration at the edges of the weld bead. For static loading, this material may indeed provide slight "reinforcement" to compensate for possible voids or inclusions in the weld metal; but for fatigue loading, strength is increased by grinding the weld bead flush with the plates. Rough approximate fatigue stress concentration factors associated with butt weld reinforcement and other weld geometries are given in Table 11.2.

SAMPLE PROBLEM 11.5 Estimate the Fatigue Strength of a Butt Weld

The tensile load on a butt weld (Fig. 11.6) fluctuates rapidly between 5000 and 15,000 lb. Plates are ½ in. thick, and the weld "reinforcement" is not ground off. An E60 series welding rod is used, giving $S_u = 62$ ksi, $S_y = 50$ ksi. A safety factor of 2.5 is to be used. What length (L) of weld is required?

Solution

1. Since the reinforcement is not ground off, there is a stress concentration at the edge of the weld metal; $K_f = 1.2$ (Table 11.2). Also the weld metal has a rough surface. In the absence of relevant experimental data, we might conservatively assume the surface to be comparable to an as-forged surface. From Fig. 8.13, $C_S = 0.55$. Assume $C_G \approx 0.8$.
2. $S_n = S'_n C_L C_G C_S = (62 \text{ ksi}/2)(1)(0.8)(0.55) = 13.6$ ksi.
3. For the *design overload* of 12,500 to 37,500 lb,

TABLE 11.2
Approximate Fatigue Stress Concentration Factors, K_f[a]

Type of Weld	K_f
Butt weld, with "reinforcing" not removed, tensile loads	1.2
Toe of fillet weld, transverse loading	1.5
End of fillet weld, parallel loading	2.7
T-butt joint weld, with sharp corners, tensile loading	2.0

[a]Originally proposed by C. H. Jennings, "Welding Design," *Trans. ASME*, **58** (1936), pp. 497–509, and widely used since then.

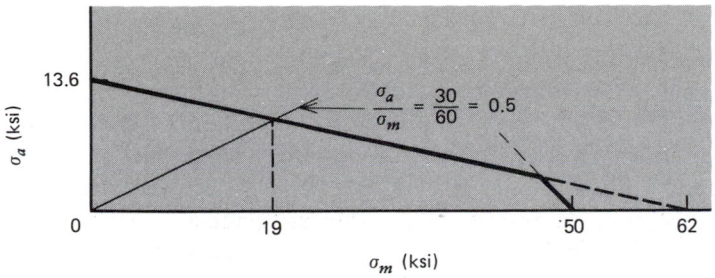

FIGURE 11.11.
$\sigma_m - \sigma_a$ curve for Sample Problem 11.5.

$$\sigma_m = K_f P_m / A = (1.2)(25{,}000)/0.5\,L = 60{,}000/L$$
$$\sigma_a = K_f P_a / A = (1.2)(12{,}500)/0.5\,L = 30{,}000/L$$

4. From Fig. 11.11, σ_m = 19,000 psi. Hence, 19,000 psi = 60,000/L, or, L = *3.16 in.* (This calculated figure might be rounded off to, say, 3 or 3¼ in., depending on circumstances and engineering judgment.) ■

11.7 BRAZING AND SOLDERING

Brazing and soldering differ fundamentally from welding in that the temperatures involved are always *below* the melting point of the workpieces. The solder or brazing filler material acts somewhat like a molten metal glue or cement that sets immediately on cooling. Thus, brazing and soldering can be classed as *bonding* processes.

Brazing involves heating the workpieces to a specified temperature above 840°F (450°C). On contact with the workpieces, the filler material melts and then flows into the space between the pieces. For capillary action to draw the molten filler metal most effectively, clearances of 0.025 to 0.050 mm are normally recommended, but special "wide-gap" filler alloys permit clearances ten times this great. It is important that the surfaces be initially clean. Either flux or an inert atmosphere is required.

Brazing is usually accomplished by heating the parts with a torch or in a furnace. Filler metals are usually alloys of copper, silver, or nickel.

Special care must be used when brazing aluminum parts because the melting point of the filler material (an aluminum-silicon alloy) is not much lower than that of the aluminum parts.

Brazing has several advantages: Dissimilar metals, cast and wrought metals, and even nonmetals (suitably coated) and metals can be brazed together. Complex assemblies can be brazed in several steps by using filler metals with progressively lower melting temperatures. Brazed joints require little or no finishing.

Soldering is similar to brazing except that the filler metal has a melting temperature below 840°F (450°C), and has relatively low strength. Most solders are tin-lead alloys, but alloys including antimony, zinc, and aluminum are also used. Nearly half of the soldering applications involve electrical and electronic parts. Another common application is sealing the seams in radiators and tin cans.

11.8 ADHESIVES

Adhesive bonding of metal parts is a rapidly growing field that influences the design of products of nearly all kinds. The advantages of adhesive are many. No holes (which weaken the part) are required, as with screws and rivets. No temperatures high enough to produce warping and residual stresses are involved, as with welding. When the joint is loaded, stresses are spread over a large area, with only minor stress concentration at the edges of contact. This often permits the use of thinner members, resulting in a weight saving. Adhesive bonding permits smooth, unbroken exterior surfaces for good appearance, easier finishing, and reduced fluid friction (in applications involving a flowing liquid or gas, as an airplane wing or helicopter rotor blade). Almost any solid materials can be bonded with a suitable adhesive. When bonding dissimilar metals, the adhesive layer can provide effective insulation against galvanic currents (the principle is illustrated in Fig. 9.7). On the other hand, the adhesive can be made electrically conducting if desired. Flexibility of the adhesive material can be made to accommodate differential thermal expansion of the bonded members. This flexibility also aids in absorbing impact loads. Furthermore, adhesive bonds can provide damping to reduce vibration and sound transmission. Adhesive bonded joints can effectively seal against leakage of any liquid or gas that does not attack the adhesive.

On the negative side, adhesives are more temperature-sensitive than mechanical fasteners. Most adhesives currently in common use are limited to the -200° to 550°F (-129° to 260°C) range. Adhesives vary greatly in temperature response, and this must always be considered when selecting an adhesive for a specific application. Inspection, disassembly, and repair of adhesives joints may not be practical. Also, long-term durability for some adhesives is questionable.

Perhaps the major factor motivating the proliferation of adhesives is cost reduction. However, the cost to be considered is *total* cost of the assembled joint, and this can be either greater or less than the cost of using mechanical fasteners, depending on a multitude of factors. As previously noted, adhesive bonding may permit thinner and lighter mating surfaces. Machining costs may be reduced by eliminating drilled holes and permitting wider tolerances (example: the "gap-filling" capability of some adhesives can eliminate the need for low-tolerance press fits). The cost of the adhesive itself can be either more or less than that of a competitive mechanical fastener. A major cost factor with bonded joints is often surface preparation. Extreme cleanliness is often required, although recent advances in adhesive research have made possible a relaxation of cleanliness standards in some applications. Automation for high production may be more costly with adhesives, particularly if elaborate fixtures and extended curing times are involved.

Bonding adhesives function as one component of a composite bonding system, with interaction between the adhesive and bonded materials influencing the properties of both. This fact, plus the tremendous variety of adhesives on the market, often makes the selection of the best adhesive for a particular application difficult. Current specialized technical literature, information from adhesive suppliers, and test results must be relied upon.

Safety and environmental factors are important considerations in the adhesives industry. Federal regulations control the use of certain solvent-based adhesives that give

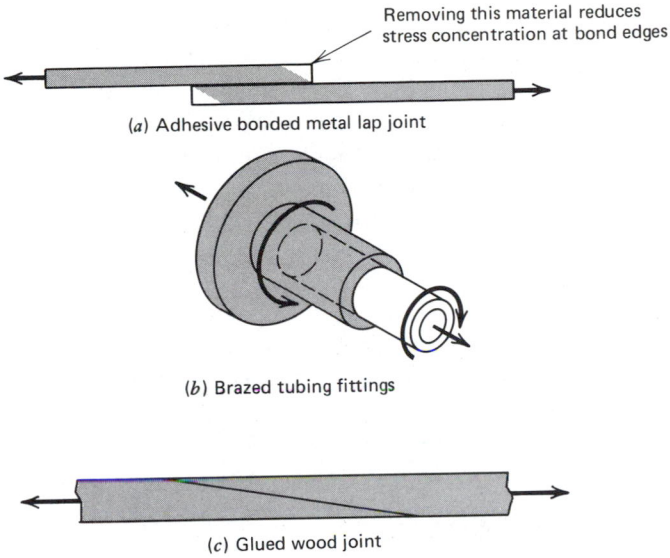

FIGURE 11.12.
Examples of bonded joints designed for loading in shear.

off flammable or toxic gases. Some adhesives are not used for certain applications because they emit an objectionable odor until cured. Petroleum is the raw material for many adhesives, and research efforts are being directed toward finding more alternatives. Several types of adhesives require heat-curing at 95° to 260°C, sometimes for several hours. Increasing energy costs are adding urgency to the development of equivalent adhesives requiring reduced curing time and temperature.

When possible, mating members should be designed so that adhesive joints are loaded in *shear,* as all adhesive bonds perform best when loaded this way. Figure 11.12 illustrates representative appropriate designs. Adhesive joints can also be loaded in tension, but "peel" and cleavage loadings should be avoided.

Most *structural* or *engineering* adhesives are *thermosetting,* as opposed to thermoplastic (heat-softening) types, such as rubber cement, airplane glue, and hot-melts. The latter are generally intended for light-duty applications requiring lower strength. The shear strength of some heat-curing one-part epoxies is as high as 70 MPa, but most structural adhesives have shear strengths in the 25 to 40 MPa range.

Epoxies, used industrially since the 1940s, are the most versatile and widely used of the structural adhesives. The two-part types cure at room temperature but require pre-mixing. One-part epoxies need no mixing but require heat-curing, usually around 150°C for at least an hour. Epoxies can be formulated to sufficient viscosity to satisfy most gap-filling requirements. Overall cost of high-strength epoxy bonds can be relatively high. *Urethanes* are similar to epoxies in that they have great versatility, relatively high strength, and are available in both two-part, room-temperature cure and one-part, heat-cure formulations. They have good toughness, flexibility, and impact resistance.

Anaerobics are one-part adhesives that are easy to apply and cure in the absence of oxygen. They are commonly used for thread-locking and machine assembly, such as retaining bearings and keys in place on shafts. Rapid curing characteristics usually restrict the use of anaerobic adhesives to the assembly of relatively small parts.

Some of the newer *acrylic* adhesives are tolerant of dirty surfaces. This can be very important in high-production assembly under ordinary industrial conditions. *Cyanoacrylates* are particularly appropriate when extremely fast set time is important. They are perhaps the easiest to apply and the fastest to cure of all industrial adhesive.

In order to facilitate production and reduce costs, most one-part (heat-curing) engineering adhesives are available as films, with removable backing. These are supplied in various thicknesses, and even in special shapes. Also, some adhesives are available in the form of a powder that is held in place electrostatically prior to assembly and heat-curing. Adhesives are sometimes used in conjunction with mechanical fasteners. The fast-moving field of modern industrial adhesives offers liberal opportunity for engineering ingenuity.

REFERENCES

1. American Welding Society Structural Welding Code, AWSD.1.77, American Welding Society, Miami, Florida.
2. Aronson, R. B., "Adhesives are Getting Stronger in Many Ways," *Machine Design* (Feb. 8, 1979), pp. 54–60.
3. Chow, W. W., *Cost Reduction in Product Design,* Van Nostrand Reinhold Co., New York, 1978.
4. Fisher, J. W., and J. H. A. Struik, *Guide to Design Criteria for Bolted and Riveted Joints,* John Wiley & Sons, Inc., New York, 1974.
5. *Machine Design,* 1980 Fastening and Joining Reference Issue, Penton/IPC, Inc., Cleveland, Nov. 13, 1980.
6. Osgood, C. C., *Fatigue Design,* Wiley-Interscience, New York, 1970.
7. Parmley, R. O. (ed.), *Standard Handbook of Fastening and Joining,* McGraw-Hill, New York, 1977.

CHAPTER TWELVE

Springs

12.1 INTRODUCTION

Springs are *elastic* members that *exert forces* (or torques) and *absorb energy* (which is usually stored and later released). Springs are usually, but not necessarily, made of metal. Plastics can be used when loads are light [3]. Modern structural composites are being introduced for some applications requiring minimum spring mass. Blocks of rubber often constitute springs, as in bumpers and in vibration-isolation mounting of various machines such as electric motors and internal combustion engines. Pneumatic springs of various types take advantage of the elastic compressibility of gases (as compressed air in automotive "air shocks," and hermetically sealed high-pressure nitrogen gas in the hydropneumatic suspensions of French Citroen automobiles). For applications requiring compact springs providing very large forces with small deflections, hydraulic springs have proven effective. These work on the basis of the slight compressibility of liquids, as indicated by their bulk modulus of elasticity. Product cost can sometimes be reduced by designing the required elasticity into other parts, rather than making those parts rigid and adding a separate spring [3].

Our concern in this book is with springs made of solids—metals or reinforced plastic composites. If energy absorption with maximum efficiency (minimum spring mass) is the objective, the ideal solution is an unnotched tensile bar, as all of the material is stressed at the same level (recall Chapter 7). Unfortunately, tensile bars of any reasonable length are too stiff for most spring applications; hence it is necessary to form the spring material so that it can be loaded in torsion or bending. (Recall Castigliano's method in Sections 5.8 and 5.9, where it was noted that deflections due to tensile and transverse shear stresses are usually negligible in comparison with deflections due to

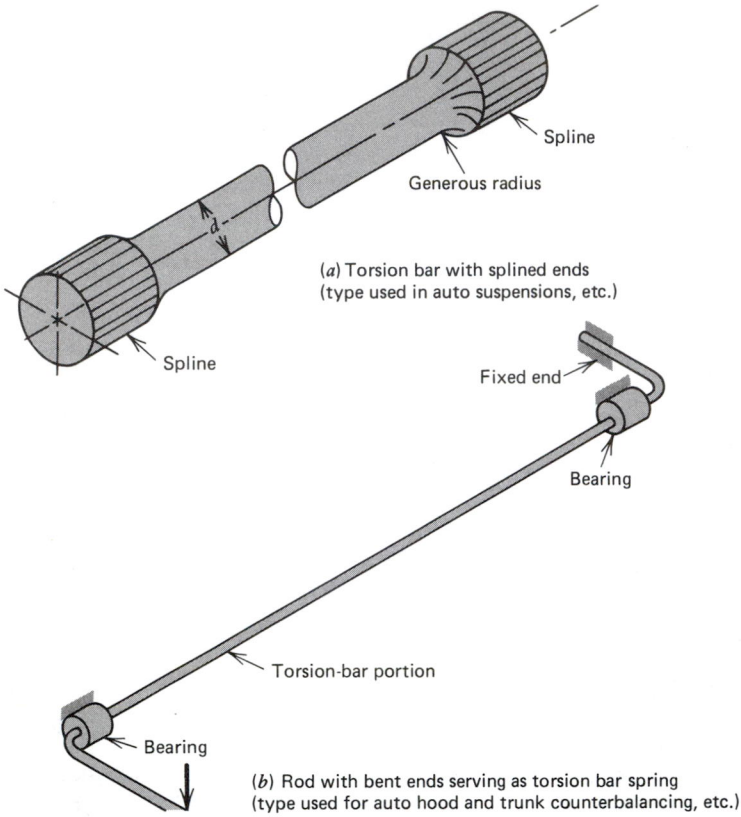

FIGURE 12.1.
Torsion-bar springs.

bending and/or torsion.) The following sections are concerned with springs of the common geometric forms that serve this purpose.

12.2 TORSION-BAR SPRINGS

Perhaps the simplest of all the spring forms is the torsion bar, illustrated in Fig. 12.1. Common applications include some automotive suspension springs, and counterbalancing springs for car hoods and trunk lids, where small torsion bars fit inconspicuously near the hinges.

The basic stress and angular deflection equations are

$$\tau = \frac{Tr}{J} \qquad \textbf{(4.3, repeated)}$$

$$\theta = \frac{TL}{JG} \quad \text{(see Table 5.1)}$$

For a solid round rod of diameter d, these become

$$\tau = \frac{16T}{\pi d^3} \qquad \textbf{(4.4, repeated)}$$

$$\theta = \frac{32TL}{\pi d^4 G} \quad \text{(see Table 5.1)}$$

12.3 COIL SPRING STRESS AND DEFLECTION EQUATIONS

Figures 12.2*a* and 12.2*c* show helical coil compression and tension springs, respectively, of relatively small helix angle, λ. Corresponding figures 12.2*b* and 12.2*d* show top por-

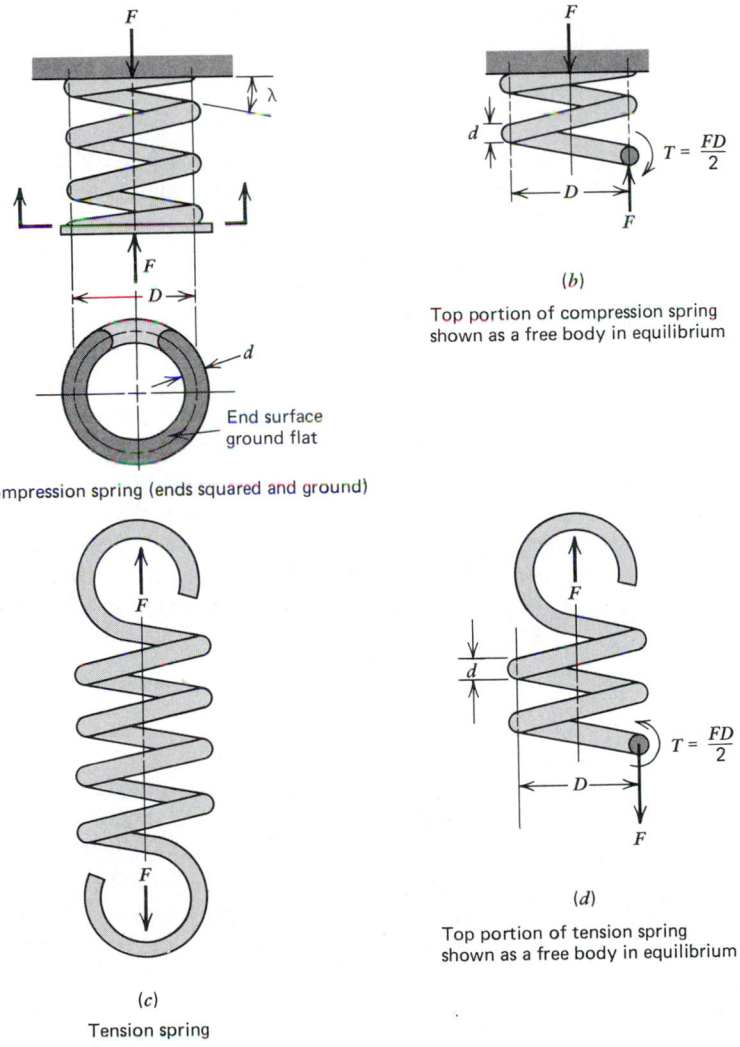

FIGURE 12.2.
Helical (coil) compression and tension springs.

tions of these springs as free bodies in equilibrium. In each case, external force F is (at least assumed to be) applied *along the axis of the helix*. In the compression spring, this is commonly achieved by winding the end coils with zero pitch and then grinding the ends flat (or by contouring the mating end plates) so that the pressure applied by the end plates is essentially uniformly distributed. In the tension spring, the end hooks are formed so as to position the load force along the spring axis.

In Figs. 12.2*b* and 12.2*d* note that regardless of where the cutting plane is taken, equilibrium considerations require that the wire be subjected to (a) a transverse shear force of F, and (b) a torque equal to $FD/2$. The shear force turns out to be of minor consequence (more about that later). The important thing is that *the entire length of active wire in the helix* (i.e., wire between the end turns that are touching the end plates, or between the end hooks) *is subjected to torque*. For springs made of solid round wire, the resulting torsional stress is

$$\tau = \frac{Tr}{J} = \frac{16T}{\pi d^3} = \frac{8FD}{\pi d^3} \tag{12.1}$$

where D is the mean coil diameter, defined as the average of the inside and outside coil diameters. Thus, *a helical compression or tension spring can be thought of as a torsion bar wound into a helix.*

In addition to the basic shear stress represented by Eq. (12.1), the inner surfaces of a coil spring are subjected to two additional shear stress components: (a) A *transverse* shear stress resulting from force F, applied to the arbitrary cutting plane in Figs. 12.2*b* and 12.2*d*. At the inner coil surface, the direction of this stress coincides with that of the torsional stress for both compressive and tensile spring loading. (b) An increase in the intensity of torsional stress because of the *curvature* of the coiled "torsion bar." The latter effect is illustrated in Fig. 12.3. Suppose the torque transmitted through the curved torsion bar (Fig. 12.3*b*) produces a 1 deg. rotation between planes m and n. This 1 deg. is distributed over short length ab at the inside of the coil, and longer length cd at the outside. This gives rise to the stress gradient shown. (Note the similarity with Fig. 4.9*c* for bending of curved beams.) The severity of this effect is obviously greatest for small values of *spring index, C*, defined as the ratio of *mean* coil diameter to wire diameter:

$$C = \frac{D}{d} \tag{12.2}$$

(i.e., the effect is greatest for springs proportioned like a "doughnut").

The first generally recognized analysis of the transverse shear and curvature effects was published by A. M. Wahl of the Westinghouse Corp. in *Transactions of the American Society of Mechanical Engineers* (May–Aug. 1929). (See [4], which is widely regarded as the standard reference on springs.) This analysis involved the derivation of the equation for a factor, K_w (now called the *Wahl factor*), by which the stress in Eq. (12.1) can be multiplied to give the total resultant shear stress on the inside of the coil:

$$K_w = \frac{4C - 1}{4C - 4} + \frac{0.615}{C} \tag{12.3}$$

When the loading on the spring is essentially *static*, the view is usually taken that the first term of Eq. (12.3), which accounts for the curvature effect, should not be used

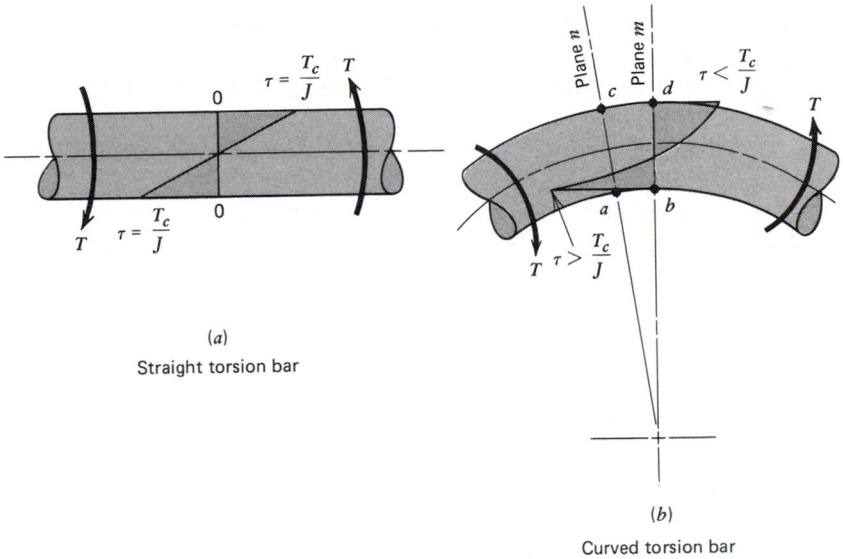

FIGURE 12.3.
Torsional stress distribution in straight vs. curved torsion bars. (Note increased stress on *inside* surface of the curved bar.)

because it is essentially a stress concentration factor. (As explained in Section 4.13, stress concentration can usually be ignored with the static loading of ductile materials.) Setting this term equal to unity gives a correction factor for transverse shear (only) of

$$K_s = 1 + \frac{0.615}{C} \tag{a}$$

where the subscript s designates static loading. Equation (a) is readily derived as follows:

An exact analysis shows the transverse shear stress acting on an element at the inside coil diameter to be $1.23P/A$. Adding this to the nominal torsional stress gives

$$\tau = \frac{8FD}{\pi d^3} + \frac{1.23F}{\pi d^2/4}$$

which reduces to

$$\tau = \frac{8FD}{\pi d^3}\left(1 + \frac{1.23(0.5)}{C}\right)$$
$$= \frac{8FD}{\pi d^3} K_s$$

where K_s is defined by Eq. (a).

After some yielding occurs (with static loading), the stresses are more uniformly distributed and the factor of 1.23 is often omitted. This gives

$$K_s = 1 + \frac{0.5}{C} \tag{12.4}$$

Actually, in applications involving static loading and elevated temperatures it is sometimes assumed that stresses become sufficiently redistributed that Eq. (12.1) can be used without correction.

The above variety of stress corrections for static loading can lead to confusion. An important point to remember is that when using allowable stress data obtained from static load tests of coil springs made of specific materials, the *same* correction factor must be used as that reflected in the test data.

In accordance with many authorities (including [1 and 2]), *it is recommended here that Eqs. (12.3) and (12.4) be used for fatigue and static loading, respectively.* These equations apply to springs of normal geometry: $C > 3$ and $\lambda < 12°$.

It follows from the above that for *fatigue* loading, the equation for *corrected* stress (i.e., including the complete Wahl factor) is

$$\tau = \frac{8FD}{\pi d^3} K_w = \frac{8F}{\pi d^2} CK_w \tag{12.5}$$

The corresponding corrected stress equation for *static* loading is

$$\tau = \frac{8FD}{\pi d^3} K_s = \frac{8F}{\pi d^2} CK_s \tag{12.6}$$

where K_s is defined in Eq. (12.4).

Values of K_w, K_s, K_wC, and K_sC are plotted in Fig. 12.4. (Use of the latter two quantities is illustrated in sample problems that follow.)

The free-body diagrams in Fig. 12.2 indicate no bending loading. For unusual cases involving pitch angle λ greater than 15 deg. and deflection of each coil greater than $D/4$, bending stresses should be considered (see [4], p. 102). Also, it should be noted that the above treatment of helical spring stresses tacitly assumed two factors to be negligible: (1) *Load eccentricity.* It is seldom possible to distribute the load on the spring ends so that the resultant coincides *exactly* with the geometric spring axis. Any eccentricity introduces bending, and changes the torsional moment arm. This causes the stresses on one side of the spring to be higher than indicated by the foregoing equations. (2) *Axial loading.* Reference to Fig. 12.2*b* indicates that in addition to creating a transverse shear stress, a small component of force F produces axial compression of the spring wire. In critical spring designs involving relatively large values of λ, this factor may warrant consideration.

Derivation of the equation for helical spring deflection is most readily accomplished using Castigliano's method (Section 5.8), as follows:

Assuming that transverse shear contributes negligibly to the deflection, only the torsional load need be considered. From Table 5.4:

$$\delta = \int_0^L \frac{T(\partial T/\partial Q)}{GK'} dx$$

where $Q = F$, and, for round wire, $K' = J = \pi d^4/32$. Letting N = number of *active* coils (i.e., not counting the end turns that do not participate in the deflection because they are in contact with the end plates):

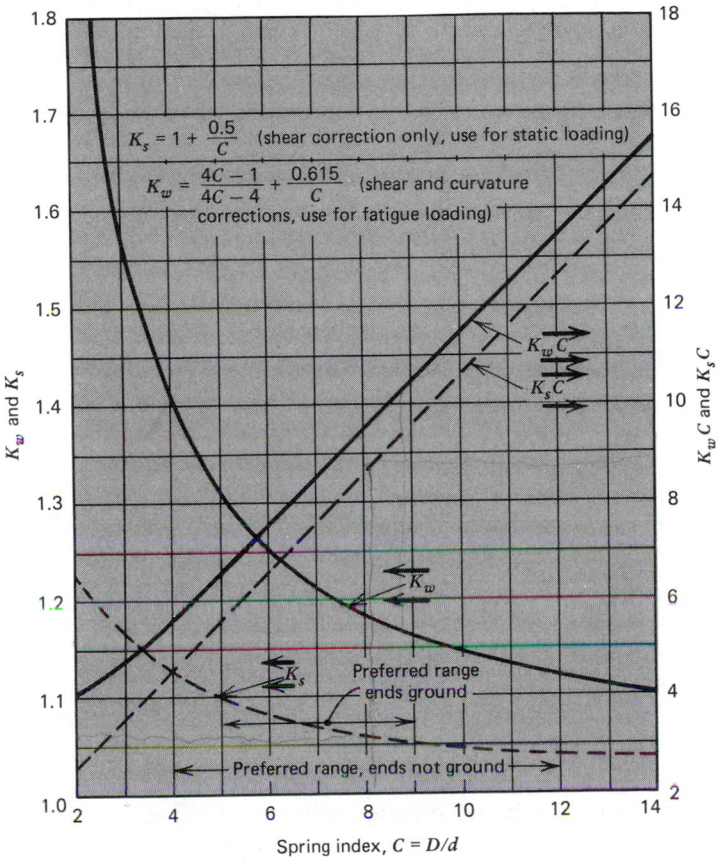

FIGURE 12.4.
Stress correction factors for helical springs.

$$\delta = \int_0^{2\pi N} \frac{(FD/2)(D/2)}{G(\pi d^4/32)} \left(\frac{D}{2}\, d\theta\right)$$

$$= \frac{4FD^3}{\pi d^4 G} \int_0^{2\pi N} d\theta$$

$$\delta = \frac{8FD^3N}{d^4G} \tag{12.7}$$

The *spring rate* (also called spring *constant* or spring *scale*—and with units of Newtons per mm, pounds per in., etc.) is commonly designated as k, where

$$k = \frac{F}{\delta} = \frac{d^4G}{8D^3N} \quad \left(\text{or, } k = \frac{dG}{8NC^3}\right) \tag{12.8}$$

Figure 12.5 shows a compression helical spring of unequal *pitch* (where *pitch* denotes axial spacing between adjacent coils). As this spring is loaded, active coils near

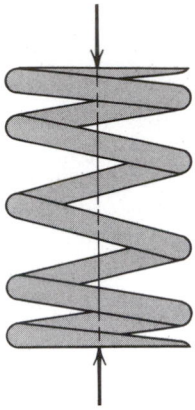

FIGURE 12.5.
Unequal pitch helical compression spring.

the ends "bottom out" first, thereby becoming "inactive" (like the end coils that are in contact with the end plates). As successive turns "bottom out," the spring becomes increasingly stiff (number of active turns, N, progressively decreases, thereby giving increased values of k in Eq. 12.8). This is desirable in some applications.

Most, but not all, coil springs are helical. A few are conical, as shown in Fig. 12.6. With a steep enough cone angle, the coils can nest together giving a *solid height* equal to the wire diameter. Since the torque $(FD/2)$ applied to the smaller turns is less than for the turns of larger diameter, the torsional stresses throughout the length of the wire vary accordingly. This means that the wire is used less efficiently than in a helical spring. By appropriately varying the pitch throughout the length of the spring, it is possible to have all turns "bottom out" simultaneously, if this is desired. The deflection and spring rate of a conical spring can be closely approximated by using Eqs. (12.7) and (12.8), where the average value of mean coil diameter is used for D.

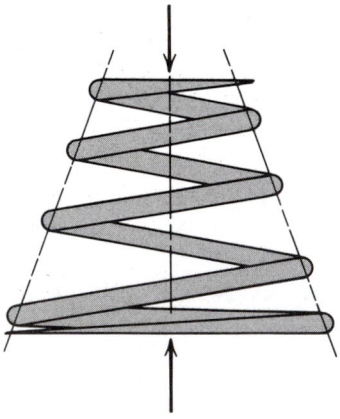

FIGURE 12.6.
Conical compression coil spring.

12.4 STRESS AND STRENGTH ANALYSIS FOR HELICAL COMPRESSION SPRINGS – STATIC LOADING

Most helical springs are wound from wire of solid round cross section. Only these will be treated in this book. Although spring wire was formerly manufactured primarily in certain standard "gauge" diameters, it is now readily available in any desired size. (Spring manufacturers may, however, carry in stock certain "preferred" sizes.)

The relative costs and minimum tensile strengths of commonly used spring wire materials are given in Table 12.1 and Fig. 12.7, respectively. (Note the tendency for smaller wires to have greater strength. For hard-drawn wire, this is mostly because the work-hardening associated with the wire drawing extends over a greater percentage of the cross section.) This information is fine as far as it goes, but for spring design allowable values of *shear* stress are needed for use with Eq. (12.5). Working with the information at hand, we proceed as follows:

1. The basic concern in designing springs for static loading is avoiding *set*, or long-term shortening (creep) of the spring under load (nicely illustrated by the springs in some old cars, which have taken sufficient set to slightly lower the car). Set is directly related to S_{ys}—and only indirectly to S_u. Experimental values of S_{ys} are seldom available. Equation (3.12) gives, for an "average" spring wire of $S_u = 220$ ksi, an *estimate* of

$$S_y = 1.05S_u - 30 \text{ ksi} = 1.05(220) - 30 = 201 \text{ ksi}$$

For such a steel, the distortion energy relationship predicts

$$S_{ys} = 0.58S_y = 0.58(201) = 116 \text{ ksi}$$

This value is $(116/220)S_u = 0.53S_u$.

TABLE 12.1
Relative Cost[a] of Common Spring Wire of 2 mm (0.079 in.) Diameter

Wire Material	ASTM Specification	Relative Cost
Patented and cold drawn steel	A227	1.0
Oil tempered steel	A229	1.3
Music (steel)	A228	2.0
Carbon steel valve spring	A230	2.5
Chrome silicon steel valve	A401	4.0
Stainless steel (Type 302)	A313 (302)	6.2
Phosphor bronze	B159	7.4
Stainless steel (Type 631)	A313 (631)	9.9
Beryllium copper	B197	22.
Inconel alloy X-750		38.

[a]Average of mill and warehouse quantities [2].

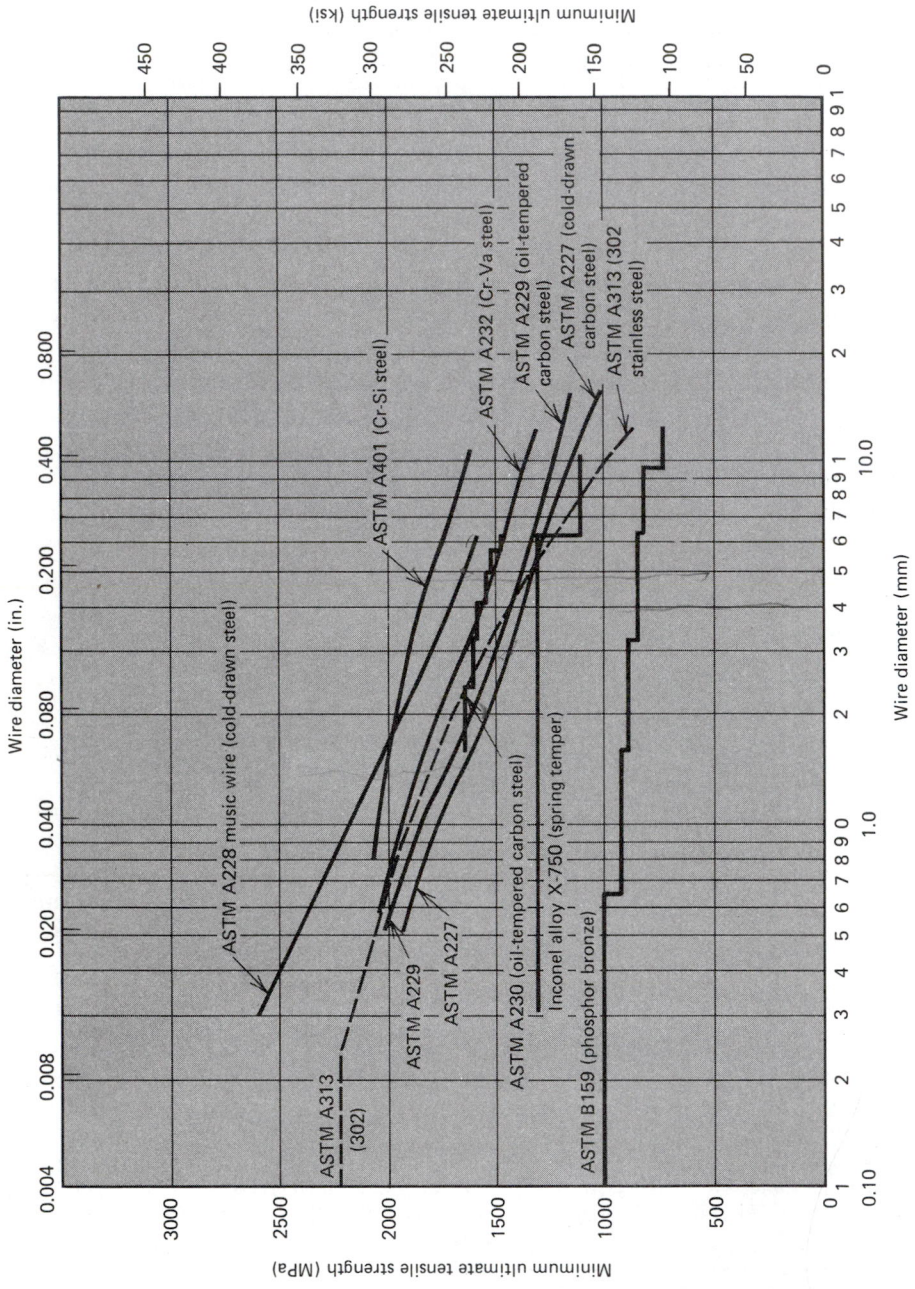

FIGURE 12.7.
Tensile strengths of various spring wire materials and diameters, minimum values [2].

2. The most severe stress to which a compression helical spring can be subjected corresponds to loading it to its *solid height* (all coils touching). Although this condition should never be experienced in service, it can be—either intentionally or accidentally—encountered when the spring is installed or removed. Typically then, τ (calculated from Eq. 12.6, with F equal to the load required to close the spring solid) should be less than S_{ys}, or, from the above, less than about $0.53S_u$.
3. Experience indicates [1] that less than 2 percent long-term "set" will occur in springs designed for τ_s (where subscript s denotes spring "solid") equal to $0.45S_u$ for ferrous spring, or $0.35S_u$ for non-ferrous and austenitic stainless steel springs.
4. Recommended stress values in step 3 thus appear to imply a safety factor, for steel springs, of about $0.53S_u/0.45S_u = 1.18$. This may seem small, but it actually illustrates the safety factor philosophy outlined in Section 6.11. For example, it was stated in that section that safety factor selection is based on five factors, three of which are (1) degree of loading uncertainty, (2) degree of material strength uncertainty, and (3) consequences of failure. For a helical compression spring there is virtually no loading uncertainty; it is impossible for the spring to be loaded beyond the point of closing solid. The manufacturing operations associated with high-quality spring production can be controlled to give a high degree of uniformity of spring yield strength. Finally, the consequences of failure (a little more than 2 percent set) are usually not serious.
5. Since stresses are limited by the "spring solid" condition, it follows that springs can be stressed most highly at their working loads if the maximum working load comes *close* to closing the spring solid. It is only necessary to provide sufficient *clash allowance* (difference in spring length between maximum load and spring solid positions) to allow for any possible combination of tolerance stack-up, differential thermal expansion, and wear of parts. Also, since all portions of the spring do not become "solid" at *precisely* the same load, the spring rate begins to build up sharply just before reaching the theoretical "spring solid" position. The usual recommendation is to *provide a clash allowance of approximately 10 percent of the total spring deflection at maximum working load.*
6. Finally, coil springs in compression are *ideal* candidates to benefit from favorable residual stresses due to yielding. Section 4.15 stated that *"An overload causing yielding produces residual stresses that are favorable to future loads in the same direction and unfavorable to future loads in the opposite direction."* Compression coil springs are loaded *only* in compression. Hence, advantage can be taken of favorable residual stresses by initially coiling the spring longer than desired, and then *yielding* it to the desired length by closing it solid. This operation, called *presetting,* is widely used.
7. According to [1], taking maximum advantage of presetting permits the design stress to be increased from the $0.45S_u$ and $0.35S_u$ values given above in step 3 to $0.65S_u$ and $0.55S_u$. This increase is actually somewhat more than can be justified theoretically on the basis of residual stresses alone, and may reflect some work-hardening (strain-strengthening) during presetting.

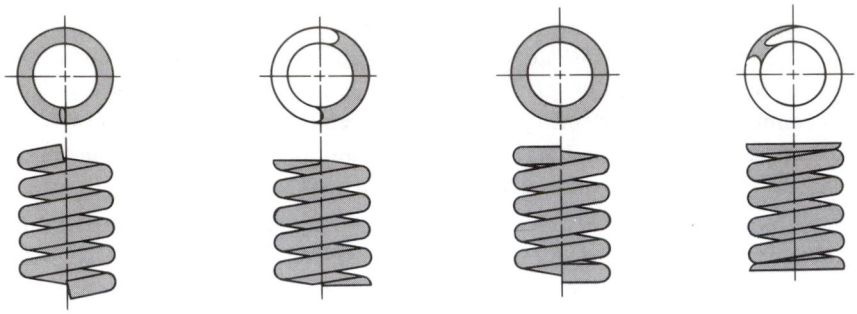

(a) $L_s = (N_t + 1)d$

(b) $L_s = N_t d$

(c) $L_s = (N_t + 1)d$

(d) $L_s = N_t d$

FIGURE 12.8.
Compression spring ends, and corresponding spring solid height equations. (Note: squared ends are wound with zero helix angle.)

Summarizing the above: to limit long-term set in compression coil springs to less than 2 percent, shear stresses calculated from Eq. (12.6) (normally with force F corresponding to spring "solid") should be:

$$\left.\begin{array}{ll} \tau_s \leqq 0.45S_u & \text{(ferrous—without presetting)} \\ \tau_s \leqq 0.35S_u & \text{(non-ferrous and austenitic stainless—without presetting)} \\ \tau_s \leqq 0.65S_u & \text{(ferrous—with presetting)} \\ \tau_s \leqq 0.55S_u & \text{(non-ferrous and austenitic stainless—with presetting)} \end{array}\right\} \quad \textbf{(12.9)}$$

12.5 END DESIGNS OF HELICAL COMPRESSION SPRINGS

The four "standard" end designs used with compression helical springs are illustrated in Fig. 12.8, together with the customarily used equations for their solid height, L_s. In all cases N = *number of active turns* (the turns that twist under load, and thereby contribute to the deflection as calculated from Eq. (12.7), and N_t = *total number of turns*. In all ordinary cases involving end plates contacting the springs on their end surfaces (i.e., *not* like the special cases shown in Figs. 12.9*b*–12.9*d*),

$$N_t \approx N + 2 \qquad \textbf{(12.10)}$$

Equation (12.10) follows from the need for both end plates to be in contact with nearly a full turn of wire in order to satisfy the requirement that the resultant load coincide with the spring axis.

The provision of essentially uniform contact pressure over the full end turns requires contoured end plates (as Fig. 12.9*a*) for all end conditions except squared and ground. The choice between (a) contoured end plates and (b) squared and ground springs (with flat end plates) is usually made on the basis of cost. Some alternative end member designs are illustrated in Fig. 12.9. Note that all of these except (a) represent

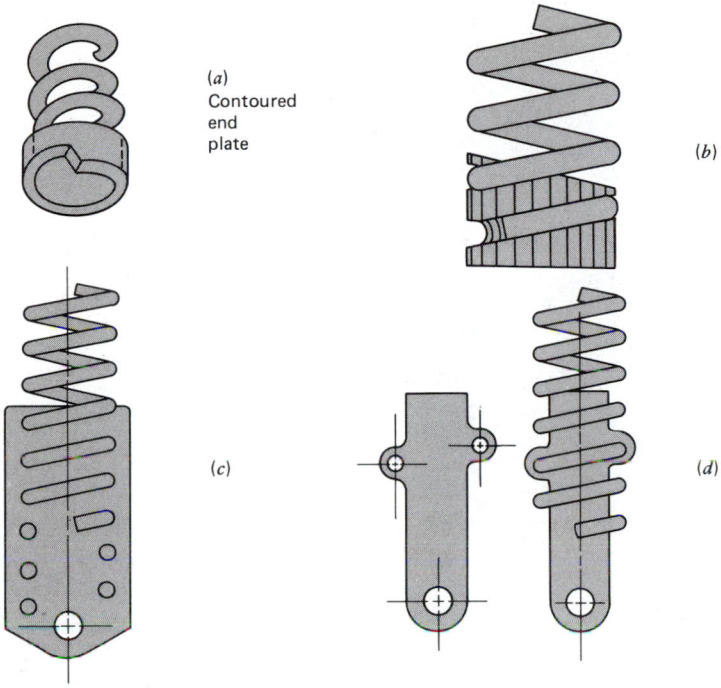

FIGURE 12.9.
Special end members for coil springs.

special designs that provide for adjusting the number of active turns, and that permit loading the spring in tension as well as in compression.

The solid height equations given in Fig. 12.8 are those normally used, but in cases (b) and (d) the actual solid height depends on the grinding operation. For springs with ground ends, the equations give a maximum solid height, associated with normal grinding practice.

For *special* applications where space is limited, ground springs can be obtained with a solid height of

$$L_s = (N_t - 0.5)(1.01\,d) \qquad \textbf{(b)}$$

12.6 BUCKLING ANALYSIS OF HELICAL COMPRESSION SPRINGS

Coil springs loaded in compression act like columns, and must be considered for possible buckling—particularly for large ratios of free length to mean diameter. The column treatment in Chapter 5 applies. Figure 12.10 gives the results for two of the end conditions illustrated in Fig. 5.25. Curve A (end plates constrained and parallel) represents the most common condition. If buckling is indicated, the preferred solution is to redesign the spring. Otherwise, the spring can be supported from buckling by placing it

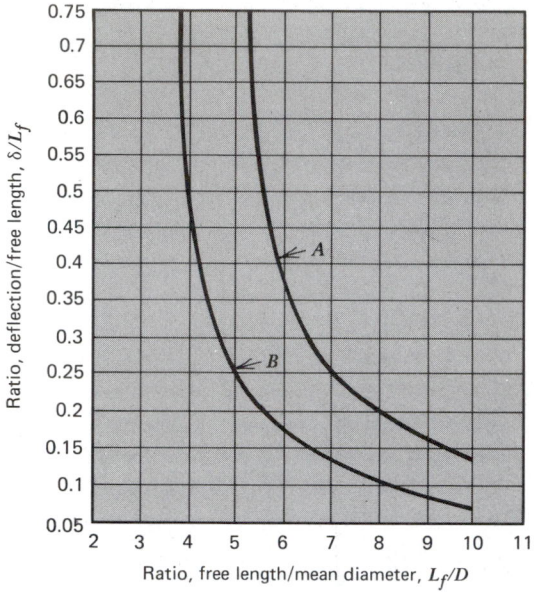

A—end plates constrained parallel (buckling pattern as in Fig. 5.25*c*)

B—one end plate free to tip (buckling pattern as in Fig. 5.25*b*)

FIGURE 12.10.
Buckling conditions for helical compression springs. (Buckling occurs for conditions above and to the right of the curves.)

either inside or outside a cylinder that provides a small clearance. Friction and wear due to rubbing on the cylindrical guide may have to be considered.

12.7 DESIGN PROCEDURE FOR HELICAL COMPRESSION SPRINGS – STATIC LOADING

The two most basic requirements of a coil spring design are (a) acceptable stress level, and (b) desired spring rate. In order to minimize weight, size, and cost, springs are usually designed to the highest stress level that will not result in significant long-term "set." In designing a spring, stress is usually considered before spring rate because stress involved D and d, but *not* N. In general, the stress requirement can be satisfied by many combinations of D and d, and the objective is to find one of these that best suits the requirements of the particular problem. With D and d at least tentatively selected, N is then determined on the basis of the required spring rate. Finally, the free length of the spring is determined so as to give the desired clash allowance. If the resulting design is prone to buckling, or if the spring does not fit into the available space, another combination of D and d may be indicated. If the spring comes out too large or too heavy, a stronger material must be considered.

SAMPLE PROBLEM 12.1

A helical spring with squared and ground ends is required to exert a force of 60 lb at a length that cannot exceed 2.5 in., and 105 lb at a length that is 0.5 in. shorter. It must fit inside a 1.5 in. diameter. Loading is essentially static. Determine a satisfactory design, using oil-tempered ASTM 229 wire, without presetting.

Solution

1. Figure 12.11 gives a convenient representation of the given information concerning spring geometry and loading. The required spring rate is:

$$k = F/\delta = \Delta F/\Delta\delta = 45 \text{ lb}/0.5 \text{ in.} = 90 \text{ lb/in.}$$

2. Using the recommended clash allowance of 10 percent of maximum working deflection,

$$\text{clash allowance} = 0.1\,\frac{105 \text{ lb}}{90 \text{ lb/in.}} = 0.12 \text{ in.}$$

3. The force when solid (i.e., maximum force that must be resisted without "set") is therefore

$$F_s = 105 + 90(0.12) = 116 \text{ lb}$$

4. We now proceed to determine a desirable combination of D and d that will satisfy the stress requirement (Eq. (12.6)). In this problem, the requirement that the spring fit inside a 1.5-in. hole permits a reasonable initial estimate of D—perhaps $D = 1.25$ in. ($D + d$ must be less than 1.5 in. by a *diametral clearance,* commonly recommended as about $0.1D$. Note that reasonable clearance is required because the outside diameter increases slightly as the spring is compressed. Since a small

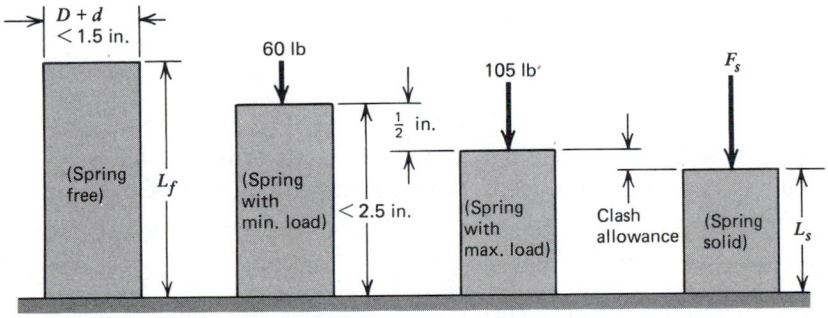

FIGURE 12.11.
Helpful representation of information given in Sample Problem 12.1.

wire size should suffice for the loads involved, D would be expected to come out in the range of 1 to 1.25 in.)

5. In order to solve Eq. (12.6) for d, preliminary values of K_s and τ_{solid} must also be determined, both of which are functions of d. Fortunately, neither quantity varies greatly over the ranges involved, so we should not be far off by estimating:

   a. $K_s = 1.05$ (Fig. 12.4 shows little variation in K_s over the normal range of C between 6 and 12.), and

   b. $\tau_{\text{solid}} = 101$ ksi. (For a "ballpark guess" of $d = 0.1$ in., Fig. 12.7 shows S_u to be about 225 ksi. The corresponding maximum acceptable value of τ_{solid} (Eq. (12.9)) is $0.45S_u$, or 101 ksi.)

6. Substituting the above into Eq. (12.6) gives

$$\tau_{\text{solid}} = \frac{8F_{\text{solid}}D}{\pi d^3} K_s$$

$$101{,}000 = \frac{8(116)(1.25)}{\pi d^3}(1.05), \quad \text{or}$$

$$d = 0.157 \text{ in.}$$

7. The estimates in steps 4 and 5 were deliberately made "rough" enough to give an unsatisfactory solution. A wire diameter of 0.157 in. has an ultimate strength of only about 210 ksi instead of the assumed 225 ksi. Furthermore, the above values of d and D provide a diametral clearance in a 1.5-in. hole of only 0.093, which is less than the desired value of $0.1D$. If we keep $d = 0.157$, and reduce D so as to subject the wire to a little less torque (hence, a little lower stress), this would also open up more diametral clearance. For a second trial, choose $d = 0.157$ in. and solve for the corresponding value of D. Both τ_{solid} and K_s will have different values than before, but this time they will be "correct" values for these quantities instead of estimates.

8. To avoid estimating K_s, use the *second* form of Eq. (12.6):

$$\tau_{\text{solid}} = \frac{8F_{\text{solid}}}{\pi d^2} CK_s$$

$$0.45(210{,}000) = \frac{8(116)}{\pi(0.157)^2} CK_s$$

$$CK_s = 7.89$$

From Fig. 12.4, $C = 7.3$,

$$D = Cd = 7.3(0.157) = 1.15 \text{ in.}$$

This combination of D and d not only conforms exactly to the desired stress criterion, but also provides a little more than the minimum desired clearance in the 1.5-in. hole.

9. From Eq. (12.8):

$$k = \frac{d^4G}{8D^3N}, \quad 90 = \frac{(0.157)^4(11.5 \times 10^6)}{8(1.15)^3N}$$

from which $N = 6.38$.

10. From Eq. (12.10), $N_t = N + 2 = 6.38 + 2 = 8.38$. From Fig. 12.8, $L_s = N_t d = 8.38(0.157) = 1.32$ in.

11. When force $F_s = 116$ lb is released, the spring will elongate a distance of 116 lb/(90 lb/in.) = 1.29 in. Thus, $L_f = L_s + 1.29 = 1.32 + 1.29 = 2.61$ in. Also, when loaded with 60 lb, the spring length will be 2.61 in. − 60 lb/(90 lb/in.) = 1.94 in. This more than satisfies the maximum length requirement of 2.5 in. at a 60-lb load.

12. Buckling is checked for the worst case of deflection approaching the solid deflection (i.e., $\delta = \delta_s = 1.29$ in.),

$$\delta_s/L_f = 1.29/2.61 = 0.49$$
$$L_f/D = 2.61/1.15 = 2.27$$

Reference to Fig. 12.10 indicates that this spring is far outside the buckling region, even if one end plate is free to tip.

13. The above solution satisfies the stress and spring rate requirements, while more than satisfying the buckling criterion and spatial limitations. (It is obvious that the requirements could also be satisfied with spring designs using a little larger or a little smaller wire, or even a wire of a little lower tensile strength.) Hence, one apparently satisfactory answer to the problem is

$$d = 0.157 \text{ in.} \quad \blacksquare$$
$$D = 1.15 \text{ in.} \quad \blacksquare$$
$$N = 6.38 \quad \blacksquare$$
$$L_f = 2.61 \text{ in.} \quad \blacksquare$$

This much information would permit a technician to draw or to make the spring. The problem is not really finished, however, without dealing with the vital matter of *tolerances*. For example, small variations in d result in large variations in stress and deflection. Imposing extremely tight tolerances can result in substantial unnecessary cost. It is best to advise the spring manufacturer of any *critical* dimensions (for example, in this problem it might be important to hold all springs to 90 ± 4 lb/in. spring rate, and to the *same* length, ±0.002 in., when loaded with 60 lb), and to allow fairly loose tolerances on all other dimensions. This would permit the manufacturer to use wire stock of *slightly* varying diameter by adjusting other dimensions as necessary in order to comply with the critical specifications.

It may be helpful to note that there are, in general, three types of problems with regard to selecting a satisfactory combination of D and d to satisfy the stress requirement:

a. Spatial restrictions placing a limit on D, as when the spring must fit inside a hole or over a rod. This situation was illustrated by the above sample problem.

b. The desire to use a fixed wire size, as, for example, to standardize on one size of wire for several similar springs. This situation is also illustrated by the sample problem, if steps 4, 5, 6, and 7 are omitted, and $d = 0.157$ in. is given.

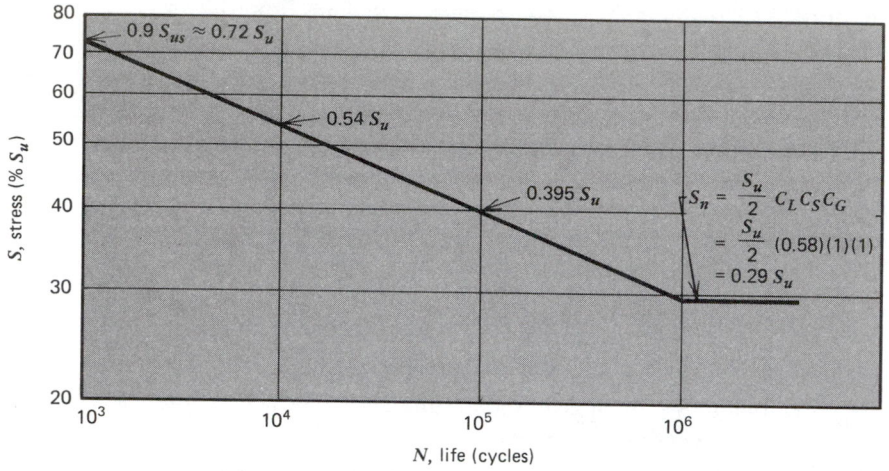

FIGURE 12.12.
Estimated *S-N* curve for round steel spring wire, $d \leq 10$ mm, $C_S = 1$ (shot-peened), reversed torsional loading.

c. No spatial restrictions are imposed, and any wire size may be selected. This completely general situation can theoretically be satisfied with an almost infinite range of D and d, but the extremes within this range would not be economical. Reference to Fig. 12.4 suggests that good *proportions* generally require values of D/d in the range of 6 to 12 (but grinding of the ends is difficult if D/d exceeds about 9). Hence, a good solution will result by selecting an appropriate value of C and then using the second form of Eq. 12.6 to solve for d. This procedure requires an estimate of S_u in order to determine the allowable value of τ_{solid}. If the resulting value of d is not consistent with the estimated value of S_u, a second trial will be necessary, as was the case in the sample problem.

12.8 DESIGN OF HELICAL COMPRESSION SPRINGS FOR FATIGUE LOADING

The design of helical springs for both static and fatigue loading can readily be computerized. A program in BASIC, applicable to microcomputers, has been published by Dietrich.[1]

Figure 12.12 shows a generalized *S-N* curve, calculated in accordance with Chapter 8, for reversed torsional loading of round steel wire of ultimate tensile strength S_u, any diameter not exceeding 10 mm, and a surface factor of unity (as with shot-peened wire—see Section 8.14). A corresponding constant-life fatigue diagram is plotted in Fig. 12.13. Since compression coil springs are always loaded in fluctuating compression (and tensile coil springs in fluctuating tension), these springs do not normally experience a stress *reversal*. In the extreme case, the load (be it tensile or compressive) drops to zero and is then reapplied in the same direction. Thus, as shown in Fig. 12.13, the *region of interest* lies between $\tau_a/\tau_m = 0$ and $\tau_a/\tau_m = 1$.

[1]Dietrich, Al, "Home Computers Aid Spring Design," *Design Engineering* (June 1981), pp. 31–35.

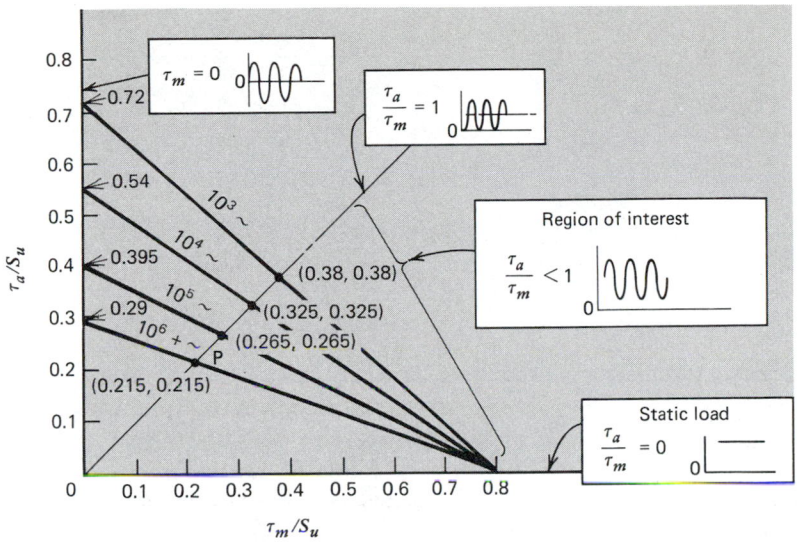

FIGURE 12.13.
Constant-life fatigue diagram corresponding to Fig. 12.12.

It is customary when working with coil springs to replot the information in Fig. 12.13 in the form used in Fig. 12.14. This alternative form of constant-life fatigue diagram contains only the "region of interest" shown in Fig. 12.13. Note, for example, that point P of Fig. 12.13 corresponds to $\tau_m = 0.215S_u$, $\tau_a = 0.215S_u$, whereas in Fig. 12.14 point P plots as $\tau_{\min} = 0$, $\tau_{\max} = 0.43S_u$.

Diagrams like Fig. 12.14 are usually based on actual torsional fatigue tests, with the specimens loaded in a zero-to-maximum fluctuation (i.e., $\tau_a/\tau_m = 1$). Figure 12.15

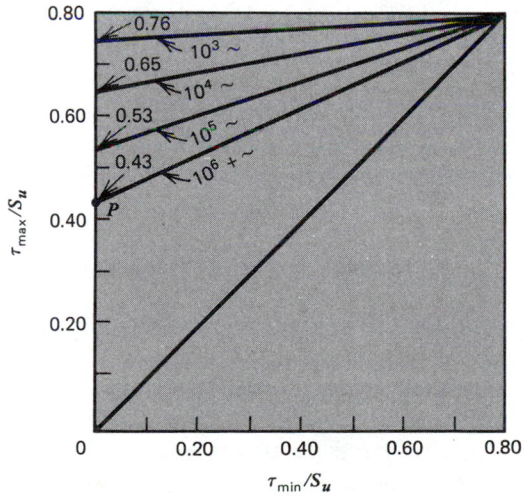

FIGURE 12.14.
Alternative form of constant-life fatigue diagram (replot of "region of interest" of Fig. 12.13).

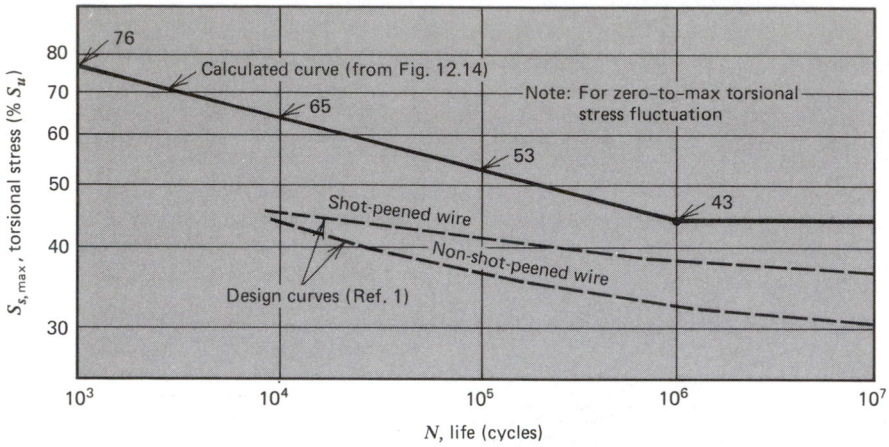

FIGURE 12.15.
$S_{s,max}$-N curves for round steel spring wire. Calculated vs. recommended maximum design values from Ref. 1.

shows S-N curves based on zero-to-maximum stress fluctuation. The top curve is drawn to agree with the values determined in Fig. 12.13. (Note that this curve shows higher values of maximum stress than does Fig. 12.12 because the fluctuating stress drops only to zero rather than completely reversing.) The lower curves in Fig. 12.15 are zero-to-maximum torsional S-N curves based on experimental data and suggested for design use by a major spring manufacturer [1]. These reflect production spring wire surface finish, rather than $C_S = 1$, as in the top curve.

Spring manufacturers customarily present design stress data of this kind in terms of the *tensile* ultimate strength (despite the fact that the loading involved is torsional) because the ultimate tensile strength is the easiest obtained and most reliable experimental measure of wire strength.

Figure 12.16 is an independently obtained empirical constant-life fatigue diagram pertaining to most grades of engine valve spring wire. It represents actual test data. Design values should be somewhat lower.

In the design of helical (or torsion bar) springs for fatigue loading, two previously mentioned manufacturing operations are particularly effective:

a. shot-peening (see Sections 8.13 and 8.14), and

b. presetting (see points 6 and 7 in Section 12.4).

Recalling that presetting always introduces surface residual stresses opposite those caused by subsequent load applications in the same direction as the presetting load, corresponding coil spring (or torsion bar) torsional stress fluctuations with and without presetting are as shown in Fig. 12.17. It can be shown that the theoretical maximum residual stress that can be introduced by presetting (assuming an idealized stress-strain curve that neglects the possibility of some strain-hardening) is $S_{ys}/3$. The practical max-

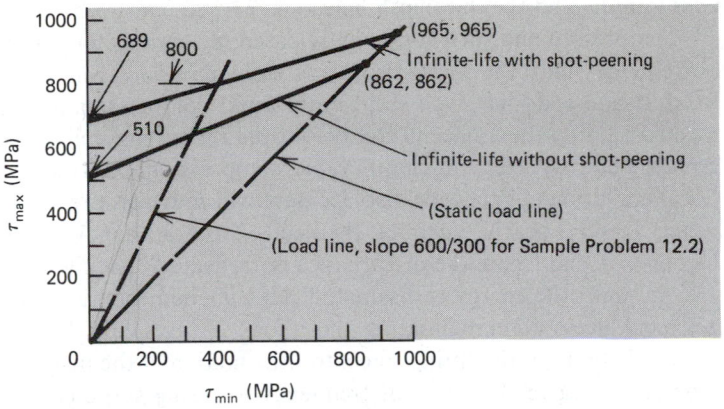

FIGURE 12.16.
Infinite-life fatigue diagram. Pretempered carbon or alloy steel high-duty spring wire, $d \leq 5$ mm (0.2 in.).

imum value is somewhat less. The fatigue improvement represented by the fluctuation with presetting in Fig. 12.17 is readily apparent when the stress fluctuations are represented on Figs. 12.13, 12.14, and 12.16. Maximum fatigue strengthening can be obtained by using *both* shot-peening *and* presetting.

Springs used in high-speed machinery must have natural frequencies of vibration well in excess of the frequency of motion they control. For example, a conventional engine valve spring goes through one cycle of shortening and elongating every two engine revolutions. The valve motion is far from sinusoidal, and a Fourier analysis of the motion typically indicates that harmonics up to about the thirteenth are of significant magnitude. Thus, at 5000 engine rpm, the *fundamental* spring motion has a fre-

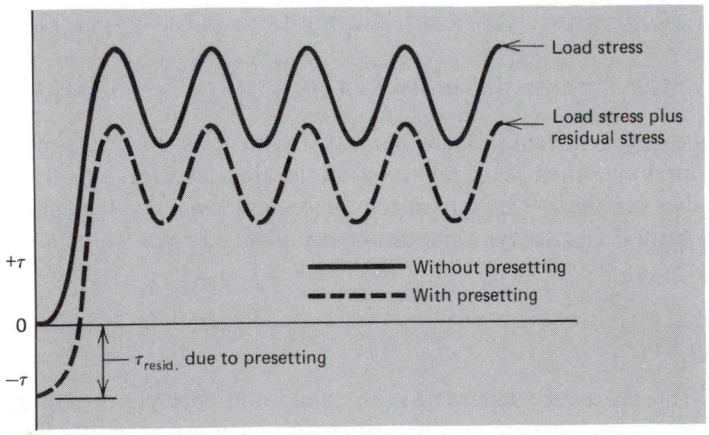

FIGURE 12.17.
Stress fluctuation in a helical (or torsion bar) spring with and without presetting.

quency of 2500 cycles per minute, and the thirteenth harmonic 32,500 cpm, or 542 Hz. When a helical spring is compressed and then suddenly released, it vibrates longitudinally at its own natural frequency until the energy is dissipated by damping. Similarly, if a helical spring is fixed at one end and given sufficiently rapid compression at the other, the end coil is pushed against the adjacent coil before the remaining coils have *time* to share in the displacement. If, after sufficiently rapid compression, the free end is then held fixed, the local condition of excessive displacement will move progressively along the spring (first, coils 1 and 2 nearly touching, then coils 2 and 3, then 3 and 4, etc.) until it reaches the opposite end where the disturbance is "reflected" back toward the displaced end, and so on, until the energy is dissipated. This phenomenon is called *spring surge,* and causes local stresses approximating those for "spring solid." Spring surge also results in decreased ability of the spring to control the motion of the machine part involved, such as the engine valve. The natural frequency of spring surge (which should be made higher than the highest significant harmonic of the motion involved—typically about the thirteenth) is

$$f_n \propto \sqrt{k/m}$$

where

$k \propto \dfrac{d^4G}{D^3N}$ (Eq. 12.8), and

$m \propto$ (volume)(density), or

$m \propto d^2DN\rho$

$$f_n \propto \sqrt{\frac{d^4G/(D^3N)}{(d^2DN\rho)}}, \quad \text{or } f_n \propto \frac{d}{D^2N}\sqrt{G/\rho}$$

For steel springs,

$$f_n = \frac{13{,}900d}{ND^2}\ \text{Hz} \quad (d \text{ and } D \text{ in inches}) \tag{12.11}$$

$$f_n = \frac{353{,}000d}{ND^2}\ \text{Hz} \quad (d \text{ and } D \text{ in mm}) \tag{12.11a}$$

Designing springs with sufficiently high natural frequencies for high-speed machinery typically requires operating at the highest possible stress level (taking advantage of presetting and shot peening). This minimizes the required *mass* of the spring, thereby maximizing its natural frequency, which is proportional to $1/\sqrt{m}$.

SAMPLE PROBLEM 12.2

A camshaft rotates 650 rpm, causing a follower to raise and lower once per revolution. The follower is to be held against the cam by a helical compression spring with a force that varies between 300 N and 600 N as the spring length varies over a range of 25 mm. Ends are to be squared and ground. The material is to be shot-peened chrome-vanadium steel valve spring wire, ASTM A232, with fatigue strength properties as represented in Fig. 12.16. Presetting is to be used. Determine a suitable combination of d, D, N, and L_f. Include in the solution a check for possible buckling and spring surge.

Solution

1. Since, at 650 rpm, a million stress cycles are accumulated in 26 operating hours, infinite fatigue life is required. To minimize possible spring surge problems, stresses should be as high as reasonable. Regardless of the spring design, the ratio τ_{max}/τ_{min} will be the same as the ratio of maximum and minimum loads—that is, 600/300. A line of this slope is drawn on Fig. 12.16, giving an intersection at τ_{max} = 800 MPa.
2. Since Fig. 12.16 represents actual test data, this value of τ_{max} makes no allowance for possible spring surge or a safety factor. To limit the amplitude of possible surge, a minimal clash allowance should be provided—say, 10 percent of the maximum working deflection. To minimize spring weight (thus minimizing weight in order to maximize natural frequency), the smallest reasonable safety factor should be selected—say 1.1 (the use of presetting will provide some additional safety factor). Thus, a design value for τ_{max} might be chosen as 800 MPa divided by 1.1 (allowance for possible surge) and divided again by 1.1 (safety factor), or 661 MPa.
3. In the absence of any restrictions on d, O.D., or I.D., let us arbitrarily select a spring *proportion* of, say, $C = 10$ (good from the standpoint of Wahl factor, but may involve extra cost for grinding the ends). Then, from Eq. (12.5)

$$d = \sqrt{\frac{8F_{max}CK_w}{\pi\tau_{max}}} = \sqrt{\frac{8(600)(10)(1.14)}{\pi(661)}} = 5.13 \text{ mm}$$

4. In the absence of any reason to stay with an odd value of d, it might be preferred to round off to $d = 5.0$ mm. Then, going back to Eq. (12.5) and solving for the value of C that gives a stress of 661 MPa (with load of 600 N) together with $d = 5.0$ mm:

$$CK_w = \frac{\pi\tau_{max}d^2}{8F_{max}} = \frac{\pi(661)(5)^2}{8(600)} = 10.82$$

From Fig. 12.4, $C = 9.4$, $D = Cd = 47.0$ mm.

5. $k = 300N/25 \text{ mm} = 12 \text{ N/mm}$.
6. From Eq. (12.8),

$$N = \frac{dG}{8C^3k} = \frac{5(79{,}000)}{8(9.4)^3(12)} = 4.95$$

7. From Fig. 12.8, $L_s = N_t d = (N + 2)d = (4.95 + 2)(5) = 34.75$ mm.

$$L_f = L_s + F_s/k$$

With 10 percent clash allowance, $F_s = 1.1F_{max} = 660N$. Then,

$$L_f = 34.75 + 660/12 = 89.75 \text{ mm}$$

8. Check for buckling (for extreme case of $\delta = \delta_s$):

$$\left.\begin{aligned} \frac{L_f}{D} &= \frac{89.75}{47} = 1.91 \\ \frac{\delta_s}{L_f} &= \frac{\dfrac{660}{12}}{89.75} = 0.61 \end{aligned}\right\} \text{Far removed from buckling per Fig. 12.10}$$

9. From Eq. (12.11a),

$$f_n = \frac{353{,}000d}{ND^2} = \frac{353{,}000(5)}{(4.95)(47)^2} = 161.4 \text{ Hz}$$

For the spring to be in resonance with the fundamental surge frequency, the camshaft would have to rotate at (161.4)(60) = 9684 rpm. For the thirteenth harmonic to be in resonance, the shaft must rotate 9684/13 = 745 rpm. Rotation at 650 rpm should not result in spring surge (unless the cam contour is highly unusual, producing significant harmonics above the thirteenth).

10. To summarize the results:

$$d = 5 \text{ mm} \quad \blacksquare$$
$$D = 47.0 \text{ mm} \quad \blacksquare$$
$$N = 4.95 \quad \blacksquare$$
$$L_f = 89.75 \text{ mm} \quad \blacksquare$$

No buckling or spring surge should occur (but allowance for possible repeated transient surge was made by appropriate selection of clash allowance and design stress).

SAMPLE PROBLEM 12.3

Repeat Sample Problem 12.2, except this time design the spring to use 5-mm wire of the same material, but with strength properties indicated in Figs. 12.7 and 12.15.

Solution

1. From Fig. 12.7, S_u = 1500 MPa for the given material and wire size.
2. From Fig. 12.15, the maximum recommended design stress for infinite-life and zero-to-maximum stress fluctuation (shot-peened wire) is $0.36S_u$ = 540 MPa.
3. From Eq. (12.9) the effective torsional yield strength associated with 2 percent long-term set is $0.65S_u$ = 975 MPa. Approximating S_{us} as $0.8S_u$ = 1200 MPa, an estimated torsional fatigue strength curve for infinite life is plotted in Fig. 12.18.
4. For τ_{max}/τ_{min} = 600/300, Fig. 12.18 shows the limiting value of τ_{max} to be 750 MPa. Because Fig. 12.15 represents *maximum* recommended values, it might be prudent to reduce this slightly. An additional "safety factor" of 1.13 would give a final design value of τ_{max} = 661 MPa, exactly as in Sample Problem 12.2.

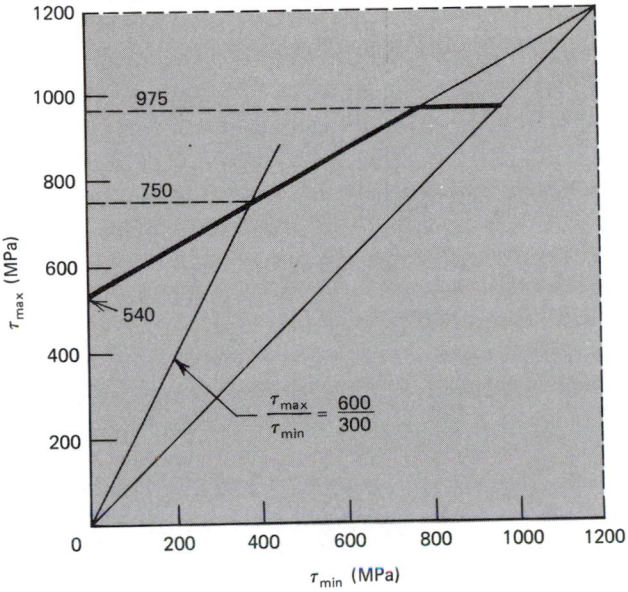

FIGURE 12.18.
Fatigue diagram for Sample Problem 12.3.

5. Using this design stress, the balance of the solution is identical to that given for Sample Problem 12.2.

It is often desirable to solve engineering problems using more than one approach (as in going through both Sample Problems 12.2 and 12.3), and the reader should be aware that the results will not always agree as well as they did in this case. (This is where "engineering judgment" comes in!)

12.9 HELICAL EXTENSION SPRINGS

Most of the foregoing treatment of helical compression springs applies equally to extension springs (illustrated in Fig. 12.2*c* and 12.2*d*), but a few points of difference should be noted. First, extension springs do not have the automatic "overload stop" feature of compression springs. A sufficient static overload can elongate any extension spring to failure. This sometimes makes them vulnerable to excessive overload stresses during installation. Furthermore, a broken compression spring may continue to provide a "stop" that holds the end plates apart. For these reasons compression springs are generally preferred to extension springs in safety-critical applications. Accordingly, some safety codes require that helical springs used for certain applications be loaded in compression.

Extension springs are usually wound while maintaining a torsional stress in the wire. This results in the coils pressing against each other, as shown in Fig. 12.19. The *initial tension* is the external force applied to the spring when the coils are just at the

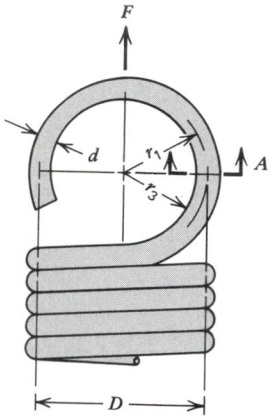

Bending stress at Sec. A:

$$\sigma = \frac{16FD}{\pi d^3}\left(\frac{r_1}{r_3}\right)$$

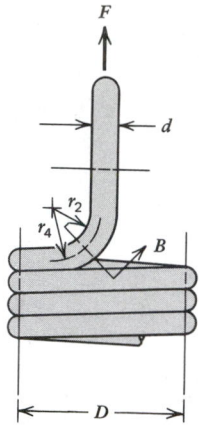

Torsional stress at Sec. B:

$$\tau = \frac{FD}{\pi d^3}\left(\frac{r_4}{r_2}\right)$$

FIGURE 12.19.
Conventional extension spring end hook.

verge of separating. Spring manufacturers recommend [1] that the initial tension be such that the resulting stress (computed from Eq. (12.6)) is

$$\tau_{\text{initial}} = (0.4 \text{ to } 0.8)\,\frac{S_u}{C} \tag{12.12}$$

An extension spring wound in this manner does not deflect until loaded with a force exceeding the initial tension. Then, the equations for stress and spring rate developed for compression springs apply.

The critical stresses in an extension spring often occur in the end hooks. Figure 12.19 shows the location and equation for critical hook bending and torsional stresses. Note that in each case the stress concentration factor is equal to the ratio of mean to inside radius. Recommended practice is to make radius r_4 greater than twice the wire diameter. Hook stresses can be further reduced by winding the last few coils with a decreasing diameter D, as shown in Fig. 12.20. This does nothing to reduce stress con-

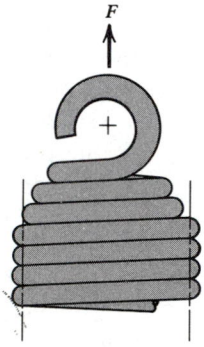

FIGURE 12.20.
Reduced diameter end turns to decrease end hook stresses.

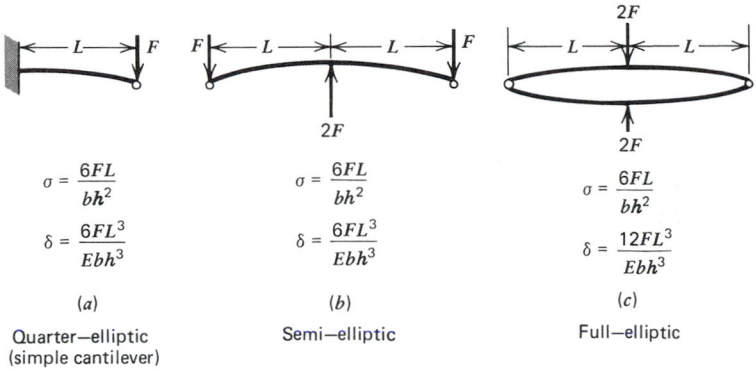

FIGURE 12.21.
Basic types of beam or leaf springs.

centration, but it does lower the nominal stresses by reducing the bending and torsional moment arms.

Depending on detailed design, each end hook typically adds the equivalent of 0.1 to 0.5 helical turn when computing spring deflection.

12.10 BEAM SPRINGS (INCLUDING LEAF SPRINGS)

Beam springs (commonly fabricated as *multi-leaf springs*) are usually arranged as cantilevers and simply supported beams in the form of a quarter, half, or full ellipse, as shown in Fig. 12.21. These springs are also called *flat* springs, although they usually have some curvature when unloaded (the curvature being a necessity in the full-elliptic spring). Note that in each case, the basic element is a *cantilever* of length L loaded by force F. The common semi-elliptic spring can be thought of as two cantilevers sharing the load in parallel. The full-elliptic spring has four cantilevers, arranged in a series-parallel arrangement. (The full-elliptic spring makes an interesting analogy with a Wheatstone bridge, with four equal resistors in a series-parallel arrangement.) It is only necessary to analyze the stress and deflection characteristics of the simple cantilever or quarter-elliptic spring, as the same equations can readily be adapted to cover the other two types.

Figure 12.22*a* shows a general cantilever beam of width w and height or thickness t, both of which vary with length, x. If the bending stresses are to be uniform over the length of a cantilever of constant *thickness,* then the width must vary linearly with x (Fig. 12.22*b*). For a uniform-stress cantilever of constant *width,* the thickness must vary parabolically with x (Fig. 12.22*c*). The triangular beam of Fig. 12.22*b* is the basic model for the design of leaf springs. The parabolic beam of Fig. 12.22*c* is the basic model for analyzing the bending strength of spur gear teeth (more about that in Chapter 15). Of course, constant strength cantilevers can be made by varying *both* w and t so that the stress, $6Fx/wt^2$, is constant for all values of x, and this is the concept behind the design of the "single-leaf" suspension springs that have been used in automobiles.

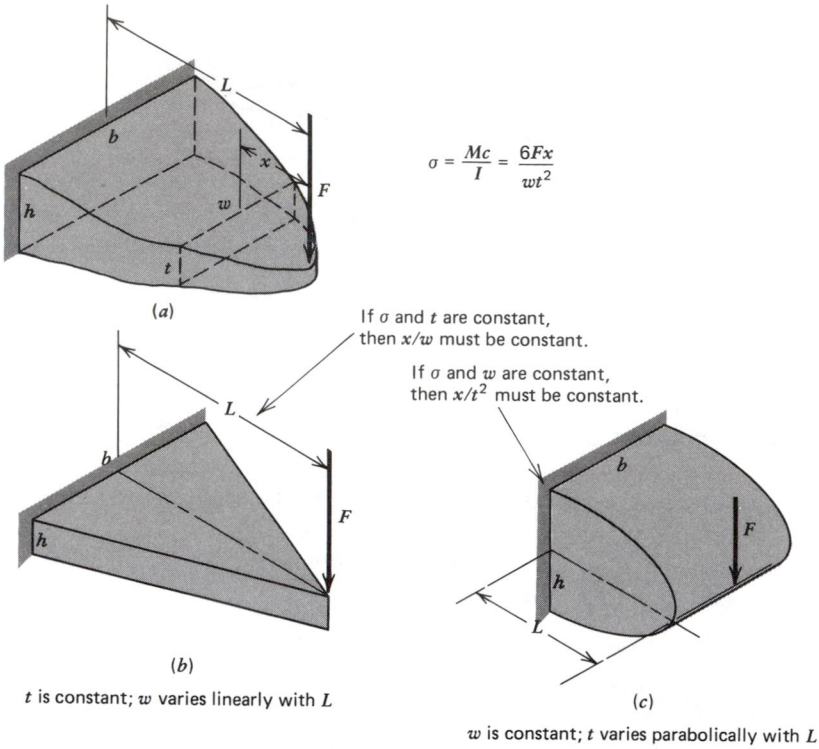

FIGURE 12.22.
Uniform-strength cantilever beams.

For *any* constant strength cantilever beam, the bending stresses throughout are equal to those at the fixed end:

$$\sigma = \frac{6FL}{bh^2} \tag{12.13}$$

Figure 12.23 shows the idea behind cutting up the constant strength triangular beam of Fig. 12.22*b* into a series of *leaves* of equal width and rearranging them in the form of a leaf spring. The triangular plate and equivalent multi-leaf spring have identical stress and deflection characteristics with two exceptions: (1) interleaf friction provides *damping* in the multi-leaf spring, and (2) the multi-leaf spring can carry a full load in only one direction (the leaves tend to separate when loaded in the opposite direction, but this is partially overcome by clips, shown in Fig. 12.24).

Because of the varying cross section, the derivation of the deflection equation for the idealized triangular leaf spring provides an excellent application of Castigliano's method (Section 5.8). It is suggested that the reader use this method to verify that:

$$\delta = \frac{FL^3}{2EI}$$

where $I = bh^3/12$ and E = Young's modulus, or

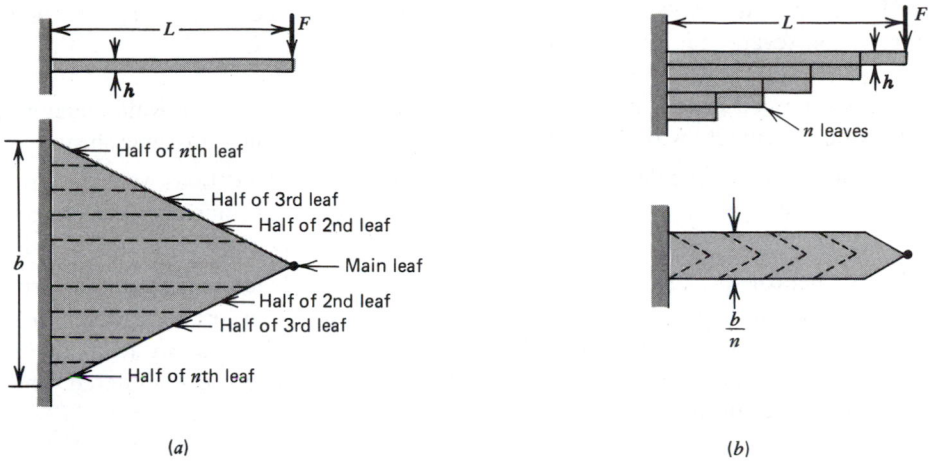

FIGURE 12.23.
Triangular plate cantilever spring and equivalent multileaf spring.

$$\delta = \frac{6FL^3}{Ebh^3} \tag{12.14}$$

The corresponding spring rate is:

$$k = \frac{F}{\delta} = \frac{Ebh^3}{6L^3} \tag{12.15}$$

Stress and deflection equations for the three basic types of leaf springs are summarized in Fig. 12.21.

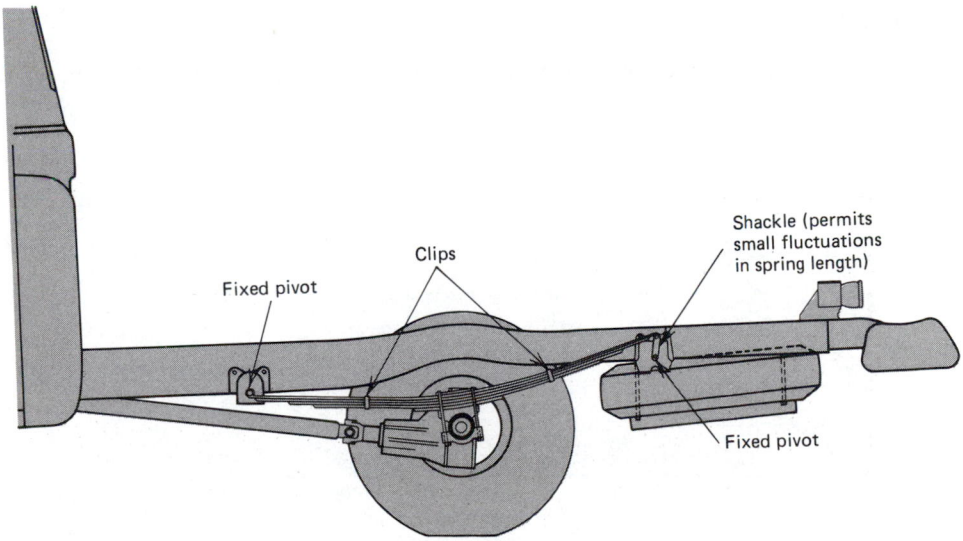

FIGURE 12.24.
Multi-leaf semi-elliptic spring installed in truck chassis.

When applying the above equations to actual springs, as the truck spring illustrated in Fig. 12.24, several additional factors must be taken into account:

1. The spring end cannot be brought to a sharp point, but must be wide enough to provide attachment with the loading member and to carry the transverse shear load.
2. The derivation of the deflection equations assumed that deflections were too small to influence significantly the geometry. For deflections exceeding about 30 percent of the cantilever length, a more precise analysis is usually needed.
3. Unlike helical springs, beam springs are able to carry structural as well as spring loads. For example, the spring in Fig. 12.24 is subjected to a torque reaction about the axle shaft axis, lateral loads due to cornering, and fore-and-aft loads due to acceleration and braking. All such loads must, of course, be considered when designing the spring.

SAMPLE PROBLEM 12.4

A semi-elliptic leaf spring is to be designed for infinite-life when subjected to a design overload (applied at the center) that varies between 2000 and 10,000 N. The spring rate is to be 30 N/mm. The material is to be shot-peened, 7-mm-thick steel having the strength characteristics represented in Fig. 12.25 (proper load, size, and surface are reflected in the given curve). Five leaves are to be used. A central bolt, used to hold the leaves together, causes a fatigue stress concentration factor of 1.3. Using the equations in this article, estimate the required overall spring length, and the width of each leaf.

Solution

1. Each half of a semi-elliptic spring acts as a cantilever carrying *half* of the total load. Thus, $F_m = 3000$ N, $F_a = 2000$ N.

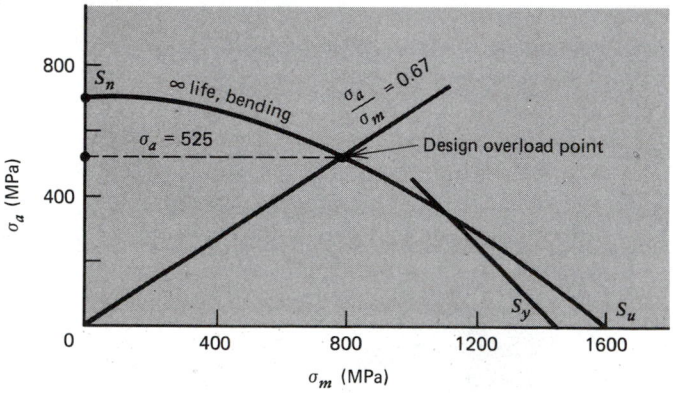

FIGURE 12.25.
Fatigue strength diagram for Sample Problem 12.4.

2. $$\sigma_a = \frac{6F_aL}{bh^2}K_f = \frac{6(2000)L}{b(7)^2}(1.3) = \frac{318L}{b}$$

Similarly,

$$\sigma_m = \frac{6F_mL}{bh^2}K_f \quad \text{and} \quad \frac{\sigma_a}{\sigma_m} = \frac{F_a}{F_m} = 0.67$$

3. Figure 12.25 shows that for infinite-life and this stress ratio, $\sigma_a = 525$ MPa. Hence,

$$525 = \frac{318L}{b}, \quad \text{or } b = 0.61L$$

4. Since the spring is loaded at the center with force $2F$, the spring rate k is $2F/\delta = Ebh^3/3L^3$. Substituting given values:

$$30 = \frac{(200{,}000)(0.61\,L)(7)^3}{3L^3}, \quad \text{or } L = 682 \text{ mm}$$

$$b = 0.61L = 0.61(682) = 416 \text{ mm}$$

5. Overall length is $2L$, or 1364 mm. Width of each of the five leaves is one fifth of 416 mm, or 83 mm. ■

SAMPLE PROBLEM 12.5

In an actual spring, the loaded end of the "triangular plate" must be widened to enable an attachment to be made. This commonly involves maintaining the main leaf at full width throughout its length, giving a spring corresponding to a trapezoidal plate, as shown in Fig. 12.26*b*. What change in spring stiffness and spring energy absorbing

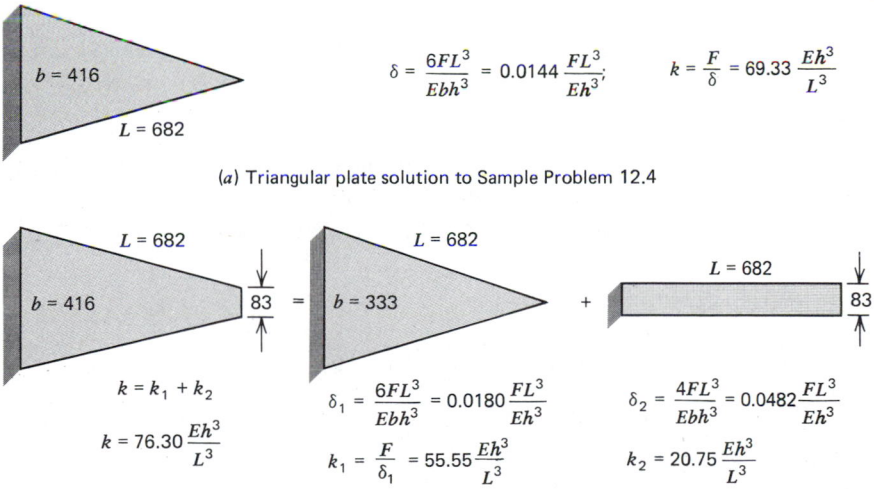

FIGURE 12.26.
Triangular versus trapezoidal spring rate comparison for Sample Problem 12.5. (Note: *k* for trapezoid is about 10 percent greater than *k* for triangle.)

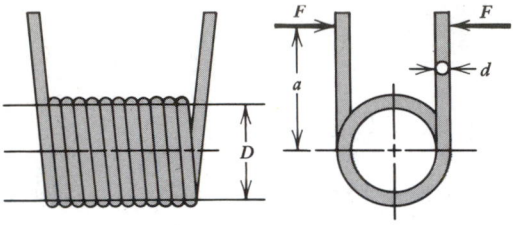

FIGURE 12.27.
Helical torsion spring.

capacity results from using the trapezoidal pattern shown in Fig. 12.26*b* instead of the original triangular pattern shown in Fig. 12.26*a*?

Solution

1. The trapezoidal plate is equivalent to the sum of a triangular plate and rectangular plate acting in parallel, as indicated in Fig. 12.26*b*.
2. Deflection and spring rate equations developed in Fig. 12.26 indicate that the added material of the trapezoid *increases the spring rate by 10 percent.*
3. Both springs will support the same force with the same level of bending stress at the built-in end of the cantilever. But since the trapezoid deflects 10 percent less, *it will absorb 10 percent less energy* (recall that energy absorbed is the integral of force times distance). This is despite the fact that that trapezoidal spring is heavier.

12.11 TORSION SPRINGS

Torsion springs are of two general types: (a) *helical,* shown in Fig. 12.27, and (b) *spiral,* shown in Fig. 12.28. The primary stress in all torsion springs is *bending,* with a bending moment of *Fa* being applied to each end of the wire. The analysis of stresses in curved

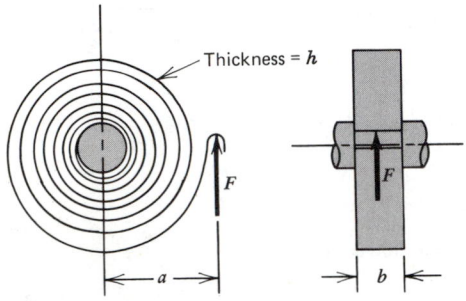

FIGURE 12.28.
Spiral torsion spring.

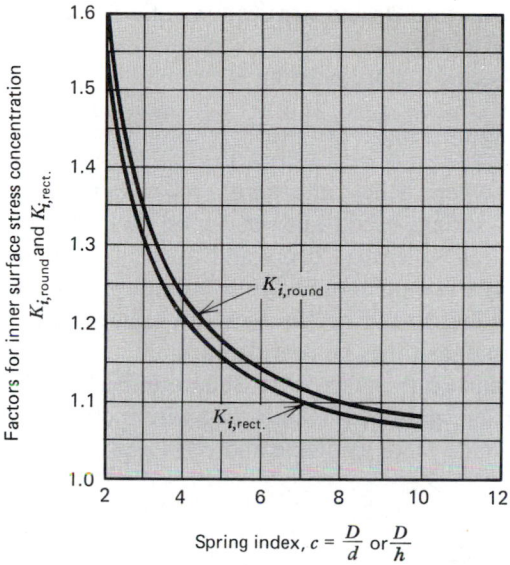

FIGURE 12.29.
Curvature-stress-concentration factors for torsion springs, round and rectangular wire [4].

beams developed in Section 4.6 is applicable. The highest stress is on the *inside* of the wire, and is equal to

$$\sigma_i = K_i Mc/I \tag{4.10}$$

where K_i is given in Fig. 4.11 for some cross sections, and given in Fig. 12.29 for the round and rectangular sections usually used in springs. Substituting the product Fa for bending moment, and also substituting the equations for section properties of round and rectangular wire gives

$$\text{round wire, } \frac{I}{c} = \frac{\pi d^3}{32}, \quad \sigma_i = \frac{32Fa}{\pi d^3} K_{i,\text{round}} \tag{12.16}$$

$$\text{rectangular wire, } \frac{I}{c} = \frac{bh^2}{6}, \quad \sigma_i = \frac{6Fa}{bh^2} K_{i,\text{rect.}} \tag{12.17}$$

For fatigue applications, special care must be taken in the design of the end portion of the wire and its engagement with the loading member, as points of stress concentration at the ends are frequent sites of fatigue failure.

Residual stresses are an important consideration in torsion springs. Recall the italicized statement in Section 4.15: *An overload causing yielding produces residual stresses that are favorable to future loads in the same direction and unfavorable to future loads in the opposite direction.* The coiling of helical and spiral springs involves obvious yielding, and the resulting residual stresses are *favorable to subsequent loads tending to coil the spring tighter* (and unfavorable to loads tending to uncoil it). It is extremely important to bear this in mind when designing applications for torsion springs. Because of the favorable residual stresses, design stresses for static service can be 100 percent of the material tensile yield strength.

The angular deflection of beams subjected to pure bending is

$$\theta = \frac{ML}{EI} \quad \text{(case 3, Table 5.1)}$$

and this equation can be applied directly to both helical and spiral torsion springs. For springs with a great many turns (as spiral springs used to power clocks, toys, etc., and helical springs used to counterbalance garage overhead doors and to roll up window shades), the angular deflection can be several complete turns. Long helical springs with many turns usually have a center rod for support. Friction with the center rod and friction between adjacent spring coils must sometimes be considered.

Spiral springs are usually made of thin rectangular wire. Square or rectangular wire is most efficient for use in helical torsion springs also, but round wire is often used in noncritical applications because it is usually more available and costs less.

12.12 MISCELLANEOUS SPRINGS

The variety of possible spring types and designs is limited only by the ingenuity and imagination of the engineer. Novel designs are regularly submitted to the Patent Office. Although space does not permit a detailed treatment of additional types here, there are at least five other kinds of springs that should be mentioned. For more information see [1, 2, 4].

Spring washers are made in tremendous variety. Six representative types are shown in Fig. 12.30. *Belleville washers* (patented in France by Julien Belleville in 1867, and also known as *coned-disk* springs) are commonly used for carrying very high loads with

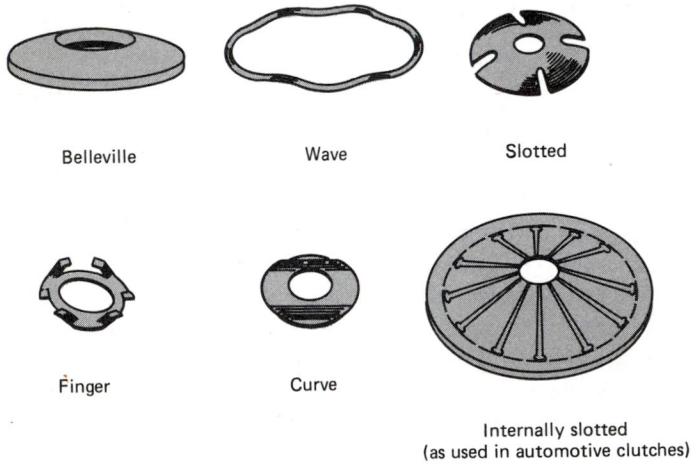

FIGURE 12.30.
Types of spring washers [1].

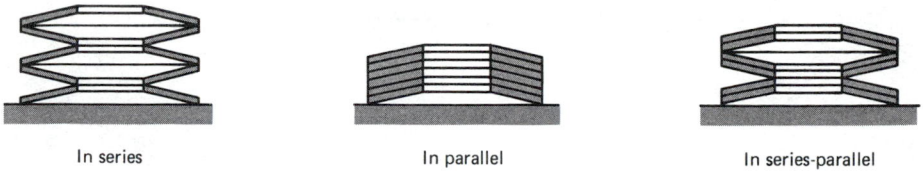

FIGURE 12.31.
Stacked Belleville washers.

small deflections. By varying the ratio of cone height to washer thickness, a constant spring rate or various progressively diminishing spring rates (which can become negative) can be obtained. Multiples of identical washers are used in combination to obtain desired characteristics, as shown in Fig. 12.31.

Volute springs, illustrated in Fig. 12.32, are wound from relatively thin strips of metal with each turn telescoping inside the preceding one. They have more lateral stability than helical compression springs, and the rubbing of adjacent turns provides damping.

Garter springs are merely helical springs with ends connected to form a circle. They are commonly used in oil seals, where they exert radial forces to hold the sealing member firmly against a shaft. Other uses include piston-ring expanders and small motor belts.

Wire forms include a variety of spring members made by bending spring wire into various shapes. Wire forms have replaced helical compression springs in many bed spring, mattress, and furniture applications. One advantage is that they can be formed flat in order to consume a minimum of thickness in the seat of a car or piece of furniture.

Constant force springs, a form originally patented and introduced under the name "Neg'ator," are shown in Fig. 12.33. They consist of rolls of prestressed strip that exert a nearly constant restraining force to resist uncoiling. The strip has a "natural" or load-free radius of curvature about 10 to 20 percent smaller than the radius of the drum (Fig. 12.33*a*). Features include long extension capability and virtual absence of intercoil friction. Figure 12.33*b* shows a form used with electric motor brushes. Other applications include typewriter carriage return springs and window sash counterbalance springs. Figure 12.33*c* shows forms used to drive mechanisms for timers, movie cameras, cable retractors, and so on.

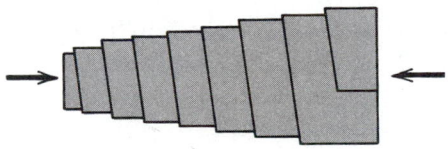

FIGURE 12.32.
Volute spring.

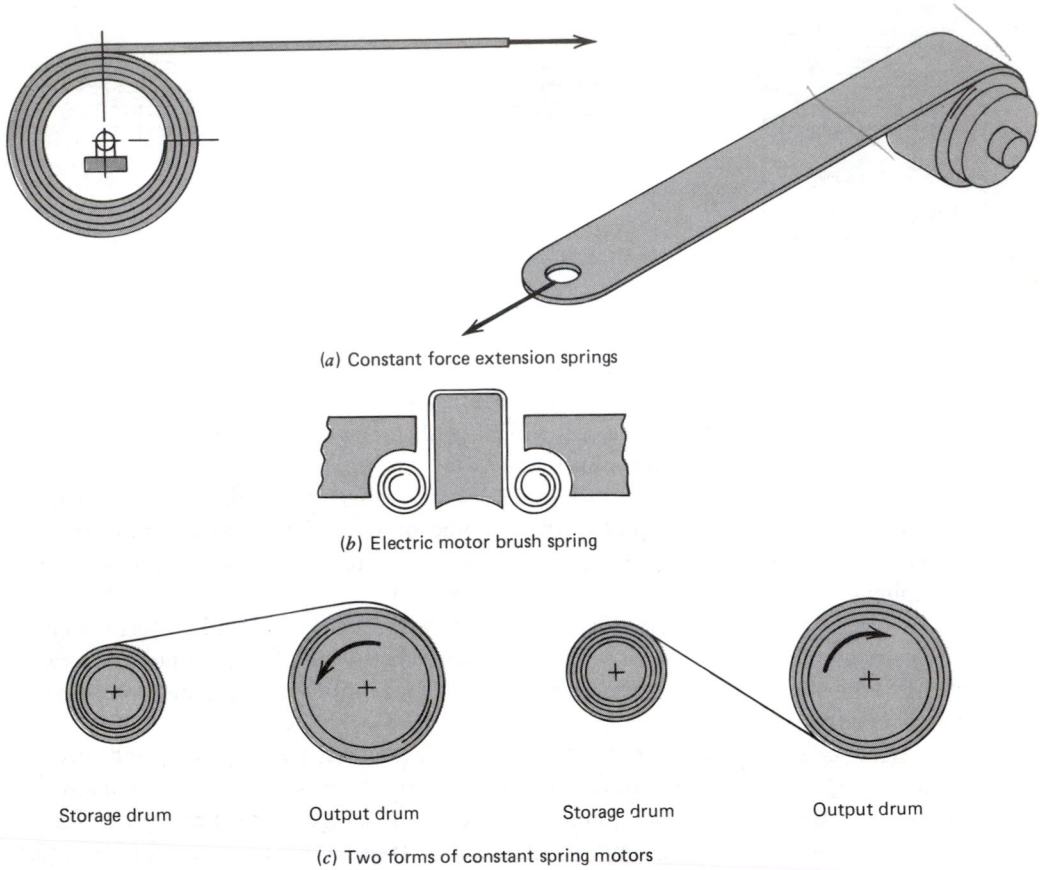

FIGURE 12.33.
Constant force springs.

REFERENCES

1. Associated Spring Corporation, "Design Handbook," Associated Spring Corporation, Bristol, Connecticut, 1970.
2. Associated Spring Corporation, "Design Handbook," Associated Spring Corporation, Bristol, Connecticut, 1981.
3. Chow, W. W., *Cost Reduction in Product Design,* Van Nostrand Reinhold, New York, 1978.
4. Wahl, A. M., *Mechanical Springs,* McGraw-Hill, New York, 1963 (also original 1944 edition).

CHAPTER THIRTEEN

Lubrication and Sliding Bearings

13.1 TYPES OF LUBRICANTS

The word *bearing,* in a machine or structure, refers to contacting surfaces through which a load is transmitted. When relative motion occurs between the surfaces it is usually desirable to minimize friction and wear. Any interposed substance that reduces friction and wear is a *lubricant.* Lubricants are usually liquid, but can be solid (as graphite, TFE[1], or molybdenum disulphide) or a gas (as pressurized air).

Liquid lubricants (oils) are characterized by their viscosity (see Section 13.5), but other properties are also important. Modern oils usually contain one or more additives designed to cause the oil to flow at lower temperatures (pour-point depressants), have less variation of viscosity with temperature (viscosity index improvers), resist foaming when agitated by high-speed machinery (defoamants), resist oxidation at high temperatures (oxidation inhibitors), prevent corrosion of metal surfaces (corrosion inhibitors), minimize the formation of engine deposits and reduce the rate at which they deposit on metal surfaces (detergents and dispersants), and reduce friction and wear when full lubricating films cannot be maintained (antiwear additives).

Greases are liquid lubricants that have been thickened in order to provide properties not available in the liquid lubricant alone. Greases are usually used where the lubricant is required to stay in position, particularly where frequent relubrication is difficult or costly. Often, by remaining in place to provide lubrication, grease also serves to prevent harmful contaminants from entering between the bearing surfaces. Unlike oils,

[1]Tetraflouroethylene, as duPont's *Teflon.*

greases cannot circulate and thereby serve a cooling and cleaning function. Except for this, greases are expected to accomplish all functions of fluid lubricants.

A detailed discussion of lubricants is contained in [13].

13.2 TYPES OF SLIDING BEARINGS

Sliding bearings involve direct sliding of the load-carrying member on its support, as distinguished from *rolling-element bearings* (Chapter 14), where balls or rollers are interposed between the sliding surfaces.

Sliding bearings (also called *plain bearings*) are of two types: (1) *journal* or *sleeve bearings,* which are cylindrical and support radial loads (i.e., loads perpendicular to the shaft axis), and (2) *thrust bearings,* which are generally flat and (in the case of a rotating shaft) support loads in the direction of the shaft axis.

Figure 13.1 shows a crankshaft supported in a housing by two *main bearings,* each consisting of a cylindrical sleeve portion plus a flanged end that serves as a thrust bearing. The cylindrical portions of the shaft in contact with the bearings are called *journals.* The flat portions bearing against the thrust bearings are called *thrust surfaces.* The bearings themselves could be integral with the housing or crankcase, but are usually thin shells that can be easily replaced and that provide bearing surfaces of an appropriate material (as babbitt or bronze).

Where the radial load on a bearing is always in one direction (as rail freight car axle bearings, which support the weight of the car), the bearing surface needs extend only part way around the periphery (usually 60 to 180 deg.), thereby providing a *partial bearing.* In this book, only the more common 360 deg. or *full bearings* will be considered.

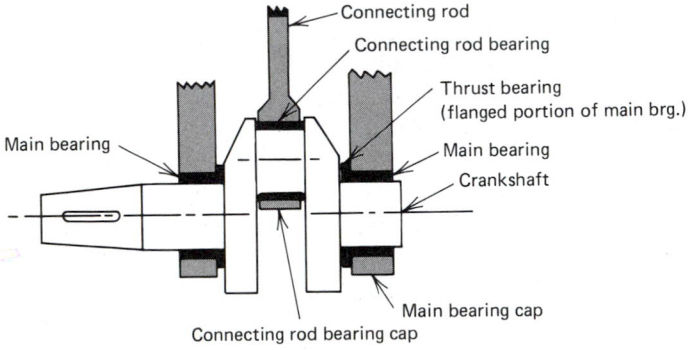

FIGURE 13.1.
Crankshaft journal and thrust bearings. The crankshaft is supported by two *main bearings,* and attaches to the connecting rod by the *connecting rod bearing.* All three are *journal* (or *sleeve*) *bearings.* Integral flanges on the main bearing inserts (commonly called merely *bearings*) serve as *thrust bearings,* which restrain axial motion of the shaft.

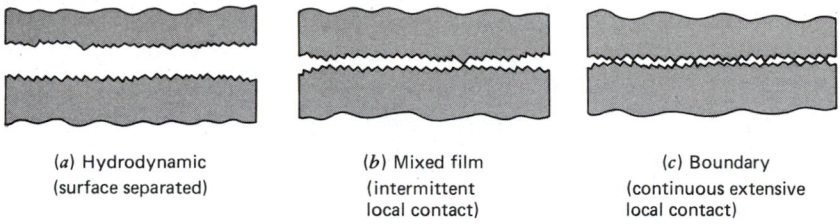

FIGURE 13.2.
Three basic types of lubrication (highly magnified surfaces).

Where assembly and disassembly operations do not require that a bearing be split, the bearing insert can be made as a one-piece cylindrical shell that is pressed into a hole in the housing. In this case the bearing insert is also known as a *bushing*.

13.3 TYPES OF LUBRICATION

Lubrication is commonly classified according to the degree with which the lubricant separates the sliding surfaces. Figure 13.2 illustrates three basic cases:

1. *Hydrodynamic lubrication* involves complete separation of the surfaces by the lubricant film. The load tending to bring the surfaces together is supported entirely by fluid pressure generated by relative motion of the surfaces (as journal rotation). Surface wear does not occur, and friction losses originate only within the lubricant film. Typical film thicknesses at the thinnest point (designated h_0) are 0.008 to 0.020 mm (0.0003 to 0.0008 in.). Typical values of coefficient of friction (f) are 0.002 to 0.010.
2. *Mixed-film lubrication* involves intermittent contact of surface peaks, and partial hydrodynamic support. With proper design, surface wear can be mild. Coefficients of friction commonly range from 0.004 to 0.10.
3. *Boundary lubrication* involves continuous and extensive surface contact, but the lubricant is continuously "smeared" over the surfaces and provides a continuously renewed adsorbed surface film which reduces friction and wear. Typical values of f are 0.05 to 0.20.

The most desirable type of lubrication is obviously hydrodynamic, and this is treated in more detail beginning with the next section. Mixed-film and boundary lubrication are further discussed in Section 13.14.

Complete surface separation (as in Fig. 13.2a) can also be achieved by *hydrostatic* lubrication. This involves introducing a highly pressurized fluid (as air, oil, or water) into the load-bearing area. Since the fluid is pressurized by external means, full surface separation can be obtained irrespective of whether or not there is relative motion between the surfaces. The principal advantage is extremely low friction at all times, including starting and low-speed operation. Disadvantages are the cost, complication,

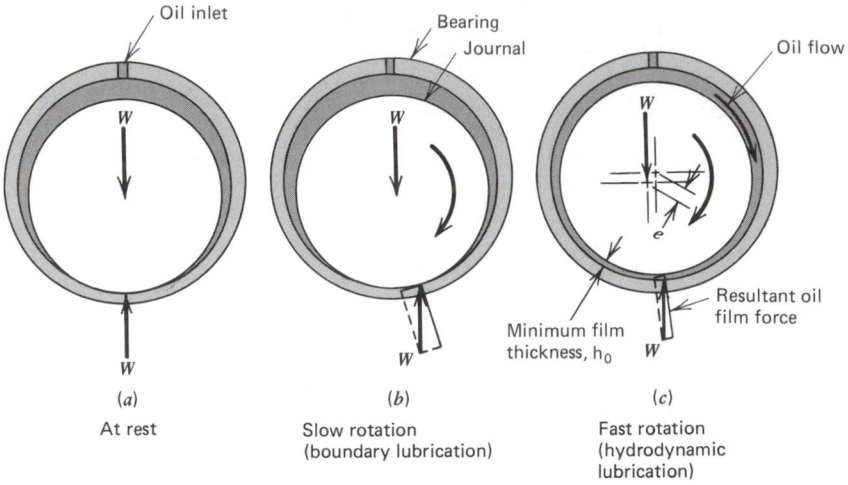

FIGURE 13.3.
Journal bearing lubrication (bearing clearance greatly exaggerated).

and bulk of the external source of fluid pressurization. Hydrostatic lubrication is used only for specialized applications. Reference 11 contains further information.

13.4 BASIC CONCEPTS OF HYDRODYNAMIC LUBRICATION

Figure 13.3*a* shows a loaded journal bearing at rest. The bearing clearance space is filled with oil, but the load (W) has squeezed out the oil film at the bottom. Slow clockwise rotation of the shaft will cause it to roll to the right, as in Fig. 13.3*b*. Continuous *slow* rotation would cause the shaft to stay in this position as it tries to "climb the wall" of the bearing surface. The result would be boundary lubrication.

If the shaft rotating speed is progressively increased, more and more oil adhering to the journal surface tries to come into the contact zone until finally enough pressure is built up just ahead of the contact zone to "float" the shaft, as shown in Fig. 13.3*c*. When this happens, the high pressure of the *converging* oil flow to the right of the minimum film thickness position (h_0) moves the shaft slightly to the left of center. Under suitable conditions, equilibrium is established with full separation of the journal and bearing surfaces. This constitutes *hydrodynamic lubrication,* also known as *full-film* or *thick-film* lubrication. The equilibrium *eccentricity* of the journal in the bearing is dimension e, shown in Fig. 13.3*c*.

Figure 13.4 illustrates the influence of three basic parameters on type of lubrication and resulting coefficient of friction:

1. *Viscosity* (μ). The higher the viscosity, the lower the rotating speed needed to "float" the journal at a given load. Increases in viscosity beyond that necessary to

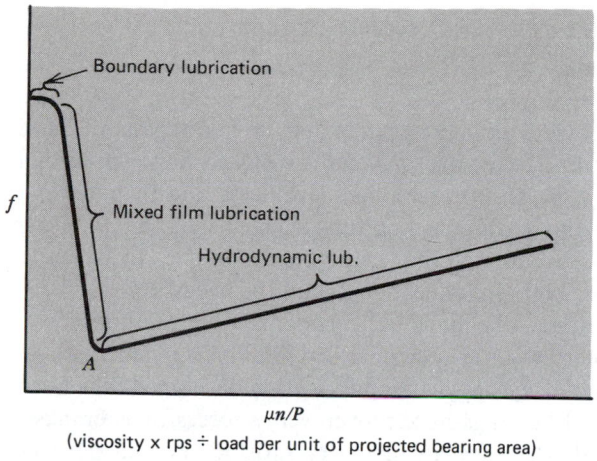

FIGURE 13.4.
Coefficient of friction (and type of lubrication) vs. dimensionless variable $\mu n/P$ (Stribeck curve).

establish full-film or hydrodynamic lubrication increase bearing friction by increasing the forces required to shear the oil film.

2. *Rotating speed* (n).[2] The higher the rotating speed, the lower the viscosity needed to "float" the journal at a given load. Once hydrodynamic lubrication is achieved, further increases in rotating speed increase bearing friction by increasing the time rate of work done in shearing the oil film.
3. *Bearing unit load* (P), defined as load W divided by bearing *projected* area (journal diameter D times bearing length L). The lower the bearing unit load, the lower the rotating speed and viscosity needed to "float" the journal. But *further* reductions in bearing load do not produce corresponding reductions in the bearing friction drag force. Thus, the bearing coefficient of friction (ratio of friction drag force to radial load W) increases.

Numerical values for the curve in Fig. 13.4 depend on details of the particular bearing design. For example, the smoother the surfaces, the thinner the oil film necessary for complete separation of surface asperities, and therefore the lower the values of $\mu n/P$ corresponding to point A. Also, the clearance or fit of the journal in the bearing has an important influence. This is to be expected when one considers that the mechanism for building up hydrodynamic fluid pressure to support the shaft requires that the journal run *eccentrically* in the bearing.

Note that the achievement of hydrodynamic lubrication requires three things:

1. relative motion of the surfaces to be separated,

[2]In lubrication calculations n is in revolutions per *second*. (This is necessary to be consistent with the units of viscosity, which involve seconds.)

2. "wedging action," as provided by the shaft eccentricity, and
3. the presence of a suitable fluid.

The sport of water-skiing provides an interesting analogy to hydrodynamic bearing lubrication, where the above three factors are provided by (1) forward speed of the skier, (2) a "wedge" formed by the ski surfaces (toes up, heels down), and (3) the presence of the water. Carrying this analogy a step further, note that the $\mu n/P$ curve of Fig. 13.4 is applicable to this case as well. For example, if one tries to ski on bare feet, the unit load P becomes very high. In order to maintain the necessary $\mu n/P$ value, the product of viscosity and speed must be made correspondingly high. Filling the lake with a more viscous fluid would help, but the more practical solution is to increase speed.

With only the understanding of hydrodynamic lubrication developed thus far, the reader can appreciate why journal bearings have proven very successful as crankshaft bearings in modern engines, and why they are not very satisfactory and have been largely replaced by roller bearings for supporting rail car axle shafts. An engine crankshaft turns slowly (producing boundary lubrication) *only* when being cranked—and then bearing loads are low because there are no combustion loads. As soon as the engine starts, combustion adds greatly to the bearing loads—but the rotating speed increases sufficiently to establish hydrodynamic lubrication despite the higher loading. Furthermore, peak combustion loads are of *momentary duration,* and inertial and transient "squeeze-film" effects prevent the oil film from being squeezed as thinly as it would be if the peak load were constant. On the other hand, rail car axle bearings must support the full weight of the car when stationary or moving slowly. A common occurrence when freight trains used journal bearings was that two locomotives were required to start a train that could easily be pulled by a single locomotive on reaching sufficient speed to establish hydrodynamic lubrication in the bearings.

13.5 VISCOSITY

Figure 13.5 illustrates the analogy between *viscosity* (also called dynamic viscosity and absolute viscosity), μ, of a fluid, and *shear modulus of elasticity,* G, of a solid. Figure 13.5*a* shows a rubber bushing bonded between a fixed shaft and an outer housing. Application of torque T to the housing subjects an element of the rubber bushing to a fixed displacement, as shown in 13.5*b*. (Note: this type of rubber bushing is commonly used at the leaf spring fixed pivot shown in Fig. 12.24.) If the material between the housing and concentric shaft is a Newtonian fluid (as are most lubricating oils), equilibrium of an element involves a fixed *velocity,* as shown in 13.5*c*. This follows from Newton's law of viscous flow, which implies a linear velocity gradient across the thin lubricant film, with molecular fluid layers adjacent to the separated surfaces having the same velocities as those surfaces.

From Fig. 13.5*c* the equation for (absolute) viscosity is written as

$$\mu = \frac{Fh}{AU} \tag{13.1}$$

The unit of viscosity in the English system is pound-second per square inch, or *reyn*

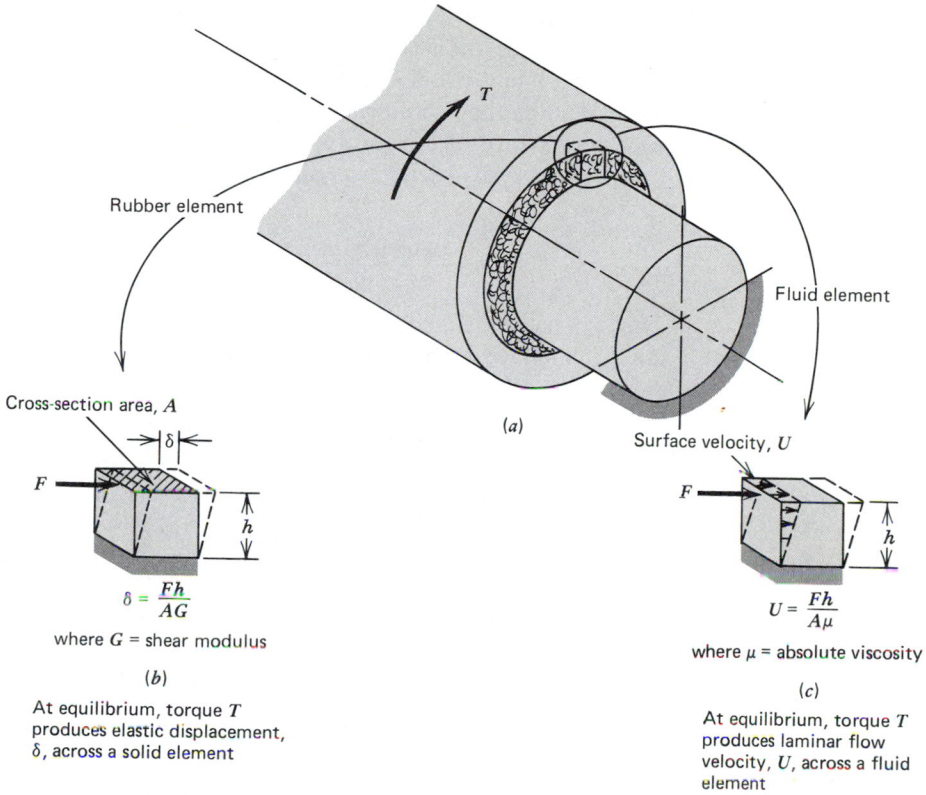

FIGURE 13.5.
Analogy between shear modulus of elasticity (of a solid) and viscosity (of a fluid).

(in honor of Osbourne Reynolds, for whom Reynold's number is also named). In SI units, viscosity is expressed as newton-seconds per square meter, or *pascal-seconds*. The conversion factor between the two is the same as for stress:

$$1 \text{ reyn} = 6890 \text{ Pa}\cdot\text{s} \tag{13.2}$$

The reyn and pascal-second are such large units that *micro reyn* (μ reyn) and *milli pascal-second* (mPa·s) are more commonly used. The former standard metric unit of viscosity, still widely found in the literature, is the *poise*.[3] Conveniently, one centipoise is equal to one milli pascal-second (1 cp = 1 mPa·s).

Fluid viscosities can be measured in several ways, including the use of an apparatus patterned after Fig. 13.5. However, liquid viscosities are usually determined by measuring the time required for a given quantity of the liquid to flow by gravity through a precision opening. In the case of lubricating oils, a commonly used instrument is the Saybolt Universal Viscosimeter, and the viscosity measurements are designated as *Saybolt seconds*, or by any of the abbreviations SUS (Saybolt Universal Seconds), SSU

[3]In the literature, Z, rather than μ, is often used to denote viscosity in terms of poises or centipoises. For convenience, μ is retained here as the symbol for absolute viscosity for all units.

(Saybolt Seconds Universal), and SUV (Saybolt Universal Viscosity). A moment's reflection reveals that these measurements are not true viscosities, since the intensity of the earth's gravitational pull on the liquid is influenced by its mass density. Thus, a liquid of high mass density flows through the viscosimeter more quickly than does a less dense liquid of the same *absolute* viscosity. The viscosity measured by a Saybolt-type viscosimeter is known as *kinematic* viscosity, defined as absolute viscosity divided by mass density:

$$\text{kinematic viscosity} = \frac{\text{absolute viscosity}}{\text{mass density}} \tag{13.3}$$

Units are length^2/time, as cm^2/s, which is named the *Stoke,* abbreviated St.

Absolute viscosities can be obtained from Saybolt viscosimeter measurements (time S, in seconds) by the equations

$$\mu\ (\text{mPa}\cdot\text{s, or cp}) = \left(0.22S - \frac{180}{S}\right)\rho \tag{13.4}$$

and

$$\mu\ (\mu\ \text{reyn}) = 0.145\left(0.22S - \frac{180}{S}\right)\rho \tag{13.5}$$

where ρ is the mass density in grams per cubic centimeter (which is numerically equal to the specific gravity). For petroleum oils, the mass density at 60°F (15.6°C) is approximately 0.89 g/cm^3. At other temperatures the mass density is

$$\rho = 0.89 - 0.00063(^\circ\text{C} - 15.6) \tag{13.6a}$$

$$\rho = 0.89 - 0.00035(^\circ\text{F} - 60) \tag{13.6b}$$

The Society of Automotive Engineers classifies oils according to viscosity. Viscosity-temperature curves for typical SAE numbered oils are given in Figs. 13.6. A particular oil may deviate significantly from these curves, as the SAE specifications define a continuous series of viscosity *bands.* For example, an SAE 30 oil may be only a trifle more viscous than the "thickest" SAE 20, or only a trifle less viscous than the "thinnest" SAE 40 oil. Furthermore, each viscosity band is specified at only one temperature. SAE 20, 30, 40, and 50 are specified at 100°C (212°F) whereas SAE 5W, 10W, and 20W are specified at −18°C (0°F). Multigrade oils must conform to a viscosity specification at both temperatures. For example, an SAE 10W-40 oil must satisfy the 10W viscosity requirement at −18°C and the SAE 40 requirement at 100°C.

Industrial fluid lubricants are commonly specified in terms of international standards, which appear as ASTM D 2422, American National Standard Z11.232, ISO (International Standards Organization) Standard 3448, and others. The various viscosity grades are designated as "ISO VG" followed by a number equal to the nominal kinematic viscosity at 40°C. Eighteen grades are specified, with kinematic viscosities at 40°C of 2, 3, 5, 7, 10, 15, 22, 32, 46, 68, 100, 150, 220, 320, 460, 680, 1000, and 1500 cSt (mm^2/s).

Grease is a non-Newtonian material that does not begin to flow until a shear stress exceeding a yield point is applied. With higher shear stresses, "viscous" flow occurs with the *apparent viscosity* decreasing with increasing shear rates. Apparent viscosities must always be reported at a specified temperature and flow rate (see ASTM D1092).

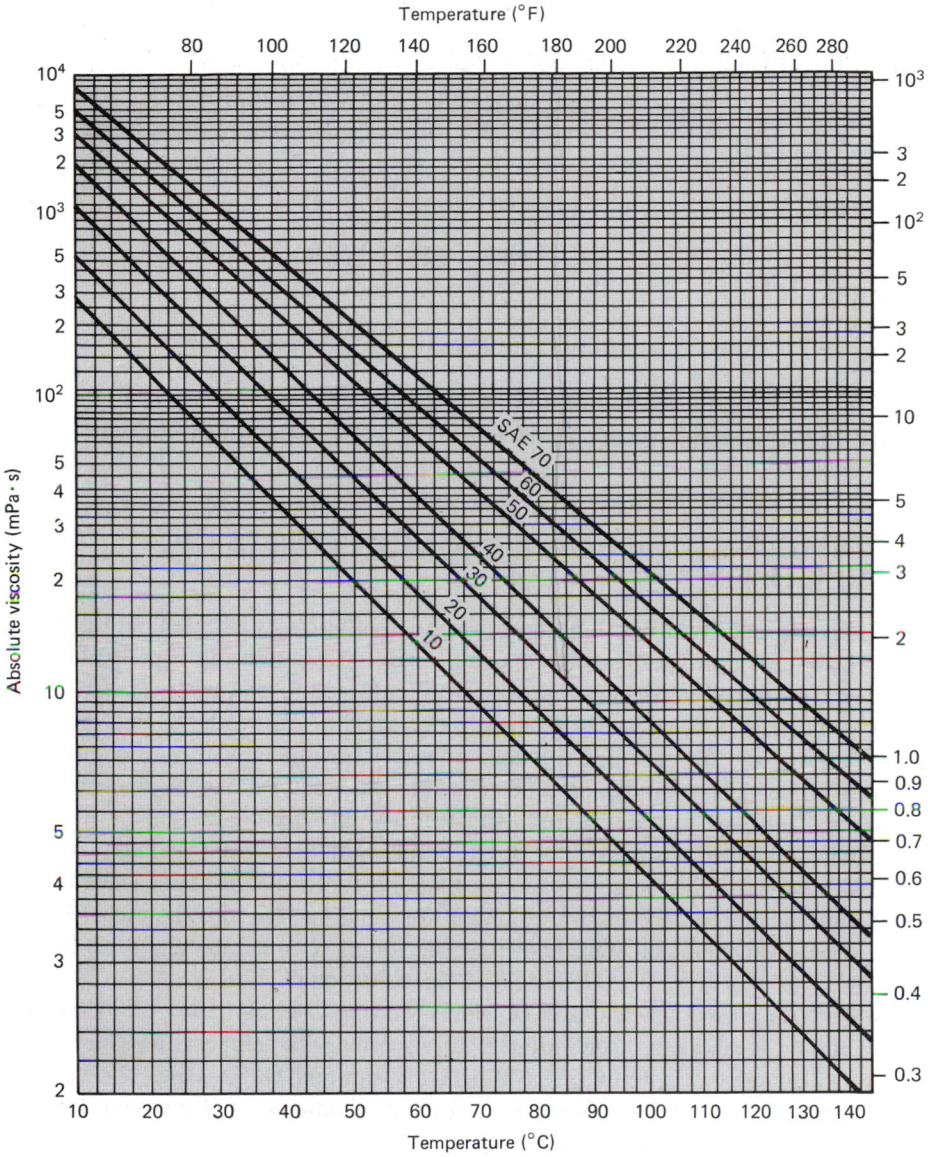

FIGURE 13.6.
Viscosity versus temperature curves for typical SAE graded oils.

SAMPLE PROBLEM 13.1

An engine oil has a kinematic viscosity at 100 °C corresponding to 58 seconds, as determined from a Saybolt viscosimeter. What is its corresponding absolute viscosity in mPa·s (or cp) and in μ reyns? To what SAE number does it correspond?

Solution

1. From Eq. (13.6a),

$$\rho = 0.89 - 0.00063(100 - 15.6) = 0.837$$

2. From Eq. (13.4),

$$\mu = \left((0.22)(58) - \frac{180}{58} \right) 0.837$$
$$= 8.08 \text{ mPa}\cdot\text{s} \quad (\text{or } 8.08 \text{ cp})$$

■

3. From Eq. (13.5),

$$\mu = 0.145 \left((0.22)(58) - \frac{180}{58} \right) 0.837$$
$$= 1.17\ \mu \text{ reyns}$$

■

4. From Fig. 13.6, the viscosity at 100°C is close to that of a typical SAE 30 oil. ■

13.6 TEMPERATURE AND PRESSURE EFFECTS ON VISCOSITY

Multigrade oils, as SAE 10W-40, have less variation of viscosity with temperature than straight-run petroleum oils that have a single grade designation (as SAE 40 or SAE 10W). The measure of variation in viscosity with temperature is *viscosity index* (abbreviated *VI*). The first generally used viscosity index scale was proposed in 1929 by Dean and Davis.[4] At that time, the least change of viscosity with temperature occurred in conventionally refined Pennsylvania crude oils, and the greatest change occurred in oils refined from Gulf Coast crudes. Pennsylvania oils were therefore assigned a VI value of 100, and Gulf Coast oils a value of 0. Other oils were rated intermediately. (Many readers will recognize this procedure as being similar to the octane rating scale of gasolines, which is based on arbitrary zero and 100 ratings for the most and least knock-prone hydrocarbon fuels that were known at the time.) A contemporary basis for viscosity index rating is given in ANSI/ASTM Specification D2270.

Nonpetroleum-base lubricants have widely varying viscosity indices. Silicone oils, for example, have relatively little variation of viscosity with temperature. Thus, their viscosity indices substantially exceed 100 on the Dean and Davis scale. The viscosity index of petroleum oils can be increased, as in the production of multigrade oils, by the use of viscosity index improvers (additives).

All lubricating oils experience an increase in viscosity with pressure. Since this effect is usually significant only at pressures higher than those encountered in sliding bearings, it will not be dealt with here. This effect is important, however, in elastohydrodynamic lubrication (see Section 13.16).

---

[4]Dean, E. W., and G. H. B. Davis, "Viscosity Variations of Oils With Temperature," *Chem. Met. Eng.*, **36** (1929), pp. 618–619.

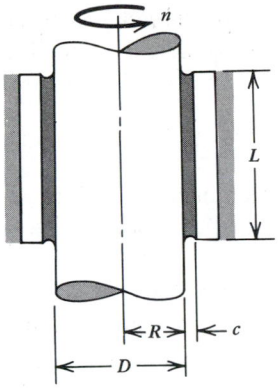

FIGURE 13.7.
Unloaded journal bearing used for Petroff's analysis.

13.7 PETROFF'S EQUATION FOR BEARING FRICTION

The original analysis of viscous friction drag in a hydrodynamic bearing (Fig. 13.7) is credited to Petroff, and was published in 1883. It applies to the simplified "ideal" case of no eccentricity between bearing and journal (hence, no "wedging action" and no ability of the oil film to support a load), and no lubricant flow in the axial direction.

With reference to Fig. 13.5, an expression for viscous friction drag torque is derived by considering the entire cylindrical oil film as the "liquid block" acted upon by force F. Solving the equation given in the figure for F gives

$$F = \frac{\mu AU}{h} \tag{a}$$

where F = friction torque/shaft radius = T_f/R

$A = 2\pi RL$

$U = 2\pi Rn$ (where n is in revolutions per *second*)

$h = c$ (where c = radial clearance = $\dfrac{\text{brg. dia.-shaft dia.)}}{2}$

Substituting and solving for friction torque:

$$T_f = \frac{4\pi^2 \mu n L R^3}{c} \tag{b}$$

If a small radial load W is applied to the shaft, the frictional drag force can be considered equal to the product fW, with friction torque expressed as

$$T_f = fWR = f(DLP)R \tag{c}$$

where P is the radial load per unit of projected bearing area.

The imposition of load W will, of course, cause the shaft to become somewhat eccentric in the bearing. *If* the effect of this on Eq. (b) can be considered negligible, Eqs. (b) and (c) can be equated to give

$$f = 2\pi^2 \frac{\mu n}{P} \cdot \frac{R}{c} \tag{13.7}$$

This is the well-known *Petroff Equation.* It provides a quick and simple means of obtaining reasonable estimates of coefficients of friction of lightly loaded bearings. More refined procedures will be presented in Section 13.9.

Note that Petroff's equation identifies two very important bearing parameters. The significance of $\mu n/P$ was discussed in Section 13.4. The ratio R/c is sometimes called the *clearance ratio,* and is commonly on the order of 500 to 1000. Its importance will become increasingly evident with the study of Section 13.8.

SAMPLE PROBLEM 13.2

A 100-mm-diameter shaft is supported by a bearing of 80 mm length with a diametral clearance of 0.10 mm. It is lubricated by oil having a viscosity (at the operating temperature) of 50 mPa·s. The shaft rotates 600 rpm and carries a radial load of 5000 N. Estimate the bearing coefficient of friction and power loss using the Petroff approach.

Solution

1. From Eq. (13.7):

$$f = 2\pi^2 \frac{(0.05\ \text{Pa}\cdot\text{s})(10\ \text{rps})}{\left(\dfrac{5000}{0.1 \times 0.08}\ \text{N/m}^2\right)} \cdot \frac{50\ \text{mm}}{0.05\ \text{mm}} = 0.0158$$

■

2. $T_f = fWD/2 = (0.0158)(5000\ \text{N})(0.1\ \text{m})/2 = 3.95\ \text{N}\cdot\text{m}$
   [Note: (1) T_f could also be obtained from Eq. (b). (2) The same *calculated* value of T_f would be obtained using any value of W, but the greater the load the greater the deviation from Petroff's assumption of zero eccentricity.]
3. Power $= 2\pi T_f n = 2\pi\,(3.95\ \text{N}\cdot\text{m})(10\ \text{rps}) = 248\ \text{N}\cdot\text{m/s} = 248\ \text{W}.$ ■

In an actual situation, one would need to verify that when dissipating 248 W, the average oil temperature in the bearing would be consistent with the value of viscosity used in the calculations.

13.8 HYDRODYNAMIC LUBRICATION THEORY

The theoretical analysis of hydrodynamic lubrication is traced to Osborne Reynolds' study of Beauchamp Tower's laboratory investigation of railroad bearings in England

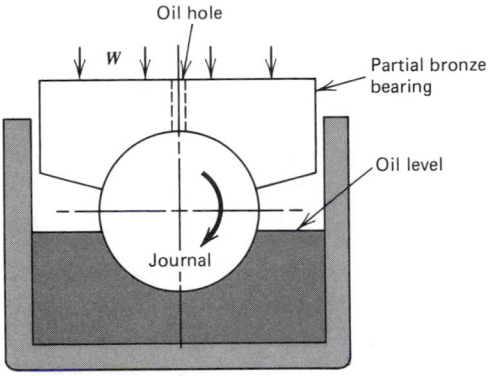

FIGURE 13.8.
Schematic representation of Beauchamp Tower's experiment.

during the early 1880s[5] (Fig. 13.8). The oil hole was drilled to test the effect of adding an oiler at this point. Tower was surprised to discover that when the test device was operated without the oiler installed, oil flowed out of the hole! He tried to block this flow by pounding cork and wooden stoppers into the hole, but the hydrodynamic pressure forced them out. At this point, Tower connected a pressure gage to the oil hole, and subsequently made experimental measurements of the oil film pressures at various locations. He then discovered that the summation of local hydrodynamic pressure times differential projected bearing area was equal to the load supported by the bearing.

Reynolds' theoretical analysis lead to his fundamental equation of hydrodynamic lubrication. The following derivation of the Reynolds equation applies to one-dimensional flow between flat plates. This analysis can also be applied to journal bearings because journal radius is so large in comparison to oil film thickness. The assumed one-dimensional flow amounts to neglecting bearing side leakage, and is approximately valid for bearings with L/D ratios greater than about 1.5. The derivation begins with the equation for equilibrium of forces in the x direction acting on the fluid element shown in Fig. 13.9.

$$p\,dy\,dz + \tau\,dx\,dz - \left(p + \frac{dp}{dx}\,dx\right) dy\,dz - \left(\tau + \frac{\partial \tau}{\partial y}\,dy\right) dx\,dz = 0 \qquad \textbf{(a)}$$

which reduces to

$$\frac{dp}{dx} = \frac{\partial \tau}{\partial y} \qquad \textbf{(b)}$$

In Eq. (13.1), the quantity F/A represents shear stress, τ, on the top of the "block." In Fig. 13.9 the "block" is shrunken to a differential element of height dy, velocity u, and top-to-bottom velocity gradient du. Making these substitutions in Eq. (13.1) gives

[5]Reynolds, O., "On the Theory of Lubrication and Its Applications to Mr. Beauchamp Tower's Experiments," Phil. Trans. Roy. Soc. (London), **177** (1886), pp. 157–234.

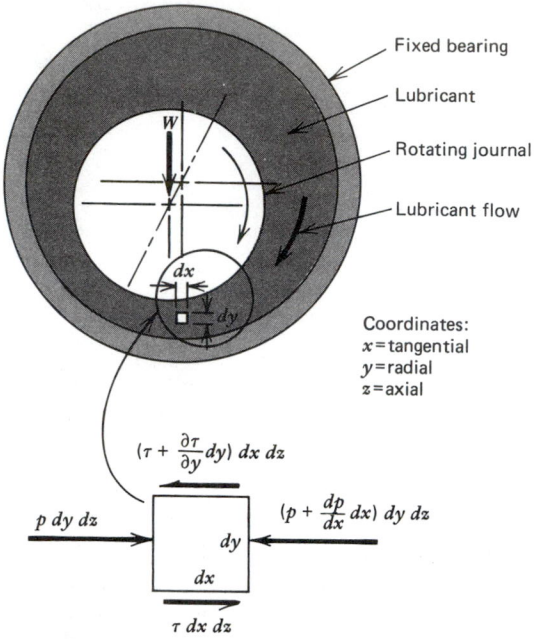

FIGURE 13.9.
Pressure and viscous forces acting on an element of lubricant (for simplicity, only X-components are shown).

$\tau = \mu\,(du/dy)$, except in this case u varies with both x and y, and so the partial derivative is used:

$$\tau = \mu \frac{\partial u}{\partial y} \tag{c}$$

Similarly, τ varies with both x and y, and so the partial derivative, $\partial\tau/\partial y$ is used in Fig. 13.9 and Eq. (a). Pressure, on the other hand, is assumed *not* to vary in the y and z directions and so the total derivative dp/dx is used.

Substitution of Eq. (c) into Eq. (b) gives

$$\frac{dp}{dx} = \mu \frac{\partial^2 u}{\partial y^2}, \quad \text{or} \quad \frac{\partial^2 u}{\partial y^2} = \frac{1}{\mu}\frac{dp}{dx}$$

Holding x constant and integrating twice with respect to y gives

$$\frac{\partial u}{\partial y} = \frac{1}{\mu}\left(\frac{dp}{dx} \cdot y + C_1\right)$$

and

$$U = \frac{1}{\mu}\left(\frac{dp}{dx}\frac{y^2}{2} + C_1 y + C_2\right)$$

The assumption of no slip between the lubricant and the boundary surfaces gives boundary conditions enabling C_1 and C_2 to be evaluated:

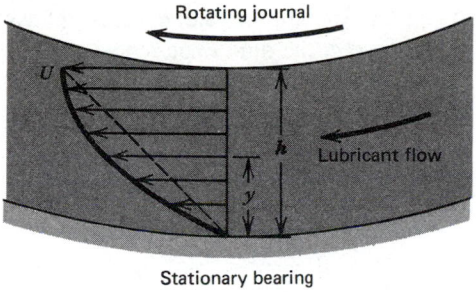

FIGURE 13.10.
Lubricant velocity gradient.

$$u = 0 \text{ at } y = 0, \quad u = U \text{ at } y = h$$

Hence,

$$C_1 = \frac{U\mu}{h} - \left(\frac{h}{2}\right)\frac{dp}{dx} \quad \text{and} \quad C_2 = 0$$

Substitution of these values in Eq. (e) gives

$$U = \frac{1}{2\mu}\frac{dp}{dx}(y^2 - hy) + \frac{U}{h}y \tag{13.8}$$

which is the equation for the velocity distribution of the lubricant film across any yz plane as a function of distance y, pressure gradient dp/dx, and surface velocity U. Note that this velocity distribution consists of two terms: (1) a linear distribution given by the second term and shown as a dotted line in Fig. 13.10, and (2) a superimposed parabolic distribution given by the first term. The parabolic term may be positive or negative, thereby adding or subtracting from the linear distribution. At the section where pressure is a maximum, $dp/dx = 0$, and the velocity gradient is linear.

Let the volume of lubricant per-unit time flowing across the section containing the element in Fig. 13.9 be Q_f. For unit width in the z direction,

$$Q_f = \int_o^h u\,dy = \frac{Uh}{2} - \frac{h^3}{12\mu}\frac{dp}{dx} \tag{d}$$

For a noncompressible lubricant, the flow rate must be the same for all cross sections, which means that

$$\frac{dQ_f}{dx} = 0$$

Hence, by differentiating Eq. (d),

$$\frac{dQ_f}{dx} = \frac{U}{2}\frac{dh}{dx} - \frac{d}{dx}\left(\frac{h^3}{12\mu}\frac{dp}{dx}\right) = 0$$

or

$$\frac{d}{dx}\left(\frac{h^3}{\mu}\frac{dp}{dx}\right) = 6U\frac{dh}{dx} \tag{13.9}$$

which is the classical Reynolds equation for one-dimensional flow. Summarizing the assumptions that were made: the fluid is Newtonian, incompressible, of constant viscosity, and experiences no inertial or gravitational forces; the fluid experiences laminar flow, with no slip at the boundary surfaces; the film is so thin that (a) it experiences negligible pressure variation over its thickness, and (b) the journal radius can be considered infinite in comparison.

When fluid flow in the z direction is included (i.e., axial flow and end leakage), a similar development gives the Reynolds equation for two-dimensional flow:

$$\frac{\partial}{\partial x}\left(\frac{h^3}{\mu}\frac{\partial p}{\partial x}\right) + \frac{\partial}{\partial z}\left(\frac{h^3}{\mu}\frac{\partial p}{\partial z}\right) = 6U\frac{\partial h}{\partial x} \tag{13.10}$$

Modern bearings tend to be shorter than those used a few decades ago. Ratios of length to diameter (L/D) are commonly in the range of 0.25 to 0.75. This results in the flow in the z direction (and the end leakage) being a major portion of the total. For these short bearings, Ocvirk [5] proposed neglecting the x term in the Reynolds equation, giving

$$\frac{\partial}{\partial z}\left(\frac{h^3}{\mu}\frac{\partial p}{\partial z}\right) = 6U\frac{\partial h}{\partial x} \tag{13.11}$$

Unlike Eqs. (13.9) and (13.10), Eq. (13.11) can be readily integrated and thus used for design and analysis purposes. The procedure is commonly known as *Ocvirk's short bearing approximation.*

13.9 DESIGN CHARTS FOR HYDRODYNAMIC BEARINGS

Solutions of Eq. (13.9) were first made in the early 1900s. Although theoretically applicable only to bearings that are "infinitely long" (i.e., have no end leakage), these solutions give reasonably good results with bearings of L/D ratios over about 1.5. At the other extreme, the Ocvirk short bearing solution, based on Eq. (13.11), is quite accurate for bearings of L/D ratios up to about 0.25, and is often used to provide reasonable approximations for bearings in the commonly encountered range of L/D between 0.25 and 0.75.

Computerized solutions of the full Reynolds equation (13.10) have been reduced to chart form by Raimondi and Boyd [7]. These provide accurate solutions for bearings of all proportions. Selected charts are reproduced in Figs. 13.11–13.17. Other Raimondi and Boyd charts apply to partial bearings (which extend around only 60 deg., 120 deg., or 180 deg. of the journal circumference), and to thrust bearings. Several quantities involved in the charts used here are illustrated in Fig. 13.18.

All of the Raimondi and Boyd charts involve plots of dimensionless bearing parameters as functions of the *bearing characteristic number,* or *Sommerfeld variable, S,* which is itself a dimensionless parameter. Note that S involves the previously discussed parameter, $\mu n/P$,[6] and the square of the clearance ratio, R/c. The S scale on the charts is logarithmic except for a linear portion between 0 and 0.01.

[6]Where n is in revolutions per second.

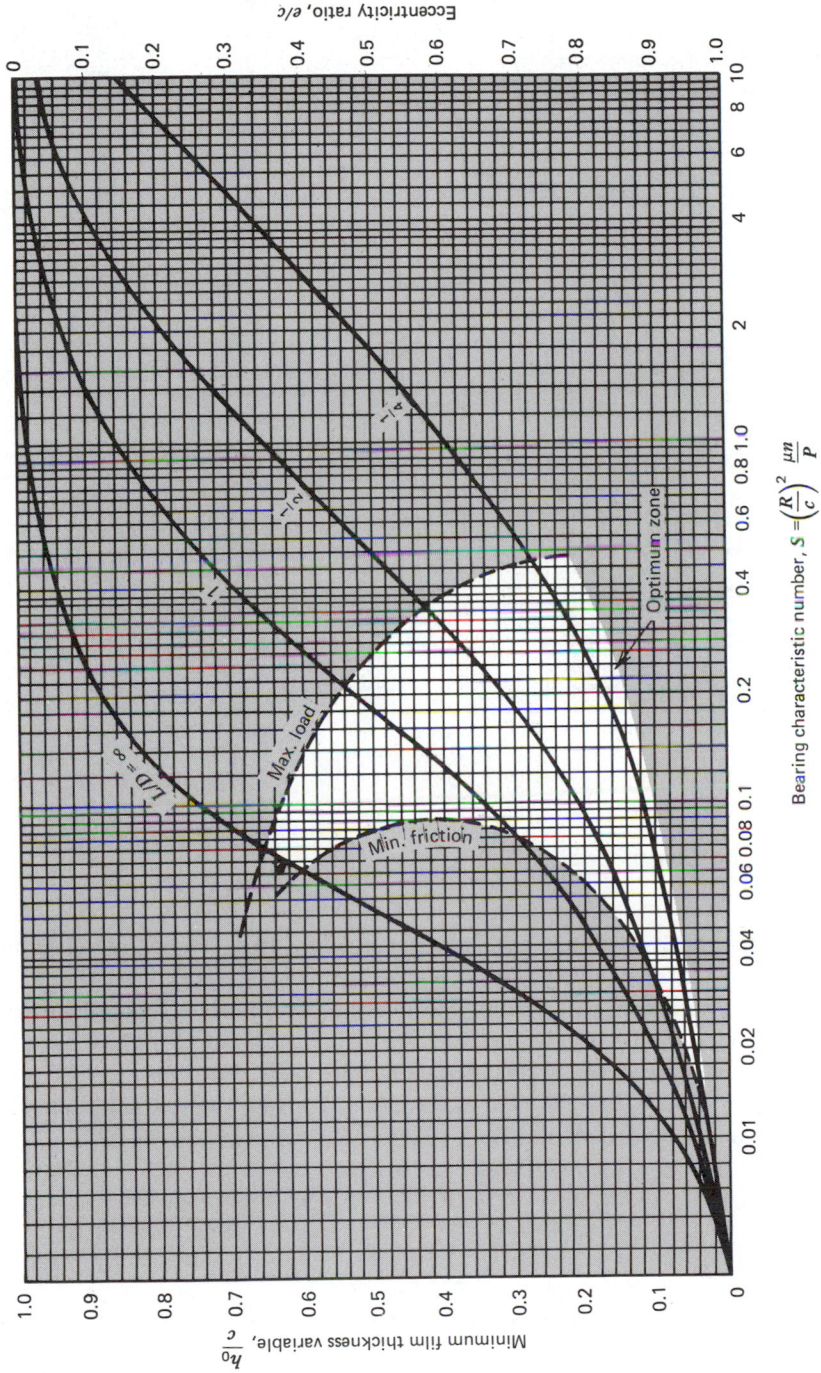

FIGURE 13.11.
Chart for minimum-film-thickness variable [7].

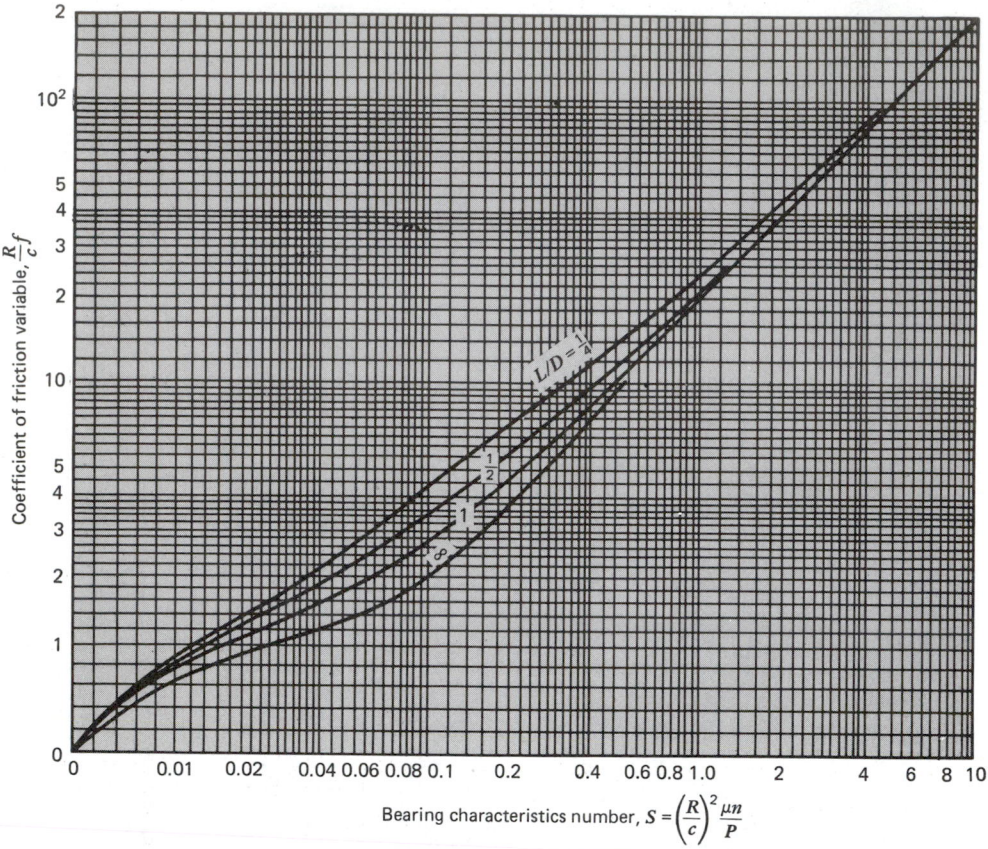

FIGURE 13.12.
Chart for coefficient-of-friction variable [7].

Figures 13.16 and 13.17 assume that lubricant is supplied to the bearing at atmospheric pressure and that the influence on flow rate of any oil holes or grooves is negligible. Viscosity is assumed to be constant, and to correspond to the average temperature of the oil flowing to and from the bearing.

Values of any of the bearing performance variables plotted in Figs. 13.12–13.17 can be determined for any ratio of L/D greater than ¼ by using the following interpolation equation given by Raimondi and Boyd [7]:

$$y = \frac{1}{(L/D)^3}\left[-\frac{1}{8}\left(1-\frac{L}{D}\right)\left(1-\frac{2L}{D}\right)\left(1-\frac{4L}{D}\right)y_\infty + \frac{1}{3}\left(1-\frac{2L}{D}\right)\left(1-\frac{4L}{D}\right)y_1 - \frac{1}{4}\left(1-\frac{L}{D}\right)\left(1-\frac{4L}{D}\right)y_{1/2} + \frac{1}{24}\left(1-\frac{L}{D}\right)\left(1-\frac{2L}{D}\right)y_{1/4}\right] \quad \textbf{(13.12)}$$

where y is the desired performance variable for any L/D ratio greater than ¼, and y_∞,

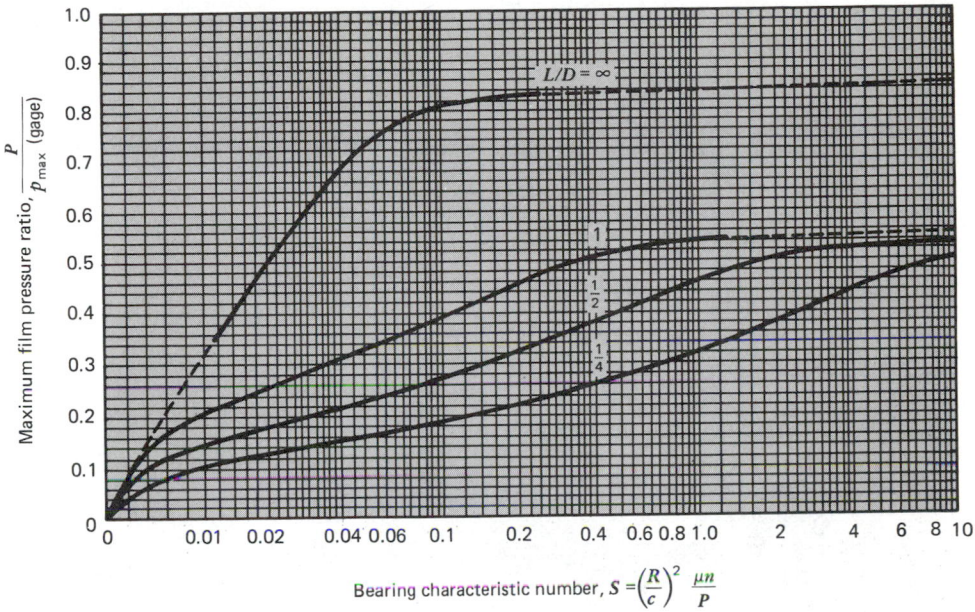

FIGURE 13.13.
Chart for determining maximum film pressure [7].

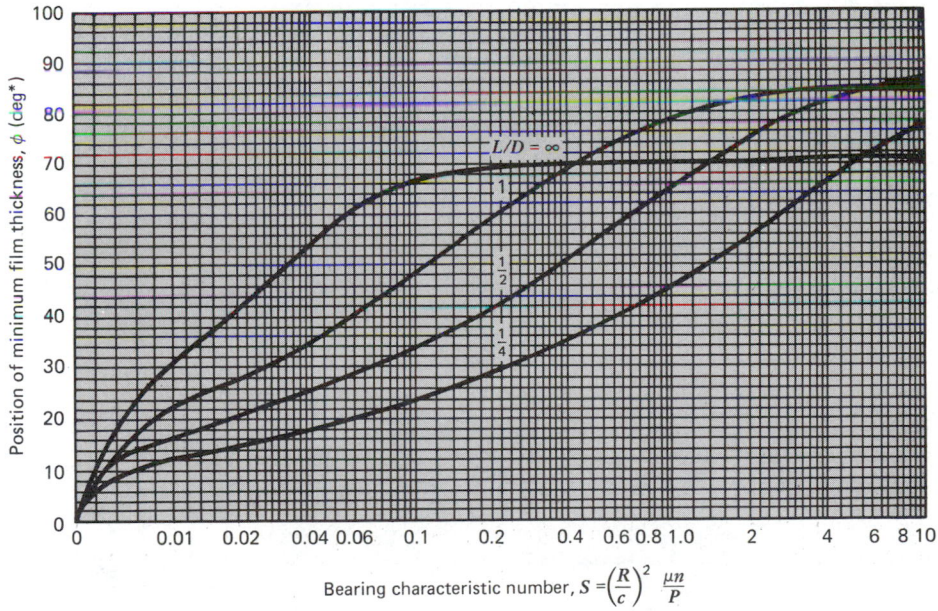

*Defined in Fig. 13.18

FIGURE 13.14.
Chart for determining position of minimum film thickness, h_0 [7].

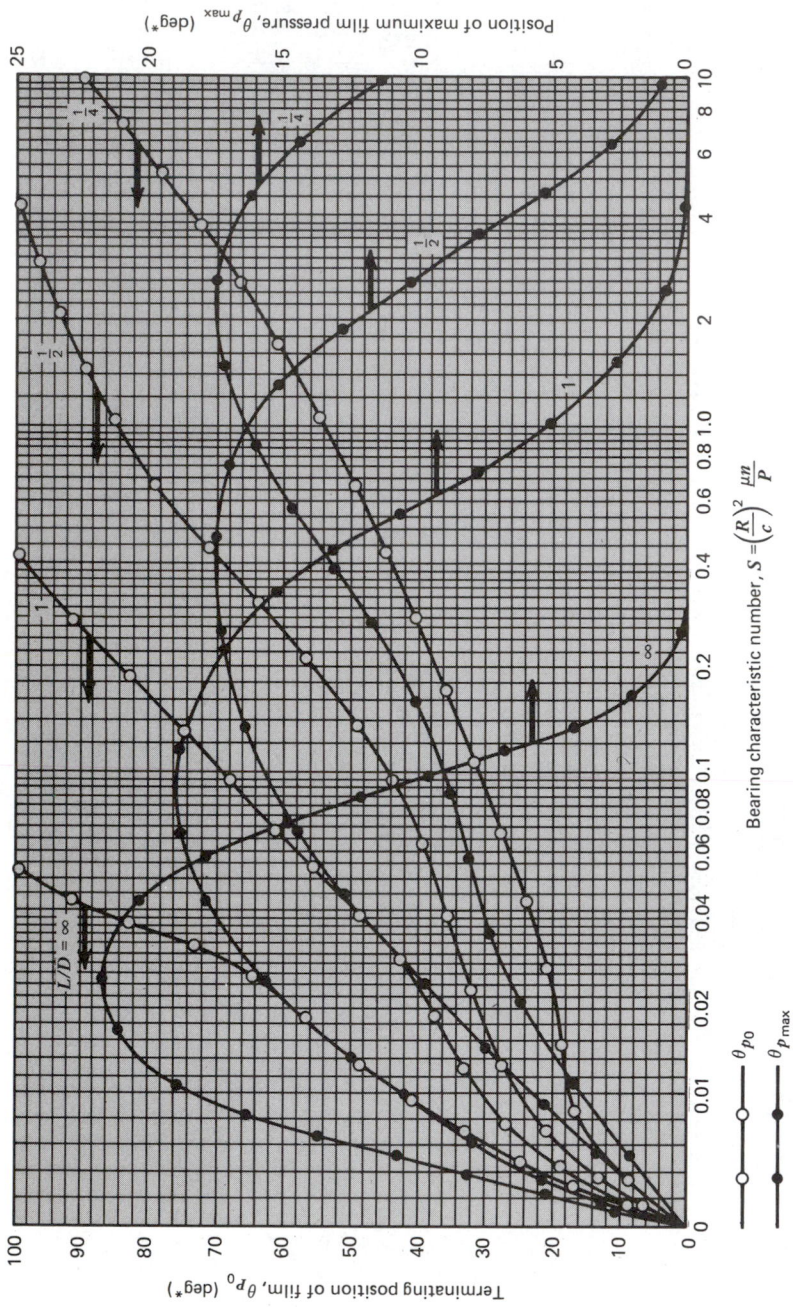

*Defined in Fig. 13.18

FIGURE 13.15.
Chart for positions of maximum film pressure and film termination [7].

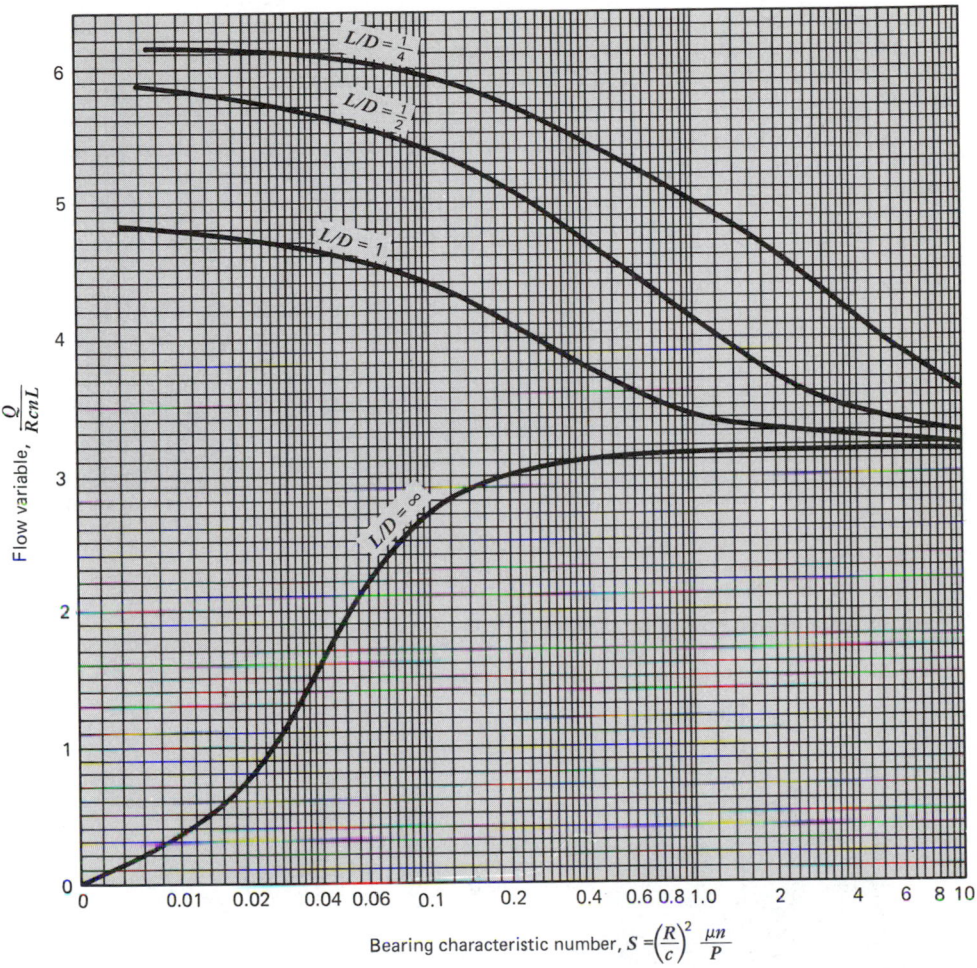

FIGURE 13.16.
Chart for flow variable [7].

y_1, $y_{1/2}$, and $y_{1/4}$ are the values of that variable for bearings having L/D ratios of ∞, 1, ½, and ¼, respectively.

SAMPLE PROBLEM 13.3

A journal bearing of 2 in. diameter, 1 in. length, and 0.0015 in. radial clearance supports a fixed load of 1000 lb when the shaft rotates 3000 rpm. It is lubricated by SAE 20 oil, supplied at atmospheric pressure. The average temperature of the oil film is estimated at 130°F.

Using the Raimondi-Boyd charts, estimate the minimum oil film thickness, bearing coefficient of friction, maximum pressure within the oil film, angles ϕ, $\theta_{p\ \max}$, and θ_{po},

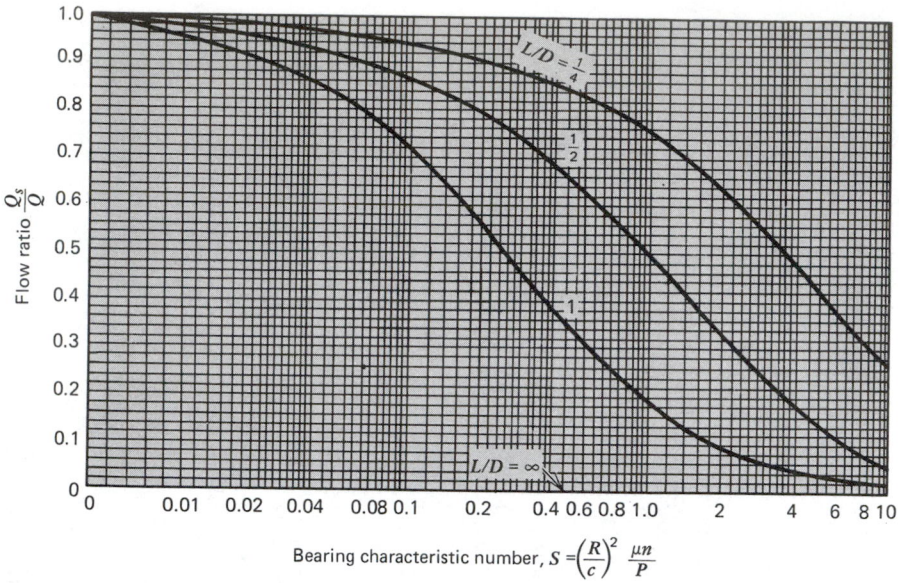

FIGURE 13.17.
Chart for ratio of side flow to total flow [7].

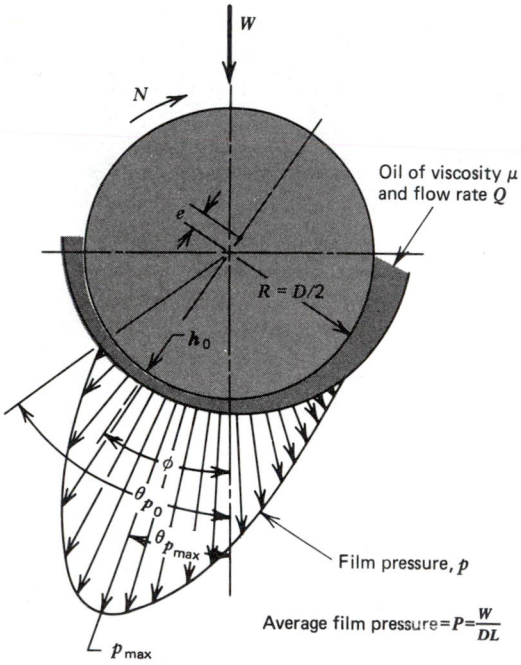

FIGURE 13.18.
Polar diagram of film-pressure distribution showing the notation used.

total oil flow rate through the bearing, the fraction of this flow rate that is recirculated oil flow, and the fraction of new flow that must be introduced to make up for side leakage.

Solution

1. From given data: $D = 2$ in., $R = 1$ in., $L = 1$ in., $c = 0.0015$ in., $n = 50$ revolutions per *second*, and $W = 1000$ lb.

2. $$P = \frac{W}{LD} = \frac{1000}{(1)(2)} = 500 \text{ psi}$$

$$\mu = 4 \times 10^{-6} \text{ reyns (Fig. 13.6)}$$

$$S = \left(\frac{R}{c}\right)^2\left(\frac{\mu n}{P}\right) = \left(\frac{1}{0.0015}\right)^2 \frac{(4 \times 10^{-6})(50)}{500} = 0.18$$

3. Use $S = 0.18$, $L/D = 0.5$ to enter all charts, and use units of inch-pound-seconds consistently: From Fig. 13.11, $h_o/c = 0.3$, hence $h_o = 0.00045$ in. (Note that this bearing is in the "optimum range"—more about that and about acceptable values of h_o in Sample Problem 13.4, Section 13.13.)

From Fig. 13.12, $(R/c)f = 5.4$, hence $f = 0.008$. ■

From Fig. 13.13, $P/p_{max} = 0.32$, hence $p_{max} = 1562$ psi. ■

From Fig. 13.14, $\phi = 40°$. ■

From Fig. 13.15, $\theta_{p_o} = 54°$, $\theta_{p_{max}} = 16.9°$. ■

From Fig. 13.16, $Q/RcnL = 5.15$, hence Q $= 0.39$ in.^{3}/sec. ■

From Fig. 13.17, $Q_s/Q = 0.81$, hence side leakage that must be made up by "new" oil represents 81 percent of the flow; the remaining 19 percent is recirculated. ■

It is important to remember that the foregoing analysis involving the Raimondi and Boyd charts applies only to steady-state operation with a load that is fixed in magnitude and direction. Bearings subjected to rapidly fluctuating loads (as engine crankshaft bearings) can carry *much* greater instantaneous peak loads than the steady-state analysis would indicate because there is not enough *time* for the oil film to be squeezed out before the load is reduced. This is sometimes called the *squeeze-film phenomenon.* It causes an apparent "stiffening" of the oil film as it is squeezed increasingly thin. The squeeze-film effect is the primary lubricating mechanism in engine wrist pin bearings (shown in Fig. 13.22) where the relative motion is oscillatory, over a small angle.

A problem sometimes encountered with high-speed, lightly loaded bearings is a *dynamic instability* that results in orbiting of the journal center with respect to the bearing center. This "whip" or "whirl" of the shaft can cause a destructive vibration, usually at a frequency of about half the rotating speed. See [2, 11, 12]. One means of

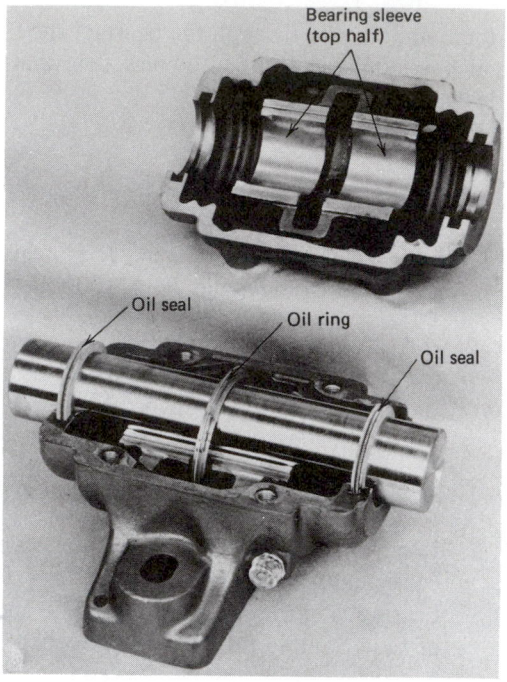

FIGURE 13.19.
Ring-oiled bearing. (Courtesy Reliance Electric Co.)

dealing with this problem is to support the shaft with *tilting-pad bearings.*[7] These are frequently used in turbomachinery.

13.10 LUBRICANT SUPPLY

The above hydrodynamic analysis assumes that oil is available to flow into the bearing at least as fast as it leaks out at the ends. The principal methods for supplying this oil are briefly described below. See [4] and others for further information.

Oil Ring

The oil ring shown in Fig. 13.19 is usually about 1.5 to 2 times the diameter of the journal from which it hangs loosely. As the shaft rotates, the ring brings oil to the top of the journal. Note that the bearing sleeve must be slotted at the top to permit the ring to bear directly on the journal. If the applied bearing load acts substantially downward, this removal of part of the bearing area at the top will not be harmful. Experience has shown ring oilers to be very effective. Sometimes a chain is substituted for the ring.

---

[7]Section 13.15 contains a brief discussion of tilting-pad thrust bearings. Tilting-pad journal bearings are similar.

Oil Collar

A somewhat similar arrangement utilizes a rigid collar attached to the shaft which dips into an oil reservoir at the bottom. The collar carries oil to the top where it is thrown off into a small upper reservoir on each side of the collar. From there it flows by gravity through holes to the bearing surface.

"Splash"

In some machines, oil splashed by rapidly moving parts can be channeled to small reservoirs above bearings. Also, small "oil scoops" on rotating parts can dip into the main oil sump or reservoir and thereby pick up oil that flows into the bearings. Some readers will recognize this as the method of lubrication used with early automobile engines.

Oil Bath

This term usually refers to oil being supplied by virtue of partially submerging the journal in the oil reservoir, as in the case of the railroad partial bearing shown in Fig. 13.8. Care must be taken with oil bath lubrication to avoid generating excessive turbulence and churning of a substantial volume of oil, thereby causing excessive viscous friction losses and possible foaming of the lubricant.

Oil Holes and Grooves

Figure 13.20 shows an axial groove used to distribute oil in the axial direction. Oil enters the groove through an oil hole, and flows either by gravity or under pressure. In general, such grooves should not be cut in the load-carrying area because the hydrodynamic pressure drops to nearly zero at the grooves. This point is illustrated in Fig. 13.21, where the circumferential groove serves to divide the bearing into two halves, each having an L/D ratio a little less than half that of the ungrooved bearing.

The development of groove patterns for adequately distributing oil over the full bearing area without objectionably disturbing the hydrodynamic pressure distribution can, for some applications, be a major problem.

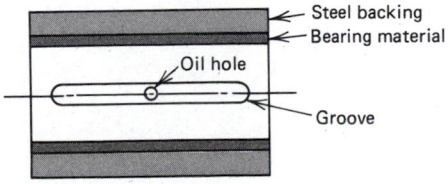

FIGURE 13.20.
Axial grooved bearing.

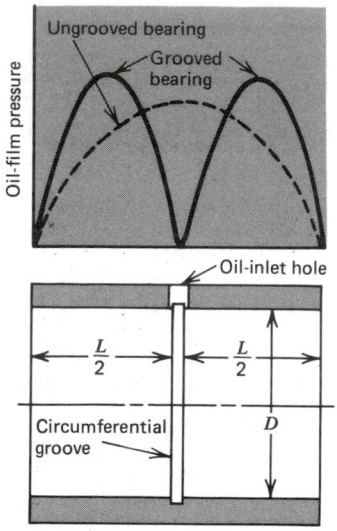

FIGURE 13.21.
Circumferential grooved bearing showing effect on pressure distribution.

Oil Pump

The most positive means of supplying oil is by an oil pump. Figure 13.22 shows the pressure-fed lubrication system of a piston-type engine or compressor. Oil supplied by the pump fills circumferential grooves in the main bearings. Holes drilled in the crankshaft carry oil from these grooves to the connecting rod bearings. Circumferential

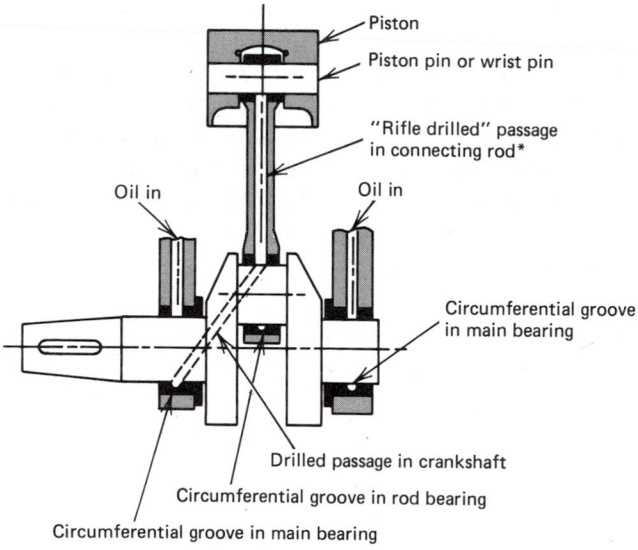

*If omitted, piston pin bearing is splash lubricated.

FIGURE 13.22.
Oil passages in a piston engine or compressor.

grooves in the connecting rod bearings connect with rifle-drilled holes in the connecting rods that carry lubricant to the wrist pins. In most automotive engines, the rifle-drilled connecting rod passages are eliminated, and the wrist pins are splash lubricated. This is less costly and has proven entirely effective.

13.11 HEAT DISSIPATION, AND EQUILIBRIUM OIL FILM TEMPERATURE

Under equilibrium conditions the rate at which heat is generated within a bearing is equal to the rate at which heat is carried away. It is essential that the oil film temperature at which this balance occurs be satisfactory. Temperatures in the range of 71 °C (160 °F) are commonly used; temperatures above 93°–121 °C (200°–250 °F) are generally unsatisfactory because of possible deterioration of petroleum-base lubricants and damage to some common bearing materials. (Remember, the maximum oil film temperature can be substantially higher than the average temperature.)

The friction power absorbed by a bearing is readily computed from the friction torque and shaft rpm (see Eqs. (1.1) and (1.2)). The portion of this power (heat) carried away by the oil is the product of the flow rate (Q_s), oil temperature rise (bearing outlet minus bearing inlet temperatures), and the volumetric specific heat, $c\rho$ (specific heat times density). For petroleum oils at normal bearing operating temperatures, approximate values are

$$c\rho = 1.36 \frac{MPa}{°C}$$

$$c\rho = 110 \frac{\text{lb}}{\text{in.}^2 \cdot °\text{F}}$$

Oil temperature rise can also be estimated by using charts devised by Raimondi and Boyd [7].

The oil temperature at thermal equilibrium depends on the effectiveness with which heat is transferred to the outside environment. For self-contained bearings, as those using an oil ring, collar, or bath, the calculation of average oil film temperature is usually a rough "ballpark" procedure at best, and experimental data under actual operating conditions must be obtained for applications close to being critical with respect to overheating. For a rough estimate, the equation

$$H = CA(t_o - t_a) \tag{13.13}$$

from which

$$t_o = t_a + \frac{H}{CA} \tag{13.14}$$

can be used, where

H = time rate of heat dissipation (as, Watts)

C = overall heat transfer coefficient (as, Watts per hour per square meter per degree Centigrade)

A = exposed housing surface area (as, square meters)

TABLE 13.1
Rough Estimates of Heat Transfer Coefficient, *C*, for Self-Contained Bearings

	C, W/m²·°C (Btu/hr·ft²·°F)[a]		
Bearing type	**Still air**	**Average air circulation**	**Air moving at 500 fpm**
Oil ring or collar	7.4 (1.3)	8.5 (1.5)	11.3 (2.0)
Oil bath	9.6 (1.7)	11.3 (2.0)	17.0 (3.0)

[a]1 Btu is equivalent to 778 ft-lb.

t_o = average oil film temperature (as, °C)

t_a = air temperature in the vicinity of the bearing housing (as, °C).

Values of C for representative conditions are given in Table 13.1. Values of A for pillow block bearings with separate shells, like the one shown in Fig. 13.19, are sometimes estimated as 20 times the bearing projected area (i.e., 20 DL). Again, the reader is cautioned that values of t_o calculated from these values may differ considerably from those actually experienced in a particular installation.

13.12 BEARING MATERIALS

During *thick-film* lubrication any material with sufficient compressive strength and a smooth surface would be an adequate bearing material. Steel, for example, would be fine. But during start-up and shut-down, these bearings experience *thin-film* lubrication and the (usually steel) journal will be damaged unless the bearing material resists seizing and welding with the journal material. Furthermore, any foreign particles larger than h_o that are present in the oil will damage the journal surface unless they can become imbedded in a relatively soft bearing material. Thus, important bearing material properties are:

1. *Mechanical Properties—Conformability* (low elastic modulus) and *deformability* (plastic flow) to relieve local high pressures due to misalignment and shaft deflection; *embeddability,* or *indentation softness,* to permit small foreign particles to become safely embedded in the material, thus protecting the shaft; and *low shear strength* for easy smoothing of surface asperities. At the same time *compressive strength* and *fatigue strength* sufficient for supporting the load and for enduring the repeated flexing that accompanies cyclic loading (as with engine bearings)—and to have these properties at the operating temperatures involved.
2. *Thermal Properties—Heat conductivity* to conduct heat away from local points of metal-to-metal contact during start-up, and from the lubricant film during normal operation; *thermal coefficient of expansion* not too different from the materials of the bearing housing and journal.

3. *Metallurgical Properties—Compatibility* with the journal material, for resistance to scoring, welding, and seizing.
4. *Chemical Properties—Corrosion resistance* to acids that may result from lubricant oxidation or from outside contamination (as engine blowby gases).

The most common bearing materials are *babbitts,* both tin-base (as 89 percent Sn, 8 percent Pb, 3 percent Cu) and lead-base (as 75 percent Pb, 15 percent Sb, 10 percent Sn), and *copper alloys,* primarily copper lead, leaded bronze, tin bronze, and aluminum bronze. Aluminum and silver are also used fairly extensively. The babbitts are unexcelled in conformability and embeddability, but have relatively low compressive and fatigue strength, particularly above about 77°C (170°F). Babbitts can seldom be used above about 121°C (250°F).

The bearing metal surface layer is usually applied to a thin steel backing shell. With babbitt bearings, the shell flexural deformation is independent of the thickness of the babbitt lining, fatigue strength is greatest when the babbitt lining is very thin—on the order of 0.5 mm (0.020 in.) for conventional bearings and 0.13 mm (0.005 in.) in automotive main and connecting rod bearings, with the latter having far greater fatigue resistance. Sometimes a thin babbitt overlay (about 0.025 mm, or 0.001 in.) is added to bearings made of other materials in order to combine the greater load-carrying capacity of the other material with the more desirable surface characteristics of the babbitt.

Rubber and other elastomers make good bearing materials for applications like ship propeller shafts that run in water. These bearings are usually fluted and are retained within a metal shell. The water flow through the bearing allows sand and grit to be flushed out with minimum damage.

13.13 HYDRODYNAMIC BEARING DESIGN

Sample Problem 13.3 in Section 13.9 illustrated an *analysis* of a given hydrodynamic bearing. The *design* of such a bearing is a considerably more involved procedure that requires all of the background in the chapter up to this point, plus empirical guidelines such as those presented below. (As with the design of most machine components, only the basic material can be included here, and there is much to be gained from a study of the detailed literature pertaining to bearing design.)

Unit loading—Table 13.2 summarizes representative values in common use. Note the dramatic influence of the squeeze film phenomenon mentioned at the conclusion of Section 13.9. Since the peak loads applied to engine bearings are of only momentary duration, the resulting bearing pressures can be of the order of ten times the values used in applications where the loads are constant.

Bearing L/D ratios—Ratios of 0.25 to 0.75 are most commonly used now, with older machinery having ratios averaging closer to unity. Greater ratios (longer bearings) result in less end leakage and reduced oil flow requirements; hence greater oil temper-

TABLE 13.2
Representative Unit Sleeve Bearing Loads in Current Practice

Application	Unit Load, $P = W_{max}/LD$ MPa	psi
Relatively steady loads		
Electric motors	0.8–1.5	120–250
Steam turbines	1.0–2.0	150–300
Gear reducers	0.8–1.5	120–250
Centrifugal pumps	0.6–1.2	100–180
Rapidly fluctuating loads		
Diesel engines		
Main bearings	6–12	900–1700
Connecting rod bearings	8–15	1150–2300
Automotive gasoline engines		
Main bearings	4–5	600–750
Connecting rod bearings	10–15	1700–2300

ature rise. Also, short bearings are less susceptible to undesirable edge loading due to shaft deflection and misalignment. Often shaft diameter is determined by strength and deflection requirements, and then bearing length is determined to provide adequate bearing capacity.

Acceptable values of h_o—The minimum acceptable oil film thickness, h_o, depends on surface finishes. Various empirical recommendations are found in the literature. For example, Trumpler [12] suggests the relationship

$$h_o \geqq 0.0002 + 0.00004D \quad (h_o \text{ and } D \text{ in in.}), \text{ or} \tag{13.15}$$
$$h_o \geqq 0.005 + 0.00004D \quad (h_o \text{ and } D \text{ in mm})$$

This equation should be used with an appropriate safety factor applied to the load. Trumpler suggests S.F. = 2 for steady loads which can be evaluated with reasonable accuracy. Also, Eq. (13.15) applies to a fine ground journal surface (peak-to-valley roughness not exceeding 0.005 mm, or 0.0002 in.), good standards of geometric accuracy (circumferential out-of-roundness, axial taper, and "waviness" both circumferential and axial), and good standards of oil cleanliness.

For bearings subjected to rapidly fluctuating loads (for example, engine bearings), simplified calculations based on the assumption that peak loads remain constant may give calculated values of h_o of the order of one-third actual values. This must be taken into account when using empirical criteria such as Eq. (13.15). More realistic calculations for such bearings take the squeeze-film phenomenon into account, and are beyond the scope of this book.

Clearance ratios (c/R, or 2c/D)—For journals of 25–150 mm diameter the c/R ratio is usually of the order of 0.001, particularly for precision bearings. For less precise bearings this ratio tends to be higher—up to about 0.002 for general machinery bearings and 0.004 for rough-service machinery. In any specific design the clearance ratio involves a range of values, depending on the tolerances assigned to the journal and bearing diameters.

The following is a list of important factors to be taken into account when designing a bearing for hydrodynamic lubrication.

1. The minimum oil film thickness must be sufficient to insure thick-film lubrication. Use Eq. (13.15) as a guide, and take into account surface finish and load fluctuation.
2. Friction should be as low as possible, consistent with adequate oil film thickness. Try to keep in the "optimum zone" of Fig. 13.11.
3. Be sure that an *adequate supply* of *clean* and *sufficiently cooled* oil is always available at the bearing inlet. This may require forced feeding and/or special cooling provisions.
4. Be sure that the maximum oil temperature is acceptable (generally below 93°–121°C or 200°–250°F).
5. Be sure that oil admitted to the bearing gets distributed over its full length. This may require grooves in the bearing. If so, they should be kept away from highly loaded areas.
6. Select a suitable bearing material to provide sufficient strength at operating temperatures, sufficient comformability and embeddability, and adequate corrosion resistance.
7. Check the overall design for shaft misalignment and deflection. If these are excessive, even a properly designed bearing will give trouble.
8. Check the bearing loads and elapsed times involved during start-up and shut-down. Bearing pressures during these periods should preferably be under 2 MPa, or 300 psi. If extended time periods of low-speed operations are involved, thin-film lubrication requirements must be considered (Section 13.14).
9. Be sure that the design is satisfactory for all reasonably anticipated combinations of clearance and oil viscosity. The operating clearance will be influenced by thermal expansion and by eventual wear. Oil temperature and therefore viscosity is influenced by thermal factors (ambient air temperature, air circulation, etc.), and by possible changes in the oil with time. Furthermore, the user may put in a lighter or heavier grade of oil than the one specified.

SAMPLE PROBLEM 13.4

A journal bearing on an 1800-rpm steam turbine rotor supports a constant gravity load of 17 kN. The journal diameter has been established as 150 mm in order to provide sufficient shaft stiffness. A forced-feed lubrication system will supply SAE 10 oil, controlled to an average film temperature of 82°C. Determine a suitable combination of bearing length and radial clearance. Also determine the corresponding values of coefficient of friction, friction power loss, oil flow rate to and from the bearing, and oil temperature rise through the bearing.

Solution

1. From the 1 to 2 MPa range given for steam turbine bearings in Table 13.2, arbitrarily select $P = 1.6$ MPa for a starting point. Then $L = 17{,}000\ \text{N}/[(1.6\ \text{mPa})(150\ \text{mm})] = 70.83$ mm. Arbitrarily round this up to $L = 75\ mm$ to give $L/D = ½$ for convenient use of the Raimondi and Boyd charts. (We note that an L/D ratio of ½ is in line with current turbine bearing practice.) With $L = 75$ mm, P is calculated to be 1.511 MPa.
2. Figure 13.11 shows that for $L/D = ½$, the optimum operating range is between $S = 0.037$ and $S = 0.35$. Figure 13.6 gives the viscosity of SAE 10 oil at 82°C as 6.3 mPa·s. Substituting known values (corresponding to the optimum zone boundaries) into the equation for S gives

$$S = \frac{\mu n}{P}\left(\frac{R}{c}\right)^2$$

$$0.037 = \frac{(6.3 \times 10^{-3}\ \text{Pa}\cdot\text{s})(30\ \text{rev/s})}{1.511 \times 10^6\ \text{Pa}}\left(\frac{75\ \text{mm}}{c\ \text{mm}}\right)^2$$

from which $c = 0.138$ mm ($c/R = 0.00184$). Similarly, for $S = 0.35$, $c = 0.0448$ mm ($c/R = 0.00060$). We note that these clearance ratios are consistent with the guidelines given in this article.

3. Before deciding on an appropriate tolerance range for the radial clearance, let us compute and plot h_o, f, Q, and Q_s as functions of c, with c extending to either side of the optimum range. Values from the table on the facing page are plotted in Fig. 13.23.
4. Figure 13.23 appears to indicate good operation over a range of radial clearances between about 0.04 and 0.15 mm, but a check should be made with Eq. (13.15):

$$h \geqq 0.005 + 0.00004(150) = 0.011\ \text{mm}$$

Let us compare this with the calculated minimum film thickness using a safety factor of 2 applied to the load, and assuming an "extreme case" of $c = 0.15$ mm:

$$S = \frac{(6.3 \times 10^{-3})(30)}{(1.511 \times 10^6)(2)}\left(\frac{75}{0.15}\right)^2 = 0.0156$$

$$h_o/c = 0.06, \quad h_o = 0.009\ \text{mm}$$

This is less than the 0.011-mm requirement of Eq. (13.15). However, the initial specification of an 82°C average oil film temperature was somewhat unrealistic in order to simplify the problem. The large oil flow rate associated with $c = 0.15$ mm would normally result in a lower oil film temperature (hence, higher viscosity) than that obtained with smaller bearing clearances. Also, a "heavier" grade of oil could be specified when wear results in increased bearing clearance. A calculation with SAE 20 oil (also at 82°C average film temperature) indicates that the minimum

Table for Sample Problem 13.4

	c (mm)	S^a	h_o/c^b	h_o (mm)	$\frac{R}{c}f^c$	f	$\frac{Q^d}{RcnL}$	Q (mm³/s)	Q_s/Q^e	Q_s (mm³/s)
	0.02	1.714	0.76	0.0152	36.	0.0096	3.8	12,800	0.37	4,700
	0.03	0.762	0.59	0.0177	16.	0.0064	4.3	21,800	0.56	12,200
←Optimum zone→ (from Fig. 13.11)	0.04	0.428	0.47	0.0188	10.	0.0053	4.65	31,600	0.68	21,500
	0.0448	0.342	0.425	0.0190	8.7	0.0052	4.8	36,300	0.72	26,100
	0.05	0.274	0.37	0.0185	7.3	0.0049	4.95	41,800	0.76	31,700
	0.07	0.140	0.26	0.0182	4.4	0.0041	5.25	62,000	0.84	52,100
	0.09	0.085	0.195	0.0176	3.1	0.0037	5.45	82,800	0.88	72,800
	0.11	0.057	0.15	0.0165	2.3	0.0034	5.55	103,000	0.91	93,800
	0.13	0.041	0.12	0.0156	1.9	0.0033	5.6	122,900	0.92	113,000
	0.138	0.036	0.11	0.0152	1.75	0.0032	5.65	131,600	0.93	122,400
	0.15	0.030	0.10	0.0150	1.6	0.0032	5.7	144,300	0.94	135,600
	0.18	0.021	0.08	0.0144	1.3	0.0031	5.75	174,700	0.95	165,900

[a] $S = \frac{\mu n}{P}\left(\frac{R}{c}\right)^2 = \frac{(6.3 \times 10^{-3}\ \text{Pa}\cdot\text{s})(30\ \text{rev/s})}{1.551 \times 10^6\ \text{Pa}}\left(\frac{75\ \text{mm}}{c\ \text{mm}}\right)^2 = \frac{6.8544 \times 10^{-4}}{c^2}$.

[b] From Fig. 13.11.

[c] From Fig. 13.12.

[d] From Fig. 13.16.

[e] From Fig. 13.17.

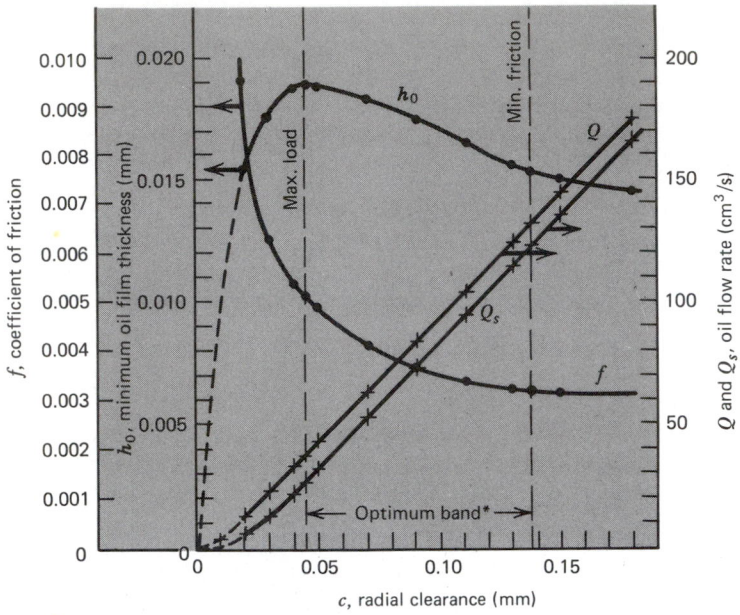

*As defined in Fig. 13.11

FIGURE 13.23.
Sample Problem 13.4. Variation of h_0, f, Q, and Q_s with c (μ, n, L, D, and W constant).

film thickness with $c = 0.15$ mm and radial load of twice 17 kN would be about 0.012 mm.

5. At this point a judgment can be made as to appropriate journal and bearing diameter tolerances. Specifying these so as to give a range of radial clearances between 0.05 and 0.07 mm would permit appreciable wear to occur without taking the bearing operation beyond the "optimum zone." Increasing the tolerances to provide a clearance range between 0.05 and 0.09 mm might permit more economical manufacture. Slightly greater initial clearances, such as 0.08 to 0.11 mm, would result in lower friction losses and tend to make the bearing run cooler.

6. Turning to the oil flow rate curves in Fig. 13.23, remember that these assume oil to be always available at the bearing inlet at *atmospheric pressure;* the flows calculated are generated by the bearing itself. The oil pump used in this forced-feed system must provide a flow rate equal to the side leakage, Q_s, just to keep up with the bearing demand. Delivery of oil to the bearing at pressures above atmospheric will cause increased flow. This means that any particular oil particle will absorb less heat as it flows through the bearing.

   Note that the difference between the two oil flow curves represents circumferential or recirculated flow, and this varies negligibly with clearance.

   The great sensitivity of oil flow rate to radial clearance suggests that wear can be monitored by checking flow rate at a constant supply pressure (or, checking supply pressure when using a constant flow rate oil pump).

7. Friction power loss for any operating clearance can be computed using values of coefficient of friction from the table or Fig. 13.23. Of particular concern is the greatest loss that will occur with the tightest bearing fit. For the clearance range of interest, this is $c = 0.04$ mm, for which:

$$\text{Friction torque, } T_f = WfD/2 = (17{,}000\text{ N})(0.0053)(0.150\text{ m})/2$$
$$= 6.76\text{ N}\cdot\text{m}$$

From Eq. (1.1)

$$\text{Friction power} = \frac{nT}{9549} = \frac{(1800)(6.76)}{9549} = 1.27\text{ kW}$$

8. Assuming that the entire 1.27 kW of heat generated in the bearing is carried away by the oil, and that oil is supplied to the bearing at atmospheric pressure (conservative for a force-feed system), the temperature rise of the oil when flowing through the bearing is:

$$\Delta t = (\text{heat rate})\left(\frac{1}{\text{flow rate}}\right)\left(\frac{1}{\text{volumetric specific heat}}\right)$$
$$= \left(1270\,\frac{\text{Nm}}{\text{s}}\right)\left(\frac{\text{s}}{21.5\times10^{-6}\text{ m}^3}\right)\left(\frac{\text{m}^2\cdot{}^\circ\text{C}}{1.36\times10^6\text{N}}\right)$$
$$= 43.4^\circ\text{C}$$

For an average temperature of 82°C, oil would have to be supplied at approximately 60° and leave at 103°. This is only marginally satisfactory. If the radial clearance is increased only slightly—say to 0.05 mm—the temperature rise will come down substantially because of reduced friction loss together with increased oil flow rate. Also, for pressurized oil flow rates exceeding the natural flow rate, the temperature rise is correspondingly lower.

9. The following appear to constitute reasonable final answers:

Bearing length = 75 mm. ■

Radial clearance = 0.05 to 0.07 mm. (The 0.07 dimension might be slightly increased due to manufacturing cost considerations.) ■

Power loss = 1.18 to 0.99 kW. (Note that power loss for various clearances is proportional to coefficient of friction.) ■

Oil flow rate (Q_s) = 31,700 to 52,100 mm^3/s. ■

Oil temperature rise = 27.3 to 13.9°C. (The brief calculations required are left to the reader.)

A few additional points relative to the above sample problem are worthy of note:

1. With the gravity force of the rotor loading the bearing only at the bottom, oil should be admitted and distributed at the top. Axial oil distribution could be

accomplished with a groove, as in Fig. 13.20. Since the entire top of the bearing is never loaded, this groove could be very wide, perhaps encompassing the entire top 180°. This would give a 180° partial bearing, with the advantage of reducing viscous drag at the top. Special Raimondi and Boyd curves for partial bearings [7] would then apply.

2. It is especially important that all oil passages be clean at the time of assembly. An appropriate oil filter should be provided.
3. It is unfortunate for the bearing that its load at rest and during starting and stopping is as great as the running load. However, since this load is under 2 MPa, and assuming neither frequent nor prolonged operation at low speed is anticipated, this should be an acceptable situation.
4. Some large turbines use hydrostatic bearings to avoid boundary lubrication during starting and stopping. In some cases the high-pressure pump can be turned off at operating speed, and hydrodynamic lubrication allowed to take over. (A low-pressure pump would normally remain on to provide a positive oil supply, as specified in the sample problem.)

13.14 BOUNDARY AND MIXED-FILM LUBRICATION

A conceptual representation of boundary and mixed-film lubrication was given in Fig. 13.2. Their representation on the f vs. $\mu n/P$ curve was shown in Fig. 13.4. Even boundary lubrication of extremely smooth surfaces involves actual contact of only a small fraction of the total area. This means that the highly localized areas of contact involve very high pressures and very high instantaneous temperatures. Were this to occur on unprotected metal surfaces, high friction and rapid surface destruction would usually result. Fortunately, even in ordinary atmosphere oxide and other protective films form on metal surfaces. The introduction of grease, oil, graphite, molybdenum disulphide, and so on results in surface films providing some "lubrication." These films are relatively weak in shear, hence the films on surface peaks and asperities tend to shear off, with new film layers forming as old ones are worn off. With mixed-film lubrication, only part of the load is carried by the film-coated solid peaks with the balance carried hydrodynamically.

Continuing research efforts are directed toward the development of new lubricants—and new combinations of surface materials and lubricants—giving improved resistance to *adhesive wear* (see Section 9.9), and reduced friction.

Improved boundary lubrication is often provided by modifying the bearing so as to promote lubrication. Common examples are "sintered" metal bearings. These are made by compressing powdered metal (usually copper and tin, or iron and copper) into the desired shape, then heating at a temperature between the melting points of the two metals. The resulting porous matrix permits the bearing to soak up oil "like a sponge" before it is put into service. In use, the entrapped oil flows to the surface in response to heat and pressure—then flows back into the porous matrix when the machine is shut down. Another example is the indentation of the surface of a solid (nonporous) bearing to give a hammered effect, with the surface indentations providing storage space for solid or semi-solid lubricants.

Some bearing materials, such as graphite and various plastics, are said to be *self-lubricating* because of their naturally low coefficients of friction with smooth metal surfaces. Plastics, such as nylon and TFE, with and without various additives and fillers, are widely used for moderate loads and speeds. Two limitations of plastic bearings should be noted: (1) "cold flow" occurs at heavy loads, and (2) they tend to run hotter than metal bearings that generate the same friction heat because of their low thermal conductivities. These disadvantages are less pronounced when the plastic bearing material is in the form of a thin *bonded coating.*

Although boundary lubrication is usually thought of in connection with bearings, the same phenomenon occurs with the relative motion of screw threads, meshing gear teeth, sliding of a piston in a cylinder, and with sliding surfaces of other machine components.

Porous metal bearings are usually designed on the basis of an allowable product of pressure times velocity, or *PV factor.* For a given coefficient of friction, this factor is proportional to the frictional heat generated per unit of bearing area. A maximum *PV* value of 50,000 (psi $\times$ fps) is commonly applied to porous metal bearings. For long-term operation at high *PV,* or for high temperatures, provision should be made for additional oil. Oil can be applied to any surface, as it will be drawn to the interior by capillary action. A grease reservoir next to the bearing can also be effective. Where long life is desired with no provision for adding lubricant, the *PV* value should be reduced by at least half. Table 13.3 gives more detailed recommendations. Except for the first three bearing materials listed in the table, hardened and ground steel shafts should be used to resist shaft wear. The tabulated values are sometimes moderately exceeded, but with a sacrifice in service life.

Table 13.4 gives corresponding design recommendations for nonmetallic bearings.

TABLE 13.3
Operating Limits of Boundary-Lubricated Porous Metal Bearings [3]

	Static P		Dynamic P		V		PV	
Material	**MPa**	**(ksi)**	**MPa**	**(ksi)**	**m/s**	**(fpm)**	**MPa·m/s**	**(ksi·fpm)**
Bronze	55	(8)	14	(2)	6.1	(1200)	1.8	(50)
Lead-bronze	24	(3.5)	5.5	(0.8)	7.6	(1500)	2.1	(60)
Copper-iron	138	(20)	28	(4)	1.1	(225)	1.2	(35)
Hardenable copper-iron	345	(50)	55	(8)	0.2	(35)	2.6	(75)
Iron	69	(10)	21	(3)	2.0	(400)	1.0	(30)
Bronze-iron	72	(10.5)	17	(2.5)	4.1	(800)	1.2	(35)
Lead-iron	28	(4)	7	(1)	4.1	(800)	1.8	(50)
Aluminum	28	(4)	14	(2)	6.1	(1200)	1.8	(50)

TABLE 13.4
Operating Limits of Boundary-Lubricated Nonmetallic Bearings [3]

	P		Temp		V		PV	
Material	**MPa**	**(ksi)**	**°C**	**(°F)**	**m/s**	**(fpm)**	**MPa·m/s**	**(ksi · fpm)**
Phenolics	41	(6)	93	(200)	13	(2500)	0.53	(15)
Nylon	14	(2)	93	(200)	3.0	(600)	0.11	(3)
TFE	3.5	(0.5)	260	(500)	0.25	(50)	0.035	(1)
Filled TFE	17	(2.5)	260	(500)	5.1	(1000)	0.35	(10)
TFE fabric	414	(60)	260	(500)	0.76	(150)	0.88	(25)
Polycarbonate	7	(1)	104	(220)	5.1	(1000)	0.11	(3)
Acetal	14	(2)	93	(200)	3.0	(600)	0.11	(3)
Carbon (graphite)	4	(0.6)	400	(750)	13	(2500)	0.53	(15)
Rubber	0.35	(.05)	66	(150)	20	(4000)	—	—
Wood	14	(2)	71	(160)	10	(2000)	0.42	(12)

13.15 THRUST BEARINGS

All rotating shafts need to be positioned axially. For example, the crankshaft shown in Fig. 13.1 is positioned axially by the flanged thrust surfaces that are integral with the main bearings which carry the radial loads. Sometimes the thrust load is carried by a separate flat "thrust washer." These flat bearing surfaces may not permit the "wedging action" required for hydrodynamic lubrication. But if the loads are light, boundary or mixed-film lubrication is adequate.

Where shaft axial loads are high (as with vertical shafts of substantial weight, and propeller shafts subjected to substantial thrust loads), *hydrodynamic* thrust bearings can be provided (Fig. 13.24). Oil supplied to the inside diameter of the rotating *collar* or *runner* flows outward by centrifugal force through the bearing interface. As the oil is dragged circumferentially through the bearing, it experiences a wedging action due

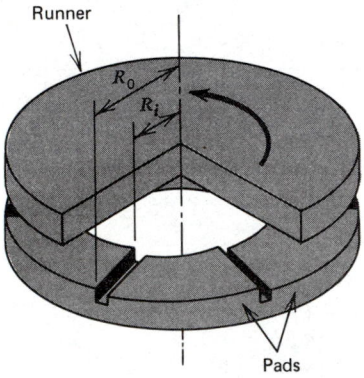

FIGURE 13.24.
Thrust bearing incorporating fixed tilted pads and rotating flat runner.

to the tapered pads on the stationary member. This is directly analogous to the wedging action produced by the eccentricity of a journal bearing (Fig. 13.3).

The fixed pads may have a fixed taper angle (as in Fig. 13.24), the pads may be pivoted and allowed to assume their own optimum tilt angle, or they may be partially constrained and permitted a small variation in tilt angle. If the pads have a fixed taper, it is obvious that a load can be supported hydrodynamically for only one direction of rotation.

Raimondi and Boyd charts are also available for thrust bearing design [4,8]. Also see [6], pp. 89–91.

13.16 ELASTOHYDRODYNAMIC LUBRICATION

Elastohydrodynamic lubrication is the term applied to the lubrication of highly loaded nonconforming surfaces, usually in at least partial rolling contact. The theory, which is not presented here, involves consideration of the elastic deflections of the mating surfaces (as discussed in Section 9.13), and the increased viscosity of the lubricant under the extreme pressures involved. (See [6], pp. 92–98 and 103–114.) Elastohydrodynamic theory is basic to any advanced study of the behavior of highly loaded curved surfaces in such machine components as gears, ball bearings, roller bearings, and cams.

REFERENCES

1. Cameron, A., *The Principles of Lubrication,* John Wiley & Sons, Inc., New York, 1966.
2. Gross, W. A., *Gas Film Lubrication,* John Wiley & Sons, Inc., New York, 1962.
3. "Mechanical Drives," *Machine Design Reference Issue,* Penton/IPC, Cleveland, June 18, 1981.
4. O'Conner, J. J., and J. Boyd, *Standard Handbook of Lubrication Engineering,* McGraw-Hill, New York, 1968.
5. Ocvirk, F. W., "Short-Bearing Approximation for Full Journal Bearings," Technical Note 2808, Nat. Advisory Comm. for Aeronautics, Washington, D.C., 1952. Also see G. B. DuBois and F. W. Ocvirk, "The Short-Bearing Approximation for Plain Journal Bearings," *Trans. ASME,* **77** (1955), pp. 1173–1178.
6. Peterson, M. B., and W. O. Winer (eds.), *Wear Control Handbook,* American Society of Mechanical Engineers, New York, 1980.
7. Raimondi, A. A., and J. Boyd, "A Solution for the Finite Journal Bearing and Its Application to Analysis and Design," Parts I, II, and III, *Trans. ASLE,* Vol. 1, No. 1, pp. 159–209, in *Lubrication Science and Technology*. Pergamon Press, New York, 1958.
8. Raimondi, A. A., and J. Boyd, "Applying Bearing Theory to the Analysis and Design of Pad-Type Bearings," *Trans. ASME,* **77** (April 1955), pp. 287–309.

9. Shaw, Milton C., and F. Macks, *Analysis and Lubrication of Bearings,* McGraw-Hill, New York, 1949.

10. Slaymaker, Robert R., *Bearing Lubrication Analysis,* John Wiley & Sons, Inc., New York, 1955.

11. Szeri, A. Z., *Tribology,* McGraw-Hill, New York, 1980.

12. Trumpler, Paul R., *Design of Film Bearings,* Macmillan, New York, 1966.

13. Wills, J. George, *Lubrication Fundamentals,* Marcel Dekker, Inc., New York, 1980.

CHAPTER FOURTEEN

Rolling-Element Bearings

14.1 COMPARISON OF ALTERNATIVE MEANS FOR SUPPORTING ROTATING SHAFTS

As a background for introducing rolling-element bearings, a brief review of sliding bearings (Chapter 13) may be helpful: The simplest possible bearings are unlubricated plain or sliding bearings—like the wooden cart wheels mounted directly on wooden axles in ancient history. Lower friction and longer life were obtained by adding a lubricant, such as animal or vegetable oil. Modern sliding bearings commonly involve steel shafts supported by bearing surfaces of a wear-compatible material such as bronze or TFE[1] (see Sections 9.9 and 9.10). Common low-speed applications (lawn mower wheels, garden carts, children's tricycles, etc.) involve the use of oil or grease, but the lubricant does not completely separate the surfaces. On the other hand, sliding bearings used with engine crankshafts experience hydrodynamic lubrication during normal operation; that is, the oil film completely separates the surfaces.

Rolling-element bearings involve separation of the shaft and outer member by balls or rollers, and thus substitute rolling friction for sliding friction. Examples are shown in Figs. 14.1–14.10. Since the contact areas are small and the stresses high (Section 9.13), the loaded parts of rolling-element bearings are normally made of hard, high-strength materials, superior to those of the shaft and outer member. These parts include inner and outer rings (also called races), and the balls or rollers. An additional component of the bearing is usually a retainer or separator, which keeps the balls or rollers separated.

Both sliding and rolling-element bearings have their places in modern machinery.

[1]Tetraflouroethylene, as duPont's Teflon.

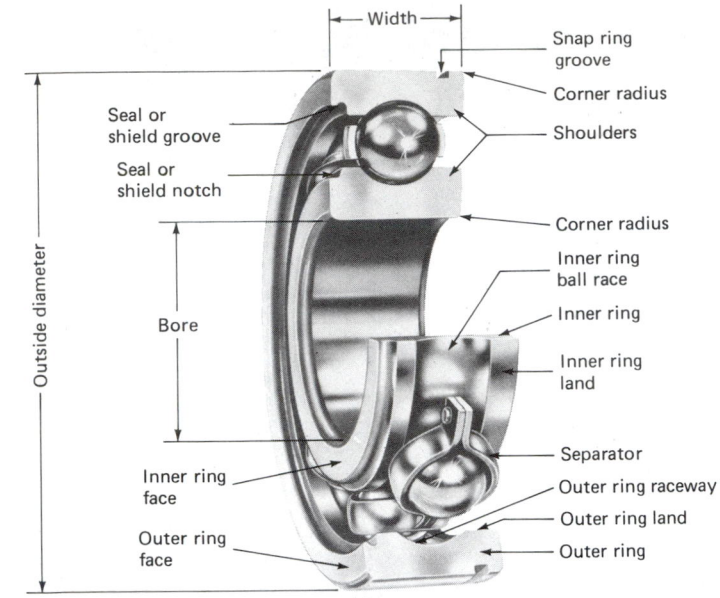

(*a*)
Construction and nomenclature

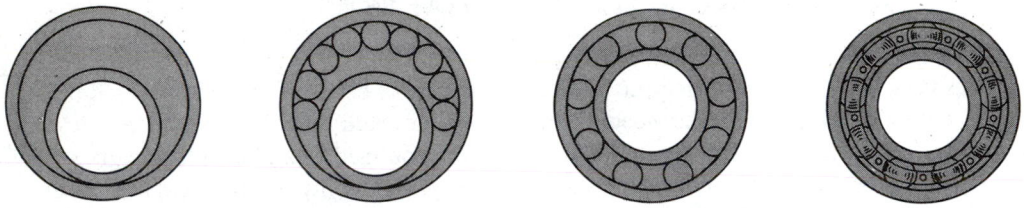

1 2 3 4

(*b*)
Steps in assembly

(*c*)
Contact surface geometry

FIGURE 14.1.
Radial ball bearing (deep-groove or "Conrad" type). (Courtesy New Departure-Hyatt Division, General Motors Corp.)

A major advantage of rolling-element bearings is low starting friction. Sliding bearings can achieve comparably low friction only with full-film lubrication (complete surface separation). This requires hydrostatic lubrication, with its attendant costly auxiliary external fluid supply system, or hydrodynamic lubrication, which cannot be achieved during starting. Rolling bearings are ideally suited for applications involving high starting loads. For example, the use of roller bearings to support rail car axles eliminates the need for an extra locomotive to get a long train started. On the other hand, fluid-film bearings are well suited for high rotating speeds with impact and momentary overloads. The higher the rotating speed the more effective the hydrodynamic pumping action. Also, the fluid film effectively "cushions" impact, as the duration involved is too short for the film to be squeezed out. High rotating speeds are generally disadvantageous to rolling-element bearings because of the rapid accumulation of fatigue cycles and the high centrifugal force on the rolling elements (see Section 9.14, Surface Fatigue Failures).

Rolling-element bearings take up more radial space around the shaft, but plain bearings usually require greater axial space. Rolling-element bearings generate and transmit a certain amount of noise, whereas fluid-film bearings do not normally generate noise and may dampen noise from other sources. Sliding bearings are cheaper than ball or roller bearings for simple applications where minimal lubrication provision is required. Where sliding bearings require a forced lubrication system, the overall cost of rolling-element bearings may be lower. Another advantage of ball and roller bearings is that they can be "preloaded" (mating bearing elements are pressed together rather than operating with a small clearance). This is important in applications requiring precise positioning of the rotating member.

Rolling-element bearings are also known as "anti-friction" bearings. This term is perhaps unfortunate because these bearings do not in all cases provide lower friction than fluid-film bearings. With normal operating loads, rolling-element bearings (without seals) typically provide coefficients of friction between 0.001 and 0.002.

14.2 HISTORY OF ROLLING-ELEMENT BEARINGS

The first recorded use of rolling elements to overcome sliding friction was for moving heavy stone slabs by Egyptian construction workers, probably before 200 B.C. [1], and possibly by the Assyrians in about 650 B.C. It is believed that some early chariot wheels used crude roller bearings made from round sticks. Around 1500 A.D., Leonardo da Vinci is considered to have invented and partially developed modern ball and roller bearings. A few ball and roller type bearings were constructed in France in the eighteenth century (the builder of a roller-bearing carriage claiming, in 1710, that his roller bearings permitted one horse to do work otherwise hardly possible for two horses). But it was not until after the invention of the Bessemer steel process in 1856 that a suitable material for rolling-element bearings was economically available. During the remainder of the nineteenth century, there was a rapid development of ball bearings in Europe for use in bicycles.

14.3 ROLLING-ELEMENT BEARING TYPES

Rolling-element bearings are either *ball bearings* or *roller bearings*. In general, ball bearings are capable of higher speeds, and roller bearings can carry greater loads. Most rolling-element bearings can be classed in one of three categories: (1) *radial* (for carrying loads that are primarily radial), (2) *thrust*, or axial-contact (for carrying loads that are primarily axial), and (3) *angular-contact* (for carrying combined axial and radial loads). Figures 14.1–14.3 illustrate ball bearings of these types: Figure 14.3*f* is a thrust bearing, Figs. 14.3*b* and 14.3*c* show angular contact bearings, and Figs. 14.1, 14.2, and the other parts of 14.3 illustrate radial bearings.

Roller bearings are also classified by roller configuration into the four types shown in Figs. 14.6–14.9: (1) *cylindrical*, (2) *spherical*, (3) *tapered*, and (4) *needle*. Needle bearings can be regarded as a special case of cylindrical roller bearings where the rollers have a length-to-diameter ratio of four or greater.

Figure 14.1*a* shows the construction and nomenclature of a typical radial ball bearing of the *deep-groove*, or *Conrad* type. Figure 14.1*b* illustrates the steps in assembling the major components, and Fig. 14.1*c* illustrates the contact between a ball and raceway. (Note that the ball and raceway curvatures involved would give a more complicated problem for contact stress analysis than the spheres, cylinders, and flat plates considered in Section 9.13.) Ball bearings are made in various proportions in order to accommodate various degrees of loading, as shown in Fig. 14.2. Although intended primarily for radial loads, it is obvious from their construction that these bearings will also carry a certain amount of thrust.

Figure 14.3*a* shows a radial ball bearing with *notches* or *loading grooves* in the race shoulders. These bearings can be assembled with races concentric, and therefore

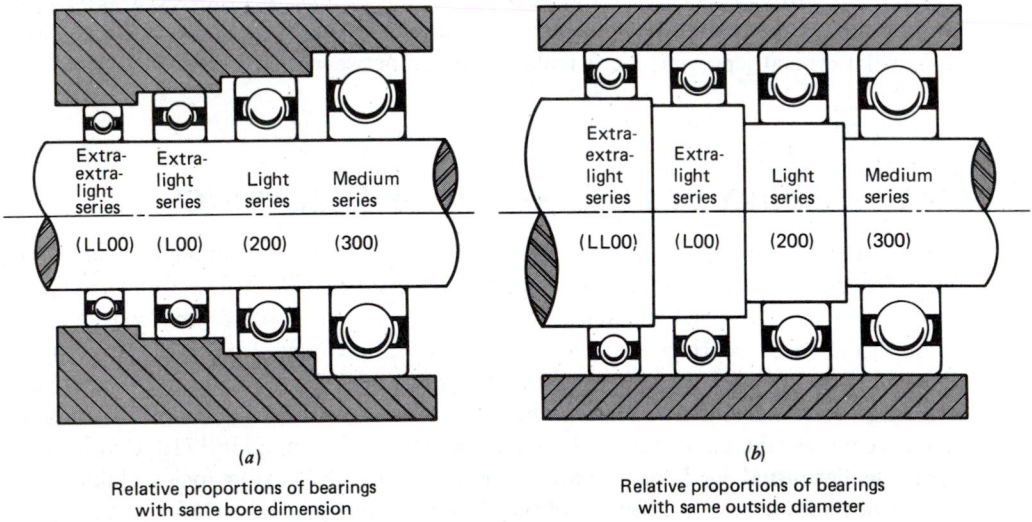

FIGURE 14.2.
Relative proportions of bearings of different series. (Courtesy New Departure-Hyatt Division, General Motors Corp.)

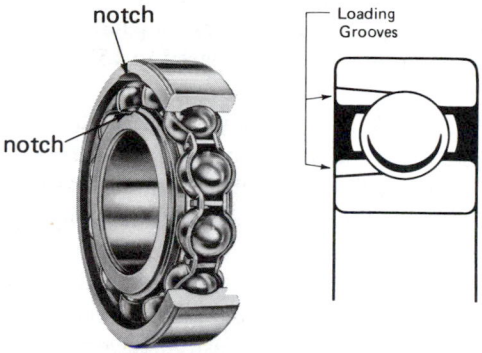

(a) Filling notch (loading groove) type

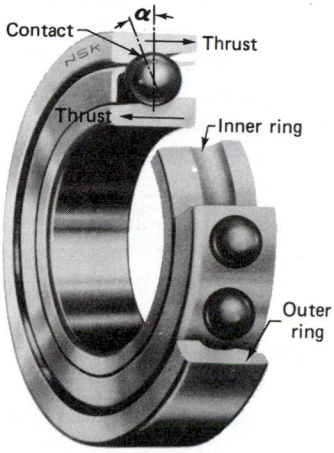

(b) Angular contact type

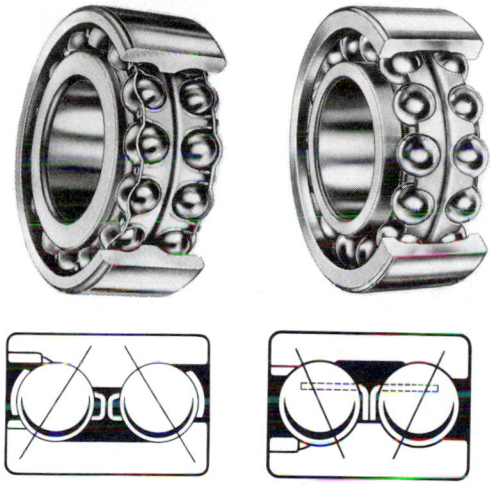

(c)
Double row

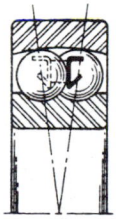

(d) Internal self-aligning

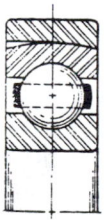

(e) External self-aligning

(f) Thrust

FIGURE 14.3.
Representative types of ball bearings (in addition to deep-groove type shown in Fig. 14.1). (*b, f* Courtesy Hoover-NSK Bearing Co.; *c* Courtesy New Departure-Hyatt Division, General Motors Corp.)

contain a larger number of balls than the deep-groove type. This gives 20 to 40 percent greater radial load capacity at the expense of sharply reduced thrust capacity. Also, these bearings tolerate only about 3′ of angular misalignment compared to 15′ for the deep-groove type.

Angular contact bearings, as shown in Fig. 14.3*b*, have substantial thrust capacity in one direction only. They are commonly installed in pairs, with each taking thrust in one direction. The double-row ball bearing shown in Fig. 14.3*c* incorporates a pair of angular contact bearings into a single unit. Figures 14.3*d* and 14.3*e* show self-aligning bearings, made to tolerate substantial shaft angular misalignment.

As a general rule ball bearings are nonseparable; that is, the two races, balls, and retainer are installed as an assembly. This makes them quite adaptable to enclosure with shields or seals, and makes self-contained grease lubrication feasible. Several types of shields and seals are shown in Fig. 14.4. Shields are close-fitting but nonrubbing thin washers that protect the bearing against all but very small foreign particles and help

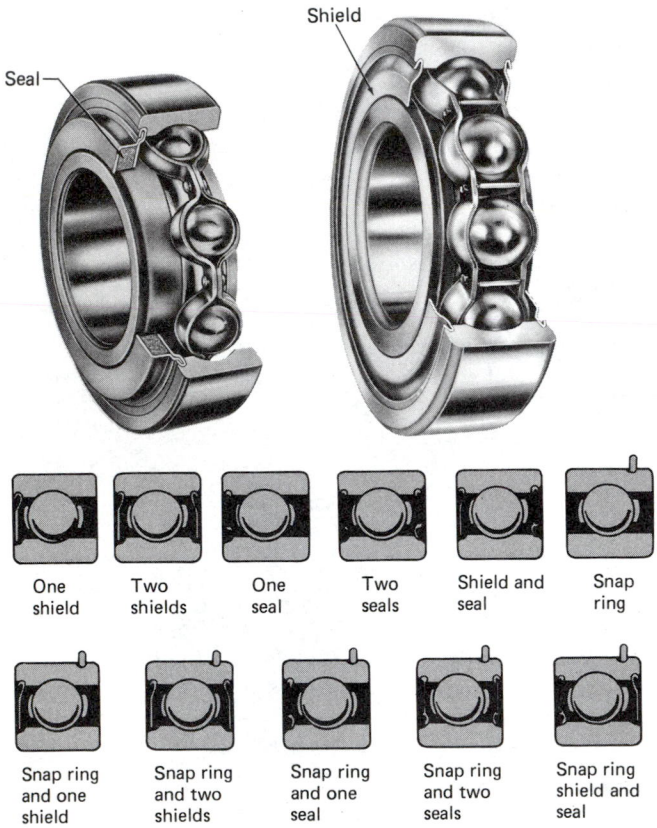

FIGURE 14.4.
Bearings with seals and shields. (Courtesy New Departure-Hyatt Division, General Motors Corp.)

retain lubricant. Seals have rubbing contact and therefore provide greater lubricant retention and protection against contamination. Their disadvantage is that they introduce some frictional drag and are subject to wear. Separate shaft seals are often used, as they can provide more effective sealing because of the greater space available for the sealing elements. With suitable seals, either integral with the bearing or separate, it is often feasible to grease-lubricate the bearing for life at the time of assembly.

Conventional ball bearings have ground rings (races). For applications with modest requirements of load capacity, life, and quietness, less costly unground ball bearings are often used. These have rings made on automatic screw machines and are hardened but not ground.

In contrast with ball bearings, roller bearings are usually made so that the rings (races) can be separated. The rollers and retainer may or may not be permanently assembled with one of the rings. This largely precludes the use of integral shields and seals, but it does facilitate the installation of the rings with the heavy press fits commonly used. Standardized dimensions often permit the use of roller bearings with rings made by different manufacturers. Also, one or both of the rings are often made integral with the mating shaft and/or housing.

Figure 14.5 illustrates four basic types of cylindrical roller bearings. In part *a* one ring has no flanges; hence, no thrust loads can be carried. In part *b* the inner ring has one flange, permitting small thrust loads in one direction. In part *c* the inner ring has an integral flange on one side and a removable flange on the other so that light thrust can be taken in either direction. Sometimes the removable flange is on the outer ring, which makes the outer ring separable. In order to relieve load concentration at the roller ends, the rollers are often slightly crowned, with the end diameters reduced typically by about 0.004 mm. Cylindrical roller bearings usually have a separator or retainer to hold the rollers in place, but can be assembled without a retainer and with a *full complement* of rollers, as illustrated for the case of needle bearings in Fig. 14.9. A roller thrust bearing is shown in Fig. 14.5d.

Figure 14.6 shows three types of spherical roller bearings. The single-row type has little thrust capacity but the double-row bearings can carry thrust up to about 30 percent of its radial load capacity. The angular contact thrust type can carry large thrust loads in one direction.

Representative tapered roller bearings are shown in Fig. 14.7. Details of the geometry are shown in Fig. 14.8. Note that conical elements (extended) of the rollers and races intersect at a common apex on the centerline of rotation. Pairs of single-row tapered roller bearings are often used for wheel bearings and other applications. The double-row and four-row types are used singly to replace a pair of single-row bearings or to carry heavier loads. Space does not permit further discussion of tapered roller bearings in this book. Selection and analysis procedures are generally similar to those presented here for other bearing types. Detailed information is available in manufacturers' catalogues and other references.

Because of their geometry, needle bearings (Fig. 14.9) have the highest load capacity for a given radial space of all rolling-element bearings.

Rolling-element bearings are available in a multitude of special forms for various purposes. Details are found in the literature published by the various bearing manufac-

(a) Nonlocating

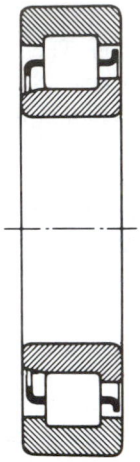

(b) One–direction locating

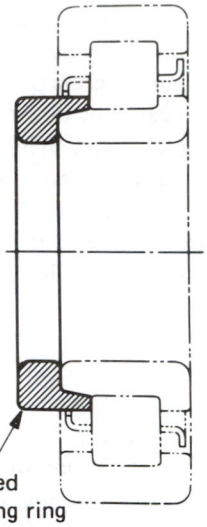

(c) Two–direction locating

(d) Thrust

FIGURE 14.5.
Cylindrical roller bearings. (Courtesy Hoover-NSK Bearing Co.)

(a) Double-row convex

(b) Thrust

FIGURE 14.6.
Spherical roller bearings. (Courtesy Hoover-NSK Bearing Co.)

(a) Single-row

(b) Double-row

(c) Four-row

FIGURE 14.7.
Tapered roller bearings. (Courtesy Hoover-NSK Bearing Co.)

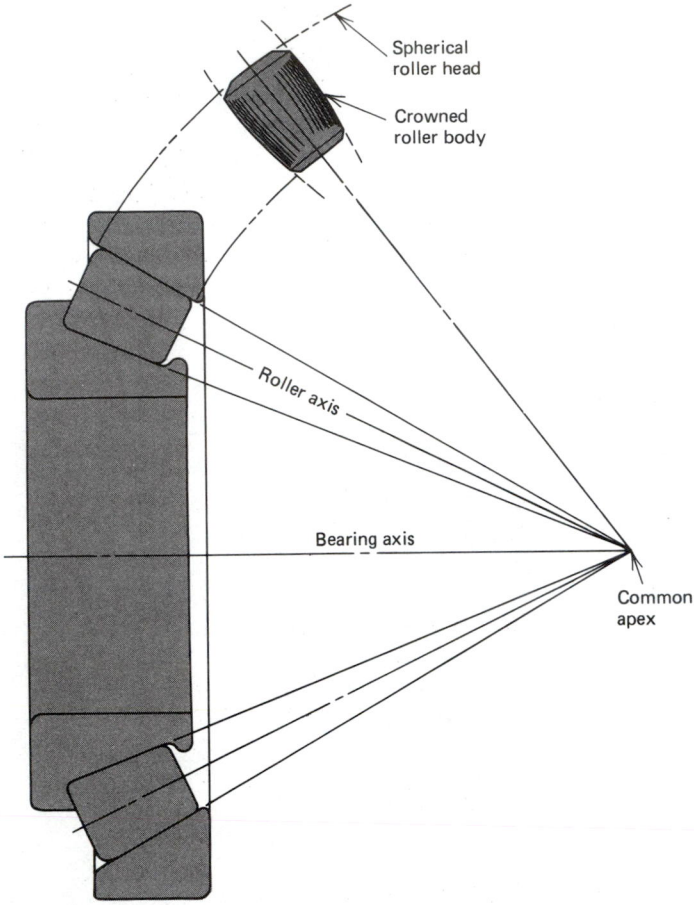

FIGURE 14.8.
Tapered roller bearing geometry. (Courtesy Federal Mogul Corp.)

turers. A few representative samples are illustrated in Fig. 14.10. The *adapter bearing* (part *a*) can be economically mounted on commercial steel shafting without machining a bearing seat. The *pillow block* mounting (part *b*) is a common arrangement used to support a rotating shaft parallel to a flat surface. The *flange bearing* (part *c*) provides for supporting a rotating shaft perpendicular to a flat surface. The *idler sheave* (part *d*) shows an inexpensive unground ball bearing built with the outer race sized and contoured to accommodate a belt. The *rod end bearing* (part *e*) is used for aircraft controls and miscellaneous machinery and mechanism applications. The needle bearing *cam follower* (part *f*) has a heavy outer ring to withstand high cam contact forces. The integral shaft bearing (part *g*) is used in automotive water pumps, lawn mowers, power saws, and other applications. The integral spindle and bearing (part *h*) represents a relatively new development for mounting wheels, pulleys, idler gears, and the like. The ball bushing (part *i*) and roller chain bearing (part *j*) illustrate less conventional applications of rolling-bearing elements. Also, recall the ball-bearing screw assembly shown in Fig. 10.10.

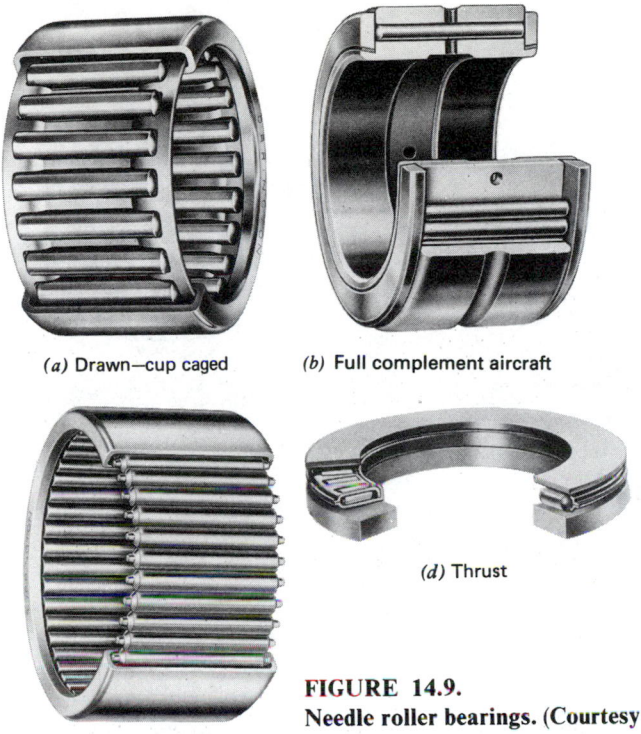

(a) Drawn–cup caged (b) Full complement aircraft

(c) Full–complement drawn–cup (d) Thrust

FIGURE 14.9.
Needle roller bearings. (Courtesy The Torrington Co.)

14.4 DESIGN OF ROLLING-ELEMENT BEARINGS

The detailed design of rolling-element bearings is a sophisticated engineering endeavor too specialized to warrant extensive treatment here. On the other hand, the selection and use of these bearings is of concern to virtually all engineers dealing with any kind of machinery. For this reason, the present section attempts only to point out a few of the fundamentals involved in bearing design so that the reader may have some appreciation of what is involved. The following sections deal in more specific detail with the selection and application of rolling-element bearings.

The underlying technology involved in rotating-element bearings is that associated with curved surface contact stresses and related fatigue failures, covered briefly in Sections 9.13 and 9.14. Figure 14.1*c* illustrates the general geometry of ball-bearing contacting surfaces. The selection of bearing race curvature is critical. A radius only a trifle larger than that of the ball gives a relatively large area of contact (on elastic deflection due to load) and low contact stresses. But different portions of this contact area are at various radii from the axis of rotation. This causes sliding, with resultant friction and wear. Thus, the radius selected (commonly about 104 percent of ball radius for the inner race; slightly more for the outer race) is a compromise between load supporting area and sliding friction.

Material selection in rolling-element bearings is also critical. Since about 1920,

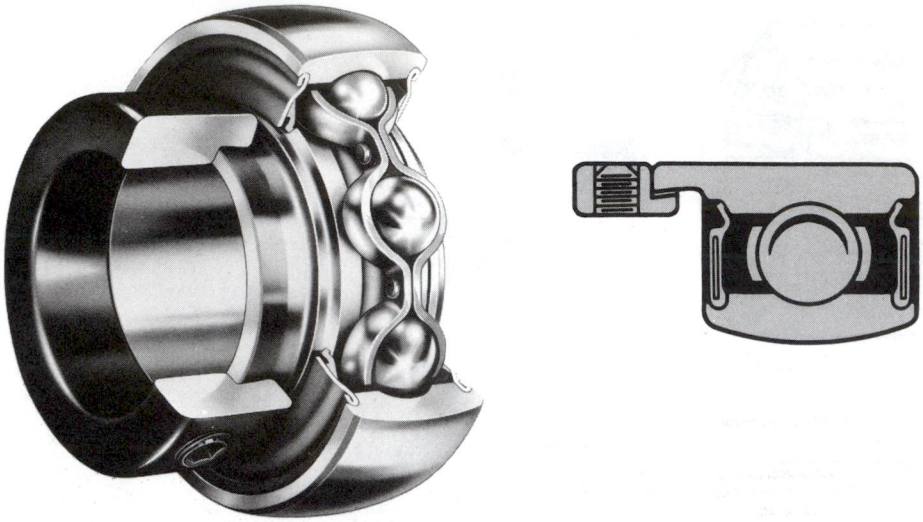

(*a*) Adapter bearing
(Courtesy New Departure—Hyatt Bearing Div.)

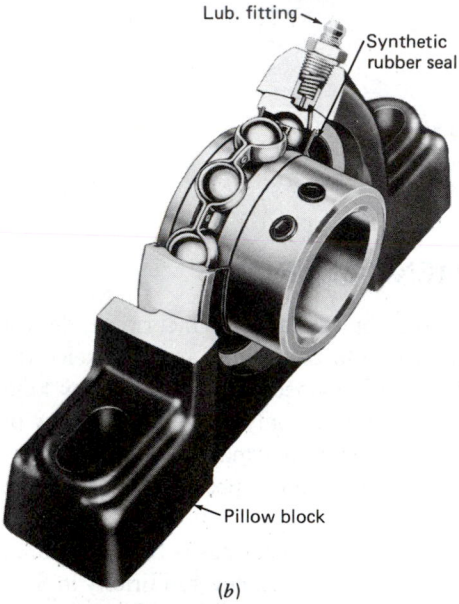

(*b*)
Pillow block mounting
(Courtesy Reliance Electric Co.)

(*c*)
Flange bearing
(Courtesy Reliance Electric Co.)

FIGURE 14.10.
Sample special bearings.

most ball bearing rings and balls have been made of high-carbon chrome steel, SAE 52100, through-hardened to 58-65 Rockwell C. Special heat-treatment is sometimes used to produce favorable residual stresses in the contacting surfaces. Roller-bearing

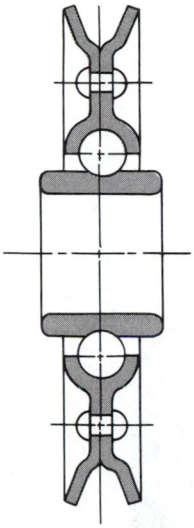

(*d*) Idler sheave
(unground bearing)

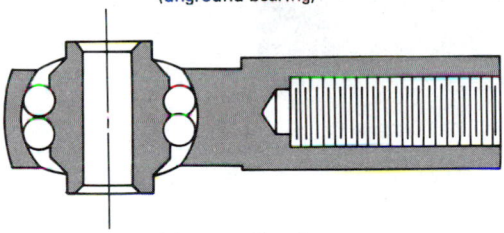

(*e*) Rod end bearing

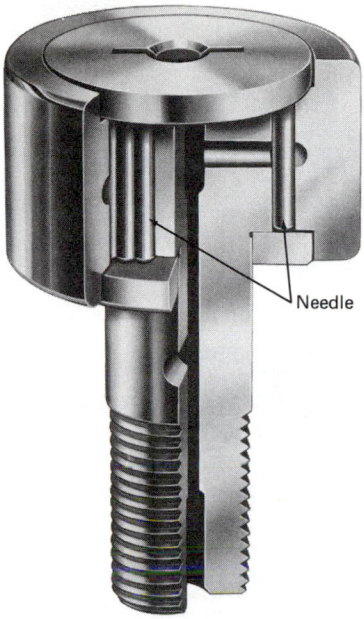

(*f*)
Needle bearing cam follower
(Courtesy The Torrington Co.)

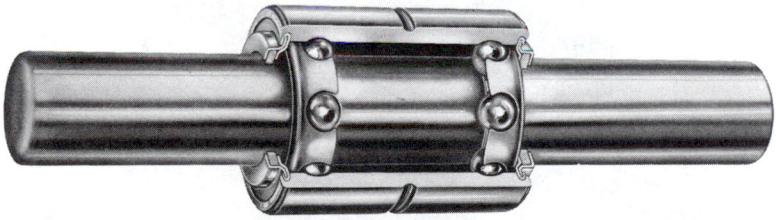

(*g*) Integral shaft bearing
(Courtesy New Departure–Hyatt Division, General Motors Corp.)

FIGURE 14.10 (*continued*)

components are more often made of carburized alloy steel. Surface compressive residual stresses are inherent with carburization. Cleanliness of the steel is of extreme importance, and for this reason nearly all bearing grade steels are vacuum-degassed.

The design of the rings for desired rigidity is very important. Ring and ball (or

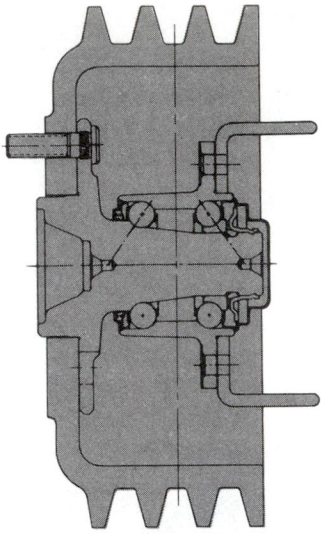

(h) Integral spindle, shown with V-belt pulley (Courtesy New Departure – Hyatt Division General Motors Corp.)

(i) Ball bushing

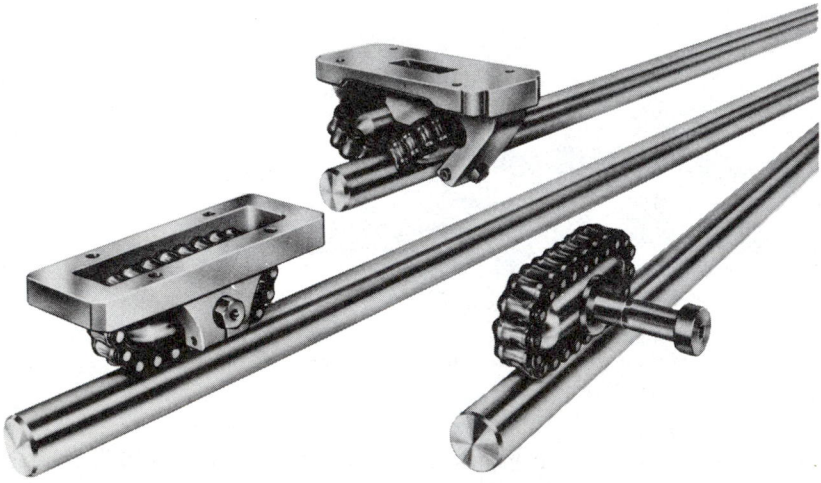

(j) Recirculating roller chain bearing

FIGURE 14.10 (*continued*) (*i, j* **Courtesy Thompson Industries, Inc.**)

roller) deflection, rotating speed, and lubricant characteristics combine to determine local stress distribution in the contact area. This takes the bearing engineer into the area of elastohydrodynamics (Section 13.16).

Manufacturing tolerances are extremely critical. In the case of ball bearings, the Annular Bearing Engineers' Committee (ABEC) of the Anti-Friction Bearing Manufacturers Association (AFBMA) has established four primary grades of precision, designated ABEC 1, 5, 7, and 9. ABEC 1 is the standard grade, and is adequate for most normal applications. The other grades have progressively finer tolerances. For example, tolerances on bearing bores between 35 and 50 mm range from (+0.0000 in. to −0.0005 in.) for ABEC Grade 1 to (+0.00000 in. to −0.00010 in.) for ABEC Grade 9. Tolerances on other dimensions are comparable. Similarly, the AFBMA Roller Bearing Engineers Committee has established RBEC standards 1 and 5 for cylindrical roller bearings. (To help visualize how small these tolerances are, compare them with the thickness of the paper in this book, as conveniently determined by measuring the thickness of 100 sheets, or 200 pages.)

For detailed information relating to the design of rolling-element bearings, see [1, 3, 4].

14.5 FITTING OF ROLLING-ELEMENT BEARINGS

Normal practice involves fitting the stationary ring with a "slip" or "tap" fit and the rotating ring with enough interference to prevent relative motion during operation. Recommended fits depend on bearing type, size, and tolerance grade. For example, a typical ABEC-1 ball-bearing fit would be in the range of 0.0005-in. clearance for the stationary ring and 0.0005-in. interference for the rotating ring. Manufacturing tolerances on the shaft and housing bearing interface dimensions are typically in the range of 0.0003 in. for ABEC-1 ball bearings. Proper fits and tolerances are influenced by the radial stiffness of the shaft and housing, and sometimes by thermal expansion.

It is important to recognize that bearing fit pressures influence the internal fits between the balls or rollers and their races. Too tight a fit can cause internal interference fits that shorten the bearing life.

Care must be taken in the installation and removal of bearings to insure that the required forces are applied directly to the bearing ring involved. If these forces are transmitted through the bearing (as when applying a force on the outer race to force the bearing into place on the shaft), damage to the bearing can result. Interference-fit installations are sometimes facilitated by heating the outer member and/or "freezing" the inner member, as by packing in "dry ice" (solidified carbon dioxide). Any heating of the bearing must not be sufficient to damage the steel or any pre-installed lubricant.

Detailed information regarding bearing fits is contained in bearing manufacturers' literature and in ANSI and AFBMA standards. Adhesives (see Section 11.8) are sometimes used together with slightly reduced tolerances on the shaft and housing bore dimensions, particularly for nonprecision applications.

14.6 "CATALOG INFORMATION" FOR ROLLING-ELEMENT BEARINGS

Bearing manufacturers' catalogues identify bearings by number, give complete dimensional information, list rated load capacities, and furnish details concerning mounting, lubrication, and operation. Dimensions of the more common series of radial ball bearings, angular ball bearings, and cylindrical roller bearings are given in Table 14.1 and illustrated in Fig. 14.11. For bearings of these types having bores of 20 mm and larger, the bore diameter is five times the last two digits in the bearing number. For example, No. L08 is an extra-light series bearing with a 40-mm bore, No. 316 is a medium series with an 80-mm bore, and so on. Actual bearing numbers include additional letters and/or numbers to provide more information. Many bearing varieties are also available in inch series.

Table 14.2 lists *rated load capacities, C.* These values correspond to a constant radial load that 90 percent of a group of presumably identical bearings can endure for 9×10^7 revolutions (as 3000 hours of 500 rpm operation) without the onset of surface fatigue failures of the kind illustrated in Fig. 9.18*b*. *Caution:* Rated capacities given by different bearing manufacturers are not always directly comparable. The basis for ratings must always be checked.

14.7 BEARING SELECTION

A specific bearing application requires the selection of bearing type, grade of precision (usually ABEC 1), lubricant, closure (i.e., open, shielded, or sealed), and basic load rating. Often, special considerations must be taken into account. For example, if the bearing will be subjected to a heavy load when not rotating, its *static load capacity* (given in bearing manufacturers' catalogues) should not be exceeded. Otherwise, the

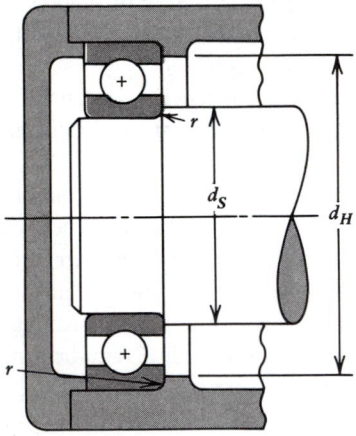

FIGURE 14.11.
Shaft and housing shoulder dimensions.

TABLE 14.1
Bearing Dimensions

		Ball Bearings					Roller Bearings				
Bearing Basic No.	**Bore (mm)**	**OD (mm)**	***w* (mm)**	***r*[a] (mm)**	**d_S (mm)**	**d_H (mm)**	***OD* (mm)**	***w* (mm)**	***r*[a] (mm)**	**d_S (mm)**	**d_H (mm)**
L00	10	26	8	0.30	12.7	23.4					
200	10	30	9	0.64	13.8	26.7					
300	10	35	11	0.64	14.8	31.2					
L01	12	28	8	0.30	14.5	25.4					
201	12	32	10	0.64	16.2	28.4					
301	12	37	12	1.02	17.7	32.0					
L02	15	32	9	0.30	17.5	29.2					
202	15	35	11	0.64	19.0	31.2					
302	15	42	13	1.02	21.2	36.6					
L03	17	35	10	0.30	19.8	32.3	35	10	0.64	20.8	32.0
203	17	40	12	0.64	22.4	34.8	40	12	0.64	20.8	36.3
303	17	47	14	1.02	23.6	41.1	47	14	1.02	22.9	41.4
L04	20	42	12	0.64	23.9	38.1	42	12	0.64	24.4	36.8
204	20	47	14	1.02	25.9	41.7	47	14	1.02	25.9	42.7
304	20	52	15	1.02	27.7	45.2	52	15	1.02	25.9	46.2
L05	25	47	12	0.64	29.0	42.9	47	12	0.64	29.2	43.4
205	25	52	15	1.02	30.5	46.7	52	15	1.02	30.5	47.0
305	25	62	17	1.02	33.0	54.9	62	17	1.02	31.5	55.9
L06	30	55	13	1.02	34.8	49.3	47	9	0.38	33.3	43.9
206	30	62	16	1.02	36.8	55.4	62	16	1.02	36.1	56.4
306	30	72	19	1.02	38.4	64.8	72	19	1.52	37.8	64.0

TABLE 14.1
Bearing Dimensions (*continued*)

		Ball Bearings					Roller Bearings				
Bearing Basic No.	**Bore (mm)**	**OD (mm)**	***w* (mm)**	***r*[a] (mm)**	**d_S (mm)**	**d_H (mm)**	***OD* (mm)**	***w* (mm)**	***r*[a] (mm)**	**d_S (mm)**	**d_H (mm)**
L07	35	62	14	1.02	40.1	56.1	55	10	0.64	39.4	50.8
207	35	72	17	1.02	42.4	65.0	72	17	1.02	41.7	65.3
307	35	80	21	1.52	45.2	70.4	80	21	1.52	43.7	71.4
L08	40	68	15	1.02	45.2	62.0	68	15	1.02	45.7	62.7
208	40	80	18	1.02	48.0	72.4	80	18	1.52	47.2	72.9
308	40	90	23	1.52	50.8	80.0	90	23	1.52	49.0	81.3
L09	45	75	16	1.02	50.8	68.6	75	16	1.02	50.8	69.3
209	45	85	19	1.02	52.8	77.5	85	19	1.52	52.8	78.2
309	45	100	25	1.52	57.2	88.9	100	25	2.03	55.9	90.4
L10	50	80	16	1.02	55.6	73.7	72	12	0.64	54.1	68.1
210	50	90	20	1.02	57.7	82.3	90	20	1.52	57.7	82.8
310	50	110	27	2.03	64.3	96.5	110	27	2.03	61.0	99.1
L11	55	90	18	1.02	61.7	83.1	90	18	1.52	62.0	83.6
211	55	100	21	1.52	65.0	90.2	100	21	2.03	64.0	91.4
311	55	120	29	2.03	69.8	106.2	120	29	2.03	66.5	108.7
L12	60	95	18	1.02	66.8	87.9	95	18	1.52	67.1	88.6
212	60	110	22	1.52	70.6	99.3	110	22	2.03	69.3	101.3
312	60	130	31	2.03	75.4	115.6	130	31	2.54	72.9	117.9
L13	65	100	18	1.02	71.9	92.7	100	18	1.52	72.1	93.7
213	65	120	23	1.52	76.5	108.7	120	23	2.54	77.0	110.0
313	65	140	33	2.03	81.3	125.0	140	33	2.54	78.7	127.0

L14	70	110	20	1.02	77.7	102.1	110	20	Not Available		
214	70	125	24	1.52	81.0	114.0	125	24	2.54	81.8	115.6
314	70	150	35	2.03	86.9	134.4	150	35	3.18	84.3	135.6
L15	75	115	20	1.02	82.3	107.2	115	20	Not Available		
215	75	130	25	1.52	86.1	118.9	130	25	2.54	85.6	120.1
315	75	160	37	2.03	92.7	143.8	160	37	3.18	90.4	145.8
L16	80	125	22	1.02	88.1	116.3	125	22	2.03	88.4	117.6
216	80	140	26	2.03	93.2	126.7	140	26	2.54	91.2	129.3
316	80	170	39	2.03	98.6	152.9	170	39	3.18	96.0	154.4
L17	85	130	22	1.02	93.2	121.4	130	22	2.03	93.5	122.7
217	85	150	28	2.03	99.1	135.6	150	28	3.18	98.0	139.2
317	85	180	41	2.54	105.7	160.8	180	41	3.96	102.9	164.3
L18	90	140	24	1.52	99.6	129.0	140	24	Not Available		
218	90	160	30	2.03	104.4	145.5	160	30	3.18	103.1	147.6
318	90	190	43	2.54	111.3	170.2	190	43	3.96	108.2	172.7
L19	95	145	24	1.52	104.4	134.1	145	24	Not Available		
219	95	170	32	2.03	110.2	154.9	170	32	3.18	109.0	157.0
319	95	200	45	2.54	117.3	179.3	200	45	3.96	115.1	181.9
L20	100	150	24	1.52	109.5	139.2	150	24	2.54	109.5	141.7
220	100	180	34	2.03	116.1	164.1	180	34	3.96	116.1	167.1
320	100	215	47	2.54	122.9	194.1	215	47	4.75	122.4	194.6
L21	105	160	26	2.03	116.1	146.8	160	26	Not Available		
221	105	190	36	2.03	121.9	173.5	190	36	3.96	121.4	175.3
321	105	225	49	2.54	128.8	203.5	225	49	4.75	128.0	203.5

TABLE 14.1
Bearing Dimensions (*continued*)

		Ball Bearings					Roller Bearings				
Bearing Basic No.	**Bore (mm)**	**OD (mm)**	***w* (mm)**	***r*[a] (mm)**	***d_S* (mm)**	***d_H* (mm)**	***OD* (mm)**	***w* (mm)**	***r*[a] (mm)**	***d_S* (mm)**	***d_H* (mm)**
L22	110	170	28	2.03	122.7	156.5	170	28	2.54	121.9	159.3
222	110	200	38	2.03	127.8	182.6	200	38	3.96	127.3	183.9
322	110	240	50	2.54	134.4	218.2	240	50	4.75	135.9	217.2
L24	120	180	28	2.03	132.6	166.6	180	28	Not Available		
224	120	215	40	2.03	138.2	197.1	215	40	4.75	139.2	198.9
324	120	Not Available					260	55	6.35	147.8	235.2
L26	130	200	33	2.03	143.8	185.4	200	33	3.18	143.0	188.2
226	130	230	40	2.54	149.9	210.1	230	40	4.75	149.1	213.9
326	130	280	58	3.05	160.0	253.0	280	58	6.35	160.3	254.5
L28	140	210	33	2.03	153.7	195.3	210	33	Not Available		
228	140	250	42	2.54	161.5	228.6	250	42	4.75	161.5	232.4
328	140	Not Available					300	62	7.92	172.0	271.3
L30	150	225	35	2.03	164.3	209.8	225	35	3.96	164.3	212.3
230	150	270	45	2.54	173.0	247.6	270	45	6.35	174.2	251.0
L32	160	240	38	2.03	175.8	223.0	240	38	Not Available		
232	160	Not Available					290	48	6.35	185.7	269.5
L36	180	280	46	2.03	196.8	261.6	280	46	4.75	199.6	262.9
236	180	Not Available					320	52	6.35	207.5	298.2

L40	200		310	51		Not Available	
240	200	Not Available	360	58	7.92	232.4	334.5
L44	220		340	56		Not Available	
244	220	Not Available	400	65	9.52	256.0	372.1
L48	240		360	56		Not Available	
248	240	Not Available	440	72	9.52	279.4	408.4

[a]Maximum fillet radius on shaft and in housing that will clear the bearing corner radius.

TABLE 14.2
Bearing Rated Capacities—90 × 10^6 Revolution Life with 90 Percent Reliability

	Radial Ball, $\alpha = 0°$			Angular Ball, $\alpha = 25°$			Roller		
Bore (mm)	L00 Xlt (kN)	200 lt (kN)	300 med (kN)	L00 Xlt (kN)	200 lt (kN)	300 med (kN)	1000 Xlt (kN)	1200 lt (kN)	1300 med (kN)
10	1.02	1.42	1.90	1.02	1.10	1.88			
12	1.12	1.42	2.46	1.10	1.54	2.05			
15	1.22	1.56	3.05	1.28	1.66	2.85			
17	1.32	2.70	3.75	1.36	2.20	3.55	2.12	3.80	4.90
20	2.25	3.35	5.30	2.20	3.05	5.80	3.30	4.40	6.20
25	2.45	3.65	5.90	2.65	3.25	7.20	3.70	5.50	8.50
30	3.35	5.40	8.80	3.60	6.00	8.80	2.40*	8.30	10.0
35	4.20	8.50	10.6	4.75	8.20	11.0	3.10*	9.30	13.1
40	4.50	9.40	12.6	4.95	9.90	13.2	7.20	11.1	16.5
45	5.80	9.10	14.8	6.30	10.4	16.4	7.40	12.2	20.9
50	6.10	9.70	15.8	6.60	11.0	19.2	5.10*	12.5	24.5
55	8.20	12.0	18.0	9.00	13.6	21.5	11.3	14.9	27.1
60	8.70	13.6	20.0	9.70	16.4	24.0	12.0	18.9	32.5
65	9.10	16.0	22.0	10.2	19.2	26.5	12.2	21.1	38.3
70	11.6	17.0	24.5	13.4	19.2	29.5		23.6	44.0
75	12.2	17.0	25.5	13.8	20.0	32.5		23.6	45.4
80	14.2	18.4	28.0	16.6	22.5	35.5	17.3	26.2	51.6
85	15.0	22.5	30.0	17.2	26.5	38.5	18.0	30.7	55.2
90	17.2	25.0	32.5	20.0	28.0	41.5		37.4	65.8
95	18.0	27.5	38.0	21.0	31.0	45.5		44.0	65.8
100	18.0	30.5	40.5	21.5	34.5		20.9	48.0	72.9
105	21.0	32.0	43.5	24.5	37.5			49.8	84.5
110	23.5	35.0	46.0	27.5	41.0	55.0	29.4	54.3	85.4
120	24.5	37.5		28.5	44.5			61.4	100.1
130	29.5	41.0		33.5	48.0	71.0	48.9	69.4	120.1
140	30.5	47.5		35.0	56.0			77.4	131.2
150	34.5			39.0	62.0		58.7	83.6	
160								113.4	
180	47.0			54.0			97.9	140.1	
200								162.4	
220								211.3	
240								258.0	

Source: New Departure Hyatt Div., General Motors Corp.

*1000 (X1t) series bearings not available in these sizes. Capacities shown are for 1900 (XX1t) series.

balls or rollers will slightly indent the rings (this is called brinelling because the indentations resemble marks produced by a Brinell hardness tester), causing noisy operation upon subsequent rotation. (If noise is not objectionable, the static capacity can often be exceeded by a factor of up to 3.) It is interesting that similar extremely slight indentation during rotation is not harmful because it leaves the ring surfaces smooth and annular.

Another special consideration is maximum speed. The limitation is one of linear surface speed rather than rotating speed; hence, small bearings can operate at higher rpm than large bearings. Lubrication is especially important in high-speed bearing applications, the best being a fine oil mist or spray. This provides the necessary lubricant film and carries away friction heat with a minimum "churning loss" within the lubricant itself. For ball bearings, nonmetallic separators permit highest speeds. ABEC 1 precision single-row ball bearings with nonmetallic separators and oil mist lubrication can run at inner ring surface speeds up to 75 m/s and have a life of 3000 hours while carrying 1/3 rated load capacity. This translates to a *DN value* (bore diameter in mm times rpm) of about 1.25×10^6. For oil drip or splash lubrication this figure is reduced by about 1/3, and for grease lubrication by about 2/3. Under the most favorable conditions, roller bearings can operate up to a *DN* value of about 450,000. For applications involving extreme rotating speeds, it is advisable to consult the bearing manufacturer.

When selecting bearings, attention should be given to possible misalignment and to sealing and lubrication. If extreme temperatures are involved, the bearing manufacturer should be consulted.

The size of bearing selected for an application is usually influenced by the size of shaft required (for strength and rigidity considerations) and by the available space. Also, the bearing must have a high enough load rating to provide an acceptable combination of life and reliability. The major factors influencing the load rating requirement are discussed below.

14.7.1 Life Requirement

Bearing applications usually require lives different from that used for the catalogue rating. Palmgren [4] determined that ball bearing life varies inversely with approximately the third power of the load. Later studies have indicated that this exponent ranges between 3 and 4 for various rolling-element bearings. Many manufacturers retain Palmgren's exponent of 3 for ball bearings and use $^{10}/_{3}$ for roller bearings. Following the recommendation of other manufacturers, this book will use the exponent $^{10}/_{3}$ for both bearing types. Thus,

$$L = L_R(C/F_r)^{3.33} \tag{14.1a}$$

or

$$C_{\text{req}} = F_r(L/L_R)^{0.3} \tag{14.1b}$$

where

C = rated capacity (as from Table 14.2) and C_{req} = the required value of C for the application,

L_R = life corresponding to rated capacity (as 9×10^7 revolutions),

F_r = radial load involved in the application,

L = life corresponding to radial load F_r, or life required by the application.

Thus, doubling the load on a bearing reduces its life by a factor of about 10.

Different manufacturers' catalogues use different values of L_R. Some use $L_R = 10^6$ revolutions. A quick calculation shows that the values in Table 14.2 must be multiplied by 3.86 to be comparable with ratings based on a life of 10^6 revolutions.

14.7.2 Reliability Requirement

Tests show that the *median* life of rolling-element bearings (ball bearings in particular) is about five times the standard 10 percent failure fatigue life. The standard life is commonly designated as the L_{10} *life* (sometimes as the B_{10} life). Since this life corresponds to 10 percent failures, it also means that this is the life for which 90 percent have *not* failed, and corresponds to *90 percent reliability*. Thus, the life for 50 percent reliability is about five times the life for 90 percent reliability.

Many designs require greater than 90 percent reliability. The distribution of fatigue lives of a group of presumably identical parts does not correspond to the normal distribution curve, discussed in Section 6.14 and illustrated in Fig. 6.15. Rather, fatigue lives characteristically have a skewed distribution, as in Fig. 14.12. This corresponds generally with a mathematical formula proposed by W. Weibull of Sweden, known as the *Weibull distribution.* Using the general Weibull equation together with extensive experimental data, the AFBMA has formulated recommended *life adjustment reliability factors,* K_r, plotted in Fig. 14.13. These factors are applicable to both ball and roller

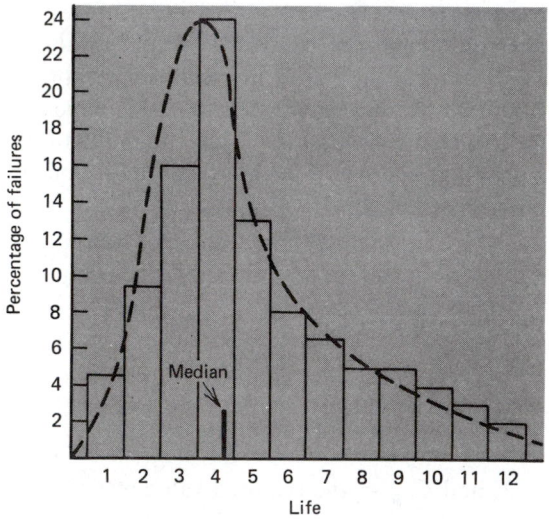

FIGURE 14.12.
General pattern of bearing fatigue-life distribution.

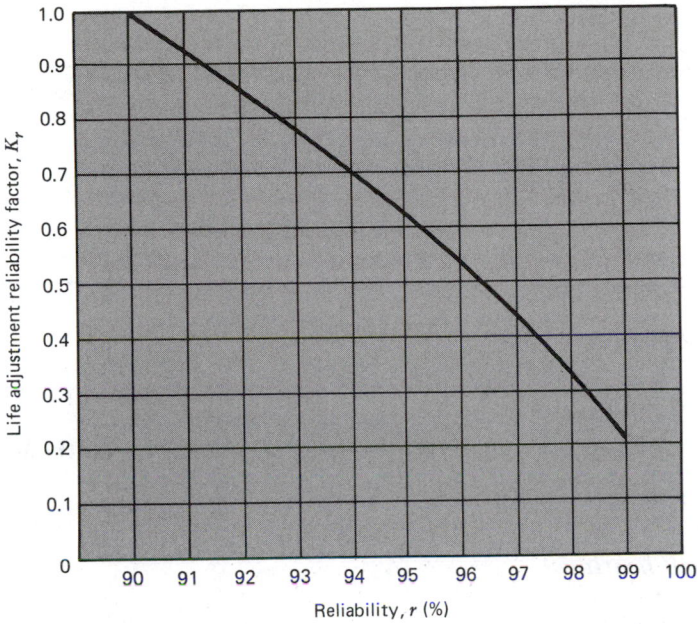

FIGURE 14.13.
Reliability factor, K_r.

bearings. The rated bearing life for any given reliability (greater than 90 percent) is thus the product, $K_r L_R$. Incorporating this factor into Eq. (14.1) gives

$$L = K_r L_R (C/F_r)^{3.33} \tag{14.2a}$$

$$C_{\text{req}} = F_r (L/K_r L_R)^{0.3} \tag{14.2b}$$

14.7.3 Influence of Axial Loading

Cylindrical roller bearings are very limited in their thrust capacity because axial loads result in sliding friction at the roller ends. Even so, when these bearings are properly aligned, radially loaded, and oil lubricated, they can carry thrust loads up to 20 percent of their rated radial capacities. This enables pairs of cylindrical roller bearings to support shafts subjected to light thrust, as by spur gears or chain sprockets. Tapered roller bearings can, of course, carry substantial thrust as well as radial loads.

For ball bearings, any combination of radial load (F_r) and thrust load (F_t) results in approximately the same life as does a pure radial *equivalent load*, F_e, calculated from the equations below. Load angle α is defined in Fig. 14.3*b*. Radial bearings have a zero load angle. Standard values of α for angular ball bearings are 15, 25, and 35 deg. Space permits including only 25 deg. angular ball bearings in this book.

TABLE 14.3
Application Factors, K_a

Type of Application	Ball Bearing	Roller Bearing
Uniform load, no impact	1.0	1.0
Gearing	1.0–1.3	1.0
Light impact	1.2–1.5	1.0–1.1
Moderate impact	1.5–2.0	1.1–1.5
Heavy impact	2.0–3.0	1.5–2.0

α = 0 deg. (radial ball bearings)

$$\left.\begin{aligned} &\text{For } 0 < F_t/F_r < 0.35, \quad F_e = F_r. \\ &\text{For } 0.35 < F_t/F_r < 10, \quad F_e = F_r\left[1 + 1.115\left(\frac{F_t}{F_r} - 0.35\right)\right]. \\ &\text{For } F_t/F_r > 10, \quad F_e = 1.176F_t. \end{aligned}\right\} \quad \textbf{(14.3)}^2$$

α = 25 deg. (angular ball bearings)

$$\left.\begin{aligned} &\text{For } 0 < F_t/F_r < 0.68, \quad F_e = F_r. \\ &\text{For } 0.68 < F_t/F_r < 10, \quad F_e = F_r\left[1 + 0.870\left(\frac{F_t}{F_r} - 0.68\right)\right]. \\ &\text{For } F_t/F_r > 10, \quad F_e = 0.911F_t. \end{aligned}\right\} \quad \textbf{(14.4)}^2$$

14.7.4 Shock Loading

The standard bearing rated capacity involves a uniform load without shock. While this desirable condition may apply in some applications (as bearings on the motor and rotor shafts of a belt-driven electric blower), other applications involve various degrees of shock loading. This has the effect of increasing the nominal load by an *application factor, K_a*. Experience within the specific industry involved is the best guide. Table 14.3 gives representative sample values.

14.7.5 Summary

By substituting F_e for F_r and adding K_a, Eq. (14.2) is modified to give

$$L = K_r L_R (C/F_e K_a)^{3.33} \quad \textbf{(14.5a)}$$

$$C_{\text{req}} = F_e K_a (L/K_r L_R)^{0.3} \quad \textbf{(14.5b)}$$

In using the above equations, the question arises as to what life, L, should be

[2]Derived from "Ball Bearing General Catalog" (BC-7, 1980), New Departure Hyatt Bearings Div. of General Motors Corp., Sandusky, Ohio.

TABLE 14.4
Representative Bearing Design Lives

Type of Application	Design life (thousands of hours)
Instruments and apparatus for infrequent use	0.1–0.5
Machines intermittently used, where service interruption is of minor importance	4–8
Machines intermittently used, where reliability is of great importance	8–14
Machines for 8-hour service, but not every day	14–20
Machines for 8-hour service, every working day	20–30
Machines for continuous 24-hour service	50–60
Machines for continuous 24-hour service where reliability is of extreme importance	100–200

required. Table 14.4 may be used as a guide when more specific information is not available. (It is worth noting that the useful life of a bearing in industrial applications where noise is not a factor may extend significantly beyond the appearance of the first small area of surface fatigue damage, which is the failure criterion in standard tests.)

Bearing manufacturers formerly reduced life ratings when the outer ring rotated relative to the load (as with a trailer wheel, rotating around a fixed spindle). As a result of more recent evidence, this is no longer done. If both rings rotate, the relative rotation between the two is used in making life calculations.

Many applications involve loads that vary with time. In such cases, the Palmgren[3] linear cumulative damage rule (Section 8.12) is applicable.

SAMPLE PROBLEM 14.1

Select a ball bearing for an industrial machine intended for continuous one-shift (8-hour day) operation at 1800 rpm. Radial and thrust loads are 1.2 and 1.5 kN, respectively, with light-to-moderate impact.

Solution

1. From Eqs. (14.3) and (14.4), the equivalent radial load for radial and angular ball bearings, respectively, are

$$F_e = 1.2\left[1 + 1.115\left(\frac{1.5}{1.2} - 0.35\right)\right] = 2.4 \text{ kN (radial bearing)}$$

$$F_e = 1.2\left[1 + 0.87\left(\frac{1.5}{1.2} - 0.68\right)\right] = 1.8 \text{ kN (angular bearing)}$$

[3]Named after Arvid Palmgren, author of Ref. 4.

2. From Table 14.3, choose $K_a = 1.5$. From Table 14.4 choose (conservatively) 30,000-hour life. Life in revolutions is $L = 1800 \text{ rpm} \times 30{,}000 \text{ hr} \times 60 \text{ min/hr} = 3240 \times 10^6$ revs.
3. For standard 90 percent reliability ($K_r = 1$), and for $L_R = 90\text{x}10^6$ revolutions (for use with Table 14.2), Eq. (14.5b) gives

$$C_{\text{req}} = (2.4)(1.5)(3240/90)^{0.3} = 10.55 \text{ kN (radial bearing)}$$
$$C_{\text{req}} = (1.8)(1.5)(3240/90)^{0.3} = 7.91 \text{ kN (angular bearing)}$$

4. From Table 14.2 (with bearing number for a given bore and series obtained from Table 14.1), appropriate choices would be radial bearings L14, 211, and 307, and angular-contact bearings L11, 207, and 306. Other factors being equal, the final selection would be made on the basis of cost of the total installation, including shaft and housing. Shaft size should be sufficient to limit bearing misalignment to no more than 15′. ■

SAMPLE PROBLEM 14.2

Suppose that radial-contact bearing 211 ($C = 12.0$ kN) is selected for the application in Sample Problem 14.1. (a) Estimate the life of this bearing, with 90 percent reliability. (b) Estimate its reliability for 30,000-hour life.

Solution

(a) From Eq. (14.5a),

$$L = (1)(90 \times 10^6)(12.0/3.6)^{3.33} = 4959 \times 10^6 \text{ revs} = 45{,}920 \text{ hr}$$ ■

(b) From Eq. (14.5a),

$$3240 \times 10^6 = K_r(90 \times 10^6)(12.0/3.6)^{3.33}$$
$$K_r = 0.65$$

From Fig. 14.13, reliability is estimated as close to 95 percent. ■

14.8 BEARING MOUNTING FOR PROPER THRUST-LOAD PROVISION

Bearing manufacturers' literature contains extensive information and illustrations pertaining to the proper application of their products. We will illustrate here only the basic principle of mounting bearings properly with respect to the thrust loading. Figures 14.14 and 14.15 illustrate two typical constructions. Even though there are no apparent axial loads on the rotating assembly, it is necessary to insure that axial movement does not take place due to gravity, vibration, and so on.

The principle that normally applies is that thrust in each direction must be carried by *one* and *only one* bearing. In Fig. 14.14, the left bearing takes thrust to the left, the

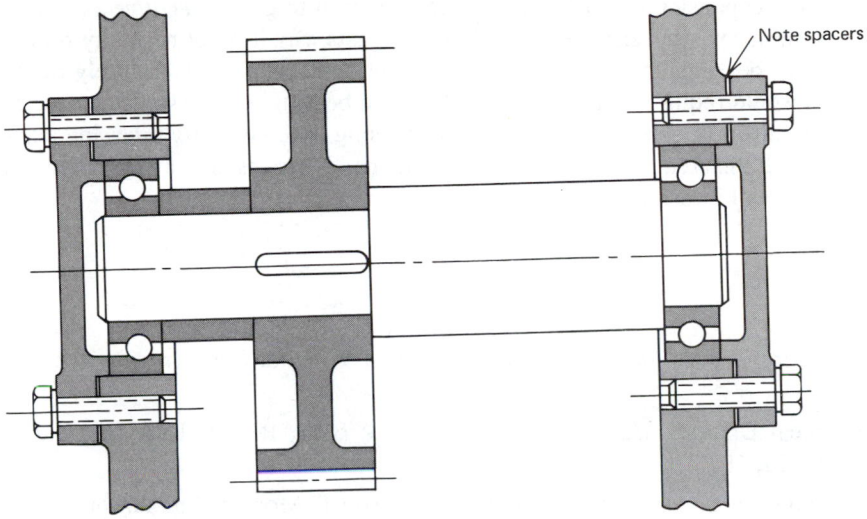

FIGURE 14.14.
Bearings mounted so that each bearing takes thrust in one direction.

right bearing thrust to the right. Neither bearing is mounted so as to be able to take thrust in the opposite direction. In Fig. 14.15, the left bearing takes thrust in both directions. The right bearing is free to slide both ways in the housing; thus, it can take thrust in neither direction.

Note in Fig. 14.14 that spacers are used to permit the axial free play of the shaft to be adjusted to the desired amount—just enough so that under no conditions of dif-

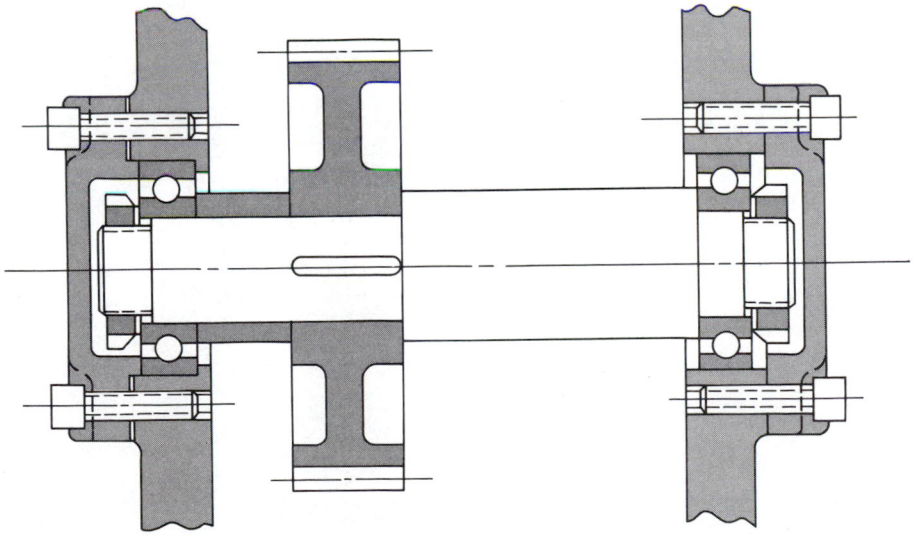

FIGURE 14.15.
Bearings mounted so that left bearing takes thrust in both directions.

ferential thermal expansion can the shaft elongate enough to give a *negative* axial free play (and resultant severe bearing loading). This suggests why it is not normally proper to design for two bearings to share thrust in the same direction. For "absolutely rigid" members, this would require that the distance between bearings on the shaft be *exactly* the same as the distance between bearings in the housing. Even if costly precision manufacture satisfied this requirement, differential thermal expansion during operation would create a small interference, thereby loading the bearings.

REFERENCES

1. Harris, Tedric A., *Rolling Bearing Analysis,* John Wiley & Sons, Inc., New York, 1966.
2. "Mechanical Drives," *Machine Design Reference Issue,* Penton/IPC, Inc. Cleveland, June 29, 1978.
3. Morton, Hudson T., *Anti-Friction Bearings,* Hudson T. Morton, Ann Arbor, Michigan, 1965.
4. Palmgren, Arvid, *Ball and Roller Bearing Engineering,* SKF Industries, Inc., Philadelphia, 1959.
5. Standards of the Anti-Friction Bearing Manufacturers Association, Arlington, Virginia.
6. Bearing Manufacturers' literature.

CHAPTER FIFTEEN

Spur Gears

15.1 INTRODUCTION AND HISTORY

Gears, defined as toothed members transmitting rotary motion from one shaft to another, are among the oldest devices and inventions of man. In about 2600 B.C., the Chinese are known to have used a chariot incorporating a complex series of gears like those illustrated in Fig. 15.1. Aristotle, in the fourth century B.C., wrote of gears as if they were commonplace. In the fifteenth century A.D., Leonardo da Vinci designed a multitude of devices incorporating many kinds of gears.

Among the various means of mechanical power transmission (including primarily gears, belts, and chains), gears are generally the most rugged and durable. Their power transmission efficiency is as high as 98 percent. On the other hand, gears are usually more costly than chains and belts. As would be expected, gear manufacturing costs increase sharply with increased precision—as required for the combination of high speeds and heavy loads, and for low noise levels. (Standard tolerances for various degrees of manufacturing precision have been established by the AGMA, American Gear Manufacturers Association.)

Spur gears are the simplest and most common type of gears. As shown in Fig. 15.2, they are used to transfer motion between parallel shafts, and have teeth that are parallel to the shaft axes. The major portion of our study of spur gears will be concerned with gear geometry and nomenclature (Sections 15.2 and 15.3), gear force analysis (Section 15.4), gear-tooth bending strength (Sections 15.6–15.8), and gear-tooth surface durability (Sections 15.9 and 15.10).

The engineer seriously involved with gears of any kind should consult the pertinent standards of the AGMA, as well as other contemporary gear literature.

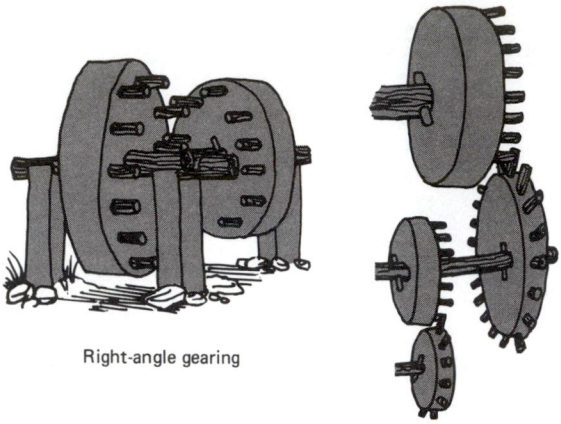

FIGURE 15.1.
Primitive gears.

15.2 GEOMETRY AND NOMENCLATURE

The basic requirement of gear-tooth geometry is the provision of angular velocity ratios that are exactly constant. For example, the angular velocity ratio between a 20-tooth and a 40-tooth gear must be precisely 2 in every position. (It must not be, for example, 1.99 as a given pair of teeth come into mesh, and then 2.01 as they go out of mesh.) Of course, manufacturing inaccuracies and tooth deflections will cause slight deviations in velocity ratio; but acceptable tooth profiles are based on theoretical curves that meet this criterion.

The action of a pair of gear teeth satisfying this requirement is termed *conjugate*

FIGURE 15.2.
Spur gears.

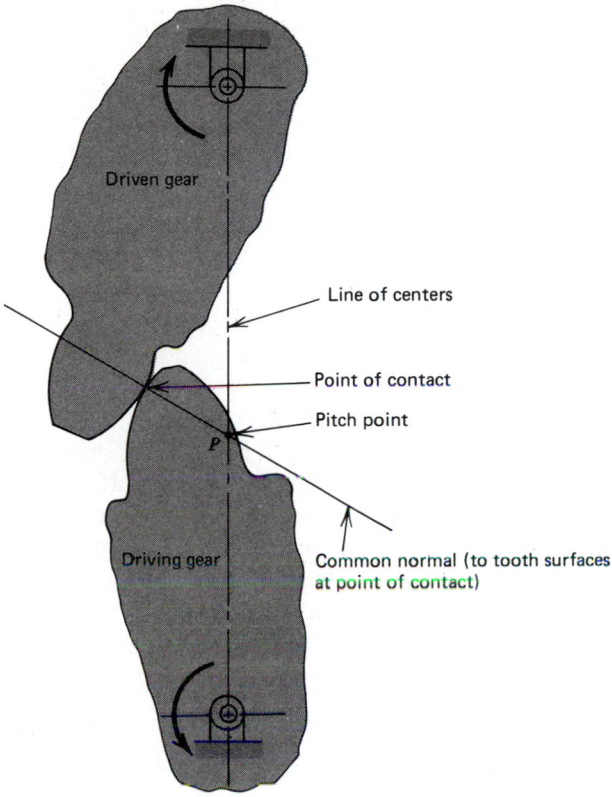

FIGURE 15.3.
Conjugate gear-tooth action.

gear-tooth action, and is illustrated in Fig. 15.3. The basic law of conjugate gear-tooth action states that as the gears rotate, *the common normal to the surfaces at the point of contact must always intersect the line of centers at the same point, P,* called the *pitch point.*

The law of conjugate gear-tooth action can be satisfied by various tooth shapes, but the only one of current importance is the *involute,* or, more precisely, the *involue of the circle.* (Its last important competitor was the cycloidal shape, used in the gears of Model T Ford transmissions.) An involute (of the circle) is the curve generated by any point on a taut thread as it unwinds from a circle, called the *base circle.* The generation of two involutes is shown in Fig. 15.4. The dotted lines show how these could correspond to the outer portions of the right sides of adjacent gear teeth. Correspondingly, involutes generated by unwinding a thread wrapped counterclockwise around the base circle would form the outer portions of the left sides of the teeth. Note that at every point, the involute is perpendicular to the taut thread. It is important to note that an involute can be developed as far as desired *outside* the base circle, but *an involute cannot exist inside its base circle.*

Let us now develop a mating pair of involute gear teeth in three steps: (1) friction

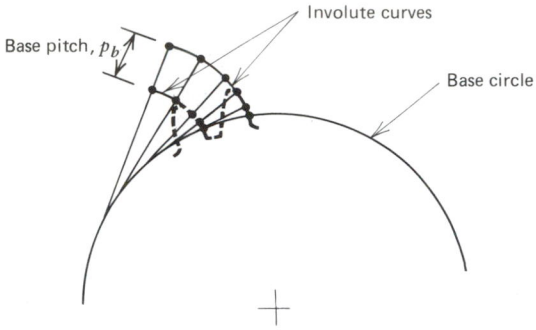

FIGURE 15.4.
Generation of an involute from its base circle.

drive, (2) belt drive, and, finally, (3) involute gear-tooth drive. Figure 15.5 shows two *pitch circles.* Imagine that they represent two cylinders pressed together. If slippage does not occur, rotation of one cylinder (pitch circle) will cause rotation of the other at an angular velocity ratio inversely proportional to their diameters. In any pair of mating gears, the smaller of the two is called the *pinion* and the larger one the *gear.* (The term "gear" is used in a general sense to indicate either of the members, and also in a specific sense to indicate the larger of the two. A bit confusing, perhaps, but life is sometimes like that!) Using subscripts p and g to denote pinion and gear, respectively,

$$\omega_p/\omega_g = -d_g/d_p \tag{15.1}$$

where the minus sign indicates that the two cylinders (gears) rotate in opposite directions.

In order to transmit more torque than is possible with friction gears alone, we now add a belt drive running between pulleys representing the *base circles,* as in Fig. 15.6.

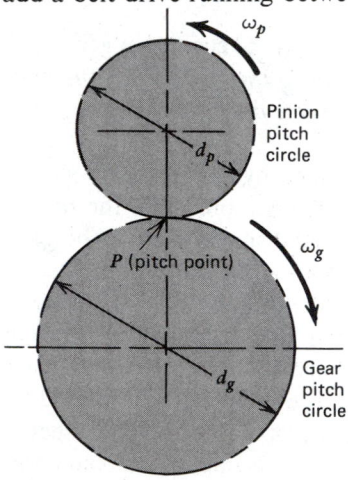

FIGURE 15.5.
Friction gears of diameter d rotating at angular velocity ω.

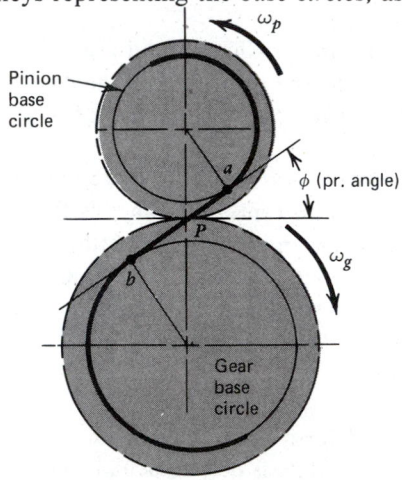

FIGURE 15.6.
Belt drive added to friction gears.

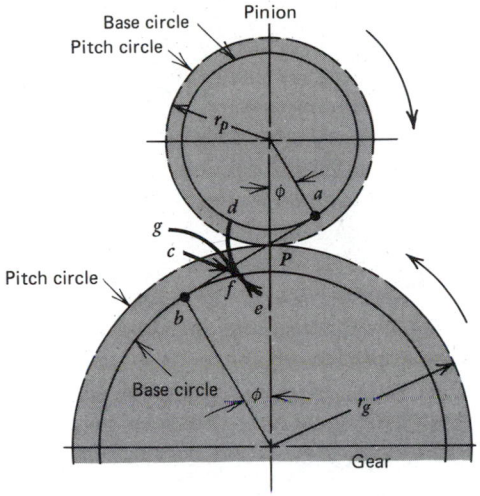

FIGURE 15.7.
Belt cut at *c* to generate conjugate involute profiles.

If the pinion is turned counterclockwise a few degrees, the belt will cause the gear to rotate in accordance with Eq. (15.1). In gear parlance, angle ϕ is called the *pressure angle*. From similar triangles, the base circles have the same ratio as the pitch circles; thus, the velocity ratios provided by the friction and belt drives are the same.

In Fig. 15.7 the belt is cut at point *c*, and the two ends are used to generate involute profiles *de* and *fg* for the pinion and gear, respectively. It should now be clear why ϕ is called the pressure angle: neglecting sliding friction, the force of one involute tooth pushing against the other is always at an angle equal to the pressure angle. A comparison of Figs. 15.7 and 15.3 shows that the involute profiles do indeed satisfy the fundamental law of conjugate gear-tooth action. Incidentally, the involute is the only geometric profile satisfying this law that maintains a constant pressure angle as the gears rotate. Note especially that conjugate involute action can take place only outside of both base circles. In Fig. 15.7, the conjugate involute profiles could be drawn only by "cutting the belt" at a point between *a* and *b*.

Figure 15.8 shows the continued development of the gear teeth. The involute profiles are extended outward beyond the pitch circle by a distance called the *addendum*. The outer circle is usually called the *addendum circle*. Similarly, the tooth profiles are extended inward from the pitch circle a distance called the *dedendum*. Of course, the involute portion can extend inward only to the base circle. The portion of the profile between the base and dedendum (root) circles cannot participate in the conjugate involute action but must clear the tip of a mating tooth as the gears rotate. This portion of the tooth profile is usually drawn as a straight radial line, but its actual shape (which depends on the manufacturing process) is usually trochoidal. A fillet at the base of the tooth blends the profile into the dedendum (root) circle. This fillet is important to reduce bending stress concentration.

An important point to keep in mind is that the "diameter" (without a qualifying

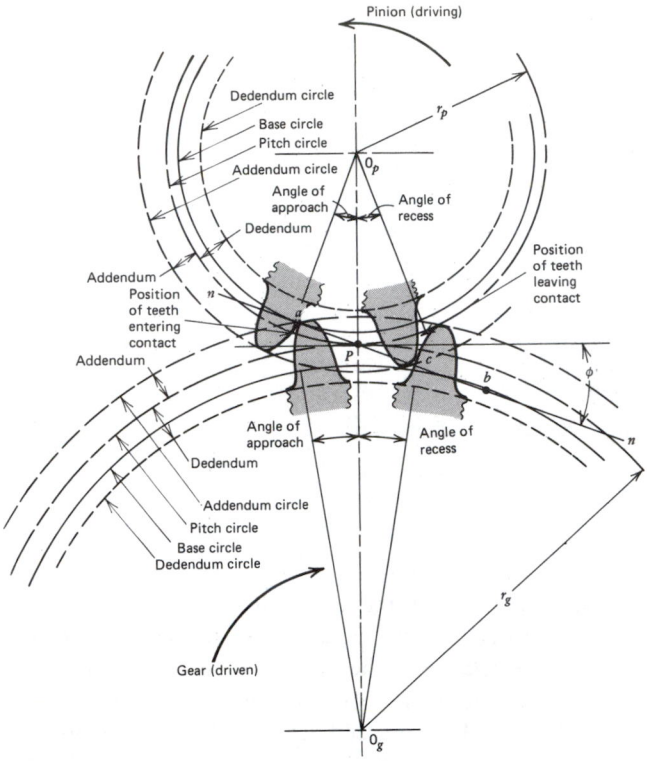

FIGURE 15.8.
Further development and nomenclature of involute gear teeth. (Note: Diagram shows special case of maximum possible gear addendum without interference [pinion addendum is far short of the theoretical limit]).

adjective) of a gear always refers to its *pitch* diameter. If other diameters (base, root, outside, etc.) are intended, these are always specified. Likewise, d, without subscripts, refers to *pitch* diameter. To distinguish between the pitch diameters of a pinion and gear, subscripts p and g are used: d_p and d_g.

Figure 15.8 shows the gear addendum extended exactly to point of tangency, a. (The pinion addendum extends to arbitrary point c, which is short of tangency point b.) This gear addendum represents the theoretical maximum without encountering "interference," which is discussed in the next article. Mating gears of standard proportions generally have shorter addenda (like the pinion in Fig. 15.8). Also, for practical reasons, the addenda of mating gears should not extend quite to the tangency points.

Figure 15.8 shows the position of a pair of mating teeth as they enter contact and again as they go out of contact. Note the corresponding *angle of approach* and *angle of recess* for both pinion and gear (measured to points on the pitch circles).

Line n-n (Fig. 15.8) is called the *line of action* (neglecting friction, the force between mating teeth always acts along this line). The *path of contact* (locus of all

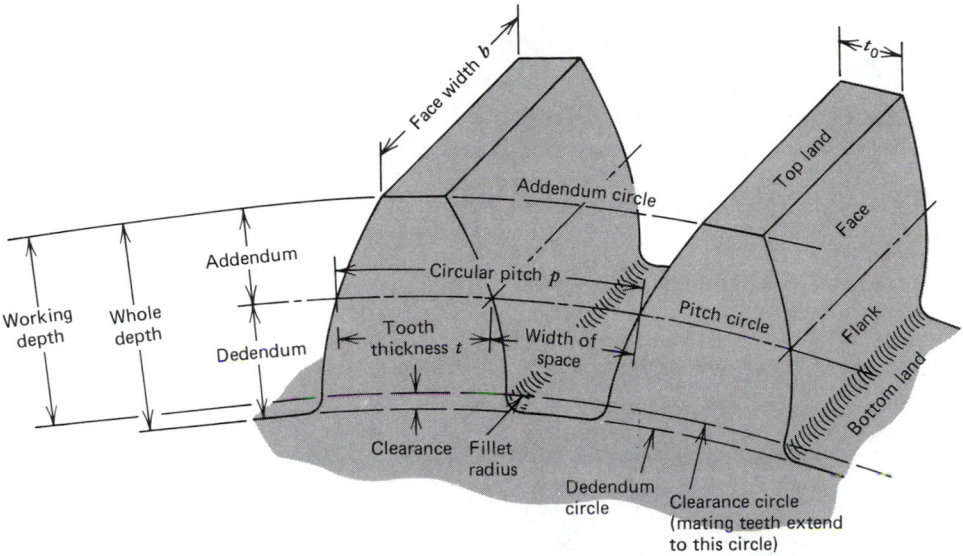

FIGURE 15.9.
Nomenclature of gear teeth.

points of tooth contact) is a segment of this line. In Fig. 15.8, the path of contact is the line segment *ac*.

Further nomenclature relating to the complete gear tooth is shown in Fig. 15.9. *Face* and *flank* portions of the tooth surface are divided by the *pitch cylinder* (which contains the pitch circle). Note particularly the *circular pitch*, designated as p, and measured in inches (English units) or mm (SI units). If N is the number of teeth in the gear (or pinion), and d the pitch diameter, then

$$p = \pi d/N, \quad p = \pi d_p/N_p, \quad p = \pi d_g/N_g \tag{15.2}$$

More commonly used indices of gear-tooth size are *diametral pitch*, P (used *only* with English units), and *module*, m (used *only* with SI or metric units). Diametral pitch is defined as the number of teeth per *inch* of pitch diameter:

$$P = N/d, \quad P = N_p/d_p, \quad P = N_g/d_g \tag{15.3}$$

Module, m, which is essentially the reciprocal of P, is defined as the pitch diameter in *millimeters* divided by the number of teeth (number of mm of pitch diameter per tooth):

$$m = d/N, \quad m = d_p/N_p, \quad m = d_g/N_g \tag{15.4}$$

The reader can readily verify that

$$pP = \pi \quad (p \text{ in in.; } P \text{ in teeth per inch}) \tag{15.5}$$

$$p/m = \pi \quad (p \text{ in mm; } m \text{ in mm per tooth}) \tag{15.6}$$

$$m = 25.4/P \tag{15.7}$$

With English units the word "pitch," without a qualifying adjective, means *diametral* pitch (a "12-pitch gear" refers to a gear with 12 teeth per inch of pitch diameter) whereas with SI units "pitch" means *circular* pitch (a "gear of pitch = 3.14"mm refers to a gear having a circular pitch of 3.14 mm).

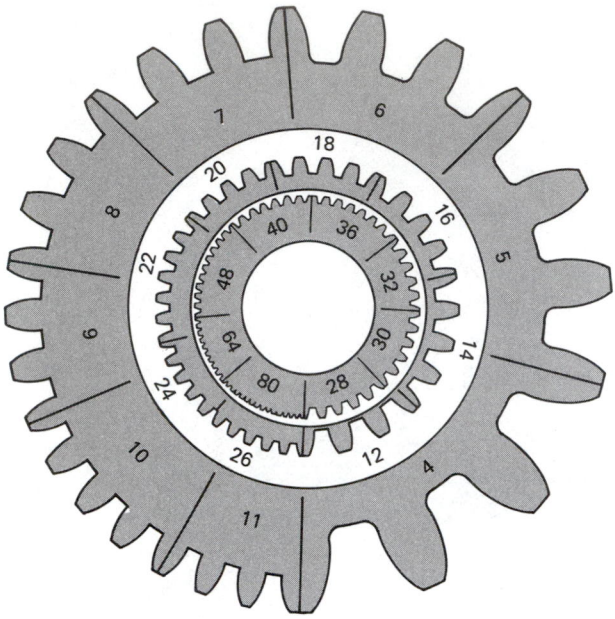

FIGURE 15.10.
Actual sizes of gear teeth of various diametral pitches. (Note: In general, fine-pitch gears have $P \geq 20$; coarse-pitch gears have $P < 20$.) (Courtesy The Barber-Colman Co.)

Gears are commonly made to an integral value of diametral pitch (English units) or standard value of module (SI units). Figure 15.10 shows the actual size of gear teeth of several standard diametral pitches. With SI units, commonly used standard values of module are:

0.2 to 1.0 by increments of 0.1,

1.0 to 4.0 by increments of 0.25,

4.0 to 5.0 by increments of 0.5.

The most commonly used pressure angle, ϕ, with both English and SI units is 20 deg. In the United States, 25 deg. is also standard, and 14½ deg. was formerly an alternative standard value.

For all systems, *the standard addendum is 1/P (in in.) or m (in mm),* and *the standard dedendum is 1.25 times the addendum.*[1] A former standard system in the United States was the 20-deg. stub system, for which the addendum was shortened to $0.8/P$. (Although 14½-deg. gears and 20-deg. stub gears are no longer standard, replacement gears are still made for these systems.) The fillet radius (at the base of the tooth) is commonly about $0.35/P$ (English units) or $m/3$ (SI units).

Face width, b (defined in Fig. 15.9), is not standardized, but generally,

$$9/P < b < 14/P, \tag{a}$$

[1]For fine pitch gears of $P \geqq 20$, the standard dedendum is $(1.20/P) + 0.002$ in.

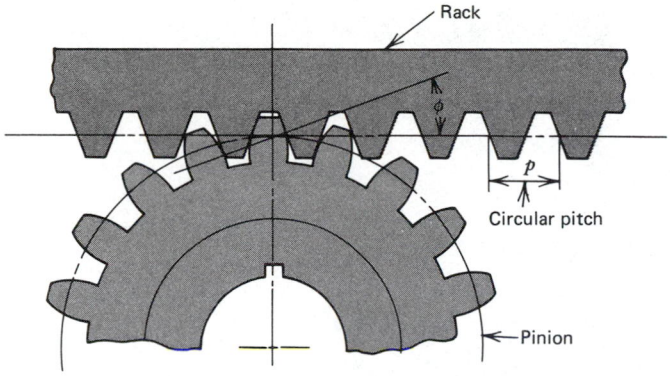

FIGURE 15.11.
Involute pinion and rack.

or

$$9m < b < 14m \tag{b}$$

The wider the face width, the more difficult it is to manufacture and mount the gears so that contact is uniform across the full face width.

Gears made to standard systems are interchangeable, and are usually available in stock. On the other hand, mass-produced gears used for particular applications (like automobile transmission gears) deviate from these standards in order to be optimal for their specific application. The present trend is toward greater use of special gears, as modern gear-cutting equipment reduces the cost penalty involved, and modern computer facilities minimize the engineering design time required.

Figure 15.11 shows a pinion in contact with a *rack,* which can be thought of as a segment of a gear of infinite diameter. Figure 15.12 shows a pinion in contact with an *internal* gear. The latter is also called an *annulus,* or *ring* gear, and is commonly used in the planetary gear trains of automotive automatic transmissions (see Section 15.13). Diameters of internal gears are considered *negative*; hence, Eq. (15.1) indicates that a pinion and internal gear rotate in the *same* direction.

An important advantage of the involute form over all others is that it provides

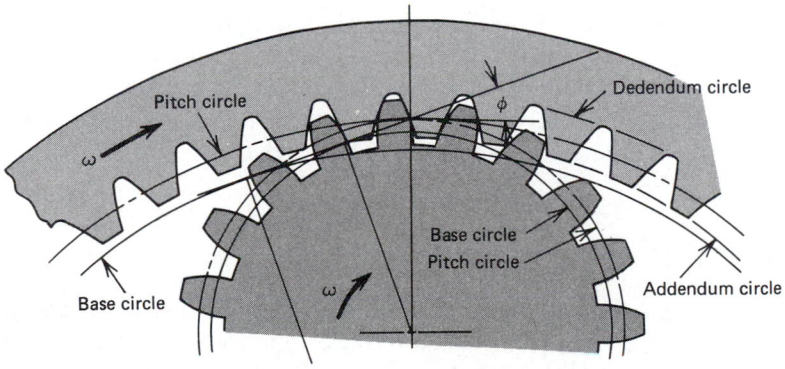

FIGURE 15.12.
Involute pinion and internal gear. (Note that both rotate in same direction.)

FIGURE 15.13.
Generating a spur gear with a pinion cutter. (Courtesy of Boston Gear)

theoretically perfect conjugate action even when the shaft center distances are not exactly correct. This fact can be verified by reviewing the basic development of involute profiles in Figs. 15.6 and 15.7. If the shafts of two mating gears are separated, proper action continues with an increased pressure angle. Of course, the *backlash* (shortest distance between the noncontacting surfaces of adjacent teeth) increases as center distance is increased. In some cases, advantage is taken of this feature by adjusting the shaft center distance to obtain desired backlash. (Some backlash is necessary to allow room for an oil film under all conditions of thermal expansion and contraction, but excessive backlash increases noise and impact loading whenever torque reversals occur.)

A second basic advantage of the involute system is that the profile for the basic rack is a straight line. This facilitates cutter manufacture and gear-tooth generation.

The cutting of gear teeth is a very highly developed engineering art and science. Two of the several methods used are shown in Figs. 15.13 and 15.14.

15.3 INTERFERENCE AND CONTACT RATIO

Interference will occur, preventing rotation of the mating gears, if either of the addendum circles extends beyond tangent points a and b (Figs. 15.6–15.8), which are called *interference points.* In Fig. 15.15 both addendum circles extend beyond the interference

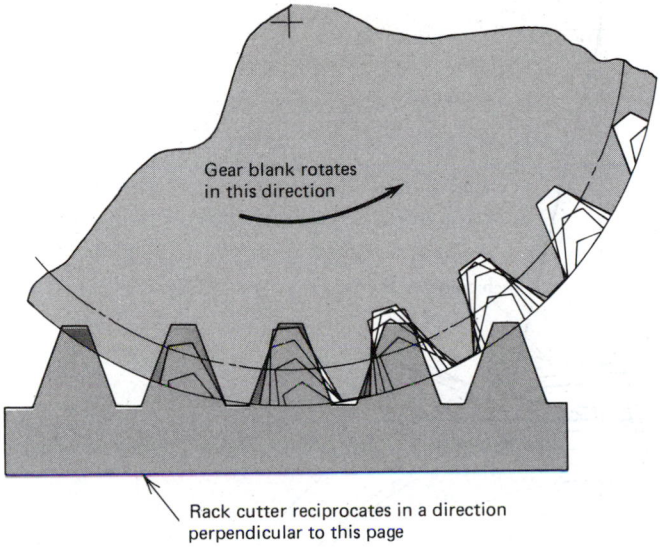

FIGURE 15.14.
Shaping teeth with a rack cutter.

points; hence these gears will not operate without modification. The preferred correction is to remove the interfering tooth tips, shown cross-hatched. Alternatively, the tooth flanks of the mating gear can be undercut in order to clear the offending tips, but this weakens the teeth. In no case is it possible to have useful contact involving the cross-hatched tips, as conjugate involute action is not possible beyond the interference points.

When teeth are generated with a rack cutter, as in Fig. 15.14, the teeth are *automatically* undercut if they would otherwise interfere with a rack. This undercutting takes place with standard 20-deg. pinions with fewer than *18* teeth, and with standard 25-deg. pinions having less than *12* teeth. For this reason, pinions with fewer than these numbers of teeth are not normally used with standard tooth proportions.

From Fig. 15.8 or 15.15 one can derive readily the equation for the maximum possible addendum circle radius without interference:

$$r_{a(\max)} = \sqrt{r_b^2 + c^2 \sin^2 \phi} \tag{15.8}$$

where

$r_{a(\max)}$ = maximum noninterfering addendum circle radius of pinion or gear

r_b = base circle radius of the same member

c = center distance, $0_1 0_2$

ϕ = pressure angle (*actual,* not nominal, value)

It is obviously necessary that the tooth profiles be proportioned so that a second pair of mating teeth come into contact before the first pair is out of contact. The average

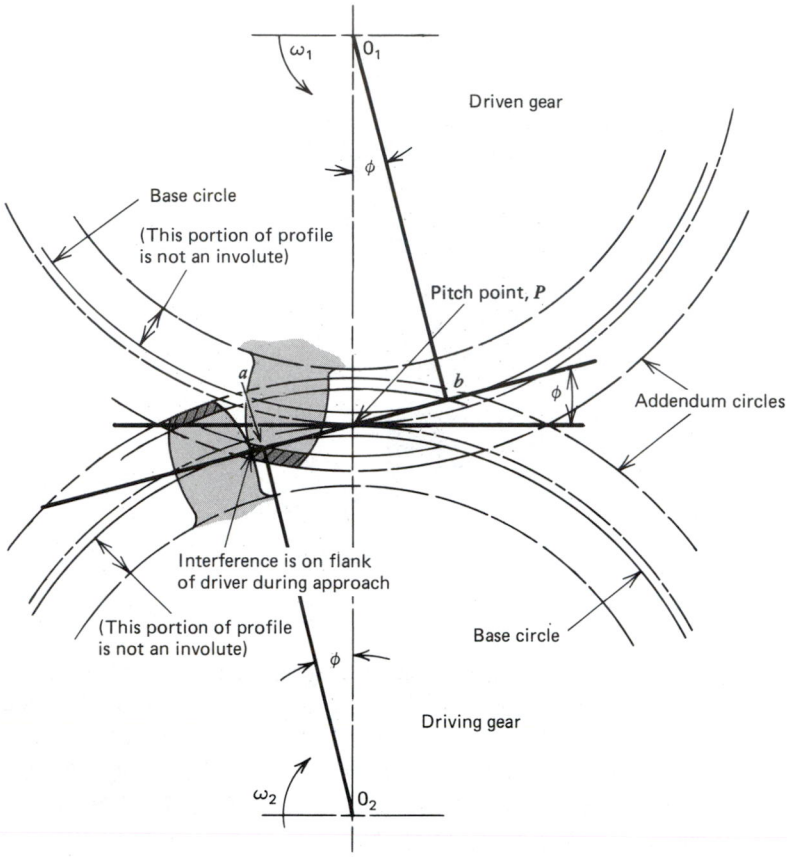

FIGURE 15.15.
Interference of spur gears (eliminated by removing the cross-hatched tooth tips).

number of teeth in contact as the gears rotate together is the *contact ratio,* which is calculated from the following equation [1]:

$$\text{C.R.} = \frac{\sqrt{r_{ap}^2 - r_{bp}^2} + \sqrt{r_{ag}^2 - r_{bg}^2} - c \sin \phi}{p_b} \tag{15.9}$$

where

r_{ap}, r_{ag} = addendum radii of the mating pinion and gear

r_{bp}, r_{bg} = base circle radii of the mating pinion and gear

p_b is the *base pitch:*

$$p_b = \pi d_b / N \tag{15.10}$$

Where N = number of teeth and d_b = diameter of base circle. From Fig. 15.7,

$$d_b = d \cos \phi, \; r_b = r \cos \phi, \text{ and } \; p_b = p \cos \phi \tag{15.11}$$

(The base pitch is like the circular pitch except that it represents an arc of the base circle rather than an arc of the pitch circle. It is illustrated in Fig. 15.4.)

In general, the greater the contact ratio the smoother and quieter the operation of the gears. A contact ratio of 2 or more means that at least two pairs of teeth are theoretically in contact at all times. (Whether or not they are *actually* in contact depends on the precision of manufacture, tooth stiffness, and applied load.)

SAMPLE PROBLEM 15.1

Two parallel shafts with 4-in. center distance are to be connected by 6-pitch, 20-deg. spur gears providing a velocity ratio of -3.0. (a) Determine the pitch diameters and numbers of teeth in the pinion and gear. (b) Determine if interference will result from the use of standard full-depth teeth. (c) Determine the contact ratio.

Solution

1. $r_p + r_g = c = 4$ in.; $r_g/r_p = -$vel. ratio $= 3$; hence, $r_p = 1$ in., $r_g = 3$ in., or $d_p = 2$ in., $d_g = 6$ in. ■
2. "6-pitch gears" means that $P = 6$ teeth per inch of pitch diameter; hence, $N_p = 12$, $N_g = 36$. ■
3. In order to use Eq. (15.8) to check for interference, we first determine the base circle radii of pinion and gear. From Eq. (15.11), $r_{bp} = 1$ in.(cos 20°), and $r_{bg} = 3$ in.(cos 20°). Substitution in Eq. (15.8) gives $r_{a(\max)} = 1.660$ in. for the pinion and 3.133 in. for the gear.
4. The limiting outer gear radius is equivalent to an addendum of only 0.133 in., whereas a standard full-depth tooth has an addendum of $1/P = 0.167$ in. Clearly, the use of standard teeth would result in interference. ■
5. Let us use unequal addenda gears (nonstandard), with somewhat arbitrarily chosen addenda of 0.060 in. for the gear and 0.290 in. for the pinion. (The reasoning being to select maximum addenda for greatest contact ratio, while at the same time limiting the gear addendum to stay safely away from interference, and limiting the pinion addendum to maintain adequate width of top land. The latter is shown as t_o in Fig. 15.9, and its minimum acceptable value is sometimes taken as $0.25/P$.)
6. Substitution in Eq. (15.11) gives $p_b = (\pi/6)\cos 20° = 0.492$ in. Substitution in Eq. (15.9) (with $r_{ap} = 1.290$ in., $r_{bp} = 1$ in.(cos 20°), $r_{ag} = 3.060$ in., $r_{bg} = 3$ in.(cos 20°)) gives $C.R. = 1.43$, which should be a suitable value. ■

A study of Eq. (15.8) and its derivation indicates that (a) interference is more likely to involve the tips of the gear teeth than the tips of the pinion teeth, and (b) interference is promoted by having a *small* number of pinion teeth, a *large* number of gear teeth, and a *small* pressure angle.

Had we wished to use standard tooth proportions in solving the above sample problem, we could have (a) increased the diametral pitch (thereby giving more teeth on the pinion—and this outweighs the influence of giving more teeth to the gear), or (b) increased the pressure angle to 25 deg. (which would be more than enough to eliminate interference).

15.4 GEAR FORCE ANALYSIS

It was noted in Figs. 15.7 and 15.8 that line *ab* was always normal to the contacting tooth surfaces, and that (neglecting sliding friction) it was the *line of action* of the forces between mating teeth.

The force between mating teeth can be resolved at the pitch point (P, in Figs. 15.15 and 15.16) into two components:

(1) tangential component, F_t, which, when multiplied by the pitch-line velocity, accounts for the power transmitted, and

(2) radial component F_r, which does no work but tends to push the gears apart.

Figure 15.16 illustrates that the relationship between these components is

$$F_r = F_t \tan \phi \tag{15.12}$$

To analyze the relationships between the gear force components and the associated shaft power and rotating speed, we note that the gear pitch-line velocity, V, in feet per minute, is equal to

$$V = \pi dn/12 \tag{15.13}$$

where d is the pitch diameter in inches of the gear rotating n rpm.

By definition, the transmitted horsepower is

$$\text{hp} = F_t V/33{,}000 \tag{15.14}$$

where F_t is in pounds and V is in feet per minute.

In SI units

$$V = \pi dn/60{,}000 \tag{15.13a}$$

where d is in mm, n in rpm, and V in m/s. Transmitted power in watts is equal to

$$W = F_t V \tag{15.14a}$$

where F_t is in newtons.

SAMPLE PROBLEM 15.2

Figure 15.17*a* shows three gears of $P = 3$, $\phi = 20$ deg. Gear *a* is the driving, or input, pinion. It rotates counterclockwise at 600 rpm and transmits 25 hp to idler gear *b*. Output gear *c* is attached to a shaft that drives a machine. Nothing is attached to the

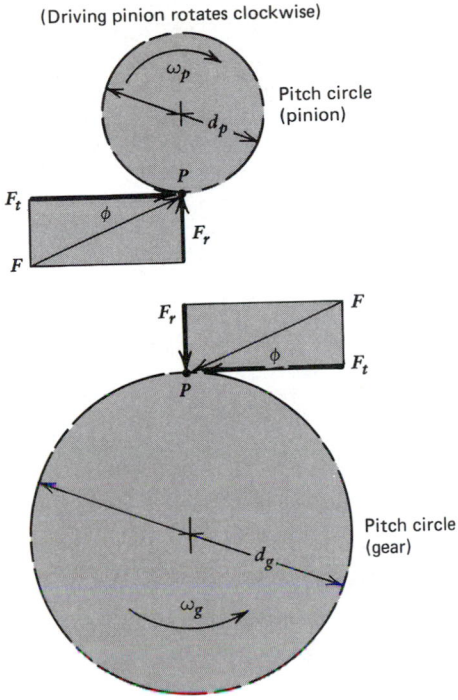

FIGURE 15.16.
Gear-tooth force, F, shown resolved at pitch point (illustrated for driving pinion and driven gear, shown separately).

idler shaft, and friction losses in the bearings and gears can be neglected. Determine the resultant load applied by the idler to its shaft.

Solution

1. Applying Eq. (15.3) to gear *a*:

$$d_a = N_a/P = (12 \text{ teeth})/(3 \text{ teeth per in.}) = 4 \text{ in.}$$

2. All three gears have the same pitch-line velocity. Applying Eq. (15.13) to gear *a*:

$$V = \pi d_a n_a/12 = \pi(4 \text{ in.})(600 \text{ rpm})/12 = 628.28 \text{ ft/min}$$

3. Applying Eq. (15.14) to gear *a*, and solving for F_t:

$$F_t = 33{,}000(25 \text{ hp})/628.28 \text{ fpm} = 1313 \text{ lb}$$

This is the horizontal force of gear *b* applied *to* gear *a*, directed to the right. Figure 15.17*b* shows the equal and opposite horizontal force of *a* applied to *b*, labeled H_{ab}, and acting to the left.

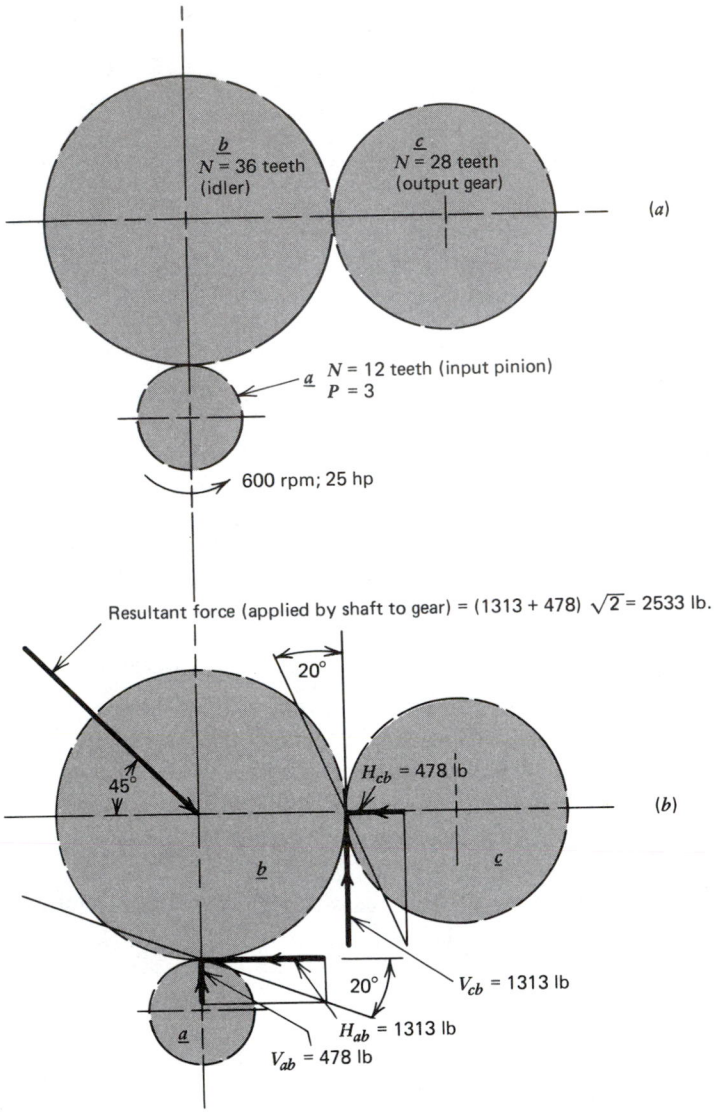

FIGURE 15.17.
Gear forces in Sample Problem 15.2. (a) Gear layout. (b) Forces acting on idler *b*.

4. From Eq. (15.12), the corresponding radial gear-tooth force is V_{ab} = 1313 tan 20° = 478 lb.
5. Forces H_{cb} and V_{cb} are shown in proper direction in Fig. 15.17*b*. (Remember, these are forces applied *by c to b*.) Since the shaft supporting idler *b* carries no torque, equilibrium of moments about its axis of rotation requires that V_{cb} = 1313 lb. From Eq. (15.12), H_{cb} = 1313 tan 20°, or 478 lb.

6. Total gear-tooth forces acting on b are $1313 + 478 = 1791$ lb both vertically and horizontally, for a vector sum of $1791\sqrt{2} = 2533$ lb acting at 45 deg. This is the resultant load applied *by* the idler *to* its shaft. ■

The equal and opposite force applied *by* the shaft *to* the idler gear is shown in Fig. 15.17*b*, where the idler is shown as a free-body in equilibrium.

15.5 GEAR-TOOTH STRENGTH

Having dealt with gear geometry and force analysis, we now turn to the question of how much power or torque a given pair of gears will transmit without tooth failure. Figure 15.18 shows a *photoelastic pattern* of gear-tooth stresses. The details of this experimental stress analysis procedure are beyond the scope of this book, and it is sufficient here to note that the highest stresses exist where the lines are bunched closest together. This occurs at two locations: (1) the point of contact with the mating gear, where force F is acting, and (2) in the fillet at the base of the tooth.

The next three articles deal with bending fatigue at the base of the tooth, and involve the principles of fatigue analysis covered in Chapter 8. The following two articles are concerned with surface durability, and make use of the information on pitting and scoring given in Chapter 9. Some of the principles of lubrication covered in Chapter 13 are involved as well. As will be seen, the load capacity and failure mode of a pair of gears are affected by their rotating speed. Altogether, the study of gear load capacity

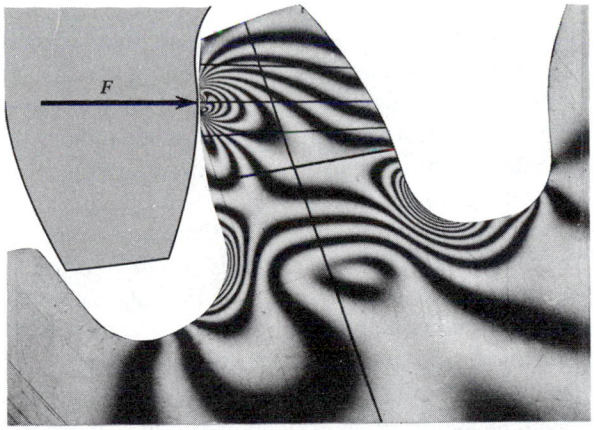

FIGURE 15.18.
Photoelastic pattern of stresses in a spur gear-tooth. (From T. J. Dolan and E. L. Broghammer, A Study of Stresses in Gear Tooth Fillets, Proc. 14th Eastern Photoelasticity Conf., PE December, 1941)

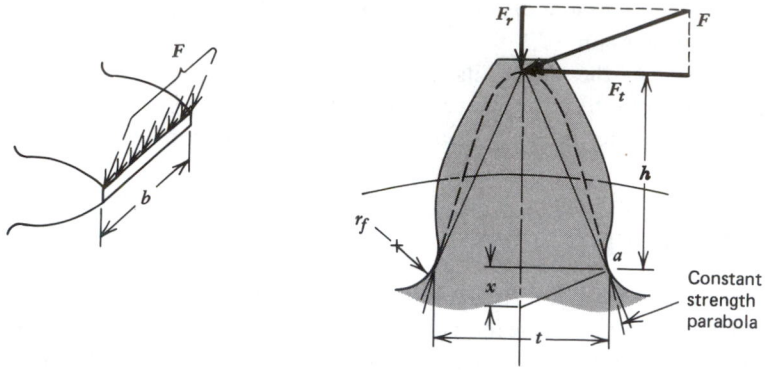

FIGURE 15.19.
Bending stresses in a spur gear-tooth (comparison with a constant-stress parabola).

affords an excellent opportunity to apply much of the basic material covered in earlier chapters.

15.6 BASIC ANALYSIS OF GEAR-TOOTH BENDING STRESS (LEWIS EQUATION)

The first recognized analysis of gear-tooth stresses was presented to the Philadelphia Engineers Club in 1892 by Wilfred Lewis. It still serves as the basis for gear-tooth bending stress analysis. Figure 15.19 shows a gear tooth loaded as a cantilever beam, with resultant force F applied to the tip. Mr. Lewis made the following simplifying assumptions:

1. *The full load is applied to the tip of a single tooth.* This is obviously the most severe condition, and is appropriate for gears of "ordinary" accuracy. For high-precision gears, however, the full load is never applied to a single tooth tip. With a contact ratio necessarily greater than unity, each new pair of teeth comes into contact while the previous pair is still engaged. After the contact point moves down some distance from the tip, the previous teeth go out of engagement and the new pair carries the full load (unless, of course, the contact ratio is greater than 2). This is the situation depicted in Fig. 15.18. Thus, with *precision gears* (not available in Mr. Lewis' time), the tooth should be regarded as carrying only part of the load at its tip, and the full load at a point on the tooth face involving a shorter bending moment arm.

2. *The radial component, F_r, is negligible.* This is a conservative assumption, as F_r produces a compressive stress that subtracts from the bending tension at point a of Fig. 15.19. (The fact that it adds to the bending compression in the opposite fillet is unimportant because fatigue failures always start on the tensile side.)

3. *The load is distributed uniformly across the full face width.* This is a nonconservative assumption, and can be instrumental in gear failures involving wide teeth and/or misaligned or deflecting shafts.

4. *Forces due to tooth sliding friction are negligible.*

5. *Stress concentration in the tooth fillet is negligible.* Stress concentration factors were unknown in Mr. Lewis' time, but are now known to be important. This will be taken into account later.

Proceeding with the development of the Lewis equation, we note from Fig. 15.19 that the gear tooth is everywhere stronger than the inscribed *constant strength parabola* (recall Fig. 12.22 *c*), except for the section at *a* where the parabola and tooth profile are tangent. At point *a*,

$$\sigma = \frac{Mc}{I} = \frac{6F_t h}{bt^2} \tag{c}$$

by similar triangles,

$$\frac{t/2}{x} = \frac{h}{t/2}, \quad \text{or } \frac{t^2}{h} = 4x \tag{d}$$

Substituting (d) into (c) gives

$$\sigma = \frac{6F_t}{4bx} \tag{e}$$

Defining the *Lewis form factor, y,* as

$$y = 2x/3p \tag{f}$$

and substituting it into (e) gives

$$\sigma = \frac{F_t}{bpy} \tag{15.15}$$

which is the basic *Lewis equation* in terms of circular pitch.

Because gears are more often made to standard values of diametral pitch, we substitute

$$p = \pi/P \tag{15.5, mod}$$

$$y = Y/\pi \tag{g}$$

into Eq. (15.14) and obtain an alternative form of the Lewis equation:

$$\sigma = \frac{F_t P}{bY} \tag{15.16}$$

or, when using SI units,

$$\sigma = \frac{F_t}{mbY} \tag{15.16a}$$

where Y is the Lewis form factor based on diametral pitch or module. Both Y and y are functions of tooth *shape* (but not size), and therefore vary with the number of teeth in the gear. Values of Y for standard gear systems are given in Fig. 15.20. For non-standard gears, the factor can be obtained by graphical layout of the tooth or by digital computation.

Note that the Lewis equation indicates that tooth bending stresses vary (1) directly with load, F_t, (2) inversely with tooth width, b, (3) inversely with tooth size, p, $1/P$, or m, and (4) inversely with tooth shape factor, Y or y.

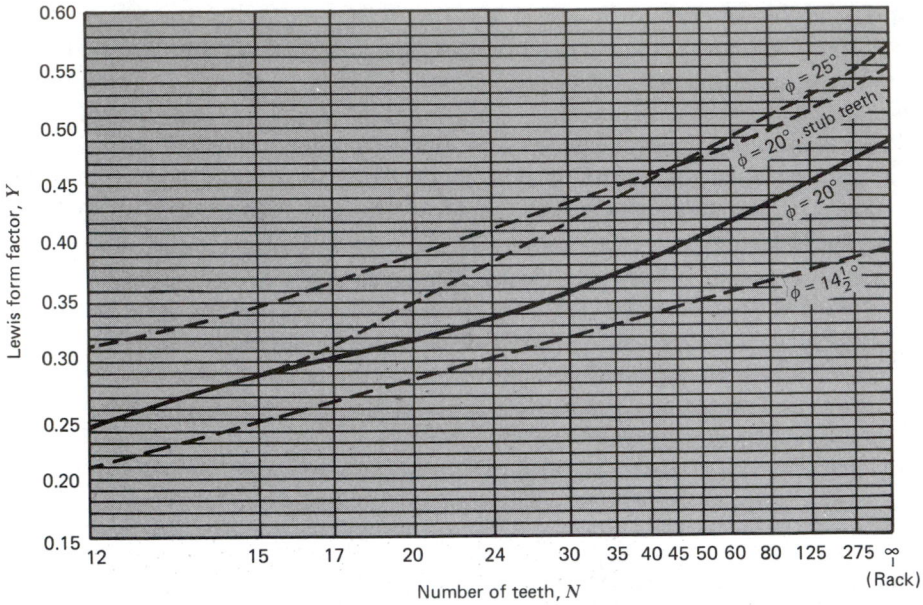

FIGURE 15.20.
Values of Lewis form factor, *Y*, for standard spur gears (load applied at tip of tooth).

15.7 REFINED ANALYSIS OF GEAR-TOOTH BENDING STRENGTH – BASIC CONCEPTS

In addition to the four basic factors included in the Lewis equation, modern gear-design procedures take into account several additional factors that influence gear-tooth bending stresses:

1. *Pitch-line velocity.* The greater the linear velocity of the gear teeth (as measured at the pitch circles), the greater the impact of successive teeth as they come into contact. These impacts arise because the tooth profiles can never be made with *absolute* perfection; and even if they were, inevitable deflections due to operating loads would cause a slight impact as each new pair of teeth came into engagement.

2. *Manufacturing accuracy.* This is also an important factor influencing impact loading. Furthermore, manufacturing accuracy is the determining factor as to whether or not teeth do in fact share the load when two or more pairs of teeth are *theoretically* in contact. (See the first assumption in Section 15.6.)

3. *Contact ratio.* For *precision* gears of "one-plus" contact ratio ($1 < C.R. < 2$), the transmitted load is divided among two pairs of teeth whenever a new tooth comes into contact at its tip. As the contact point moves down the face of the new tooth, the meshing teeth ahead go out of contact at the highest point of single tooth contact of the new tooth. Thus, there are two loading conditions to be considered: (a) carrying part of the load (often assumed to be half) at the tooth tip, and (b) carrying the full load at

the point of highest single tooth contact. For gears of "two-plus" contact ratio ($2 < C.R. < 3$), one should consider a three-way division of load at tooth-tip contact, and a two-way division at the highest point of double tooth contact.

4. *Stress concentration* at the base of the tooth, as mentioned in assumption 5 of Section 15.6.

5. *Degree of shock loading* involved in the application. (This is similar to the "application factor" given for ball bearings in Section 14.7.4.)

6. *Accuracy and rigidity of mounting.* (See assumption 3, Section 15.6.)

7. *Moment of inertia of the gears and attached rotating members.* Slight tooth inaccuracies tend to cause momentary angular accelerations and decelerations of the rotating members. If the rotating inertias are small, the members easily accelerate without imposing high momentary tooth loads. With large inertias, the rotating members tend strongly to resist acceleration, thereby causing large momentary tooth loads. Significant torsional elasticity between the gear teeth and the major sources of inertia may tend to isolate the gear teeth from the harmful inertial effect. (This situation sometimes provides a fruitful area for dynamic analysis.)

The problem of gear-tooth bending fatigue involves an evaluation of (a) the fluctuating stresses in the tooth fillet, and (b) the fatigue strength of the material *at this same highly localized location.* So far only stresses have been considered; now consider the strength aspect of the problem.

The important strength property is usually the bending fatigue strength, as represented by the endurance limit. From Eq. (8.1)

$$S_n = S'_n C_L C_G C_S$$

which, for steel members, is usually

$$S = (S_u/2) C_L C_G C_S$$

Most gear teeth are loaded in *only one direction.* However, the teeth of idler gears (Figs. 15.17 and 15.21*a*) and planet pinions (Fig. 15.28, and described in Section 15.13) are loaded in both directions. Although ideally one might prefer to make a mean stress-alternating stress diagram for each particular case, reference to Fig. 15.21*b* shows the basis for the common generalization:

> *For infinite-life, peak stresses must be below the reversed bending endurance limit for an idler gear, but peak stresses can be 40 percent higher for a driving or driven gear.*

For a reliability of other than 50 percent, gear bending strength calculations are commonly based on the assumption that the tooth bending fatigue strength has a normal distribution (recall Figs. 6.15–6.17) with one standard deviation being about 8 percent of the nominal endurance limit.

If gear teeth operate at elevated temperatures, the fatigue properties of the material at the temperatures involved must be used.

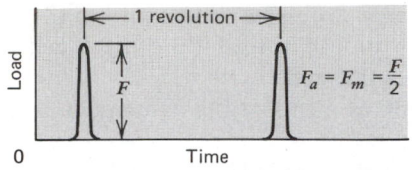

Driving and driven gears

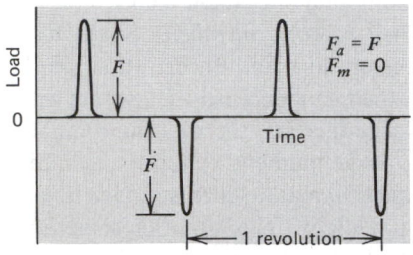

Idler gear

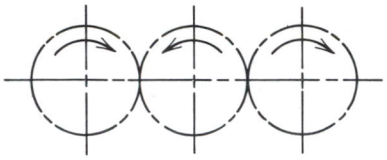

(*a*) Load fluctuation

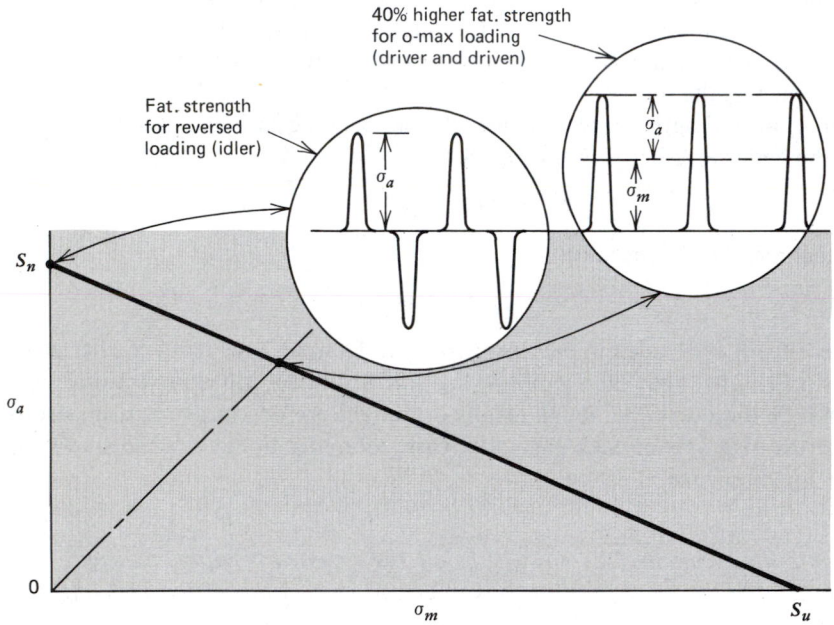

(*b*) Stress fluctuation

FIGURE 15.21.
Load and stress fluctuations in driving, driven, and idler gears.

15.8 REFINED ANALYSIS OF GEAR-TOOTH BENDING STRENGTH – RECOMMENDED PROCEDURE

The engineer seriously concerned with gear design and analysis should consult the latest standards of the American Gear Manufacturers Association and the relevant current literature. The procedures given here are representative of current practice.

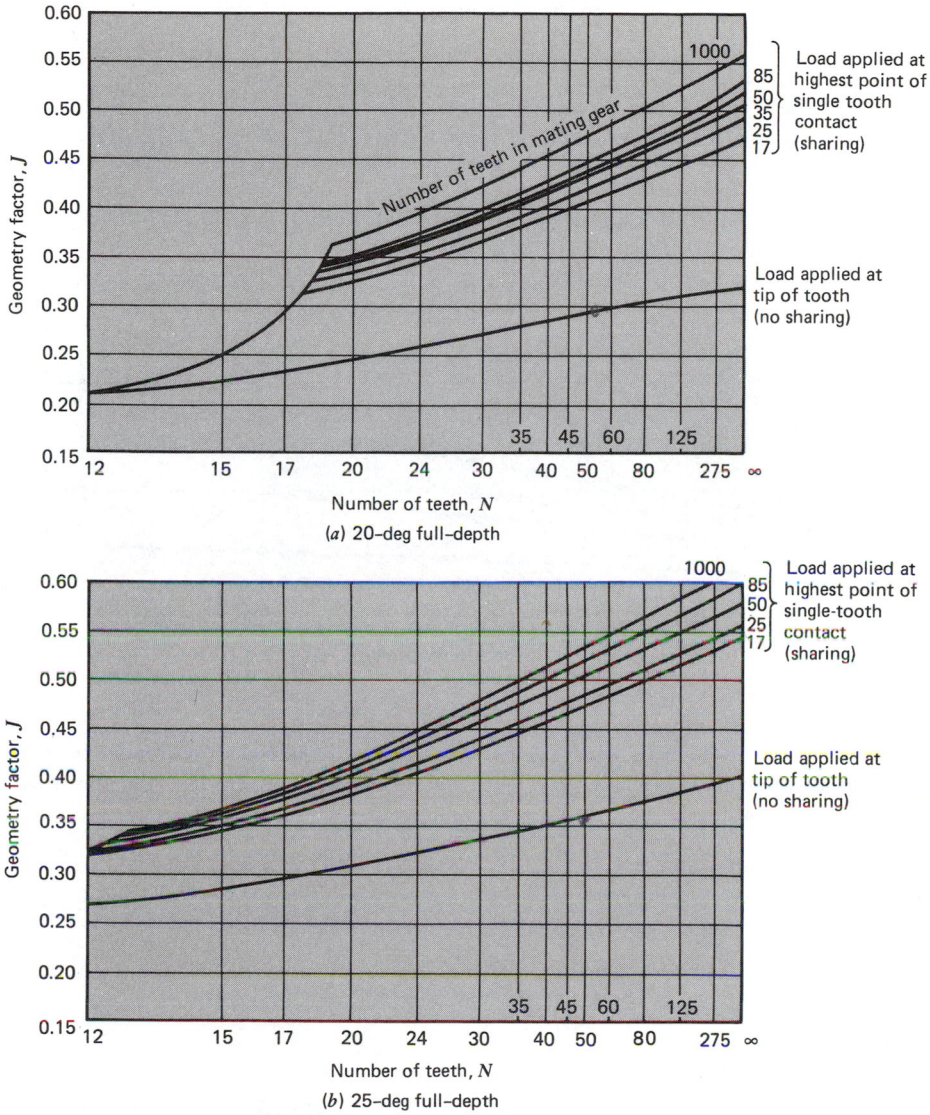

FIGURE 15.22.
Geometry factor, *J*, for standard spur gears (based on tooth fillet radius = 0.35/P.) (From AGMA Information Sheet 225.01)

In the absence of more specific information, the factors affecting gear-tooth bending stress can be taken into account by embellishing the Lewis equation to the following form:

$$\sigma = \frac{F_t P}{bJ} K_v K_o K_m \tag{15.17}$$

where

J = spur gear *geometry factor* from Fig. 15.22. This factor includes the *Lewis form*

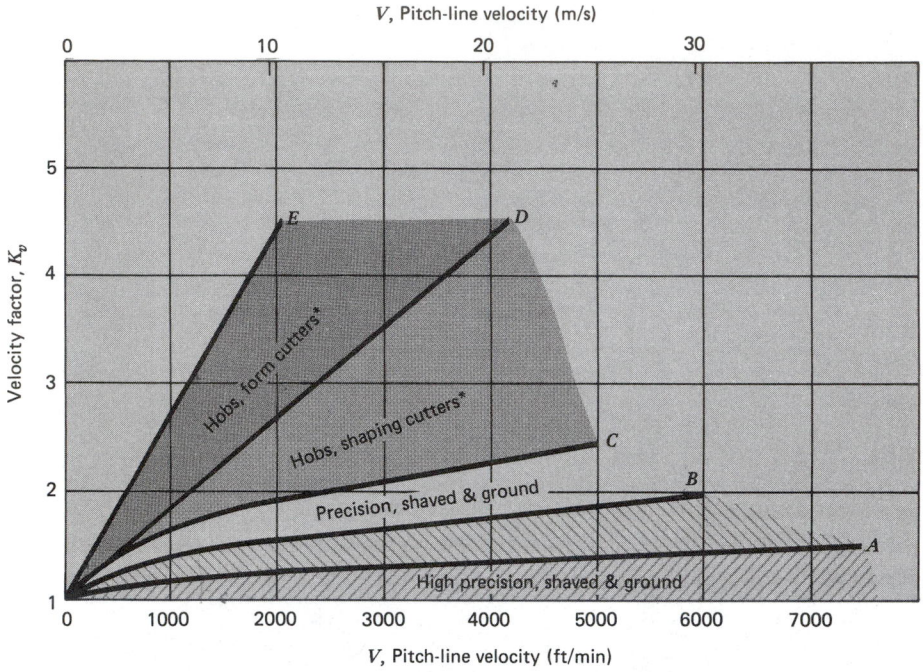

*Limited to about 350 Bhn

FIGURE 15.23.
Velocity factor, K_v. (Note: This figure, in a very rough way, is intended to account for the effects of tooth spacing and profile errors, tooth stiffness, and the velocity, inertia, and stiffness of the rotating parts.)

$$A{:}K_v = \sqrt{\frac{78 + \sqrt{V}}{78}} \qquad D{:}K_v = \frac{1200 + V}{1200}$$

$$B{:}K_v = \frac{78 + \sqrt{V}}{78} \qquad E{:}K_v = \frac{600 + V}{600}$$

$$C{:}K_v = \frac{50 + \sqrt{V}}{50}$$

Note: V is in ft/min

factor, Y, and also a *stress concentration factor* based on a tooth fillet radius of $0.35/P$. Note that values are given for the case of no load sharing (nonprecision gears) and also for load sharing (high-precision gears). In the latter case, the J factor depends on the number of teeth in the mating gear, as this controls the contact ratio, which in turn determines the highest point of single tooth contact.

K_v = *velocity* or *dynamic factor,* indicating the severity of impact as successive pairs of teeth engage. This is a function of pitch-line velocity and manufacturing accuracy. Figure 15.23 gives guidelines pertaining to representative gear manufacturing processes.

K_o = *overload factor,* reflecting the degree of nonuniformity of driving and load

TABLE 15.1
Overload Correction Factor, K_o

	Driven Machinery		
Source of Power	**Uniform**	**Moderate Shock**	**Heavy Shock**
Uniform	1.00	1.25	1.75
Light shock	1.25	1.50	2.00
Medium shock	1.50	1.75	2.25

torques. In the absence of better information, the values in Table 15.1 have long been used as a basis for rough estimates.

K_m = *mounting factor,* reflecting the accuracy of mating gear alignment. Table 15.2 is used as a basis for rough estimates.

The effective fatigue stress from Eq. (15.17) must be compared with the corresponding fatigue strength. For infinite life, the appropriate endurance limit is estimated from the equation

$$S_n = S'_n C_L C_G C_S k_r k_t k_{ms} \tag{15.18}$$

where

S'_n = standard R. R. Moore endurance limit

C_L = load factor = 1.0 for bending loads

C_G = gradient factor = 1.0 for $P > 5$, and 0.85 for $P \leq 5$

C_S = surface factor from Fig. 8.13. Be sure that this pertains to the surface *in the fillet,* where a fatigue crack would likely start. (In the absence of specific information, assume this to be equivalent to a machined surface.)

k_r = reliability factor, determined from Fig. 6.17. For convenience, values corresponding to an endurance limit standard deviation of 8 percent are given in Table 15.3.

Table 15.2
Mounting Correction Factor, K_m

	Face width (in.)			
Characteristics of Support	**0 to 2**	**6**	**9**	**16 up**
Accurate mountings, small bearing clearances, minimum deflection, precision gears	1.3	1.4	1.5	1.8
Less rigid mountings, less accurate gears, contact across full face	1.6	1.7	1.8	2.2
Accuracy and mounting such that less than full-face contact exists		Over 2.2		

TABLE 15.3
Reliability Correction Factor, k_r (From Fig. 6.17 with Assumed Standard Deviation of 8 Percent)

Reliability (%)	50	90	99	99.9	99.99	99.999
Factor k_r	1.000	0.897	0.814	0.753	0.702	0.659

k_t = temperature factor. For steel gears, use $k_t = 1.0$ if the temperature (usually estimated on the basis of lubricant temperature) is less than 160°F. If not, and in the absence of better information, use

$$k_t = \frac{620}{460 + T} \quad (\text{for } T > 160°\text{F}) \tag{15.19}$$

k_{ms} = mean stress factor. In accordance with Section 15.7, use 1.0 for idler gears (subjected to two-way bending) and 1.4 for input and output gears (one-way bending).

The safety factor for bending fatigue can be taken as the ratio of fatigue strength (Eq. (15.18)) to fatigue stress (Eq. (15.17)). Its numerical value should be chosen in accordance with Section 6.12. Since factors K_o, K_m, and k_r have been taken into account separately, the "safety factor" need not be as large as would otherwise be necessary. Typically, a safety factor of 1.5 might be selected, together with a reliability factor corresponding to 99.9 percent reliability.

SAMPLE PROBLEM 15.3

Figure 15.24 shows a specific application of a pair of spur gears. Estimate the maximum horsepower that the gears can transmit continuously with only a 1 percent chance of encountering tooth bending fatigue failure.

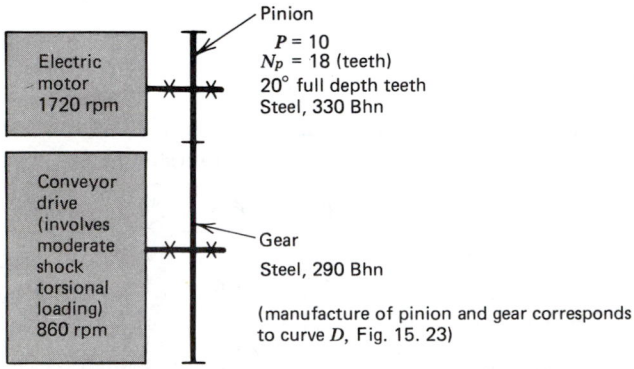

FIGURE 15.24.
Data for Sample Problem 15.3.

Solution

1. The bending endurance strength is estimated from Eq. (15.18) as:

$$S_n = S'_n C_L C_G C_S k_r k_t k_{ms}$$

where

$S'_n = 290/4 = 72.5$ ksi (gear)

$= 330/4 = 82.5$ ksi (pinion)

$C_L = 1$ (for bending loads)

$C_G = 1$ (since $P > 5$)

$C_S = 0.68$ (pinion) (from Fig. 8.13, machined surfaces)

$= 0.70$ (gear)

$k_r = .814$ (from Table 15.3; 99 percent reliability)

$k_t = 1.0$ (temperature should be $<160°$F)

$k_{ms} = 1.4$ (for one-way bending)

$S_n = 63.9$ ksi (pinion); $S_n = 57.8$ ksi (gear)

2. The bending fatigue stress is estimated from Eq. (15.17) as:

$$\sigma = \frac{F_t P}{bJ} K_v K_o K_m$$

where

$P = 10$ and $b = 1.25$ in. (given)

$J = 0.235$ (pinion) (for $N = 18$, no load sharing because of inadequate precision of manufacture)

$= 0.28$ (gear) (for $N = 36$, which is needed to provide the given speed ratio)

K_v involves pitch-line velocity, V, calculated as

$$V = \frac{\pi d_p n_p}{12}$$

$$= \frac{\pi(18 \text{ teeth}/10 \text{ teeth per inch})(1720 \text{ rpm})}{12}$$

$$= 811 \text{ fpm}$$

$K_v = 1.68$ (from Fig. 15.23)

$K_o = 1.25$ (from Table 15.1)

$K_m = 1.6$ (from Table 15.2)

$\sigma = 114F_t$ psi (pinion), $\sigma = 96F_t$ psi (gear)

3. Equating bending fatigue strength and bending fatigue stress:

$$63{,}900 \text{ psi} = 114F_t \text{ psi}, \quad F_t = 561 \text{ (pinion)}$$
$$57{,}800 \text{ psi} = 96F_t \text{ psi}, \quad F_t = 602 \text{ (gear)}$$

4. In this case the pinion is the weaker member, and the power that can be transmitted is (561 lb) (811 fpm) = 456,000 ft-lb/min. Dividing by 33,000 to convert to horsepower gives *13.8 hp* (without provision for a safety factor). ■

15.9 GEAR-TOOTH SURFACE DURABILITY – BASIC CONCEPTS

Gear teeth are vulnerable to the various types of surface damage discussed in Chapter 9. As was the case with rolling-element bearings (Chapter 14), gear teeth are subjected to *Hertz contact stresses,* and the lubrication is often *elastohydrodynamic* (Section 13.14). Excessive loading and/or lubrication breakdown can cause various combinations of *abrasion, pitting,* and *scoring.* In this and the next section, it will become evident that gear-tooth surface durability is a more complex matter than tooth bending fatigue capacity.

Previous sections dealt with the determination of the compressive force, F, acting between the gear teeth, and it was noted that the contacting surfaces are cylindrical in nature with involute profiles. Nothing has previously been said about the *rubbing velocity* of the contacting surfaces. Figure 15.25*a* shows the same conjugate gear teeth as in Fig. 15.3, with vectors added to show velocities, V_p and V_g, of the instantaneous contact points on the pinion and gear teeth, respectively. These velocities are *tangential* with respect to their centers of rotation. If the teeth do not separate or crush together, the components of V_{pn} and V_{gn} normal to the surface must be the same. This results in their components tangential to the surface (V_{pt} and V_{gt}) being different. The sliding velocity is the difference between V_{pt} and V_{gt}.

Figure 15.25*b* shows that when contact of the mating teeth is at the *pitch point, P,* (i.e., on the line of centers), the sliding velocity is zero, and the tooth relative motion is one of *pure rolling.* For contact at all other points, the relative motion is one of *rolling plus sliding,* with the sliding velocity being directly proportional to the distance between the point of contact and the pitch point. The maximum sliding velocity occurs with contact at the tooth tips. This means that teeth with long addenda (as in Fig. 15.8) experience higher maximum sliding velocities than do corresponding gears with shorter addenda. (But, of course, the gears with shorter addenda have a smaller contact ratio.)

Note that the relative sliding velocity *reverses direction* as a pair of teeth roll through the pitch point. During approach (see "angle of approach," Fig. 15.8), the sliding friction forces tend to compress the teeth; during recess, friction forces tend to elongate the teeth. The latter tends to give smoother action. For this reason, special teeth are sometimes designed to have most or all of the contact to lie within the angle of recess (for the direction of rotation involved). Gear-tooth sliding can be readily demonstrated by spreading and extending the fingers of the two hands, meshing them as gears, and then rotating them.

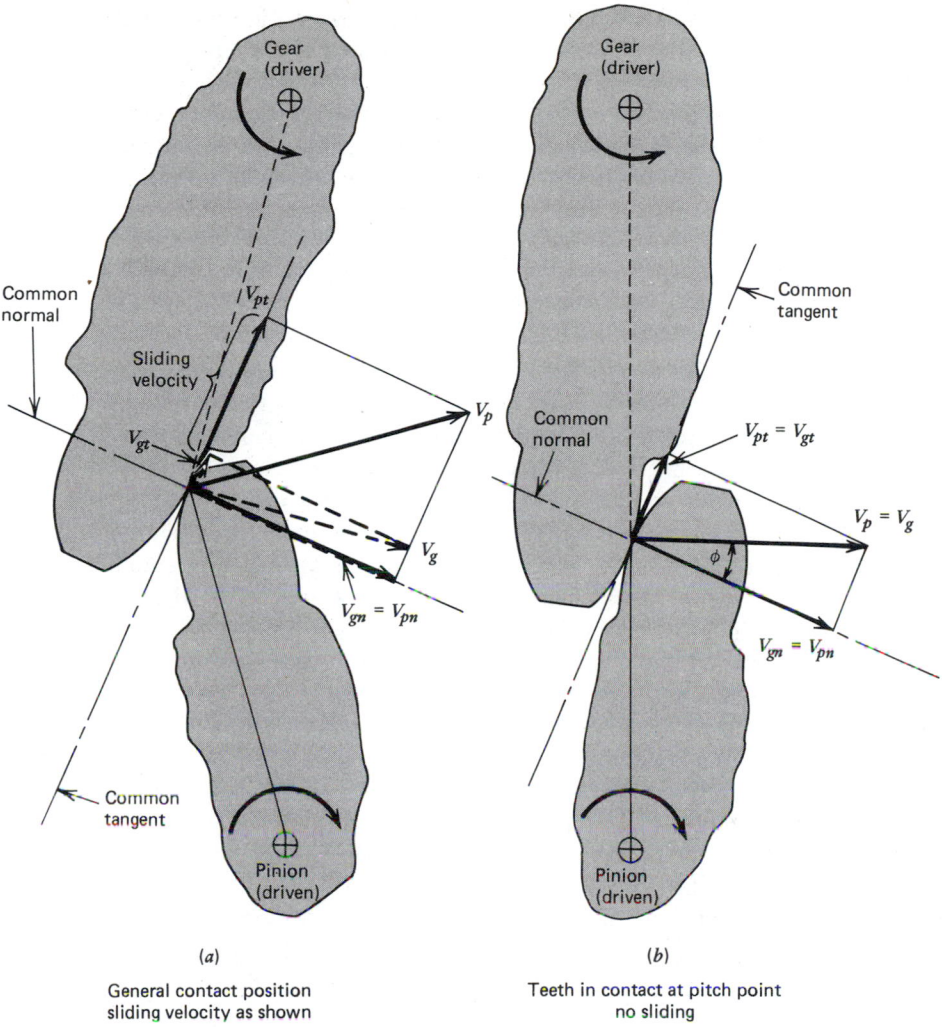

FIGURE 15.25.
Gear-tooth sliding velocity.

We now consider briefly the three basic types of surface deterioration that occur in gear teeth:

1. *Abrasive wear* (treated in Section 9.10), which requires the presence of foreign particles, such as with gears that are not enclosed, with enclosed gears assembled with abrasive particles present, or with gears lubricated by an oil supply with inadequate filtration.

2. *Scoring* (a form of adhesive wear described in Section 9.9), which occurs, usually at high speeds, when adequate lubrication is not provided by elastohydrodynamic action (Section 13.16) [or, possibly in some cases, by boundary and mixed film lubrication (Section 13.14)]. This causes a high coefficient of sliding friction that, together with high tooth loading and high sliding velocities, produces a high rate of heat gener-

ation in the *localized* regions of contact. This results in temperatures and pressures that cause welding and tearing apart. Scoring can often be prevented by directing an *adequate flow* (to provide cooling) of *appropriate* lubricant (generally, one sufficiently resistant to extreme pressures to maintain hydrodynamic lubrication) to the teeth as they come into mesh. Surface finish is also important, with finishes as fine as 20 microinches being desirable where scoring is a factor. Also, allowing the gears to smooth themselves during an initial "break-in" period of moderate load usually increases their scoring resistance.

3. *Pitting and spalling,* which represent, respectively, surface and sub-surface fatigue failure due to the complex stresses within the contact zone. These failure types are discussed in Section 9.14.

With proper care, gears should not fail because of abrasive wear. With proper lubrication and cooling, they will not fail because of scoring. (If, with the best heat transfer provision available, an adequate lubricant cannot be found, then loads and/or speeds must be reduced, more score-resistant materials used, or the gears made larger.) Unlike scoring, which is not time-dependent and occurs early in the operating life if at all, pitting is typical of fatigue failures because it occurs only after accumulating a sufficient number of load cycles. Furthermore, since contact stress *S-N* curves do not level off after 10^6 or 10^7 cycles even with steel members, this type of potential surface failure must be considered in every gear design.

Generally good correlation has been observed between spur gear surface fatigue failure and the computed elastic surface stress (Hertz stress). Just as the Lewis equation serves as the basis for gear-tooth bending strength analysis, the Hertz stress (Eq. (9.5)) is the basis for gear-tooth surface durability analysis.

The classic work of adapting the Hertz equation to spur gear teeth was done by Professor Earle Buckingham of MIT [1]. Professor Buckingham noted that gear-tooth pitting occurs predominantly in the vicinity of the pitch line where, because of zero sliding velocity, the oil film (elastohydrodynamic) breaks down. Hence, he treated a pair of gear teeth as two cylinders of radii equal to the radii of curvature of the mating involutes at the pitch point. From basic involute geometry, these radii are

$$R_p = (d_p \sin \phi)/2 \quad \text{and} \quad R_g = (d_g \sin \phi)/2 \tag{15.20}$$

(Refer to Fig. 15.7 and imagine the belt to be cut at P for generating the involutes.)

To adapt Eq. (9.5) (and Eq. (9.2)) for convenient use with spur gears, we make the following substitutions:

Eq. (9.5) Quantity	**Equivalent Notation for Spur Gears**
F	F (which is equal to $F_t/\cos \phi$)
p_o	σ_H
L	b
R_1	$(d_p \sin \phi)/2$
R_2	$(d_g \sin \phi)/2$

This gives

$$\sigma_H = 0.564 \sqrt{\frac{F_t[2/(d_p \sin\phi) + 2/(d_g \sin\phi)]}{\cos\phi\, b\left(\dfrac{1-\nu_p^2}{E_p} + \dfrac{1-\nu_g^2}{E_g}\right)}} \tag{15.21}$$

Several fundamental relationships are evident from this equation. Because of the increased contact area with load, stress increases only as the square root of load F_t (or square root of load per inch of face width, F_t/b). Similarly, contact area increases (and stress decreases) with decreased moduli of elasticity, E_p and E_g. Also, larger gears have greater radii of curvature, hence lower stress.

In much the same way as tooth bending stresses, contact stresses are influenced by manufacturing accuracy, pitch-line velocity, shock loading, shaft misalignment and deflection, and moment of inertia and torsional elasticity of the connected rotating members. Similarly, surface fatigue strength of the material is affected by the reliability requirement and by possible temperature extremes.

15.10 GEAR-TOOTH SURFACE FATIGUE ANALYSIS – RECOMMENDED PROCEDURE

Equation (15.21) becomes more manageable when we (a) combine terms relating to the elastic properties of the materials into a single factor, C_p, commonly called the *elastic coefficient,* and (b) combine terms relating to tooth shape into a second factor, I, commonly called the *geometry factor:*

$$C_p = 0.564 \sqrt{\frac{1}{\dfrac{1-\nu_p^2}{E_p} + \dfrac{1-\nu_g^2}{E_g}}} \tag{15.22}$$

$$I = \frac{\sin\phi\cos\phi}{2}\,\frac{R}{R+1} \tag{15.23}$$

where R is the ratio of gear and pinion diameters

$$R = \frac{d_g}{d_p} \tag{f}$$

Note that R is positive for a pair of external gears (Fig. 15.2). Because diameters of internal gears are considered negative, R is negative for a pinion and internal gear (Fig. 15.12).

Substituting C_p and I into Eq. (15.21), and also introducing factors K_v, K_o, and K_m, which were used with the bending fatigue analysis, gives

$$\sigma_H = C_p \sqrt{\frac{F_t}{bd_pI} K_vK_oK_m} \tag{15.24}$$

Note that I is a dimensionless constant that is readily calculated from Eq. (15.23), whereas C_p has units of $\sqrt{\text{ksi}}$ or $\sqrt{\text{MPa}}$ depending upon the system of units used. For convenience, values of C_p are given in Tables 15.4*a* and 15.4*b*.

TABLE 15.4a
Values of Elastic Coefficient C_P for Spur Gears, in $\sqrt{\text{psi}}$ (Values Rounded Off)

Pinion Material (ν = 0.30 in All Cases)	Gear Material: Steel	Cast Iron	Aluminum Bronze	Tin Bronze
Steel, E = 30,000 ksi	2300	2000	1950	1900
Cast iron, E = 19,000 ksi	2000	1800	1800	1750
Aluminum bronze, E = 17,500 ksi	1950	1800	1750	1700
Tin bronze, E = 16,000 ksi	1900	1750	1700	1650

As noted in Sections 9.13 and 9.14, the actual stress state at the point of contact is influenced by several factors not considered in the simple Hertz equation (Eqs. (9.5), (15.21), and (15.24)). These include thermal stresses, pressure distribution changes due to the presence of a lubricant, stresses due to sliding friction, and so on. For this reason, stresses calculated from Eq. (15.24) must be compared with surface fatigue strength *S-N* curves that have been obtained experimentally from *tests where these additional factors were at least roughly comparable with those for the situation under study*. Thus, the surface fatigue strength curve for spur gears in Fig. 9.19 is appropriate for use now, whereas the other curves in the same figure are not.

In the absence of surface fatigue strength information more directly pertinent to the specific application being considered, Table 15.5 gives representative values.

As noted in Section 9.14, it is often desirable for one of the contacting members to be harder than the other. In the case of steel gears, the pinion is invariably made the harder (if they are different) because pinion teeth are subjected to a greater number of fatigue cycles, and because it is generally more economical to manufacture the smaller member to the higher hardness. Typically, the hardness differential ranges from about 30 Bhn for gears in the 200 Bhn range to about 100 Bhn for the 500 Bhn range and 2 Rockwell C for the 60 R_C range. For hardness differentials not exceeding these values, it has been found that the average hardness can be used for checking both pinion and gear.

For surface-hardened steel gears, the hardness used with Table 15.5 is the surface hardness, but the depth of the hardened case should extend down to the peak shear

TABLE 15.4b.
Values of Elastic Coefficient C_p for Spur Gears, in $\sqrt{\text{MPa}}$ (Values Converted from Table 15.4*a*)

Pinion Material (ν = 0.30 in All Cases)	Gear Material: Steel	Cast Iron	Aluminum Bronze	Tin Bronze
Steel, E = 207 GPa	191	166	162	158
Cast iron, E = 131 GPa	166	149	149	145
Aluminum bronze, E = 121 GPa	162	149	145	141
Tin bronze, E = 110 GPa	158	145	141	137

TABLE 15.5
Surface Fatigue Strengths, S_{fe}, for Use with Metallic Spur Gears (10^7 Cycle Life, 99 Percent Reliability, Temperature <250°F)

Material	S_{fe} (ksi)	S_{fe} (MPa)
Steel	0.4 (Bhn)−10 ksi	28 (Bhn)−69 MPA
Nodular iron	0.95[0.4 (Bhn)−10 ksi]	0.95[28 (Bhn)−69 MPa]
Cast Iron, grade 20	55	379
grade 30	70	482
grade 40	80	551
Tin bronze AGMA 2C (11 percent tin)	30	207
Aluminum bronze (ASTM B 148—52 (Alloy 9C—H.T.)	65	448

stresses shown in Figs. 9.14*b* and 9.16. This would normally be at least 1 mm, or 0.040 in.

For fatigue lives other than 10^7 cycles, multiply the values of S_{fe} (from Table 15.5) by a life factor, C_{Li}, from Fig. 15.26. The latter represents somewhat of an average shape of surface fatigue *S-N* curve for steel. The alert reader will note a discrepancy between this shape (slope) and that of the *S-N* curve for steel gears in Fig. 9.19. No attempt was made to "fudge" these curves to bring them into consistency, because this would mask a significant "fact of life" regarding published materials strength data: Independent studies presenting generalizations of different sets of data are apt to be at some variance, and should be used with appropriate caution. If possible, it is always best to obtain good test data applying closely to the case at hand. Incidentally, the spur gear curve in Fig. 9.17 is about the highest that is normally obtainable for steel gears.

For reliabilities other than 99 percent, an appropriate reliability factor, C_R, should be used (as in Eq. (15.25), below). Relevant data are scarce, but a rough guide is to use C_R equal to about 1.25 and 0.8 for 50 and 99.9 percent reliability, respectively.

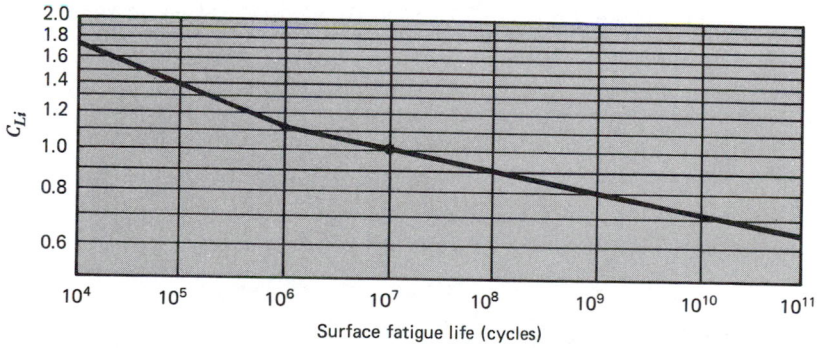

FIGURE 15.26.
Values of C_{Li} for steel gears (general shape of surface fatigue *S-N* curve).

For cases involving extreme gear-tooth surface temperatures (above about 120°C, or 250°F), one must determine the appropriate surface fatigue strength for the material and temperature involved. (A temperature correction factor for surface fatigue strength has not been included in Eq. (15.25).)

Applying the above, the resulting equation for gear-tooth surface fatigue strength, which should be compared to the gear-tooth surface fatigue stress from Eq. (15.24), is

$$S_H = S_{fe} C_{Li} C_R \tag{15.25}$$

In accordance with the philosophy presented in Section 6.12, the safety factor (defined as the multiplier of F_t needed to make σ_H equal to S_H) can be small. Many of the factors often included in the "safety factor" are already taken into account by various multiplying factors in Eqs. (15.24) and (15.25). Furthermore, the consequences of failure are mitigated by the fact that pitting failures develop slowly, and give warning by gradually increasing gear noise. Also, the extent of surface fatigue damage constituting "failure" is arbitrary, and the gears will continue to operate for some period of time after their surface endurance "life" has expired. Accordingly, safety factors of 1.1 to 1.5 are usually appropriate.

SAMPLE PROBLEM 15.4

For the gears in Sample Problem 15.3, estimate the maximum horsepower that the gears can transmit with only a 1 percent chance of a surface fatigue failure during 5 years of 40 hr/week, 50 week/year operation.

Solution

1. The surface endurance strength is estimated from Eq. (15.25) as:

$$S_H = S_{fe} C_{Li} C_R$$

where

$S_{fe} = 0.4\ (\text{Bhn}) - 10\ \text{ksi} = 0.4\ (310) - 10 = 114\ \text{ksi}$

$C_{Li} = 0.8$ (from Fig. 15.26, life $= (1720)(60)(40)(50)(5) = 1.03 \times 10^9$ cycles)

$C_R = 1$

$S_H = (114)(0.8)(1) = 91.2\ \text{ksi}$

2. The surface (Hertz) fatigue stress is estimated from Eq. (15.24) as:

$$\sigma_H = C_p \sqrt{\frac{F_t}{b d_p I} K_v K_o K_m}$$

where

$C_p = 2300\ \sqrt{\text{psi}}$ (from Table 15.4)

$b = 1.25$ in., $d_p = 1.8$ in., $K_v = 1.68$, $K_o = 1.25$, and $K_m = 1.6$ (all are the same as in Sample Problem 15.3)

$$I = \frac{\sin\phi\cos\phi}{2}\frac{R}{R+1} = 0.107 \text{ (from Eq. (15.23))}$$

$$\sigma_H = 2300\sqrt{\frac{F_t}{(1.25)(1.8)(0.107)}(1.68)(1.25)(1.6)} = 8592\sqrt{F_t}$$

3. Equating surface fatigue strength and surface fatigue stress:

$$8592\sqrt{F_t} = 91{,}200 \text{ psi} \quad \text{or } F_t = 113 \text{ lb}$$

(This single value applies to the combination of the two surfaces acting together.)

4. The corresponding power is (113 lb)(811 fpm) = 91,600 ft-lb/min, or 2.8 hp. ■

This compares with a bending fatigue-limited power of nearly 14 hp and illustrates the usual situation of steel gears being stronger in bending fatigue. Although much of the 14-hp bending fatigue capacity is obviously wasted, a moderate excess of bending capacity is desirable because bending fatigue failures are sudden and total, whereas surface failures are gradual and cause increasing noise levels to warn of gear deterioration.

15.11 SPUR GEAR DESIGN PROCEDURES

Sample Problems 15.3 and 15.4 illustrated the *analysis* of estimated capacity of a given pair of gears. As is generally the case with machine components, it is a more challenging task to *design* a suitable (hopefully, near optimal) pair of gears for a given application. Before illustrating this procedure with a sample problem, let us make a few general observations:

1. Increasing the surface hardness of steel gears pays off handsomely in terms of surface endurance. Table 15.5 indicates that doubling the hardness *more* than doubles surface fatigue strength (allowable Hertz stress); Eq. (15.24) shows that doubling the allowable Hertz stress *quadruples* the load capacity, F_t.
2. Increases in steel hardness also increase bending fatigue strength, but the increase is far less. For example, doubling the hardness will likely *not* double the basic endurance limit, S'_n (note flattening of curves in Fig. 8.6). Furthermore, doubling hardness substantially reduces C_S (see Fig. 8.13). An additional factor in the case of surface hardened gears is that the hardened case may effectively increase surface fatigue strength, yet be too shallow to contribute much to bending fatigue strength (recall Fig. 8.29 and the related discussion in Section 8.13).
3. Increasing tooth size (using a coarser pitch) increases bending strength more than surface strength. This fact, together with points 1 and 2, above, correlates with the observations that (a) a balance between bending and surface strengths occurs typically in the region of $P = 8$ for high-hardness steel gears (above about 500 Bhn, or 50 R_C), with coarser teeth failing in bending fatigue and finer teeth failing in surface fatigue, and (b) with progressively softer steel teeth, surface fatigue

becomes critical at increasingly fine pitches. Other materials have properties giving different gear-tooth strength characteristics. Further information is given in the next section.

4. In general, the harder the gears the more costly they are to manufacture. On the other hand, harder gears can be smaller and still do the same job. And if the gears are smaller, the housing and other associated parts may also be smaller and lighter. Furthermore, if the gears are smaller, pitch-line velocities are lower, and this reduces the dynamic loading and rubbing velocities. Thus, overall cost can often be reduced by using harder gears.
5. If minimum size gears are desired (for any given gear materials and application), it is best in general to start by choosing the minimum acceptable number of teeth for the pinion (usually 18 teeth for 20-deg. pinions, and 12 for 25-deg. pinions), and then solving for the pitch (or module) required.

SAMPLE PROBLEM 15.5

Design, using a standard gear system, a pair of spur gears to connect a 100-hp, 3600-rpm motor to a 900-rpm load shaft. Shock loading from the motor and driven machine is negligible. The center distance is to be as small as reasonably possible. A life of 5 years of 2000 hr/year operation is desired, but full power will be transmitted only about 10 percent of the time, with half power the other 90 percent. Likelihood of failure during the 5 years should not exceed 10 percent.

Solution

1. For minimum center distance, choose hardened gears corresponding to the spur gear curve in Fig. 9.19. Specify surface hardness of 660 Bhn and 600 Bhn for pinion and gear, respectively.
2. For these hardnesses (which are too hard for normal machining), specify a ground finish, and precision manufacture corresponding to the average of curves A and B in Fig. 15.23.
3. Choose the more common 20-deg. full-depth involute tooth form, with 18 teeth in the pinion.
4. Total life required = 3600 rev/min $\times$ 60 min/hr $\times$ 2000 hr/yr $\times$ 5 yr = 2.16 $\times 10^9$ revolutions of the pinion. Only 2.16×10^8 cycles are at full power. Looking at the spur gear curve in Fig. 9.19, we note that if the stresses for 2×10^8 cycles of full power are on the curve, stresses for 50 percent power would correspond to over 10^{10} cycle life. Considering the Palmgren-Miner cumulative damage rule (Section 8.12), and recognizing the approximate nature of our solution, it appears justified to design for the full-load cycles only, and ignore the half-load cycles.
5. For minimum center distance (i.e., minimum gear diameters), tentatively choose width, b, at the maximum of the normal range, $14/P$.

6. Anticipating that surface fatigue will likely be more critical than bending fatigue, we solve for the value of P that will balance σ_H and S_H with a small safety factor, S.F., of say 1.25:

$$\sigma_H \text{ (from Eq. (15.24))} = S_H \text{ (from Eq. (15.25))}$$

$$C_p \sqrt{\frac{F_t(\text{S.F.})}{bd_pI} K_vK_oK_m} = S_{fe}C_{Li}C_R$$

A few auxiliary calculations are required:

$$V = \pi d_p(3600 \text{ rpm})/12 = 942d_p = 942(18/P) = 16{,}960/P$$

$K_v \approx 1.4$ (this is a rough estimate from Fig. 15.23 that must be confirmed or modified after P is determined)

$K_m = 1.3$ (must be increased if b comes out >2 in.)

$$F_t = 100 \text{ hp } (33{,}000)/V = 195P$$

$$I = ((\sin 20^\circ \cos 20^\circ)/2)(4/5) = 0.128$$

$S_{fe}C_{Li}C_R = 165{,}000$ psi (directly from Fig. 9.19)
Substituting:

$$2300 \sqrt{\frac{(195P)(1.25)}{(14/P)(18/P)(0.128)} (1.4)(1)(1.3)} = 165{,}000$$

from which

$$P = 7.21$$

7. Tentatively choose a standard pitch of 7, compute the corresponding value of V, refine the estimate of K_v, and compute the value of b required to balance σ_H and S_H. (Note that if $P = 8$ were chosen, b would have to exceed $14/P$ to balance σ_H and S_H.)

$$V = \frac{\pi d_p n_p}{12} = \frac{\pi(18/7)(3600)}{12} = 2424 \text{ fpm}$$

From Fig. 15.23, $K_v = 1.5$,

$$2300 \sqrt{\frac{(195 \times 7)(1.25)}{b(18/7)(0.128)} (1.5)(1)(1.3)} = 165{,}000$$

from which $b = 1.96$ in. Round off to $b = 2$ in. (For this value of b, $K_m = 1.3$ is satisfactory. Also, note that b remained at $14/P$ because decreasing P from 7.21 to 7 offset increasing K_v from 1.4 to 1.5)

8. Check contact ratio, using Eq. (15.9).

The pitch radii are $r_p = 9/7$ and $r_g = 36/7$.

Adding addendum $= 1/P$: $r_{ap} = 10/7$, $r_{ag} = 37/7$.

Center distance, $c = r_p + r_g = 45/7$.

From Eq. (15.11), $r_{bp} = (9/7) \cos 20^\circ$, $r_{bg} = (36/7) \cos 20^\circ$.

From Eq. (15.10), $p_b = \pi(18/7)(\cos 20°)/18$.

Substituting in Eq. (15.9) gives $C.R. = 1.67$.

This is satisfactory, but it means that a single pair of teeth carries the load in the vicinity of the pitch line, where pitting is most apt to occur. Thus, there can be no sharing of the surface fatigue load, regardless of manufacturing precision. (Note that no sharing was assumed in the above calculations.)

9. Provide for adequate bending fatigue strength. Detailed consideration of gear-tooth bending fatigue for case hardened gears involves an analysis of stress and strength gradients, as represented in Fig. 8.29. Since we anticipate no problem in satisfying this requirement, let us make the conservative assumption that the fatigue strength of the *core* material (Eq. (15.18)) must be equal to the bending fatigue stresses at the surface (Eq. (15.17)):

$$S'_n\, C_L C_G C_S k_r k_t k_{ms} = \frac{F_t P}{bJ} K_v K_o K_m$$

The manufacturing accuracy involved is in a "gray area" with respect to load sharing. There will likely be at least a partial sharing, meriting a value of J at least intermediate between the "sharing" and "not sharing" curves (i.e., between $J = 0.235$ and 0.32). But let us conservatively assume no sharing—if it is easy to provide adequate fatigue strength for this possibility, there is no need to consider the matter further. In choosing C_S, remember that we are considering fatigue strength *under* the surface, where surface roughness would not be involved:

$$S'_n(1)(1)(1)(0.897)(1)(1.4) = \frac{1365(7)}{2(0.235)}(1.5)(1)(1.3)$$

from which $S'_n = 31{,}600$ psi, which requires a (core) hardness of 126 Bhn, a value that will be satisfied or exceeded by any steel selected to meet the case-hardened surface requirement.

10. In summary, our tentatively proposed design involves 20-deg. full-depth teeth, precision manufactured with ground finish (between curves A and B of Fig. 15.23) from case hardening steel, surface hardened to 660 and 600 Bhn for pinion and gear, respectively, and with core hardness at least 126 Bhn. In accordance with good practice, we shall specify a case hardening procedure leaving compressive residual stresses in the surfaces. $P = 7$, $N_p = 18$, $N_g = 72$, $b = 2$ in. ($D_p = 2.57$ in., $D_g = 10.29$ in., $c = 6.43$ in.) ■

The above sample problem represents but one of a great many situations and approaches encountered in the practical design of spur gears. The important thing for the student is to gain a clear understanding of the basic concepts involved, and to understand how these may be brought to bear in handling specific situations. In addition to the fundamentals, it has been seen that a great deal of empirical data are needed. It is always important to seek out the best and most directly relevant empirical data for use in any given situation. Textbooks such as this can include only sample empirical information. Better values for actual use are often found in company files, contemporary specialized technical literature, and current publications of the AGMA.

15.12 GEAR MATERIALS

The least expensive gear material is usually ordinary cast iron, ASTM (or AGMA) grade 20. Grades 30, 40, 50, and 60 are progressively stronger and more expensive. Cast iron gears typically have greater surface fatigue strength than bending fatigue strength. Their internal damping tends to make them quieter than steel gears. Nodular cast iron gears have substantially greater bending strength, together with good surface durability. A good combination is often a steel pinion mated to a cast iron gear.

Unheat-treated steel gears are relatively inexpensive, but have low surface endurance capacity. Heat-treated steel gears must be designed to resist warpage; hence, alloy steels and oil quenching are usually preferred. For hardnesses over 250–350 Bhn, machining must usually be done before hardening. Greater profile accuracy is obtained if the surfaces are finished after heat-treating, as by grinding. (But if grinding is done, care must be taken to avoid residual tensile stresses at the surface.) Through-hardened gears generally have 0.35 to 0.6 percent carbon. Surface or case-hardened gears are usually processed by flame hardening, induction hardening, carburizing, or nitriding.

Of the non-ferrous metals, bronzes are most often used for making gears.

Nonmetallic gears (rawhide, nylon, and other plastics) are generally quiet, durable, reasonably priced, and often can operate under light loads without lubrication. Their teeth deflect more easily than those of corresponding metal gears. This promotes effective load sharing among teeth in simultaneous contact, but results in substantial hysteresis heating if the gears are rotating at high speed. Since nonmetallic materials have low thermal conductivity, special cooling provisions may be required. Furthermore, these materials have relatively high coefficients of thermal expansion, and this may require installation with greater backlash than metal gears.

Often the base plastics used for gears are formulated with fillers, such as glass fibers, for strength, and with lubricants. Nonmetallic gears are usually mated with cast iron or steel pinions. For best wear resistance, the hardness of the mating metal pinion should be at least 300 Bhn. Design procedures for gears made of plastics are similar to those for gears made of metals, but are not yet as reliable. Hence, prototype testing is even more important than for metal gears.

15.13 GEAR TRAINS

The speed ratio (or "gear ratio") of a single pair of *external* spur gears is expressed by the simple equation:

$$\frac{\omega_p}{\omega_g} = \frac{n_p}{n_g} = -\frac{d_g}{d_p} = -\frac{N_g}{N_p} \tag{15.28}$$

(an expanded version of Eq. (15.1)) where ω and n are rotating speed in radians per second and rpm, respectively, d represents pitch diameter, and N the number of teeth. The minus sign indicates that an ordinary pinion and gear (both with external teeth) rotate in *opposite* directions. If the gear has internal teeth (as in Fig. 15.12), its diameter is negative and the members rotate in the *same* direction. In most applications, the pinion is the driver and the gear the output, which provides a *reduction ratio* (reduction

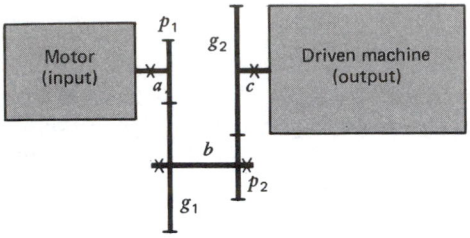

FIGURE 15.27.
Double-reduction gear train.

in speed, but increase in torque). This is because power sources (engines, motors, turbines) usually rotate relatively fast in order to provide a large amount of power from a given size unit. The machinery being driven usually runs slower. (There are exceptions: for example, engine-driven superchargers, and large centifugal compressors for refrigeration and air conditioning.)

Figure 15.27 shows a double reduction gear train involving countershaft b as well as input shaft a and output shaft c. The overall speed ratio is

$$\frac{\omega_a}{\omega_c} = \frac{\omega_a}{\omega_b} \cdot \frac{\omega_b}{\omega_c} = -\frac{d_{gl}}{d_{pl}} \cdot -\frac{d_{g2}}{d_{p2}} = + \frac{d_{gl}d_{g2}}{d_{pl}d_{p2}} = \frac{N_{gl}N_{g2}}{N_{pl}N_{p2}} \tag{15.29}$$

Note that if the two gear pairs have the same center distance, the input and output shafts can be in exact alignment, which may facilitate economical manufacture of the housing.

Figure 15.27 and Eq. (15.29) can be extended to three, four, or any number of gear pairs, with the overall ratio being the product of the ratios of the individual pairs. Familiar examples are the gear trains in odometers and in mechanical watches and clocks.

Planetary (or *epicyclic*) gear trains are more complicated to analyze because some of the gears rotate about axes that are *themselves* rotating. Figure 15.28*a* illustrates a typical planetary train, which includes a sun gear, S, at the center, surrounded by planets, P, that rotate freely on shafts mounted in arm (also called "carrier"), A. Also meshing with the planets is a ring or annulus gear, R, that has internal teeth. Figure 15.28*b* is a simplified version in that only a single planet is shown. Actual planetary trains incorporate two or more planets, equally spaced, to balance the forces acting on the sun, ring, and arm. Also, by dividing the load between multiple planets, the torque and power capacity of the train are correspondingly increased. When analyzing planetary train speed ratios, it may be more convenient to refer to the single planet drawing (Fig. 15.28*b*).

The three members S, A, and R are normally assigned three functions: input, output, and fixed reaction member. Let us examine three alternative arrangements: (1) With A as the reaction member, we have a simple gear train (all of the axes are fixed), and members S and R rotate in opposite directions, providing a reverse gear. (2) With

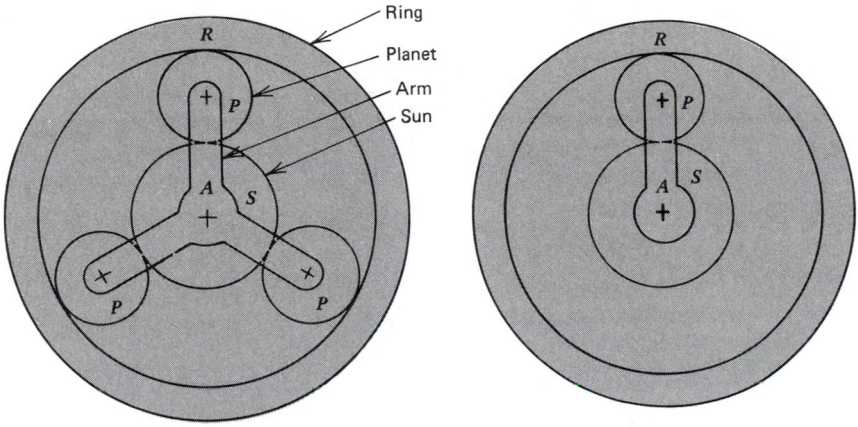

(a)
With three planets (typical)

(b)
With one planet (for analysis only)

FIGURE 15.28.
Typical planetary gear train.

R held fixed, S and A rotate in the same direction but at different speeds. (3) With S fixed, R and A also rotate in the same direction, but with different speed ratios than when R is fixed. Regardless of the arrangement chosen, one can install a clutch enabling any two of members S, A, and R to be locked together. This causes the entire planetary train to rotate as a single member, and provides a *direct drive* (gear ratio = 1) from input to output. Automotive automatic transmissions use combinations of planetary gear trains, with a clutch for direct drive and with brakes or one-way clutches to hold various members fixed to obtain the different ratios.

We now present three methods for determining planetary gear ratios, illustrated for the train shown in Fig. 15.28 with R as the input, A as the output, and S as the fixed reaction member. In order to simplify the notation, letters R, S, and P are used to represent either the diameters or the numbers of teeth in the ring, sun, and planet, respectively.

1. *Free-body force analysis.* Figure 15.29 shows an exploded diagram of three components. Using the above notation, the arm radius is equal to:

$$\frac{S+P}{2} = \frac{S+(R/2-S/2)}{2} = \frac{R+S}{4}$$

(as shown). Starting with input torque T_i applied to the ring, loads are placed on each of the members to put them in equilibrium. This leads to the conclusion that:

$$\frac{\omega_i}{\omega_o} = \frac{T_o}{T_i} = 1 + \frac{S}{R} \tag{g}$$

2. *Velocity vector analysis.* Figure 15.30 shows arbitrary velocity vector V drawn at the meshing of R and P. The linear velocity is zero at the meshing of S and P because

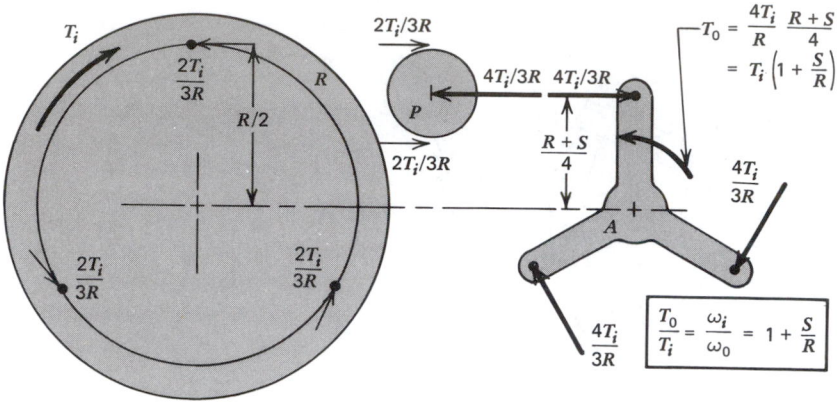

(R = input; A = output; S = fixed member)

FIGURE 15.29.
Torque ratio (= 1/speed ratio) determined by free-body diagrams.

the sun is fixed. Angular velocities of R and A are determined as the ratios of linear velocities to corresponding radii. This analysis again leads to Eq. (g).

3. *General planetary train equation.* In Fig. 15.28 (and with no member necessarily held fixed), the angular velocity of the ring with respect to the arm, and of the sun with respect to the arm are, by definition:

$$\omega_{R/A} = \omega_R - \omega_A \quad \text{and} \quad \omega_{S/A} = \omega_S - \omega_A$$

from which

$$\frac{\omega_{R/A}}{\omega_{S/A}} = \frac{\omega_R - \omega_A}{\omega_S - \omega_A} \tag{h}$$

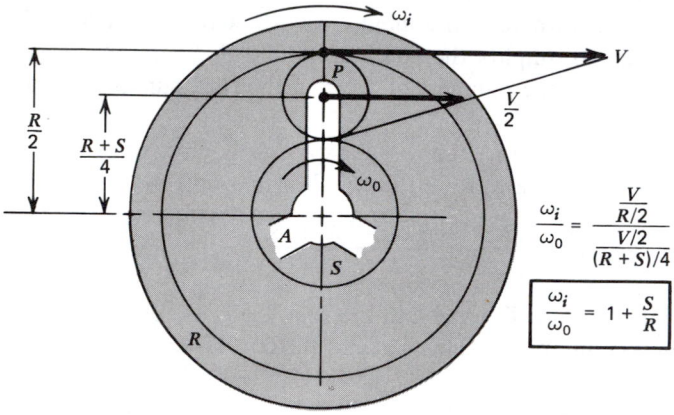

(R = input; A = output; S = fixed member)

FIGURE 15.30.
Speed ratio determined by velocity vector diagram.

Equation (h) is true for *any* angular velocity of the arm, including zero. With the arm fixed, angular velocity ratios are computed from Eq. (15.29), and the result is known as the *train value, e*. Thus,

$$\frac{\omega_{R/A}}{\omega_{S/A}} = \frac{\omega_R}{\omega_S} = e = \left(-\frac{S}{P}\right)\left(+\frac{P}{R}\right) = -\frac{S}{R} \tag{i}$$

Combining (h) and (i) gives

$$e = -\frac{S}{R} = \frac{\omega_R - \omega_A}{\omega_S - \omega_A} \tag{15.30}$$

where R and S again represent either the pitch diameters or numbers of teeth in the ring and sun gears. Applying Eq. (15.30) to Fig. 15.28 with S the fixed member, we once more obtain Eq. (g).

To adapt Eq. (15.30) to a complex planetary train, first identify the three members providing the input, output, and reaction functions. One of these will be the arm. Call the other two X and Y. Then the train value is

$$e = \frac{\omega_X}{\omega_Y} = \frac{\omega_X - \omega_A}{\omega_Y - \omega_A} \tag{15.31}$$

The Fig. 15.29 arrangement with S fixed is perhaps the most commonly used planetary train. Depending on the relative gear sizes, the value of the ratio determined from Eq. (15.30) can be any value between 1 and 2. With R as the input and A the output, this train is often used as a reduction gear for propeller-driven aircraft. With A as the input and R the output, it is the basis of the conventional automotive overdrive. Perhaps the most familiar application of all is the Sturmy-Archer 3-speed bicycle hub, which is shifted between (1) low gear, connected as in the aircraft reduction gear, (2) intermediate gear, with all parts rotating together for direct drive, and (3) high gear, connected as in the automotive overdrive—a very ingenious arrangement.

In Fig. 15.29 it was assumed that the load is divided equally among all planets. Actually, this happens only if (1) the parts are made with sufficient precision, or (2) special construction features are employed to automatically equalize the loading.

There are two basic factors that control the number and spacing of the planets employed: (1) the maximum number of planets is limited by the space available—that is, the tooth tips of any planet must clear those of the adjacent planets—and (2) the teeth of each planet must align simultaneously with teeth of the sun and annulus. Figure 15.31 shows an example where the latter condition is satisfied for two equally spaced planets but not for four. (The reader is invited to continue this study and show that with the tooth numbers used in Fig. 15.31, the number of equally spaced planets can only be two or five.)

REFERENCES

1. Buckingham, Earle, *Analytical Mechanics of Gears,* McGraw-Hill, New York, 1949.
2. Dudley, D. W. (ed.), *Gear Handbook,* McGraw-Hill, New York, 1962.

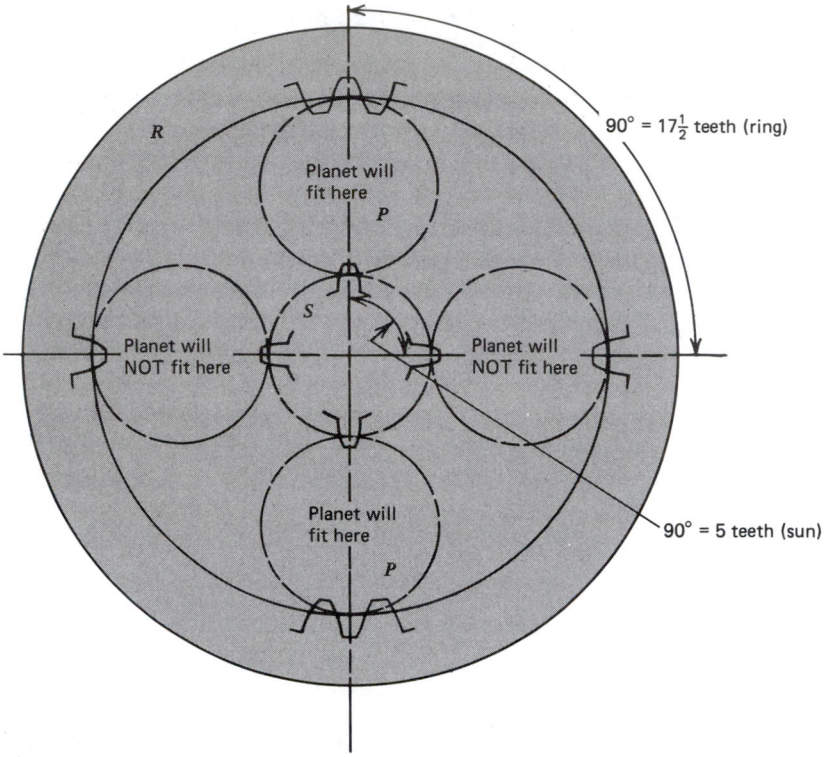

FIGURE 15.31.
Geometric study of 2 vs. 4 equally spaced planets for 20-tooth sun and 70-tooth ring.
1. Planets must have $(70 - 20)/2 = 25$ teeth. Being an odd number, the sun and ring *must* be aligned with a sun tooth opposite a ring space between teeth, as shown at top and bottom position.
2. With sun and ring properly indexed for top and bottom planets, the two side planets will *not* fit.
3. Conclusion: Two equally spaced planets can be used; four cannot.

3. Dudley, W. W., *The Evolution of the Gear Art,* American Gear Manufacturers Association, Washington, D.C., 1969.

4. Kelley, O. K., "Design of Planetary Gear Trains," Chapter 9 of *Design Practices—Passenger Car Automatic Transmissions,* Society of Automotive Engineers, New York, 1973.

5. "1981 Mechanical Drives Reference Issue," *Machine Design,* Penton/IPC, Cleveland, June 18, 1981.

6. Merritt, H. E., *Gear Engineering,* Pitman Publishing, Inc., Marshfield, Mass., 1971.

7. Standards of the American Gear Manufacturers Association, Arlington, Virginia.

CHAPTER SIXTEEN

Helical, Bevel, and Worm Gears

16.1 INTRODUCTION

The treatment here of the principal types of nonspur gears will be relatively brief. The basic principles and many of the equations from the previous chapter apply to these gear types as well.

A *helical gear* (Fig. 16.1*a*) can be thought of as an ordinary spur gear machined from a stack of thin shim stock where each lamination is rotated slightly with respect to its neighbors (Fig. 16.1*b*). When power is transmitted by a pair of helical gears, both shafts are subjected to a thrust load. This can be eliminated by using *double helical* or *herringbone* gears (Fig. 16.1*c*), but this significantly increases gear manufacturing and mounting cost.

As helical (or herringbone) gears rotate, each tooth comes first into engagement on one side, with contact spreading across the tooth as rotation continues. Thus, teeth come into engagement *gradually*, which makes for smoother and quieter operation than with spur gears. The gradual engagement also results in a lower dynamic factor, K_v, and often permits higher rotating speeds. A common application of helical gears is in automotive transmissions, where quietness is a priority consideration.

Although helical gears usually operate on parallel shafts, they can be made to operate on nonparallel, nonintersecting shafts (Fig. 16.1*d*). In this case they are called *crossed helical* gears (previously called "spiral," or "skew," gears). Because crossed helical gears theoretically have point contact, they can carry only light loads. A common application is driving the distributor and oil pump from the camshaft in automotive engines.

Bevel gears (Fig. 16.2) have teeth shaped generally like ordinary spur gears, except

(a) Mounted on parallel shafts (most common type). Gears have helices of opposite hand

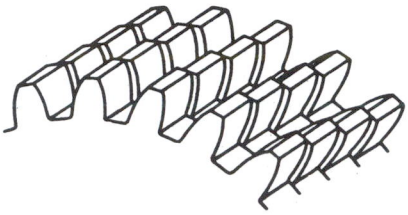

(b) Rotated spur gear laminations approach a helical gear as laminations approach zero thickness.

(c) *Double helical* or *herringbone* gears may or may not have a center space, depending on manufacturing method.

(d) When mounted on non–parallel shafts, they are *crossed helical* gears, and usually have the same hand.

FIGURE 16.1.
Types of helical gears. (*a* and *d* Courtesy Boston Bear; *c* Courtesy Horsburgh & Scott)

that the tooth surfaces are made up of conical elements. The teeth may be straight (Figs. 16.2*a* and 16.2*b*), or spiral (Fig. 16.2*c*). Spiral teeth engage *gradually* (starting at one side, as with helical gears), a feature enabling them to operate smoother and quieter. Except for hypoid gears (Fig. 16.2*e*), bevel gears are mounted on shafts having intersecting axes. The shaft axes are usually, but not necessarily, perpendicular. Figure

(*a*) Straight-tooth bevel gears

(*b*) Straight-tooth bevel gears
Special case of miter gears (1:1 ratio)

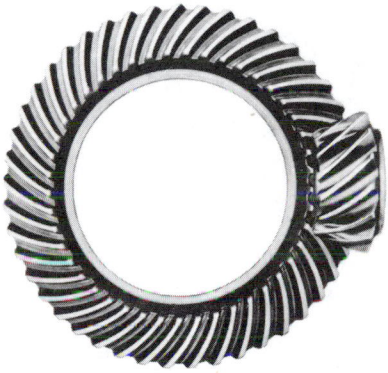

(*c*) Spiral bevel gears

(*d*) Bevel gears mounted on nonperpendicular shafts

(*e*) Hypoid gears

FIGURE 16.2.
Types of bevel gears. (*a, c, d,* and *e* Courtesy Gleason Machine Division; *b* Courtesy Horsburgh & Scott)

16.2*d* shows a rather extreme case of nonperpendicular shaft axes. The nonintersecting shaft feature of hypoid gears is desirable for automobile rear-axle applications as this permits the propeller shaft to be lowered, resulting in a lower floor, roof, and center of gravity.

A *worm and worm gear* set (Fig. 16.3) is essentially a screw meshing with a special

(*a*)
Single enveloping
(Courtesy Horsburg & Scott)

(*b*)
Double enveloping
(Courtesy Ex-Cell-O Corp., Cone Drive Operations)

FIGURE 16.3.
Worm and worm gear sets.

helical gear. Like a screw, the worm can have one or more threads (as illustrated in Fig. 10.1). As will be seen in Section 16.11, the analysis of forces acting on a worm is essentially the same as for a screw (Fig. 10.6).

Two things immediately characterize worm gearing: (a) large velocity ratios (up to 300 or more), and (b) high sliding velocities. The latter means that heat generation and power transmission efficiency are of greater concern than with other types of gears.

See Ref. 3 for other variations of these basic gear types.

16.2 HELICAL GEAR GEOMETRY AND NOMENCLATURE

Spur gears, treated in the preceding chapter, are merely helical gears with a zero helix angle. Figure 16.4 shows a portion of a helical rack (with nonzero helix angle). The helix angle, ψ, is always measured on the cylindrical pitch surface. Values of ψ are not standardized, but commonly range between 15 and 30 degrees. Lower values give less end thrust, but higher values tend to give smoother operation. Helical gears are right- or left-hand, defined the same as for screw threads (see Fig. 10.1). In Fig. 16.1*a* the pinion is right-hand and the gear left-hand. Note that mating helical gears (on parallel shafts) must have the *same helix angle* but *opposite hand.*

Figure 16.4 illustrates that circular pitch (p) and pressure angle (ϕ) are measured in the *plane of rotation*, as with spur gears. New quantities for circular pitch and pressure angle measured in a *plane normal to the teeth* are shown in the figure as p_n and ϕ_n. From simple geometry

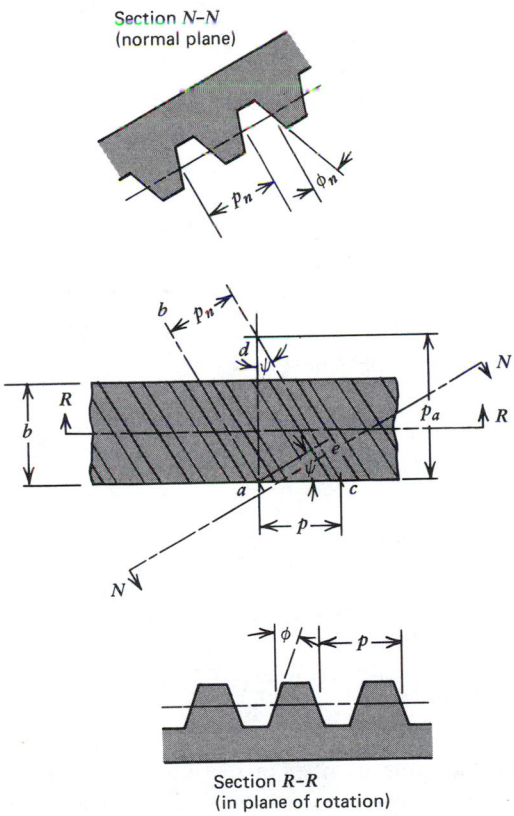

FIGURE 16.4.
Portion of a helical rack.

$$p_n = p \cos \psi \tag{16.1}$$

In the next article it will be shown that

$$\tan \phi_n = \tan \phi \cos \psi \tag{16.2}$$

Whether ϕ or ϕ_n is taken as a standard value depends on the method used to cut the teeth. Standard values for addendum and dedendum are $1/P_n$ and $1.25/P_n$, respectively (in inches), but special gears with nonstandard addendum and dedendum are common.

Since the product of circular and diametral pitch is π for the normal as well as the rotational plane,

$$P_n = P/\cos \psi \tag{16.3}$$

The pitch diameter of a helical gear is

$$d = N/P = N/(P_n \cos \psi) \tag{16.4}$$

Note that axial pitch, p_a, is defined in Fig. 16.4 as the distance between corresponding points on adjacent teeth measured on the pitch surface in the axial direction. Thus

$$p_a = p/\tan \psi \tag{16.5}$$

To achieve axial overlap of adjacent teeth, $b \geqq p_a$. In practice, it is usually considered desirable to make $b \geqq 1.15p_a$, and in many cases, $b \geqq 2p_a$.

Neglecting sliding friction, the resultant load between mating teeth is always perpendicular to the tooth surface. Thus, with helical gears, the load is in the *normal* plane. Hence, bending stresses are computed in the normal plane, and the strength of the tooth as a cantilever beam depends on its profile in the normal plane. Since this is different than the profile in the plane of rotation, the appropriate Lewis form factor (Y) and geometry factor (J) must be based on the tooth profile in the normal plane.

Figure 16.5 shows the pitch cylinder and one tooth of a helical gear. The normal plane intersects the pitch cylinder in an ellipse. The shape of the tooth in the normal plane is nearly (not exactly) the same as the shape of a spur gear tooth having a pitch radius equal to radius R_e of the ellipse. From analytical geometry

$$R_e = (d/2) \cos^2 \psi \tag{a}$$

The equivalent number of teeth (also called the formative or virtual number of teeth), N_e, is defined as the number of teeth in a gear of radius R_e:

$$N_e = \frac{2\pi R_e}{p_n} = \frac{\pi d}{p_n \cos^2 \psi} \tag{b}$$

From Eq. (16.1)

$$p_n = p \cos \psi = \pi d(\cos \psi)/N, \quad \text{or } \pi d/p_n = N/\cos \psi \tag{c}$$

Substituting this into (b) gives

$$N_e = N/\cos^3 \psi \tag{16.6}$$

When computing bending strength of helical teeth, values of the Lewis form factor Y are the same as for spur gears having the same number of teeth as the formative number of teeth in the helical gear, and a pressure angle equal to ϕ_n. This is taken into account in the determination of the appropriate values of geometry factor J for helical gears, which are plotted in Figs. 16.8, and discussed in Section 16.4.

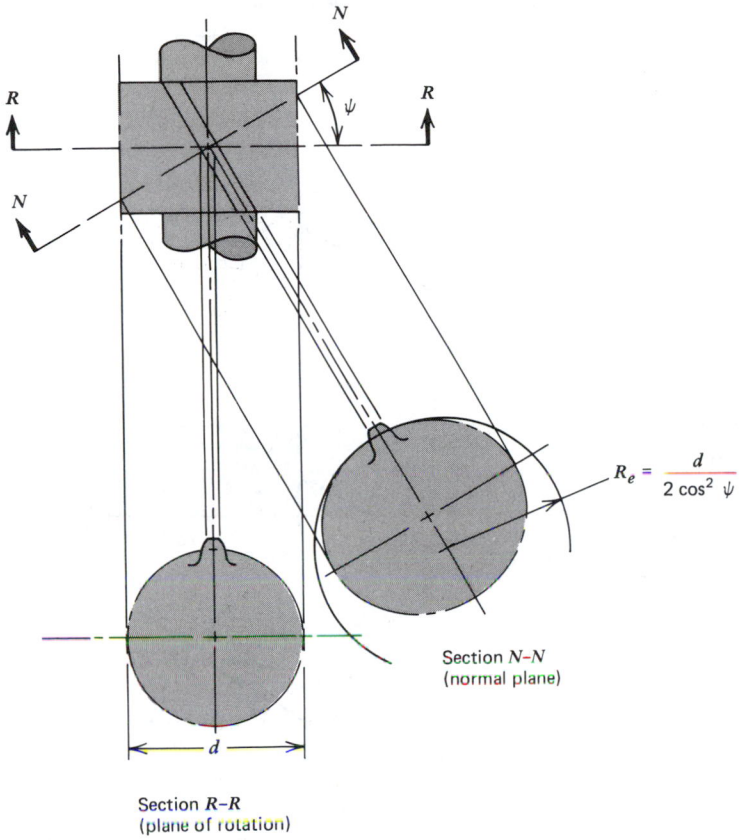

FIGURE 16.5.
Helical gear pitch cylinder and one tooth.

16.3 HELICAL GEAR FORCE ANALYSIS

Figure 16.6 illustrates the comparison of force components on spur and helical gears. For spur gears, the total tooth force F consists of components F_t and F_r. For helical gears, component F_a is added, and section N-N is needed to show a true view of total tooth force, F. Force $F_t \mapsto F_a$ is labeled F_b, the subscript b being chosen because F_b is the bending force on the helical tooth (just as F_t is the bending force on the spur tooth).

The force component associated with power transmission is, of course, F_t, where

$$F_t \text{ (lb)} = 33{,}000\ hp/V \tag{16.7}$$

where V is the pitch-line velocity in feet per minute; or

$$F_t \text{ (newtons)} = w \text{ (watts)}/V \tag{16.7a}$$

where V is measured in meters per second.

From basic geometric relationships, the other forces shown in Fig. 16.6 can be expressed in terms of F_t:

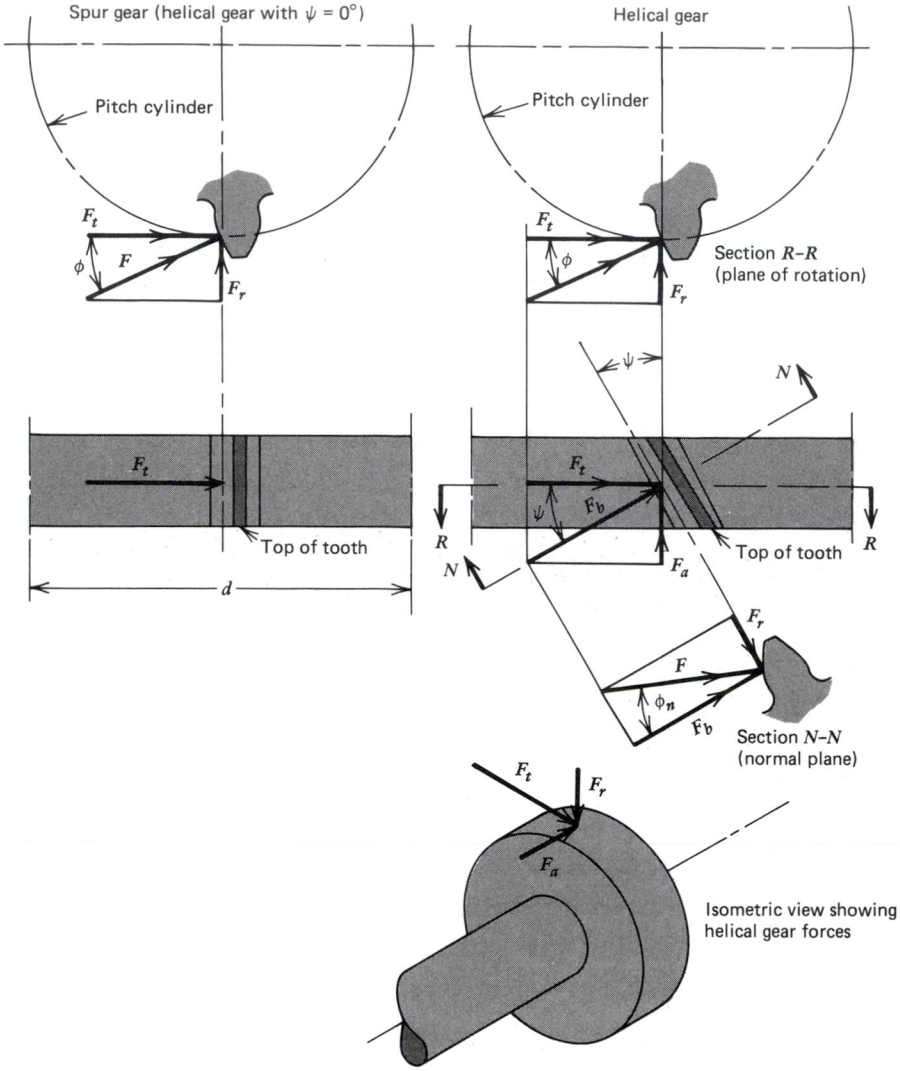

FIGURE 16.6.
Force components in spur and helical gears.

$$F_r = F_t \tan \phi \tag{16.8}$$

$$F_a = F_t \tan \psi \tag{16.9}$$

$$F_b = F_t / \cos \psi \tag{16.10}$$

$$F = F_b / \cos \phi_n = F_t / \cos \psi \cos \phi_n \tag{16.11}$$

In order to derive Eq. (16.2) (given in the previous article) note from Fig. 16.6 that

$$F_r = F_b \tan \phi_n \tag{d}$$

Combining Eqs. (d) and (16.10) gives

$$F_r = F_t \tan \phi_n / \cos \psi \tag{e}$$

Combining Eqs. (e) and (16.8) gives

$$\tan \phi_n = \tan \phi \cos \psi \tag{16.2}$$

SAMPLE PROBLEM 16.1

Figure 16.7*a* shows a motor driving a machine through a helical-gear speed reducer. From information given in the drawing, determine gear dimensions ϕ, P, d_p, d_g, and

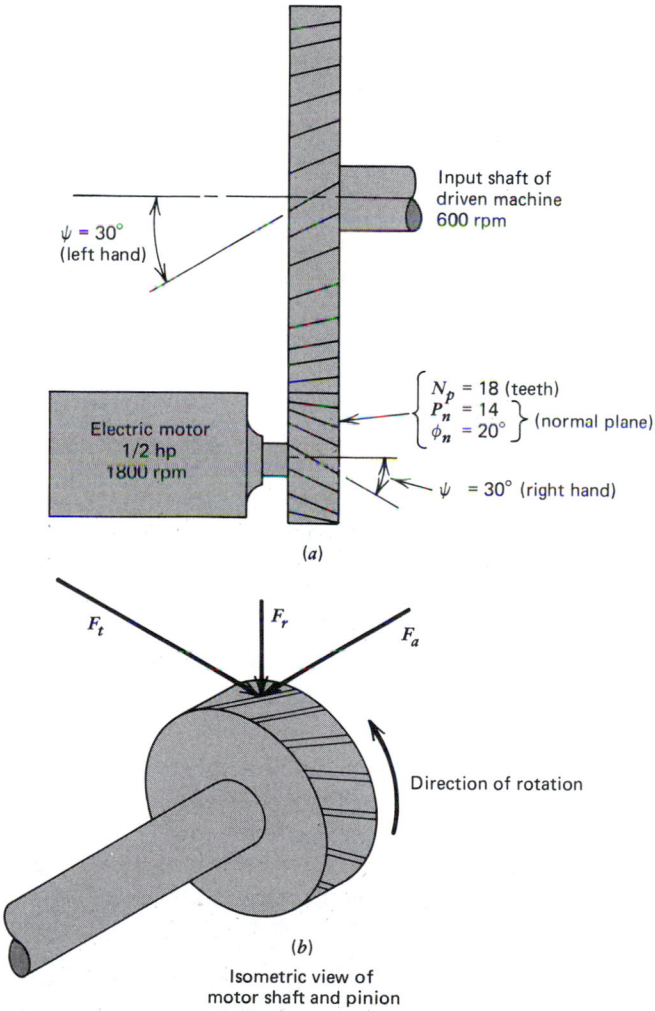

FIGURE 16.7.
Drawings for Sample Problem 16.1.

N_g; pitch-line velocity V; gear-tooth forces F_t, F_r, and F_a; and gear face width, b, to give $b = 1.5p_a$.

Solution

1. From Eq. (16.2): $\phi = \tan^{-1}(\tan \phi_n/\cos \psi)$
$= \tan^{-1}(\tan 20°/\cos 30°) = 22.8°$ ■
2. From Eq. (16.3): $P = P_n \cos \psi = 14 \cos 30° = 12.12$ teeth/in. ■
3. $d_p = N_p/P = 18/12.12 = 1.48$ in. ■
4. $N_g = N_p(n_p/n_g) = 18\,(1800 \text{ rpm}/600 \text{ rpm}) = 54$ teeth; ■
$d_g = N_g/P = 54/12.12 = 4.45$ in. ■
5. $V = \pi d_p n_p/12 = \pi(1.49)(1800)/12 = 702$ fpm. ■
6. $F_t = 33{,}000 \text{ hp}/V = 33{,}000(0.5)/702 = 23.6$ lb. ■
7. $F_r = F_t \tan \phi = 23.5 \tan 22.8° = 9.9$ lb. ■
8. $F_a = F_t \tan \psi = 23.5 \tan 30° = 13.6$ lb. ■

   Directions of the three force components acting on the pinion are shown in Fig. 16.7*b*. (Forces acting on the gear are, of course, equal and opposite.)

9. $p_a = p/\tan \psi = \pi/P \tan \psi = \pi/12.2 \tan 30° = 0.45$ in.
10. For $b = 1.5p_a$,

$$b = 1.5(0.45) = 0.67 \text{ in.}$$ ■

16.4 HELICAL GEAR-TOOTH BENDING AND SURFACE FATIGUE STRENGTHS

The bending stress equation for straight spur gear teeth (Eq. (15.17)) applies with only slight modification to helical gear teeth:

$$\sigma = \frac{F_t P}{bJ} K_v K_o (0.93 K_m) \tag{16.12}$$

where J is obtained from Fig. 16.8 and K_v is normally taken from curves A or B of Fig. 15.23. Introduction of the constant 0.93 with the mounting factor reflects the slightly lower sensitivity of helical gears to mounting conditions.

Bending stresses computed from Eq. (16.12) are compared with fatigue strengths computed from Eq. (15.18), repeated below, exactly as with spur gears:

$$S_n = S_n' C_L C_G C_S k_r k_t k_{ms} \tag{15.18}$$

In modifying the surface fatigue stress equation for spur gears (Eq. (15.24)) to make it applicable to helical gears, a more fundamental difference between the two types of gears is encountered: At the pitch surface (where, because of zero sliding velocity, the oil film gets squeezed out and surface pitting is most likely to occur), spur gears

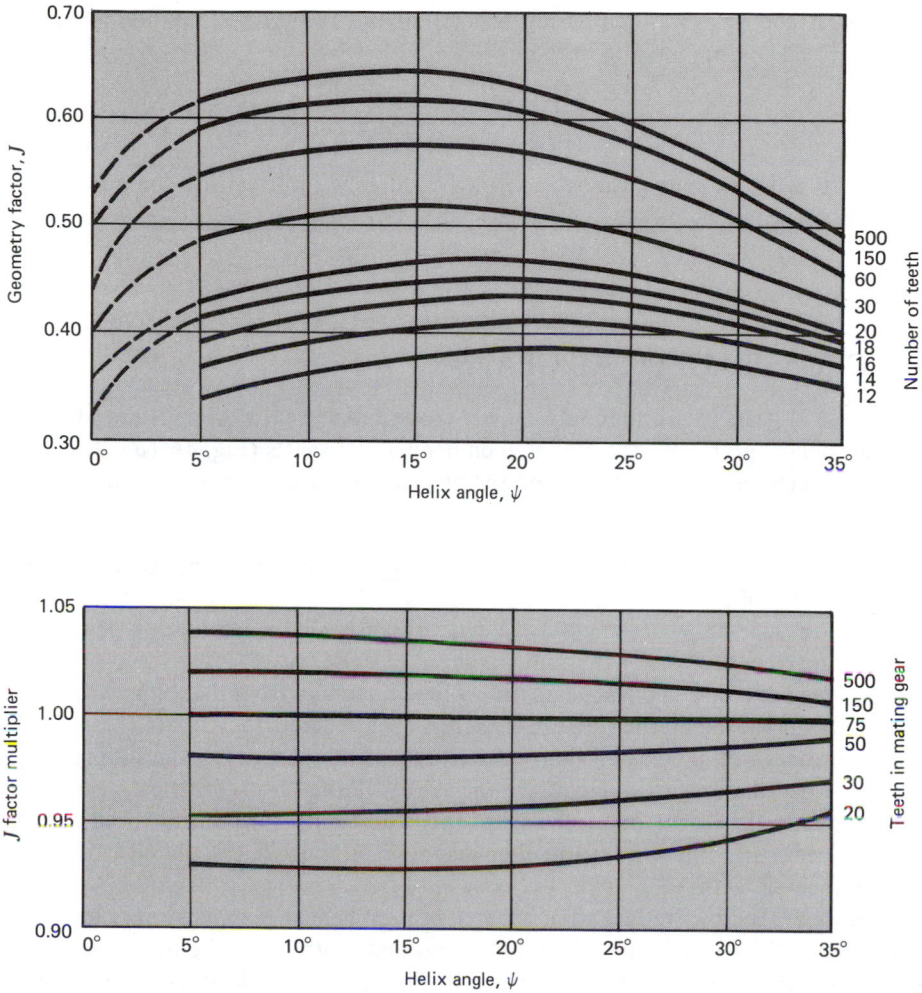

FIGURE 16.8.
Geometry factor (J) for helical gears having $\phi_n = 20$ deg., standard addendum of $1/P_n$, and shaved teeth (a) when used with a 75-tooth mating gear. (Spur gear values from Fig. 15.22 are shown at $\psi = 0$ deg. for comparison). (b). J factor multipliers for use when mating gear has other than 75 teeth. (From AGMA Information Sheet 226.01, which also gives J factors for $\phi_n = 14.5$, 15, and 20 deg., for unequal gear and pinion addendum, and for ground and hobbed tooth surfaces.)

of contact ratio less than 2 have a theoretical length of tooth contact of $1.0b$. With helical gears, the length of contact per tooth is $b/\cos\psi$, and the helical action causes the total length of tooth contact to be approximately $b/\cos\psi$ times the contact ratio ($C.R.$) at all times. The AGMA[1] recommends that 95 percent of this value be taken as

[1]See AGMA Standard 211.02.

the length of contact when computing contact stress. Thus, when applied to helical gears, Eq. (15.24) is modified to become

$$\sigma_H = C_p \sqrt{\frac{F_t}{bd_pI}\left(\frac{\cos\psi}{0.95\ C.R.}\right) K_vK_o(0.93K_m)} \tag{16.13}$$

As with ordinary spur gears, the surface endurance limit can be computed from Eq. (15.25), repeated below:

$$S_H = S_{fe}C_{Li}C_R \tag{15.25}$$

16.5 CROSSED HELICAL GEARS

Crossed helical gears (more precisely called "crossed-axes helical gears") are identical with other helical gears, but are mounted on nonparallel shafts (Fig. 16.1*d*). The relationship between the shaft angle, Σ, and the helix angles of mating gears 1 and 2 is:

$$\Sigma = \psi_1 + \psi_2 \tag{16.14}$$

Mating gears usually have helices of the same hand, but if not, use a negative sign with the smaller value of ψ.

The most common shaft angle is 90 deg., which results when mating gears have complementary helix angles of the same hand.

The action of crossed helical gears differs fundamentally from that of parallel-shaft helical gears in that the mating teeth *slide* across each other as they rotate. This sliding velocity increases with increasing shaft angle. For a given shaft angle, the sliding velocity is least when the two helix angles are the same. Mating crossed helical gears must have the same p_n and ϕ_n, but not necessarily the same p and ϕ. Furthermore, the velocity ratio is not necessarily the ratio of the pitch diameters; it must be calculated as the ratio of the numbers of teeth.

Because of their theoretical point contact, crossed helical gears have very low load-carrying capacities—usually less than 400*N* resultant tooth load. The limitation is one of surface deterioration, not bending strength. To increase the load capacity, contact ratios of 2 or more are usually used. To increase contact ratio, low values of pressure angle and relatively large values of tooth depth are commonly specified.

16.6 BEVEL GEAR GEOMETRY AND NOMENCLATURE

When intersecting shafts are connected by gears, the *pitch cones* (analogous to the pitch cylinders of spur and helical gears) are tangent along an element, with their apexes at the intersection of the shafts. Figure 16.9 shows the basic geometry and terminology. The size and shape of the teeth are defined at the *large end*, where they intersect the back cones. Note that the pitch cone and back cone elements are perpendicular. Figure 16.9 shows the tooth profiles at the back cones. These profiles resemble those of spur gears having pitch radii equal to the developed back cone radii r_{bg} (gear) and r_{bp} (pinion). The numbers of teeth in these imaginary spur gears are

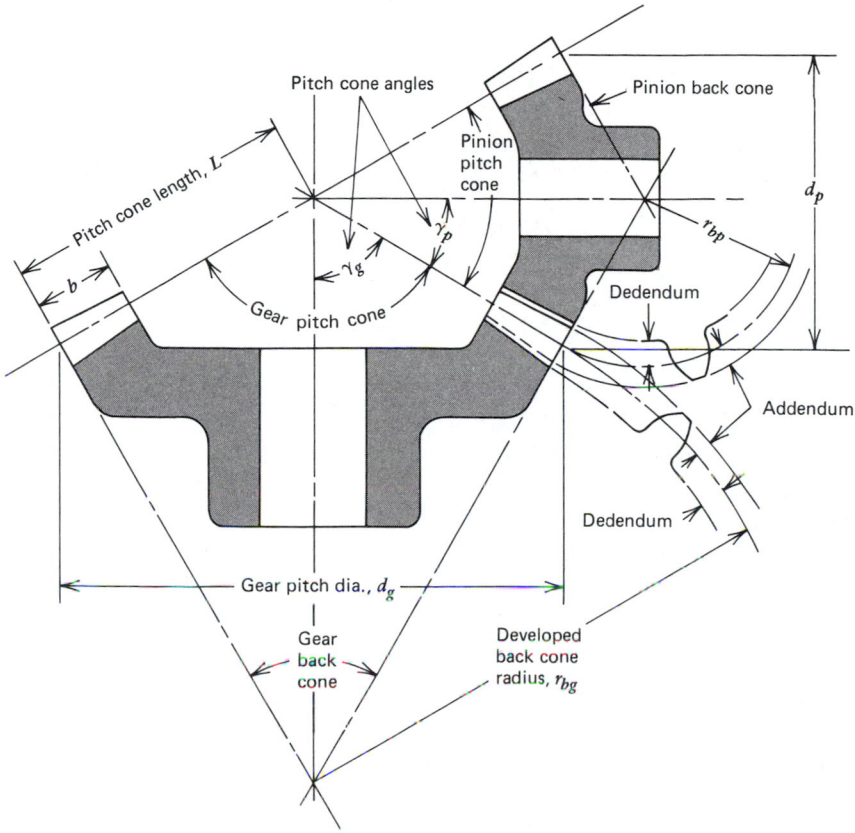

FIGURE 16.9.
Bevel gear terminology.

$$N'_p = 2\pi r_{bp}/p \quad \text{and} \quad N'_g = 2\pi r_{bg}/p \tag{16.15}$$

where N' is called the *virtual number of teeth,* and p is the circular pitch of both the imaginary spur gears and the bevel gears. In terms of diametral pitch, P (of both the imaginary spur and bevel gears),

$$N'_p = 2r_{bp}P \quad \text{and} \quad N'_g = 2r_{bg}P \tag{16.15a}$$

The practice of characterizing the size and shape of bevel gear teeth as those of an imaginary spur gear appearing on the developed back cone corresponds to what is known as *Tredgold's approximation.*

Bevel gear teeth are inherently noninterchangeable. The working depth of the teeth (gear addendum plus pinion addendum) is usually $2/P$, the same as for standard spur and helical gears, but the bevel pinion is designed with the larger addendum. This avoids interference and results in stronger pinion teeth. The gear addendum varies from $1/P$ for a gear ratio of 1, to $0.54/P$ for ratios of 6.8 or greater.

The gear ratio can be determined from the numbers of teeth, the pitch diameters, or the pitch cone angles:

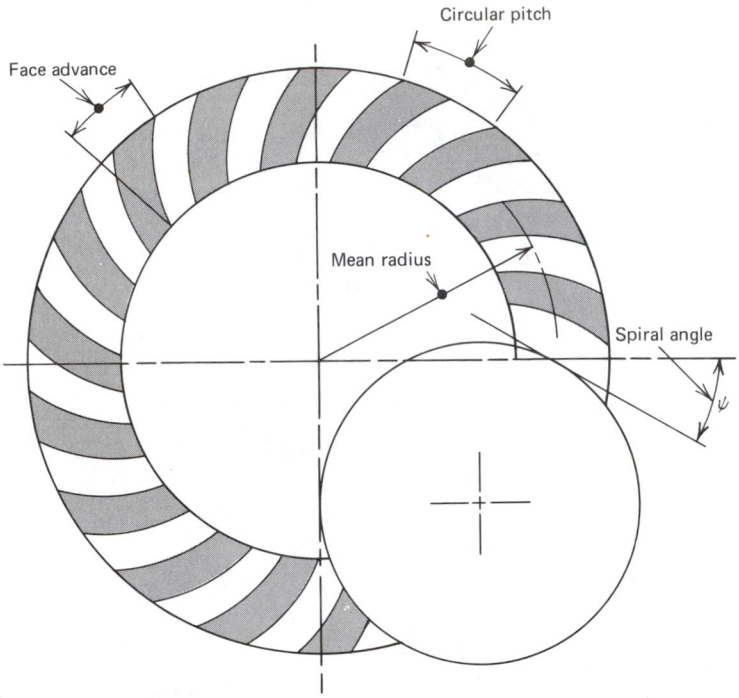

FIGURE 16.10.
Measurement of spiral angle at mean radius.

$$\text{Gear ratio} = \omega_p/\omega_g = N_g/N_p = d_g/d_p = \tan\gamma_g = \cot\gamma_p \tag{16.16}$$

Accepted practice usually imposes two limits on the face width:

$$b \leqq 10/P \quad \text{and} \quad b \leqq L/3 \quad (L \text{ is defined in Fig. 16.9}) \tag{16.17}$$

Figure 16.10 illustrates the measurement of the spiral angle, ψ, of a spiral bevel gear. Bevel gears most commonly have a pressure angle, ϕ, of 20 deg., and spiral bevels usually have a spiral angle, ψ, of 35 deg.

Figure 16.11 illustrates *Zerol*® bevel gears, developed by the Gleason Machine Division. These have curved teeth like spiral bevels, but have a zero spiral angle.

16.7 BEVEL GEAR FORCE ANALYSIS

Figure 16.12 shows the resolution of resultant tooth force F into its tangential (torque producing), radial (separating), and axial (thrust) components, designated F_t, F_r, and F_a, respectively. Note that an auxiliary view is needed to show the true length of the vector representing resultant force F (which is normal to the tooth profile).

Resultant force F is shown applied to the tooth at the pitch cone surface, and midway along tooth width b. This conforms to the usual assumption that the load is uniformly distributed along the tooth width, despite the fact that the tooth is largest at its

FIGURE 16.11.
Zerol® bevel gears (Courtesy Gleason Machine Div.)

outer end. The following relationships are derived directly from the geometry shown in Fig. 16.12:

$$d_{av} = d - b \sin \gamma \tag{16.18}$$

$$V_{av} = \pi d_{av} n \text{ (where } n \text{ is in rpm or rps)} \tag{16.19}$$

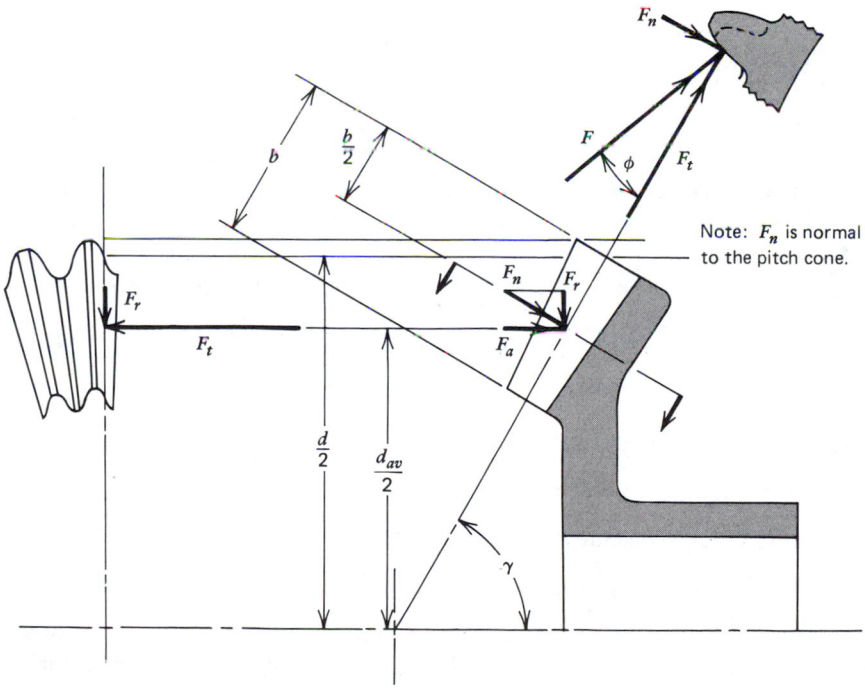

FIGURE 16.12.
Resolution of resultant bevel gear-tooth force, *F*.

$$F_t \text{ (lb)} = 33{,}000 \text{ hp}/V_{av} \text{ (} V_{av} \text{ in ft/min.), or} \tag{16.20}$$

$$F_t \text{ (newtons)} = w \text{ (watts)}/V_{av} \text{ (} V_{av} \text{ in m/s)} \tag{16.20a}$$

$$F = F_t/\cos\phi \tag{16.21}$$

$$F_n = F\sin\phi = F_t\tan\phi \tag{f}$$

$$F_a = F_n\sin\gamma = F_t\tan\phi\sin\gamma \tag{16.22}$$

$$F_r = F_n\cos\gamma = F_t\tan\phi\cos\gamma \tag{16.23}$$

For a spiral bevel gear, the axial and radial force components are functions of the spiral angle ψ:

$$F_a = \frac{F_t}{\cos\psi}(\tan\phi_n\sin\gamma \mp \sin\psi\cos\gamma) \tag{16.24}$$

$$F_r = \frac{F_t}{\cos\psi}(\tan\phi_n\cos\gamma \pm \sin\psi\sin\gamma) \tag{16.25}$$

Where $\pm$ or $\mp$ is used in the above equations, the upper sign applies to a driving pinion with right-hand spiral rotating clockwise as viewed from its large end, and to a driving pinion with left-hand spiral rotating counterclockwise when viewed from its large end. The lower sign applies to a left-hand driving pinion rotating clockwise, and to a right-hand driving pinion rotating counterclockwise. As was the case with helical gears, ϕ_n is the pressure angle measured in the plane normal to the tooth.

16.8 BEVEL GEAR-TOOTH BENDING AND SURFACE FATIGUE STRENGTHS

The calculation of bevel gear-tooth bending and surface fatigue strengths is even more complex than for spur and helical gears. The treatment given here is very brief. The serious student wishing to pursue this subject further should consult appropriate AGMA publications and other specialized literature, such as that published by the Gleason Machine Division.

The equation for bevel gear bending stress is the same as for spur gears:

$$\sigma = \frac{F_t P}{bJ} K_v K_o K_m \tag{15.16}$$

where

F_t = tangential load in pounds, from Eq. (16.20)

P = diametral pitch at large end of tooth

b = face width in inches (should be in accordance with Eq. (16.17))

J = geometry factor from Fig. 16.13 (straight bevel) or Fig. 16.14 (spiral bevel)[2]

K_v = velocity factor (When better information is not available, use a value between unity and curve C of Fig. 15.23, depending on degree of manufacturing precision.)

[2]See AGMA 226.01 for J values corresponding to other helix angles and pressure angles.

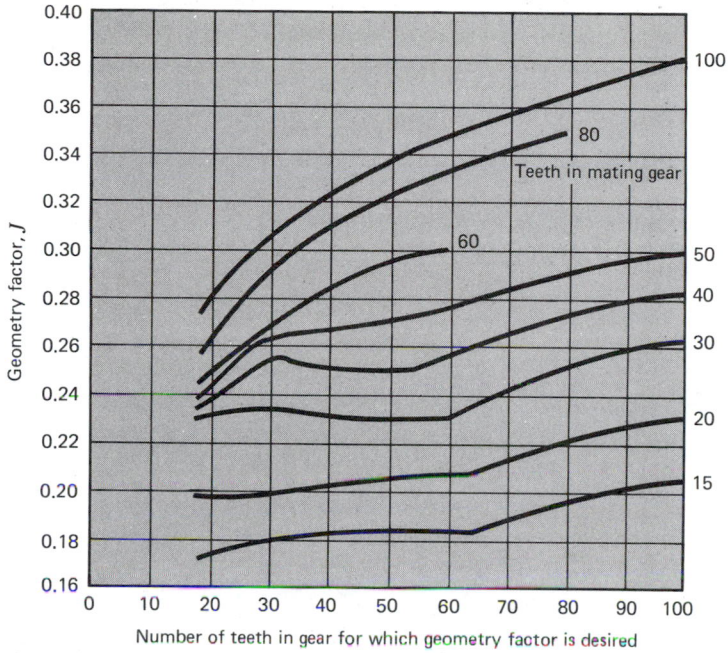

FIGURE 16.13.
Geometry factors, *J*, for straight bevel gears. Pressure angle = 20°, shaft angle = 90°. (From AGMA Information Sheet 226.01)

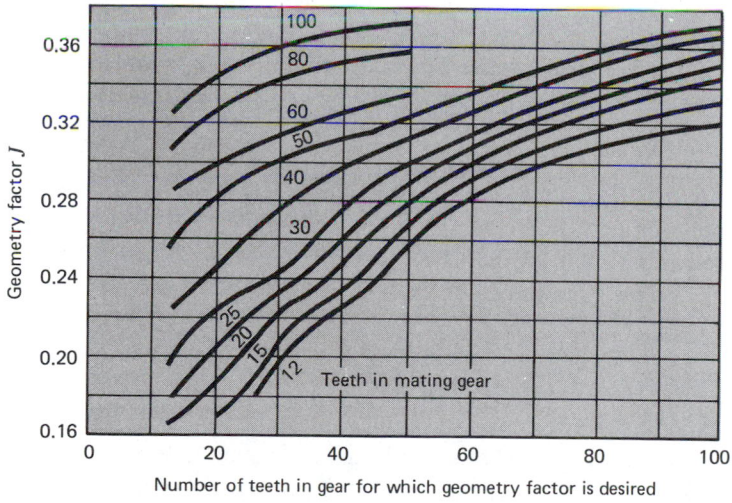

FIGURE 16.14.
Geometry factors, *J*, for spiral bevel gears. Pressure angle = 20°, spiral angle = 35°, shaft angle = 90°. (From AGMA Information Sheet 226.01)

TABLE 16.1
Mounting Factor, K_m, for Bevel Gears

Mounting Type	Mounting Rigidity: Maximum	Mounting Rigidity: Questionable
Both gears straddle mounted	1.0	to 1.25
One gear straddle mounted; the other overhung	1.1	to 1.4
Both gears overhung	1.25	to 1.5

K_o = overload factor, from Table 15.1

K_m = mounting factor, depending on whether gears are straddle mounted (between two bearings) or overhung (outboard of both bearings), and on the degree of mounting rigidity (See Table 16.1.)

The bending endurance limit is calculated from Eq. (15.18), exactly as for spur gears. The safety factor relating bending fatigue stress and strength (Eqs. (15.17) and (15.18)) is also handled as with spur gears.

Bevel gear surface fatigue stresses can be calculated as for spur gears:

$$\sigma_H = C_p \sqrt{\frac{F_t}{bd_pI} K_v K_o K_m} \tag{15.24}$$

with only two modifications: (a) Values of C_p are 1.23 times the values given in Table 15.4. This modification reflects a somewhat more localized contact area than for spur gears. (b) Values of geometry factor, I, are taken from Fig. 16.15 (straight teeth) and 16.16 (spiral teeth). (See AGMA 215.01 for calculation of I values for other tooth shapes.)

Bevel gear surface endurance strength is found from Eq. 15.25, exactly as for spur gears.

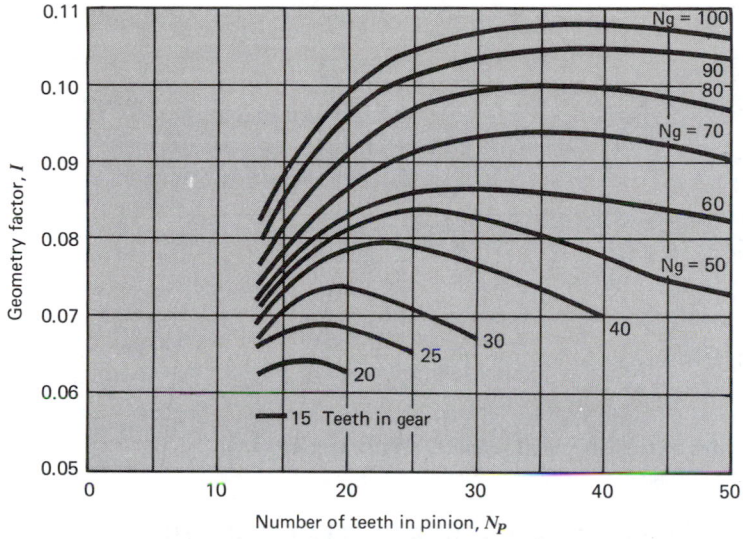

FIGURE 16.15.
Geometry factors, *I*, for straight bevel gears. Pressure angle = 20°, shaft angle = 90°. (From AGMA Information Sheet 215.91)

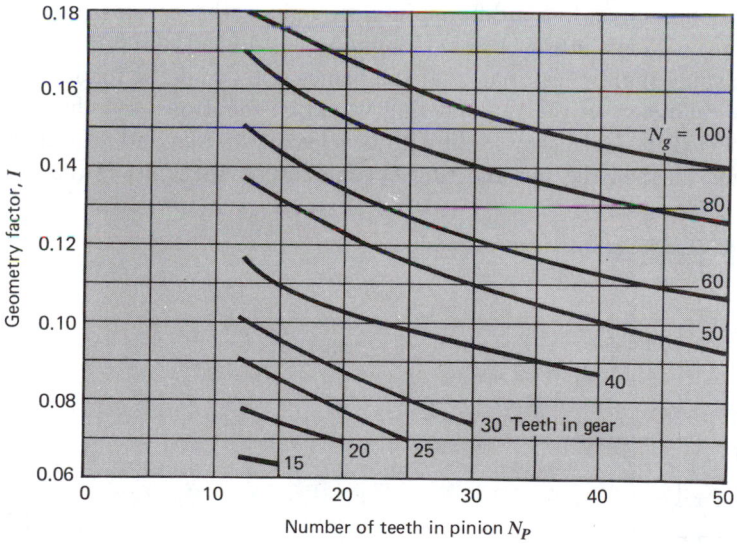

FIGURE 16.16.
Geometry factors, *I*, for spiral bevel gears. Pressure angle = 20°, spiral angle = 35°, shaft angle = 90°. (From AGMA Information Sheet 215.91)

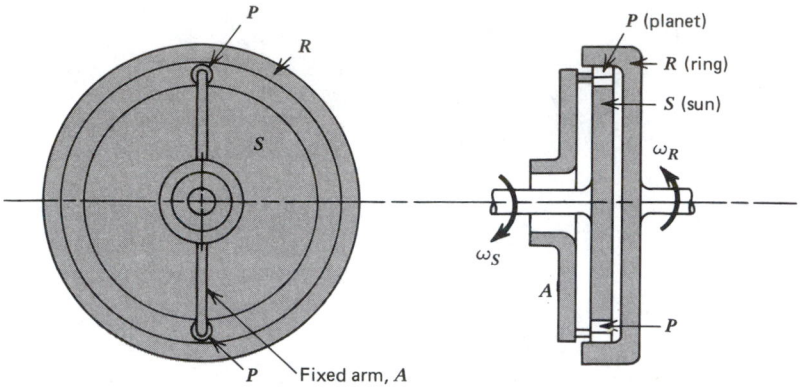

FIGURE 16.17.
Planetary gear train with extremely small planets. (With arm fixed, ω_R/ω_S is close to -1.)

16.9 BEVEL GEAR TRAINS; DIFFERENTIAL GEARS

For the usual gear trains, with all gears rotating about axes that are fixed with respect to each other, train values are readily determined from Eq. (16.16). If one or more bevel gears are mounted with their axes on a rotating arm, the result is a planetary gear train, much like the planetary spur gear trains discussed in Section 15.13. A planetary bevel gear train of particular interest is the differential gear train used in automobiles. Its purpose is to divide the torque equally between the left and right driving wheels, and permit these two wheels to rotate at different speeds when the car turns.

To gain an understanding of how a differential gear train works, consider first the spur gear planetary train shown in Fig. 16.17. Holding the arm fixed, the ratio ω_R/ω_S would be a negative value slightly less than 1. If the planet gears could be made infinitesimally small, the diameters of the sun and ring would be the same, and the ratio would be exactly -1. Figure 16.18 shows that with bevel gears the sun and ring diameters can be the same; in fact, the sun and ring become interchangeable. If the arm

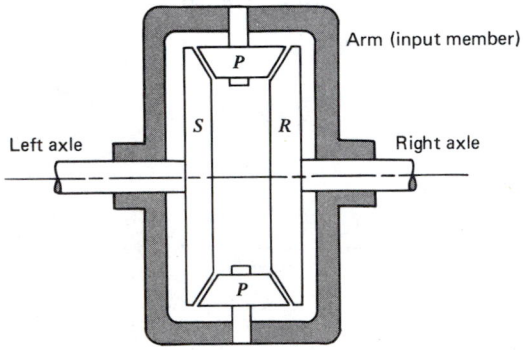

FIGURE 16.18.
Differential (planetary bevel gear) train. (With arm fixed, $\omega_R/\omega_S = -1$ exactly.)

(which is the input member, driven by the car engine) is held fixed, the two axle shafts rotate at equal speeds in opposite directions. If the "sun" (call it the gear attached to the left axle shaft) is held fixed, the "ring" (attached to the right axle) rotates at twice arm speed. Similarly, holding the right axle fixed causes the left axle to rotate at twice arm speed. If friction in the planet bearings is negligible, the differential gear train applies equal torque to the right and left shafts at all times. Also, the *average* of the axle speeds is equal to the arm speed. If, when rounding a curve, the outer wheel turns at 101 percent of arm speed, the inner wheel must turn at 99 percent. As discussed above, if one side turns at zero speed, the other turns at 200 percent of arm speed.

16.10 WORM GEAR GEOMETRY AND NOMENCLATURE

Figure 16.19 illustrates a typical worm and worm gear set. The worm shown has two threads, but any number up to six or more may be used. The geometry of a worm is similar to that of a power screw (recall Sections 10.2 and 10.3). Rotation of the worm simulates a linearly advancing involute rack. The geometry of a worm gear (sometimes called a worm wheel) is similar to that of a helical gear, except that the teeth are curved

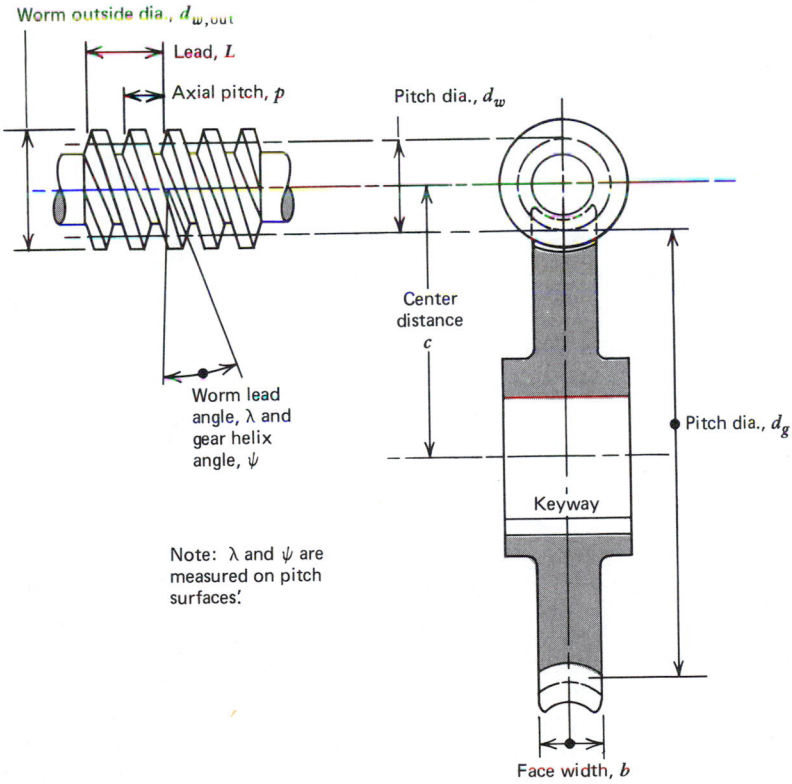

FIGURE 16.19.
Worm gearset nomenclature shown for double threaded worm, enveloping gear.

to envelop the worm. Sometimes the worm is modified to envelop the gear, as shown in Fig. 16.3*b*. This gives a greater area of contact, but requires extremely precise mounting. (Note that the axial positioning of a conventional nonenveloping worm is not critical.)

Figure 16.19 shows the usual case of a 90 deg. angle between the nonintersecting shafts. In this case the worm lead angle, λ (which corresponds to the screw lead angle illustrated in Fig. 10.1), is equal to the gear helix angle, ψ (which is also shown in Figs. 16.4–16.7). Also, angles λ and ψ have the same "hand."

As with a spur or helical gear, the pitch diameter of a worm gear is related to its circular pitch and number of teeth by the formula

$$d_g = N_g p/\pi \qquad \textbf{(15.2, modified)}$$

The pitch diameter of a worm is *not* a function of its number of threads, N_w. This means that the velocity ratio of a worm gearset is determined by the ratio of gear teeth to worm threads; it is *not* equal to the ratio of gear and worm diameters:

$$\omega_w/\omega_g = N_g/N_w \qquad \textbf{(16.26)}$$

Worm gears usually have at least 24 teeth, and the number of gear teeth plus worm threads should be more than 40:

$$N_w + N_g > 40 \qquad \textbf{(16.27)}$$

A worm of any pitch diameter can be made with any number of threads and any axial pitch. For maximum power transmitting capacity, the pitch diameter of the worm should normally be related to the shaft center distance by the following equation:

$$c^{0.875}/3.0 \leq d_w \leq c^{0.875}/1.7 \qquad \textbf{(16.28)}$$

Integral worms, cut directly on the shaft can, of course, be made of smaller diameter than shell worms, which are made separately. The latter are bored to slip over the shaft, and are driven by splines, key, or pin. Strength considerations seldom permit a shell worm to have a pitch diameter less than

$$d_w = 2.4p + 1.1 \text{ (in.)} \qquad \textbf{(16.29)}$$

The face width of the gear should not exceed half the worm outside diameter:

$$b \leq 0.5 d_{w,\text{out}} \qquad \textbf{(16.30)}$$

Lead angle, lead, and worm pitch diameter have the relationship previously noted in connection with screw threads:

$$\tan \lambda = L/\pi d_w \qquad \textbf{(10.1, modified)}$$

Largely to avoid interference, pressure angles are commonly related to worm lead angle as indicated in Table 16.2. The following are frequently used standard values of p (axial pitch of worm, or circular pitch of gear): ¼, 5/16, ⅜, ½, ⅝, ¾, 1, 1¼, 1½, and 2 in. Values of addendum and tooth depth often conform generally to helical gear practice, but may be strongly influenced by manufacturing considerations. The specialized literature should be consulted for this and other design details.

The load capacity and durability of worm gears can be significantly increased by modifying the design to give predominantly "recess action." (With reference to Fig. 15.8, the angle of approach would be made small or zero, and the angle of recess larger.) See [2] for details.

TABLE 16.2
Maximum Worm Lead Angle, and Worm Gear Lewis Form Factor for Various Pressure Angles

Pressure Angle ϕ_n (deg.)	Maximum lead angle λ (deg.)	Lewis form factor y
14½	15	0.100
20	25	0.125
25	35	0.150
30	45	0.175

16.11 WORM GEAR FORCE AND EFFICIENCY ANALYSIS

Figure 16.20 illustrates the tangential, axial, and radial force components acting on a worm and gear. For the usual 90 deg. shaft angle, note that the worm tangential force is equal to the gear axial force and vice versa ($F_{wt} = F_{ga}$, and $F_{gt} = F_{wa}$). Also, the worm and gear radial or separating forces are equal ($F_{wr} = F_{gr}$). If the power and speed of either the input (almost always the worm) or output (normally the gear) are known, the tangential force acting on that member can be determined from Eq. (15.14) or (15.14a).

In Fig. 16.20, the driving member is a clockwise-rotating right-hand worm. The force directions shown can readily be visualized by thinking of the worm as a right-hand screw being turned so as to pull the "nut" (worm gear tooth) toward the "screw head." Force directions for other combinations of worm-hand and rotation-direction can be similarly visualized.

The analysis of power screw force components given in Section 10.3 is also applicable to worm gears. By noting that the thread angle, α_n, of a screw thread corresponds to the pressure angle, ϕ_n, of the worm (illustrated in Fig. 16.21*a*), the force, efficiency, and self-locking equations of Section 10.3 can be applied directly to a worm and gear set. In the interest of making these equations more meaningful, they are derived below with reference to worm and gear geometry.

Figure 16.21*a* shows in detail the forces acting on the gear of Fig. 16.20. Components of the normal tooth force, F_n, are shown solid. Components of the friction force, fF_n, are shown dotted. Note that the friction force is always directed so as to *oppose the sliding motion.* In the case illustrated in Fig. 16.21*a*, the driving worm is rotating clockwise:

$$F_{gt} = F_{wa} = F_n \cos\phi_n \cos\lambda - fF_n \sin\lambda \tag{g}$$

$$F_{wt} = F_{ga} = F_n \cos\phi_n \sin\lambda + fF_n \cos\lambda \tag{h}$$

$$F_{gr} = F_{wr} = F_n \sin\phi_n \tag{i}$$

Combining (g) and (h) gives

$$\frac{F_{gt}}{F_{wt}} = \frac{\cos\phi_n \cos\lambda - f\sin\lambda}{\cos\phi_n \sin\lambda + f\cos\lambda} \tag{16.31}$$

Combining (i) with (g), and combining (i) with (h) gives

$$F_{gr} = F_{wr} = F_{gt}\frac{\sin\phi_n}{\cos\phi_n \cos\lambda - f\sin\lambda} = F_{wt}\frac{\sin\phi_n}{\cos\phi_n \sin\lambda + f\cos\lambda} \tag{16.32}$$

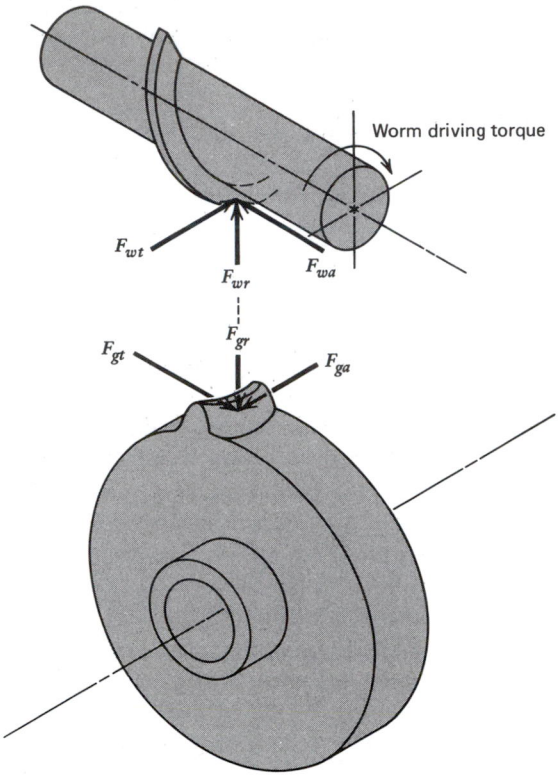

FIGURE 16.20.
Worm and gear force directions illustrated for clockwise-driven right-hand worm.

Figure 16.22 shows the relationship between worm tangential velocity, gear tangential velocity, and sliding velocity.

$$V_g/V_w = \tan \lambda \tag{16.33}$$

Efficiency, e, is the ratio of work out to work in. For the usual case of the worm serving as input member,

$$e = \frac{F_{gt}V_g}{F_{wt}V_w}$$

$$= \frac{\cos \phi_n \cos \lambda - f \sin \lambda}{\cos \phi_n \sin \lambda + f \cos \lambda} \tan \lambda$$

$$e = \frac{\cos \phi_n - f \tan \lambda}{\cos \phi_n + f \cot \lambda} \tag{16.34}$$

This corresponds to Eq. (10.9) and Fig. 10.8. It is important to remember that the overall efficiency of a worm gear reducer is a little lower because of friction losses in the bearings and shaft seals, and because of "churning" of the lubricating oil.

The coefficient of friction, f, varies widely depending on variables such as the gear

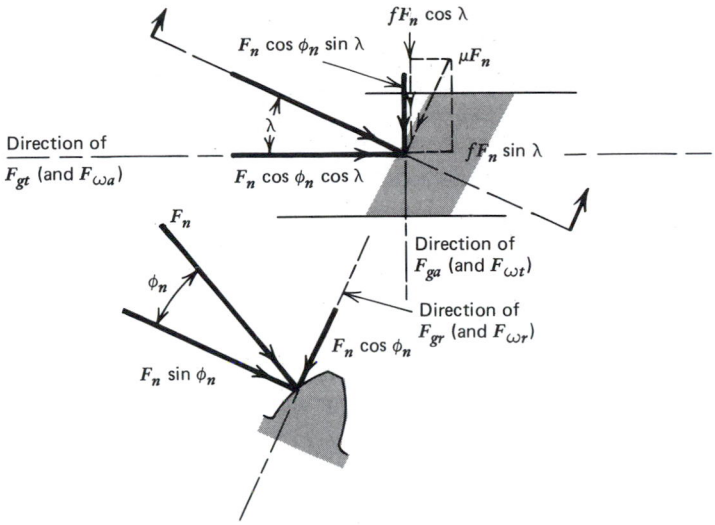

(*a*) Worm driving (as in Fig. 16.20)

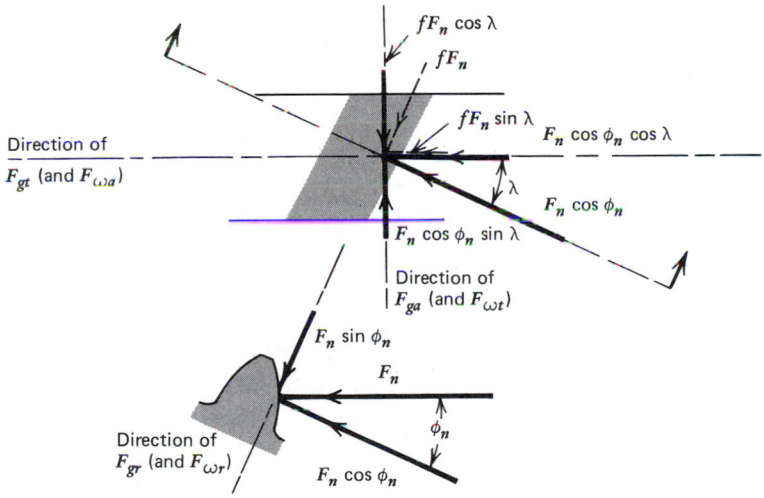

(*b*) Gear driving (same direction of rotation)

FIGURE 16.21.
Forces acting on the gear tooth shown in Fig. 16.20.

materials, lubricant, temperature, surface finishes, accuracy of mounting, and sliding velocity. Values reported in the literature cover a wide range. Figure 16.23 gives values recommended for use by the American Gear Manufacturers Association.

Figure 16.22 shows that sliding velocity, V_s, is related to the worm and gear pitch-line velocities and to the worm lead angle by

$$V_s = V_w/\cos\lambda = V_g/\sin\lambda \tag{16.35}$$

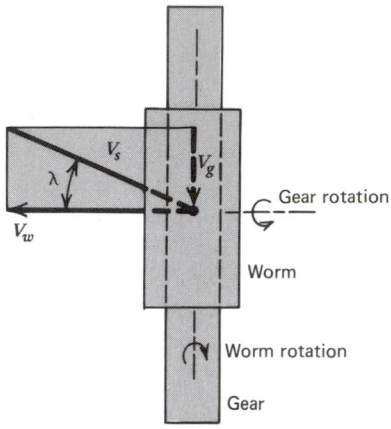

FIGURE 16.22.
Vector relationship between worm velocity, gear velocity, and sliding velocity.

Equation (g), above, indicates that with a sufficiently high coefficient of friction the gear tangential force becomes zero, and the gear set "self-locks," or does not "overhaul." (See Section 10.3.3 for discussion of self-locking and overhauling in power screws.) With this condition, no amount of worm torque can produce motion. Non-

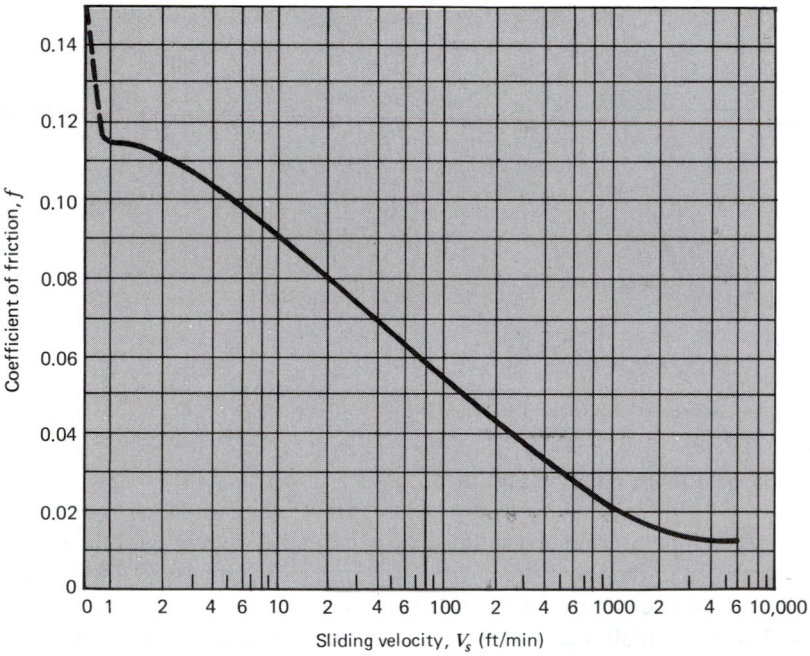

FIGURE 16.23.
Worm gear coefficient of friction. (From AGMA Standard 440.04)

overhauling (self-locking) occurs, if at all, with the gear driving. In many cases this is desirable and serves to hold the load from reversing, much like a self-locking power screw. In other cases, nonoverhauling is undesirable and can be destructive, as in a worm gear truck axle[3] where the torque direction reverses to provide engine braking.

Figure 16.21*b* illustrates the same directions of rotation as Fig. 16.21*a*, but with the torque directions reversed (i.e., gear driving). In this case contact shifts to the other side of the gear tooth, and the normal load reverses. Since the sliding velocity has the same direction regardless of which member is driving, the friction force has the same direction in Figs. 16.21*a* and 16.21*b*. In Fig. 16.21*b*, the tangential force tending to drive the worm is

$$F_{wt} = F_n \cos \phi_n \sin \lambda - fF_n \cos \lambda \tag{j}$$

Nonoverhauling occurs if this force goes to zero, which happens if

$$f \geqq \cos \phi_n \tan \lambda \tag{16.36}$$

If a worm gear set is to be *always* overhauling or *never* overhauling, it is necessary to take variation in coefficient of friction into account when selecting the value of λ (and to a lesser extent when selecting ϕ_n).

SAMPLE PROBLEM 16.2

A 2-hp 1200-rpm motor drives a 60-rpm machine by means of a worm gear reducer with a 5-in. center distance. The right-hand worm has two threads, an axial pitch of ⅝ in., and a normal pressure angle of 14½ deg. The worm is made of steel, hardened and ground, and the gear is made of bronze. Determine (a) all force components corresponding to rated motor power, (b) the power delivered to the driven machine, and (c) whether or not the drive is self-locking.

Solution

1. For a gear reduction ratio of 1200 rpm/60 rpm = 20, together with a double threaded worm, the gear must have 40 teeth.
2. With p = ⅝ in., $d_g = (\frac{5}{8})(40)/\pi = 7.96$ in.
3. For c = 5 in., $d_g + d_w = 10$ in. Hence, $d_w = 2.04$ in.
4. Lead angle $(\lambda) = \tan^{-1} L/\pi d_w = \tan^{-1} 1.25/\pi(2.04) = 11.04°$.
5. $V_w = \pi d_w n_w/12 = \pi(2.04)(1200)/12 = 640$ ft/min.
6. $F_{wt}\ (= F_{ga}) = 2$ hp $(33{,}000)/(640$ fpm$) = 103$ lb. ■
7. The other force components are functions of the coefficient of friction. To estimate f from Fig. 16.23, the sliding velocity must first be determined. From Eq. (16.35), $V_s = V_w/\cos \lambda = 640/\cos 11.04°. = 652$ fpm. From Fig. 16.23, f is estimated

[3]Truck axles normally use spiral bevel or hypoid gear drives.

as approximately 0.026. Note that all answers beyond this point are only as good as the estimated value of f.

8. From Eq. (16.31),

$$\frac{F_{gt}}{F_{wt}} = \frac{\cos 14.5^\circ \cos 11.04^\circ - 0.026 \sin 11.04^\circ}{\cos 14.5^\circ \sin 11.04^\circ + 0.026 \cos 11.04^\circ} = 4.48$$

Hence, F_{gt} $(= F_{wa})$ = 103 lb. (4.48) = 461 lb. ■

9. From Eq. (16.32),

$$F_{gr} = F_{wr} = 103 \text{ lb.} \frac{\sin 14.5^\circ}{\cos 14.5^\circ \sin 11.04^\circ + 0.026 \cos 11.04^\circ} = 122 \text{ lb.}$$

All force components are shown with their proper directions in Fig. 16.20. ■

10. From Eq. (16.34),

$$e = \frac{\cos 14.5^\circ - 0.026 \tan 11.04^\circ}{\cos 14.5^\circ + 0.026 \cot 11.04^\circ} = 87 \text{ percent}$$

or

$$e = \frac{F_{gt}V_g}{F_{wt}V_w} = \frac{F_{gt}}{F_{wt}} \tan \lambda = 4.48 \tan 11.04^\circ = 87 \text{ percent}$$

(Note that this value also checks with Fig. 10.8.)

11. To allow for small friction losses due to bearings, shaft seals, and oil "churning," assume an overall efficiency of about 85 percent. On this basis, output hp = input hp times efficiency = 2(0.85) = 1.7 hp. ■

12. Cos ϕ_n tan λ = cos 14.5° tan 11.04° = 0.19. Since this is greater than f, Eq. (16.31) indicates that the drive is *not* self-locking; rather, it is reversible. ■

16.12 WORM GEAR BENDING AND SURFACE FATIGUE STRENGTHS

The determination of load capacity is more involved with worm gears than with other gear types. This results in a wider variation in procedures used for estimating bending and surface strengths. Furthermore, worm gear capacity is often limited not by fatigue strength but by cooling capacity. The latter is discussed in the next section.

Investigations of bending and surface fatigue strengths for gear types previously considered have been made by comparing estimated bending and surface stresses with the corresponding estimated material fatigue strengths. The same thing can be accomplished by comparing the estimated total gear-tooth tangential *load* (nominal load multiplied by factors accounting for impact due to tooth inaccuracies and deflections, misalignment, etc.) with limiting values of total tooth *load* based on bending and surface fatigue strengths. The total tooth load is called the *dynamic load*, F_d, the bending-fatigue limiting load is called the *strength capacity*, F_s, and the surface-fatigue limiting load is called (somewhat inappropriately), the *wear capacity*, F_w. For satisfactory gear performance it is necessary that

$$F_s \geqq F_d \tag{16.37}$$

and

$$F_w \geqq F_d \tag{16.38}$$

This "dynamic load" approach was worked out in considerable detail for all gear types in Buckingham's classic treatise [1]. The following is a simplified version applied to worm gears.

The dynamic load is estimated by multiplying the nominal value of gear tangential force (as determined from Eq. (15.14) or (15.14a)) by velocity factor "D" of Fig. 15.23:

$$F_d = F_{gt}K_v = F_{gt}\frac{1200 + V_g}{1200} \tag{16.39}$$

where V_g is the pitch-line velocity of the gear in feet per minute.

Bending stresses are much higher in the gear than in the worm. Adapting the Lewis equation (Eq. (15.15)) to the worm gear teeth:

$$F_s = S_n bpy \tag{16.40}$$

where

F_s = maximum allowable value of dynamic load with respect to bending fatigue

S_n = zero-to-maximum bending fatigue strength of the gear material (usually taken as 24 ksi for gear bronze; see below)

b = gear face width

p = gear circular pitch

y = Lewis form factor, usually taken as a function only of the normal pressure angle (See Table 16.2.)

Although other materials are occasionally used (as aluminum, cast iron, and plastics), most worm gears are made of a special gear bronze (as SAE 65). Instead of estimating the zero-to-maximum fatigue strength from Eq. (15.18), a value of 24 ksi, originally proposed by Buckingham [1], has been found satisfactory over long experience.

The "wear" capacity, F_w, is a function of the materials, radii of curvature, and length of theoretical line of contact. Because of the high sliding velocities and associated heat generated, lubrication is extremely important. Important also are smooth surfaces, particularly on the worm. Assuming the presence of an adequate supply of appropriate lubricant, the following equation may be used for rough estimates:

$$F_w = d_g b K_w \tag{16.41}$$

where

F_w = maximum allowable value of dynamic load with respect to surface fatigue

d_g = pitch diameter of the gear

b = face width of the gear

K_w = a material and geometry factor with values empirically determined (See Table 16.3.)

TABLE 16.3
Worm Gear Wear Factors, K_w

Material		K_w (lb/in.^{2})		
Worm	**Gear**	$\lambda < 10$ **deg.**	$\lambda < 25$ **deg.**	$\lambda > 25$ **deg.**
Steel, 250 Bhn	Bronze[a]	60	75	90
Hardened steel (surface 500 Bhn)	Bronze[a]	80	100	120
Hardened steel	Chill-cast bronze	120	150	180
Cast iron	Bronze[a]	150	185	225

[a] Sand-cast

The combination of large loads and high sliding velocities encountered with worm gear sets makes them somewhat analogous to shafts and journal bearings. Bronze and hardened steel can be a good material combination for both applications. The bronze member is able to "wear-in" and increase contact area.

As mentioned earlier, the above constitutes a simplified treatment of a complex subject. For example, factors corresponding to those given for spur gears in Tables 15.1, 15.2, and 15.3 were not mentioned, although they obviously have an influence on worm gear capacity.

16.13 WORM GEAR THERMAL CAPACITY

The continuous rated capacity of a worm gear set is often limited by the ability of the housing to dissipate friction heat without developing excessive gear and lubricant temperatures. Normally, oil temperatures must not exceed about 200°F (93°C) for satisfactory operation. The fundamental relationship between temperature rise and rate of heat dissipation was previously applied to journal bearings:

$$H = CA(t_o - t_a) \qquad \textbf{(13.13, repeated)}$$

where

H = time rate of heat dissipation (as ft·lb per minute)

C = heat transfer coefficient (as ft·lb per minute per sq. ft of housing surface area per degree F)

A = housing external surface area (as sq. ft)

t_o = oil temperature (allowable values are commonly 160°–200°F)

t_a = ambient air temperature (as degrees F)

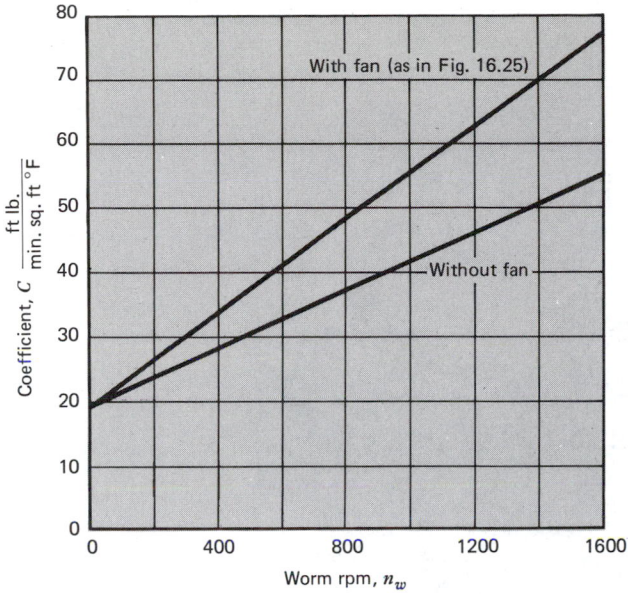

FIGURE 16.24.
Estimated heat transfer coefficient, *C*, for worm gear housings. (Based on Walker, H., "Thermal Rating of Worm Gear Boxes," Proc. Inst. Mech. Engrs., 151, 1944.)

Values of A for conventional housing designs may be roughly estimated from the equation[4]

$$A = 0.3c^{1.7} \tag{16.42}$$

where A is in sq. ft and c (distance between shafts) is in inches.

Rough estimates of C can be taken from Fig. 16.24. Figure 16.25 shows an example of a fan installed on a worm shaft for cooling.

The approximate nature of Eq. (16.42) and the curves in Fig. 16.24 should be emphasized. Housing surface area can be made far greater than the Eq. (16.42) value by incorporating cooling fins. Almost any desired amount of cooling can be provided by using an external heat exchanger to cool the oil, and directing jets of cooled oil at the gear mesh.

Where thermal considerations may be critical, it is important to obtain reliable test data under actual operating conditions.

SAMPLE PROBLEM 16.3

An 11:1 ratio worm gear reducer is to be designed using a hardened steel worm and a chill-cast bronze gear. Center distance should be approximately 6 in. The worm will be

[4]The AGMA recommends that housings be designed to provide at least this much area, exclusive of base, flanges, and fins.

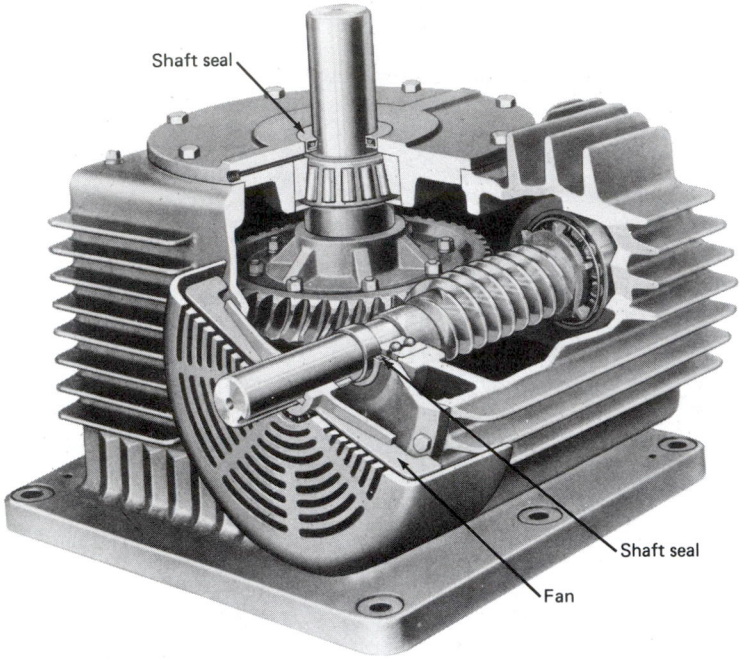

FIGURE 16.25.
Worm-gear speed reducer with fan and fins to increase heat transfer. (Courtesy Cleveland Gear Co.)

driven by a 1200 rpm motor. Determine appropriate values of d_w, d_g, N_w, N_g, p, λ, and ϕ_n. Estimate the horsepower capacity and efficiency of the reducer. Can the worm be bored for mounting on a separate shaft?

Solution

1. From Eq. (16.26), $N_g/N_w = 11$; from Eq. (16.27), $N_g + N_w > 40$; hence, choose $N_w = 4$, $N_g = 44$. ■
2. For high efficiency, Fig. 10.8 indicates that λ should be as large as possible—hopefully close to 35 deg. From Table 16.2, select $\phi_n = 25$ deg. ■
3. To obtain a large value of λ, d_w must be small. For a 6 in. center distance, the smallest value of d_w permitted by Eq. (16.28) is $6^{0.875}/3 = 1.60$ in. This leaves 10.4 in. for the gear diameter, with a circular pitch of $p = 10.4\pi/44 = 0.7425$ in. Select a standard pitch of $p = 0.75$ in. ■
4. This slight increase in p gives a larger gear, and requires us to choose between making the worm slightly smaller than the normally recommended range, or increasing the center distance. Let us choose the latter. $d_g = 44(0.75)/\pi = 10.50$

in. Using $d_w = 1.60$ in. would result in a center distance of 6.05 in. Let us choose an even fractional dimension of $c = 6\frac{1}{8}$ in. The corresponding worm diameter is $12.25 - 10.50 = 1.75$ in. (note that this is slightly larger than the minimum of $6.125^{0.875}/3 = 1.63$ in.). Thus, $d_g = 10.50$ in., $d_w = 1.75$ in., $c = 6.125$ in. ■

5. From Eq. (16.29) the smallest worm diameter normally suitable for boring (to fit over a separate shaft) is

$$d_w = 2.4(0.75) + 1.1 = 2.9 \text{ in.}$$

Obviously, the chosen worm diameter of 1.75 in. requires that the worm be cut directly on the shaft. ■

6. From Eq. (10.1), $\tan \lambda = (4)(0.75)/(1.75\,\pi)$, or $\lambda = 28.62$ deg. ■

7. To estimate the efficiency we must first estimate the friction coefficient. For this we must find V_s, which requires a knowledge of V_g.

$$V_g = \pi d_g n_g = \pi(10.5/12)(1200/11) = 300 \text{ fpm}$$

From Eq. (16.35), $V_s = 300/\sin 28.62$ deg. $= 626$ fpm. From Fig. 16.23, f is estimated as 0.027. From Eq. (16.34),

$$e = \frac{\cos 25^\circ - 0.027 \tan 28.62^\circ}{\cos 25^\circ + 0.027 \cot 28.62^\circ} = 93.3 \text{ percent}$$

Note that this agrees with Fig. 10.8. Note also that bearing, shaft seal, and oil-churning losses would reduce this figure slightly—perhaps to about 92 percent. ■

8. Gear face width should be close to—but not greater than—half the worm outside diameter. The latter is equal to d_w plus twice the addendum. Although the tooth form would not necessarily correspond to the standard addendum of $1/P = p/\pi = 0.75/\pi = 0.24$ in., this is accurate enough for our present purpose. Thus, $d_{w,\text{out}} \approx 2.23$ in., which limits the face width to 1.11 in. Choose an even dimension: $b = 1$ in. ■

9. The velocity factor for a gear tangential velocity of 300 fpm is determined from curve D of Fig. 15.23 as 1.25. From Eq. (16.39) the dynamic load is $F_d = 1.25F_{gt}$.

10. From Eq. (16.40), the strength capacity is

$$F_s = (24{,}000 \text{ psi})(1 \text{ in.})(0.75 \text{ in.})(0.150) = 2700 \text{ lb}$$

Equating F_d with F_s, we obtain the strength-limiting value of F_{gt} as $2700/1.25 = 2160$ lb. The corresponding gear power is (300 fpm)(2600 lb)/33,000 = 19.6 hp.

11. From Eq. (16.41), the wear capacity is

$$F_w = (10.5 \text{ in.})(1 \text{ in.})(180 \text{ lb/in.}) = 1890 \text{ lb}$$

Equating F_d with F_w, we obtain the wear-limiting value of F_{gt} as $1890/1.25 = 1512$ lb. The corresponding gear power is 13.7 hp.

12. Estimating the heat dissipation capacity of the housing for a limiting 100°F temperature rise: From Fig. 16.24, $C = 45$ (assuming no fan). From Eq. 16.42, $A = 0.3(6.125)^{1.7} = 6.53$ sq. ft. From Eq. (13.13) (repeated in Section 16.13),

$$H = (45)(6.53)(100) = 29{,}385 \text{ ft·lb/min.} = 0.89 \text{ hp}$$

13. Accepting the estimated overall efficiency as 92 percent (from step 7), the 0.89 heat dissipation hp represents 8 percent of the input or worm power. Thus, input power = 0.89/0.08 = 11.1 hp, and output or gear power = (0.92)(11.1) = 10.2 hp.

14. With no special cooling provisions, the gear reducer has an estimated input capacity of approximately 11 hp (output capacity approximately 10 hp). With sufficient cooling, the capacity would be limited by "wear" to 13.7 output or gear hp, and 14.9 input or worm hp. This might reasonably be rounded off to 15 input hp. The cooling capacity required for this rating would be 15(0.08) = 1.2 hp, an increase of (1.2 − 0.89)/0.89 = 35 percent. Figure 16.24 indicates that the addition of a fan on the worm shaft increases cooling capacity by 36 percent (C increases from 45 to 61). Thus, with the fan an input power rating of 15 hp would appear to be justified. In view of the many empirical approximations involved, the final rating, based on test results, might be a little different. ■

Important details of the reducer design not mentioned above include (1) making sure that the worm root diameter and both shaft diameters are adequate for carrying the torsional, bending, and axial loads involved, (2) insuring that housing rigidity, bearing positioning, and shaft diameters provide for sufficient rigidity of worm and gear mounting, (3) providing an appropriate grade and quantity of clean lubricant in the housing, and (4) insuring that the shaft oil seals are adequate to prevent lubricant leakage.

REFERENCES

1. Buckingham, Earle, *Analytical Mechanics of Gears,* McGraw-Hill, New York, 1949.
2. Buckingham, Earle, and H. H. Ryffel, *Design of Worm and Spiral Gears,* Industrial Press, New York, 1960.
3. "1979 Mechanical Drives Reference Issue," *Machine Design,* Penton/IPC, Cleveland, June 29, 1979.
4. Standards of the American Gear Manufacturers Association, Arlington, Virginia.

CHAPTER SEVENTEEN

Shafts and Associated Parts

17.1 INTRODUCTION

The term *shaft* usually refers to a relatively long member of round cross section that rotates and transmits power. One or more members such as gears, sprockets, pulleys, and cams are usually attached to the shaft by means of pins, keys, splines, snap rings, and other devices. These latter members are among the "associated parts" considered in this chapter, as are couplings and universal joints, which serve to connect the shaft to its source of power or load.

A shaft can have a nonround cross section, and it need not necessarily rotate. It can be stationary and serve to support a rotating member, such as the short shafts (also called *spindles*) that support the nondriving wheels of an automobile. The shafts supporting idler gears (Figs. 15.17 and 15.21) can be either rotating or stationary depending on whether the gear is attached to the shaft or supported by it through bearings. Shafts supporting and/or driving vehicle wheels are also called *axles*.

It is apparent that shafts can be subjected to various combinations of axial, bending, and torsional loads, and that these loads may be static or fluctuating. Typically, a rotating shaft transmitting power is subjected to a constant torque (producing a mean torsional stress) together with a completely reversed bending load (producing an alternating bending stress). Sample Problems 8.3 and 8.4 (Section 8.11) illustrate the application of fatigue analysis procedures to shafts subjected to combinations of static and fluctuating loads.

In addition to satisfying strength requirements, shafts must be designed so that deflections are within acceptable limits. Excessive lateral shaft deflection can hamper gear performance and cause objectionable noise. The associated angular deflection can be very destructive to non-self-aligning bearings (either plain or rolling). Torsional deflection can affect the accuracy of a cam or gear driven mechanism. Furthermore, the greater the flexibility—either lateral or torsional—the lower the corresponding critical speed.

17.2 PROVISION FOR SHAFT BEARINGS

Rotating shafts carrying gears, pulleys, cams, and so on must be supported by bearings (Chapters 13 and 14). If two bearings can provide sufficient radial support to limit shaft bending and deflection to acceptable values, this is highly desirable and simplifies manufacturing. If three or more bearings must be used to provide adequate support and rigidity, precise alignment of the bearings in the supporting structure must be maintained (as, for example, with the three or more main bearings supporting an engine crankshaft).

Shaft axial positioning and provision for carrying thrust loads normally require that *one and only one* bearing take thrust in each direction. The reason for this is given in Section 14.8, and examples of proper thrust provision are illustrated in Figs. 14.14, 14.15, and 13.1. Sometimes a thrust load is shared among two or more plain thrust bearings (Fig. 13.1-type). In this case there must be sufficient axial clearance to insure against "binding" under any operating conditions. Production tolerances may be such that only one bearing will carry the thrust until after initial "wearing-in."

It is important that the members *supporting* the shaft bearings be sufficiently strong and rigid.

17.3 MOUNTING PARTS ONTO ROTATING SHAFTS

Sometimes members like gears and cams are made integral with the shaft, but more often such members (which also include pulleys, sprockets, etc.) are made separately and then mounted onto the shaft. This is shown in Figs. 14.14 and 14.15 for the case of gears. The portion of the mounted member in contact with the shaft is the *hub*. Attachment of the hub to the shaft is made in a variety of ways. In Figs. 14.14 and 14.15 the gear is gripped axially between a shoulder on the shaft and a spacer, with torque being transmitted through a *key*. Figure 17.1 illustrates a variety of keys. The grooves in the shaft and hub into which the key fits are called *keyways* or *keyseats*.

A simpler attachment for transmitting relatively light loads is provided by *pins*, such as the types illustrated in Fig. 17.2. Pins provide a relatively inexpensive means of transmitting both axial and circumferential loads.

Radially tapped holes in the hub permit *set screws* to bear on the shaft, thereby tending to prevent relative motion. (Local flats or grooves are usually machined into the shaft where the set screw(s) bear, so that "burrs" caused by jamming the screws tightly against the shaft do not inhibit subsequent removal and reinstallation of the hub.) The screw diameter is typically about ¼ the shaft diameter. Two screws are commonly used, spaced 90 deg. apart. Set screws are inexpensive and sometimes adequate for relatively

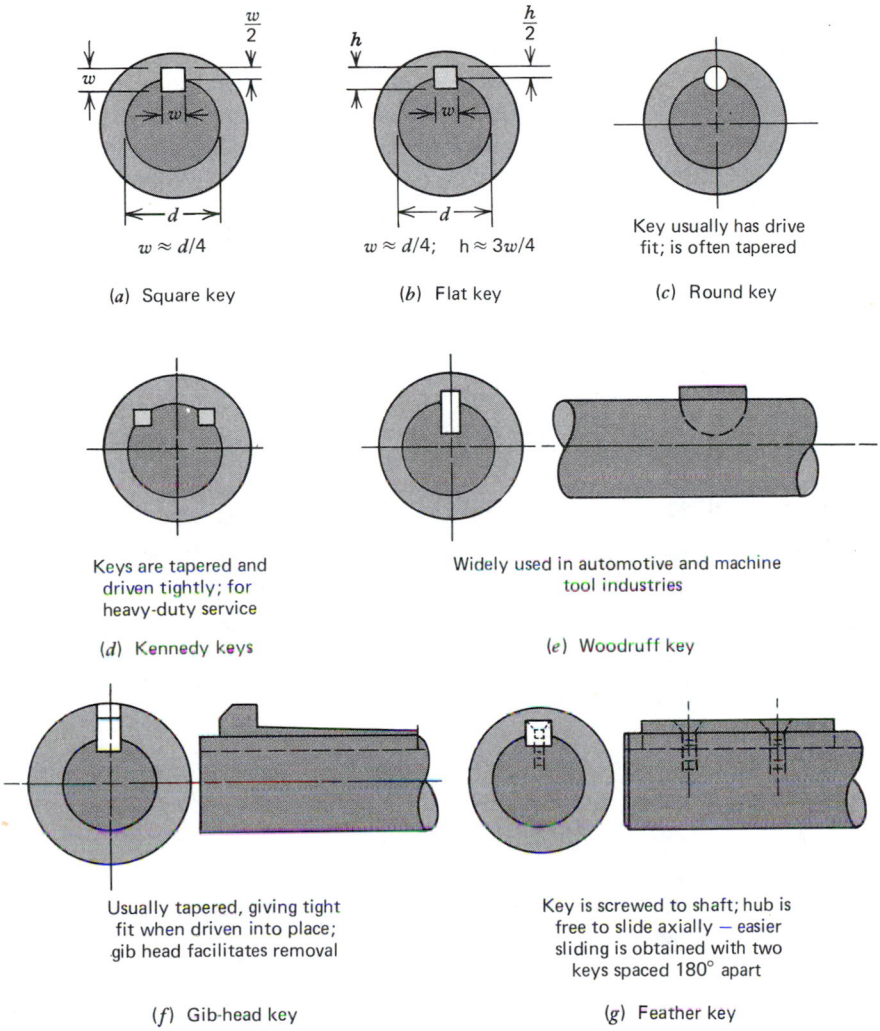

FIGURE 17.1.
Common types of shaft keys.

light service. Although special designs that provide increased protection against loosening in service are available, set screws should not be counted on in applications where loosening would impose a safety hazard. (Recall safety example (c) in Section 1.2.) Set screws are sometimes used in conjunction with keys. Typically, one screw bearing on the key and another bearing directly on the shaft are used to prevent axial motion.

An excellent and inexpensive method of axially positioning and retaining hubs and bearings onto shafts is by means of *retaining rings,* commonly called *snap rings.* Figure 17.3*a* illustrates a few of the numerous varieties which are available. Figure 17.6*b* shows snap rings used to retain the bearing at *A* to both the shaft and housing. Shaft snap rings require grooves that weaken the shaft, but this is no disadvantage if they are located where stresses are low, as in Fig. 17.6*b*. The cost of the shaft in Fig. 17.6*b* could

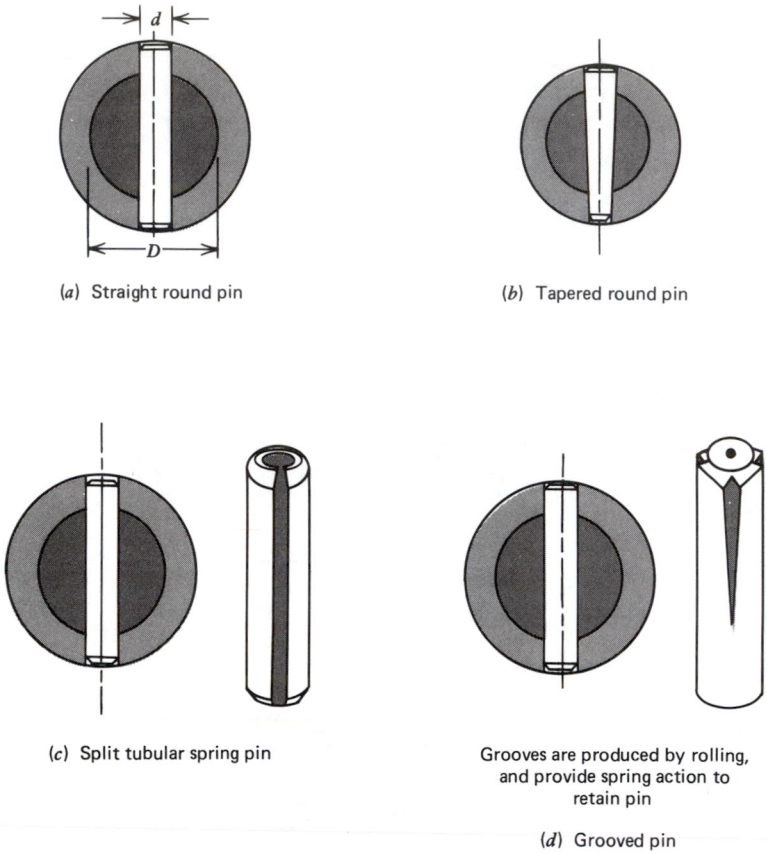

FIGURE 17.2.
Common types of shaft pins. (All pins are driven into place: for safety, pins should not project beyond hub.)

have been reduced by positioning hubs T_1 and T_2 with snap rings instead of with the shaft shoulders that require a large diameter between the two hubs. This was not done because the grooves would have weakened the shaft in a highly stressed region. Figure 17.3*b* illustrates "push on" retaining rings that do not require grooves. These provide a low cost, compact means of assembling parts; but they do not provide the positive, precision positioning and retention of parts that is possible with conventional retaining rings.

Perhaps the simplest of all hub-to-shaft attachments is by means of an *interference fit* (hub bore is slightly smaller than shaft diameter, with assembly being made with a press, or by thermally expanding the hub—sometimes also thermally contracting the shaft, as with dry ice—prior to making a quick assembly before the temperatures equalize). Sometimes a combination of pin and interference fit is used.

The strongest provision for torque transmission is usually by means of mating *splines* cut in the shaft and hub (Fig. 17.4). Both splines and keys can be fitted to permit the hub to slide axially along the shaft.

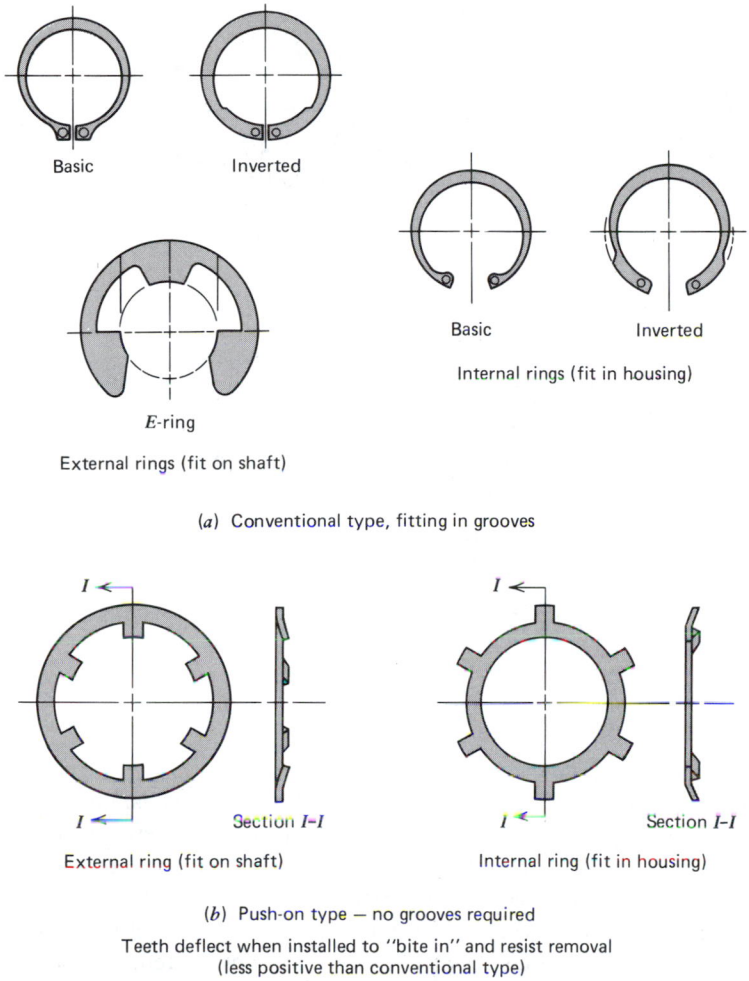

FIGURE 17.3.
Common types of retaining (or snap) rings. (Courtesy Waldes Kohinoor, Inc.)

More information on the design of keys, pins, and splines is given in Section 17.6.

17.4 ROTATING SHAFT DYNAMICS

Rotating shafts, particularly those that run at high speeds, must be designed to avoid operation at *critical speeds*. This usually means providing sufficient lateral rigidity to place the lowest critical speed significantly above the operating range. When torsional fluctuations are encountered (as with camshafts, engine and compressor crankshafts, etc.) an additional dynamic requirement is imposed: the *torsional natural frequencies* of the shaft must be well removed from the frequency of torsional input. This usually

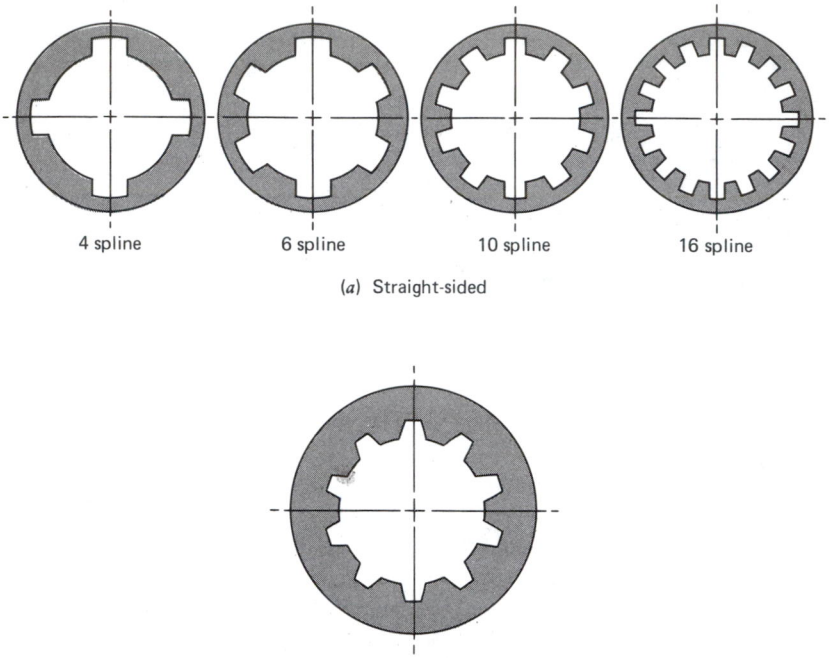

FIGURE 17.4.
Common types of splines.

means providing sufficient torsional stiffness (and low enough torsional inertias) to make the lowest torsional natural frequency significantly above that of the highest torsional disturbing frequency. These important subjects are treated in detail in books on mechanical vibrations, and can be touched on only briefly here.

With respect to lateral vibration and critical speeds, the practicalities of manufacturing and operation are such that the center of mass of a rotating system can never coincide *exactly* with the center of rotation. Hence as the shaft rotating speed is increased, centrifugal force acting at the mass center tends increasingly to bow the shaft. The more the shaft bows, the greater the eccentricity and the larger the centrifugal force. Below the lowest (or *fundamental*) critical speed of rotation, the centrifugal and shaft elastic forces balance at a finite shaft deflection. At the critical speed, equilibrium theoretically requires an infinite displacement of the mass center. Damping (from the shaft bearings, windage, and internal hysteresis within the rotating member) causes equilibrium to occur at a finite displacement. However, this displacement is often large enough to break the shaft or to cause rotating bearing forces of a highly objectionable, if not destructive, magnitude. Rotation sufficiently above the critical speed results in a satisfactory equilibrium position by moving the mass center *toward* the center of rotation. In unusual situations (as some high-speed turbines), satisfactory operation is provided by quickly going through the critical speed (not allowing sufficient *time* for an equilibrium deflection to be reached) and then running well above the critical speed.

Configuration	Critical speed equation
(*a*) Single mass 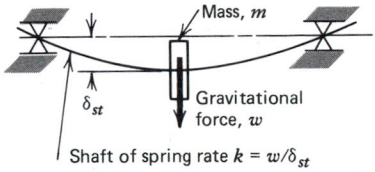	$\omega_n = \sqrt{\frac{k}{m}} = \sqrt{\frac{kg}{w}} = \sqrt{\frac{g}{\delta_{st}}}$ (17.1) $n_c = \frac{30}{\pi}\sqrt{\frac{k}{m}} = \frac{30}{\pi}\sqrt{\frac{kg}{w}} = \frac{30}{\pi}\sqrt{\frac{g}{\delta_{st}}}$ (17.1a)
(*b*) Multiple masses 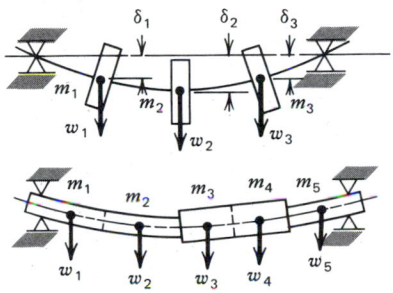	$n_c \approx \frac{30}{\pi}\sqrt{\frac{g(w_1\delta_1 + w_2\delta_2 + \ldots)}{w_1\delta_1^2 + w_2\delta_2^2 + \ldots}}$ $n_c \approx \frac{30}{\pi}\sqrt{\frac{g\Sigma w\delta}{\Sigma w\delta^2}}$ (17.2)
(*c*) Shaft mass only 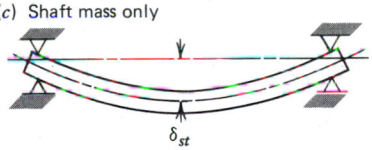	$n_c \approx \sqrt{\frac{5g}{4\delta_{st}}}$ (17.3)

Quantity	Symbol	SI	English
Mass	m	kg	lb-sec^2/in.
Gravitational force	w	N	lb
Static deflection	δ_{st}	m	in.
Shaft spring rate	k	N/m	lb/in.
Accel. of gravity	g	m/s^2	in./sec^2
Natural frequency	ω_n	rad/s	rad/sec
Critical speed	n_c	rpm	rpm

FIGURE 17.5.
Shaft critical speeds (lowest, or fundamental).

The critical speed of rotation is numerically the same as the lateral natural frequency of vibration resulting from stopping the rotation, displacing the shaft center laterally, and then suddenly releasing it. For all except the simple "ideal" case of a massless shaft supporting a single concentrated mass, additional critical speeds at higher frequencies are also present. Equations for the lowest or fundamental critical speed are summarized in Fig. 17.5, which contains Eqs. (17.1)–(17.3). The derivations are given in standard elementary vibration texts.

Since the estimation of shaft critical speeds involves the calculation of static deflections at several points, an extension of the BASIC program given in Table 5.3 enables shaft critical speeds to be calculated with a microcomputer.

17.5 OVERALL SHAFT DESIGN

A few general principles that should be kept in mind are:

1. Keep shafts as short as possible, with bearings close to the applied loads. This reduces deflections and bending moments, and increases critical speeds.
2. Place necessary stress raisers away from highly stressed shaft regions if possible. If not possible, use generous radii and good surface finishes. Consider local surface-strengthening processes (as shot-peening or cold-rolling).
3. Use inexpensive steels for deflection-critical shafts, as all steels have essentially the same elastic modulus.
4. When weight is critical, consider hollow shafts. (Example: propeller shafts on rear-wheel-drive cars are made of tubing in order to obtain the low weight/stiffness ratio needed to keep critical speeds above the operating range.)

The maximum allowable deflection of a shaft is usually determined by critical speed, gear, or bearing requirements. Critical speed requirements vary greatly with the specific application. Allowable shaft deflections for satisfactory gear and bearing performance vary with the gear or bearing design and with the application, but the following can be used as a general guide:

1. Deflections should not cause mating gear teeth to separate more than about 0.13 mm (0.005 in.), nor should they cause the relative slope of the gear axes to change more than about 0.03 deg.
2. The shaft (journal) deflection across a plain bearing must be small compared to the oil film thickness. (If the angular deflection of the shaft at the bearing is excessive, the shaft will bind unless the bearings are self-aligning.)
3. The shaft angular deflection at a ball or roller bearing should generally not exceed 0.04 deg. unless the bearing is self-aligning.

Shaft deflections can be computed by the methods discussed in Section 5.7 and illustrated by Sample Problem 5.2. Additionally, torsional deflections must often be considered because of (a) torsional natural frequency requirements, and (b) necessary torsional deflection limitations.

Determination of the fatigue strength of a rotating shaft usually involves an analysis for the general case of biaxial loading, as summarized in Fig. 8.17, and illustrated by Sample Problem 17.1, below.

Early in the design of any given shaft, an estimate is usually made of whether strength or deflection will be the critical factor. A preliminary design is based on that criterion; then, the remaining factor (deflection or strength) is checked.

SAMPLE PROBLEM 17.1 Snowmobile track drive shaft

Figure 17.6*a* shows a drive shaft, supported in a snowmobile frame by bearings *A* and *B*, which is chain-driven by sprocket *C*. (The engine and transmission are above and forward of the shaft; hence the 30 deg. chain angle.) Track sprockets T_1 and T_2 drive the snowmobile track. Basic dimensions are given in Fig. 17.6*a*. Determine an appropriate design for the shaft, based on a maximum engine output of 20 kw at a vehicle speed of 72 km/hr. Because the track and chain do not impose stringent deflection requirements and the bearings can be self-aligning if necessary, the preliminary design should be based on fatigue strength.

Solution

1. Figure 17.6*b* shows a proposed shaft layout. Note that the chain sprocket is mounted outboard to provide easy access to the chain for servicing, and that the retaining nut on the end tightens directly against the shaft. (If the nut were to tighten against the hub of sprocket *C*, the nut initial tightening load would impose a static tensile stress in the shaft between the nut and shoulder *S*. This would be undesirable from the standpoint of shaft fatigue strength.) Since bearing *B* will carry by far the greater load, provision is shown for bearing *A* to carry thrust in both directions. (Note snap rings in the housing and on the shaft to retain bearing *A*.) Thrust loads will be small (due to cornering maneuvers only), and this arrangement permits the use of a straight roller bearing at *B* if desired. Torque is shown transmitted from the chain sprocket by splines, and to the track sprockets by keys. Because of the large shaft bending moment in the vicinity of sprocket T_2, a tentative dimension locating the shaft shoulder is selected as shown in Fig. 17.6*b*. (This dimension will be needed to compute loads at this point of stress concentration.)

2. Conservatively assume that full engine power reaches the track. Track tension F_T is then computed as

$$F_T = \frac{\text{engine power}}{\text{vehicle velocity}} = \frac{20{,}000 \text{ w}}{20 \text{ m/s}} = 1000 \text{ N}$$

Half of this force is assumed to be transmitted to each of the track sprockets, T_1 and T_2.

3. By setting the summation of moments about the shaft axis equal to zero, the chain sprocket tension, F_C, is determined as

$$F_C = 1000 \text{ N } (125 \text{ mm}/50 \text{ mm}) = 2500 \text{ N}$$

4. Force, shear, and moment diagrams for the vertical and horizontal planes are determined in the usual manner (as in Fig. 2.11) and drawn in Fig. 17.6*c*. Note that the only vertical applied load is a component of the chain tension ($F_C \cos 30°$). Applied horizontal loads include the chain tension component ($F_C \sin 30°$) and also the two track sprocket forces (each of $F_T/2$). Bearings *A* and *B* are assumed to be self-aligning within the range of angular shaft deflection involved.

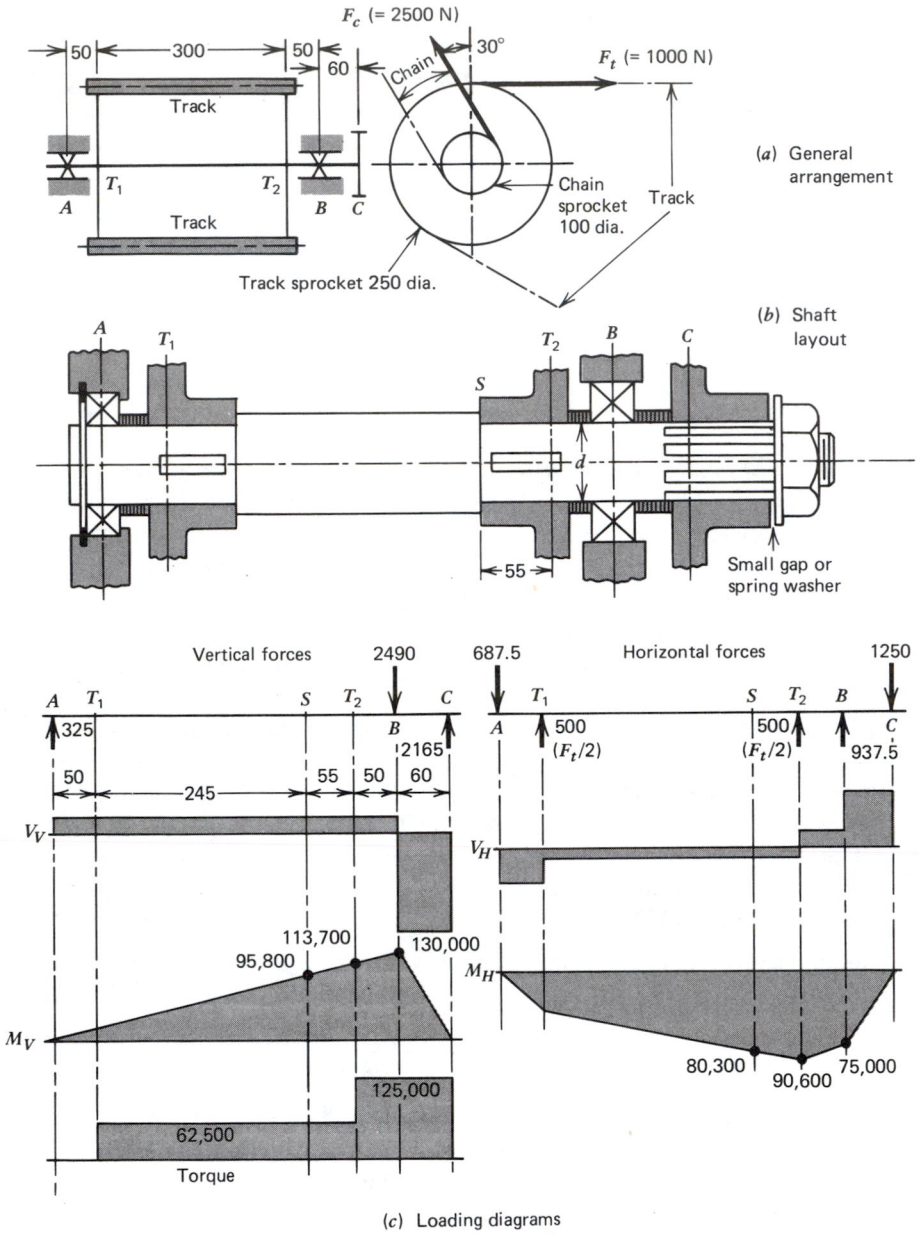

FIGURE 17.6.
(Above and on facing page)

5. The torque diagram assumes an even division of torque between the two track sprockets.
6. From an inspection of the loading diagrams and shaft layout it is clear that the critical location for determining the value of d will either be at S, or near B or C. Failure could occur at S due to the stress concentration. Failure at precisely B or

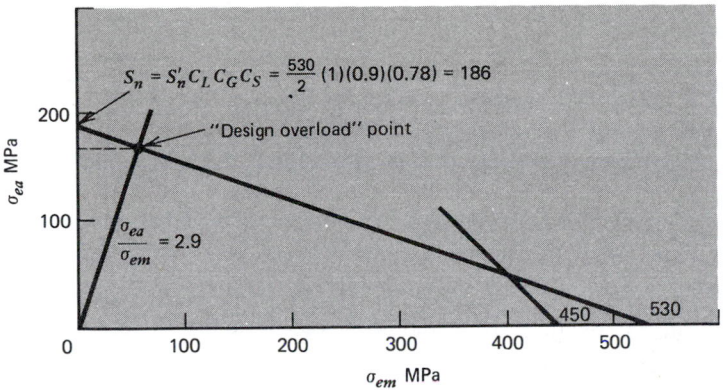

(*d*) Fatigue diagram

FIGURE 17.6.
Sample Problem 17.1—Snowmobile track drive shaft (Note: dimensions are in mm).

C is unlikely because the shaft is reinforced by the bearing race and sprocket hub at these points; but at the edges of the sprocket hub and bearing inner race a stress concentration exists not unlike that at S. Conservatively, select shaft diameter d on the basis of the loads at B (where the resultant bending moment is slightly higher than at T_2) and the estimated stress concentration factors at S.

7. To estimate K_f at shoulder S, decisions need to be made concerning material, surface finish, and geometric proportions of the shoulder. On the basis of cost, tentatively select cold-drawn 1020 steel having $S_u = 530$ MPa, $S_y = 450$ MPa, and machined surfaces. Ratios of $D/d = 1.25$ and $r/d = 0.03$ should give conservatively high values of K_f. From Fig. 4.34, $K_t = 2.25$ and 1.8 for bending and torsional loads, respectively. From Fig. 8.23 and an assumption that r will be about 1 mm, q is estimated as 0.7. Applying Eq. (8.2), values of K_f are 1.9 and 1.6 for bending and torsional loading.

8. Following the procedure specified for general biaxial loads in Fig. 8.17, the equivalent alternating stress is due only to bending:

$$\sigma_{ea} = \sigma = \frac{32M}{\pi d^3} K_{f(b)} = \frac{32\sqrt{130{,}000^2 + 75{,}000^2}}{\pi d^3}(1.9) = \frac{2.9 \times 10^6}{d^3}$$

The equivalent mean stress is due only to torsion:

$$\sigma_{em} = \tau = \frac{16T}{\pi d^3} K_{f(t)} = \frac{16(125{,}000)}{\pi d^3}(1.6) = \frac{1.0 \times 10^6}{d^3}$$

Thus, regardless of the value of d,

$$\sigma_{ea}/\sigma_{em} = 2.9$$

9. Figure 17.6*d* shows a fatigue strength diagram for this case, with "load line" at a slope of 2.9. The diagram indicates that for infinite life, σ_{ea} is limited to 165 MPa; but this is at the design overload that incorporates a safety factor. Reference

to Section 6.12 indicates that 2.5 might be an appropriate safety factor for this case. Hence,

$$\frac{2.9 \times 10^6}{d^3}(2.5) = 165 \text{ MPa}, \quad \text{or } d = 35.3 \text{ mm}$$

10. With the shaft layout shown, dimension d must correspond to a standard bearing bore. Selection of $d = 35$ mm should be satisfactory; to be a little more conservative, $d = 40$ mm might be preferred. In accordance with assumptions in step 7, the fillet radius must be at least $0.03d$. A more generous radius would be preferred. Specify, say, $r = 2$ mm.

11. A check of the angular deflections at A and B should be made to determine if self-aligning bearings are necessary.

Shafts supporting helical or bevel gears are subjected to mean loads that include tension or compression as well as torsion. Since axial stresses are functions of d^2 rather than d^3, a simple ratio between σ_{ea} and σ_{em} for all values of d does not exist for these cases. The most expedient procedure is first to ignore the axial stress when solving for d, and then to check the influence of the axial stress for the diameter obtained. A slight change in diameter may or may not be indicated. If the diameter selected must correspond to a standard size (as in Sample Problem 17.1), it is likely that a consideration of the axial stress will not change the final choice.

17.6 KEYS, PINS, AND SPLINES

Perhaps the most common of these torque-transmitting shaft-to-hub connections are *keys* (Fig. 17.1), of which the most common type is the square key (Fig. 17.1*a*). Standard proportions call for the key width to be approximately one-fourth the shaft diameter (see various handbooks and ANSI Standard B17.1 for details). Keys are usually made of cold-finished low-carbon steel (as SAE or AISI 1020), but heat-treated alloy steels are used when greater strength is required.

The loading of a key is a complex function of the clearances and elasticities involved. Figure 17.7*a* shows the loading of a loosely fitted square key. The primary loading involves the heavy horizontal force vectors; but these tend to rotate the key counterclockwise until one or both pairs of its diagonally opposite corners make contact with the keyway sides and/or bottoms.

Figure 17.7*b* shows a key that is tightly fitted at the top and bottom (sometimes radial setscrews in the hub are used to hold the key tightly against the bottom of the shaft keyway). The horizontal forces shown are commonly assumed to be uniformly distributed over the key surfaces, and to be equal to shaft torque divided by shaft radius. (Neither assumption is strictly correct, but in view of the complexities and uncertainties involved, they provide a reasonable basis for design and analysis.)

As an illustration of key sizing, let us estimate the length of key required in Fig. 17.7*b* to transmit a torque equal to the elastic torque capacity of the shaft. Assume that

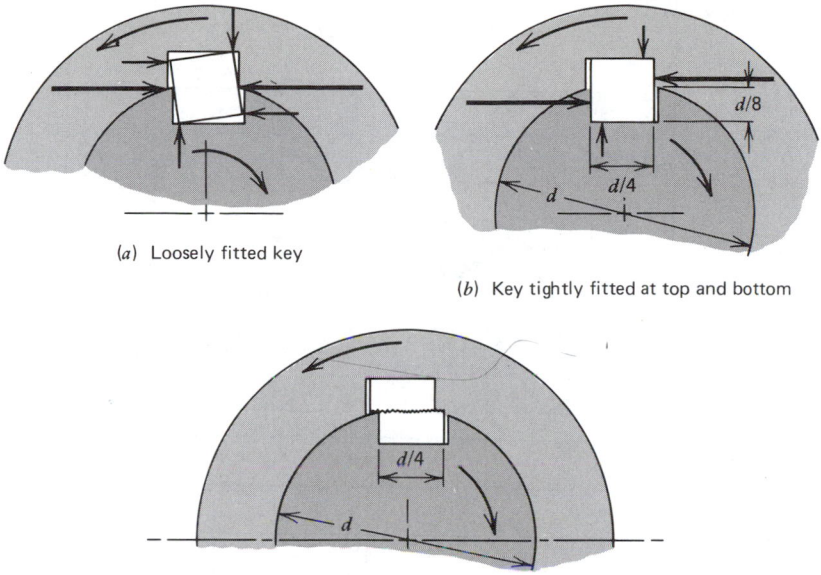

(*a*) Loosely fitted key

(*b*) Key tightly fitted at top and bottom

(*c*) Shear failure of a tightly fitted key

FIGURE 17.7.
Square key loading and stresses.

the shaft and key are made of ductile materials having the same strength, and that (in accordance with the distortion energy theory—Section 6.8), $S_{ys} = 0.58S_y$.

From Eq. (4.4), the shaft torque capacity is

$$T = \frac{\pi d^3}{16}(0.58S_y) \tag{a}$$

The torque that can be transmitted by compressive forces acting on the sides of the key is the product of limiting stress, contact area, and radius:

$$T = S_y \frac{Ld}{8}\left(\frac{d}{2}\right) = S_y Ld^2/16 \tag{b}$$

The torque that can be transmitted by key shear (Fig. 17.7*c*) is likewise the product of limiting stress, area, and radius:

$$T = (0.58S_y)\left(\frac{Ld}{4}\right)\left(\frac{d}{2}\right) = 0.58S_y Ld^2/8 \tag{c}$$

Equating (a) and (b) gives $L = 1.82d$; equating (a) and (c) gives $L = 1.57d$. Thus, on the basis of the simplifying assumptions that were made, a balanced design requires the key to be about $1.8d$ in length. Also, note that a key designed for theoretically balanced compression and shear strength would require a key depth a little greater than the key width. Keys normally extend along the full width of the hub, and for good stability, hub widths are commonly $1.5d$ to $2d$.

If the shaft diameter is based on deflection rather than strength, a shorter key may

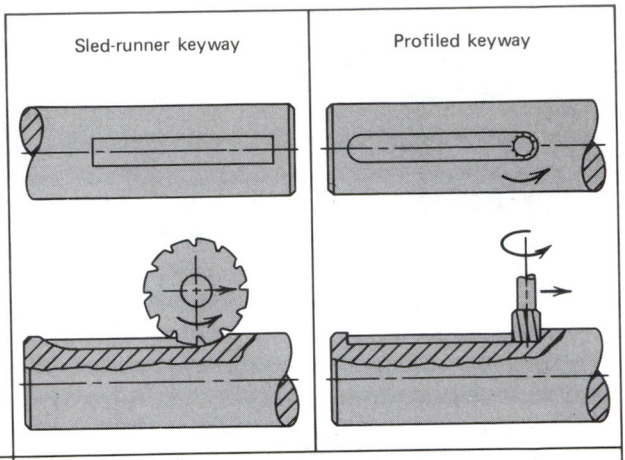

Steel	Fatigue stress concentration factor, K_f^*			
	Bending	Torsion	Bending	Torsion
Annealed (less than 200 Bhn)	1.3	1.3	1.6	1.3
Quenched and drawn (over 200 Bhn)	1.6	1.6	2.0	1.6

*From [6] of Chap. 8. Base nominal stress on total shaft section.

FIGURE 17.8.
Keyway types, and corresponding fatigue stress concentration factors, K_f.

be entirely adequate. If the shaft diameter is based on strength with shock or fatigue loading present, stress concentration at the keyway must be taken into account when estimating shaft strength. Figure 17.8 illustrates the two usual ways of cutting keyways, with associated approximate values of K_f (*fatigue* stress concentration factor).

Figure 17.2*a* shows a round *pin* attaching a hub and shaft. The torque capacity of the connection is limited by the strength of the pin in double shear (i.e., failure involves shearing both pin cross sections at the shaft-hub interface). For a solid pin of diameter d and shear yield strength S_{ys}, the reader can readily show that the torque capacity (based on pin yielding) is

$$T = \pi d^2 D S_{ys}/4 \tag{17.4}$$

Sometimes shear-loaded, torque-transmitting pins (as Fig. 17.2*a*) are made small and of relatively weak material to limit their capacity to the safe torque that can be carried by the shaft. In this case the *shear pin* serves as a safety or protective device. A common example is the shear pin used to attach the propeller of an outboard motor, If the propeller strikes an obstruction, the shear pin fails, thereby preventing possible damage to the more costly members of the drive train.

Splines (Fig. 17.4) act like multiple keys. They have either involute or straight-

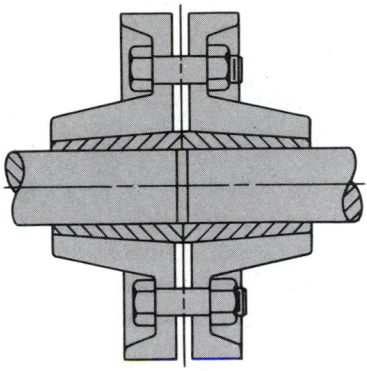

FIGURE 17.9.
Rigid shaft coupling.

sided profiles, the former being the usual type in modern machinery. Involute splines usually have a 30 deg. pressure angle, and one half the depth of standard gear teeth (other standard pressure angles are 37.5 deg. and 45 deg.). The fit between mating splines is characterized as sliding, close, or press. (See ANSI Standard B92.1, B92.1M, and B92.2M; the SAE Handbook, or other handbooks for details.) Figure 17.13 shows a sliding-fit spline which permits the length of an automotive propeller shaft to change slightly with rear wheel jounce. Splines can be cut or rolled onto a shaft (much like screw threads). The strength of a splined shaft is usually taken as the strength of a round shaft of diameter equal to the minor spline diameter. However, in the case of rolled splines, the favorable effects of cold-working and residual stresses may make the strength more nearly equal to that of the original unsplined shaft.

17.7 COUPLINGS AND UNIVERSAL JOINTS

Collinear shafts can be joined by *rigid couplings*, like the one shown in Fig. 17.9. The coupling halves can be attached to the shaft ends with keys, but the one shown transmits torque by friction through slotted tapered sleeves. (The sleeves are wedged tightly in place by tightening the bolts connecting the two halves.) Note that the flanged portions at the outside diameter serve a safety function by shielding the bolt heads and nuts. In designing such a coupling, the force-flow concept (Section 2.4) is appropriate. This leads one to consider, in order, (1) the torque capacity of the key or wedged friction connection with the shaft, (2) the strength of the relatively thin web portions that are drilled to accommodate the bolts, and (3) the strength of the bolts.

Rigid couplings are limited in their application to the *unusual* cases where the shafts are collinear within extremely close tolerances, and where they can be counted on to stay that way. If the shafts are laterally misaligned (i.e., parallel but offset) or angularly misaligned (axes intersect at an angle), the installation of a rigid coupling *forces* them into alignment. This subjects the coupling, shafts, and shaft bearings to unnecessary loads that may lead to early failures.

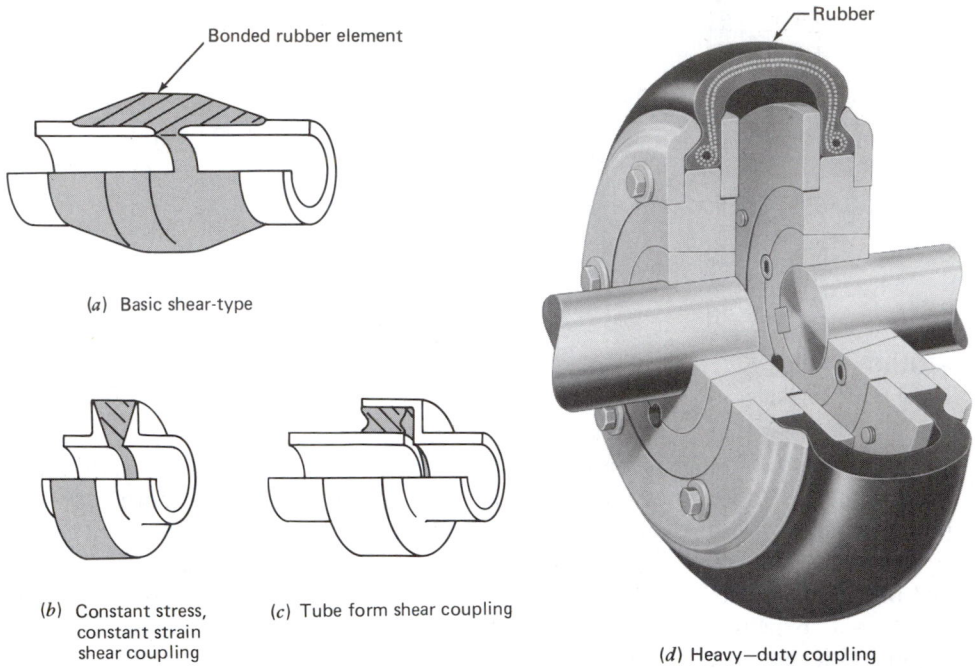

FIGURE 17.10.
Rubber-element flexible couplings. (*a, b, c* Courtesy Lord Corporation; *d* Courtesy Reliance Electric Co.)

The problems of a small amount of shaft misalignment can be eliminated by using a *flexible* coupling. A large variety of ingenious designs are on the market. Figure 17.10 shows a few of the many designs that use a flexible material like rubber. These can be designed to provide appropriate elasticity and damping for control of torsional vibration as well as providing for misalignment. Other flexible couplings use all metallic components (two are shown in Fig. 17.11), and these tend to have greater torque capacity for a given size.

One ingenious design of fairly ancient origin is the *Oldham coupling,* shown in Fig. 17.12. The sliding of the center block permits a substantial amount of shaft lateral offset, while built-in axial clearance permits some angular misalignment.

Universal joints permit substantial angular misalignment of shafts having intersecting axes. Figure 17.13 shows the common *cross-type* (known as *Cardan joint* or *Hooke's joint*) that is usually used at the ends of the drive shaft on rear-wheel-drive automobiles. Plain bushings or needle bearings are used at the yoke-to-cross connections. If the input yoke rotates at constant angular velocity, the output yoke velocity will have a speed fluctuation at twice rotating speed, the magnitude of which increases with the misalignment angle. If two joints are used, *with yokes aligned as shown in Fig. 17.13,* speed fluctuations across the two joints cancel to give uniform output yoke rota-

(*a*) Roller chain coupling

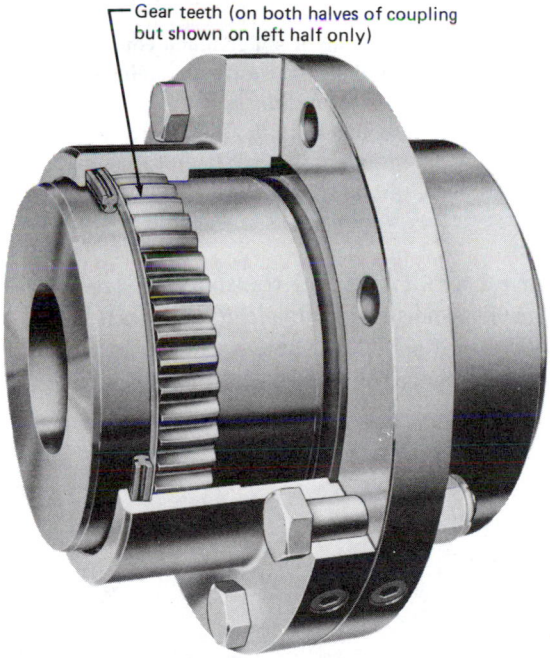

(*b*) Gear coupling

FIGURE 17.11.
Metallic-element flexible couplings. (Courtesy Reliance Electric Co.)

tion *if all three shafts between and at each end of joints are in the same plane, and if the misalignment angles at the two joints are equal.*

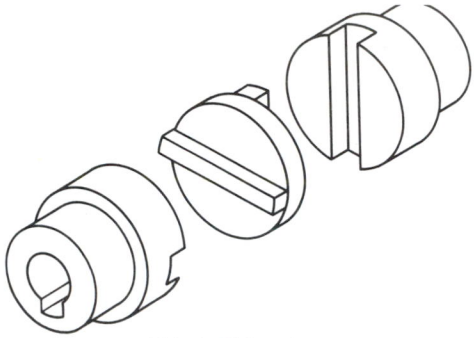

(*a*) basic Oldham type

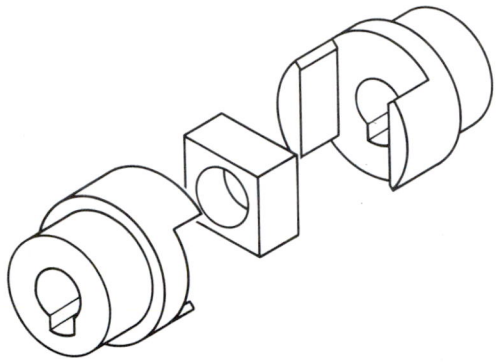

(*b*) modified type

FIGURE 17.12.
Oldham or slider-block couplings. Both versions have a freely sliding center slider-block that provides pairs of sliding surfaces at 90 deg. orientation. The greater the shaft misalignment, the greater the sliding. Lubrication and wear must be considered.

Other types of universal joints have been devised that transmit angular velocity uniformly across a single joint. These are known as *constant velocity* universal joints. A common application is in front-wheel-drive automobiles, where drive shafts are short, and shaft angles (due to steering and wheel jounce) can be large.

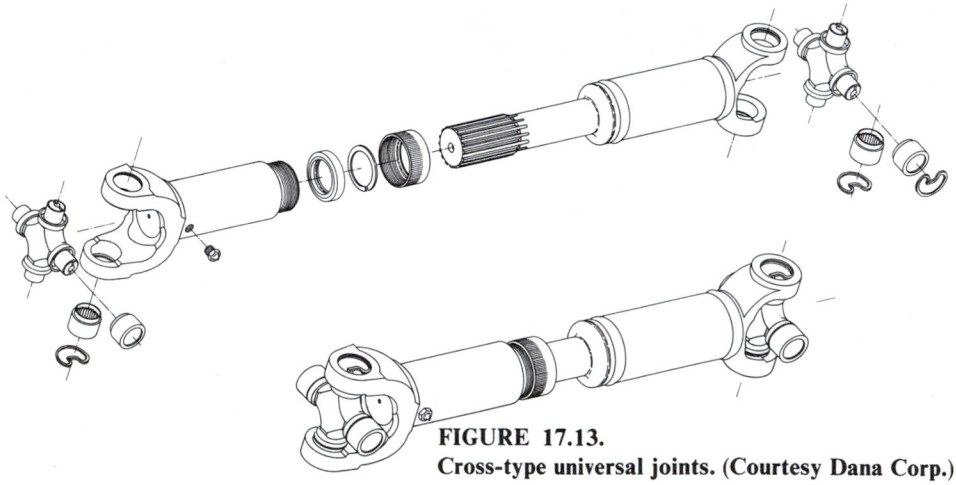

FIGURE 17.13.
Cross-type universal joints. (Courtesy Dana Corp.)

CHAPTER EIGHTEEN

Clutches and Brakes

18.1 INTRODUCTION

With sliding surfaces encountered in most machine components (as bearings, gears, cams, etc.) it is desirable to minimize interface friction in order to reduce energy loss and wear. In contrast, clutches and brakes depend on friction in order to function. With these members, one objective is to *maximize* the friction coefficient (and to keep it uniform over a wide range of operating conditions) and at the same time to minimize wear.

The function of a *clutch* is to permit the smooth, gradual connection and disconnection of two members having a common axis of rotation. A brake acts similarly except that one of the members is fixed. The clutches and brakes considered here are all of the *friction* type, involving sliding friction between solid surfaces. Other types use magnetic, eddy current, and hydrodynamic forces (as fluid couplings and torque converters, treated in Chapter 19).

Several types of friction clutches and brakes are considered in this chapter. All must be designed to satisfy three basic requirements: (1) the required friction torque must be produced by an acceptable actuating force, (2) the energy converted to friction heat (during braking or during clutch engagement) must be dissipated without producing destructively high temperatures, and (3) the wear characteristics of the friction surfaces must be such as to give acceptable life. Sections 9.8, 9.9, 9.10, 9.12, and 9.15 provide appropriate background with respect to clutch and brake wear characteristics.

18.2 DISK CLUTCHES

Figure 18.1 shows a simple disk clutch with one driving and one driven surface. Driving friction between the two is developed when they are forced together. Practical embodiments of this principle are illustrated in Figs. 18.2 and 18.3.

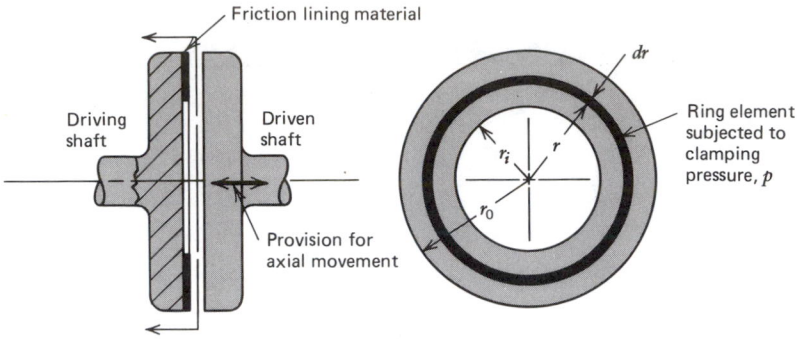

FIGURE 18.1.
Basic disk clutch.

Figure 18.2 shows an automotive-type clutch, used with "standard" transmissions. The flywheel, clutch cover, and pressure plate rotate with the crankshaft. A series of circumferentially distributed springs (or a single internally slotted coned-disk spring—Fig. 12.30) force the pressure plate toward the flywheel, clamping the clutch plate (driven disk) between them. The hub of the clutch plate is spline-connected to the transmission input shaft. The clutch is disengaged by depressing the clutch pedal, which rotates the lever marked "to release." This pushes the clutch release bearing against a series of radially oriented release levers that pull the pressure plate away from the flywheel. Note that the clutch release bearing is a *thrust* bearing—the right side bears against the release mechanism, which does not rotate; the left side bears against the release levers, which rotate with the crankshaft. This clutch has two driving surfaces (one on the flywheel and one on the pressure plate) and two driven surfaces (the two sides of the clutch plate).

Figure 18.3 illustrates the principle of multiple-disk clutches. Disks *a* are constrained (as with splines) to rotate with the input shaft; disks *b* are similarly constrained to rotate with the output shaft. When the clutch is disengaged, the disks are free to slide axially to separate themselves. When the clutch is engaged, they are clamped tightly together to provide (in the case illustrated) six driving and six driven surfaces. The two end disks (which have only their inner sides serving as friction surfaces) should be members of the same set in order to avoid transmitting the clamping force through a thrust bearing. (Note that in Figs. 18.2 and 18.3 the clutch clamping force is localized to the region of the disks, whereas in Fig. 18.1 it would have to be transmitted through a thrust bearing.)

As with other friction members, disk clutches can be designed to operate either "dry" or "wet" (in oil). Automotive clutches of the type shown in Fig. 18.2 are "dry"; most multiple-disk clutches (including those used in automotive automatic transmissions) operate "wet." The oil serves as an effective coolant during clutch engagement, and the use of multiple disks compensates for the reduced coefficient of friction.

Equations are developed below relating clutch size, friction coefficient, torque capacity, axial clamping force, and interface pressure, using each of two basic assumptions. Throughout the development it is assumed that the coefficient of friction, f, is a constant.

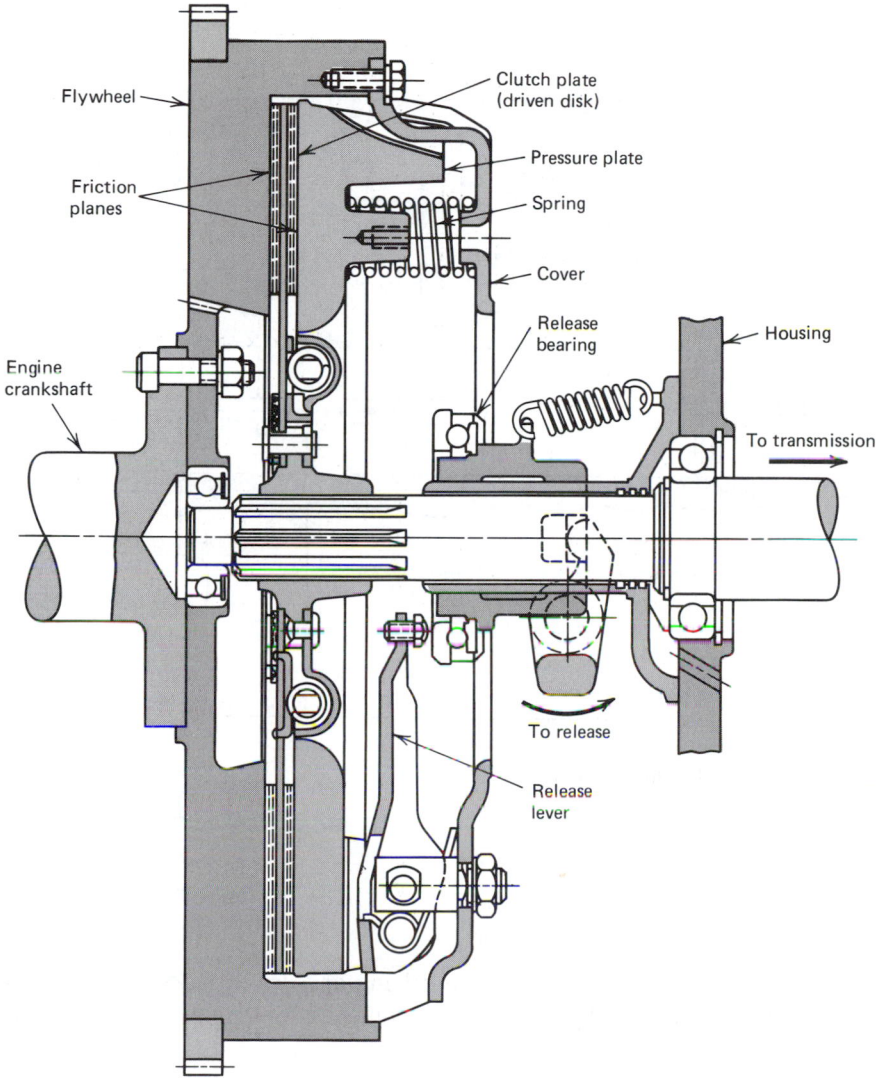

FIGURE 18.2.
Automotive-type disk clutch. (Courtesy Borg-Warner Corporation)

(a) Assume Uniform Distribution of Interface Pressure This assumption is valid for an unworn (new), accurately manufactured clutch, with rigid outer disks. With reference to Fig. 18.1, the normal force acting on a differential ring element of radius r is

$$dF = (2\pi r\, dr)p, \tag{a}$$

where p is the uniform level of interface pressure. The total normal force acting on the area of contact is

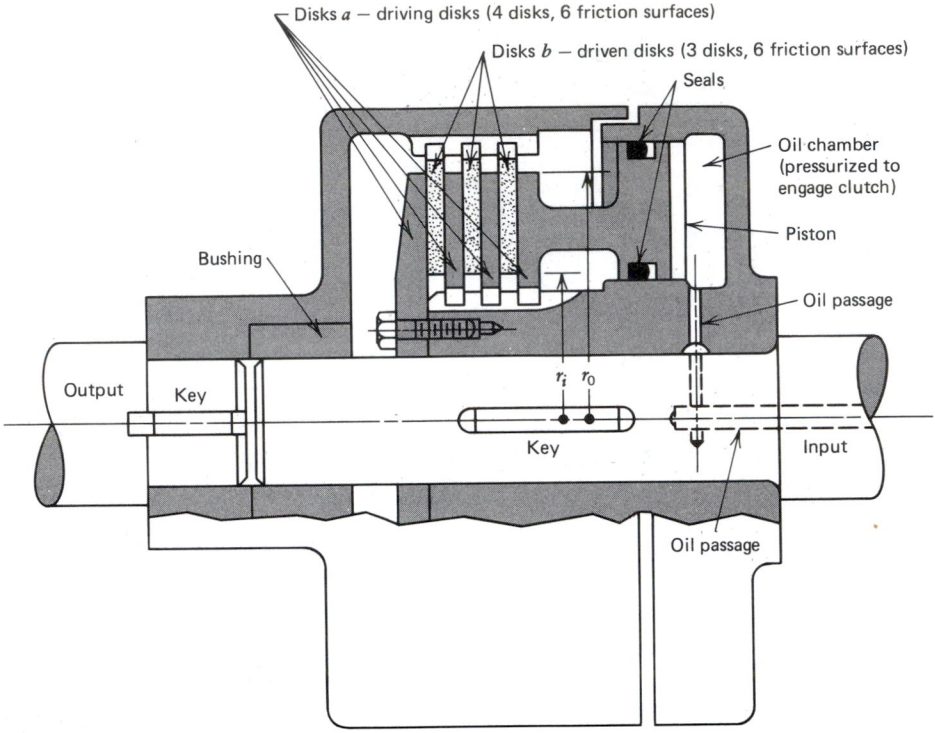

FIGURE 18.3.
Multiple-disk clutch, hydraulically operated.

$$F = \int_{r_i}^{r_0} 2\pi p r \, dr$$

$$F = \pi p(r_0^2 - r_i^2) \tag{18.1}$$

where F is also the axial force clamping the driving and driven disks together. The friction torque that can be developed on a ring element is the product of the normal force, coefficient of friction, and radius:

$$dT = (2\pi r \, dr)pfr$$

and the total torque that can be developed over the entire interface is

$$T = \int_{r_i}^{r_0} 2\pi pfr^2 \, dr = \tfrac{2}{3}\pi pf(r_0^3 - r_i^3) \tag{b}$$

Equation (b) represents the torque capacity of a clutch with one friction interface (one driving disk mating with one driven disk, as in Fig. 18.1). Actual clutches (as shown in Figs. 18.2 and 18.3) employ N friction interfaces transmitting torque in parallel, where N is an *even* number (two in Fig. 18.2; six in Fig. 18.3). For a clutch with N friction interfaces, Eq. (b) is modified to give

$$T = \tfrac{2}{3}\pi pf(r_0^3 - r_i^3)N \tag{18.2}$$

Solving Eq. (18.1) for p, and substituting its value into Eq. (18.2) gives an equation for torque capacity as a function of axial clamping force

$$T = \frac{2Ff(r_0^3 - r_i^3)}{3(r_0^2 - r_i^2)} N \tag{18.3}$$

(b) Assume Uniform Rate of Wear at Interface As implied in Eq. (9.1), wear rate is generally proportional to the rate of friction work—that is, friction force times rubbing velocity. With a uniform coefficient of friction, wear rate is proportional to the product of pressure times sliding velocity. (Hence, the common experience of wearing down a piece of wood with sandpaper at a rate proportional to both pressure and rubbing velocity.) On the clutch face, velocity is proportional to radius; hence, rate of work is proportional to the product of pressure and radius. On this basis, a new clutch (considered in assumption (a), above) would have its greatest initial wear at the outer radius. After this initial "break-in" wear, the friction lining tends to wear at a uniform rate, being "ground down" between assumed rigid and parallel outer plates. This uniform wear rate is assumed to result from a uniform rate of friction work—that is, a constant product of pressure and velocity, or a constant product of pressure and radius. Thus:

$$pr = C \quad \text{(where } C \text{ is a constant)}$$

The greatest pressure, p_{max}, obviously occurs at the inside radius, and this has an allowable value determined by the characteristics of the friction lining material. Hence, for a clutch of inside radius r_i and friction lining with allowable pressure p_{max}, the clutch design is based on

$$pr = C = p_{max} r_i \tag{18.4}$$

Using Eq. (18.4) and proceeding as in the derivation of Eqs. (18.1)–(18.3), we have:

$$F = \int_{r_i}^{r_0} 2\pi p_{max} r_i \, dr$$

$$F = 2\pi p_{max} r_i (r_0 - r_i) \tag{18.5}$$

$$T = \int_{r_i}^{r_0} 2\pi p_{max} r_i f r \, dr \, N$$

$$T = \pi p_{max} r_i f (r_0^2 - r_i^2) N \tag{18.6}$$

$$T = Ff\left(\frac{r_0 + r_i}{2}\right) N \tag{18.7}$$

Note the simple physical interpretation of Eq. (18.7).

The assumption of uniform wear rate gives a lower calculated clutch capacity than the assumption of uniform pressure. (This must be so because the higher initial wear toward the outside diameter shifts the center of pressure toward the inside, giving a smaller torque arm.) Hence, clutches are usually designed on the basis of uniform wear, and have a little extra torque capacity when new.

Approximate data relative to coefficients of friction and allowable pressures for various friction lining materials are given in Tables 18.1 and 18.2.

A parameter in the design of clutches is the ratio of inside to outside radius. It is

TABLE 18.1
Representative Properties of Friction Materials, Operating Dry

Friction Material[a]	Dynamic Friction Coefficient, f[b]	Maximum Pressure[c]		Max. Bulk Temp.	
		psi	kPa	°F	°C
Molded	0.25–0.45	150–300	1030–2070	400–500	204–260
Woven	0.25–0.45	50–100	345–690	400–500	204–260
Sintered metal	0.15–0.45	150–300	1030–2070	450–1250	232–677
Cork	0.30–0.50	8–14	55–95	180	82
Wood	0.20–0.30	50–90	345–620	200	93
Cast iron, hard steel	0.15–0.25	100–250	690–1720	500	260

[a]When rubbing against smooth cast iron or steel.
[b]Experimental values of f vary with detailed composition, rubbing speed, pressure, temperature, humidity, and composition. Consult manufacturer or obtain test data. For design, commonly use 50 to 75 percent of test values to provide a factor of safety.
[c]Use of lower values will give longer life. Consult the manufacturer or obtain test data. Compute average pressure on cylindrical surfaces on the basis of projected area of contact (as for bearing stresses and journal bearing pressures—see Section 13.3).

left to the reader to show, from Eq. (18.6), that maximum torque for a given outside radius is obtained when

$$r_i = \sqrt{\tfrac{1}{3}}\,r_0 = 0.58r_0 \tag{18.8}$$

Proportions commonly used range from $r_i = 0.45r_0$ to $r_i = 0.80r_0$.

SAMPLE PROBLEM 18.1

A multiple-disk wet clutch is to be designed for transmitting a torque of 85 N·m. Space restrictions limit the outside disk diameter to 100 mm. Design values for the molded friction material and steel disks to be used are $f = 0.06$ ("wet") and $p_{max} = 1400$ kPa. Determine appropriate values for disk inside diameter, total number of disks, and clamping force.

Solution

1. Based on Eq. (18.8), tentatively select $r_i = 29$ mm.
2. The only unknown in Eq. (18.6) is N, which is solved for to give $N = 6.69$.
3. Since N must be an *even* integer, choose $N = 8$. Reference to Fig. 18.3 shows that this requires a total of 4 + 5, or 9, disks (remember that the two outer disks have friction surfaces on one side only).
4. With no other changes, this will give a clutch that is overdesigned by a factor of 8/6.69 = 1.19. Possible alternatives include (a) accepting the 19 percent overde-

TABLE 18.2
Representative Values of Friction Coefficient for Friction Materials Operating in Oil

Friction Material[a]	Dynamic Friction Coefficient, f
Molded	0.06–0.09
Woven	0.08–0.10
Sintered metal	0.05–0.08
Paper	0.10–0.14
Graphitic	0.12 (avg.)
Polymeric	0.11 (avg.)
Cork	0.15–0.25
Wood	0.12–0.16
Cast iron, hard steel	0.03–0.06

[a]When rubbing against smooth cast iron or steel.

sign, (b) increasing r_i, (c) decreasing r_0, and (d) leaving both radii unchanged and reducing both p_{max} and F by a factor of 1.19.

5. Arbitrarily choosing alternative (d), the clamping force is computed from Eq. (18.7) to be just sufficient to produce the desired torque:

$$85\ \text{N}\cdot\text{m} = F(0.06)\left(\frac{0.050 + 0.029}{2}\ \text{m}\right)8, \qquad F = 4483\ \text{N}$$

6. Rounding up the calculated value of F to an even number, the final proposed answers are (a) inside diameter = 58 mm, (b) clamping force = 4500 N, and (c) a total of nine disks. ■

18.3 DISK BRAKES

As noted previously, a brake is similar to a clutch except that one of the shafts is replaced by a fixed member. Hence, with minor modification, the designs illustrated in Figs. 18.2 and 18.3 can be converted to disk brakes. Such brakes would be unsatisfactory for general use because of inadequate cooling. For this reason, *caliper disk brakes* are commonly used. Bicycle brakes are undoubtedly the best known examples. The wheel rim constitutes the disk. Friction lining on the caliper contacts only a small portion of the disk surface, leaving the remainder exposed to dissipate heat. Figure 18.4 shows a hydraulically actuated caliper disk brake that uses a ventilated disk. Air circulation through the interior passages provides substantial additional cooling. Disk brakes can conveniently be examined on the front wheel of most larger motorcycles.

The torque capacity and clamping force requirements of caliper disk brakes can be determined using the procedures of the previous article. The cooling or heat-dissipating characteristics of brakes are discussed below.

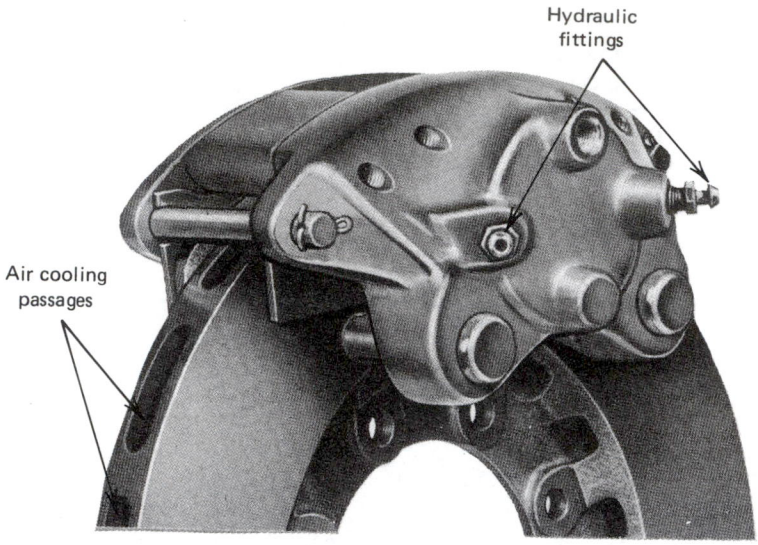

FIGURE 18.4.
Caliper disk brake, hydraulically operated. (Courtesy Auto Specialities Manufacturing Co.)

18.4 ENERGY ABSORPTION AND COOLING

The basic function of a brake is to absorb energy (i.e., to convert kinetic and potential energy into friction heat) and to dissipate the resulting heat without developing destructively high temperatures. Clutches also absorb energy and dissipate heat, but usually at a lower rate. Where brakes (or clutches) are used more or less continuously for extended periods of time, provision must be made for rapid transfer of heat to the surrounding atmosphere. For intermittent operation, the thermal capacity of the parts may permit much of the heat to be stored, and then dissipated over a longer period of time. Brake and clutch parts must be designed to avoid objectionable thermal stresses and thermal distortion (Section 4.16).

The basic heat transfer equation involved is Eq. (13.13), previously encountered in connection with journal bearings and worm gears. With slightly modified notation this is:

$$H = CA(t_s - t_a) \tag{18.9}$$

where

H = time rate of heat dissipation (W or hp)

C = overall heat transfer coefficient (W per m^2 per °C, or hp per in.^{2} per °F)

A = exposed heat-dissipating surface area (m^2 or in.^{2})

t_s = average temperature of heat-dissipating surfaces (°C or °F)

t_a = air temperature in the vicinity of the heat-dissipating surfaces (°C or °F)

The ability of brakes to absorb large amounts of energy without reaching destructive temperatures can be increased by (a) increasing exposed surface areas, as by fins and ribs, (b) increasing air flow past these surfaces by minimizing air flow restrictions and maximizing the air pumping action of the rotating parts, and (c) increasing the mass and specific heat of parts in immediate contact with the friction surfaces, thereby providing increased heat storage capacity during short time periods of peak braking load.

The sources of energy to be absorbed are primarily three:

(1) kinetic energy of translation:

$$KE = \tfrac{1}{2}MV^2 \tag{18.10}$$

(2) kinetic energy of rotation:

$$KE = \tfrac{1}{2}I\omega^2 \tag{18.11}$$

(3) potential (gravitational) energy, as in an elevator being lowered, or an automobile descending a hill:

$$PE = Wd \text{ (weight times vertical distance)} \tag{18.12}$$

Some appreciation of the magnitude of the rate of energy that brakes are sometimes called on to absorb can be gained by considering a "panic stop" of a high-speed automobile: the instantaneous braking power is the same as the engine power that would be required to "squeal" all wheels during 4-wheel-drive acceleration from the high speed!

A dramatic example of the power capacity of vehicle brakes is the stopping of a modern jet airliner during a "rejected take-off" (fully loaded plane is at speed for lift-off when, at the last minute, the brakes are applied for an emergency stop). In the case of a Boeing 707, this involves stopping a 260,000 pound vehicle traveling 185 mph. Eighty percent of the kinetic energy is absorbed by the brakes, and they are designed to do this *once*. In the unlikely event that this occurs, the brakes reach a destructive temperature and must be replaced before the aircraft is used again. (This represents good engineering design, as it is more economical to replace the brakes than it is to carry the extra weight of brakes that could make this extreme emergency stop without damage.)

The rate at which heat is generated on a unit area of friction interface is equal to the product of the normal (clamping) pressure, coefficient of friction, and rubbing velocity. Manufacturers of brakes and of brake lining materials have conducted tests and accumulated experience enabling them to arrive at empirical values of pV (normal pressure times rubbing velocity) and of power per unit area of friction surface (as hp per square inch or kW per square mm) that are appropriate for specific types of brake design, brake lining material, and service conditions. Table 18.3 lists typical values of pV in industrial use.

18.5 CONE CLUTCHES AND BRAKES

Figure 18.5 shows a cone clutch. It is similar to a disk clutch, and can be regarded as the *general case* of which the disk clutch is a special case involving a cone angle, α, of

TABLE 18.3
Typical Values of Pressure Times Rubbing Velocity Used in Industrial Shoe Brakes

Operating Conditions	pV (psi) (ft/min.)	pV (kPa) (m/s)
Continuous, poor heat dissip.	30,000	1050
Occasional, poor heat dissip.	60,000	2100
Continuous, good heat dissip. as in an oil bath	85,000	3000

90 deg. The construction of a cone clutch makes it impractical to have more than one friction interface; hence, $N = 1$. As previously noted, this requires that the shaft bearings take a thrust load equal to the clamping force. This is acceptable because the inherent wedging action of a typical cone clutch enables the clamping force to be reduced to only about one-fifth that of a corresponding disk clutch of $N = 1$.

Figure 18.5*b* shows that the surface area of a ring element is

$$dA = 2\pi r\, dr/\sin\alpha$$

The normal force on the element is

$$dN = (2\pi r\, dr)p/\sin\alpha$$

The corresponding clamping force is

$$dF = dN \sin\alpha = (2\pi r\, dr)p$$

which is exactly the same as for a disk clutch ring element (Eq. (a)). The torque that can be transmitted by the element is

$$dT = dNfr = 2\pi pfr^2\, dr/\sin\alpha$$

From this point the equations for clamping force and torque capacity are derived exactly as for a disk clutch. Equation (18.4) applies for the case of assumed uniform wear rate. The resulting equations indicate that: Eqs. (18.1) and (18.5) are also valid for cone clutch clamping force, and the torque capacity of a cone clutch is given by the disk clutch equations divided by $\sin\alpha$. Thus, for assumed uniform pressure:

$$T = \tfrac{2}{3}\pi pf(r_0^3 - r_i^3)/\sin\alpha \tag{18.2a}$$

$$T = \frac{2Ff(r_0^3 - r_i^3)}{3(r_0^2 - r_i^2)} \Big/ \sin\alpha \tag{18.3a}$$

and for assumed uniform wear rate:

$$T = \pi p_{\max} r_i f(r_0^2 - r_i^2)/\sin\alpha \tag{18.6a}$$

$$T = Ff\left(\frac{r_0 + r_i}{2}\right) \Big/ \sin\alpha \tag{18.7a}$$

The above equations have been numbered after the corresponding disk clutch equations because they are merely the general case of the disk clutch equations, with $N = 1$.

The smaller the cone angle, α, the lower the theoretical clamping force required. This angle cannot be made less than about 8 deg. or the clutch may tend to "grab"

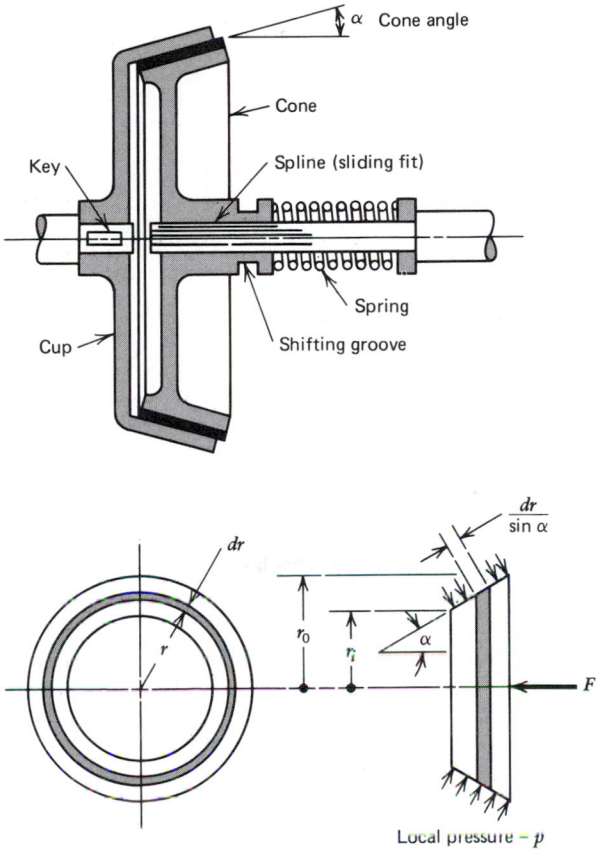

FIGURE 18.5.
Cone clutch.

when engaged. Also, for $\alpha < 8$ deg. cone clutches tend to be difficult to disengage. An angle of 12 deg. is usually considered about optimum, with values of α between 8 deg. and 15 deg. being commonly used.

The clutch in Fig. 18.5 is disengaged by a fork fitting in the shifting groove. Since the fork does not rotate, a thrust bearing must be used (as with disk clutches—Fig. 18.2).

18.6 SHORT-SHOE DRUM BRAKES

Drum brakes are of two types: (1) those with *external* shoes that contract to bear against the outer (cylindrical) drum surface, and (2) those with *internal* shoes that expand to contact the inner drum surface.Figure 18.6 shows a schematic representation of a simple external drum brake with a "short shoe"—that is, a shoe that contacts only a small segment of the drum periphery. Force F at the end of the lever applies the brake. Although the normal force (N) and the friction force (fN) acting between the drum

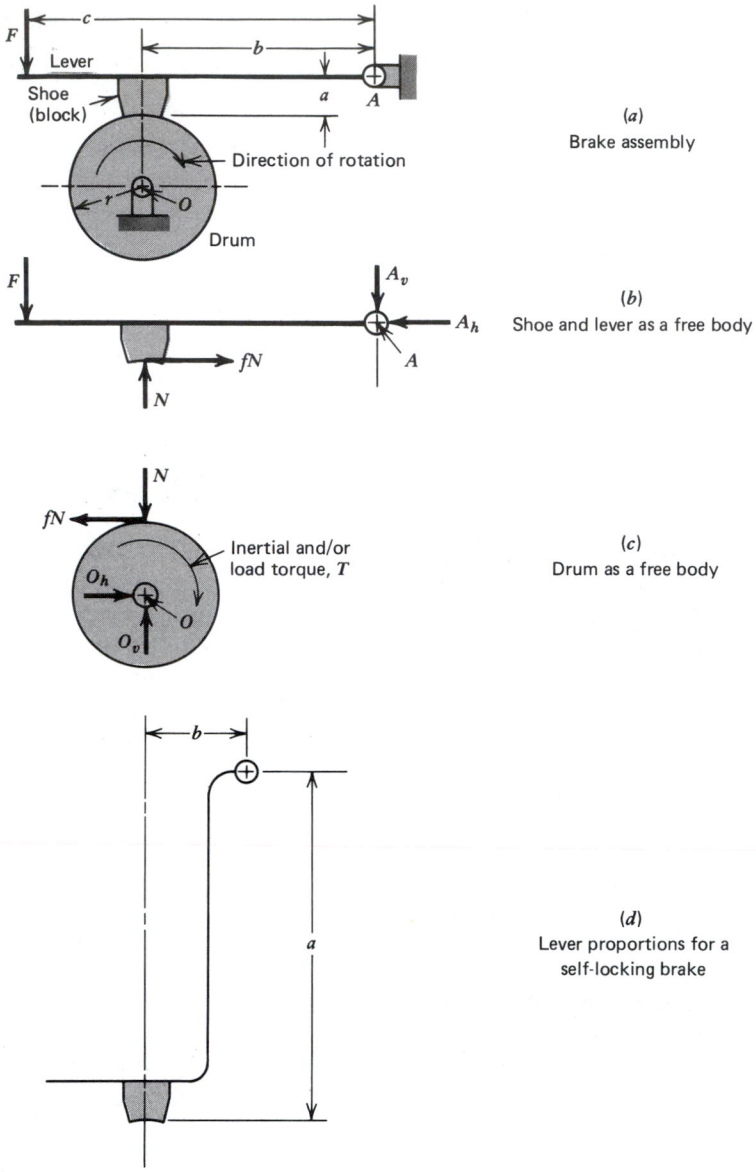

FIGURE 18.6.
"Short-shoe" drum brake.

and shoe are distributed continuously over the contacting surfaces, the short-shoe analysis assumes these forces to be concentrated at the center of contact. The complete brake assembly is shown in Fig. 18.6*a*. Free-body diagrams of the basic components are shown in 18.6*b* and 18.6*c*. The case shown is for *clockwise* drum rotation.

Taking moments about pivot A for the shoe and lever assembly:

$$Fc + fNa - bN = 0 \tag{c}$$

From summation of moments about 0 for the drum:

$$T = fNr \tag{d}$$

Solving (c) for N and substituting in (d):

$$N = Fc/(b - fa) \tag{18.13}$$
$$T = fFcr/(b - fa) \quad \text{(self-energizing case)}$$

Torque T is the inertial and/or load torque required for equilibrium, and it is numerically equal to the friction torque developed by the brake.

Equation (18.13) is labeled "self-energizing case" because the moment of the friction force (fNa) *assists* the applied force (F) in applying the brake. For counterclockwise drum rotation, the direction of the friction force would be reversed. This would cause it to *oppose* the application of the brake, giving the self-*de-energizing* case. The derivation of the equation for the de-energizing case is the same as for the self-energizing case except that the signs of the friction force terms are reversed. This leads to:

$$T = fFcr/(b + fa) \quad \text{(self-de-energizing case)} \tag{18.14}$$

Returning now to the self-energizing case (clockwise drum rotation), note that the brake is *self-locking* if the denominator of Eq. (18.13) is zero or negative. Thus, for self-locking:

$$b \leq fa \tag{18.15}$$

For example, if $f = 0.3$, self-locking (for clockwise drum rotation) is obtained if $b \leq 0.3a$. This is illustrated in Fig. 18.6*d*. A self-locking brake requires only that the shoe be brought in contact with the drum (with $F = 0$) for the drum to be "locked" against rotation in one direction.

Because the brake in Fig. 18.6 has only one shoe (block), the entire force exerted on the drum by the shoe must be reacted by the shaft bearings. Partly for this reason, two opposing shoes are almost always used, as in Fig. 18.7. Although the opposing shoe forces are seldom in complete balance, the resultant shaft bearing loads are usually small.

SAMPLE PROBLEM 18.2

The two-shoe external drum brake shown in Fig. 18.7 has shoes 80 mm wide that contact 90 deg. of drum surface. For a coefficient of friction of 0.20 and an allowable contact pressure of 400 kN per square meter of projected area, estimate (a) the maximum lever force, F, which can be used, (b) the resulting braking torque, and (c) the radial load imposed on the shaft bearings. Use the "short-shoe" equations derived above.

Solution

1. Figures 18.7*b*–18.7*f* show each component of the brake as a free body. The force analysis begins with floating lever 5 because it receives the applied force, F. Note the convention used in designating the other forces: H_{45} is a horizontal force of 4

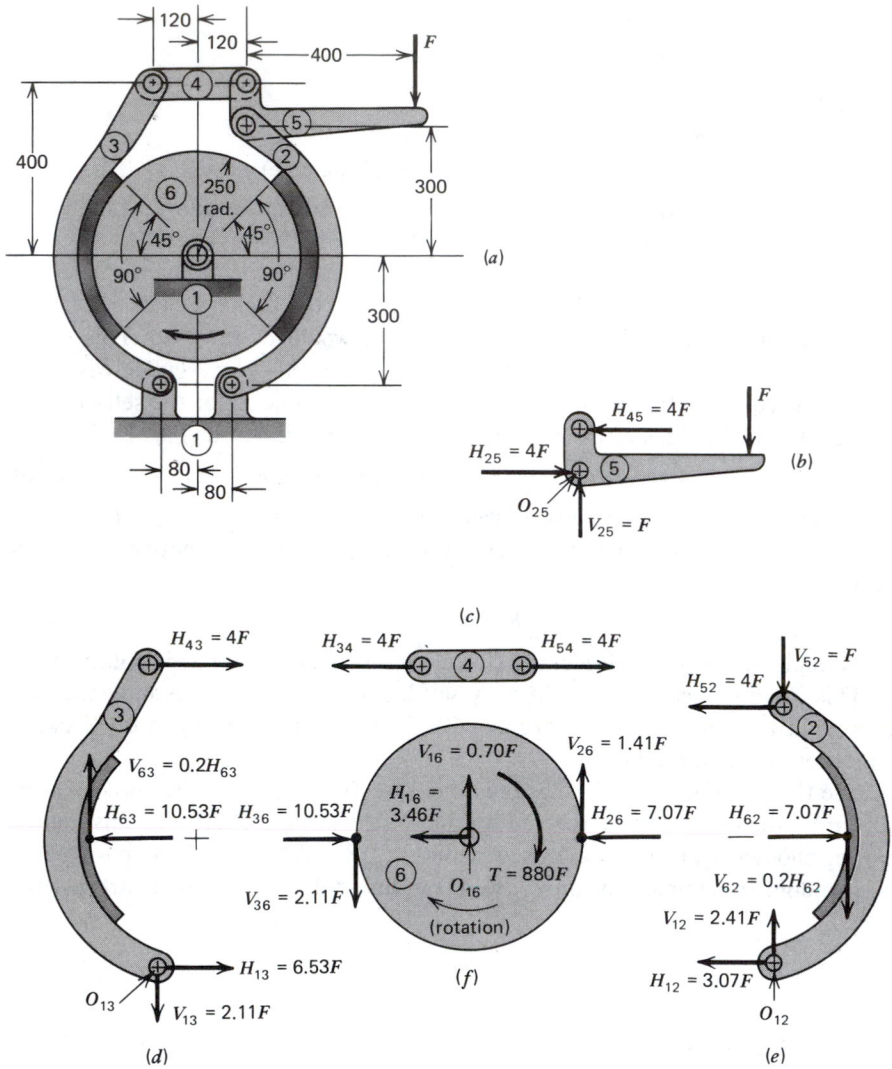

FIGURE 18.7.
Two-shoe external drum brake (Sample Problem 18.2).

applied to 5 (since link 4 can carry only simple tension or compression, there can be no vertical force V_{45}). The pin connection at floating pivot O_{25} provides both horizontal and vertical forces to link 5 (H_{25} and V_{25}). Taking moments about O_{25} establishes $H_{45} = 4F$; force summations establish $V_{25} = F$ and $H_{25} = 4F$.

2. Turning to link 4, H_{54} (horizontal force of 5 on 4) is equal and opposite to H_{45}; summation of forces requires that H_{34} be equal and opposite to H_{54}.

3. On the left shoe (link 3), the applied force is H_{43} (which is equal and opposite to H_{34}). The "short-shoe" analysis assumes that the normal and friction forces applied by drum 6 act at the center of the shoe as shown. The normal force is H_{63} and the

friction force is H_{63} multiplied by the given friction coefficient of 0.2. Summation of moments about O_{13} gives

$$4F(700) + 0.2H_{63}(170) - H_{63}(300) = 0, \quad \text{or } H_{63} = 10.53F$$

Forces H_{13} and V_{13} acting at fixed pivot O_{13} are determined from the force equilibrium equations.

4. Normal and friction forces H_{62} and V_{62} acting on shoe 2 are determined in the same manner. The moment equation has an additional term because both horizontal and vertical forces are applied by link 2:

$$4F(600) - F(40) - H_{62}(300) - 0.2H_{62}(170) = 0, \text{ or } H_{62} = 7.07F$$

5. Horizontal and vertical forces applied to drum 6 are equal and opposite to corresponding forces applied to the shoes. Load torque T (which tends to continue the clockwise direction of rotation) is equal to $(2.11F + 1.41F)$ times the drum radius, or $880F$ N·mm. Forces applied at fixed pivot O_{16} are $H_{16} = 3.46F$ and $V_{16} = 0.70F$.

6. Note that the greater loads applied to the drum by the left shoe (3) result from its being self-energizing, while right shoe (2) is de-energizing. If the direction of drum rotation is reversed, shoe 2 becomes self-energizing. (The self-energizing shoe wears more rapidly, but if the brake is used about equally for both directions of drum rotation, the shoes will wear about equally.)

7. The allowable value of F is governed by the allowable pressure on the self-energized shoe. The projected area of the shoe is the 80 mm width multiplied by the chord length subtended by a 90-deg. arc of the 250-mm-radius drum:

$$A = 80[2(250 \sin 45°)] = 28{,}284 \text{ mm}^2$$

The normal pressure on shoe 3 is

$$p = 10.53F/28284 = 0.0003723F \text{ N/mm}^2$$

Equating this to the allowable value of 0.40 N/mm^2 gives

$$F = 1074 \text{ N}$$ ■

8. The corresponding brake torque is

$$T = 880(1074) = 945 \times 10^3 \text{ N·mm, or } 945 \text{ N·m}$$ ■

9. The resultant radial load transmitted to the bearings is

$$\sqrt{(0.70)^2 + (3.46)^2}\, F = 3.53F = 3791 \text{ N}$$ ■

18.7 LONG-SHOE DRUM BRAKES WITH EXTERNAL SHOES

If a brake shoe or block contacts the drum over an arc of about 45 deg. or more, errors introduced by the "short-shoe" equations are usually significant. In that case the following "long-shoe" analysis is appropriate.

Figure 18.8 shows a brake shoe in contact with its drum, represented by a solid circle. As the shoe wears, it pivots about O_2. However, the wear pattern can be more

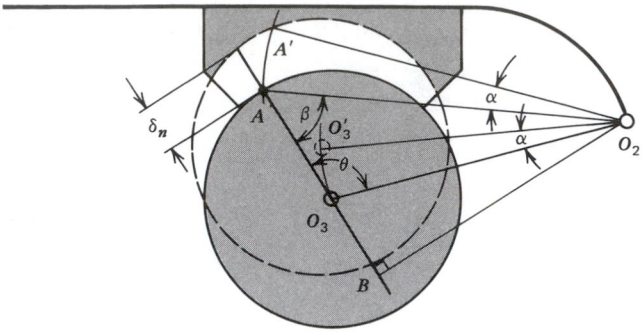

FIGURE 18.8.
Wear distribution on a brake shoe.

simply represented by keeping the shoe fixed and showing the drum pivoting over an angle α about O_2, as represented by the dotted circle. For the rather extreme amount of wear shown, center O_3 moves to O_3', and arbitrary contact point A moves to A'. The wear at point A normal to the contacting surface is the distance δ_n, where

$$\delta_n = AA' \sin \beta = O_2A\alpha \sin \beta \tag{e}$$

From the geometry of the figure,

$$\left.\begin{aligned} \sin \beta &= O_2B/O_2A \\ O_2B &= O_2O_3 \sin(180^\circ - \theta) = O_2O_3 \sin \theta \end{aligned}\right\} \tag{f}$$

Substituting (f) into (e):

$$\delta_n = O_2O_3\alpha \sin \theta \tag{g}$$

Assuming normal wear to be proportional to friction work, which, for various locations on the shoe contact surface, is proportional to local pressure:

$$p \propto \delta_n \propto \sin \theta \tag{h}$$

and

$$p = p_{\max} \sin \theta/(\sin \theta)_{\max} \tag{18.16}$$

The maximum value of $\sin \theta$ is, of course, unity, with $\theta = 90$ deg. Hence, the maximum pressure and the maximum wear occur at $\theta = 90$ deg. If, as is usually the case, the geometry is such that the angles of contact include $\theta = 90$ deg.,

$$p = p_{\max} \sin \theta \tag{18.17}$$

Note that the above development was based on the assumptions that no deflection occurs in either the shoe or drum, the drum does not wear, and shoe wear is proportional to friction work, which is in turn proportional to local pressure.

Figure 18.9 shows the forces applied to a brake shoe (including its associated lever). Considering the shoe as a free-body, $\Sigma M_{O_2} = 0$:

$$Fc + M_n + M_f = 0 \tag{18.18}$$

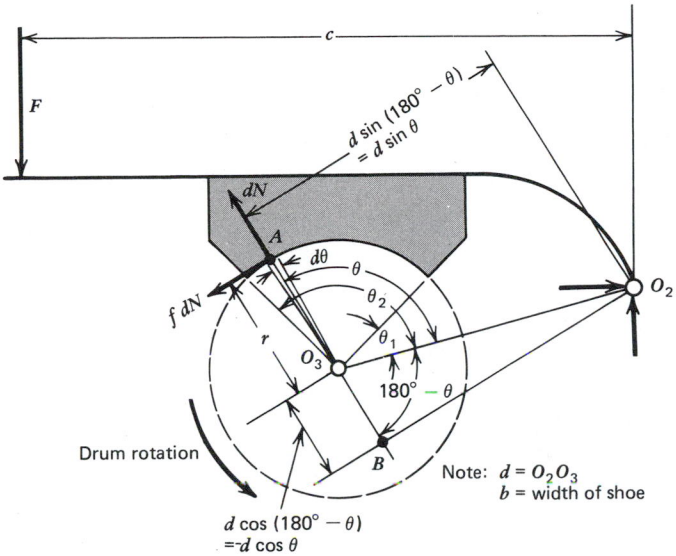

FIGURE 18.9.
Forces acting on a brake shoe.

where M_n is the moment of the normal forces:

$$M_n = -\int_{\theta_1}^{\theta_2} dN(d \sin \theta) \tag{i}$$

and

$$dN = p(r\, d\theta)b$$

Applying Eq. (18.16):

$$dN = p_{\max} br \sin \theta\, d\theta/(\sin \theta)_{\max} \tag{j}$$

Combining (i) and (j):

$$M_n = -\int_{\theta_1}^{\theta_2} \frac{p_{\max} brd \sin^2 \theta}{(\sin \theta)_{\max}}\, d\theta$$

$$= -\frac{p_{\max} brd}{(\sin \theta)_{\max}} \int_{\theta_1}^{\theta_2} \sin^2 \theta\, d\theta$$

$$M_n = -\frac{p_{\max} brd}{4(\sin \theta)_{\max}} [2(\theta_2 - \theta_1) - \sin 2\theta_2 + \sin 2\theta_1] \tag{18.19}$$

Similarly, for the moment of friction forces, M_f:

$$M_f = \int_{\theta_1}^{\theta_2} f\, dN(r - d \cos \theta)$$

$$= \int_{\theta_1}^{\theta_2} \frac{f p_{\max} \sin \theta r\, d\theta\, b(r - d \cos \theta)}{(\sin \theta)_{\max}}$$

Substituting the relationship: $\sin \theta \cos \theta = \frac{1}{2} \sin 2\theta$:

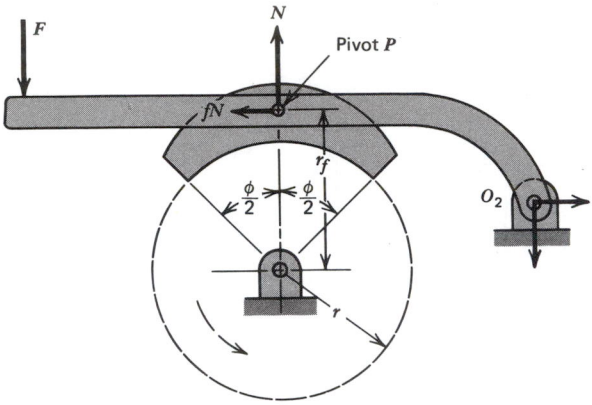

FIGURE 18.10.
Pivoted-shoe brake.

$$M_f = \frac{fp_{\max}rb}{(\sin\theta)_{\max}} \int_{\theta_1}^{\theta_2} \left(r\sin\theta - \frac{d}{2}\sin 2\theta \right) d\theta$$

$$M_f{}^{1} = \frac{fp_{\max}br}{(\sin\theta)_{\max}} \left(r(\cos\theta_1 - \cos\theta_2) + \frac{d}{4}(\cos 2\theta_2 - \cos 2\theta_1) \right) \tag{18.20}$$

For equilibrium of moments acting on the drum:

$$T + \int_{\theta_1}^{\theta_2} rf\,dN = 0$$

$$T = -\int_{\theta_1}^{\theta_2} \frac{r^2 fbp_{\max}\sin\theta\,d\theta}{(\sin\theta)_{\max}}$$

$$T = -\frac{r^2 fbp_{\max}}{(\sin\theta)_{\max}} \int_{\theta_1}^{\theta_2} \sin\theta\,d\theta$$

$$T = -\frac{r^2 fbp_{\max}}{(\sin\theta)_{\max}} (\cos\theta_1 - \cos\theta_2) \tag{18.21}$$

Reaction forces at O_2 and O_3 can readily be obtained from horizontal and vertical force equilibrium equations.

With reference to Eq. (18.18), it is evident that a self-energizing brake is self locking if $M_f \geqq M_n$. It is often desired to make a brake shoe strongly self-energizing while staying safely away from a self-locking condition. This can be done by designing the brake so that the value of M_f, as calculated using a value of f 25 to 50 percent greater than the true value, is equal to the value of M_n.

Figure 18.10 shows a *pivoted* long-shoe brake. If pivot P is located at the intersection of the resultant normal and frictional forces acting on the shoe (designated as N and fN, respectively), there is no tendency of the shoe to rotate about the pivot. This is desirable to equalize wear. Pivoted shoes are not generally practical because the pivot moves progressively closer to the drum as wear occurs. This results in an increasing

[1]When substituting in Eq. (18.18) use the negative of this expression if the drum surface in contact with the shoe is moving toward pivot O_2 (as would be the case in Fig. 18.9 for clockwise drum rotation).

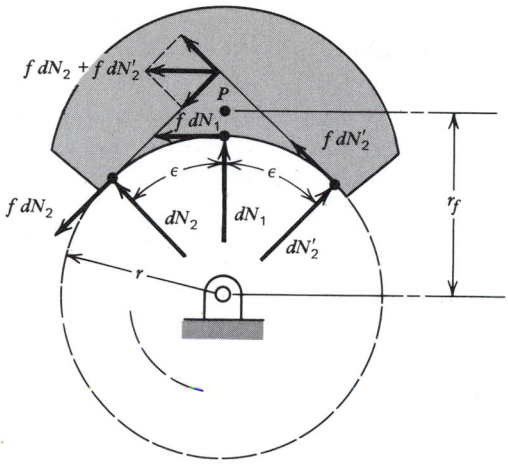

FIGURE 18.11.
Resultant friction force vectors illustrating why $r_f > r$.

tendency of the shoe to pivot about P, with resultant rapid wear on either the *toe* of the shoe (edge farthest from pivot O_2) or *heel* (edge closest to pivot O_2).

By integrating over the surface of a symmetrically loaded shoe, it can be shown that the intersection of the resultant normal and frictional forces occurs at "friction radius" r_f, where

$$r_f = r\frac{4\sin(\theta/2)}{\theta + \sin\theta} \tag{18.22}$$

and that the resulting braking torque is

$$T = fNr_f \tag{18.23}$$

Figure 18.11 shows why $r_f > r$: the resultant of friction forces acting on any symmetrical pair of elements is located above the drum surface.

SAMPLE PROBLEM 18.3

Figure 18.12 represents a double-shoe drum brake with spring force F applied at distance $c = 500$ mm to both shoes. (The brake is released by a solenoid, not shown.) Design values of coefficient of friction and allowable pressure are 0.3 and 600 kPa, respectively. For occasional industrial usage, determine an appropriate value of spring force, and the resulting brake torque and power absorption for 300-rpm drum rotation in either direction.

Solution

1. With reference to the right shoe, as shown in Fig. 18.12*b*: $\phi = \tan^{-1}(200/150) = 53.13$ deg.; hence, $\theta_1 = 8.13$ deg., $\theta_2 = 98.13$ deg.

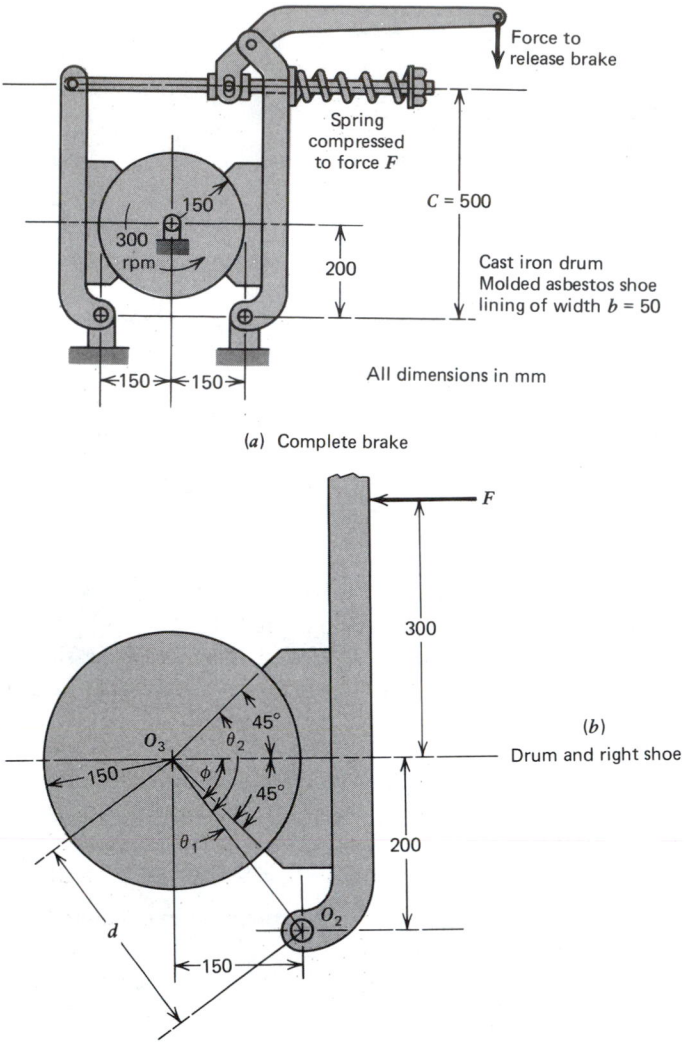

FIGURE 18.12.
Double-shoe drum brake (Sample Problem 18.3).

2. $d = \sqrt{200^2 + 150^2} = 250$ mm.
3. Since $\theta_2 > 90$ deg., $(\sin\theta)_{\max} = 1$.
4. From Eq. (18.19),

$$M_n = -(p_{\max}/4)(50)(150)(250)[2(\pi/2) - \sin 196.26^\circ + \sin 16.26^\circ]$$
$$= -1735 \times 10^3 p_{\max}$$

5. From Eq. (18.20),

$$M_f = fp_{\max}(50)(150)[150(\cos 8.13^\circ - \cos 98.13^\circ) + (250/4)(\cos 196.26^\circ - \cos 16.26^\circ)]$$
$$= 373 \times 10^3 fp_{\max}$$

6. From Eq. (18.21),

$$T = -(150)^2(50)fp_{max}(\cos 8.13^\circ - \cos 98.13^\circ)$$

The absolute value of brake torque, T, is $1273 \times 10^3 fp_{max}$.

7. From Table 18.3, assume a reasonable design value for $p_{max}V$ to be 2.1 MPa m/s. $V = 0.3\pi(300/60) = 4.71$ m/s. Hence, $p_{max} = 2.1/4.71 = 0.446$ MPa (good—substantially below the given allowable value of 0.600 MPa).

8. From Eq. (18.18),

$$500F - 1735 \times 10^3(0.446) + 373 \times 10^3(0.3)(0.446) = 0, \quad \text{or}$$
$$F = 1448 \text{ N (spring force)}$$ ■

9. The appropriate spring force was determined above on the basis of allowable pressure on the *right* shoe, which proved to be mildly self-energizing. This same spring force will produce a lower value of p_{max} on the left shoe. Applying Eq. (18.18) to the left shoe, the only change is the sign of the friction moment term:

$$500(1448) - 1735 \times 10^3 p_{max} - 373 \times 10^3(0.3)p_{max} = 0, \quad \text{or}$$
$$p_{max} = 0.392 \text{ MPa}$$

10. From the equation in step 6, the total brake torque (right shoe plus left shoe) is:

$$1273 \times 10^3(0.3)(0.446 + 0.392) = 320{,}032 \text{ N}\cdot\text{mm} \approx 320 \text{ N}\cdot\text{m}$$ ■

11. From Eq. (1.1), the corresponding power at 300 rpm is

$$nT/9549 = 300(320)/9549 = 10.1 \approx 10 \text{ kw}$$ ■

Three points in the above solution are worthy of note:

(a) The above answers apply to either direction of drum rotation; reversing the direction merely reverses the values of p_{max} and T for the two shoes.

(b) If the short-shoe analysis is used, neither shoe appears self-energizing or de-energizing. For $F = 1448$ N, the short-shoe equations estimate the torque for each shoe as

$$T = Nfr = [1448(500/200)](0.3)(0.150) = 162.9 \text{ Nm}$$

or a total brake torque of 325.8 Nm. [The actual self-energizing and de-energizing action is determined by the radius of the resultant friction force (see Figs. 18.10 and 18.11).]

(c) The long-shoe equations indicate that the friction material at the heel of the shoes (end closest to pivot O_2) is contributing very little; thus, it might be desirable to increase θ_1. Similarly, considerable additional capacity could be added by increasing the value of θ_2.

18.8 LONG-SHOE DRUM BRAKES WITH INTERNAL SHOES

This type of brake is typified by conventional automotive drum brakes. Figure 18.13 shows the basic construction. Both shoes pivot about anchor pins and are forced against the inner surface of the drum by a piston in each end of the hydraulic wheel cylinder.

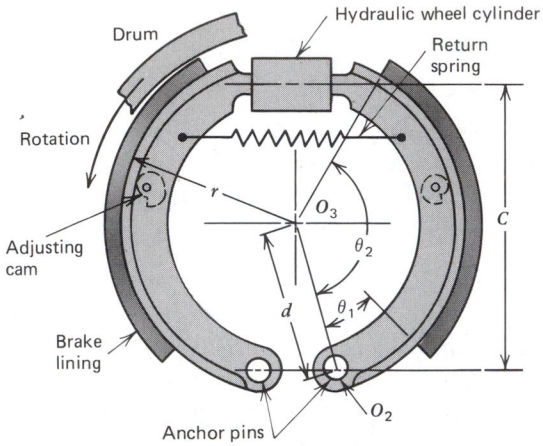

FIGURE 18.13.
Automotive-type drum brake (internal shoes).

The return spring exerts only enough force to retract the shoes against the adjusting cams, which serve as "stops." By adjusting the cams to minimize the clearance between the shoes and drum, a minimum of hydraulic piston movement is consumed in bringing the shoes into contact with the drum. Equations (18.16)–(18.23) apply also to internal-shoe drum brakes. Notation is given on Fig. 18.13.

The energizing or de-energizing character of any shoe and direction of drum rotation can readily be visualized by (1) noting the direction of the friction force acting on a portion of the shoe surface, and (2) determining if this force tends to bring the shoe into or out of contact with the drum. In Fig. 8.13 the left shoe is self-energizing and the right shoe de-energizing for the direction of drum rotation shown.

In automotive brakes, substantial self-energizing action is desired in order to reduce pedal pressure, but self-locking (for any coefficient of friction that might be encountered in service) must, of course, be avoided. One method that has been used to obtain greater self-energizing action for forward car motion is shown in Fig. 18.14. By using two hydraulic wheel cylinders (each with a single piston), both shoes are made self-energizing for forward drum rotation. This arrangement has been used on front brakes, with the Fig. 18.13-type used in the rear. The result is a car with six shoes that are self-energizing for forward motion, and two for reverse. Primarily due to cost considerations, the two-cylinder front brake was largely replaced by a single-cylinder design that achieves forward self-energizing action with both shoes by mounting the anchor pins on a plate which is itself free to pivot through a small arc.

In recent years, caliper-type disk brakes have replaced front drum brakes on most passenger cars because of their greater cooling capacity and resultant resistance to fade. Also, brake drums tend inherently to distort more with heat and large interface forces than do disk brake rotors.

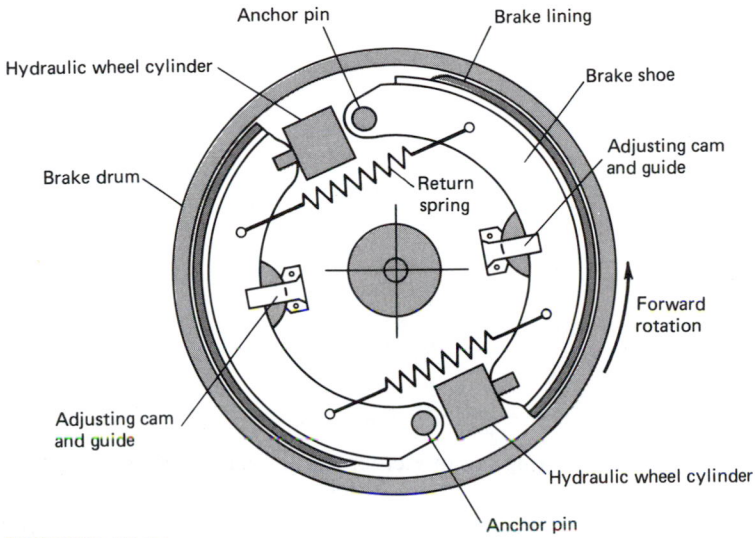

FIGURE 18.14.
Automotive brake with two hydraulic wheel cylinders (both shoes are self-energizing for forward car motion). (Courtesy Chrysler Corporation)

18.9 BAND BRAKES

Perhaps the simplest of the many braking devices is the band brake, shown in Fig. 18.15. The band itself is usually made of steel, lined with a woven friction material (for flexibility). For the clockwise drum rotation shown, friction forces acting on the band serve to increase P_1 and decrease P_2. Considering the drum and band portion above the cutting plane (Fig. 18.15) as a free body, brake torque, T, is equal to

$$T = (P_1 - P_2)r \tag{18.24}$$

Considering the lever and band portions below the cutting plane as a free-body, the applied lever force, F, is

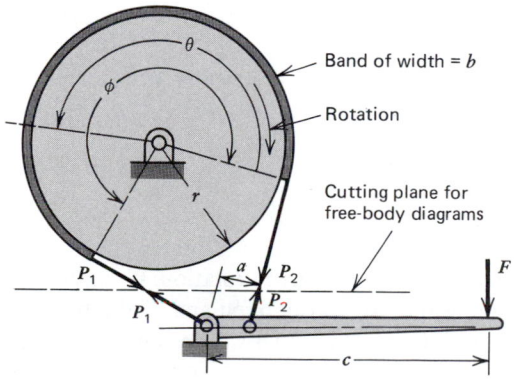

FIGURE 18.15.
Band brake.

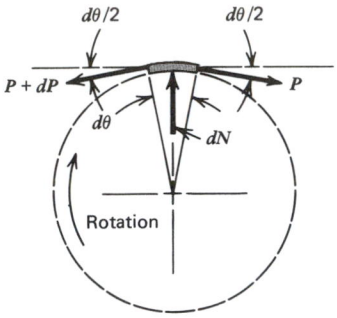

FIGURE 18.16.
Forces on a band element (width = *b*.).

$$F = P_2 a/c \tag{18.25}$$

Figure 18.16 shows the forces acting on an element of the band. For small angle $d\theta$,

$$dP = f\,dN \tag{k}$$

and

$$dN = 2(P\,d\theta/2) = P\,d\theta \tag{l}$$

Also by definition,

$$dN = pbr\,d\theta \tag{m}$$

where p is the local contact pressure between drum and band. Substituting (l) into (k),

$$dP = fP\,d\theta, \quad \text{or } \frac{dP}{P} = f\,d\theta \tag{n}$$

Band force P varies from P_1 to P_2 over the band portion between $\theta = 0$ and $\theta = \phi$. Hence, integrating Eq. (n) over the length of contact:

$$\int_{P_2}^{P_1} \frac{dP}{P} = f \int_0^{\phi} d\theta$$

$$\ln P_1 - \ln P_2 = \ln \frac{P_1}{P_2} = f\phi$$

$$\frac{P_1}{P_2} = e^{f\phi} \tag{18.26}$$

The maximum normal pressure, p_{max}, acting on the band occurs at $\theta = \phi$, where $P = P_1$. Applying Eqs. (l) and (m) to this point:

$$dN = P_1\,d\theta \quad \text{and} \quad dN = p_{max} br\,d\theta$$

Hence,

$$P_1 = p_{max} rb \tag{18.27}$$

The brake in Fig. 18.15 is self-energized for clockwise drum rotation. Greater self-energizing action can be obtained by attaching the previously fixed end of the band to the lever at the opposite side of the pivot, as shown in Fig. 18.17. The tensile force of

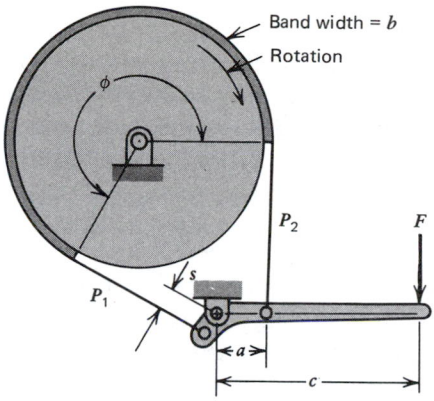

FIGURE 18.17.
Differential band brake.

this attachment now serves to *assist* in the application of the brake. Note also that distance s must be less than distance a so that rotating the lever with force F tightens the end of the band attached at distance a more than it loosens the end attached at distance s. A study of the motion and forces involved at the two band attachment points shows why the name *differential band brake* is appropriate. For a differential band brake, Eq. (18.25) is replaced by

$$F = (P_2 a - P_1 s)/c \tag{18.28}$$

SAMPLE PROBLEM 18.4

A differential band brake similar to Fig. 18.17 uses a woven lining having a design value of $f = 0.20$. Dimensions are: $b = 80$ mm, $r = 250$ mm, $c = 700$ mm, $a = 150$ mm, $s = 35$ mm, and $\phi = 270$ deg. Find (a) the brake torque if the maximum lining pressure is 0.5 MPa, (b) the corresponding actuating force, F, and (c) the values of dimension s that would cause the brake to be self-locking.

Solution

1. From Eq. (18.27): $P_1 = (0.5)(250)(80) = 10{,}000$ N (tight side).
2. From Eq. (18.26): $P_2 = \dfrac{P_1}{e^{(0.2)(1.5\pi)}} = 3897$ N (slack side).
3. From Eq. (18.24): $T = (10{,}000 - 3897)(0.250) = 1526$ Nm. ■
4. From Eq. (18.28): $F = \dfrac{3897(150) - 10{,}000(35)}{700} = 335$ N. ■
5. From Eq. (18.28): $F = 0$ for $s = 3897(150)/10{,}000 = 58.5$ mm.
   The brake is self-locking (for $f = 0.2$) if $s \geq 58.5$ mm. ■

18.10 MATERIALS

The critical material selections in the design of brakes and clutches involve the friction interface surfaces. One of the mating surfaces is normally metal—usually cast iron or steel. It must have good friction characteristics (which are relatively stable over the temperature range involved) when interfaced with the mating material. Other requirements include good thermal conductivity, and good resistance to wear and thermal fatigue. Surface finish of the metal member must be sufficiently smooth to minimize wear of the mating friction material.

Thermal fatigue is due to internal stresses that result from the expansion and contraction of the surface material relative to the subsurface during use (see Section 4.16). One fatigue cycle is accumulated each time the brake (or clutch) is used and subsequently cooled. Yielding and associated residual stresses can occur. These accelerate fatigue damage and cause warpage. Initial fatigue cracks resulting from thermal cycling are often called "heat checks" or "temperature checks." Resistance to thermal fatigue is improved by using a material with greater thermal conductivity (thereby reducing surface temperature gradients), a lower thermal coefficient of expansion, and greater high-temperature yield and fatigue strengths.

The primary requirements of friction materials are: a high dynamic coefficient of friction that is relatively stable over a usable temperature range and affected relatively little by moisture and small amounts of dirt and oil; a static coefficient of friction that exceeds the dynamic value by as little as possible (to avoid "slip-stick" chatter and noise problems); high resistance to abrasive and adhesive wear (see Sections 9.9 and 9.19); good thermal conductivity; and enough resilience to promote a good distribution of interface pressure. The requirement of temperature stability of the coefficient of friction is often expressed as resistance to *fade* (reduction in friction coefficient at elevated temperatures).

Table 18.1 lists the common friction materials used "dry" in brakes and clutches. Some industrial brakes have used wood as a friction element, and many rail car brakes use cast iron shoes bearing on cast iron or steel wheels.

Most applications use friction materials classed as either molded, woven, or sintered metal. *Molded* materials are the most common and the least costly. They consist primarily of a binder (usually a thermosetting resin or rubber that serves to bond the other ingredients together into a heat-resisting compound), reinforcing fibers (almost always asbestos in the past, but other materials are now being used increasingly), friction modifying additives, and fillers. *Woven* materials have more flexibility (as required by band brakes), and generally perform better, particularly under adverse conditions where contaminants such as mud, grease, and dirt are present. They are produced by spinning asbestos or other fibers into yarn, twisting the yarn onto zinc, copper, or brass wire for added strength and thermal conductivity, weaving the yarn into cloth or tape, saturating it with resins and friction modifiers, and then curing it under heat and pressure. *Sintered metal* friction materials are the most costly but also the best suited for heavy-duty applications, particularly when continuous operation at temperatures above 260°C (500°F) is involved. These materials are composed of metal powder and inorganic fillers that are molded under high pressure and then "sintered," which involves heating to thermally fuse the metal particles without completely melting them. Sintered metal-ceramic friction materials are similar except that ceramic particles are added before sintering.

Table 18.2 lists friction coefficients for "wet" (in oil) operation of several friction materials. The so-called "paper" materials are the least costly. They are made from fibrous sheets, saturated with resin (with fillers and friction modifiers added), cured at high temperatures, and bonded to a carrier (usually steel). Because of their high coefficient of friction and low cost, they are used extensively in multiple-disk clutches, as in automotive automatic transmissions. The high friction coefficient permits the use of fewer disks. Graphitic materials are molded compounds of graphite and resin binders. They have good thermal capacities for high-energy applications. Polymerics are a relatively new class of friction materials that are highly resilient and have high thermal storage capacities.

The recognition of health hazards associated with asbestos (and the resulting government regulations) has resulted in unusual efforts to develop alternative friction materials. This is a good example of the influence of health, ecological, and legal factors on modern engineering design.

CHAPTER NINETEEN

Miscellaneous Power Transmission Components

19.1 INTRODUCTION

Power transmission between shafts can be accomplished in a variety of ways. In addition to gears (Chapters 15 and 16), *flexible elements* such as belts and chains are in common use. These permit power to be transmitted between shafts that are separated by a considerable distance, thus providing the engineer with considerable flexibility in the relative placement of driving and driven machinery.

Belts are relatively quiet in operation. Except for timing belts (Fig. 19.5), slippage between belt and pulleys causes speed ratios to be inexact. This slippage characteristic is sometimes used to advantage by permitting the pulleys to be moved closer together in order to disengage the drive (as in some snowblowers and self-propelled lawn mowers). This may save substantial cost, weight, and bulk of providing a separate clutch. The flexibility and inherent damping in belts (and to a lesser extent, chains) serves to reduce the transmission of shock and vibration.

The design of chains illustrates the general proposition that if a component of desired characteristics is not already available, an engineer should consider the possibility of inventing something new. For example, the conventional roller and inverted-tooth chains discussed in Sections 19.5 and 19.6 require that all sprockets engaging a single chain lie in a common plane. Suppose a positive flexible drive is needed between sprockets lying in different planes. If power requirements are low, a "beaded chain" (similar to the pull cord on a plain light fixture) can be used. A stronger type of flexible chain incorporates parallel steel cables bonded to the sides of plastic cylindrical "buttons" that simulate the rollers of a conventional roller chain. A chain embodying the latter concept was used between the pedal and propeller shafts of the Gossamer Albatross, the man-powered airplane that flew across the English Channel.

For transmitting small amounts of torque, flexible shafts often offer inexpensive solutions. The common automotive speedometer drive is a familiar example.

For transmitting power between nominally collinear shafts, flexible couplings, universal joints, and friction clutches have already been discussed. Another important general class of collinear power transmission members involves *hydrodynamic* action. These are fluid couplings (also called fluid clutches) and hydrodynamic torque converters.

Other types of power transmission devices use rope or cable, and involve moving or lifting a weight using power delivered to a rotating shaft. Examples include hoists, elevator drives, and capstans.

19.2 FLAT BELTS

Large flat leather belts were in common use a few decades ago when one large motor or engine was often used to drive several pieces of machinery. Today's more limited use of flat belts usually involves driving high-speed machines with thin, light belts. Often, the vibration-isolating characteristic of the belt is an important consideration.

The basic equations for the limiting torque that can be transmitted by a flat belt are the same as for band brake torque:

$$T = (P_1 - P_2)r \tag{18.24}$$

and

$$P_1/P_2 = e^{f\phi} \tag{18.26}$$

where P_1 and P_2 are the tight and slack side belt tensions, f is the coefficient of friction, and ϕ is the angle of contact with the pulley (see Fig. 18.15). These two equations enable P_1 and P_2 to be determined for any combination of T, f, and ϕ. The required initial belt tension, P_i, depends upon the elastic characteristics of the belt, but it is usually satisfactory to assume that

$$P_i = (P_1 + P_2)/2 \tag{19.1}$$

Note that the capacity of the belt drive is determined by the angle of wrap, ϕ, on the *smaller* pulley, and that this is particularly critical for drives involving pulleys of greatly differing size, spaced closely together. Also, an important practical consideration is that the required belt initial tension must not be lost when the belt stretches slightly over a period of time. Of course, one solution might be to make the initial installation with an excessive initial tension, but this would overload the bearings and shafts, as well as shorten belt life. Three methods of maintaining belt tension are illustrated in Fig. 19.1. Note that all of these show the slack side of the belt on top, so that its tendency to sag acts to increase the angle of wrap.

The coefficient of friction between belt and pulley varies with the usual list of environmental factors, and with the extent of slippage involved. In addition to ordinary "torque-transmission slippage," belts experience slip (commonly called "creep") due to

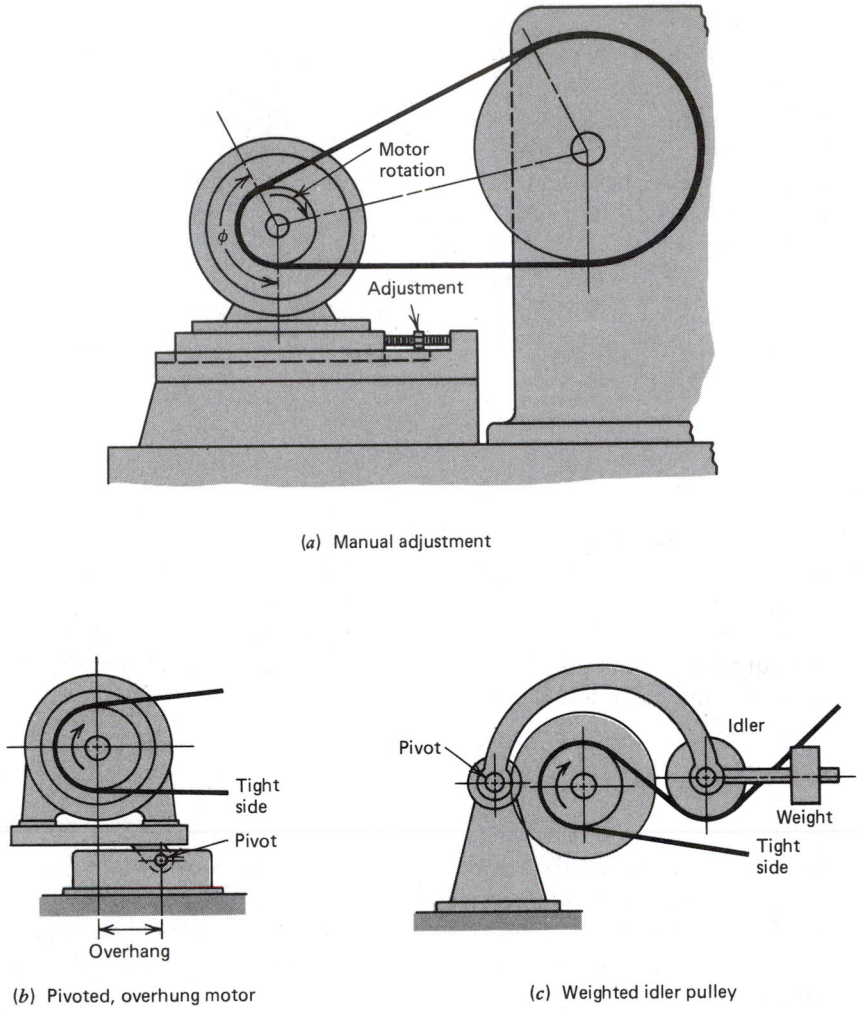

(*a*) Manual adjustment

(*b*) Pivoted, overhung motor

(*c*) Weighted idler pulley

FIGURE 19.1.
Alternative means of maintaining desired belt tension.

slight stretch or contraction of the belt as its tension varies between P_1 and P_2 while going through angles ϕ in contact with the pulleys. For leather belting and cast iron or steel pulleys, $f = 0.3$ is often used for design purposes. Rubber coated belting usually gives a lower value (perhaps $f = 0.25$), whereas running on plastic pulleys usually results in a slightly higher value. It is always best to obtain friction values from tests or from the belt manufacturer.

The allowable value of tight-side tension, P_1, depends on the belt cross section and material strength. As the belt makes one revolution, it goes through a rather complex cycle of fatigue loading: In addition to the tensile fluctuation between P_1 and P_2, the belt is subjected to bending stresses when in contact with the pulleys. The greater bending stress occurs with the smaller pulley, and for this reason there are minimum pulley diameters that should be used with any particular belting. For leather belting, a tight-side tensile stress (P_1/A) of 250–400 psi is usually specified.

The above discussion pertains to belts that run slowly enough that centrifugal loading can be neglected. For greater power transmitting capacity, most belt drives operate at relatively high speeds. The centrifugal force acting on the belt creates a tension, P_c, of

$$P_c = m' V^2 = m' \omega^2 r^2 \tag{19.2}$$

where m' is the mass per unit length of belt, V is the belt velocity, and r the pulley radius. Force P_c (when large enough to be considered) should be added to both P_1 and P_2 in Eqs. (18.24) and (18.26). The result is that Eq. (18.24) is unchanged and (18.26) becomes

$$\frac{P_1 - P_c}{P_2 - P_c} = e^{f\phi} \tag{19.3}$$

It should also be noted that centrifugal force tends to reduce the angles of wrap, ϕ.

Were it not for centrifugal force, the friction-limited torque transmission would be constant, and the power transmitted would increase linearly with speed. On the other hand, if an unloaded belt drive is driven at sufficient speed, centrifugal force alone can load the belt to its tensile capacity. It follows, then, that there is some speed at which the power transmitting capacity is a maximum. With leather belting, this is commonly in the vicinity of 30 m/s (6000 ft/min.), with about 20 m/s being regarded as an "ideal" operating speed, all factors (including noise, life, etc.) being considered.

As was the case in determining appropriate sizes of bearings and gears, a variety of "experience factors" should be taken into account when selecting belt sizes. These include torque fluctuations in the driving and driven shaft, starting overloads, pulley diameters, and environmental contamination (moisture, dirt, oil, etc.).

19.3 V-BELTS

V-belts are used with electric motors to drive blowers, compressors, appliances, machine tools, farm and industrial machinery, and so on. One or more V-belts are used to drive the accessories on automotive and most other internal combustion engines. They are made to standard lengths, and with the standard cross-sectional sizes shown in Fig. 19.2. The grooved pulleys that V-belts run in are called *sheaves*. They are usually made of cast iron, pressed steel, or die-cast metal. V-belts work well with short center distances. Because of the stretch resistance of their interior tension cords, V-belts do not require frequent adjustment of initial tension.

When a single V-belt has insufficient capacity, multiple belts may be used, as shown in Fig. 19.3. As many as 12 or more belts are commonly used in heavy-duty applications. It is important that these be obtained as matched sets, so that the load is divided equally. Also, when one belt needs replacement, a complete new set should be installed.

Figure 19.4*a* shows how a V-belt rides in the sheave groove with contact on the sides and clearance at the bottom. This "wedging action" increases the normal force on

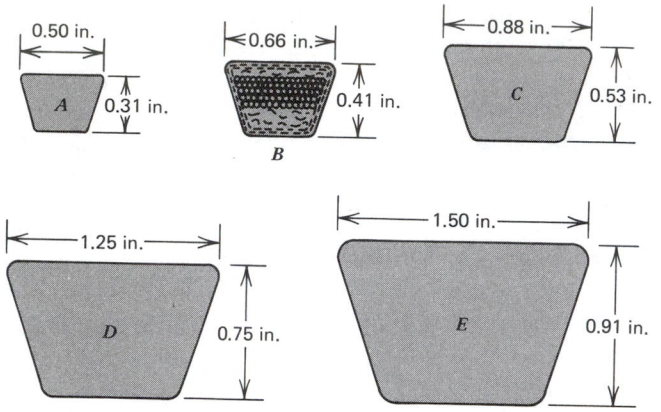

(*a*) Standard sizes A, B, C, D, and E

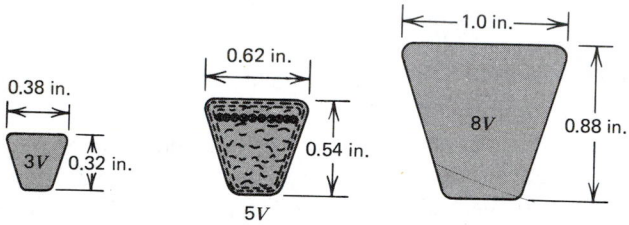

(*b*) High-capacity sizes 3V, 5V, and 8V

FIGURE 19.2.
Standard cross sections of V-belts. (All belts have rubber-impregnated fabric jacket with interior tension cords above a rubber cushion.)

a belt element from dN (as in Figs. 18.16 or 19.4*b*) to $dN/\sin\beta$, which is approximately equal to 3.25 dN. Since the friction force available for torque transmission is assumed proportional to normal force, the torque capacity is thus increased more than threefold. The flat-belt equations can be modified to take this into account by merely replacing the coefficient of friction, f, with the quantity $f/\sin\beta$.

Since belt-drive capacity is normally limited by slippage of the *smaller* pulley, V-belt drives can sometimes be used with a flat larger pulley (as in Fig. 19.4*b*) with no sacrifice in capacity. For example, the V-belt(s) driving the drum of a domestic clothes drier or the flywheel of a large punch press normally rides directly on the flat drum or flywheel surface.

Some sheaves are made with provision for adjusting the groove width. This varies the effective pitch diameter, and permits moderate changes in speed ratio. (A familiar example is the V-belt drive of domestic furnace blowers that have this feature to permit an adjustment of the discharge air velocity.) An extension of this principle is used in

FIGURE 19.3.
Multiple V-belt drive. (Courtesy Reliance Electric Co.)

some continuously variable adjustable speed drives. These employ special extra-wide V-belts with matched pairs of variable width sheaves that adjust simultaneously (with the machine running) to accommodate the fixed belt length.

There is a sufficient variation in the strength and friction properties of commercial V-belts that the detailed selection for a specific application is best made using manufacturer's literature that reflects test data and service experience. In general, it is recommended that belt speeds in the range of 20 m/s (4000 ft/min.) be used where feasible.

V-belt life is markedly affected by temperature. Where elevated belt temperatures (say, above 200°F or 93°C for conventional belts) are encountered, belt life can often be substantially improved by increasing air circulation, as with fins on the sheaves.

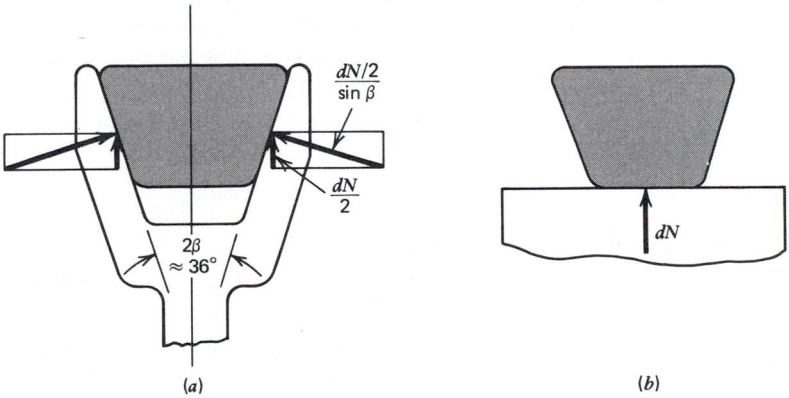

FIGURE 19.4.
V-belt in sheave and on flat pulley.

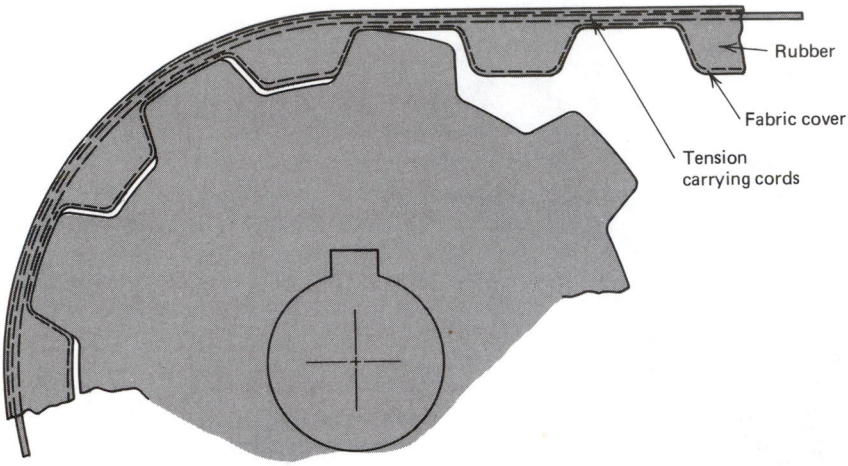

FIGURE 19.5.
Toothed (timing) belt.

19.4 TOOTHED BELTS

Figure 19.5 illustrates toothed belts, also known as *timing belts.* Since the drive is through teeth rather than friction, there is no slippage and the driving and driven shafts remain synchronized. This permits toothed belts to be used for many applications like driving an engine camshaft (from the crankshaft) where the use of other types of belts would be impossible. The toothed drive, together with the use of tension-carrying cords with minimum stretch, permits installation with minimal initial tension. This reduces bearing loading and shaft bending loads.

Toothed belts permit the use of small pulleys and small arcs of contact. Contact on only six teeth is sufficient to develop full rated capacity. Toothed belts are relatively lightweight, and can give efficient operation at speeds up to at least 80 m/s (16,000 ft/min.). Their main disadvantage is the higher cost of both the belt and the toothed pulleys. As with other belts, long service life can be obtained, but not as long as with metallic power transmission members (gears and chains). For example, automotive engines using timing belts for the camshaft drive usually require belt replacement at around 60,000 miles (100,000 km), whereas gear and chain camshaft drives usually last for the life of the engine.

19.5 ROLLER CHAINS

There are several types of power transmission chains, but the most widely used is the *roller chain.* Of its many applications, the most familiar is the chain drive on a bicycle. Figure 19.6 illustrates the construction of the chain. Note the alternation of pin links and roller links. In analyzing the load that can be carried by a given chain, the force-flow concept of Section 2.4 is appropriate. The procedure involves beginning with the portion of the load (depending on its distribution among the driving sprocket teeth in

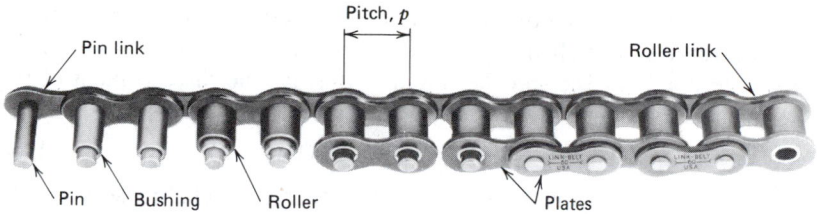

FIGURE 19.6.
Construction of a roller chain. (Courtesy PT Components, Inc., Link-Belt Chain Div.)

contact) applied to a chain roller by a driving sprocket tooth. From the roller, the load is transmitted, in turn, to a bushing, pin, and pair of link plates. Moving along the chain, this load is added to loads from other sprocket teeth. Finally, successive pins, bushings, and link plates transmit the full load along the tight side of the chain. At chain speeds above about 3000 feet per minute, centrifugal force adds significantly to the tensile loading of the plates and to the bearing loading between the pins and bushings.

Along the force path are several potentially critical locations. At the sprocket tooth-chain roller interface there is a Hertz contact stress, as with gear teeth. Also like gears, there is an impact involved when each new tooth comes into contact, and the intensity of this impact increases markedly with speed. Since the roller turns freely on its bushing, there is negligible sliding between the sprocket tooth and roller. Lubrication and wear considerations are involved at both bushing interfaces—with the roller on the outside and the pin on the inside. Wear at the pin interface is the more critical because the load-supporting area is much smaller, and superimposed centrifugal loading is involved. The link plates are subjected to essentially a zero-to-maximum tensile fatigue loading, with stress concentration at the pin holes.

An important factor affecting the operating smoothness of a roller chain drive, particularly at high speeds, is *chordal action,* depicted in Fig. 19.7. In Fig. 19.7*a*, roller

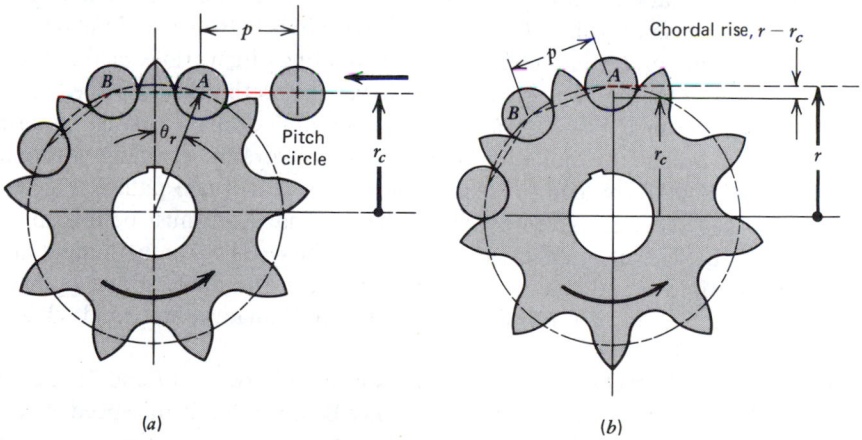

FIGURE 19.7.
Chordal action of a roller chain.

A has just seated on the sprocket, and the centerline of the chain is at the chordal radius, r_c. After the sprocket rotates through angle θ, the chain is in the position shown in 19.7*b*. Here, the chain centerline is at the sprocket pitch radius, r. The displacement of the chain centerline (amount of chain rise and fall) is

$$\Delta r = r_c - r = r(1 - \cos\theta) = r[1 - \cos(180^\circ/N_t)] \tag{19.4}$$

where N_t is the number of teeth in the sprocket. Together with chain rise and fall, chordal action results in a nonuniformity of speed ratio by effectively varying the sprocket pitch radius between r and r_c. Fortunately, the elasticity of the chain absorbs most of this small speed fluctuation in a properly designed drive. Because of chordal action, a chain drive is analogous to a belt drive using polygonal (flat-sided) pulleys. If the number of sides in the polygon is sufficiently large, the effect may be hardly noticeable.

Roller chains are designed so that they seldom break, but eventually require replacement because of wear between the pins and bushings. This causes the pitch (distance between centers of adjacent rollers) to increase. Some wear can be compensated for by increasing the distance between sprockets, or by adjusting or spring loading an idler sprocket. When wear elongates the chain by about 3 percent, the increased pitch causes the rollers to ride objectionably high on the sprocket teeth and the chain (also the sprockets, if they are worn too) should be replaced. A conservatively designed chain, properly lubricated, normally has an operating life of about 15,000 hours. If a shorter life is adequate, a lighter and less costly chain drive can be used.

At low speeds, the chain tensile loading necessary to produce objectionable wear in 15,000 hours may be high enough to cause *prior* fatigue failure of the plates. At a sufficiently high speed, pin bearing loading due to centrifugal plus shock loading can be so high that no useful load can be transmitted.

The American National Standards Institute has issued ANSI B29.1 (last revised in 1975), which gives standard dimensions of chains and sprockets (with pitches ranging from ½ to 3 in.) so that these parts made by various manufacturers are interchangeable. Standard load capacity data are also given. Details of materials and construction vary, however, and for this reason it is best to check load capacities with the chain manufacturer. All manufacturers give tables of basic capacity ratings for various chain speeds and numbers of teeth in the smaller sprocket (the fewer the teeth, the lower the capacity). These ratings consider both wear and fatigue failure. They are based on uniform input power (as from an electric motor, or an engine with hydraulic coupling or torque converter drive) and a uniform load (as from a blower or centrifugal pump). For less favorable conditions, factors are given by which the nominal load must be multiplied when selecting a chain. These factors range up to 1.7 for heavy shock in both input and output shafts. If multiple strand chains are used (Fig. 19.8), the single strand rating is multiplied by 1.7, 2.5, and 3.3 for double, triple, and quadruple strand chains, respectively.

The number of teeth on the smaller sprocket is usually between 17 and 25. Fewer may be acceptable if the speed is very low; more may be desirable if the speed is very high. The larger sprocket is usually limited to about 120 teeth (speed ratios are usually limited to 12, even with slow speed).

FIGURE 19.8.
Multiple (quadruple) strand roller chain and sprockets. (Courtesy PT Components, Inc., Link-Belt Div.)

19.6 INVERTED-TOOTH CHAINS

Inverted-tooth chains (Fig. 19.9), also called *silent chains* because of their relatively quiet operation, consist of a series of toothed link plates that are pin connected to permit articulation. The link "teeth" are usually straight-sided. Corresponding sprocket teeth are also straight-sided, with the included angle between the tooth sides increasing with the number of teeth in the sprocket. The most critical part of the chain is the pinned connection. Different manufacturers have devised a variety of detailed joint designs in order to improve wear life.

Provision must be made to prevent the chain from sliding off the sprockets. Figure 19.9*b* shows a chain with center guide links that engage with center grooves in the sprockets. Figure 19.9*c* shows a chain with side guide links that straddle the sprocket face. Fig. 19.9*d* shows a *duplex chain,* used for "serpentine" drives where sprockets are driven from both sides of the chain.

Inverted-tooth chains and sprockets are standardized (ANSI B-29.2 (1957, reaffirmed 1971)). Most of the remarks in the previous section also apply to inverted-tooth chains and sprockets. Standard pitches range from ⅜ to 2 in., and capacities are rated per inch of chain width. Because of their smoother (quieter) operation, inverted-tooth chains can operate at somewhat higher speeds than roller chains. Lubrication is very important, and the chains should usually operate in an enclosure.

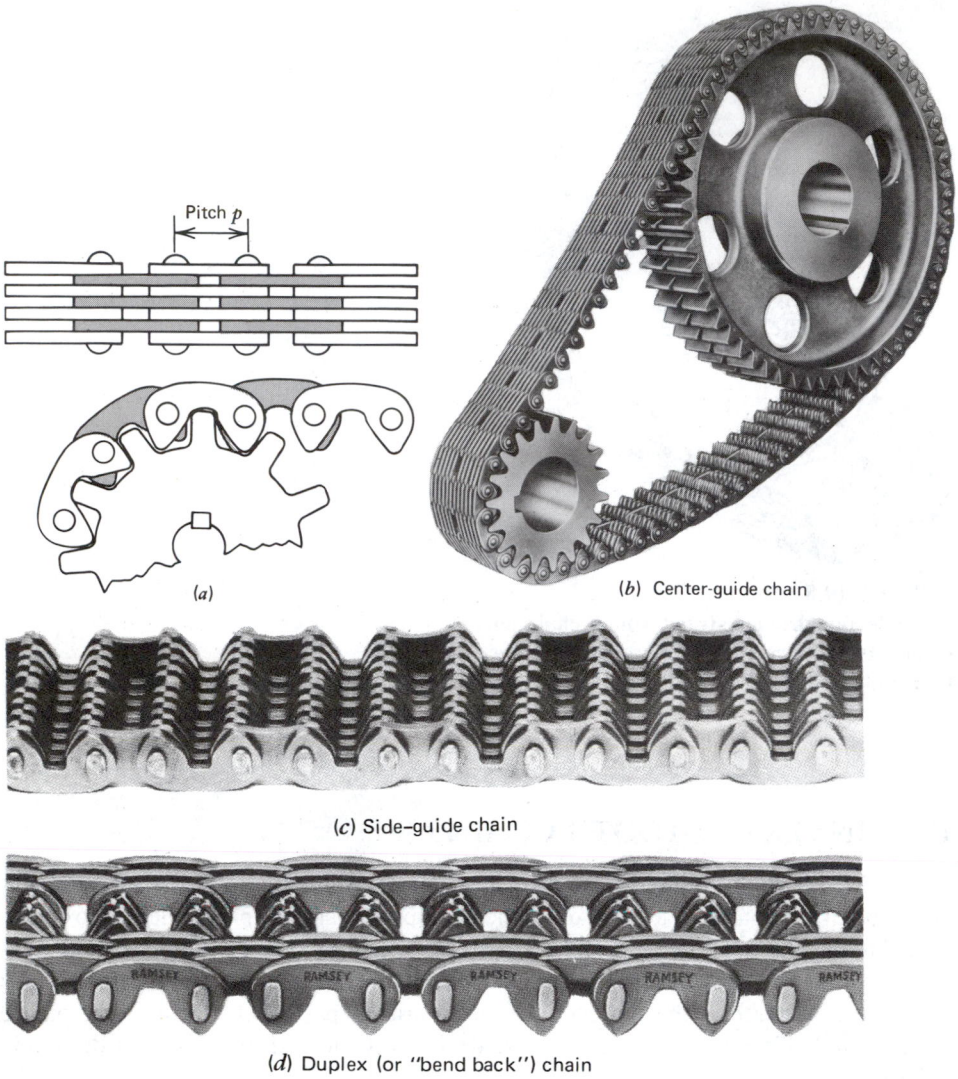

FIGURE 19.9.
Inverted tooth ("silent") chains. (Courtesy Ramsey Products Corp.)

19.7 HISTORY OF HYDRODYNAMIC DRIVES

In contrast with virtually every other means of power transmission, hydrodynamic drives (i.e., *fluid couplings* and *torque converters*) are strictly twentieth century devices. Dr. H. Föttinger of Germany built the first hydrodynamic drive in 1905. At that time, steam turbines had just been introduced for marine propulsion, and no practical means of providing a reduction gear were available. Direct coupling of the engine and propeller caused both to operate inefficiently—the turbine ran too slowly and the propeller too fast. Föttinger was associated with the Vulcan-Werke A.G., where his invention was developed to provide a 5:1 reduction ratio for drives up to 15,000 hp. The increased turbine and propeller efficiencies more than made up for the 15 percent or so hydraulic loss in the converter. Föttinger's converters were used quite successfully until helical gears were developed for use with ship turbines. The much greater efficiency of the gear drive (plus its lower cost, availability of larger reduction ratios, and compactness) enabled it quickly to displace the hydraulic torque converter.

The next development in hydrodynamic drives did not occur until after World War I, when, in connection with efforts to improve the efficiency of hydraulic converters, it was found that the simpler fluid coupling (which provides no torque multiplication) could operate at efficiencies above 95 percent. This suggested to Dr. Bauer, then a director of Vulcan-Werke, that newly developed high-speed diesel engines could be adapted to marine service by using helical gearing for speed reduction together with a fluid coupling to isolate the substantial torsional shock of the engine from the gearing and propeller shaft. (The only diesels previously used for marine propulsion were low-speed and directly coupled.) This worked very successfully.

Hydrodynamic drive development continued in England as well as Germany. In 1926, Harold Sinclair of the Hydraulic Coupling and Engineering Co. was disturbed by the "jerky" clutch engagement of the London busses, which he frequently rode. This reportedly motivated his development of the first automotive fluid couplings. These were used on several British cars, beginning a few years later.

In about 1930 the Chrysler Corporation began experimentation with fluid couplings, and purchased patent rights from Harold Sinclair. This lead to the introduction of "Fluid Drive" on the Chrysler Custom Imperial in 1939. Also in the early 1930s, the American Blower Co. developed and manufactured hydrodynamic drives for use with induced-draft fans. In the meantime, General Motors introduced a four-speed semiautomatic transmission on the 1937 Oldsmobile, but dropped it in 1939. This transmission was modified and integrated with a fluid coupling to become the original "Hydra-Matic Drive," introduced on the 1940 Oldsmobile.

Since the mid-1950s, all American car manufacturers have used automatic transmissions incorporating hydraulic torque converters. Also, hydrodynamic drives have gained world-wide acceptance in a great variety of industrial and marine applications.

19.8 FLUID COUPLINGS

Figure 19.10 shows the essential parts of a fluid coupling. The rotor attached to the input shaft is called the *impeller,* which is also bolted to the case. These two units form the housing containing the hydraulic fluid (usually a low-viscosity mineral oil). The output shaft is supported in the housing by two bearings. Attached to it is the *turbine,* which is driven by oil discharged from the impeller. The oil seal between the housing and output shaft represents extensive development work. It must allow only the microscopic leakage needed for lubrication of the seal itself. Each rotor encloses a semitoroidal space that is divided into compartments by uniformly spaced, flat, radial *blades* (also called plates or vanes). An optional core or inner shroud, attached to the vanes of both rotors, is usually used. This guides the circulating fluid into a more nearly circular

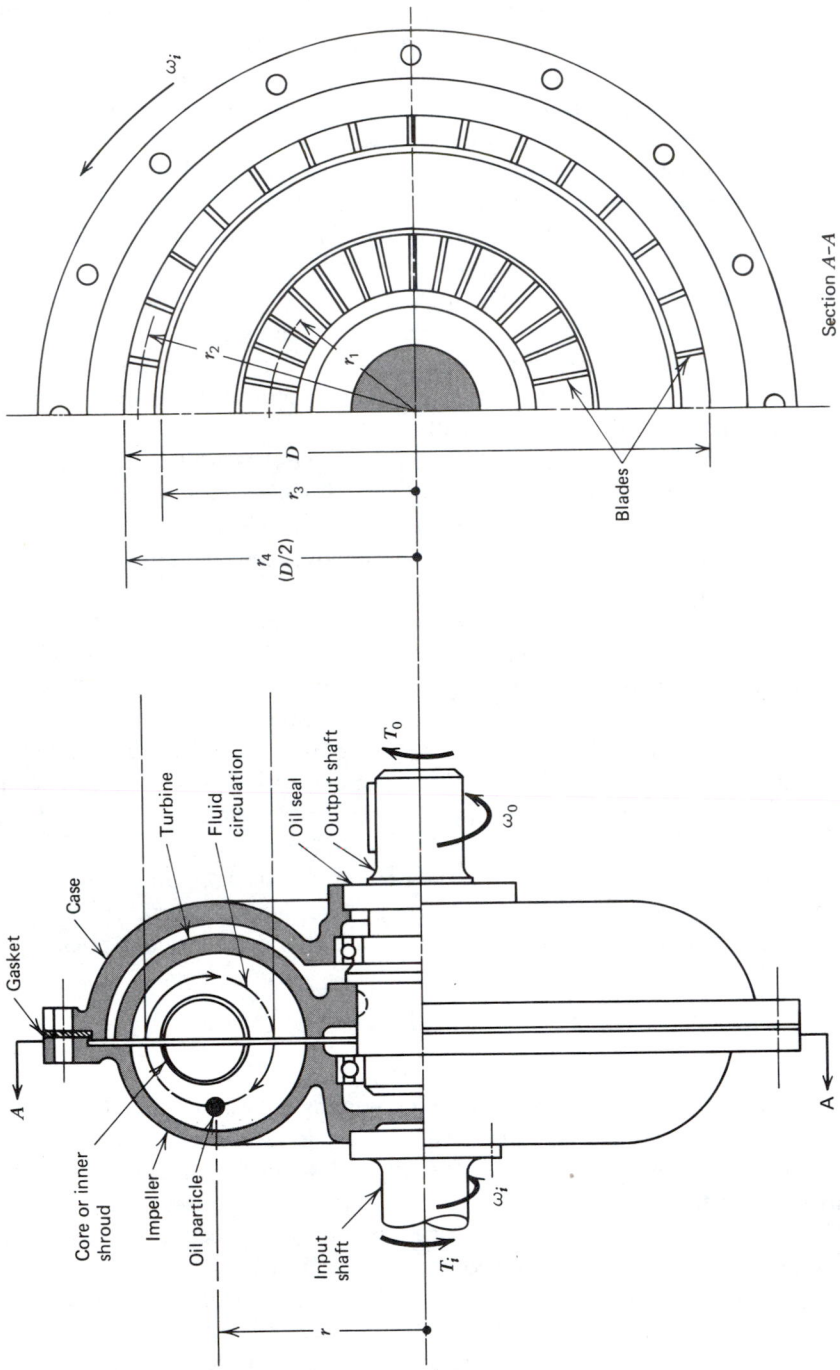

FIGURE 19.10.
Fluid coupling.

path. As will be shown, *no hydrodynamic drive can transmit power with 100 percent efficiency*. To assist in dissipating the resulting heat, fins are often added to the housing to improve air circulation. In many heavy-duty applications the fluid is continuously removed, cooled by an external heat exchanger, and returned.

Rotation of the impeller creates centrifugal forces within the oil trapped between adjacent blades, causing it to flow radially outward (note direction of arrows on "fluid circulation" path in Fig. 19.10). When the flow crosses over to the turbine, the oil impinges upon the turbine blades, giving up much of its kinetic energy (in the plane of impeller and turbine rotation). Pressure from the "oil behind" forces the flow to proceed radially inward through the spaces between adjacent turbine blades, where it gives up additional kinetic energy. During normal operation the turbine rotation may be only slightly slower than that of the impeller. The centrifugal forces developed within the oil on the turbine side oppose those on the impeller side, thereby reducing the velocity of flow around the transverse "fluid circulation" paths. If the direction of the drive reverses and the output shaft runs faster than the input, the direction of fluid flow also reverses. This is what happens when a car with a fluid coupling (or torque converter) experiences "engine braking." Normal operation involves the output shaft rotating slower than the input shaft by a factor called *slip*. Representing the input and output rotating velocities by ω_i and ω_o, respectively, slip, S, is defined as:

$$S = (\omega_i - \omega_o)/\omega_i \tag{19.5}$$

A fundamental concept is that *without slip, there can be no fluid circulation and therefore no transmission of power.*

A second fundamental concept follows from considering the fluid coupling as a free-body in equilibrium. The only external attachments are to the input and output shafts. For the summation of moments about the axis of rotation to be zero, *the input and output torques must be exactly equal.* Thus, every fluid coupling always transmits 100 percent of the input torque. This is true even if the coupling has no oil. In this case, both input and output torques are zero (i.e., the coupling provides no load to the driving motor). This suggests the possibility of draining and refilling the coupling oil while the motor is running as a means of providing a smooth declutching-clutching function.

In order to develop fundamental relationships between the parameters of a fluid coupling, consider a typical particle of oil at radius r (Fig. 19.10) that has unit volume, a density of ρ, and negligible viscosity. Assume the input shaft to be rotating at speed ω and the output shaft to be stationary (as during "start-up"). Centrifugal force acting on the oil particle is

$$F = mr\omega^2 = \rho r\omega^2 \tag{a}$$

Since fluid couplings are, in general, geometrically similar, it is desirable to develop equations in terms of the outside diameter of the fluid cavity, D, as shown in Fig. 19.10. All other dimensions can be expressed as k_nD, where k_n represents numerical constants $k_1, k_2, k_3 \ldots$. Accordingly, Eq. (a) can be written as

$$F = \rho k_1 D\omega^2 \tag{b}$$

From $F = ma$, and the fact that the oil particle has a mass of ρ,

$$a = k_1 D\omega^2 \tag{c}$$

With the output shaft stationary, an oil particle enters the impeller at radius r_1 with essentially zero tangential and radial velocity. Assuming, as an approximation, that the particle experiences uniform acceleration over distance $(r_2 - r_1) = k_2D$, its terminal velocity (from the elementary equations of uniformly accelerated motion) will be

$$V = \sqrt{k_2Da} = \sqrt{k_2Dk_1D\omega^2} = k_3D\omega \tag{d}$$

Assuming that the fluid velocity when moving across the small gap from the impeller to the turbine is given by (d), it follows that the mass flow rate, Q, of oil entering the turbine is

$$Q = \rho VA$$

where area A is $\pi(r_4^2 - r_3^2) = k_4D^2$. Hence

$$Q = \rho k_3D\omega k_4D^2 = k_5D^3\rho\omega \tag{e}$$

The momentum of the oil in the plane of fluid coupling rotation when it enters the turbine is the product of mass times velocity in the rotating plane, or $mr_2\omega$, and the moment of its momentum is $mr_2^2\omega = m(k_6D)^2\omega$.

The torque imparted to the turbine is equal to the time rate of change of angular momentum:

$$T = Q(k_6D)^2\omega = (k_5D^3\rho\omega)(k_6D)^2\omega = k_7\rho\omega^2D^5 \tag{f}$$

Since the densities of hydraulic fluids vary relatively little, ρ can be combined with the other constants to give a single constant, k, with the result:

$$T = k\omega^2D^5 \tag{19.6}$$

The above equation gives the *stall torque* of the coupling, for which slip, S, is unity. When *power* is transmitted the slip must, of course, be less than unity. In this case the transmitted torque is approximately proportional to slip. With the slip factor included, Eq. (19.6) becomes the equation for the *running torque* of the coupling

$$T = k\omega^2D^5S \tag{19.7}$$

For torque in lb·ft, ω in rpm, and D in inches, tests of automotive-type couplings in the 30 to 100 percent slip range indicate that as a rough approximation

$$T \approx 5 \times 10^{-10}\omega^2D^5S \tag{19.7a}$$

This equation contains two important conclusions:

1. The torque and horsepower capacity of a fluid coupling vary with the *fifth power of the diameter.* This is a rather remarkable fact—doubling the diameter increases the capacity by a factor of 32.
2. Torque capacity varies as the *square of the speed;* power capacity varies with the *cube of the speed.* This explains why a fluid coupling can transmit the full output of an engine at operating speeds, yet transmit an almost negligible torque (essentially declutching) at idle speed.

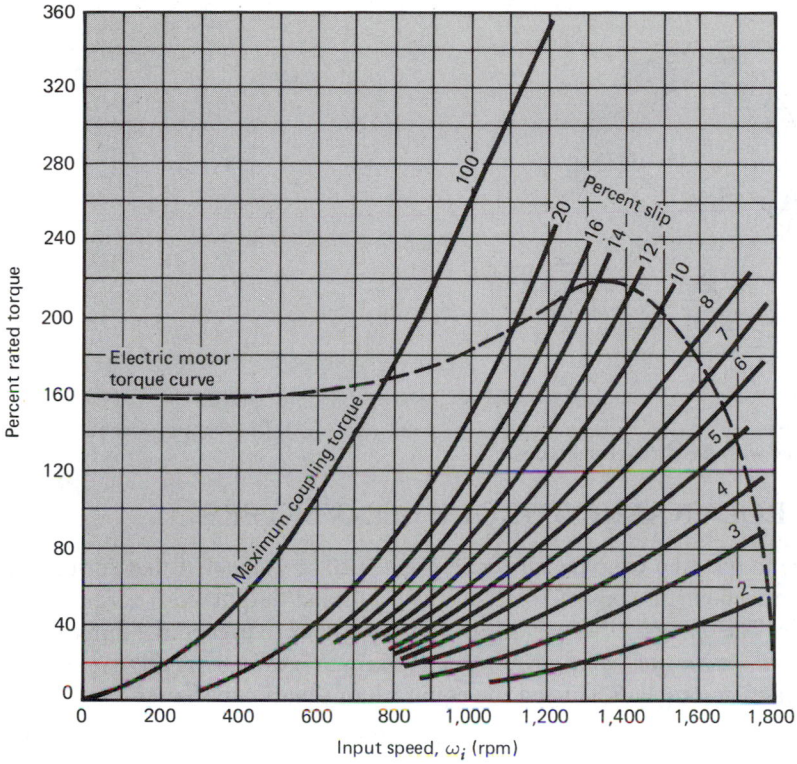

FIGURE 19.11.
Typical fluid coupling torque-speed-slip curves.

From Eq. (19.5), the coupling output shaft speed is

$$\omega_o = \omega_i(1 - S) \qquad \textbf{(19.5, mod.)}$$

Since input and output torques are equal, the efficiency of power transmission by the coupling is

$$e = \frac{\omega_o}{\omega_i} = 1 - S \qquad \textbf{(19.8)}$$

During normal operation, a well-designed and properly applied coupling usually operates at 95 to 98 percent efficiency.

Representative fluid coupling performance curves are illustrated in Fig. 19.11. Note that at the rated speed and torque of the motor, coupling slip is only 3½ percent (efficiency = 96.5 percent).

Fluid couplings are particularly useful in isolating the torsional impulses of diesel engines. When a fluid coupling is interposed between the engine and a gear reducer, cheaper and lighter gears and shafts can be used. Another important use is in heavy-duty hoisting equipment, where the smooth, gradually increasing output torque obtainable from the coupling enables heavy loads to be picked up with minimal shock.

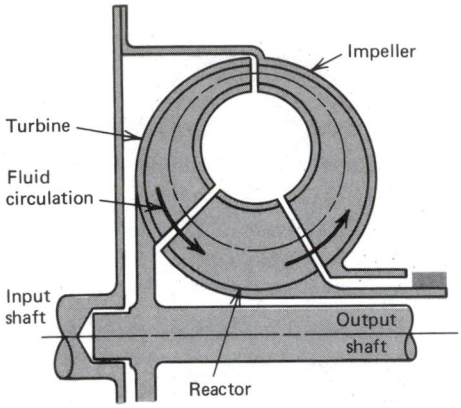

FIGURE 19.12.
Torque converter with fixed reactor blades.

19.9 HYDRODYNAMIC TORQUE CONVERTERS

With reference to the fluid coupling shown in Fig. 19.10, it was noted that for equilibrium of moments about the axis of rotation, the input and output shaft torques must exactly balance. The only possible way for the output torque to exceed the input torque is to *add a third torque-contributing member* (usually a *stationary reaction member* contributing reaction torque, T_r). Then the equilibrium equation becomes

$$T_i + T_o + T_r = 0 \tag{19.9}$$

Figure 19.12 shows one way to add a reaction member to a fluid coupling to make it into a torque converter. The blades of the stationary reactor as well as those of the impeller and turbine are curved and set at an angle. When the circulating fluid strikes the reactor blades, external torque T_r must be applied to keep the reactor from rotating.

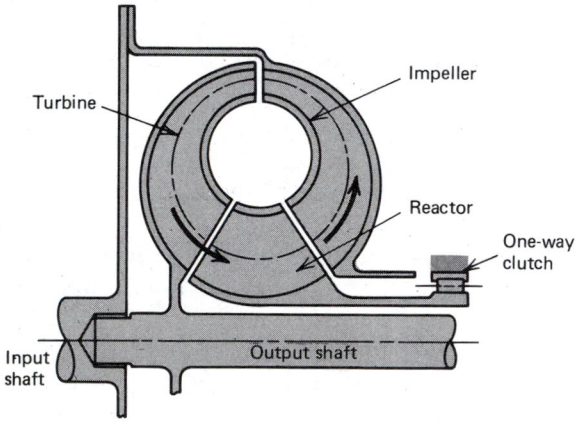

FIGURE 19.13.
Torque converter with reactor controlled by a one-way clutch.

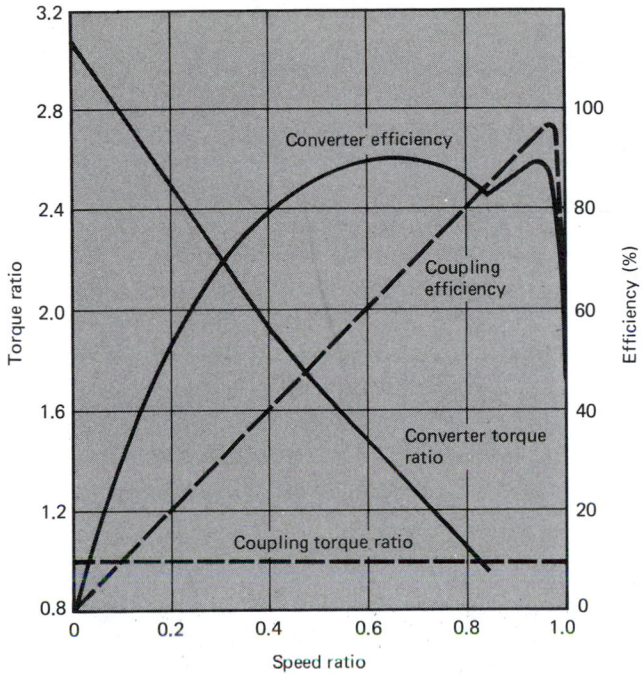

FIGURE 19.14.
Representative torque converter and fluid coupling performance curves.

The reactor blades *redirect* the circulating fluid so as to cause an increased torque to be delivered to the turbine in accordance with Eq. (19.9).

Since the reactor blades are stationary and do no work, the increased turbine torque must be accompanied by a proportional decrease in output speed, ω_o.

In applications requiring a torque multiplication only to start and bring a load up to speed, the "fixed" reactor is mounted through a "free-wheeling" or "one-way" clutch, as shown in Fig. 19.13. When normal operating speed is approached, the reactor "free-wheels" (turns without restraint in the direction permitted by the free-wheeling clutch), and the torque converter behaves as a fluid coupling. This is the case, for example, with torque converters incorporated in automotive automatic transmissions. These converters typically provide a torque multiplication ratio of the order of 3 with a stalled output shaft. As the output shaft accelerates, the torque ratio decreases rapidly. When the ratio reaches unity, the free-wheeling clutch allows the reactor to begin turning freely (in the direction opposite that which it was previously restrained from turning by the one-way clutch). Performance characteristics of a typical torque converter, compared with those of a fluid coupling of the same diameter, are shown in Fig. 19.14.

CHAPTER TWENTY

Machine Component Interrelationships – A Case Study

20.1 INTRODUCTION

The ultimate purpose of machine components is, of course, to function as parts of a complete machine that serves a useful purpose. The design of a complete machine is a creative process, involving the use of imagination and ingenuity to conceive of effective combinations of components, and the optimization of selected combinations, taking into account the multitude of factors that generally apply to a given specific situation. Advancing from the design of individual components to the design of a complete machine is obviously a major step. The purpose of this chapter is to help prepare the student for undertaking this step.

In order to develop skill in conceiving effective combinations and arrangements of components, it is helpful to study the design of selected existing machines and to observe carefully the various interrelationships between them. The way one component is designed may influence both the loads carried by other parts and the space available for those parts.

The machine chosen here for a "case study" is the original "Hydra-Matic" transmission, introduced on the 1940 Oldsmobile (see Section 19.7). It was the first modern fully automatic transmission used in an automobile. It represents an ingenious combination of components, all worked out to give a well-balanced design (i.e., all parts of adequate strength with none substantially "over-designed"). The parts fit together into a compact "package," and they are designed so that the transmission can be mass-produced at reasonable cost. Examples showing similar ingenuity and sound engineering design are found in a wide range of industrial and consumer products.

20.2 DESCRIPTION OF ORIGINAL HYDRA-MATIC TRANSMISSION

Figure 20.1 shows a highly simplified, diagrammatic representation of the general layout of the transmission. Its three basic components are: (a) front planetary gear train, (b) fluid coupling, and (c) rear planetary gear train. Shifting is accomplished by means of hydraulically actuated band brakes and multi-disk clutches. Torque ratios of the planetary trains can be determined by using the procedures given in Section 15.13.

Figure 20.2 gives a diagram of the power flow through the transmission. The engine drives the front planetary train, which provides a torque multiplication of either 1.44 or 1.00. The output of the front train passes through a fluid coupling to the rear planetary train, which provides a torque multiplication of 2.53, 1.00, or −2.99. In third

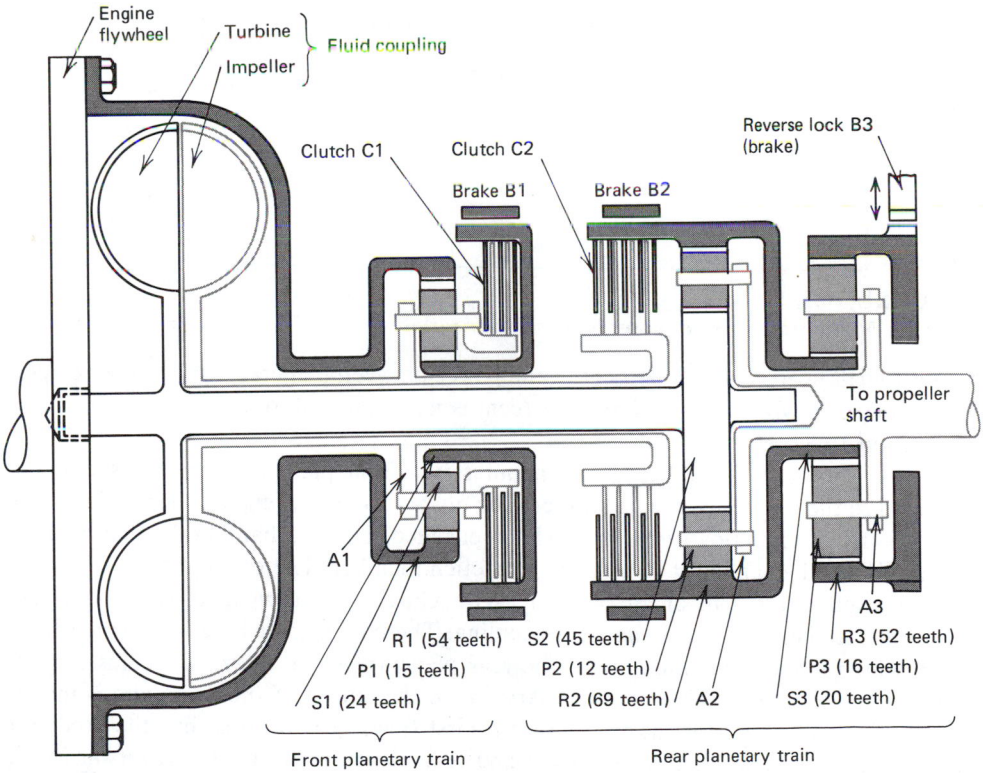

FIGURE 20.1.
Schematic drawing of Hydra-Matic transmission.

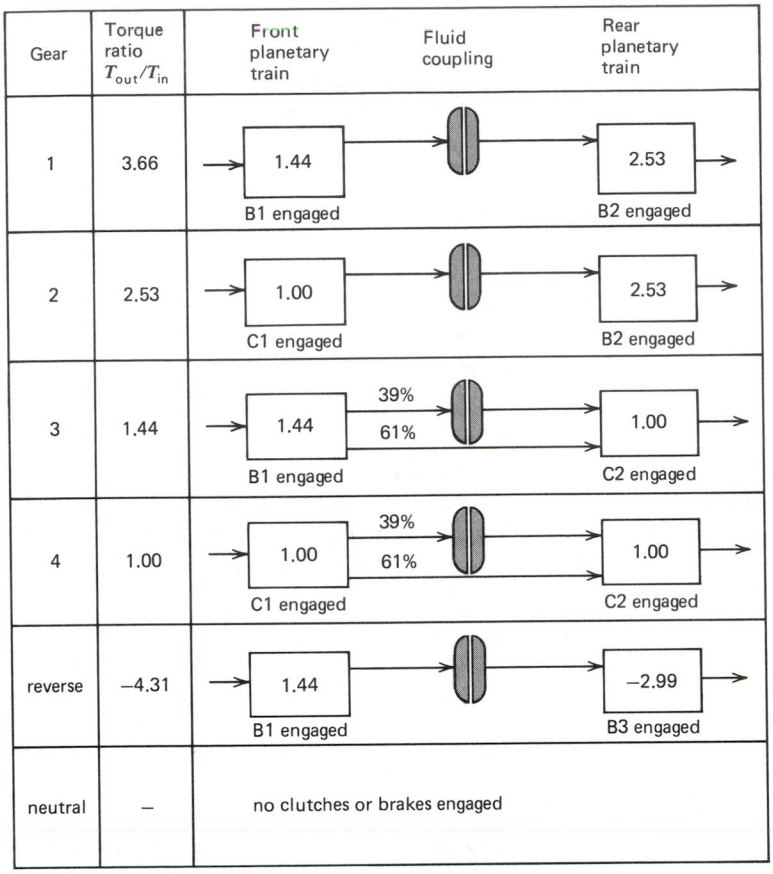

FIGURE 20.2.
Hydra-Matic transmission power flow and gear ratios.

and fourth gears, only 39 percent of the power between planetary trains flows through the coupling, which means that 61 percent is not subjected to coupling slippage losses (Section 19.8).

As shown in Fig. 20.1, ring or annulus R1 (front planetary train) is solidly connected to the engine flywheel and therefore *always* rotates at engine speed. Figure 20.2 shows that for first (low) gear, brake B1 is engaged. This is a band brake that holds sun gear S1 fixed, enabling it to serve as a reaction member. The output of the front planetary train is the arm or carrier, A1. With clutch C2 disengaged, this output goes entirely to the impeller of the fluid coupling. The coupling turbine is attached to sun S2, which is the input member of the rear planetary train. Brake B2 holds ring R2, which serves as a reaction member. Arm A2 is the output of the train, and is integral with the output shaft of the transmission. First (low) gear ratio is thus the product of the speed reduction ratios of the front and rear planetary trains. In all forward ratios the third planetary train (S3, P3, A3) is inactive—these parts rotate, but carry no loads.

Whereas brakes B1 and B2 provide reaction members for the front and rear planetary trains, clutches C1 and C2 provide a direct-drive torque ratio of unity through these trains. Thus, in second gear, clutch C1 forces ring R1 and arm A1 to rotate together. This causes the entire front planetary train to rotate as one solid member, transmitting power without change in speed. In third gear, it is the rear planetary train that rotates essentially as a solid member. As far as torque ratios are concerned, coupling slip can be neglected. Hence, with clutch C2 actuated, S2 and A2 can be considered as rotating together.

From Fig. 20.2, it can be seen that the control system provides for the product of the two planetary train ratios for first gear, the larger ratio alone for second gear, the smaller ratio alone for third gear, and a direct 1:1 drive for fourth gear.

For reverse gear, the "reverse lock" B3 holds ring R3 as a fixed reaction member. Planetary gears with labels 2 and 3 act together and constitute a single compound planetary train with S2 as input and A2-A3 (the two arms are integral) as output.

One of the most interesting and ingenious features of the original Hydra-Matic design is the action of the fluid coupling. Note that in first gear, the coupling turns at engine speed divided by 1.44. This means that with the engine idling, the stall torque of the coupling is so low that chassis friction (most of which is tire rolling resistance) will usually keep the car from "creeping" forward.[1] Thus, when stopped at a traffic light with transmission in "drive," it is not normally necessary to keep a foot on the brake pedal.

When the transmission is in third or fourth gear, the output of the front planetary train is transmitted directly to R2 (via clutch C2), and through the fluid coupling to S2. These parallel power paths are shown in Fig. 20.2. It will be shown later that for the numbers of gear teeth involved, the division of torque is in the ratio of 39:61. An advantage of the split power feature is that for high (fourth) gear cruising, only 39 percent of the power is subject to fluid coupling losses. When cruising above about 33 mph, power loss due to coupling slippage in the Hydra-Matic is less than 1 percent. More modern automatic transmissions have *all* of the power going through a torque converter, thereby sustaining about three times the fluid slippage loss of the original Hydra-Matic. To eliminate this loss, the torque converter is sometimes mechanically "locked" after shifting into high gear, but this has the disadvantage of eliminating the isolation of torsional shock and vibration which is a feature of hydrodynamic drives.

Although the original Hydra-Matic transmission was eminently successful, there was some problem in maintaining a smooth shift between second and third gears. Note in Fig. 20.2 that this shift involves the simultaneous engagement or release of all four friction components, B1, C1, B2, and C2. Another minor problem was that the gears turned with the transmission in "neutral" or "park," thereby causing some noise.

20.3 FREE-BODY DIAGRAM DETERMINATION OF GEAR RATIOS AND COMPONENT LOADS

The various transmission gear ratios can be determined by any of the methods given in Section 15.13, as well as others. The *free-body diagram method* provides a convenient

[1]See Section 19.8.

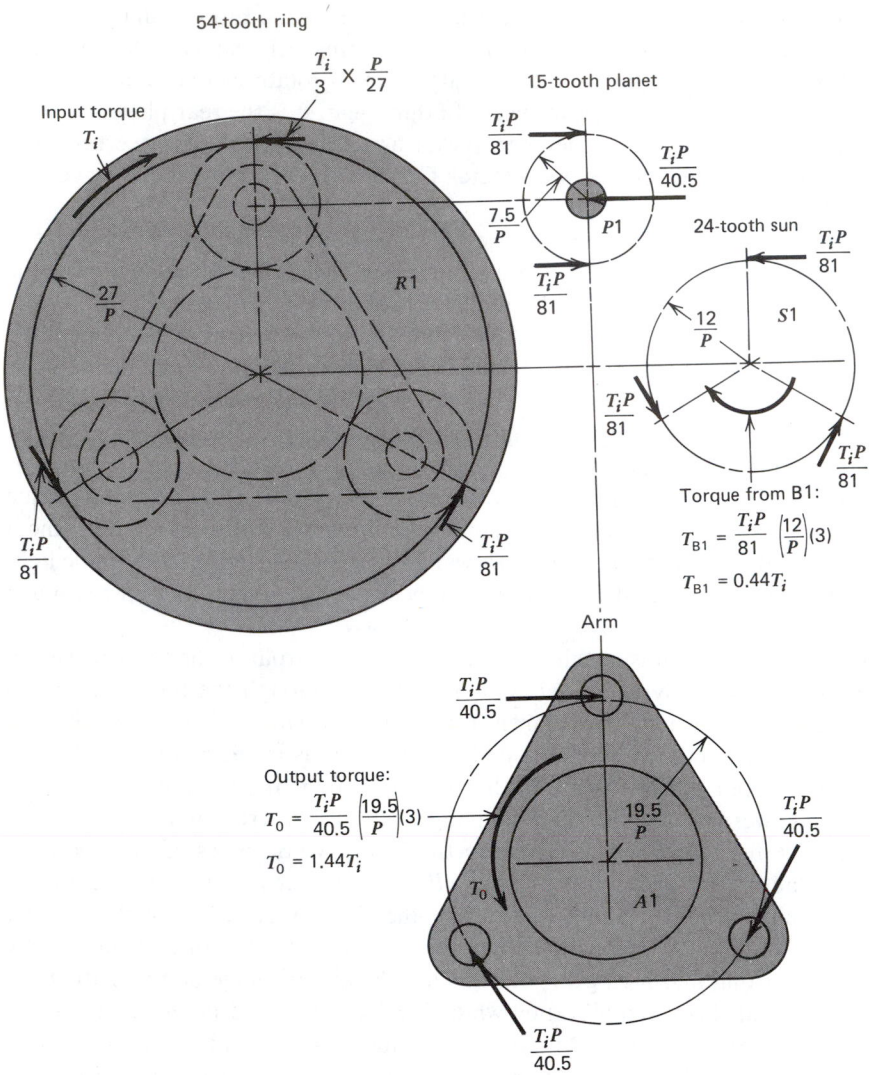

FIGURE 20.3.
Front planetary train with brake B1 engaged. (Note: Radial and axial gear forces are not shown. P = gear diametral pitch, teeth per inch of diameter.)

means of determining not only the gear ratios but the loads applied to the various components as well.

Figure 20.3 shows free-body diagrams for each component of the front planetary train when brake B1 is engaged and clutch C1 is disengaged (as in first, third, and reverse gears). We begin with ring R1, as it receives input torque T_i (pound inches) from the engine. The transmission has three equally spaced planets, P1, which apply tangential forces to the ring teeth. The ring has 54 teeth of pitch P teeth per inch. Hence the radius of the tangential forces is $27/P$ inches. For equilibrium of moments about the axis of the ring, the tangential gear-tooth forces must be $T_iP/81$ lb. Equal and opposite forces are applied *by* the annulus *to* the planet teeth. For a planet to be in equilibrium with respect to moments about its axis, the sun teeth must apply forces to the planets which are also $T_iP/81$ lb. in magnitude. Equilibrium of forces acting on a planet requires that a force of $T_iP/40.5$ be applied at the center spindle, which is part of the arm or carrier. Equal and opposite forces are applied by the planets to the carrier. Equilibrium of the arm requires that the output reaction torque be $1.44T_i$. Equilibrium of the sun involves forces applied to it by the planet teeth, and a torque applied by brake B1. From the diagrams it is concluded that:

1. The torque ratio (output torque/input torque) of the front planetary train is $+1.44$.
2. Brake B1 must provide a torque of $0.44T_i$.
3. The bearings supporting each of the members carry negligible loads except for the planet. Planet bearings must carry a load of $T_iP/40.5$ lb.

Figure 20.4 shows a similar set of free-body diagrams for the rear planetary train when brake B2 is engaged, clutch C2 disengaged, and reverse lock B3 disengaged (as in first and second gears). The diagrams begin with applying input torque T_i to the sun (note that the input torque to the rear planetary unit is the output torque from the front gear train). Free-body diagrams for the sun, planet, ring, and arm are determined as shown, with the conclusions that:

1. The torque ratio of the rear unit is $T_o/T_i = +2.53$.
2. Brake B2 must provide a torque of $1.53\ T_i$.
3. The planet bearings must carry a load of $T_iP/33.75$ pounds.

Figure 20.5 shows the set of free-body diagrams for the rear planetary train with reverse lock B3 engaged (as in reverse gear). Brake B2 and clutch C2 are disengaged. The analysis is slightly more complicated because planetary gears S2, P2, A2 and S3, P3, A3 act together as a compound planetary train. As in Fig. 20.4, we begin by applying the input torque to sun S2 (this is the output torque of the front unit that, for reverse gear, is 1.44 times engine torque). Sun S2 and planet P2 are placed in equilibrium, and loads are applied to ring R2. As shown in Fig. 20.1, ring R2 is integral with sun S3; hence, R2 and S3 are shown as a single free-body. For equilibrium of moments about the axis of this member, the forces applied *to* S3 *by* planets P3 must be as shown. Planet P3 and ring R3 are next placed in equilibrium. Finally, integral arm A2 and A3 is placed in equilibrium. Since the torque applied to the arm by planets P3 exceeds that applied by planets P2, the reaction torque applied to the arm by the transmission output shaft is in the direction shown. The torque applied *to* the output shaft is, of course, in the opposite direction. Since this is also opposite in direction to the engine torque, the torque ratio is negative, as it must be for a reverse gear. From the above, it is concluded that:

1. The ratio of the rear unit with B3 engaged is -2.99.

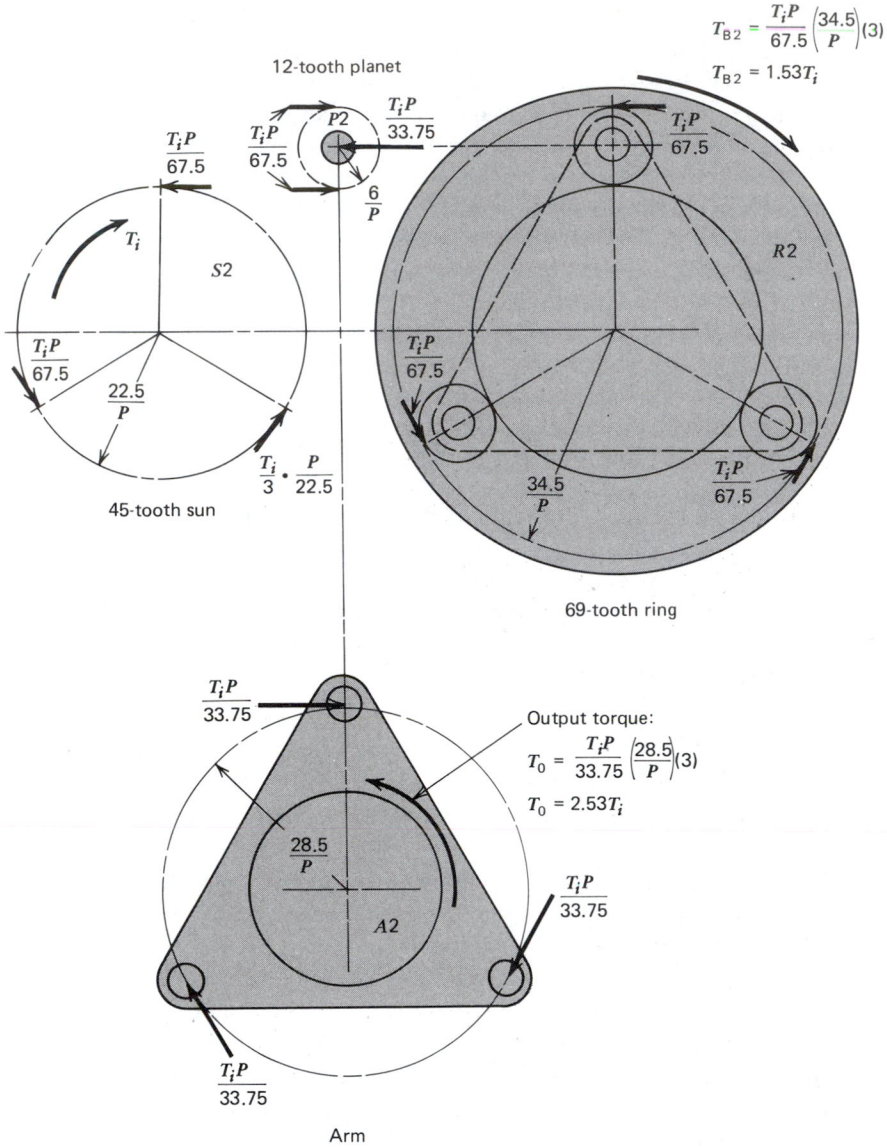

FIGURE 20.4.
Rear planetary train with brake B2 engaged.

2. The reverse lock must provide a brake torque of 3.99 times rear train input torque, or 5.75 times engine torque.
3. Bearing loads for planets P2 are $T_iP/33.75$, as in the Fig. 20.4 analysis. Bearing loads for planets P3 are over three times as high: $69T_iP/675$.

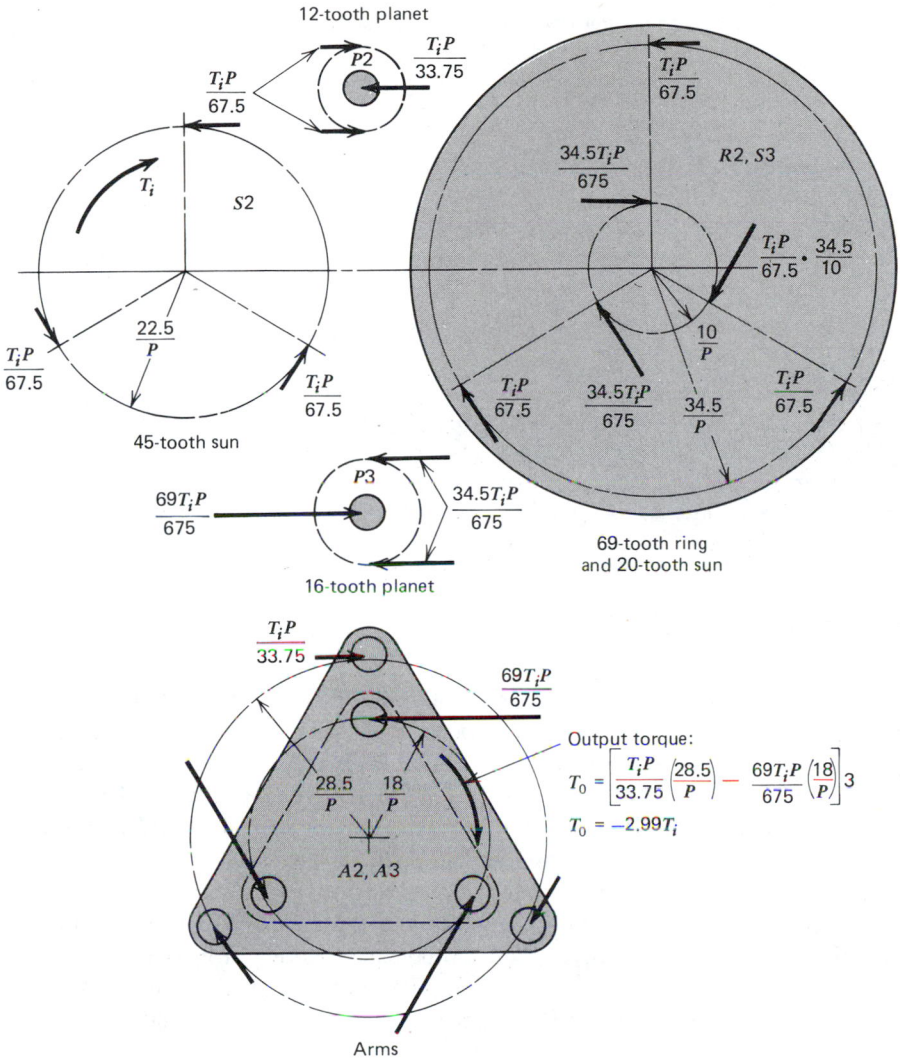

FIGURE 20.5.
Rear planetary train with reverse lock B3 engaged. [See next page for ring R3.]

20.4 GEAR DESIGN CONSIDERATIONS

The last paragraph of Section 15.13 gave two requirements that must be satisfied when choosing the number of planets to be mounted on an arm: (1) the number must be small enough to give clearance between the planets, and (2) the teeth of each planet must align simultaneously with teeth of the sun and ring. Applying these criteria to the front planets, P1: (1) fewer than eight planets must be used, and (2) two, three, or six equally spaced planets can be used; four cannot.

Note from the free-body diagrams that the tangential loads applied to the teeth of

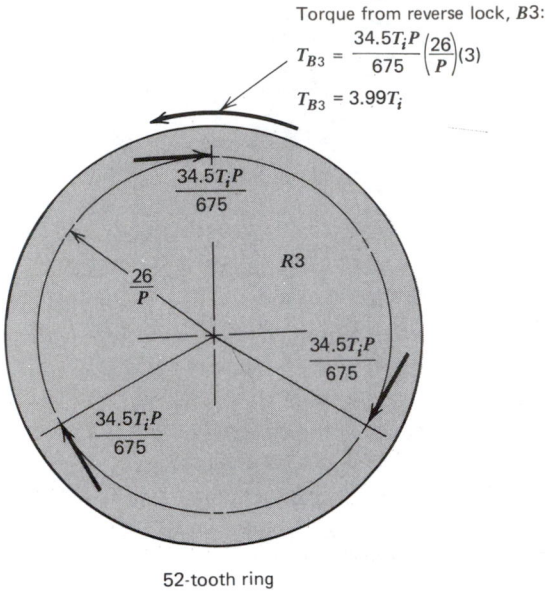

FIGURE 20.5 (*continued*)

the sun, planets, and ring (of the same gear set) are all the same. Although it may seem surprising at first, the effective pitch-line velocity is also the same for all gears in the set, *regardless of which member is held fixed.* For each gear, the effective pitch-line velocity reflects the frequency with which successive teeth come into contact. To accomplish this, the pitch-line velocity is computed *relative to the arm.* The effective pitch-line velocity appears correctly to an observer stationed on the arm.

Helical gears are usually used in automotive transmissions because of their relatively quiet operation. For the free-body diagrams in Figs. 20.3–20.7 to be complete, the gear forces should include both an axial component (function of helix angle) and a radial component (function of pressure angle). The existence of radial forces has no bearing on the conclusions drawn from the diagrams. The existence of axial gear forces results in thrust loads being applied to the bearings.

The basic arrangement of the transmission is such that the size of the gears is a major factor controlling the size of the transmission. There are three favorable factors to be taken into account when sizing the gears (i.e., determining an appropriate value of pitch, P): (1) modern gear cutting machines, together with the high volume of production involved, make it economically feasible to specify gears of excellent quality; (2) use of the fluid coupling protects the gears from much of the shock loading that they would otherwise encounter; and (3) while the transmission must be designed for a life of well over 100,000 miles of reasonably severe service, the gears encounter little cyclic load in high gear (when the front planetary train is locked and the rear train moves together except for coupling slip), and experience a relatively few miles of full engine torque operation in the other gears.

20.5 BRAKE AND CLUTCH DESIGN CONSIDERATIONS

The free-body analysis in Fig. 20.3 indicated that brake B1 must be designed for 0.44 times engine torque. Note that this value can also be obtained by considering the entire front unit of the transmission as a single free-body: With an input torque of T_i and an output torque of $1.44T_i$, there *must* be a reaction torque of $0.44T_i$. The only provision for a reaction torque to be transmitted to the transmission housing and its external support is through brake B1.

Similarly, brake B2 must be designed for 1.53 times the input torque to the rear planetary unit. For first and reverse gears, this brake torque is 2.20 (1.53×1.44) times engine torque. This value can readily be checked by considering the overall equilibrium of the rear section of the transmission: the input shaft receives 1.44 times engine torque, the brake adds 2.20 times engine torque, and this balances the output reaction torque of 3.64 times engine torque. B1 and B2 are both band brakes operating in oil (transmission fluid). Clearly, B2 is the more critical.

Reverse lock B3 must take 3.99×1.44 or 5.75 times engine torque. As a check, equilibrium of the rear unit for unit engine torque involves an input torque of 1.44, a reaction torque of -5.75, and an output torque of -4.31.

In order to determine the load applied to clutch C1, free-body diagrams for the front planetary unit with C1 engaged (and B1 disengaged) are drawn as in Fig. 20.6. These apply to operation in second and fourth gears. Loads on the ring and planet gears are the same as in Fig. 20.3. For the sun, the brake torque of $0.44T_i$ shown in Fig. 20.3 is replaced by a clutch torque of $0.44T_i$. This torque is transmitted through the clutch to the arm, where it causes the output torque to be reduced from $1.44T_i$ (shown in Fig. 20.3) to $1.00T_i$.

In order to determine the load carried by clutch C2, free-body diagrams for the rear planetary unit in third and fourth gears (C2 engaged, B2 disengaged) are drawn in Fig. 20.7. First, however, note in Fig. 20.1 that with C2 engaged, the output torque from the arm of the front planetary train is *split*. A portion is transmitted through the fluid coupling to sun S2; the remainder is transmitted through clutch C2 to ring R2. The fluid-transmitted and clutch-transmitted torques are designated as T_f and T_{C2}, respectively.

Turning to the free-body diagrams of Fig. 20.7, the loads on the sun gear are the same as in Fig. 20.4, except that T_f replaces T_i. The clutch torque is applied to the ring gear, causing loading on the teeth as shown. Planet gear-tooth loads come from the sun and from the ring. For equilibrium of moments, these loads must be equal. This leads to the conclusion that *61 percent of the torque is transmitted through the clutch and 39 percent through the fluid coupling,* as indicated in Fig. 20.2. The maximum loading on clutch C2 occurs in third gear, when the clutch carries 61 percent of 1.44 times engine torque, or 88 percent of engine torque.

Space limitations restrict both clutches to about the same outside diameter, as indicated diagrammatically in Fig. 20.1. Since clutch C2 carries twice the maximum torque of clutch C1, it will require about twice as many plates. Both clutches operate in oil.

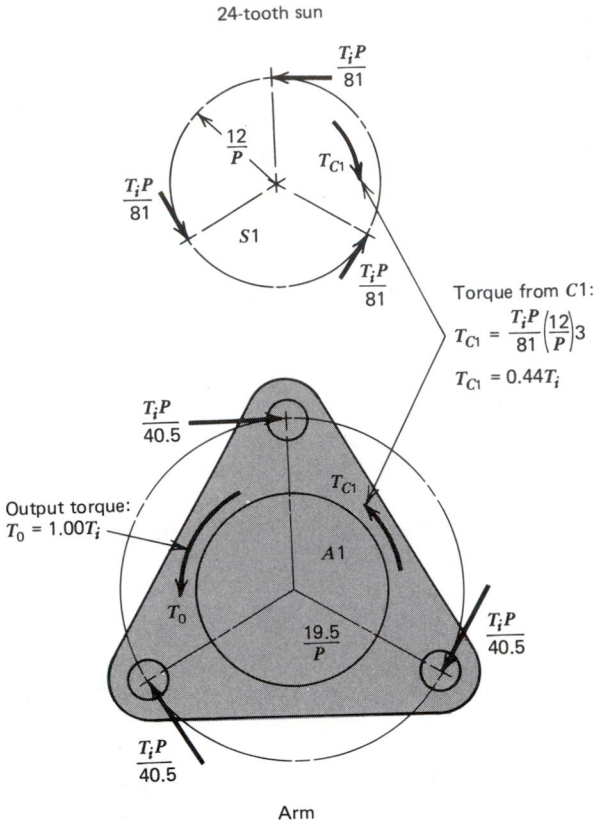

FIGURE 20.6.
Front planetary train with clutch C1 engaged. (Note: Loads on R1 and P1 are as shown in Fig. 20.3.)

20.6 MISCELLANEOUS DESIGN CONSIDERATIONS

The general transmission layout shown in Fig. 20.1 includes both solid and hollow shafts. The outer hollow shaft between the fluid coupling and front planetary unit carries engine torque to ring gear R1. Inside that is another hollow shaft that, in first and reverse gears, carries 1.44 times engine torque from arm A1 to the fluid coupling. Inside that is a solid shaft that, in first and reverse gears, carries 1.44 times engine torque from the coupling to sun gear S2. At the rear of the transmission, the hollow shaft connecting arm A2 to the output carries 3.66 times engine torque in first gear. The output shaft must carry 4.31 times engine torque in reverse gear.

The concentric shaft construction involved here provides a good illustration of the relatively high strength of hollow shafts. For example, the torsional stresses in a solid shaft of diameter d are the same as those in a hollow shaft of inside diameter d and outside diameter of only $1.26d$.

As previously noted, the planet bearings are the only bearings in the transmission subjected to significant radial loads. Unfortunately, these are also the bearings for

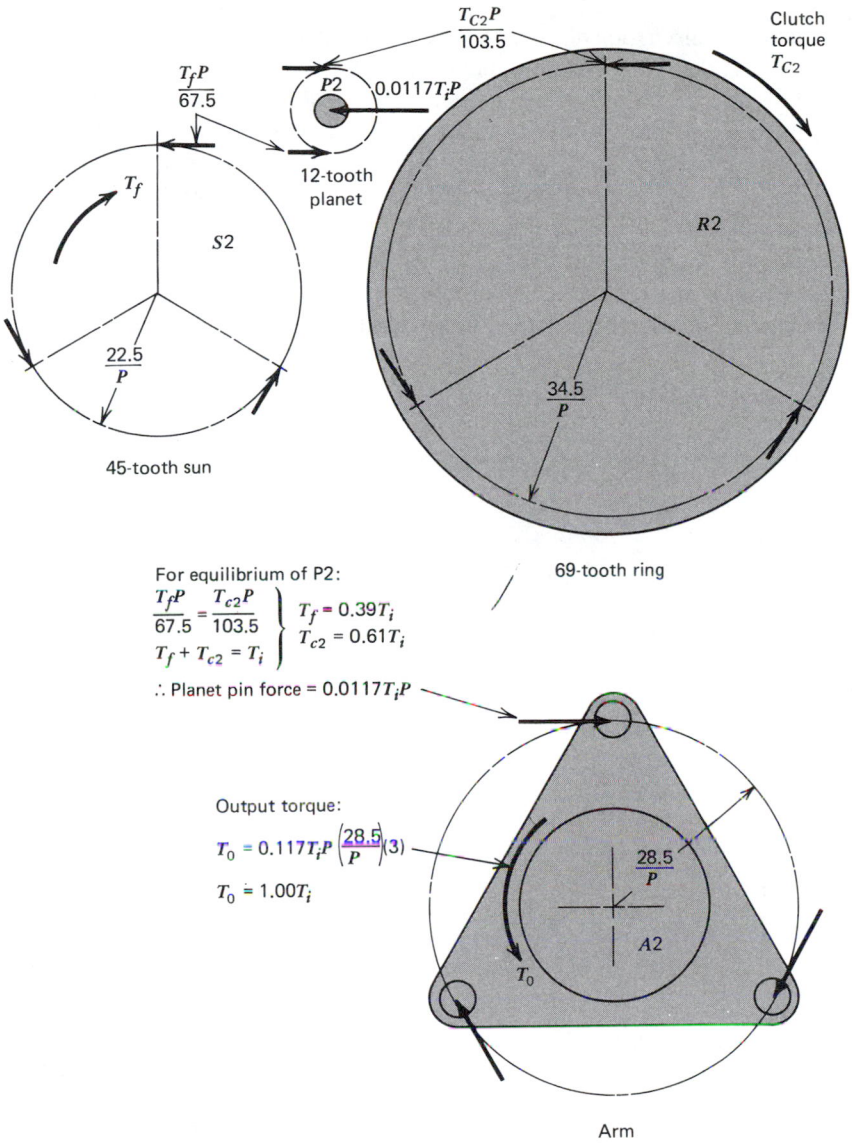

FIGURE 20.7.
Rear planetary train with clutch C2 engaged.

which there is the least available space. This is so because the planets are of relatively small outside diameter, and the center supporting shafts (which are part of the arm) must be sufficiently large to carry the bearing load. The radial space left for the bearing is sufficient only for a plain bearing (Chapter 13) or needle bearing (Section 14.3, Fig. 14.9). Needle bearings are usually chosen because of their ability to carry load with a limited oil supply. To carry the thrust load of the helical gears, numerous thrust washers (Section 13.13) are used.

We conclude this discussion of the design of an automotive automatic transmission with a reminder that there is much more involved than the points covered here. Figure 20.1 is a simplified diagrammatic representation—not an actual assembly drawing. For example, there are hydraulic pistons for actuating the clutches and brakes, and an elaborate hydraulic control system for timing the shifts and actuating them smoothly. Also, liberties were taken in representing the structure, as in showing the planet pinion shafts cantilevered. Nevertheless, it is hoped that this short "case study" will help to aid and inspire the student to continue the study of mechanical engineering design, and eventually to contribute his or her own imagination and ingenuity to the design of improved products that make a positive contribution to the quality of life in our society.

APPENDIX A

Units

APPENDIX A-1 Conversion Between English and SI Units

Quantity	English Unit[a]	SI Unit[a]	Ratio, SI/English: Approximate	Ratio, SI/English: Accurate	Ratio, English/SI: Approximate	Ratio, English/SI: Accurate
Length	inch (in. or ″)	meter (m)	0.0254	2.540 000 E − 2	39.4	3.937 008 E + 1
	foot (ft or ′)	meter (m)	0.305	3.048 000 E − 1	3.28	3.280 840 E + 0
	mile (mi U.S. Statute)	kilometer (km)	1.61	1.609 344 E + 0	0.621	6.213 712 E − 1
Volume	gallon (gal U.S.)	meter^3 (m^3)	0.00379	3.785 412 E − 3	264.	2.641 720 E + 2
Force, (weight)	pound (lb)	newton[d] (N)	4.45	4.448 222 E + 0	0.225	2.248 089 E − 1
Torque	pound-foot (lb-ft)	newton-meter (N·m)	1.36	1.355 818 E + 0	0.737	7.375 620 E − 1
Work, energy	foot-pound (ft-lb)	joule[c] (J)	1.36	1.355 818 E + 0	0.737	7.375 620 E − 1
Power	foot-pound/second (ft-lb/s)	watt[g] (W)	1.36	1.355 818 E + 0	0.737	7.375 620 E − 1
	horsepower[b] (hp)	kilowatt (kW)	0.747	7.456 999 E − 1	1.34	1.341 022 E + 0
Stress, pressure	pounds/in.^{2} (psi)	pascal[e] (Pa)	6890.	6.894 757 E + 3	0.000145	1.450 370 E − 4
	thousand pounds/in.^{2} (ksi)	megapascal (MPa)	6.890	6.894 757 E + 0	0.145	1.450 370 E − 1
Mass	slug[f]	kilogram (kg)	14.6	1.459 390 E + 1	0.0685	6.852 178 E − 2

Examples:
A door 36 in. wide is 36 in. (0.0254 m/in.) = 0.914 m wide;
a door 1 m wide is 1 m (39.4 in./m) = 39.4 in. wide.

[a]The *larger* unit is underlined.
[b]1 hp = 550 ft·lb/s [c]1 J = 1 N·m [d]1 N = 1 kg·m/s^2 [e]1 Pa = 1 N/m^2 [f]1 slug = 1 lb·s^2/ft [g]1 W = 1 J/s

APPENDIX A-2
Standard SI Prefixes

Category	Name	Symbol	Factor
Recommended, important for this course.	giga	G	1 000 000 000 = 10^9
	mega	M	1 000 000 = 10^6
	kilo	k	1 000 = 10^3
	milli	m	0.001 = 10^{-3}
	micro	μ	0.000 001 = 10^{-6}
Not recommended but sometimes encountered.	hecto	h	100 = 10^2
	deca	da	10 = 10^1
	deci	d	0.1 = 10^{-1}
	centi	c	0.01 = 10^{-2}
Not encountered in this course.	tera	T	1 000 000 000 000 = 10^{12}
	nano	n	0.000 000 000 = 10^{-9}
	pico	p	0.000 000 000 000 = 10^{-12}
	femto	f	0.000 000 000 000 000 = 10^{-15}
	atto	a	0.000 000 000 000 000 000 = 10^{-18}

Note: Spaces are sometimes used instead of commas to group numbers. This is to avoid confusion with the practice in some European countries of using commas for decimal points.

APPENDIX A-3
Suggested SI Prefixes for Stress Calculations

$\sigma = P/A,\ Mc/I,\ M/Z;\quad \tau = P/A,\ V/A,\ Tr/J,\ T/Z',\ (V/Ib)\int y\,dA$

σ,τ	P,V	M,T	A	I,J	c,r,b,y	Z,Z'
Pa	N	N·m	m^2	m^4	m	m^3
kPa	kN	kN·m	m^2	m^4	m	m^3
MPa	N	N·mm	mm^2	mm^4	mm	mm^3
GPa	kN	N·m	mm^2	mm^4	mm	mm^3

APPENDIX A-4
Suggested SI Prefixes for Linear Deflection Calculations

$\delta = PL/AE^a;\ \delta \propto PL^3/EI^a,\ wL^4/EI,\ ML^2/EI^a$

δ	P	w	M	L	A	E	I
μm	N	N/m	N·m	m	m^2	MPa	m^4
μm	N	N/mm	N·mm	mm	mm^2	GPa	mm^4
μm	kN	N/m	kN·m	m	m^2	GPa	m^4

[a]Illustrated in Table 5.1.

APPENDIX A-5
Suggested SI Prefixes for Angular Deflection Calculations

$\theta = TL/K'G^a, ML/IE^a$				
θ	T,M	L	K',I	E,G
rad	N·m	m	m^4	Pa
μrad	N·m	m	m^4	MPa
mrad	N·mm	mm	mm^4	GPa
μrad	kN·m	m	m^4	GPa

[a]Illustrated in Table 5.1.

APPENDIX B

Properties of Sections and Solids

B-1 PROPERTIES OF SECTIONS

A = area, in.^{2}

I = moment of inertia, in.^{4}

J = polar moment of inertia, in.^{4}

Z = section modulus, in.^{3}

ρ = radius of gyration, in.

$\bar{y}$ = centroidal distance, in.

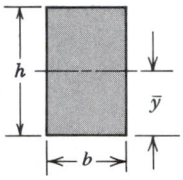

Rectangle

$$A = bh \qquad I = \frac{bh^3}{12} \qquad Z = \frac{bh^2}{6}$$

$$\rho = 0.289h \qquad \bar{y} = \frac{h}{2}$$

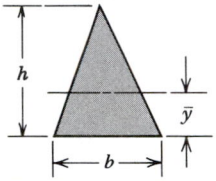

Triangle

$$A = \frac{bh}{2} \qquad I = \frac{bh^3}{36} \qquad Z = \frac{bh^2}{24}$$

$$\rho = 0.236h \qquad \bar{y} = \frac{h}{3}$$

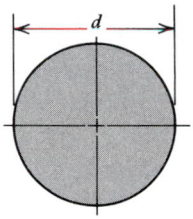

Circle

$$A = \frac{\pi d^2}{4} \qquad I = \frac{\pi d^4}{64} \qquad Z = \frac{\pi d^3}{32}$$

$$J = \frac{\pi d^4}{32} \qquad \rho = \frac{d}{4}$$

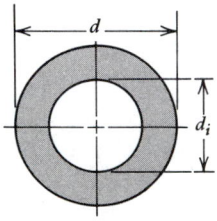

Hollow circle

$$A = \frac{\pi}{4}(d^2 - d_i^2) \qquad I = \frac{\pi}{64}(d^4 - d_i^4) \qquad Z = \frac{\pi}{32d}(d^4 - d_i^4)$$

$$J = \frac{\pi}{32}(d^4 - d_i^4) \qquad \rho = \sqrt{\frac{d^2 + d_i^2}{16}}$$

B-2 MASS AND MASS MOMENTS OF INERTIA OF HOMOGENEOUS SOLIDS

ρ = mass density

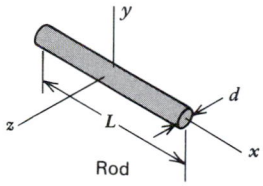

Rod

$$m = \frac{\pi d^2 L \rho}{4}$$

$$I_y = I_z = \frac{mL^2}{12}$$

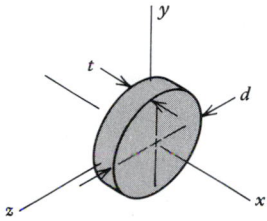

Disk

$$m = \frac{\pi d^2 t \rho}{4}$$

$$I_x = \frac{md^2}{8}$$

$$I_y = I_z = \frac{md^2}{16}$$

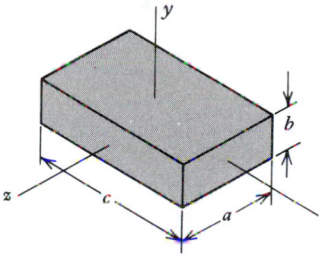

Rectangular prism

$$m = abc\rho$$

$$I_x = \frac{m}{12}(a^2 + b^2)$$

$$I_y = \frac{m}{12}(a^2 + c^2)$$

$$I_z = \frac{m}{12}(b^2 + c^2)$$

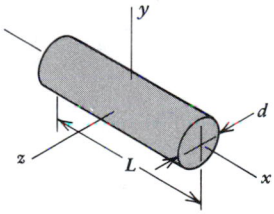

Cylinder

$$m = \frac{\pi d^2 L \rho}{4}$$

$$I_x = \frac{md^2}{8}$$

$$I_y = I_z = \frac{m}{48}(3d^2 + 4L^2)$$

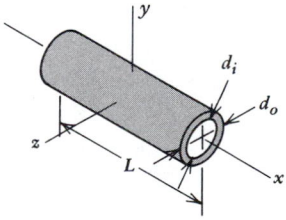

Hollow cylinder

$$m = \frac{\pi L \rho}{4}(d_o^2 - d_i^2)$$

$$I_x = \frac{m}{8}(d_o^2 - d_i^2)$$

$$I_y = I_z = \frac{m}{48}(3d_o^2 + 3d_i^2 + 4L^2)$$

APPENDIX C

Material Properties and Uses

APPENDIX C-1
Physical Properties of Common Metals[a]

Metal	Modulus of Elasticity, E		Modulus of Rigidity, G		Poisson's ratio, ν	Unit wt., w	Density, ρ	Coefficient of Therm. Expan., α		Thermal Conductivity		Specific Heat	
	Mpsi	GPa	MPsi	GPa		(lb/in.^{3})	(Mg/m^3)	10^{-6}/°F	10^{-6}/°C	Btu/hr-ft-°F	W/m-°C	Btu/lb-°F	J/kg-°C
Aluminum alloy	10.4	72	3.9	27	0.32	0.10	2.8	12.	22	100	173	0.22	920
Beryl. copper	18.5	127	7.2	50	0.29	0.30	8.3	9.3	17	85	147	0.10	420
Brass, Bronze	16	110	6.0	41	0.33	0.31	8.7	10.5	19	45	78	0.10	420
Copper	17.5	121	6.6	46	0.33	0.32	8.9	9.4	17	220	381	0.10	420
Iron, gray cast[b]	15	103	6.0	41	0.26	0.26	7.2	6.4	12	29	50	0.13	540
Magnesium alloy	6.5	45	2.4	17	0.35	0.065	1.8	14.5	26	55	95	0.28	1170
Nickel alloy	30	207	11.5	79	0.30	0.30	8.3	7.0	13	12	21	0.12	500
Steel, carbon	30	207	11.5	79	0.30	0.28	7.7	6.7	12	27	47	0.11	460
Steel, alloy	30	207	11.5	79	0.30	0.28	7.7	6.3	11	22	38	0.11	460
Steel, stainless	27.5	190	10.6	73	0.30	0.28	7.7	8.0	14	12	21	0.11	460
Titanium alloy	16.5	114	6.2	43	0.33	0.16	4.4	4.9	9	7	12	0.12	500
Zinc alloy	12	83	4.5	31	0.33	0.24	6.6	15.0	27	64	111	0.11	460

[a]Values given are representative. Exact values may vary with composition and processing, sometimes greatly.
[b]See Appendix C-3 for more detailed elastic properties of cast irons.

Note: See Appendix C-18 for physical properties of some plastics.

APPENDIX C-2
Tensile Properties of Some Metals

Material	Ultimate Strength, S_u		Yield Strength, S_y		σ_0[a]		m[a]	ϵ_{Tf}[a]
	ksi	Mpa	ksi	Mpa	ksi	Mpa		
Carbon and Alloy Steels								
1002 A[b]	42	290	19	131	78	538	0.27	1.25
1010 A	44	303	29	200	82	565	0.23	1.20
1018 A	49.5	341	32	221	90	621	0.25	1.05
1020 HR	66	455	42	290	115	793	0.22	0.92
1045 HR	92.5	638	60	414	140	965	0.14	0.58
1212 HR	61.5	424	28	193	110	758	0.24	0.85
4340 HR	151	1041	132	910	210	1448	0.09	0.45
52100 A	167	1151	131	903	210	1448	0.07	0.40
Stainless Steels								
302 A	92	634	34	234	210	1448	0.48	1.20
303 A	87	600	35	241	205	1413	0.51	1.16
304 A	83	572	40	276	185	1276	0.45	1.67
440C A	117	807	67	462	180	1241	0.14	0.12
Aluminum Alloys								
1100-0	12	83	4.5	31	22	152	0.25	2.30
2024-T4	65	448	43	296	100	690	0.15	0.18
7075-0	34	234	14.3	99	61	421	0.22	0.53
7075-T6	86	593	78	538	128	883	0.13	0.18
Magnesium Alloys								
HK31XA-0	25.5	176	19	131	49.5	341	0.22	0.33
HK31XA-H24	36.2	250	31	214	48	331	0.08	0.20
Copper Alloys								
90-10 Brass A	36.4	251	8.4	58	83	572	0.46	—
80-20 Brass A	35.8	247	7.2	50	84	579	0.48	—
70-30 Brass A	44	303	10.5	72	105	724	0.52	1.55
Naval brass A	54.5	376	17	117	125	862	0.48	1.00

[a]Defined in Section 3.4.
[b]A = annealed, HR = hot-rolled.

Note: Values are from single tests and believed typical. Actual values may vary due to small differences in composition and processing; hence, some values here do not agree with values in other Appendix C tables.
Source: Datsko, J., *Materials in Design and Manufacturing,* Mallory, Inc., Ann Arbor, Michigan, 1977.

APPENDIX C-3a
Typical Mechanical Properties and Uses of Gray Cast Iron[a]

ASTM class[a]	Tensile Strength		Torsional Shear Strength		Compressive Strength		Reversed Bending Fatigue Limit		Brinell Hardness, H_B	Tensile Modulus		Torsional Modulus		Typical Uses
	MPa	ksi[a]	MPa	ksi	MPa	ksi	MPa	ksi		GPa	10^6 psi	GPa	10^6 psi	
20	152	22	179	26	572	83	69	10	156	66 to 97	9.6 to 14.0	27 to 39	3.9 to 5.6	Miscellaneous soft iron castings
25	179	26	220	32	669	97	79	11.5	174	79 to 102	11.5 to 14.8	32 to 41	4.6 to 6.0	Cylinder heads and blocks, housings
30	214	31	276	40	752	109	97	14	210	90 to 113	13.0 to 16.4	36 to 45	5.2 to 6.6	Brake drums, clutch plates, flywheels
35	252	36.5	334	48.5	855	124	110	16	212	100 to 119	14.5 to 17.2	40 to 48	5.8 to 6.9	Heavy-duty brake drums, clutch plates
40	293	42.5	393	57	965	140	128	18.5	235	110 to 138	16.0 to 20.0	44 to 54	6.4 to 7.8	Cylinder liners, camshafts
50	362	52.5	503	73	1130	164	148	21.5	262	130 to 157	18.8 to 22.8	50 to 55	7.2 to 8.0	Special high-strength castings
60	431	62.5	610	88.5	1293	187.5	169	24.5	302	141 to 162	20.4 to 23.5	54 to 59	7.8 to 8.5	Special high-strength castings

[a]Minimum values of S_u (in ksi) are given by the class number.

APPENDIX C-3b
Mechanical Properties and Typical Uses of Malleable Cast Iron[a]

Specification No.	Class or Grade	Tensile Strength MPa	Tensile Strength ksi	Yield Strength MPa	Yield Strength ksi	Brinell Hardness, H_B	Elongation[b] (%)	Typical Uses
Ferritic								
ASTM A47, A338; ANSI G48.1; FED QQ-I-666c	32510	345	50	224	32	156 max	10	General purpose at normal and elevated temperatures; good machinability, excellent shock resistance
	35018	365	53	241	35	156 max	18	
ASTM A197	—	276	40	207	30	156 max	5	Pipe flanges, valve parts
Pearlitic and Martensitic								
ASTM A220; ANSI G48.2; MIL-I-11444B	40010	414	60	276	40	149-197	10	General engineering service at normal and elevated temperatures
	45008	448	65	310	45	156-197	8	
	45006	448	65	310	45	156-207	6	
	50005	483	70	345	50	179-229	5	
	60004	552	80	414	60	197-241	4	
	70003	586	85	483	70	217-269	3	
	80002	655	95	552	80	241-285	2	
	90001	724	105	621	90	269-321	1	
Automotive								
ASTM A602; SAE J158	M3210[c]	345	50	224	32	156 max	10	Steering gear housing, mounting brackets
	M4504[d]	448	65	310	45	163-217	4	Compressor crankshafts and hubs
	M5003[d]	517	75	345	50	187-241	3	Parts requiring selective hardening, as gears
	M5503[e]	517	75	379	55	187-241	3	For machinability and improved induction hardening
	M7002[e]	621	90	483	70	229-269	2	Connecting rods, universal joint yokes
	M8501[e]	724	105	586	85	269-302	1	Gears with high-strength and good wear resistance

[a]Condensed from *ASM Metals Reference Book*, American Society for Metals, Metals Park, Ohio, 1981.
[b]Minimum in 50 mm (2 in.).
[c]Annealed.
[d]Air quenched and tempered.
[e]Liquid quenched and tempered.

APPENDIX C-3c
Average Mechanical Properties and Typical Uses of Ductile (Nodular) Iron

Grade[a]	Brinell Hardness	Elongation (%) (in 50 mm)	Poisson's Ratio	Tensile Modulus GPa	Tensile Modulus 10^6 psi	Typical Uses
60-40-18	167	15.0	0.29	169	24.5	Valves and fittings for steam and chemicals
65-45-12	167	15.0	0.29	168	24.4	Machine components subject to shock and fatigue
80-55-06	192	11.2	0.31	168	24.4	Crankshafts, gears, rollers
120-90-02	331	1.5	0.28	164	23.8	Pinions, gears, rollers, slides

	Tensile Strength				Compr. Strength: Ultimate		Torsional Strength			
	Ultimate		Yield				Ultimate		Yield	
Grade	MPa	10^6 psi	MPa	10^6 psi	MPa	10^6 psi	MPa	10^6 psi	MPa	10^6 psi
60-40-18	461	66.9	329	47.7	359	52.0	472	68.5	195	28.3
65-45-12	464	67.3	332	48.2	362	52.5	475	68.9	297	30.0
80-55-06	559	81.8	362	52.5	386	56.0	504	73.1	193	28.0
120-90-02	974	141.3	864	125.3	920	133.5	875	126.9	492	71.3

[a]First two sections of grade number indicate minimum values (in ksi) of tensile ultimate and yield strengths.
Source: ASM Metals Reference Book, American Society for Metals, Metals Park, Ohio, 1981.

APPENDIX C-4a
Mechanical Properties of Selected Carbon and Alloy Steels

I:AISI No.[a]	Treatment	Tensile Strength MPa	Tensile Strength ksi	Yield Strength MPa	Yield Strength ksi	Elongation (%)	Reduction in Area (%)	Brinell Hardness, H_B	Izod Impact Strength J	Izod Impact Strength ft-lb
1015	As-rolled	420.6	61.0	313.7	45.5	39.0	61.0	126	110.5	81.5
	Normalized	424.0	61.5	324.1	47.0	37.0	69.6	121	115.5	85.2
	Annealed	386.1	56.0	284.4	41.3	37.0	69.7	111	115.0	84.8
1020	As-rolled	448.2	65.0	330.9	48.0	36.0	59.0	143	86.8	64.0
	Normalized	441.3	64.0	346.5	50.3	35.8	67.9	131	117.7	86.8
	Annealed	394.7	57.3	294.8	42.8	36.5	66.0	111	123.4	91.0
1030	As-rolled	551.6	80.0	344.7	50.0	32.0	57.0	179	74.6	55.0
	Normalized	520.6	75.5	344.7	50.0	32.0	60.8	149	93.6	69.0
	Annealed	463.7	67.3	341.3	49.5	31.2	57.9	126	69.4	51.2
1040	As-rolled	620.5	90.0	413.7	60.0	25.0	50.0	201	48.8	36.0
	Normalized	589.5	85.5	374.0	54.3	28.0	54.9	170	65.1	48.0
	Annealed	518.8	75.3	353.4	51.3	30.2	57.2	149	44.3	32.7
1050	As-rolled	723.9	105.0	413.7	60.0	20.0	40.0	229	31.2	23.0
	Normalized	748.1	108.5	427.5	62.0	20.0	39.4	217	27.1	20.0
	Annealed	636.0	92.3	365.4	53.0	23.7	39.9	187	16.9	12.5
1095	As-rolled	965.3	140.0	572.3	83.0	9.0	18.0	293	4.1	3.0
	Normalized	1013.5	147.0	499.9	72.5	9.5	13.5	293	5.4	4.0
	Annealed	656.7	95.3	379.2	55.0	13.0	20.6	192	2.7	2.0
1118	As-rolled	521.2	75.6	316.5	45.9	32.0	70.0	149	108.5	80.0
	Normalized	477.8	69.3	319.2	46.3	33.5	65.9	143	103.4	76.3
	Annealed	450.2	65.3	284.8	41.3	34.5	66.8	131	106.4	78.5
3140	Normalized	891.5	129.3	599.8	87.0	19.7	57.3	262	53.6	39.5
	Annealed	689.5	100.0	422.6	61.3	24.5	50.8	197	46.4	34.2
4130	Normalized	668.8	97.0	436.4	63.3	25.5	59.5	197	86.4	63.7
	Annealed	560.5	81.3	360.6	52.3	28.2	55.6	156	61.7	45.5

4140	Normalized	1020.4	148.0	655.0	95.0	17.7	46.8	302	22.6	16.7
	Annealed	655.0	95.0	417.1	60.5	25.7	56.9	197	54.5	40.2
4340	Normalized	1279.0	185.5	861.8	125.0	12.2	36.3	363	15.9	11.7
	Annealed	744.6	108.0	472.3	68.5	22.0	49.9	217	51.1	37.7
6150	Normalized	939.8	136.3	615.7	89.3	21.8	61.0	269	35.5	26.2
	Annealed	667.4	96.8	412.3	59.8	23.0	48.4	197	27.4	20.2
8650	Normalized	1023.9	148.5	688.1	99.8	14.0	40.4	302	13.6	10.0
	Annealed	715.7	103.8	386.1	56.0	22.5	46.4	212	29.4	21.7
8740	Normalized	929.4	134.8	606.7	88.0	16.0	47.9	269	17.6	13.0
	Annealed	695.0	100.8	415.8	60.3	22.2	46.4	201	40.0	29.5
9255	Normalized	932.9	135.3	579.2	84.0	19.7	43.4	269	13.6	10.0
	Annealed	774.3	112.3	486.1	70.5	21.7	41.1	229	8.8	6.5

[a]All grades are fine-grained except for those in the 1100 series, which are coarse-grained. Heat-treated specimens were oil quenched unless otherwise indicated.

Note: Values tabulated are approximate median expectations for 1-in. round sections. Individual test results may differ considerably.

Source: ASM Metals Reference Book, American Society for Metals, Metals Park, Ohio, 1981.

APPENDIX C-4b
Typical Uses of Plain Carbon Steels

Carbon (%)	Typical Uses
0.05–0.10	Stampings, rivets, wire, cold-drawn parts
0.10–0.20	Structural shapes, machine parts, carburized parts
0.20–0.30	Gears, shafts, levers, cold-forged parts, welded tubing, carburized parts
0.30–0.40	Shafts, gears, connecting rods, crane hooks, seamless tubing (This and higher hardnesses can be heat-treated.)
0.40–0.50	Gears, shafts, screws, forgings
0.60–0.70	Hard-drawn spring wire, lock washers, locomotive tires
0.70–0.90	Plow shares, shovels, leaf springs, hand tools
0.90–1.20	Springs, knives, drills, taps, milling cutters
1.20–1.40	Files, knives, razors, saws, wire-drawing dies

APPENDIX C-5a
Properties of Some Water-Quenched and Tempered Steels

Steel	Dia. Treated (in.)	Dia. Tested (in.)	Normaliz. Temp. (°F)	Reheat Temp. (°F)	As Quenched, H_B
1030	1.0	0.505	1700	1600	514
1040	1.0	0.505	1650	1550	534
1050	1.0	0.505	1650	1525	601
1095	1.0	0.505	1650	1450	601
4130	0.53	0.505	1600	1575	495

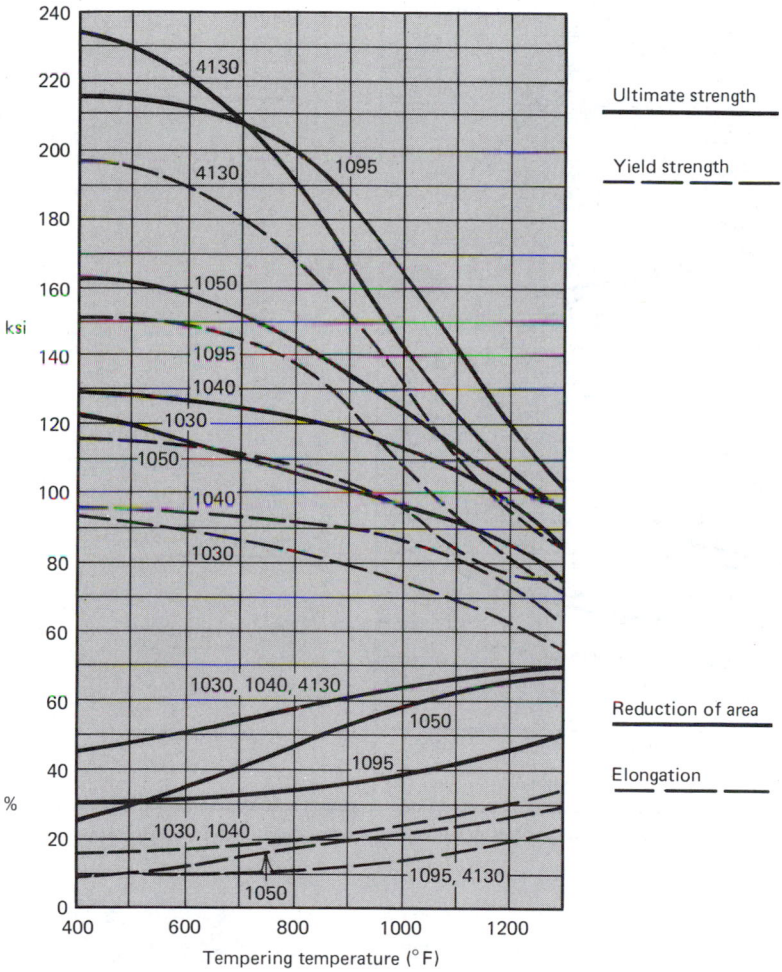

Source: Modern Steels and Their Properties, Bethlehem Steel Corp., Bethlehem, Pennsylvania, 1972.

APPENDIX C-5b
Properties of Some Oil-Quenched and Tempered Carbon Steels

Steel	Dia. Treated (in.)	Dia. Tested (in.)	Normaliz. Temp. (°F)	Reheat Temp. (°F)	As Quenched, H_B
1040	1.0	0.505	1650	1575	269
1050	1.0	0.505	1650	1550	321
1095	1.0	0.505	1650	1475	401

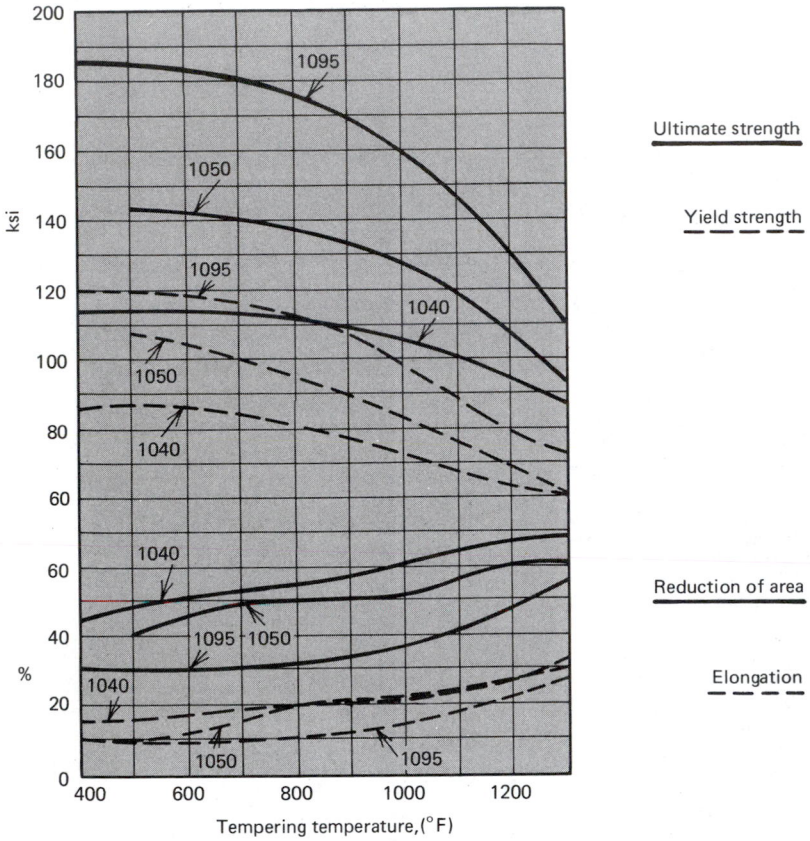

Source: Modern Steels and Their Properties, Bethlehem Steel Corp., Bethlehem, Pennsylvania, 1972.

APPENDIX C-5c
Properties of Some Oil-Quenched and Tempered Alloy Steels

Steel	Dia. Treated (in.)	Dia. Tested (in.)	Normaliz. Temp. (°F)	Reheat Temp. (°F)	As Quenched, H_B
4140	0.53	0.505	1600	1525	555
4340	0.53	0.505	1600	1550	601
9255	1.0	0.505	1650	1625	653

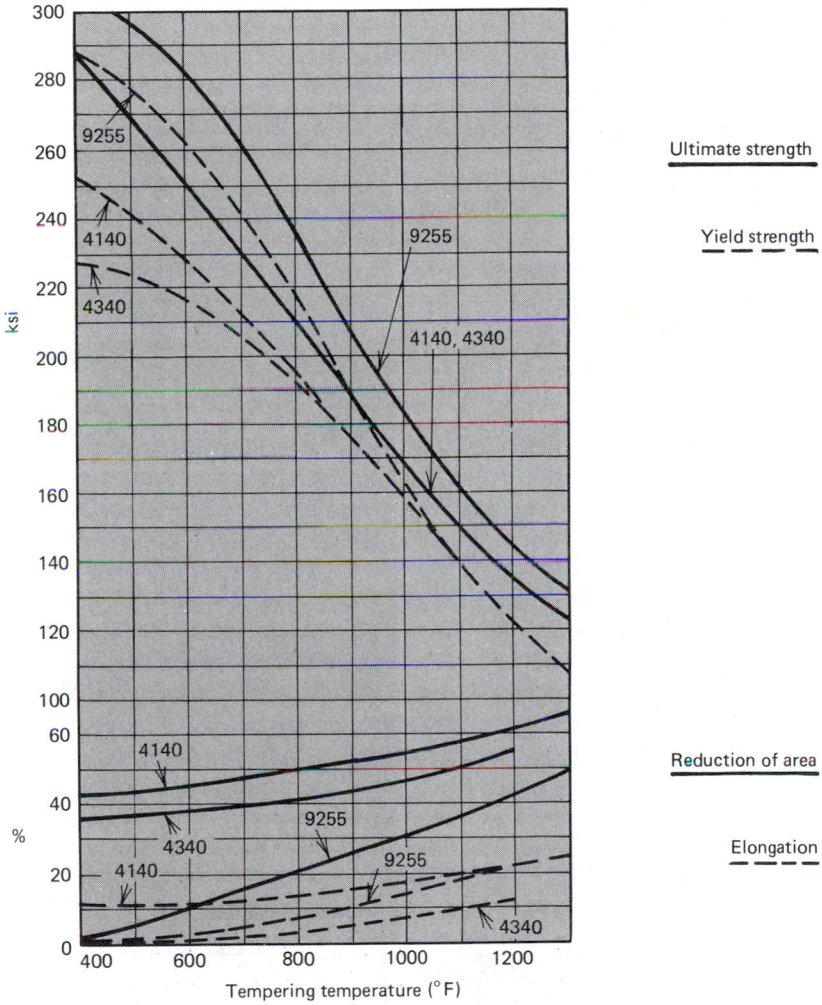

Source: Modern Steels and Their Properties, Bethlehem Steel Corp., Bethlehem, Pennsylvania, 1972.

APPENDIX C-6
Effect of Mass on Strength Properties of Steel

All specimens oil-quenched and tempered at 1000°F (538°C)

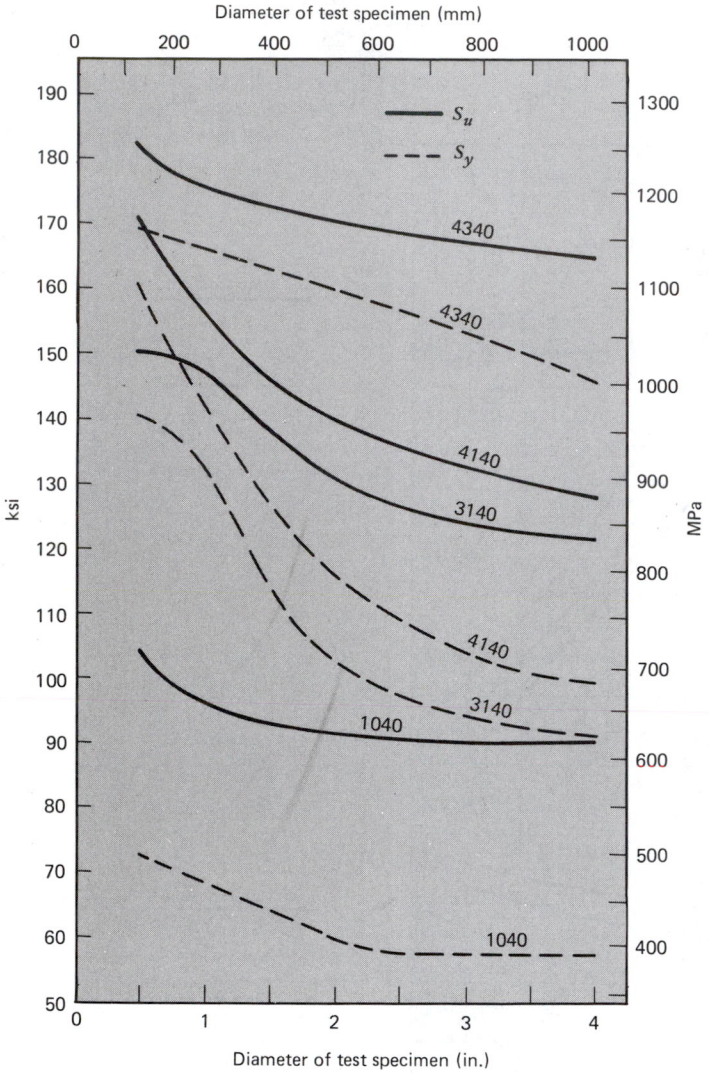

Source: Modern Steels and Their Properties, Bethlehem Steel Corp., Bethlehem, Pennsylvania, 1972.

APPENDIX C-7
Mechanical Properties of Some Carburizing Steels

	Core									Case		
		Tensile Strength				Ductility		Impact Strength Izod			Thickness	
		Ultimate, S_u		Yield, S_y								
Steel AISI	Hardness, H_B	ksi	Mpa	ksi	Mpa	Elongation in 2 in. (%)	Reduction of Area (%)	ft-lb	J	Hardness, R_C	in.	mm
1015[a]	149	73	503	46	317	32	71	91	123	62	.048	1.22
1022[a]	163	82	565	47	324	27	66	81	110	62	.046	1.17
1117[a]	192	96	662	59	407	23	53	33	45	65	.045	1.14
1118[a]	229	113	779	76	524	17	45	16	22	61	.065	1.65
4320[b]	293	146	1006	94	648	22	56	48	65	59	.075	1.91
4620[b]	235	115	793	77	531	22	62	78	106	59	.060	1.52
8620[b]	262	130	896	77	531	22	52	66	89	61	.070	1.78
E9310[b]	352	169	1165	138	952	15	62	63	85	58	.055	1.40

[a]1 -in. round section treated, 0.505-in. round section tested. Single quench in water, tempered 350°F (177°C).
[b]0.565-in. round section treated, 0.505-in. round section tested. Double quench in oil, tempered 450°F (232°C). (Tempering at 300°F gives higher case hardness but less core toughness.)

Note: Values tabulated are approximate median expectations.
Source: Modern Steels and Their Properties, Bethlehem Steel Corp., 4th ed., 1958, and 7th ed., 1972.

APPENDIX C-8 Mechanical Properties of Some Wrought Stainless Steels (Approximate Median Expectations)

AISI Type	Ult. Strength, S_u (ksi)			Yield Strength, S_y (ksi)			Elongation (%)			Izod Impact (ft-lb)			Drawability	Machinability	Weldability	Typical Uses
	An.	CW	H&T	An.	CW	H&T	An.	CW	H&T	An.	CW	H&T				
Austenitic																
302	85	110		35	75		60	35		110	90		VG	P	G	General purpose; springs
303	90	110		35	80		50	22		85	35		G	G	P	Bolts, nuts, rivets, aircraft fittings
304	85	110		35	75		60	55		110	90		VG	P	G	General purpose; welded construction
310, 310S	95			45			50			110			G	P	G	Turbine, furnace, heat exchanger parts
347, 348	90	110		35	65		50	40		110			VG	P	G	Jet engine, nuclear energy parts
384 (wire)	75			35			55						E			Severely cold-worked parts; fasteners
Martensitic																
410	75	105	115	40	85	85	35	17	23	90	75	80	F	F−	F	Machine parts, shafts, bolts, cutlery
414	115	130[a]	160	90	110[a]	125	20	15[a]	17	50		45		F	F	Machine parts, springs, bolts, cutlery
416, 416Se	75	100[b]	110	40	85[b]	85	30	13[b]	18	70	20[b]	25	P	G	P	Cutlery, fasteners, tools, screw mach. parts
431	125	130[a]	165	95	110[a]	125	20	15[a]	17	50		40		P−	F	High strength bolts, aircraft fittings
440 A,B,C	105	115[a]	260	60	90[a]	240	14	7[a]	3	2	2[a]	2		VP	P	Balls, bearing parts, nozzles, cutlery (highest H&T hardness of any stainless)
Ferritic																
430, 430F	75	83		43	63		27	20					G	F–G	F	Decorative trim, mufflers, screw mach. parts
446	83	85		53	70		23	20		2			P	F	F	Parts subjected to high-temperature corrosion

[a]Annealed and cold-drawn. [b]Tempered and cold-drawn.

Note: An., CW, H&T mean annealed, cold-worked, and hardened & tempered, respectively.
E, VG, G, F, P, VP mean excellent, very good, good, fair, poor, very poor, respectively.
Sources: Metal Progress Databook 1980, Am. Society for Metals, Metals Park, Ohio, Vol. 118, No. 1 (mid-June 1980); *ASME Handbook Metal Properties,* McGraw-Hill, New York, 1954; 1981 Materials Selector Issue, *Materials Engineering,* Penton/IPC, Cleveland, Vol. 92, No. 6 (Dec. 1980); *Machine Design,* 1981 Matls. Ref. Issue, Penton/IPC, Cleveland, Ohio, Vol. 53, No. 6 (March 19, 1981).

APPENDIX C-9
Mechanical Properties of Some Iron-Based Superalloys

AISI Grade	Ult. Strength, S_u (ksi)		Yield Strength, S_y (ksi)		Elongation (%)		Rupture Strength 100 hr @ 1000°F (ksi)	Creep Strength 0.0001%/hr @ 1000°F (ksi)	Charpy Impact Strength @ 70°F (ft-lb)
	70°F	1000°F	70°F	1000°F	70°F	1000°F			
Martensitic									
604 (Chromalloy)	125–138	110	95–108	85	7	—	75	—	—
610 (H-11)	135–310	180	100–240	140	3–17	11	95–115	—	10–32
Austenitic									
635 (Stainless W)	220–225	75–80	215–290	37–50	1–5	47–58	32	—	4–106
650 (16-12-G)	110–140	90	50–100	33	20–45	58	78	26	15
653 (17-24 CuMo)	86–112	65	40–90	29	30–45	37	48	10	8–26
665 (W-545)	176–187	154	123–142	120	19	13	120	—	—

Note: Values tabulated are approximate median expectations.
Source: Machine Design, 1981 Materials Reference Issue, Penton/IPC, Cleveland, Ohio, Vol. 53, No. 6 (March 19, 1981).

APPENDIX C-10
Mechanical Properties, Characteristics, and Typical Uses of Some Wrought Aluminum Alloys

Alloy	Hard., H_B	Tensile Strength: Ultimate, S_u ksi	Ultimate, S_u Mpa	Yield, S_y ksi	Yield, S_y Mpa	Elong. in 2 in. (%)	Corr. Resist.	Cold Work	Mach-ine	Braze	Gas Weld	Arc Weld	Resist. Weld	Typical Uses
1100-0	23	13	90	5	34	45	A	A	E	A	A	A	B	Spinnings, drawn shapes, heat exchangers, cooking utensils, tanks
-H14	32	18	125	17	115	20	A	A	D	A	A	A	A	
-H18	44	24	165	22	150	15	A	B	D	A	A	A	A	
2011-T3	95	55	380	43	295	15	D	C	A	D	D	D	D	Screw machine products
-T8	100	59	405	45	310	15	D	D	A	D	D	D	D	
2014-0	45	27	185	14	97	18	—	—	D	D	D	D	B	Heavy duty forgings, aircraft structures and fittings, truck frames
-T4	105	62	425	42	290	20	D	C	B	D	D	B	B	
-T6	135	70	485	60	415	13	D	D	B	D	D	B	B	
2024-0	47	27	185	11	76	22	—	—	D	D	D	D	D	Aircraft structures, truck wheels, screw mach. products
-T4	120	68	470	47	325	19	D	C	B	D	C	B	B	
6061-0	30	18	125	8	55	30	B	A	D	A	A	A	B	Boats, rail cars, pipe, flanges, trailers
-T6	95	45	310	40	275	17	B	C	C	A	A	A	A	
6063-0	25	13	90	7	48	—	A	A	—	A	A	A	A	Furniture tube, doors, windows, pipe, fuel tanks
-T6	73	35	240	31	215	12	A	C	C	A	A	A	A	
7075-0	60	38	230	15	105	16	—	—	D	D	D	C	B	Aircraft structures and skins, skis, railings
-T6	150	83	570	73	505	11	C	D	B	D	D	C	B	

Note: Values are approximate median expectations for sizes about ½ in. H_B values were obtained from 500-kg load and 10-mm ball. Letters A, B, C, D indicate relative ratings in decreasing order of merit.

Source: ASM Metals Reference Book, American Society for Metals, Metals Park, Ohio, 1981.

APPENDIX C-11
Tensile Properties, Characteristics, and Typical Uses of Some Cast Aluminum Alloys

Alloy	Casting Type	Tensile Strength: Ultimate, S_u MPa	Tensile Strength: Ultimate, S_u ksi	Tensile Strength: Yield, S_y MPa	Tensile Strength: Yield, S_y ksi	Elongation (%)	Corrosion resistance	Machining	Weldability	Anodized appearance	Typical Uses
201-T4	sand	365	53	215	31	20.					Aircraft components
-T6	sand	485	70	435	63	7.					
208-F	sand	145	21	97	14	2.5	4	3	2	3	Manifolds, valve bodies, pressure tight parts
295-T4	sand	220	32	110	16	8.5	4	3	2	2	Crankcases, wheels, housings, spring hangers, fittings
-T6	sand	250	36	165	24	5.0					
355-T6	sand	240	35	175	25	3.0	3	3	1	4	Cylinder heads, water jackets, housings, impellers, timing gears, meter parts
-T6	perm mold	290	42	190	27	4.0					
356-T6	sand	230	33	165	24	3.5	2	3	1	4	Automotive housings, aircraft and marine fittings, general purpose castings
-T6	perm mold	265	38	185	27	5.0					
A390-F	sand	180	26	180	26	<1.0					Automotive engine blocks, pumps, pulleys, brake shoes
-T6	sand	280	40	280	40	<1.0					
-F	perm mold	200	29	200	29	<1.0					
-T6	perm mold	310	45	310	45	<1.0					
520-T4	sand	330	48	180	26	16.	1	1	4	1	Aircraft fittings, levers, brackets, parts requiring shock resistance

Note: Values are approximate median expectations for sizes about ½ in. Characteristics are comparably rated from 1 to 5; 1 is the highest or best possible rating.
Sources: ASM Metals Reference Book, American Society for Metals, Metals Park, Ohio, 1981. *1981 Materials Selector,* Materials Engineering, Penton/IPC, Cleveland, Vol. 92, No. 6 (Dec. 1980).

APPENDIX C-12
Temper Designations for Aluminum and Magnesium Alloys

Temper	Process
F	As cast.
0	Annealed.
Hxx	Strain-hardened. First digit indicates the specific combination of operations, second digit indicates the degree of strain-hardening; thus, H18 indicates a greater degree of hardening than does H14 or H24.
T3	Solution-heat-treated, cold-worked, and naturally aged
T4	Solution-heat-treated and naturally aged
T5	Cooled from an elevated-temperature shaping process and artificially aged
T6	Solution-heat-treated and artificially aged
T8	Solution-heat-treated, cold-worked, and artificially aged

APPENDIX C-13
Mechanical Properties of Some Copper Alloys

	Alloy	UNS Designation	Composition	Tensile Strength: Ultimate, S_u ksi	Tensile Strength: Ultimate, S_u Mpa	Tensile Strength: Yield, S_y ksi	Tensile Strength: Yield, S_y Mpa	Elongation in 2 in. (%)
←Wrought Alloys→	Leaded beryllium copper	C17300		68–200	469–1379	25–178	172–1227	43–3
	Med leaded brass	C34000	(65Cu-34Zn)	50–55	345–379	19–42	131–290	60–40
	Free cutting brass	C36000		49–68	338–469	18–45	124–310	53–18
	Leaded phos bronze	C54400	(88Cu-4Zn)	68–75	469–517	57–63	393–434	20–15
	Aluminum silicon-bronze	C64200	(91Cu-7Al-2Si)	75–102	517–703	35–68	241–469	32–22
	Silicon bronze	C65500	(97Cu-3Si)	58–108	400–745	22–60	152–414	60–13
	Manganese bronze	C67500		65–84	448–579	30–60	207–414	33–19
←Cast Alloys→	Leaded red brass	C83600	(85Cu-5Zn-5Sn-5Pb)	37	255	17	117	30
	Leaded yellow brass	C85200		38	262	13	90	35
	Manganese bronze	C86200		95	655	48	331	20
	Navy M bronze	C92200		40	276	20	138	30
	Leaded Ni-Sn bronze	C92900		47	324	26	179	20
	Bearing bronze	C93200		35	241	18	124	20
	Aluminum bronze	C95400		85–105	586–724	35–54	241–372	18–8
	Copper nickel	C96200	(90Cu-10Ni)	45	310	25	172	20

Note: Values tabulated are approximate median expectations.
Source: Machine Design, 1981 Materials Reference Issue, Penton/IPC, Cleveland, Ohio, Vol. 53, No. 6 (March 19, 1981).

APPENDIX C-14
Mechanical Properties of Some Magnesium Alloys

		Tensile Strength				
		Ultimate, S_u		Yield, S_y		Elongation in 2 in.
Alloy	**Form**	**ksi**	**Mpa**	**ksi**	**Mpa**	(%)
AZ91B-F	Die casting	34	234	23	159	3
AZ31B-F ZK60A-T5	Extrusion	38–53	262–365	28–44	193–303	11–15
AZ31B-F HM21A-T5 AZ80A-T5 ZK60A-T6	Forging	34–50	234–345	22–39	152–269	6–11
AZ31B-H24 HK31A-H24 HM21A-T8	Sheet, plate	33–42	228–290	21–32	145–221	9–21

Note: Values tabulated are approximate median expectations.
Source: Machine Design, 1981 Materials Reference Issue, Penton/IPC, Cleveland, Ohio, Vol. 53, No. 6 (March 19, 1981).

APPENDIX C-15
Mechanical Properties of Some Nickel Alloys

		Tensile Strength				Creep Strength 0.0001 %/hr		Elongation in 2 in.	Impact Strength Notched Charpy	
		Ultimate, S_u		Yield, S_y						
Alloy	Form	ksi	Mpa	ksi	Mpa	ksi	Mpa	(%)	ft-lb	J
Wrought nickel	CD annealed bar	55–80	379–552	15–30	103–207	12	83	55–40	228	309
Duranickel 301	CD annealed bar	90–120	621–827	30–60	207–414			55–35		
	CD aged bar	170–210	1172–1448	125–175	862–1207			25–15		
Monel 400	Annealed bar	70–90	483–621	25–50	173–345	24	165	60–35	216	293
	Hot-rolled bar	80–110	552–758	40–100	276–690	25	172	60–30	219	297
Monel K-500	Aged bar	140–190		110–150		87		30–20	39	53
Hastelloy B[a]	As-cast bar	134	924	67	462			52		
Udimet HX[a]	Sheet (0.109 in.)	114 (70°F)	786	52 (70°F)	359			43 (70°F)		
Unitemp HK[a]		13 (2000°F)	89	8 (2000°F)	55			50 (2000°F)		
Hastelloy X[a]										
Rene 95[a]	Forging	235 (70°F)	1620	190 (70°F)	1310			15 (70°F)		
		225 (1000°F)	1551	182 (1000°F)	1255			13 (1000°F)		
Inconel 600[a]	Annealed bar	96 (70°F)	662	41 (70°F)	283	40 (800°F)	276	45 (70°F)	180	244
		37 (1400°F)	255	25 (1400°F)	172	2.0 (1600°F)	14	68 (1400°F)		
Inconel 625[a]	Annealed bar	140 (70°F)	965	71 (70°F)	490	12 (1400°F)	83	50 (70°F)	49	66
		78 (1400°F)	538	61 (1400°F)	421	3.9 (1600°F)	27	45 (1400°F)		
Inconel X-750[a]	Aged bar	184 (70°F)	1269	126 (70°F)	869	63 (1200°F)	434	25 (70°F)	37	50
		143 (1200°F)	986	110 (1200°F)	758			7 (1200°F)		
Incoloy 800[a]	Annealed bar	87 (70°F)	600	43 (70°F)	296	6.0 (1400°F)	41	44 (70°F)	107	145
		33 (1400°F)	228	23 (1400°F)	159	3.5 (1600°F)	24	84 (1400°F)		

[a]"Superalloys," noted for hi temp strength and corrosion resistance. Used in jet engines, turbines, and furnaces.

Note: Values tabulated are approximate median expectations.

Source: Machine Design, 1981 Materials Reference Issue, Penton/IPC, Cleveland, Ohio, Vol. 53, No. 6 (March 19, 1981).

APPENDIX C-16
Mechanical Properties of Some Wrought Titanium Alloys

		Tensile Strength					Charpy Impact Strength	
		Ultimate, S_u		Yield, S_y		Elongation in 2 in.		
Alloy	**Designation**	**ksi**	**Mpa**	**ksi**	**Mpa**	(%)	**ft-lb**	**J**
Comml. pure alpha Ti	Ti-35A	35	241	25	172	24	11–40	15–54
Comml. pure alpha Ti	Ti-50A	50	345	40	276	20	11–40	15–54
Comml. pure alpha Ti	Ti-65A	65	448	55	379	18	11–40	15–54
Alpha alloy	Ti-0.2Pd	50	345	40	276	20	—	—
Alpha-beta alloy	Ti-6Al-4V	130–160[a]	896–1103[a]	120–150[a]	827–1034[a]	10-7	10–20	14–27
Beta alloy	Ti-3Al-13V-11Cr	135–188[a]	931–1296[a]	130–175[a]	896–1207[a]	16-6	5–15	7–20

[a]Depending on heat treatment.

Note: Values tabulated are approximate median expectations.
Source: Machine Design, 1981 Materials Reference Issue, Penton/IPC, Cleveland, Ohio, Vol. 53, No. 6 (March 19, 1981).

APPENDIX C-17
Mechanical Properties of Some Zinc Casting Alloys

Alloy Designation			Tensile Strength				Elongation in 2 in.	Charpy Impact Strength		Hardness
			Ultimate, S_u		Yield, S_y					
ASTM	SAE	ADCI	ksi	Mpa	ksi	Mpa	(%)	ft-lb	J	H_B
AG40A[a]	903	No. 3	41	283			10	43	58	82
AC41A[a]	925	No. 5	47	324			7	48	65	91
ZA-12										
sand cast			40–45	276–310	30	207	1–3			105–120
perm mold			45–50	310–345	31	214	1–3			105–125
die cast			57	393	46	317	2			110–125

[a]Die cast.

Note: Values tabulated are approximate median expectations.

Sources: Machine Design, 1981 Materials Reference Issue, Penton/IPC, Cleveland, Ohio, Vol. 53, No. 6 (March 19, 1981); *Metal Progress, Databook 1980,* American Society for Metals, Metals Park, Ohio, Vol. 118, No. 1 (mid-June, 1980).

APPENDIX C-18a
Representative Mechanical Properties of a Few Plastics

Plastic	Tensile Strength, S_u ksi	Tensile Strength, S_u Mpa	Elongation in 2 in. (%)	Izod Impact Strength ft-lb	Izod Impact Strength J	Friction Coefficient With Self	Friction Coefficient With Steel
ABS (general purpose)	6	41	5–20	6.5	8.8		
Acrylic (standard molding)	10.5	72	6	0.4	0.5		
Cellulosic (cellulose acetate)	2–7	14–48		1–7	1.4–9.5		
Epoxy (glass filled)	10–20	69–138	4	2–30	2.7–41		
Fluorocarbon (PTFE)	3.4	23	300	3	4.1		0.05
Nylon (6/6)	12	83	60	1	1.4	0.04–0.13	
Phenolic (wood flour filled)	7	48	0.4–0.8	0.3	0.4		
Polycarbonate (general purpose)	9–10.5	62–72	110–125	12–16	16–22	0.52	0.39
Polyester (20–30 percent glass fill)	16–23	110–90	1–3	1.0–1.9	1.4–2.6	0.12–0.22	0.12–0.13
Polypropylene (unmodified resin)	5	34	10-20	0.5–2.2	0.7–3.0		

Note: Values shown are typical; both higher and lower values may be commercially obtainable. Also see Appendix C-19.
Source: Machine Design, 1981 Materials Reference Issue, Penton/IPC, Cleveland, Ohio, Vol. 53, No. 6 (March, 19, 1981); *1981 Materials Selector,* Materials Engineering, Penton/IPC, Cleveland, Ohio, Vol. 92, No. 6 (Dec. 1980).

APPENDIX C-18b
Properties of a Few Unreinforced vs. Glass-Reinforced Thermoplastic Resins

Base Resin ASTM Test ⟶	Tensile Strength (ksi) D638	Flexural Modulus (Mpsi) D790	Izod Impact Strength ft-lb/in. Notched	Izod Impact Strength ft-lb/in. Unnotched D256	Specific Gravity D792	Mold Shrinkage (%) D955	Water Absorption (in 24 hr) D570	Thermal Expansion $10^{-5}/°F$ D696	Deflection Temperature °F (264 psi) D648
ABS	14.5 (6.0)	1.10 (0.32)	1.4 (4.4)	6–7	1.28 (1.05)	0.1 (0.6)	0.14 (0.30)	1.6 (5.3)	220 (195)
Acetal	19.5 (8.8)	1.40 (0.40)	1.8 (1.3)	8–10 (20)	1.63 (1.42)	0.3 (2.0)	0.30 (0.22)	2.2 (4.5)	325 (230)
Fluorocarbon ETFE	14.0 (6.5)	1.10 (0.20)	7.5 (>40)	17–18	1.89 (1.70)	0.3 (1.8)	0.20 (0.02)	1.6 (4.0)	460 (160)
Nylon 6/12	22.0 (8.8)	1.20 (0.295)	2.4 (1.0)	20	1.30 (1.06)	0.4 (1.1)	0.21 (0.25)	1.5 (5.0)	415 (194)
Polycarbonate	18.5 (9.0)	1.20 (0.33)	3.7 (2.7)	17 (60)	1.43 (1.20)	0.1 (0.6)	0.07 (0.15)	1.3 (3.7)	300 (265)
Polyester[a]	19.5 (8.5)	1.40 (0.34)	2.5 (1.2)	16–18	1.52 (1.31)	0.3 (2.0)	0.06 (0.08)	1.2 (5.3)	430 (130)
Polyethylene[b]	10.0 (2.6)	0.90 (0.20)	1.1 (0.4)	8–9	1.17 (0.95)	0.3 (2.0)	0.02 (0.02)	2.7 (6.0)	260 (120)
Polypropylene[c]	9.7 (4.9)	0.55 (0.18)	3.0 (0.4)	11–12	1.12 (0.91)	0.4 (1.8)	0.03 (0.01)	2.0 (4.0)	295 (135)
Polystyrene	13.5 (7.0)	1.30 (0.45)	1.0 (0.45)	2–3	1.28 (1.07)	0.1 (0.4)	0.05 (0.10)	1.9 (3.6)	215 (180)

[a]Polybutylene terephthalate (PBT) resin. [b]High-density (HD). [c]Impact-modified grade.

Note: Values in parentheses pertain to unreinforced resins. Other values are typical of 30% glass-reinforcement formulas. All values shown are typical; both higher and lower values may be commercially obtainable.

Source: Machine Design, 1981 Materials Reference Issue, Penton/IPC, Cleveland, Ohio, Vol. 53, No. 6 (March 19, 1981).

APPENDIX C-18c
Typical Applications of Common Plastics

Application \ Plastic	Thermoplastic																	Thermoset		
	ABS	Acetal	Acrylic	Cellulosics	Fluoroplastics	Nylon	Phenylene oxide	Polycarbonate	Polyester	Polyethylene	Polyimide	Polyphenylene sulfide	Polypropylene	Polystyrene	Polysulfone	Polyurethane	Polyvinyl chloride	Phenolic	Polyester	Polyurethane
Structural, mechanical gears, cams, pistons, rollers, valves, pump impellers, fan blades, rotors, wash-machine agitators		X				X		X	X				X					X		
Light-duty mechanical and decorative knobs, handles, camera cases, pipe fittings, battery cases, auto steering wheels, trim moldings, eyeglass frames, tool handles	X		X	X						X				X	X		X	X		
Small housings and hollow shapes phone and flashlight cases, helmets; housings for power tools, pumps, small appliances	X			X			X	X	X	X				X	X			X	X	
Large housings and hollow shapes boat hulls, large appliance housings, tanks, tubs, ducts, refrigerator liners	(Foam)						(Foam)		(Foam)	(H.D. Foam)				(Foam)	(Foam)		(Foam)		(Glass filled) (Foam)	
Optical and transparent parts safety glasses, lenses, safety and vandal-resistant glazing, snowmobile windshields, signs, refrig. shelves			X	X				X							X	X				
Parts for wear applications gears, bushings, bearings, tracks, chute liners, roller-skate wheels, wear strips		X			X	X			X	(UHMW)			X				X		X	

Source: Machine Design, 1981 Materials Reference Issue, Penton/IPC, Cleveland, Ohio, Vol. 53, No. 6. (March 19, 1981).

APPENDIX D

Shear, Moment, and Deflection Equations for Beams

APPENDIX D-1
Shear, Moment, and Deflection Equations for Cantilever Beams

	Slope at Free End	Maximum Deflection	Deflection at Any Point x
1. Concentrated load at end	$\theta = \dfrac{PL^2}{2EI}$	$\delta_{max} = \dfrac{PL^3}{3EI}$	$\delta = \dfrac{Px^2}{6EI}(3L - x)$
2. Concentrated load at any point	$\theta = \dfrac{Pa^2}{2EI}$	$\delta_{max} = \dfrac{Pa^2}{6EI}(3L - a)$	For $0 \leq x \leq a$: $\delta = \dfrac{Px^2}{6EI}(3a - x)$ For $a \leq x \leq L$: $\delta = \dfrac{Pa^2}{6EI}(3x - a)$

3. Uniform load

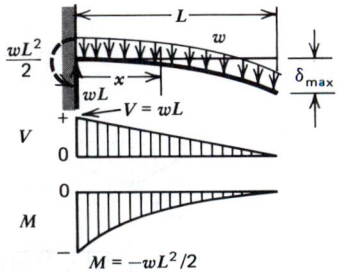

$$\theta = \frac{wL^3}{6EI} \qquad \delta_{max} = \frac{wL^4}{8EI} \qquad \delta = \frac{wx^2}{24EI}(x^2 + 6L^2 - 4Lx)$$

4. Moment load at free end

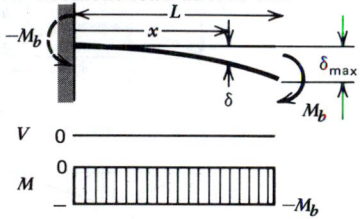

$$\theta = \frac{M_b L}{EI} \qquad \delta_{max} = \frac{M_b L^2}{2EI} \qquad \delta = \frac{M_b x^2}{2EI}$$

APPENDIX D-2
Shear, Moment, and Deflection Equations for Simply Supported Beams

	Slope at Ends, θ	Maximum Deflection, δ_{max}	Deflection at Any Point x, δ
1. Concentrated center load 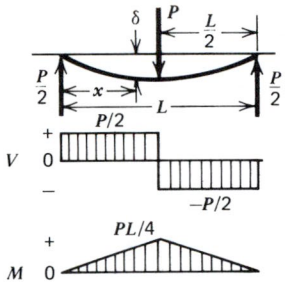	$\dfrac{PL^2}{16EI}$	At center: $\dfrac{PL^3}{48EI}$	For $0 \leq x \leq L/2$: $\dfrac{Px}{12EI}\left(\dfrac{3L^2}{4} - x^2\right)$
2. Concentrated load at any point 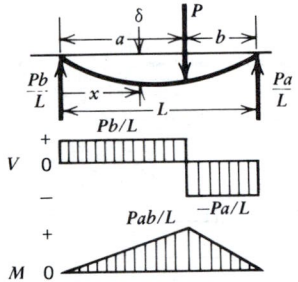	At left end: $\dfrac{Pb(L^2 - b^2)}{6LEI}$	At $x = \sqrt{\dfrac{L^2 - b^2}{3}}$: $\dfrac{Pb(L^2 - b^2)^{3/2}}{9\sqrt{3}LEI}$	For $0 \leq x \leq a$: $\dfrac{Pbx}{6LEI}(L^2 - x^2 - b^2)$

	Slope at Ends, θ	Maximum Deflection, δ_{max}	Deflection at Any Point x, δ
3. Uniform load 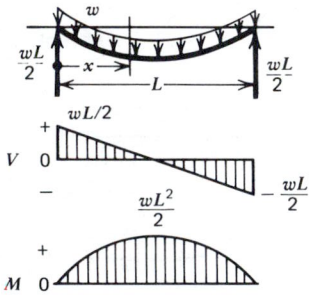	$\frac{wL^3}{24EI}$	$\frac{5wL^4}{384EI}$	$\frac{wx}{24EI}(L^3 - 2Lx^2 + x^3)$
4. Overhung load 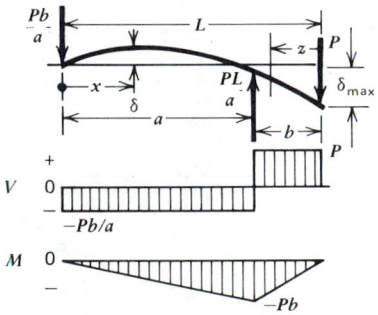	At left supt: $\frac{Pab}{6EI}$ At right supt: $\frac{Pab}{3EI}$ At load: $\frac{Pb}{6EI}(2L + b)$	$\delta_{max} = \frac{Pb^2L}{3EI}$	For $0 \leqq x \leqq a$: $\frac{Pbx}{6aEI}(x^2 - a^2)$ For $0 \leqq z \leqq b$: $\frac{P}{6EI}[z^3 - b(2L + b)z + 2b^2L]$

(*continued*)

APPENDIX D-2 *(continued)*
Shear, Moment, and Deflection Equations for Simply Supported Beams

	Slope at Ends, θ	Maximum Deflection, δ_{max}	Deflection at Any Point x, δ
5. Moment load between supports 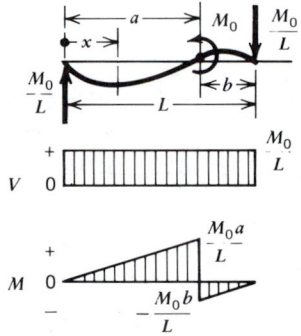	At left supt: $\frac{-M_0}{6EIL}(2L^2 - 6aL + 3a^2)$ At load: $\frac{M_0}{EI}\left(\frac{L}{3} + \frac{a^2}{L} - a\right)$ At right supt: $\frac{M_0}{6EIL}(L^2 - 3a^2)$	At load: $\frac{M_{0a}}{3EIL}(2a^2 - 3aL + L^2)$	For $0 \leqq x \leqq a$: $\frac{M_0 x}{6EIL}(x^2 + 3a^2 - 6aL + 2L^2)$
6. Overhung moment load 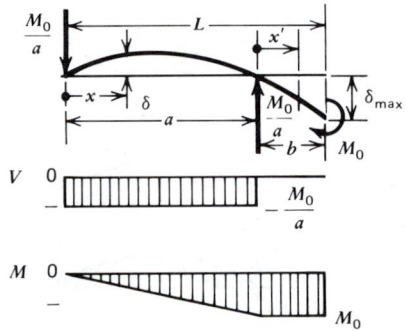	At left supt: $\frac{M_0 a}{6EI}$ At right supt: $\frac{M_0 a}{3EI}$ At load: $\frac{M_0(a + 3b)}{3EI}$	$\delta_{max} = \frac{M_0 b}{6EI}(2L + b)$	For $0 \leqq x \leqq a$: $-\frac{M_0 x}{6aEI}(a^2 - x^2)$ For $0 \leqq x' \leqq b$: $\frac{M_0}{6EI}(2ax' + 3x'^2)$

APPENDIX D-3
Shear, Moment, and Deflection Equations for Beams with Fixed Ends

	Deflection, δ	Deflection, δ, at Any Point, x
1. Concentrated center load	At center: $\delta_{max} = \dfrac{PL^3}{192EI}$	For $0 \leqq x \leqq L/2$: $\dfrac{Px^2}{48EI}(3L - 4x)$
2. Concentrated load at any point	At load: $\delta = \dfrac{Pb^3a^3}{3EIL^3}$	For $0 \leqq a$: $\dfrac{Pb^2x^2}{6EIL^3}[3aL - (3a + b)x]$
3. Uniform load	At center: $\delta_{max} = \dfrac{wL^4}{384EI}$	$\dfrac{wx^2}{24EI}(L - x)^2$

APPENDIX E

Fits and Tolerances

Fits between parts (as a cylindrical member fitting in a cylindrical hole) influence the accuracy of relative positioning of the members, the ease with which the members can be assembled and disassembled, the ease with which they can slide with respect to each other (with clearance fits), and the load they can sustain without relative motion (with interference fits). Differential thermal expansion is often a factor to be considered in determining appropriate fits. The *tolerance* (permitted variation) applied to each dimension influences both function and cost. Unnecessarily small tolerances are an important factor causing excessive cost.

Fits and tolerances are usually specified on the basis of experience with the specific application involved. USAS (ANSI) Standard B4.1-1967 contains detailed recommendations that serve as a valuable guide. Its original predecessor, ASA B4a-1925, provides a simpler guide that is helpful for obtaining an introductory understanding of the subject. It is summarized in the table below. Classes 1–4 are *clearance fits,* Classes 7 and 8 are *interference fits,* and Classes 5 and 6 are known as *transitional fits* because they can be either clearance or interference fits, depending on the random assembly of the parts.

The table illustrates the *basic hole system,* wherein the minimum hole size is chosen as the standard nominal dimension for all classes of fits.

d = nominal diameter
h = hole diameter tolerance = $C_h \sqrt[3]{d}$
s = shaft diameter tolerance = $C_s \sqrt[3]{d}$
a = allowance (minimum diametral clearance, obtained with maximum shaft and minimum bore dimensions) = $C_a \sqrt[3]{d^2}$
i = average interference, obtained with average shaft and bore diameters = $C_i d$

The bar graphs are to scale for d = 25 mm or 1 in.

APPENDIX E

Class of fit	1	2	3	4	5	6	7	8
Bar graph (basic hole system)	*h* hole; *a*; *s* shaft	*h*; *a*; *s*	*h*; *a*; *s*	*h*; *s*	*h*; *s*	*h*; *s*	*h*; *s*	*h*; *s*
	Loose fit	Free fit	Medium fit	Snug fit	Wringing fit	Tight fit	Medium force fit	Heavy force and shrink fit
C_h	.0216 (.0025)	.0112 (.0013)	.0069 (.0008)	.0052 (.0006)	.0052 (.0006)	.0052 (.0006)	.0052 (.0006)	.0052 (.0006)
C_s	.0216 (.0025)	.0112 (.0013)	.0069 (.0008)	.0035 (.0004)	.0035 (.0004)	.0052 (.0006)	.0052 (.0006)	.0052 (.0006)
C_a	.0073 (.0025)	.0041 (.0014)	.0026 (.0009)	0 (0)				
C_f					0 (0)	.00025 (.00025)	.0005 (.0005)	.0010 (.0010)

Note: Numbers in the table are for use with all dimensions in *millimeters*, except for those in parentheses, which are for use with *inches*.

Problems

CHAPTER 1 – MECHANICAL ENGINEERING DESIGN IN BROAD PERSPECTIVE

Sections 1.2-1.5

1.1 From your own experience and observation, describe briefly (perhaps 1 or 2 typed pages, double spaced) a specific example of mechanical engineering design that you regard as *excellent* from a *safety* standpoint. (Preferably, choose an example reflecting your own observation and safety-consciousness rather than one featured in the news media.) Your write-up should reflect the professional appearance expected of an engineer. Use illustrations if and where appropriate.

1.2 Repeat Problem 1.1 except describe an example of *poor* mechanical engineering design from the standpoint of safety.

1.3 Repeat Problem 1.1 except describe a *good* example of mechanical engineering design from an *ecological* standpoint.

1.4 Repeat Problem 1.2 except describe a *poor* example of mechanical engineering design from an ecological standpoint.

1.5 Repeat Problem 1.1 except describe a *good* example of mechanical engineering design from a *sociological* standpoint. (This should represent the kind of engineering activity that you would feel proud to be associated with.)

1.6 Repeat Problem 1.5 except describe what you regard as a *questionable* mechanical engineering design from a sociological standpoint.

1.7 Describe briefly a mechanical engineering design wherein you recognize compromises that had to be made among the various considerations discussed in Sections 1.2–1.5.

Sections 1.6-1.8

1.8 What is the rate of work output of a press that delivers 100 strokes per minute, each stroke providing a force of 2000 lb throughout a distance of ½ in.? If the press efficiency is 90%, what average torque must be provided by a 1750 rpm driving motor?

1.9 The crankshaft of a single-cylinder air compressor rotates 1800 rpm. The piston area is 2000 mm^2, and the piston stroke is 50 mm. Assume a simple "idealized" case where the average gas pressure acting on the piston during the compression stroke is 1 MPa, and pressure during the intake stroke is negligible. The compressor is 80% efficient. A flywheel provides adequate control of speed fluctuation.

(a) What motor power (kW) is required to drive the crankshaft?
(b) What torque is transmitted through the crankshaft?
(Ans.: 3.75 kW, 19.9 N·m)

1.10 A punch press with flywheel adequate to minimize speed fluctuations produces 120 punching strokes per minute, each providing an average force of 2000 N over a stroke of 50 mm. The press is driven through a gear reducer by a shaft rotating 300 rpm. Overall efficiency is 80%.

(a) What power (watts) is transmitted through the shaft?
(b) What average torque is applied to the shaft?
(Ans.: 250 W, 8.0 N·m)

1.11 Repeat Problem 1.10 except change the punch force to 10,000 N, stroke to 50 mm, and drive shaft speed to 900 rpm.

1.12 An 1800 rpm motor drives a camshaft 360 rpm by means of a belt drive. During each revolution of the cam, a follower rises and falls 20 mm. During each follower up-stroke, the follower resists a constant force of 500 N. During the down-strokes, the force is negligible. The inertia of the rotating parts (including a small flywheel) provides adequate speed uniformity. Neglecting friction, what motor power is required? You should be able to get the answer in three ways: by evaluating power at the (1) motor shaft, (2) camshaft, and (3) follower.
(Ans.: 60 W)

1.13 Repeat Problem 1.12 except work with English units and use a follower displacement of 1 in. and a follower force (during the up-stroke) of 100 lb.

1.14 The crankshaft of a small punch press rotates 100 rpm, with the shaft torque fluctuating between 0 and 1000 N·m in accordance with curve A of Fig. P1.14. The press is driven (through a gear reducer) by a 1200 rpm motor. Neglecting friction losses, what motor power would theoretically be required:

(a) With a flywheel adequate to minimize speed fluctuations.
(b) With no flywheel.
[Ans.: (a) 2618 W, (b) 10,472 W]

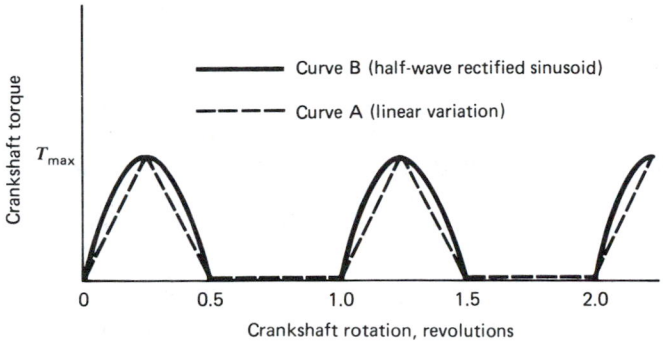

FIGURE P1.14

1.15 Repeat Problem 1.14 except use curve B of Fig. P1.14.

1.16 A piston-type air compressor with an inlet pressure of 1 atmosphere (100 kPa) rotates 1800 rpm and delivers 0.1 m^3/min of air compressed to a gage pressure of 4 atmospheres. For simplicity, assume that the compression is isothermal, and that the compressor efficiency is 50%.

(a) What average torque must be supplied to the crankshaft?
(b) Assuming an adequate flywheel, what motor power is required?
(Ans.: 14.24 N·m, 2.62 kW)

1.17 Assume in Problem 1.16 that the rotating inertia of the crankshaft, piston, connecting rod, and flywheel all taken together is equivalent to a steel flywheel of 0.3 m diameter (rotating at crankshaft speed) with proportions as shown in Fig. 1.7, and assume that the instantaneous torque requirement varies as in curve A of Fig. P1.14. If 1800 rpm is the maximum rotating speed, what is the minimum rotating speed?

1.18 Repeat Problem 1.17 except assume that instantaneous torque varies as in curve B of Fig. P1.14.
(Ans.: 1789 rpm)

1.19 A steel flywheel has the proportions shown in Fig. 1.7. The hub and arms add 10% to the inertia of the rim. How much energy does the flywheel give up in slowing from 1800 to 1700 rpm:

(a) If diameter, d, is 500 mm (use SI units)?
(b) If diameter, d, is 12 in. (use English units)?

1.20 How great a reduction in the Fig. 1.8 "road-load" horsepower requirement would be necessary to enable a car with an "ideal" transmission to get 30 miles per gallon at 70 mph? (Assume that the engine would have a minimum brake specific fuel consumption of 0.45 lb/hp-hr, as shown in Fig. 1.10.)
(Ans.: About 25%)

1.21 How steep a grade can be climbed by the automobile in Sample Problem 1.6 (with a 2.64 axle ratio) while maintaining a constant 55 mph:

(a) With transmission in direct drive?
(b) With a transmission reduction ratio of 1.6?

1.22 How great a speed can the car in Sample Problem 1.6 (with 2.64 axle ratio) maintain when going up a continuous 10% grade (1-ft rise in 10-ft horizontal travel)?

CHAPTER 2 – LOAD ANALYSIS

Section 2.2

2.1 The drawing (Fig. P2.1) shows an exploded view of a 1200 rpm motor, a gear box, and a 6000 rpm blower. The gear box weighs 20 pounds, with center of gravity midway between the two mountings. All shafts rotate counterclockwise, viewed from the blower. Neglecting friction losses, determine all loads acting on the gear box when the motor output is 1 hp. Sketch the gear box as a free-body in equilibrium.

2.2 The drawing (Fig. P2.2) shows an electric fan supported by mountings at A and B. The motor delivers a torque 1 N·m to the fan blades. They, in turn, push the air forward with a force of 10 N. Neglecting gravity forces, determine all loads acting on the fan (complete assembly). Sketch it as a free-body in equilibrium.

2.3 Figure P2.3 shows an exploded drawing of a pump driven by a 1-kW, 1800-rpm motor integrally attached to a 4:1 ratio gear reducer. Reducer shaft C is directly connected to pump shaft C' through a flexible coupling (not shown). Face A of the reducer housing is bolted to flange A' of the connecting tube (a one-piece solid unit). Pump face B is similarly attached to flange B'. Sketch the connecting tube and show all loads acting on it. (Neglect gravity.)

2.4 Figure P2.4 shows an exploded view of an airplane engine, reduction gear, and propeller. The engine and propeller rotate clockwise, viewed from the propeller end. The reduction gear housing is bolted to the engine housing through the bolt holes shown. Neglect friction losses in the reduction gear. When the engine develops 137 hp at 3600 rpm:

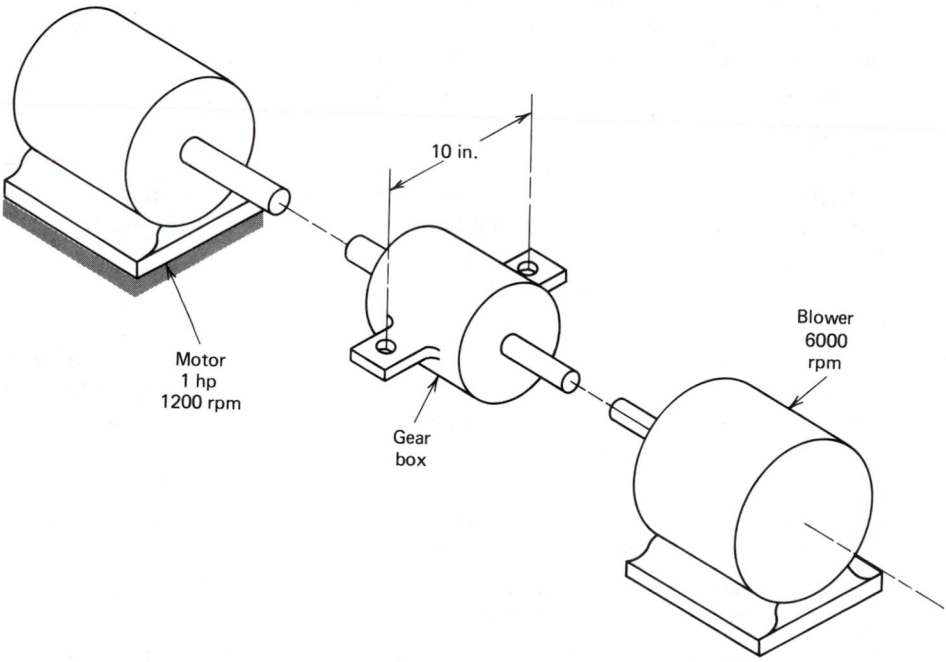

FIGURE P2.1

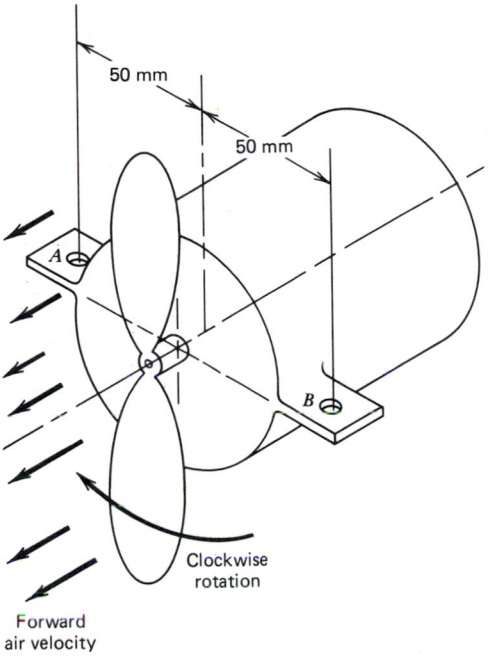

FIGURE P2.2

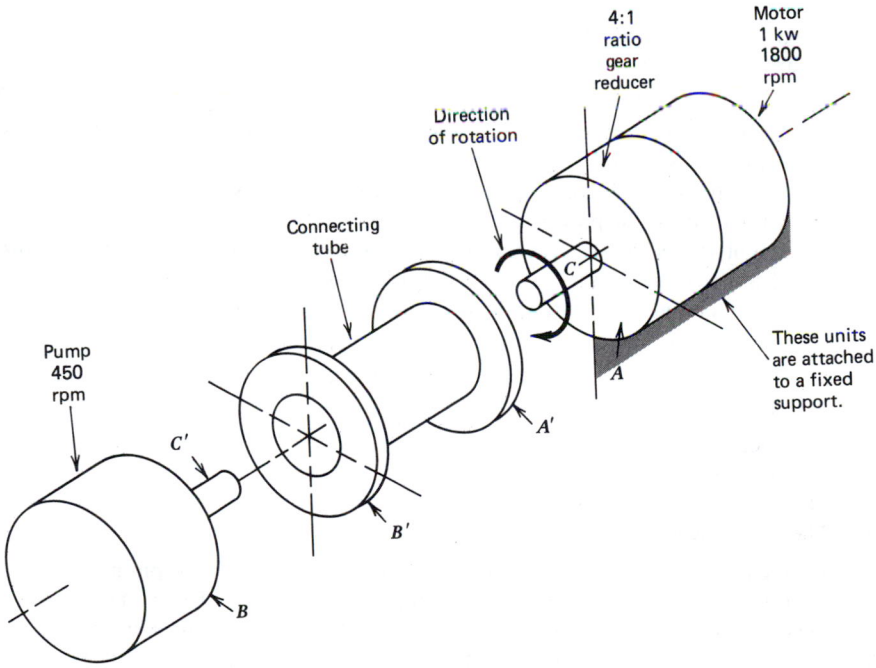

FIGURE P2.3

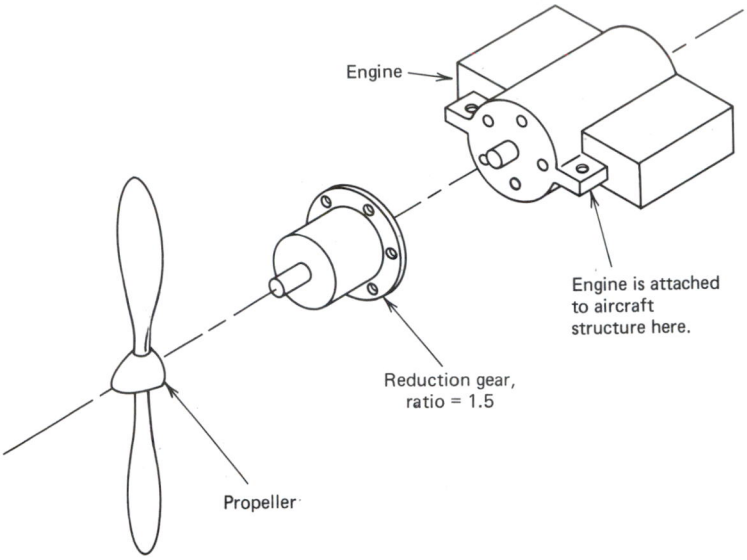

FIGURE P2.4

(a) What is the direction and magnitude of the torque applied *to* the engine housing *by* the reduction gear housing?
(b) What is the magnitude and direction of the torque reaction tending to rotate (roll) the aircraft?
(c) What is an advantage of using opposite-rotating engines with twin-engine propeller-driven aircraft?

[Ans.: (a) 100 lb-ft ccw, (b) 300 lb-ft ccw]

2.5 Figure P2.5 shows the gear reduction unit and propeller of an outboard motor boat. It is attached to the boat structure at the mounting flange at the top. The motor is mounted above this unit, and turns the vertical shaft with a torque of 15 N·m. By means of bevel gearing, this shaft turns the propeller at half the vertical shaft speed. The propeller provides a thrust of 400 N to drive the boat forward. Neglecting gravity and friction, show all external loads acting on the assembly shown. (Make a sketch, and show moments applied to the mounting flange using the notation suggested in the drawing.)

2.6 The drawing (Fig. P2.6) represents a bicycle with a 700-N rider applying full weight to one pedal. Treat this as a two-dimensional problem, with all components in the plane of the paper. Draw as free-bodies in equilibrium:

(a) The pedal, crank, and pedal sprocket assembly.
(b) The rear wheel and sprocket assembly.
(c) The front wheel.
(d) The entire bicycle and rider assembly.

2.7 The drawing (Fig. P2.7) shows a bevel gear reducer driven by an 1800-rpm motor delivering a torque of 10 N·m. The output drives a 600-rpm load. The reducer is held in place by vertical forces applied at mountings *A*, *B*, *C*, and *D*. Torque reaction about the motor shaft is reacted at *A* and *B*; torque reaction about the output shaft is reacted at *C* and *D*. Determine the forces applied to the reducer at each of the mountings:

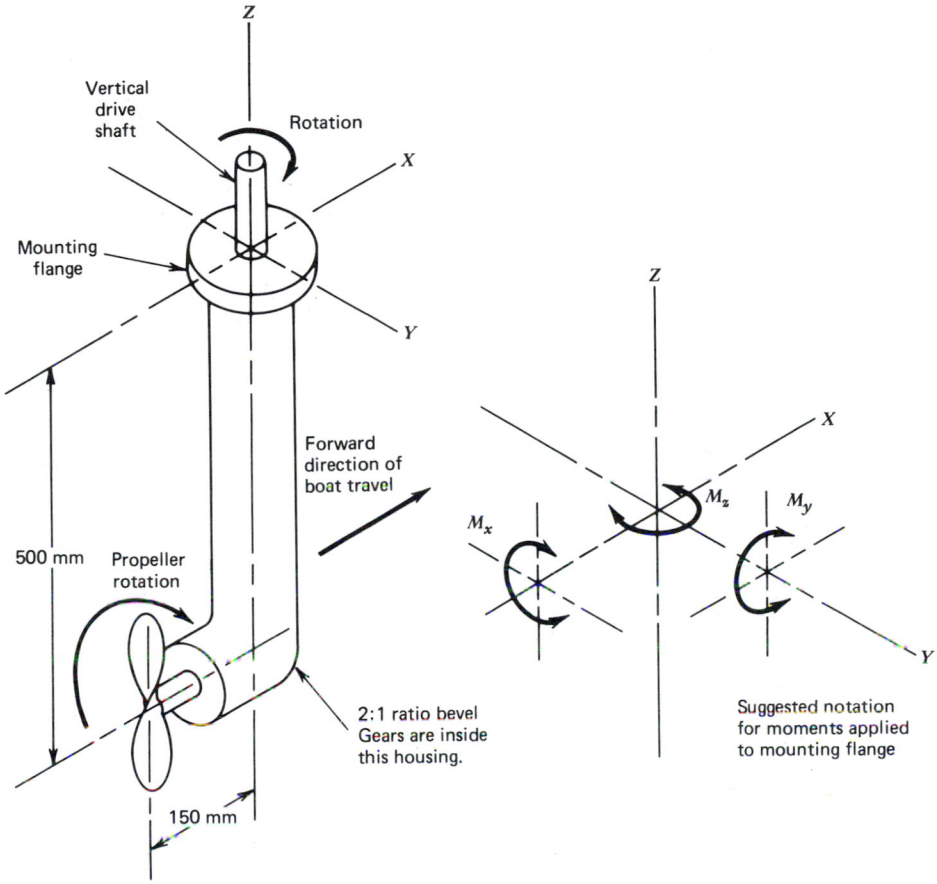

FIGURE P2.5

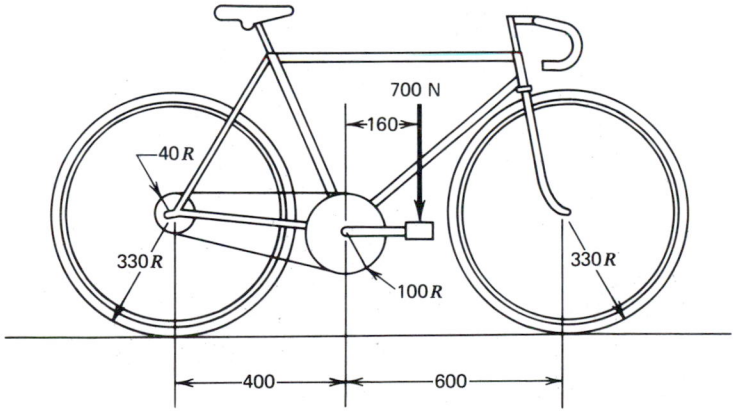

FIGURE P2.6

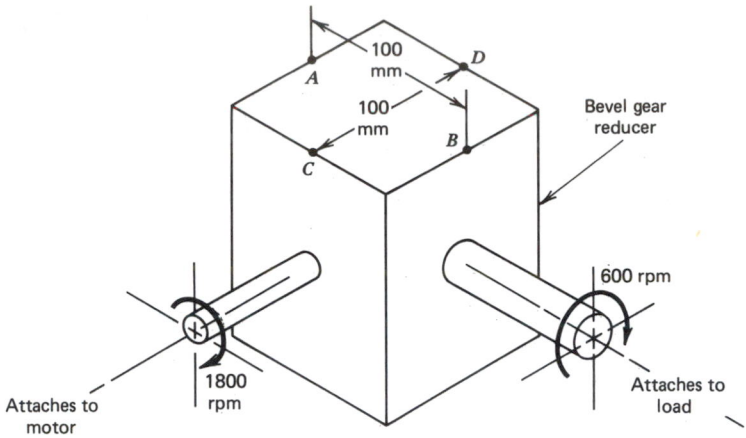

FIGURE P2.7

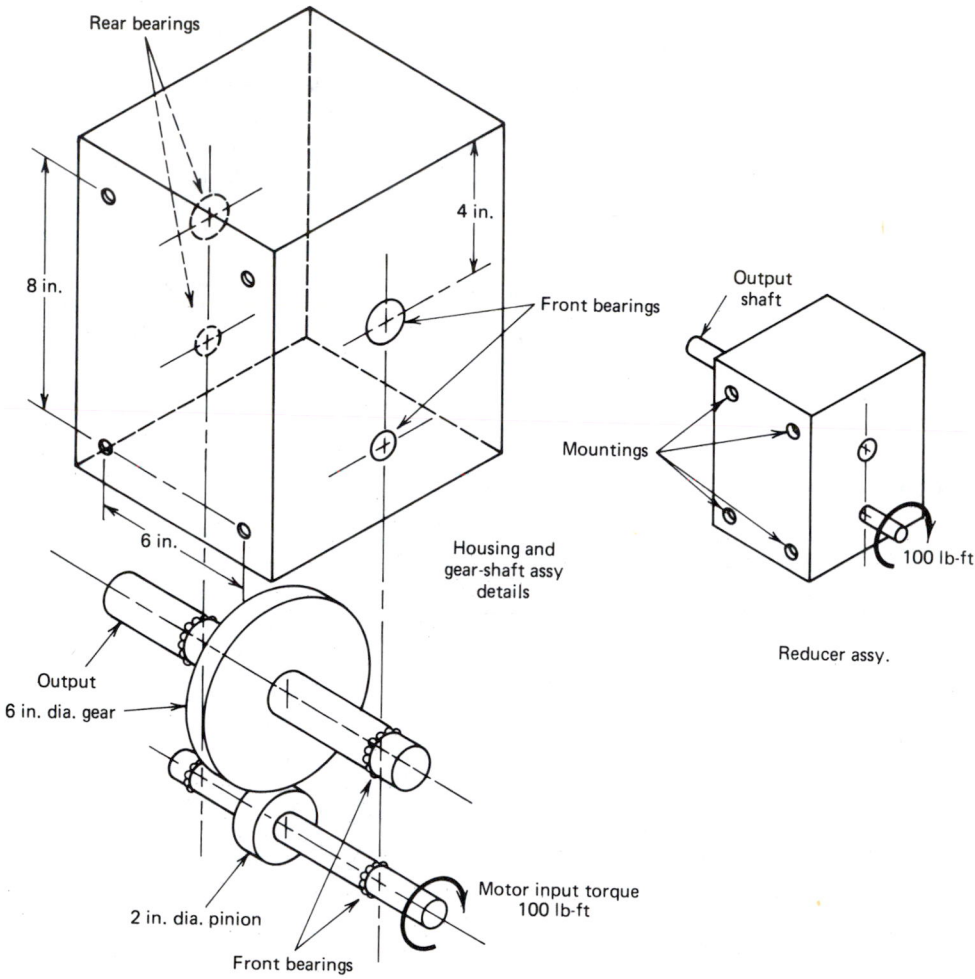

FIGURE P2.8

(a) Assuming 100% reducer efficiency.
(b) Assuming 95% reducer efficiency.

2.8 The drawings (Fig. P2.8) pertain to a spur gear reducer. A motor applies a torque of 100 lb-ft to the pinion shaft, as shown. The gear shaft drives the output load. Both shafts are connected with flexible couplings (which transmit only torque). The gears are mounted on their shafts mid-way between the bearings. The reducer is supported by four identical mountings on the side of the housing, symmetrically spaced on 6-in. and 8-in. centers, as shown. For simplicity, neglect gravity and assume that forces between the gears (i.e., between gear and pinion) act tangentially. Sketch, as free-bodies in equilibrium:

(a) The pinion and shaft assembly.
(b) The gear and shaft assembly.
(c) The housing.
(d) The entire reducer assembly.

2.9 The drawing (Fig. P2.9) is a highly simplified diagrammatic representation of the engine, transmission, drive shafts, and front axle of a 4-wheel-drive car. All members shown may be treated as a single free-body supported by mountings at *A*, *B*, *C*, and *D*. The engine rotates 2400 rpm and delivers a torque of 200 lb-ft. The transmission ratio is 2.0 (drive shafts rotate 1200 rpm); the front and rear axle ratios are both 3.0 (wheels rotate 400 rpm). Neglect friction and gravity, and assume that the mountings exert only vertical forces. Determine the forces applied to the free-body at *A*, *B*, *C*, and *D*.
(Ans.: 300 lb down, 300 lb up, 200 lb down, and 200 lb up, respectively)

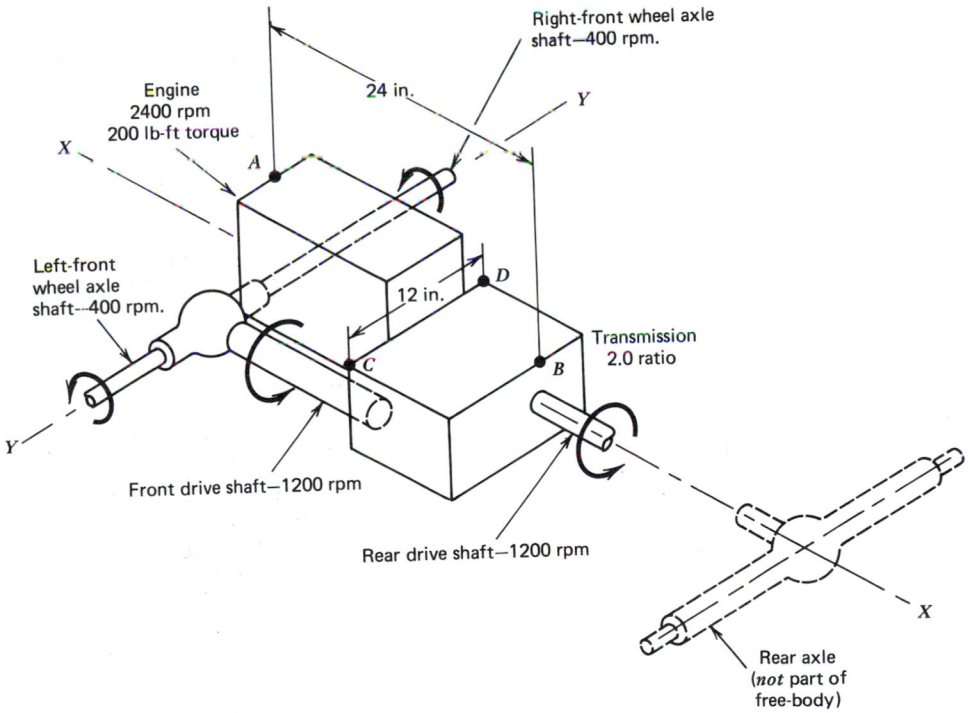

FIGURE P2.9

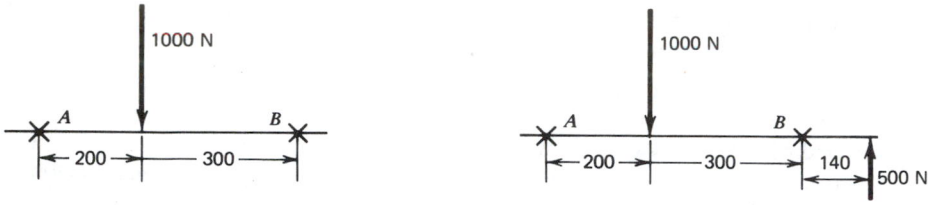

FIGURE P2.10

Section 2.3

2.10 The drawings (Fig. P2.10) show steel shafts supported by self-aligning bearings (which can provide radial but not bending loads to the shaft) at *A* and *B*. A gear (or a pulley or sprocket) causes each force to be applied as shown. Draw shear and bending moment diagrams neatly and to scale for each case. (Given dimensions are in mm.)

2.11 With reference to Fig. P2.11:

(a) Draw a free-body diagram of the structure supporting the pulley.

(b) Draw shear and bending moment diagrams for both the vertical and horizontal portions of the structure.

2.12 The drawing (Fig. P2.12) shows a bevel gear attached to a shaft supported by self-align-

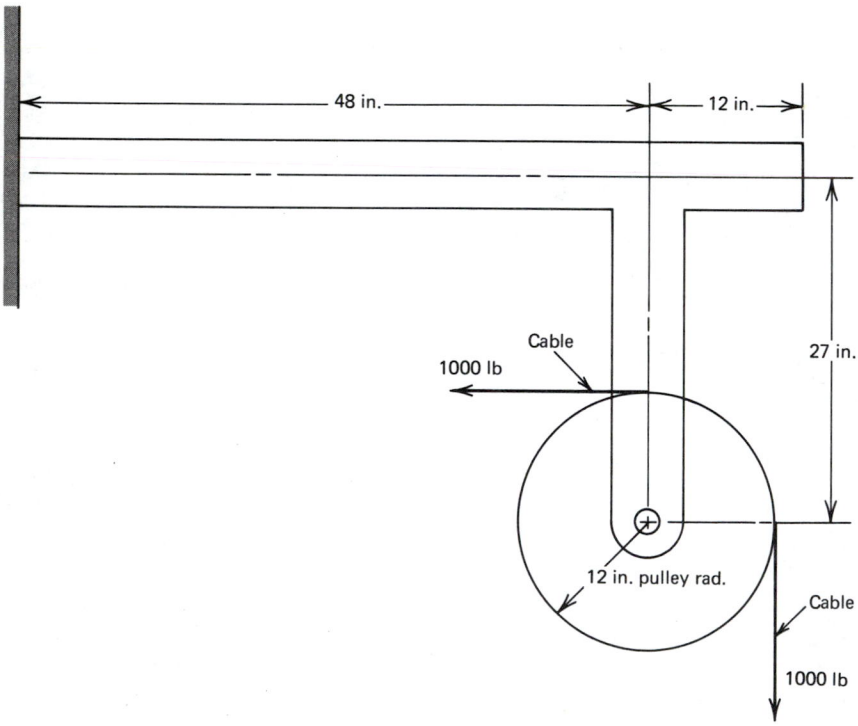

FIGURE P2.11

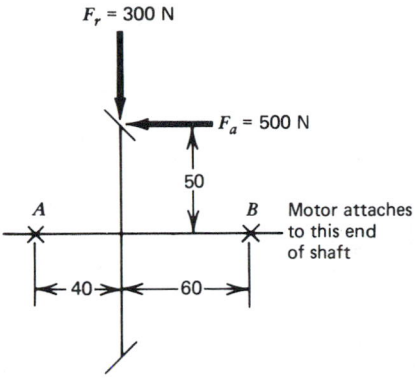

FIGURE P2.12

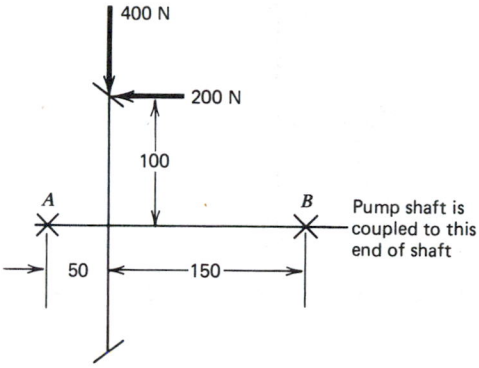

FIGURE P2.13

ing bearings at A and B, and driven by a motor. Axial and radial components of the gear force are shown. The tangential or torque producing component is perpendicular to the plane of the paper and has a magnitude of 1000 N. Bearing A takes thrust; B does not. Dimensions are in mm.

(a) Draw (to scale) shear and bending moment diagrams.
(b) To what values of axial load and torque is the shaft subjected, and what portion(s) of the shaft experience these loads?

2.13 The shaft with bevel gear shown in Fig. P2.13 is supported by self-aligning bearings A and B. (Given dimensions are in mm.) Only bearing A takes thrust. Gear loads in the plane of the paper are shown (the tangential or torque-producing force component is perpendicular to the paper and hence does not enter into this problem). Draw shear and bending moment diagrams for the shaft.

2.14 Same as Problem 2.13 except the shaft in this drawing (Fig. P2.14) has one bevel and one spur gear, and neither end of the shaft is connected to a motor or load.

2.15 Same as Problem 2.13 except the shaft in this drawing (Fig. P2.15) has two bevel gears, and neither end of the shaft is connected to a motor or load.

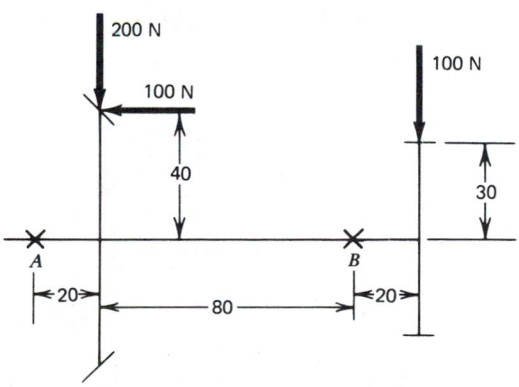

FIGURE P2.14

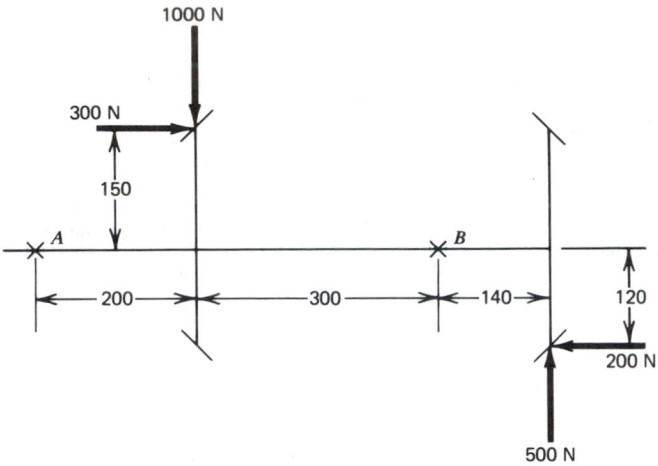

FIGURE P2.15

Section 2.4

2.16 Figure P2.16 shows a static force, F, applied to the tooth of a gear that is keyed to a shaft. Making appropriate simplifying assumptions, identify the stresses in the key, and write an equation for each.* State the assumptions made, and discuss briefly their effect.

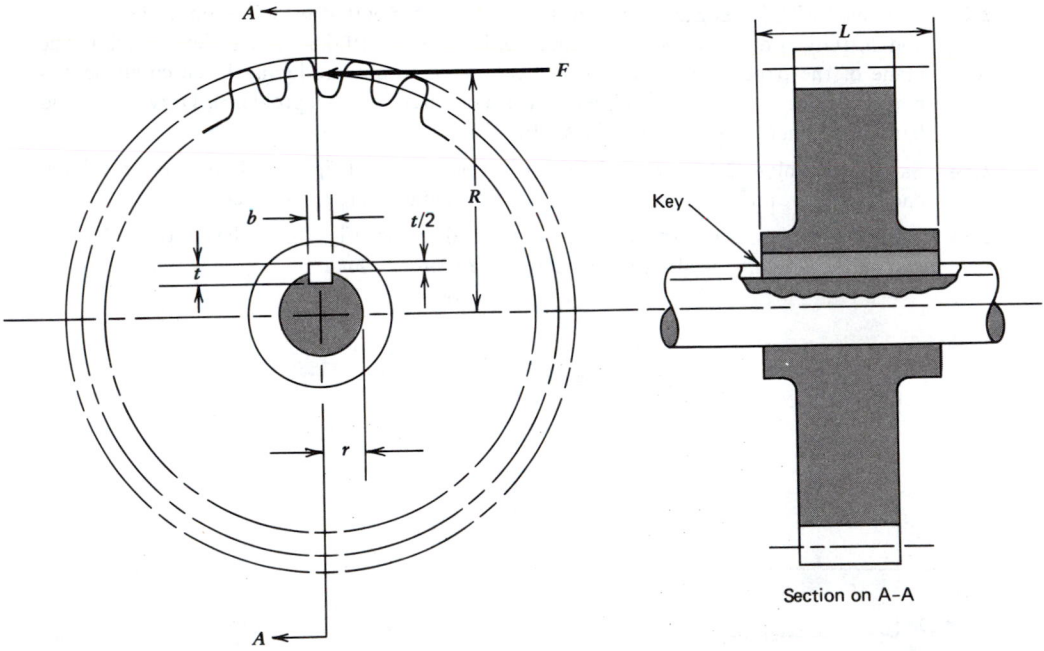

FIGURE P2.16

*The first five sections of Chapter 4 review simple stress equations.

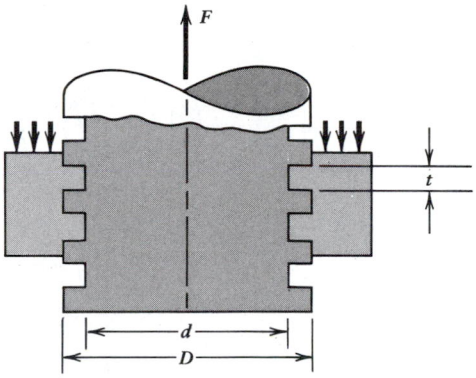

FIGURE P2.17

2.17 Figure P2.17 shows a screw with a square thread transmitting axial force F through a nut with n threads engaged (the drawing illustrates $n = 2$). Making appropriate simplifying assumptions, identify the stresses in the threaded portion of the screw and write an equation for each.* State the assumptions made, and discuss briefly their effect.

2.18 Figure P2.18 shows a total gas force F applied to the top of a piston.

(a) Copy the drawing and sketch the force paths through the piston, through the piston pin, and into the connecting rod.

(b) Making appropriate simplifying assumptions, identify the stresses in the piston pin and write an equation for each.* State the assumptions made, and discuss briefly their effect.

2.19 Figure P2.19 shows force P applied to an engine crankshaft by a connecting rod. The shaft is supported by main bearings A and B. Torque is transmitted to an external load through flange F.

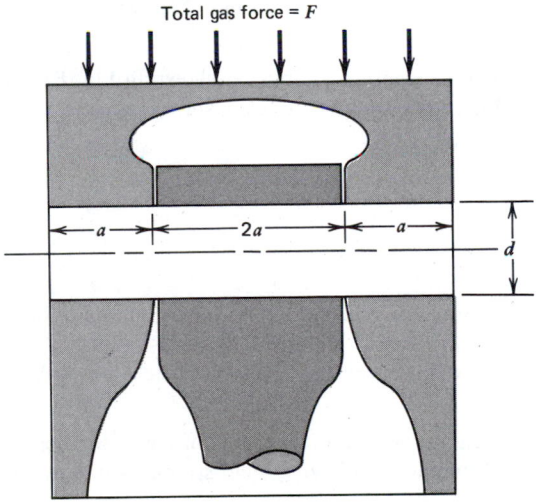

FIGURE P2.18

*The first five sections of Chapter 4 review simple stress equations.

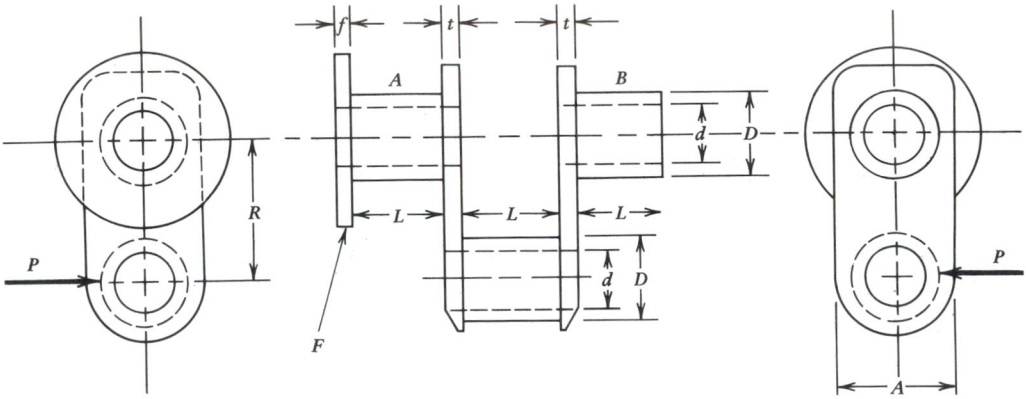

FIGURE P2.19

(a) Draw the shaft, and show all loads necessary to place it in equilibrium as a free-body.
(b) Starting with P and following the force paths through the shaft to the flange, identify the locations of potentially critical stresses.
(c) Making appropriate simplifying assumptions, write an equation for each.* State the assumptions made.

Section 2.5

2.20 A "T" bracket, attached to a fixed surface by four bolts, is loaded at point E as shown in Fig. P2.20.

(a) Copy the drawing and sketch paths of force flow going to each bolt.
(b) If the stiffness between point E and the plate through bolts B and C is twice the stiffness between point E and the plate through bolts A and D, how is the load divided between the four bolts?

2.21 A very stiff horizontal bar, supported by four identical springs, as shown in Fig. P2.21, is subjected to a center load of 50 N. What load is applied to each spring?
(Ans.: lower springs, 20 N; upper springs, 10 N)

Section 2.6

2.22 With reference to the bolts in Problem 2.20:

(a) If they are brittle and each one fractures at a load of 5000 N, what maximum force F can be applied to the bracket?
(b) What load can be applied if they are ductile, and each bolt has a yield strength of 5000 N.

2.23 Figure P2.23 shows two plates joined with straps and a single row of rivets (or bolts). Plates, straps, and rivets are all made of ductile steel having yield strengths in tension,

*The first five sections of Chapter 4 review simple stress equations.

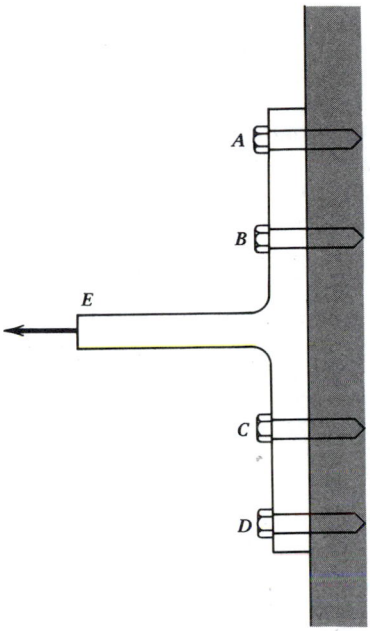

FIGURE P2.20

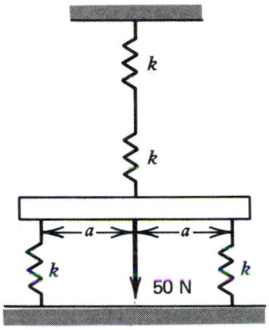

FIGURE P2.21

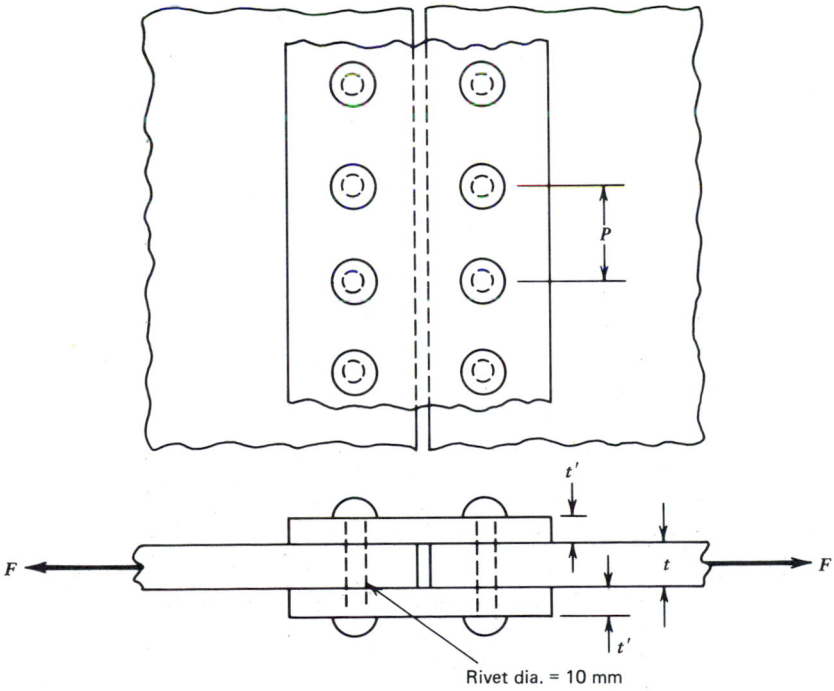

FIGURE P2.23

compression, and shear of 300, 300, and 170 MPa, respectively. Neglect frictional forces between the plates and straps.

(a) What force F can be transmitted across the joint per pitch, P, of joint width, based on rivet shear strength?
(b) Determine minimum values of t, t', and P that will permit the total joint to transmit this same force (thus giving a "balanced" design).
(c) Using these values, what is the "efficiency" of the joint (ratio of joint strength to strength of a continuous plate)?

2.24 Plates 20 mm thick are butted together and spliced using straps 10 mm thick and rivets (or bolts) of 40 mm diameter. A double-riveted joint is used, and this is exactly as shown in Fig. 2.19 except that the inner row of rivets is eliminated on both sides. All materials have tensile, compression, and shear yield strengths of 200, 200, and 120 MPa, respectively. Neglect friction between plates and straps. Determine the pitch, P, giving the greatest joint strength. How does this compare with the strength of a continuous plate?

CHAPTER 3 – MATERIALS

Section 3.3

3.1 The critical location of a part made from hot-rolled AISI 1020 steel is cold worked during its fabrication to an extent corresponding to point C of Fig. 3.2. What values of S_u, S_y, and ductility (in terms of ϵ, R, and A_r at fracture) are applicable to this location?

3.2 A tensile specimen made of hot-rolled AISI 1020 steel is loaded to point C of Fig. 3.2. What are the values of σ, ϵ, σ_T, and ϵ_T involved? Next, the specimen is unloaded. Treating it then as a *new* specimen, it is reloaded to point C. What are the values of these same quantities for the "new" specimen?

Section 3.10

3.3 An AISI 4340 steel part is heat-treated to 300 Bhn. Estimate the corresponding values of S_u and S_y.

3.4 If the curve in Fig. P3.4 represents the results of a Jominy end-quench test of AISI 4340 steel, draw corresponding curves (roughly) for low-alloy steel and for plain carbon steel, each having 0.40%C and heat-treated to the same surface hardness as the 4340 steel.

3.5 For each of the following applications calling for steel, choose between (a) 0.1%C and 0.4%C, and between (b) plain carbon and alloy steel:

(a) A machine frame requiring extreme rigidity (when massive enough to satisfy this requirement, stresses will be very low).
(b) A small round rod subjected to high bending and torsional stresses.
(c) A large irregularly shaped part subjected to high stresses.
(d) A rail car wheel (interior stresses low but surface must be carburized to resist wear).

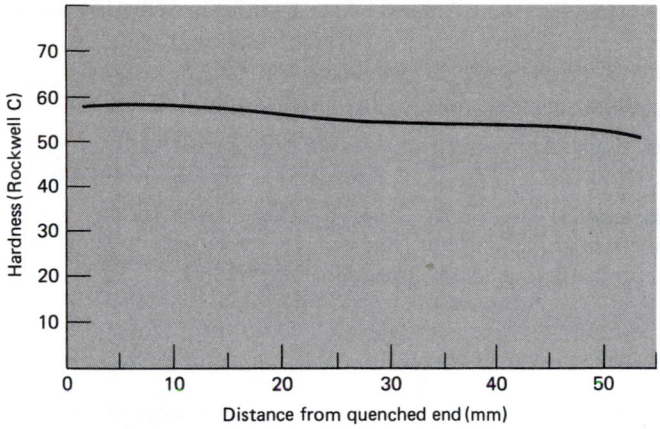

FIGURE P3.4

CHAPTER 4 – STATIC BODY STRESSES

Section 4.2

4.1 For the axially loaded shaft in Fig. P4.1, at which of the lettered sections is the average compressive stress equal to P/A? At which is the maximum stress equal to P/A?

Section 4.3

4.2 What force, P, is required to produce shear failure in a 20-mm-diameter bolt or pin made of a ductile metal having $S_{us} = 200$ MPa:

(a) With the configuration shown in Fig. 4.3?
(b) With the configuration shown in Fig. 4.4?

Section 4.4

4.3 A 2-in.-diameter steel propeller shaft of an experimental high-speed boat transmits 3000 hp at 2000 rpm. Bending and axial loads are negligible.

(a) What is the nominal shear stress at the surface?

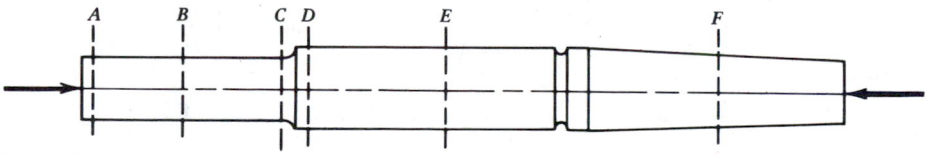

FIGURE P4.1

(b) If a hollow shaft of inside diameter 0.9 times outside diameter is used, what outside diameter would be required to give the same outer-surface stress?
(c) How do the weights of the solid and hollow shafts compare?

4.4 What torque is required to produce a maximum shear stress of 300 MPa:
(a) In a round shaft of 40 mm diameter?
(b) In a square shaft, 40 mm on a side?

4.5 Compare the torque-transmitting strength of a solid round shaft with that of a solid square shaft of the same size (circle diameter equal to side of square). Compare the weight of the two shafts, and also the ratio of strength to weight.

Section 4.5

4.6 A 2-in. diameter straight round shaft is subjected to a bending moment of 3000 ft-lb.
(a) What is the nominal bending stress at the surface?
(b) If a hollow shaft of inside diameter 0.5 times outside diameter is used, what outside diameter would be required to give the same outer-surface stress? (*Note:* If the hollow shaft is too thin, buckling will occur. See Section 5.15.)

4.7 What bending moment is required to produce a maximum normal stress of 450 MPa:
(a) In a straight round rod of 40 mm diameter?
(b) In a straight square rod, 40 mm on a side (with bending about the X-axis as shown for a rectangular section in Appendix B-2)?

Section 4.6

4.8 The rectangular beam shown in Fig. P4.8 has an initial curvature, $\bar{r}$, equal to twice the section depth, h. How do its extreme-fiber bending stresses compare with those of an otherwise identical straight beam?

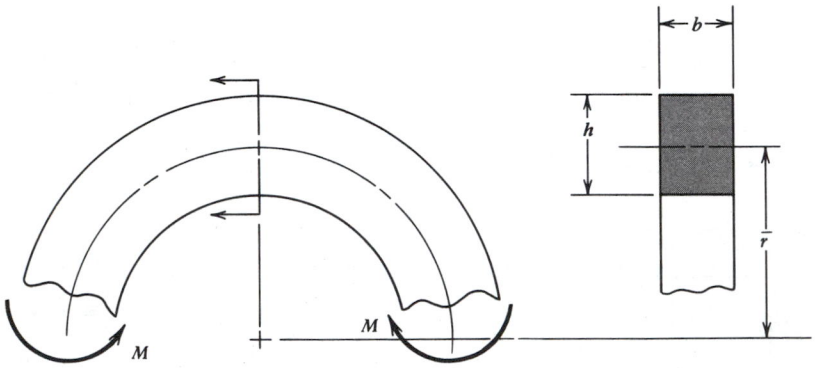

FIGURE P4.8

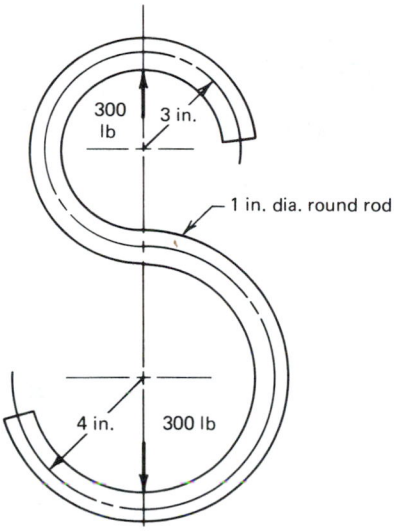

FIGURE P4.9

4.9 Determine the location and magnitude of the maximum tensile stress in the S-hook shown in Fig. P4.9. (Note: The lower portion experiences the larger bending moment, but the upper part has a smaller radius of curvature; hence, both locations must be investigated.)

4.10 Critical section A-A of a crane hook (Fig. P4.10) is considered, for purposes of analysis, to be trapezoidal with dimensions as shown. Determine the resultant stress (bending plus direct tension) at points P and Q.

4.11 Figure P4.11 shows a portion of a C-clamp. What force F can be exerted by the screw if the maximum tensile stress in the clamp is to be limited to 25 ksi?

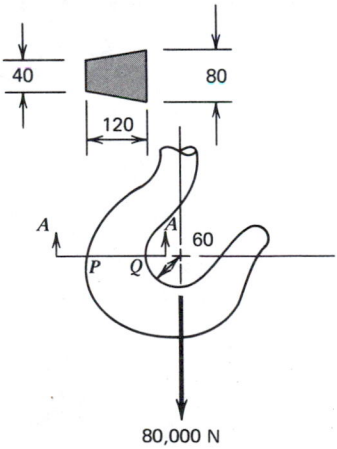

FIGURE P4.10

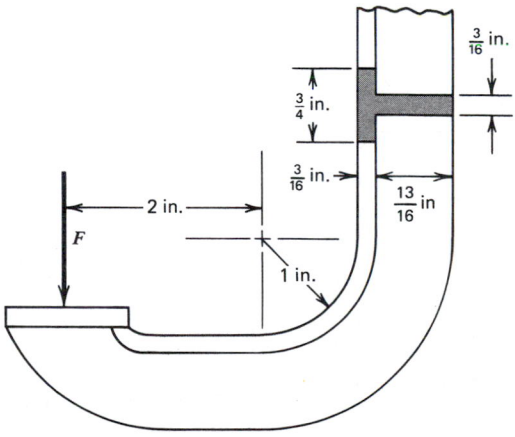

FIGURE P4.11

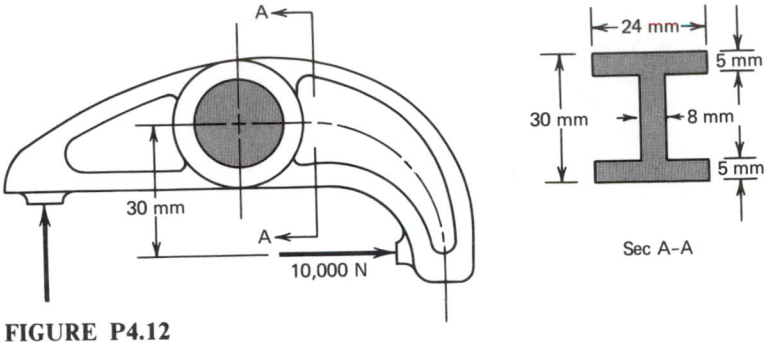

FIGURE P4.12

4.12 For the rocker arm shown in Fig. P4.12, determine the maximum tensile stress in section A-A.

Section 4.7

4.13 A solid square-section beam, 60 mm on a side, is used in place of the beam in Sample Problem 4.2. What is the location and magnitude of the maximum shear stress? Use Eq. (4.12) and check the result with Eq. (4.14).

4.14 Using Eq. (4.12), derive Eq. (4.13).

4.15 Using Eq. (4.12), derive Eq. (4.14).

4.16 For the 8-in. I-beam shown (Fig. P4.16), compute the maximum transverse shear stress when the beam is simply supported at each end and subjected to a load of 2000 lb in the center. Compare your answer with the approximation obtained by dividing the shear load by the area of the webb (only), with the webb considered to extend for the full 8-in. depth.

4.17 Figure P4.17 shows a plastic beam having a box section, where the top plate is cemented in place, as indicated. For the 6-kN load shown, what is the shear stress acting on the cemented joint?

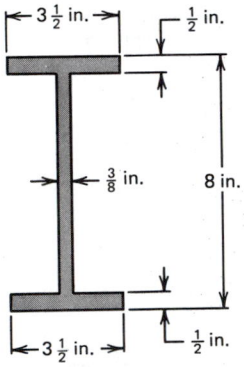

FIGURE P4.16

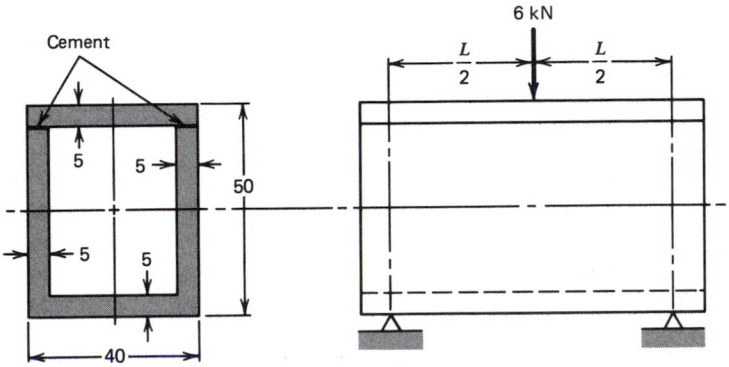

FIGURE P4.17

Sections 4.9 and 4.11

4.18 The shaft shown in Fig. P4.18 is 200 mm long between self-aligning bearings A and B. Belt forces are applied to a sheave in the center, as shown. The left end of the shaft is connected to a clutch by means of a flexible coupling. Nothing is attached to the right end.

(a) Determine and make a sketch showing the stresses acting on the top and side elements, T and S, located adjacent to the sheave. (Neglect stress concentration.)

(b) Represent the states of stress at T and S with three-dimensional Mohr circles.

(c) At location S, show the orientation and stresses acting on a principal element, and on a maximum shear element.

4.19 We wish to analyze the stresses in a bicycle crankshaft. (This is the horizontal shaft, supported in the frame by two ball bearings, which connects the two pedal crank arms.)

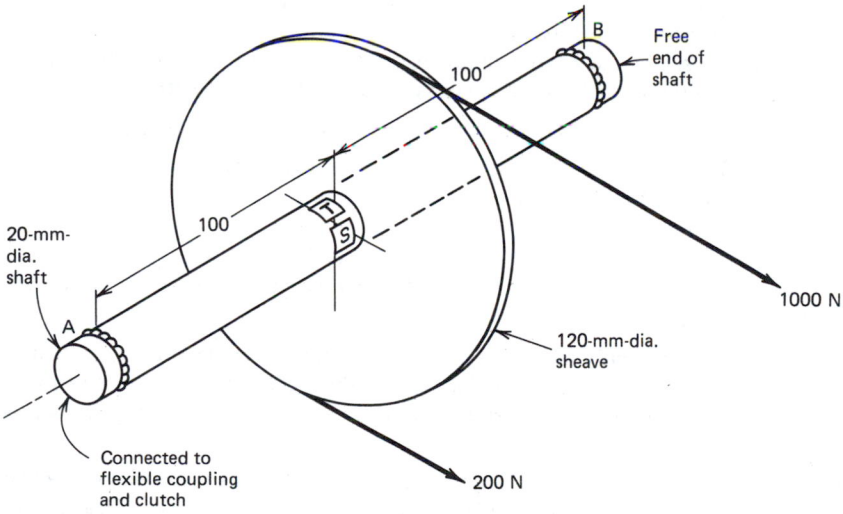

FIGURE P4.18

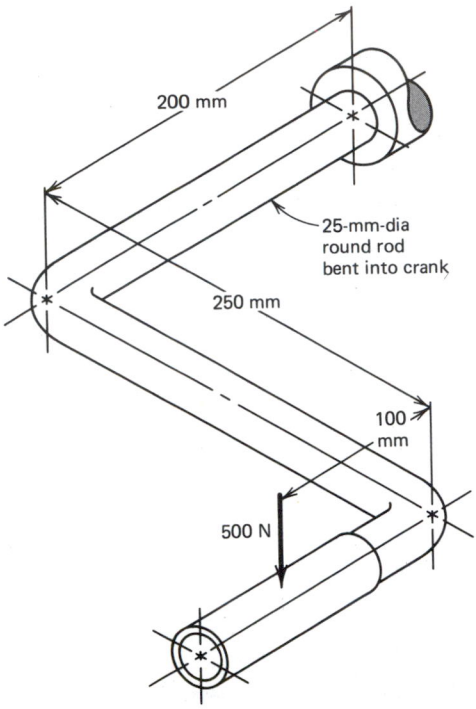

FIGURE P4.20

Obtain whatever dimensions you need by measuring an actual bicycle of standard adult size.

(a) Show, with the aid of a simple sketch, the most severe loading condition normally encountered by this shaft. Show all important dimensions, and state any assumptions made concerning the loading.

(b) Show on your sketch the location of greatest stress in this shaft, and make a Mohr-circle representation of this state of stress. (Neglect stress concentration.)

4.20 Figure P4.20 shows a hand crank with static vertical load applied to the handle.

(a) Copy the drawing and mark on it the location of highest bending stress. Make a three-dimensional Mohr-circle representation of the stresses at this point. (Neglect stress concentration.)

(b) Mark on the drawing the location of highest combined torsional and transverse shear stress. Make a three-dimensional Mohr-circle representation of the stresses at this point, again neglecting stress concentration.

4.21 Figure P4.21 shows an electric motor loaded by a belt drive. Copy the drawing and show on both views the location or locations on the shaft of the highest stress. Make a complete Mohr-circle representation of the stress state at this location. (Neglect stress concentration.)

4.22 Figure P4.22 shows a 1-in. solid round shaft supported by self-aligning bearings at *A* and *B*. Attached to the shaft are two chain sprockets that are loaded as shown. Treat this as a static loading problem, ignoring fatigue and stress concentration. Identify the specific

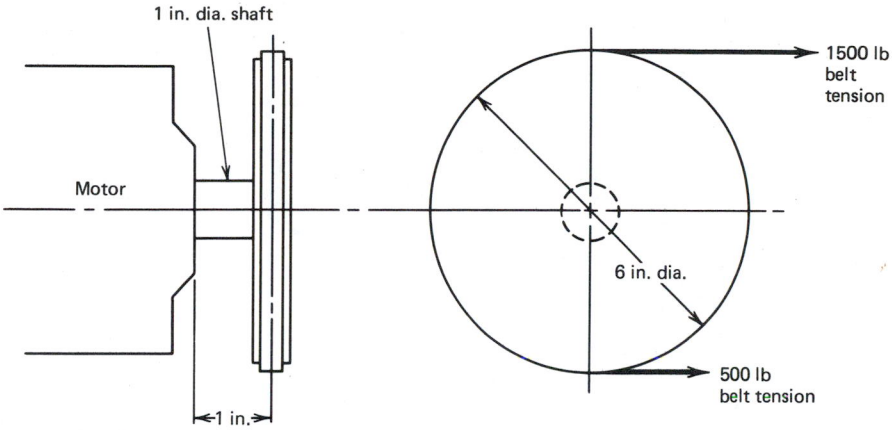

FIGURE P4.21

shaft location subjected to the most severe state of stress, and make a Mohr-circle representation of this stress state.

4.23 Repeat Problem 4.22, except use Fig. P4.23.

4.24 Figure P4.24 shows a small pressurized cylinder, attached at the one end and loaded with a pipe wrench at the other. Due to the internal pressure, there is a tangential stress of 200 MPa and an axial stress of 100 MPa acting on an element at point A. The pipe wrench superimposes a bending stress of 50 MPa and a torsional stress of 100 MPa.

(a) Make a Mohr-circle representation of the state of stress at point A.

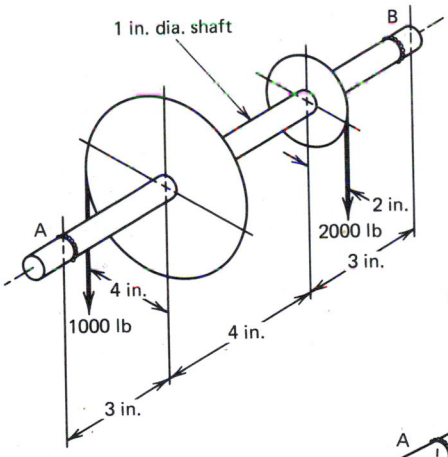

FIGURE P4.22

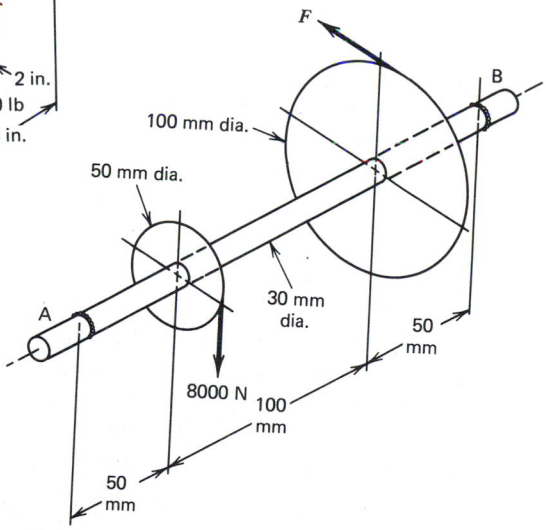

FIGURE P4.23

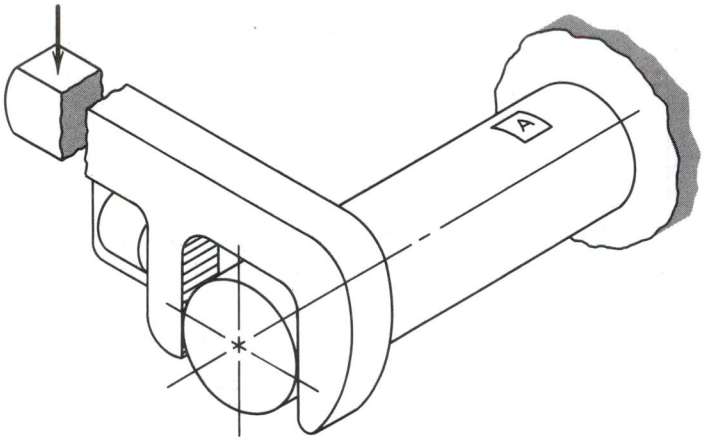

FIGURE P4.24

(b) What is the magnitude of the maximum shear stress at A?
(c) Make a sketch showing the orientation of a principal element (with respect to the original element drawn at A), and show all stresses acting on it.

[Ans.: (b) 140 MPa]

4.25 An internally pressurized section of round steel tubing is subjected to tangential and axial stresses at the surface of 400 and 200 MPa, respectively. Superimposed on this is a torsional stress of 100 MPa. Make a Mohr-circle representation of the surface stresses.

4.26 A cylinder is internally pressurized to a pressure of 50 MPa. This causes tangential and axial stresses in the outer surface of 200 and 100 MPa, respectively. Make a Mohr-circle representation of the stresses in the outer surface. What maximum shear stress is experienced by the outer surface?
(Ans.: 100 MPa)

4.27 The inner surface of the cylinder in Problem 4.26 experiences tangential and axial stresses of 300 and 100 MPa, respectively. Make a Mohr-circle representation of the stresses in the inner surface. What maximum shear stress exists at the inner surface?
(Ans.: 175 MPa)

Section 4.14

4.28 A notched flat bar (as shown in Fig. 4.39) has a stress concentration factor for tensile loads of 2. Its cross-sectional area in the plane of the notches is 0.5 sq. in. The material is steel, with tensile and compressive yield strengths of 60 ksi. Assume an idealized stress-strain curve. The bar is initially free of residual stress.

(a) Make a drawing showing the approximate shape of the stress distribution curve when the bar is loaded to 10,000 lb tension, and also after the load is removed.
(b) Repeat for a 20,000 lb load.
(c) Repeat for a 30,000 lb load.

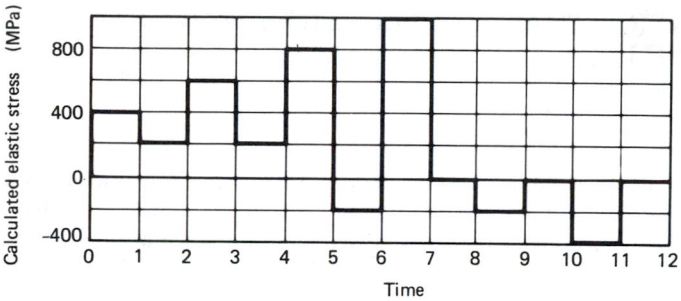

FIGURE P4.34

4.29 Repeat Problem 4.28 except for loading in compression.

4.30 Repeat Problem 4.28 except for a stress concentration factor of 3.

4.31 Repeat Problem 4.28 except for a stress concentration factor of 3, and loading in compression.

4.32 A 20-mm x 60-mm rectangular bar with 10-mm-dia. central hole (as shown in Fig. 4.40) is made of steel having tensile and compressive yield strengths of 300 MPa. Assume an idealized stress-strain curve. The bar is initially free of residual stress. Make a drawing showing the approximate stress distribution in the plane of the hole:

(a) When a tensile force of 200 kN is applied to each end of the bar.
(b) After the load is removed.

4.33 Repeat Problem 4.32 for loading in compression.

4.34 A notched bar (illustrated in Fig. 4.39) has a stress concentration factor for tensile loading of 2.5. It is made of ductile steel (assume an idealized stress-strain curve) with tensile and compressive yield strengths of 400 MPa. The bar is loaded in tension with *calculated* notch-root stresses varying with time as shown in Fig. P4.34. Copy the drawing and add to it a curve showing the variation with time of *actual* notch-root stresses.

Section 4.15

4.35 Two rectangular beams are made of steel having a tensile yield strength of 40 ksi and an assumed idealized stress-strain curve. Beam A has a uniform 1-in. x 0.5-in. section. Beam B has a 1-in. x 0.5-in. section that blends symmetrically into a 1.5-in. x 0.5-in. section with fillets giving a stress concentration factor of 3. The beams are loaded in bending such that $Z = I/c = bh^2/6 = 0.5(1)^2/6 = 1/12$ in^3.

(a) For each beam, what moment, M, causes (1) initial yielding, and (2) complete yielding?
(b) Beam A is loaded to cause yielding to a depth of ¼ in. Determine and plot the distribution of residual stresses which remain after the load is removed.

[Ans.: (a1) beam A, 3333 in.-lb, beam B, 1111 in.-lb; (a2) 5000 in.-lb for both beams]

CHAPTER 5 – ELASTIC STRAIN, DEFLECTION, AND STABILITY

Sections 5.2-5.4

5.1 The following readings are obtained from an equiangular strain gage rosette mounted on a free and unloaded surface of a part: $\epsilon_0 = +1900$, $\epsilon_{120} = +1250$, and $\epsilon_{240} = +600$. Gage orientation angles are measured counterclockwise, and strain values are in micrometers per meter (or microinches per inch). Determine the magnitude of the principal strains and their orientation with respect to the 0° gage. Check the results with a Mohr circle.
(Ans.: $\epsilon_1 = 0.0020$, $\epsilon_2 = 0.0005$, ϵ_1 is 15° cw from 0° gage)

5.2 The following readings are obtained from a rectangular strain gage rosette mounted on a free and unloaded surface: $\epsilon_0 = +1000$, $\epsilon_{90} = -600$, $\epsilon_{225} = -200$. Gage orientation angles are measured counterclockwise, and strain values are in micrometers per meter. Determine the magnitude of the principal strains and their orientation with respect to the 0° gage. Check the results with a Mohr circle.

5.3 Repeat Problem 5.2 except use gage orientations and readings of: $\epsilon_0 = -1200$, $\epsilon_{135} = -1520$, and $\epsilon_{270} = -800$.

Section 5.5

5.4 Determine values of principal stresses and maximum shear stress for the following:

(a) Sample Problem 5.1, if the material is steel.
(b) Sample Problem 5.2, if the material is aluminum.
(c) Problem 5.1 (above), if the material is titanium.
(d) Problem 5.2, if the material is steel.
(e) Problem 5.3, if the material is aluminum.

5.5 An equiangular strain gage rosette is mounted on a free and unloaded surface of a part made of steel. When loaded the gages read: $\epsilon_0 = +950$, $\epsilon_{120} = -350$, and $\epsilon_{240} = +150$ micrometers per meter. Determine all principal stresses and strains, and draw Mohr circles for both stress and strain.

5.6 Repeat Problem 5.5 except use a rectangular rosette mounted on a magnesium part, and gage readings of $\epsilon_0 = -1250$, $\epsilon_{90} = 3150$, and $\epsilon_{135} = -780$ micrometers per meter. (Gage angles are measured counterclockwise.)

Section 5.6

5.7 What is the torsional spring constant (N·m torque per degree of angular deflection) of a solid round steel shaft 300 mm long if its diameter is:

(a) 30 mm?
(b) 20 mm?
(c) 30 mm for half its length and 20 mm for the other half?

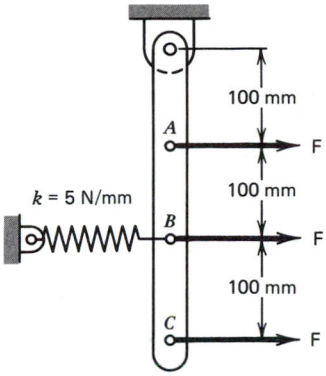

FIGURE P5.8

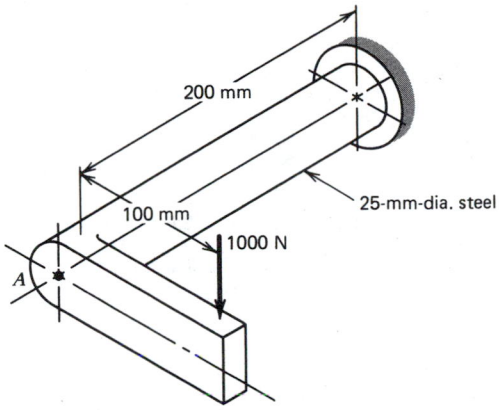

FIGURE P5.9

5.8 Figure P5.8 shows one end of a spring attached to a lever. What is the spring constant (newtons force per mm deflection):

(a) with respect to a horizontal force applied at A?
(b) with respect to a horizontal force applied at B?
(c) with respcet to a horizontal force applied at C?

5.9 What are the angular and linear displacements of point A of Figure P5.9?

Section 5.7

5.10 Figure P5.10 shows a simply supported steel shaft subjected to two loads. Determine the deflection at all points along the shaft?

Section 5.8

5.11 The bracket in Figure P5.11 is loaded with a force in the Y-direction, as shown. Derive an expression for the deflection of the free end in the Y direction.

(Ans.: $\dfrac{Fb^3}{3EI} + \dfrac{Fa^3}{3EI} + \dfrac{Fb^2a}{GJ}$ plus generally negligible transverse shear terms)

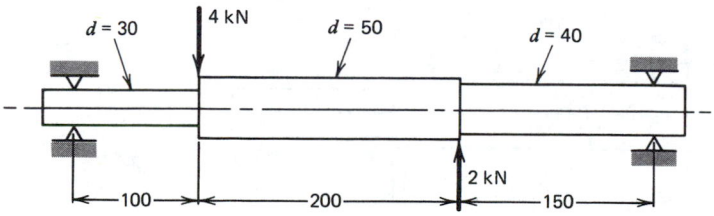

FIGURE P5.10

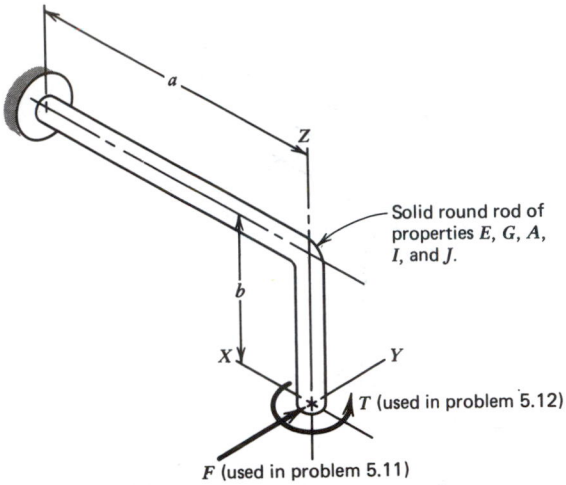

FIGURE P5.11

5.12 The bracket in Figure P5.11 is loaded with a torque about the Z-axis, as shown. Derive an expression for the resulting deflection of the free end in the Y-direction. (Ans.: $Ta^2/2EI$)

5.13 Figure P5.13 shows a steel shaft supported by self-aligning bearings and subjected to a uniformly distributed load. Using Castigliano's method, determine the required diameter, d, to limit the deflection to 0.2 mm. (You may assume that transverse shear is negligible.)

5.14 The structure shown (Fig. P5.14) is fabricated by welding together three pieces of square rod or tubing, each having the same cross-sectional area A, moment of inertia I, and modulus of elasticity E. Derive an expression for the deflection between the points of force application. Omit terms that are likely to be negligible, but enumerate any such terms.

5.15 The helical lock washer in Figure P5.15 has material elastic properties of E and G, and cross-sectional properties of A, I, and K' (J if the section is circular). What is its spring rate with respect to the force P, which tends to flatten it? You may neglect terms expected to be unimportant, but enumerate any terms neglected.

$$\left(\text{Partial ans.:}\quad \delta = \frac{\pi PR^3}{EI} + \frac{3\pi PR^3}{K'G} + \frac{12\pi PR}{5GA}\right)$$

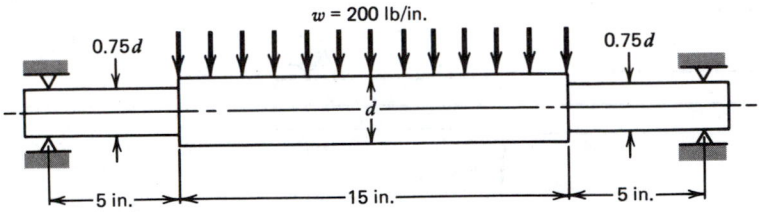

FIGURE P5.13

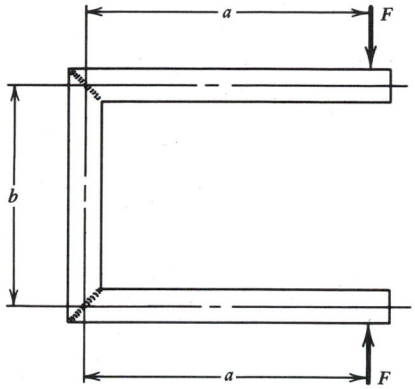

FIGURE P5.14

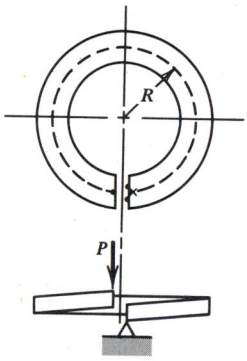

FIGURE P5.15

5.16 The triangular plate cantilever beam shown (Fig. P5.16) represents an idealization of a leaf spring (more about these in Chapter 12). Using Castigliano's method, derive an expression for the deflection of the loaded end, assuming that transverse shear will contribute negligibly.

$$\left(\text{Ans.:}\quad \delta = \frac{6FL^3}{Ebh^3}\right)$$

Section 5.9

5.17 In order to reduce the deflection of the I-beam cantilever shown (Fig. P5.17), a support is to be added at S:

(a) What vertical force at S is needed to reduce the deflection at that point to zero?

(b) What force is needed to cause an upward deflection at S of 5 mm (thereby perhaps reducing the end deflection to a desired value)?

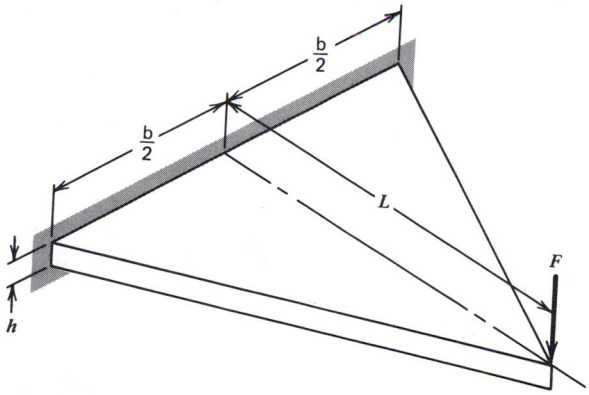

FIGURE P5.16

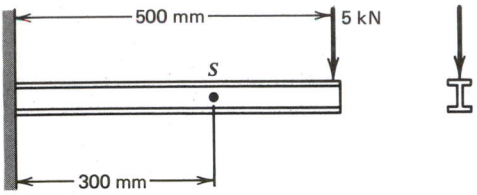

FIGURE P5.17

(c) What can you say about the effect of these forces at S on the bending stresses at the point of beam attachment?

Sections 5.10–5.12

5.18 A solid round steel rod with $S_y = 350$ MPa is 1 m long and 70 mm in diameter. With a safety factor of 4, what axial compressive load can be applied if:

(a) Both ends are hinged?
(b) Both ends are built in, as in Fig. 5.25*c*?

5.19 A 1 in. x 2 in. bar is 20 in. long and made of aluminum having $S_y = 25$ ksi. With a safety factor of 4, what axial compressive load can be applied if:

(a) Both ends are hinged?
(b) One end is built in and the other is unsupported, as in Fig. 5.25*e*?

5.20 A steel angle iron, loaded in compression, is added to a structure in order to increase its rigidity. Although the two ends are attached by rivets, the end fixity is sufficiently questionable that pinned ends (Fig. 5.25*a*) are to be assumed. The length is 1.2 m, and the yield strength 350 MPa. The radius of gyration about centroidal axis parallel to either side is 8 mm, but the minimum radius of gyration (about a centroidal axis at 45° to the sides) is only 5 mm. What compressive load can be carried with a safety factor of 3?

5.21 The 3-in. I-beam shown in Fig. P5.21 has cross-sectional properties of $A = 1.64$ in.^{2}, I_{11}

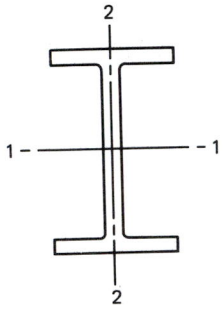

FIGURE P5.21

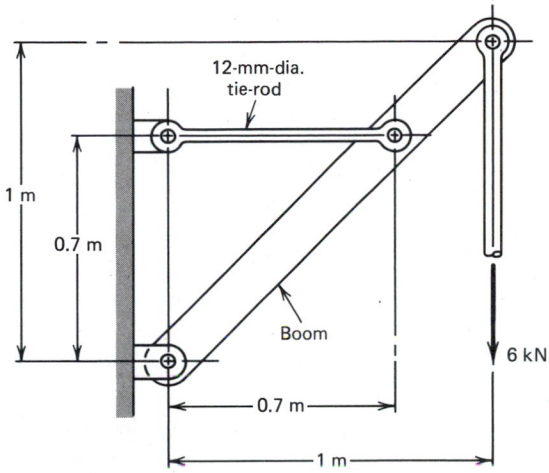

FIGURE P5.23

$= 2.5$ in.^{4}, and $I_{22} = 0.46$ in.^{4}. It is made of steel having $S_y = 42$ ksi. Find the safe axial compressive load based on a safety factor of 3 for pinned ends and unsupported lengths of (a) 10 in., (b) 50 in., (c) 100 in., and (d) 200 in.

5.22 A 20-mm-diameter steel rod of $S_y = 350$ MPa is loaded as a column with pinned ends. If sufficiently short, it can carry a limiting load of $S_yA = 110$ kN. How long can the rod be and still carry the following percentages of this 110 kN load: (a) 90%, (b) 50%, (c) 10%, (d) 2%?

5.23 Figure P5.23 drawing shows a boom and tie-rod arrangement supporting a load of 6 kN. The tie-rod is made of steel having a tensile yield strength of 400 MPa.

(a) What is the safety factor of the tie-rod with respect to static yielding?
(b) What is the safety factor of the tie-rod if the vertical rod is rotated 180° so that the 6-kN load acts upward?
(c) What conclusion do you draw with respect to the relative desirability of designing machines with column members loaded in tension vs. loaded in compression?

[Ans.: (a) 5.3, (b) none, it will buckle]

CHAPTER 6 – FAILURE THEORIES, SAFETY FACTORS, AND RELIABILITY

Sections 6.3 and 6.4

6.1 Repeat Sample Problem 6.1 using a material of Ti-6Al-4V annealed titanium alloy, with properties of $S_u = 130$ ksi, $S_y = 120$ ksi, and a plane stress $K_{Ic} = 110$ ksi$\sqrt{\text{in.}}$

6.2 Repeat Sample Problem 6.2 using materials of (a) 7075-T651 aluminum, (b) D6AC steel at room temperature, (c) D6AC steel at −40°F, and (d) 4340 steel at room temperature.

Sections 6.5-6.10

6.3 A machine component is loaded so that stresses at the critical location are $\sigma_1 = 20$ ksi, $\sigma_2 = -15$ ksi, and $\sigma_3 = 0$. The material is ductile, with yield strengths in tension and compression of 60 ksi. What is the safety factor according to (a) the maximum-normal-stress theory, (b) the maximum-shear-stress theory, and (c) the maximum-distortion-energy theory? Which theory would be expected to agree most closely with an actual test?
[Partial ans.: (a) 3.0, (b) 1.71, (c) 1.97]

6.4 What tensile yield strength must a ductile material have in order to provide a safety factor of 2 with respect to initial yielding at the location(s) investigated in problems (a) 4.18, (b) 4.20, (c) 4.21, (d) 4.22, (e) 4.23, (f) 4.24, (g) 4.25, (h) 4.26, and (i) 4.27. In each case, determine the answer using both the maximum-shear-stress theory and the maximum-distortion-energy theory.

6.5 What ultimate tensile strength would be required of a brittle material in order to provide a safety factor of 4 to a member subjected to the same state(s) of stress as those determined in problems (a) 4.18, (b) 4.20, (c) 4.21, (d) 4.22, (e) 4.23, (f) 4.24, (g) 4.25, (h) 4.26, and (i) 4.27. Use the modified Mohr theory, and assume a compressive ultimate strength of 3.5 times the tensile strength. If overloaded to failure, what would be the orientation of the brittle crack in each case?

6.6 The surface of a steel machine member is subjected to principle stresses of 200 MPa and 100 MPa. What tensile yield strength is required to provide a safety factor of 2 with respect to initial yielding: (a) According to the maximum-shear-stress theory? (b) According to the maximum-distortion-energy theory?
[Ans.: (a) 400 MPa, (b) 346 MPa]

6.7 A round steel rod is subjected to axial tension of 50 MPa with superimposed torsion of 100 MPa. What is your best estimate of the safety factor with respect to initial yielding if the material has a tensile yield strength of 500 MPa?
(Ans.: 2.77)

6.8 A straight round shaft is subjected to a torque of 5000 lb-in. Determine the required diameter, using steel with a tensile yield strength of 60 ksi and a safety factor of 2 based on initial yielding: (a) According to the maximum-normal-stress theory. (b) According to the maximum-shear-stress theory. (c) According to the maximum-distortion-energy theory. Discuss briefly the relative validity of the three predictions.

6.9 A machine frame is made of steel having $S_y = 400$ MPa and $S_{ys} = 250$ MPa. When loaded in a test fixture, the stresses were found to vary linearly with load. Two points on the surface were found to be most critical. With a 4-kN test load, stresses at these points were: point a, $\sigma_1 = 200$ MPa, $\sigma_2 = 100$ MPa; point b, $\sigma_1 = 150$ MPa, $\sigma_2 = -100$ MPa. Compute the test load at which the frame will experience initial yielding according to the (a) maximum-normal-stress theory, (b) maximum-shear-stress theory, and (c) maximum-distortion-energy theory. Discuss briefly the relative validity of each theory for this application. Considering all of the information available, what is your best judgment of the value of test load at which yielding would actually be expected to begin?

6.10 Consider the following states of biaxial stress: (1) $\sigma_1 = 30$, $\sigma_2 = 0$, (2) $\sigma_1 = 30$, $\sigma_2 = -15$, (3) $\sigma_1 = 30$, $\sigma_2 = -30$, (4) $\sigma_1 = 30$, $\sigma_2 = 15$, (5) pure shear, $\tau = 30$. With the aid of a σ_1-σ_2 plot, list these stress states in order of increasing likelihood of causing failure according to (a) the maximum-normal-stress theory, (b) the maximum-shear-stress theory, and (c) the maximum-distortion-energy theory.

6.11 A round steel bar having $S_y = 800$ MPa is subjected to loads producing calculated stresses of $P/A = 70$ MPa, $Tc/J = 200$ MPa, $Mc/I = 300$ MPa, and $4V/3A = 170$ MPa.

(a) Make a sketch showing the relative locations of maximum normal stress and maximum shear stress.

(b) Determine the safety factor with respect to initial yielding according to the maximum-shear-stress theory and according to the maximum-distortion-energy theory.

Section 6.15

6.12 A particular machine part is subjected in service to a maximum load of 10 kN. With the thought of providing a safety factor of 1.5, it is designed to withstand a load of 15 kN. If the maximum load encountered in various applications is normally distributed with a standard deviation of 2 kN, and if part strength is normally distributed with a standard deviation of 1.5 kN, what failure percentage would be expected in service?
(Ans.: 2.3%)

6.13 Assume that the failure rate calculated in Problem 6.12 is unacceptable. (a) To what value would the standard deviation of part strength have to be reduced in order to give a failure rate of only 1%, with no other changes? (b) To what value would the nominal part strength have to be increased in order to give a failure rate of only 1%, with no other changes?

CHAPTER 7 – IMPACT

Section 7.2

7.1 A tensile impact bar, similar to the one in Fig. 7.4*a*, fractured in service. Because the failure happened to occur near the center, a naive technician makes a new bar exactly like the old one except that the middle third is enlarged to twice the diameter of the ends. Assuming that stress concentration can be neglected (not very realistic), how do the impact capacities of the new and old bars compare?

7.2 A car skidded off an icy road and became stuck in deep snow at the road shoulder. Another car, of 1400 kg mass, attempted to jerk the stuck vehicle back onto the road using a 5-m steel tow cable of stiffness $k = 5000$ N/mm. The traction available to the rescue car prevented it from exerting any significant force on the cable. With the aid of a push from bystanders, the rescue car was able to back against the stuck car and then go forward and reach a speed of 4 km/hr at the instant the cable became taut. If the cable is attached rigidly to the masses of the cars, estimate the maximum impact force that can be developed in the cable, and the resulting cable elongation.

7.3 The rescue attempt in Problem 7.2 resulted in only slight movement of the stuck car because the cable force decayed so quickly to zero. Besides, concern was felt about possible damage to the car attachment points because of the high "instantaneous" force developed. One witness to the proceedings brought a 12-m elastic cable of overall stiffness only 2.4 N/mm and suggested that it be tried. Because of the longer length of the elastic cable, its use enabled the rescue car to reach 12 km/hr at the point of becoming taut. Estimate the impact force developed and the resulting cable elongation. If the stuck vehicle does not move significantly until the rescue car has just come to a stop, how much energy is stored in the cable? (Think of this in terms of the height from which a 100 kg mass would have to be dropped to represent an equivalent amount of energy, and consider the potential hazard if the cable should break or come loose from either car.) What warnings would you suggest be provided with elastic cables sold for this purpose?

7.4 A 5-ton elevator is supported by a stranded steel cable of 2.5 in.^{2} cross section and an effective modulus of elasticity of 12 x 10^6 psi. As the elevator is descending at a constant 400 fpm, an accident causes the top of the cable, 60 ft above the elevator, to suddenly stop. Estimate the maximum elongation and maximum tensile stress developed in the cable.

Section 7.3

7.5 The vertical drive shaft in Fig. P2.5 is 20 mm in diameter, 750 mm long, and made of steel. The motor to which it is attached at the top is equivalent to a steel flywheel 300 mm in diameter and 25 mm thick. When the vertical shaft is rotating 3000 rpm, the propeller strikes a heavy obstruction, bringing it to a virtually instantaneous stop. Assume that the short horizontal propeller shaft and the bevel gears have negligible flexibility. Calculate the elastic torsional shear stress in the vertical shaft. (Since this stress far exceeds any possible torsional elastic strength, a shear pin or slipping clutch would be used to protect the shaft and associated costly parts.)

Section 7.4

7.6 For the tensile impact bar shown in Fig. P7.6, estimate the ratio of impact energy that can be absorbed with and without the notch which reduces the diameter to 24 mm. (Ans.: 0.06)

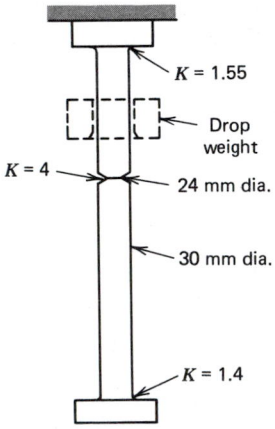

FIGURE P7.6

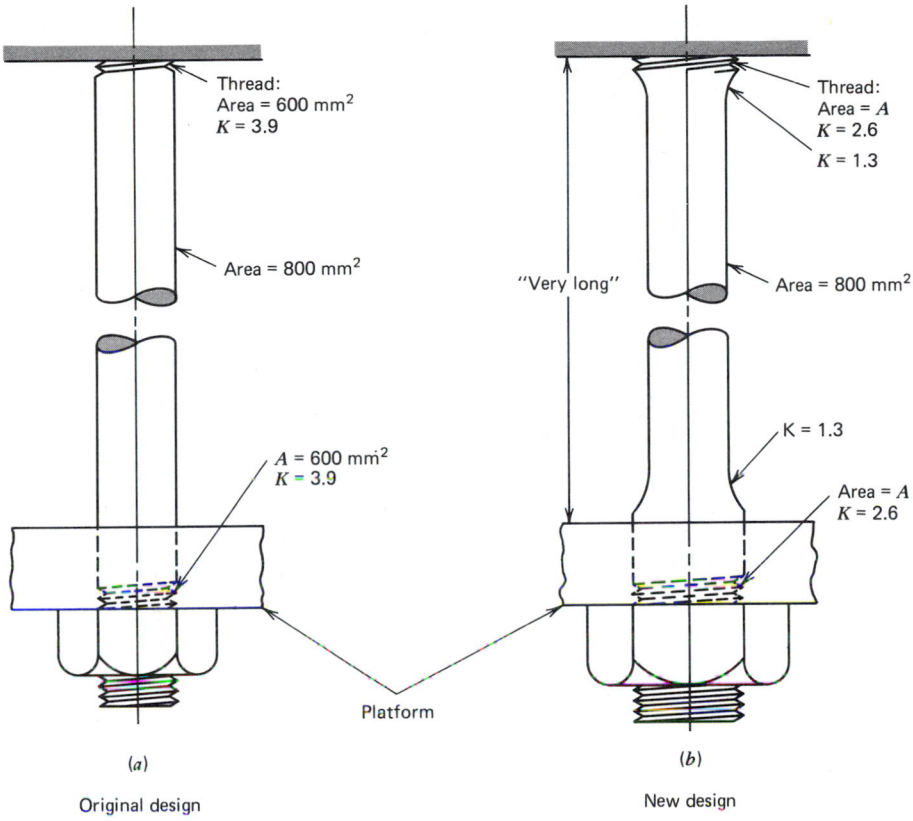

FIGURE P7.7

7.7 A platform is suspended by long steel rods as shown in Fig. P7.7*a*. Because heavy items are sometimes dropped on the platform, it is decided to modify the rods as shown in Fig. P7.7*b* to obtain greater energy absorbing capacity. The new design features enlarged ends, blended into the main portion with generous fillets, and special threads giving less stress concentration.

(a) What is the smallest effective threaded section area A (Fig. P7.7*b*) that would provide maximum energy absorbing capability?

(b) Using this value of A (or that of the next larger standard thread size), what increase in energy absorbing capacity would be provided by the new design?

7.8 The initial design of a bolt loaded in tensile impact is shown in Fig. P7.8*a*. The bolt fractures next to the nut, as shown. A proposed redesign, Fig. P7.8*b*, involves drilling an axial hole in the unthreaded portion, and incorporating a larger fillet radius under the bolt head.

(a) What is the theoretically optimum diameter of the drilled hole?

(b) Using this hole size, by what approximate factor do the modifications increase the energy absorbing capacity of the bolt?

7.9 Figure P7.9 shows a tensile impact bar with a small transverse hole. By what factor does the hole reduce the impact energy absorbing capacity of the bar?

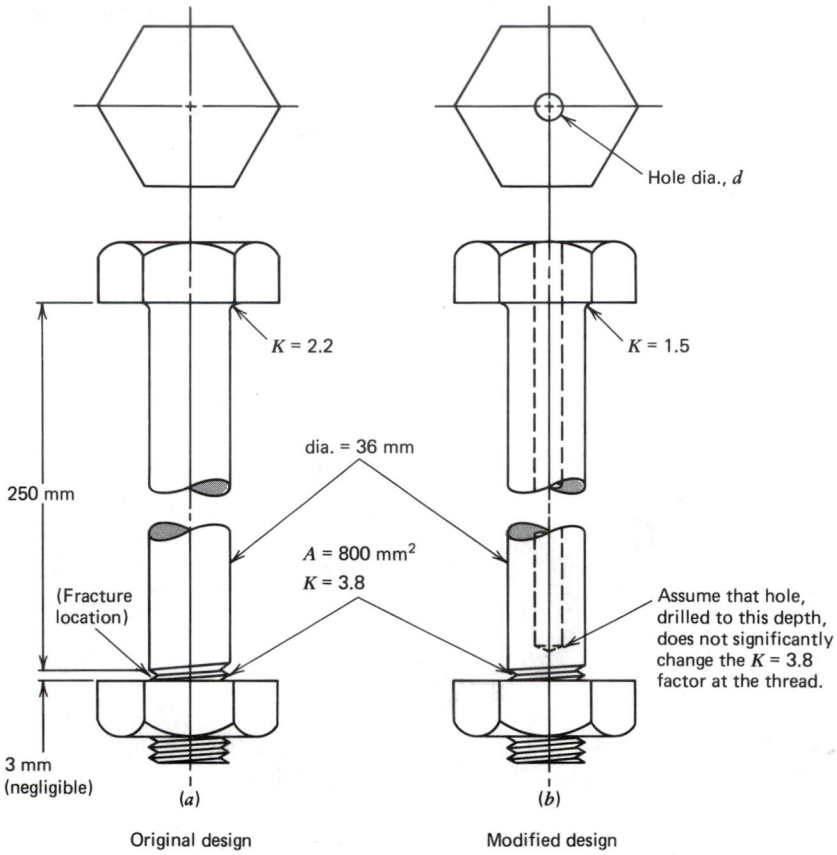

FIGURE P7.8

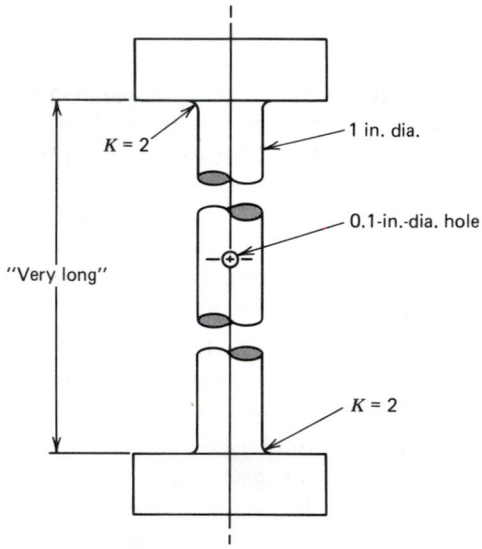

FIGURE P7.9

CHAPTER 8 – FATIGUE

Section 8.3

8.1 Estimate the rotating bending endurance limit and also the 10^3 cycle fatigue strength for standard R. R. Moore test specimens made of steels having Brinell hardnesses of 200, 400, and 500.

8.2 Estimate the long-life rotating bending fatigue strength (state whether it is for 10^8 or 5×10^8 cycles) for standard R. R. Moore specimens made of (a) wrought aluminum, $S_u = 200$ MPa, (b) wrought aluminum, $S_u = 500$ MPa, (c) average grade cast aluminum, and (d) average grade forged magnesium.

Section 8.4

8.3 How would the answers to Problems 8.1 and 8.2 change if the loading is reversed bending rather than rotating bending?

8.4 How would the answers to Problems 8.1 and 8.2 change if the loading is reversed axial loading rather than rotating bending? Assume that the concentricity of load and geometric axes reflects an intermediate condition between precision ground parts and non-precision as-cast or as-forged parts.

Section 8.5

8.5 Repeat Problem 8.1 for reversed torsional loading.

8.6 Repeat Problem 8.2 for reversed torsional loading.

Sections 8.7 and 8.8

8.7 Plot on log-log coordinates estimated S-N curves for (a) bending, (b) axial, and (c) torsional loading of a 1-in.-dia. steel bar having $S_u = 103$ ksi, $S_y = 72$ ksi, and average machined surfaces. For each of the three types of loading, what is the fatigue strength corresponding to (1) 10^6 or more cycles, and (2) 6×10^4 cycles?
(Partial ans.: for bending, 34.8 ksi, 51 ksi)

8.8 Repeat Problem 8.7 for a 20-mm-dia. steel bar having $S_u = 1000$ MPa, $S_y = 650$ MPa, and average ground surfaces.

Section 8.9

8.9 Repeat the determination of the six fatigue strengths in Problem 8.7 for the case of zero-to-maximum (rather than completely reversed) load fluctuation.
(Partial ans.: for bending, 0–52 ksi, 0–68 ksi)

8.10 Repeat the determination of the six fatigue strengths in Problem 8.8 for the case of zero-to-maximum (rather than completely reversed) load fluctuation.

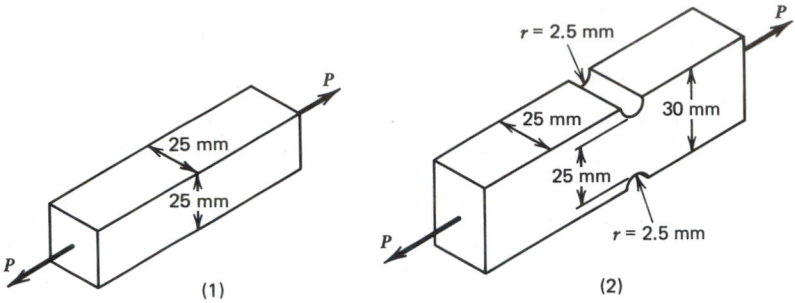

FIGURE P8.11

Section 8.10

8.11 Figure P8.11 shows (1) an unnotched bar and (2) a notched bar of the same minimum cross section. Both bars were machined from AISI 1050 normalized steel. For each bar, estimate (a) the value of static tensile load P causing fracture, and (b) the value of alternating axial load $\pm P$ that would be just on the verge of producing eventual fatigue fracture (after perhaps 1–5 million cycles).
[Ans.: (a) 470 kN for both, (b) 138 kN, 63 kN]

8.12 A stepped shaft, as illustrated in Fig. 4.35, has dimensions of $D = 2$ in., $d = 1$ in., and $r = 0.05$ in. It was machined from steel having tensile properties of $S_u = 80$ ksi and $S_y = 67$ ksi.

(a) Estimate the torque T required to produce static yielding. (Note: For static loading of a ductile material, assume that the very first yielding at the notch-root is not significant; hence, ignore stress concentration.)

(b) Estimate the value of reversed torque, $+T$, required to produce eventual fatigue failure.

(Ans.: 7670 lb-in.; 2050 lb-in.)

8.13 The shaft illustrated in Fig. P8.13 rotates at high speed while the imposed loads remain static. The shaft is machined from AISI 1040 steel, oil quenched and tempered at

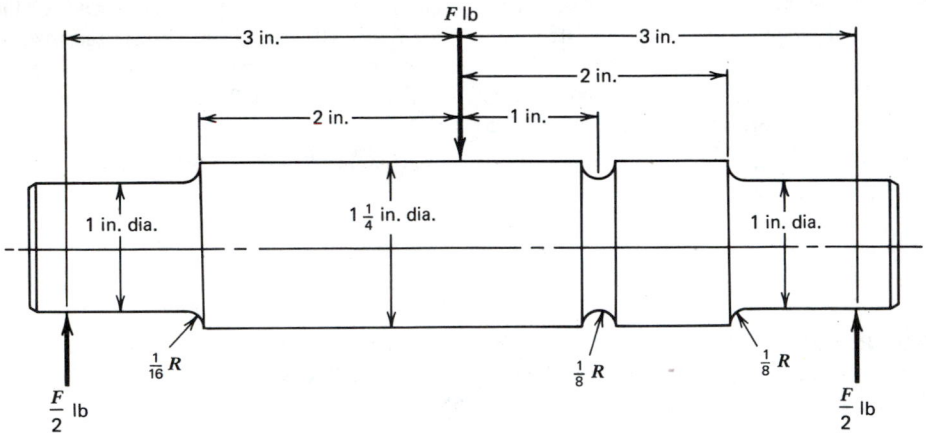

FIGURE P8.13

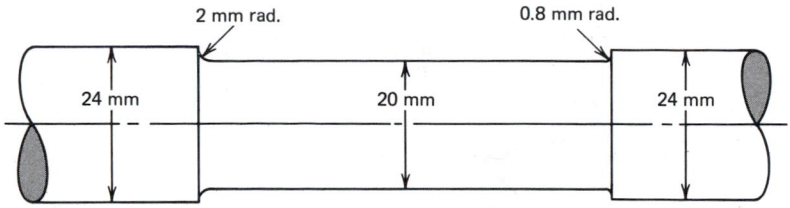

FIGURE P8.14

1000°F. If the loading is sufficiently great to produce a fatigue failure (after perhaps 10^6 cycles), where would the failure most likely occur? (Show all necessary computations and reasoning, but do not do unnecessary computations.)

8.14 When in use, the shaft shown in Fig. P8.14 experiences completely reversed torsion. It is machined from steel having a hardness of 160 Bhn. With a safety factor of 2, estimate the value of reversed torque that can be applied without causing eventual fatigue failure. (Ans.: 59.5 N·m)

Section 8.11

8.15 A grooved shaft like the one shown in Fig. 4.36 is machined from steel of 160 Bhn and $S_y = 55$ ksi. Dimensions are $D = 1.1$ in., $d = 1.0$ in., and $r = 0.05$ in. A commercial polish is given only to the surface of the groove. With a safety factor of 2, estimate the maximum value of torque T that can be applied for infinite life when the fluctuating torsional load consists of: (a) completely reversed torsion, with the torque varying between $+T$ and $-T$, (b) a steady torque of T with superimposed alternating torque of $2T$.
(Ans.: 1240 lb-in., 530 lb-in.)

8.16 Figure P8.16 shows a cantilever beam serving as a spring for a latching mechanism. When assembled, the free end is deflected 0.075 in., which corresponds to a force F of 8.65 lb. When the latch operates the end deflects an additional 0.15 in. Would you expect eventual fatigue failure?
(Ans.: It is marginal.)

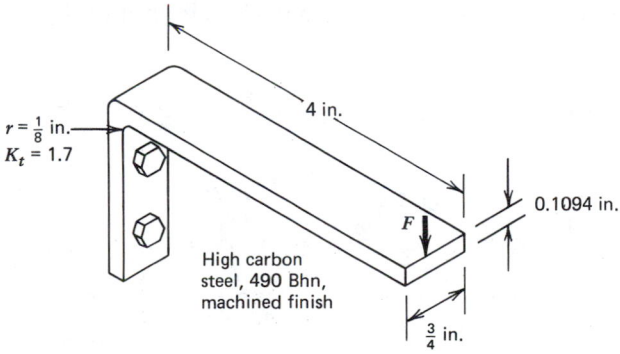

FIGURE P8.16

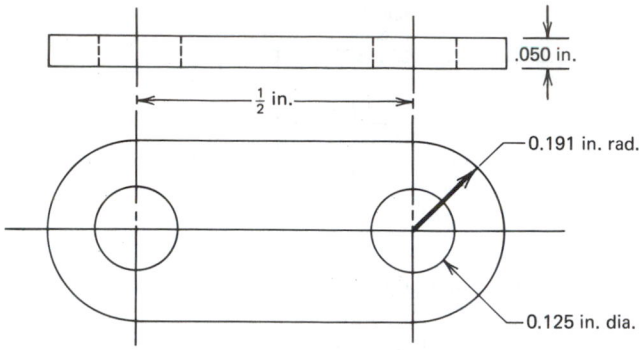

FIGURE P8.20

8.17 A cold-drawn rectangular steel bar of 128 Bhn is 10 mm thick, 60 mm wide, and has a central hole of 12 mm diameter (as in Fig. 4.40). Estimate the maximum tensile force that can be applied to the ends and have infinite life with 90% reliability and a safety factor of 1.3: (a) if the force is completely reversed, (b) if the force varies between zero and a maximum value.
(Ans.: 20 kN, 32 kN)

8.18 A 20-mm-dia. shaft with a 6-mm-dia. transverse hole is made of cold-drawn steel having S_u = 550 MPa and S_y = 462 MPa. Surfaces in the vicinity of the hole have a machined finish. Estimate the safety factor with respect to infinite fatigue life for (a) torque fluctuations between 0 and 90 N·m, (b) a completely reversed torque of 40 N·m, (c) a mean torque of 50 N·m plus a superimposed alternating torque of 35 N·m.
(Ans.: 1.7, 2.4, 2.0)

8.19 For the shaft and loading involved in Problem 8.18, estimate the safety factors with respect to *static* yielding. (Note: For these calculations stress concentration is usually neglected. Why?)
(Ans.: 3.5, 7.8, 3.7)

8.20 Figure P8.20 shows a ½-in. pitch roller chain plate, as used on a bicycle chain. It is made of carbon steel, heat-treated to give S_u = 150 ksi and S_y = 120 ksi. All surfaces are comparable to the "machined" category. Since a roller chain cannot transmit compression, the link is loaded in repeated axial tension (load fluctuates between 0 and a maximum force as the link goes from the slack side to the tight side of the chain) by pins that go through the two holes. Estimate the value of maximum tensile force that would give infinite fatigue life with a safety factor of 1.2.
(Ans.: 240 lb)

8.21 Figure P8.21 shows a shaft and the fluctuating nominal stress (in the center of the 50-

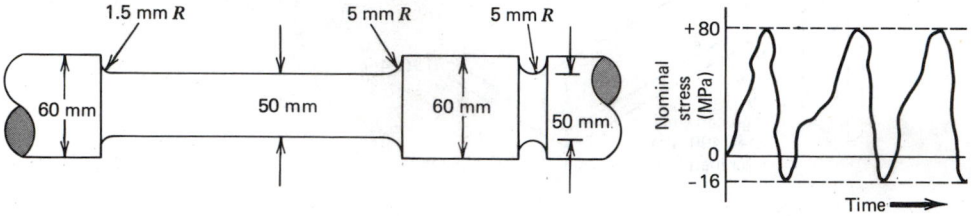

FIGURE P8.21

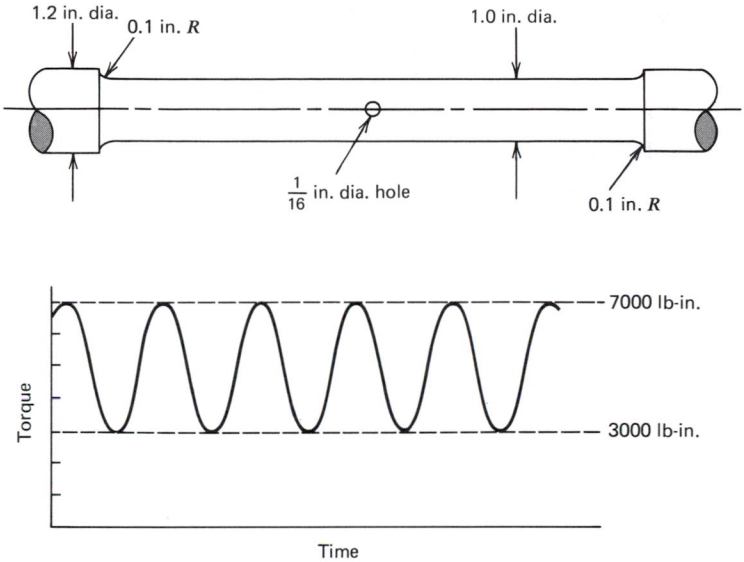

FIGURE P8.22

mm section) to which it is subjected. The shaft is made of steel having $S_u = 700$ MPa and $S_y = 450$ MPa. Estimate the safety factor with respect to eventual fatigue failure if (a) the stresses are bending, (b) the stresses are torsional.

8.22 Figure P8.22 shows a round shaft and a torque fluctuation to which it is subjected. The material is steel, with $S_u = 152$ ksi and $S_y = 135$ ksi. All critical surfaces are ground. Estimate the safety factor for infinite fatigue life with respect to an overload that (a) increases both mean and alternating torque by the same factor, (b) an overload that increases only the alternating torque.

8.23 A stepped shaft, as shown in Fig. 4.35, has dimensions of $D = 10$ mm, $d = 8$ mm, and $r = 0.8$ mm. It is made of steel having $S_u = 1200$ MPa and is finished with a grinding operation. In service, it is loaded with a fluctuating zero-to-maximum torque. Estimate the magnitude of maximum torque which would provide a safety factor of 1.3 with respect to a 75,000 cycle fatigue life.

8.24 The critical portion of a machine part is shaped like the bar in Fig. 4.38 with $H = 35$ mm, $h = 25$ mm, $b = 20$ mm, and $r = 2$ mm. The material is steel, of 160 Bhn hardness. All surfaces are machined. The part is loaded in zero-to-maximum cyclic bending. Estimate the value of maximum bending moment that would give infinite fatigue life with 99% reliability (and a safety factor of 1).
(Ans.: 300 N·m)

8.25 A solid round shaft has a shoulder (as in Fig. 4.35) with $D = 1$ in., $d = 0.5$ in., and r to be determined. The shaft is made of steel, with $S_u = 150$ ksi and $S_y = 120$ ksi. All surfaces are machined. In service the shaft is subjected to a torsional load which fluctuates between 82 and 123 lb-ft. Estimate the smallest fillet radius that would permit infinite life (with safety factor = 1).
(Ans.: about 0.040 in.)

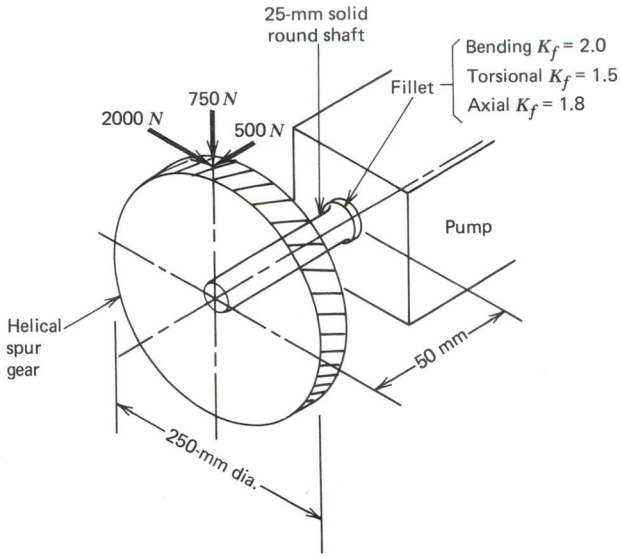

FIGURE P8.27

8.26 A steel shaft used in a spur gear reducer is subjected to a constant torque together with lateral forces that tend always to bend it downward in the center. These result in calculated stresses of 80 MPa torsion and 60 MPa bending. However, these are nominal values and do not take into account stress concentration caused by a shoulder (as in Fig. 4.35), where dimensions are D = 36 mm, d = 30 mm, and r = 3 mm. All surfaces are machined, and the steel has strength values of S_u = 700 MPa and S_y = 550 MPa. Hardness is 200 Bhn. Estimate the safety factor with respect to infinite fatigue life.
(Ans.: 1.9)

8.27 Figure P8.27 shows a portion of a pump which is gear-driven at uniform load and speed. The shaft is supported by bearings mounted in the pump housing. The shaft is made of steel having S_u = 1000 MPa, S_y = 800 MPa. The tangential, axial, and radial components of force applied to the gear are shown. The surface of the shaft fillet has been shot-peened, which is estimated to be equivalent to a laboratory mirror-polished surface. Fatigue stress concentration factors for the fillet have been determined and are shown on the drawing. Estimate the safety factor with respect to eventual fatigue failure at the fillet.
(Ans.: 1.9)

8.28 Drawing 1 of Figure P8.28 shows a countershaft with helical gear (B), bevel gear (D), and two supporting bearings (A and C). Loads acting on the bevel gear are shown. Forces on the helical gear can be determined from equilibrium of moments about the shaft axis plus the given proportions of the helical gear force components. Shaft dimensions are given in drawing 2. All shoulder fillets (at points where diameter changes) have a radius of 5 mm. Note that the shaft is designed so that only bearing A takes thrust. The shaft is made of hardened steel, S_u = 1069 MPa, S_y = 896 MPa. All important surfaces are finished by grinding.

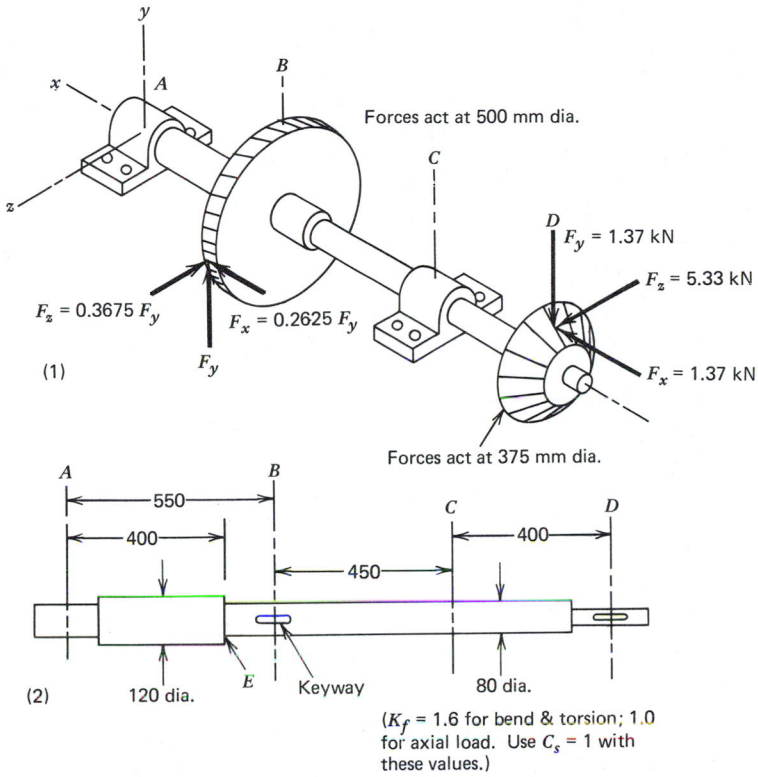

FIGURE P8.28

(a) Draw load, shear force, and bending moment diagrams for the shaft in the *xy*- and *xz*-planes. Also draw diagrams showing the intensity of the axial force and torque along the length of the shaft.

(b) At points *B*, *C*, and *E* of the shaft, calculate the equivalent stresses in preparation for making a fatigue safety factor determination. (Note: Refer to Fig. 8.17.)

(c) For a reliability of 99% (and assuming $\sigma = 0.08\ S_n$), estimate the safety factor of the shaft at points *B*, *C*, and *E*.

[Ans. to (c): 5.0, 6.8, and 5.8, respectively]

Section 8.12

8.29 A stepped shaft, as shown in Fig. 4.35, has dimensions of $D = 2$ in., $d = 1$ in., and $r = 0.1$ in. It was machined from AISI steel of 200 Bhn hardness. The loading is one of completely reversed torsion. During a typical 30 seconds of operation under overload conditions the nominal (Tc/J) stress in the 1-in.-dia. section was measured to be as shown in Fig. P8.29. Estimate the life of the shaft when operating continuously under these conditions.

(Ans.: Roughly 43 hours)

17.2

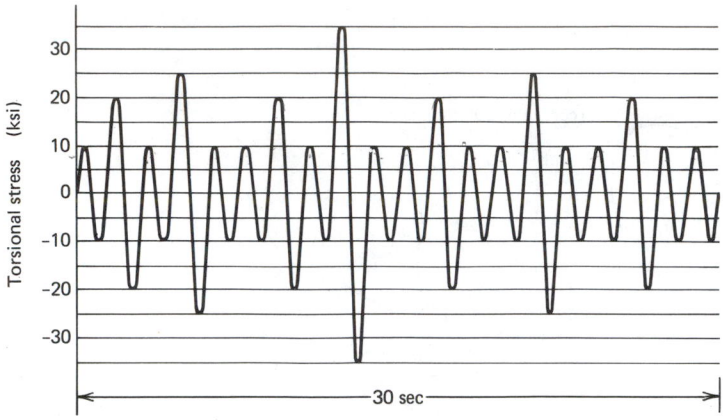

FIGURE P8.29

CHAPTER 9 – SURFACE DAMAGE

Sections 9.2-9.4

9.1 Metal plates having a total exposed area of 1 m^2 are fastened together with rivets having a total exposed area of 100 cm^2. The environment involves moisture, and possibly some salt. Consider two cases: (1) steel plates with copper rivets, and (2) copper plates with steel rivets. (a) For each case, which metal will corrode? (b) How do the rates of corrosion compare for the two cases? (c) If twice as many rivets are used, what influence would this have on the total rate of corrosion?

9.2 Some plumbing codes require that an electrical insulator be used when copper tubing is connected to steel pipe. With the aid of a simple sketch, explain the rationale behind this requirement.

9.3 Corrosion of the internal steel surfaces of the crankcase of a particular engine is a problem. Someone suggests replacing the steel oil drain plug with one made of magnesium. You recommend retaining the steel plug, but that one end of a small magnesium rod be embedded in the inside surface of the plug. Explain, briefly.

Section 9.12

9.4 A latching mechanism involves steel mating surfaces of 100 and 300 Bhn rubbing back and forth over a distance of 20 mm each time the latch is operated. Lubrication is questionable (the surfaces are supposed to get a drop of oil every few months). The latch is operated an average of 30 times per day, every day. Estimate the volume of metal that will wear away from the softer steel member during one year of use if the compressive load between the surfaces is 100N.

Sections 9.13 and 9.14

9.5 The ball and socket joint at the end of a rocker arm involves a hardened steel spherical surface of 10-mm-dia. fitting in a hard-bronze bearing-alloy spherical seat of 10.1 mm diameter. What maximum contact stress will result from a load of 1800 N?

9.6 Figure P9.6 shows a Geneva indexing mechanism used, for example, on the indexing heads of machine tools. Each time the driving arm makes one revolution the Geneva wheel (for the 4-slot design illustrated) turns 90°. The arm supports a rolling cylindrical hardened steel pin that fits into the Geneva wheel slots. The pin is to have the same length and diameter. The wheel is made of heat-treated alloy cast iron (E = 140 GPa, ν = 0.25). For a design contact stress of 700 MPa, determine the smallest acceptable pin diameter if the design overload torque (normal torque times safety factor) applied to the arm is 50 N·m.

9.7 Two mating steel spur gears are 20 mm wide, and the tooth profiles have radii of curvature at the line of contact of 10 mm and 15 mm. A force of 200 N is transmitted between them. (a) Compute the maximum contact pressure and the width of contact. (b) How deep below the surface is the maximum shear stress, and what is its value?
[Ans.: (a) 246 MPa, 0.052 mm; (b) 0.021 mm, 74 MPa]

9.8 Power is transmitted between two steel rollers pressed together, as in Fig. 9.16. Loading is such that the maximum contact pressure is 1 GPa and the width of contact 1 mm. There is slight slippage, and the coefficient of friction is estimated to be 0.3.

(a) What is the maximum completely reversed shear stress, τ_{yz}, and how far on either side of the load line does it occur?
(b) What is the maximum value of reversed tensile stress developed at the surface?
(c) What is the maximum value of shear stress developed at the surface?
(d) Explain briefly the kinds of surface deterioration that might occur?
[Ans.: (a) 0.23 GPa, 0.45 mm; (b) 0.6 GPa, (c) 0.36 GPa]

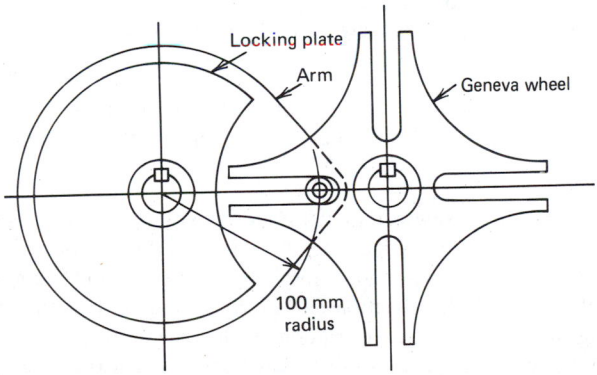

FIGURE P9.6

CHAPTER 10 – THREADED FASTENERS AND POWER SCREWS

Section 10.3

10.1 A square-threaded power screw with single thread is used to raise a load of 20,000 lb. The screw has a mean diameter of 1 in., and four threads per inch. The collar mean diameter is 1.5 in. The coefficient of friction is estimated as 0.1 for both the thread and the collar.

(a) What is the major diameter of the screw?
(b) Estimate the screw torque required to raise the load.
(c) If collar friction is eliminated, what minimum value of thread coefficient of friction is needed to prevent the screw from overhauling?

[Ans.: (a) 1.125 in., (b) 3310 lb-in., (c) 0.08]

10.2 A double-threaded Acme stub screw of 2 in. major diameter is used in a jack having a plain thrust collar of 2.5 in. mean diameter. Coefficients of running friction are estimated as 0.10 for the collar and 0.11 for the screw.

(a) Determine the pitch, lead, thread depth, mean pitch diameter, and helix angle of the screw.
(b) Estimate the starting torque for raising and for lowering a 6000-lb load.
(c) If the screw is lifting a 6000-lb load at the rate of 4 ft/min, what is the screw rpm? What is the efficiency of the jack under this steady-state condition?
(d) Would the screw overhaul if a ball thrust bearing (of negligible friction) were used in place of the plain thrust collar?

10.3 A jack similar to the ones shown in Fig. 10.5 uses a single square-thread screw to raise a load of 60 kN. The screw has a major diameter of 36 mm and a pitch of 6 mm. The thrust collar mean diameter is 80 mm. Running friction coefficients are estimated to be 0.15 for the screw and 0.12 for the collar:

(a) Determine the thread depth and helix angle.
(b) Estimate the starting torque for raising and lowering the load.
(c) Estimate the efficiency of the jack for raising the load.
(d) Estimate the power required to drive the screw at a constant 1 revolution per second?

10.4 An ordinary C-clamp uses a ½ in. Acme thread and a collar of ⅝ in. mean diameter. Estimate the force required at the end of a 5-in. handle to develop a 300 lb clamping force.

10.5 Two identical 3 in. major diameter power screws (single threaded) with modified square threads are used to raise and lower a 50-ton sluice gate of a dam. Quality of construction and maintenance (including lubrication) are good, resulting in an estimated friction coefficient of only 0.1 for the screw. Collar friction can be neglected, as ball thrust bearings are used. Assuming that, because of gate friction, each screw must provide a lifting force of 26 tons, what power is required to drive each screw when the gate is being raised at the rate of 3 ft/min? What is the corresponding rotating speed of the screws?

Section 10.4

10.6 Calculate the nominal values of torsional, axial, thread bearing, and thread shear stresses under starting conditions for the power screws in Problems (a) 10.1, (b) 10.2, (c) 10.3, (d) 10.4, and (e) 10.5. Assume in each case that the length of thread engagement is 1.5 times the outside diameter of the screw.

Section 10.9

10.7 The cylinder head of a piston-type air compressor is held in place by ten bolts. Total joint stiffness is four times total bolt stiffness. Each bolt is tightened to an initial tension of 6000 N. The total external force acting to separate the joint fluctuates between 0 and 30,000 N. Assume that the bolt size and material are such that the bolt load remains within the elastic range. Make a graph (plotting force vs. time) showing three or four external load fluctuations, and corresponding curves showing the fluctuations in total bolt load and total joint clamping force.

10.8 Repeat Problem 10.7 except for an external joint separating force that varies between 20,000 and 30,000 N.

10.9 Two parts of a machine are held together by bolts that are initially tightened to provide a total initial clamping force of 10,000 N. The elasticities are such that $k_c = 2k_b$:

(a) What external separating force would cause the clamping force to be reduced to 2000 N (assume that the bolt remains within its elastic range)?

(b) If this separating force is repeatedly applied and removed, what will be the values of mean and alternating force acting on the bolts?

10.10 Drawings 1 and 2 of Fig. P10.10 are identical except for placement of the spring washer. The bolt and the clamped members are "infinitely" rigid in comparison with the spring washer. In each case the bolt is initially tightened to a force of 10,000 N before the two 1000-N loads are applied.

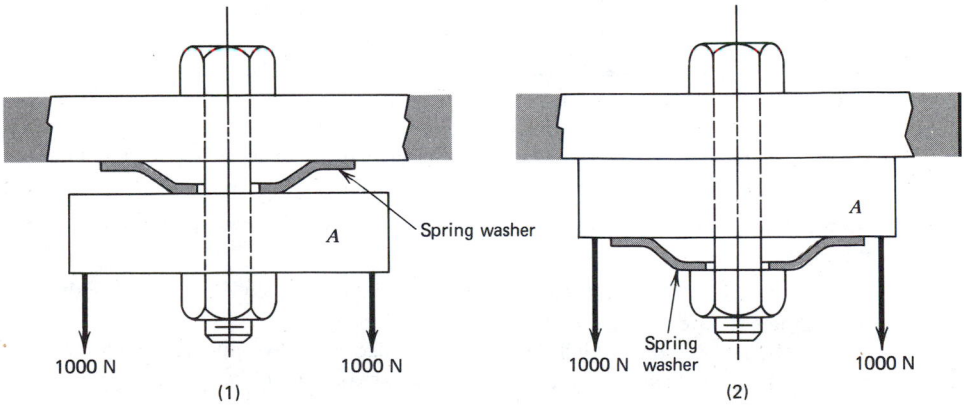

FIGURE P10.10

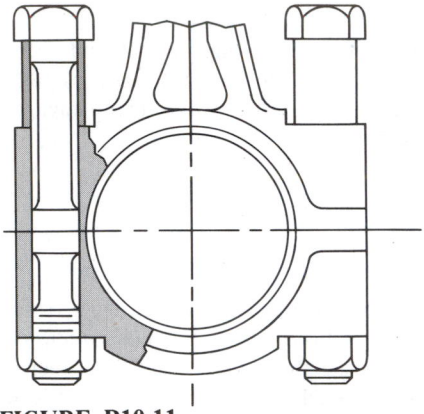

FIGURE P10.11

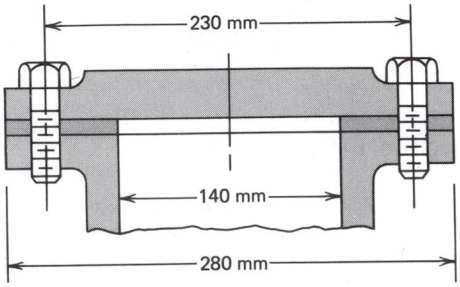

FIGURE P10.15

(a) For both arrangements, draw block A as a free-body in equilibrium.
(b) For both arrangements, draw a bolt-force-vs-time plot for the case involving repeated application and removal of the 1000-N external loads.

10.11 Figure P10.11 shows an idea for reducing the fluctuating load in a connecting rod bolt. Will it work? Explain, briefly.

Section 10.10

10.12 The bolts that attach a bracket to an industrial machine must each carry a static tensile load of 4 kN.

(a) Using a safety factor of 5 (based on proof strength), what size of class 5.8 coarse-thread metric bolts is required?
(b) If the nuts are made of a steel with ⅔ the yield strength and proof strength of the bolt steel, what is the least number of threads that must be engaged for the thread shear strength to be equal to the bolt tensile strength?

[Ans.: (a) M10 x 1.5, (b) 4.3]

10.13 What size of UNF bolt made from SAE grade 5 steel is needed to carry a static tensile load of 3000 lb with a safety factor of 4 based on proof strength? If this bolt is used with a nut made of steel corresponding to SAE grade 2 specifications, what is the least number of threads that must be engaged for the thread shear strength to be equal to the bolt tensile strength?

10.14 An industrial motor weighing 22 kN is to be provided with a steel eye bolt for use when it is lifted.

(a) What size of class 8.8 steel bolt would you recommend? Explain briefly your choice of safety factor.
(b) If the housing into which the bolt is threaded has only half the yield strength of the bolt steel, what minimum number of threads should be engaged?

10.15 Figure P10.15 shows a pressure vessel with a gasketed end plate. The internal pressure is sufficiently uniform that the bolt loading can be considered static. The gasket supplier advises a gasket clamping pressure of at least 13 MPa (this includes an appropriate safety factor) to insure a leakproof joint. To simplify the problem, you may neglect the bolt holes when calculating gasket area.

(a) For 12-, 16-, and 20-mm bolts with coarse threads and made of SAE class 8.8 or 9.8 steel (whichever is appropriate), determine the number of bolts needed.

(b) If the ratio of bolt spacing to bolt diameter should not exceed 10 in order to maintain adequate flange pressure between bolts, and if this ratio should not be less than 5 in order to provide convenient clearance for standard wrenches, which of the bolt sizes considered gives a satisfactory bolt spacing?

[Ans.: (a) 13, 7, and 5; (b) 16 and 20 mm]

Section 10.11

10.16 A bolt made of steel with assumed idealized stress-strain curve is initially tightened to its full yield strength (i.e., $F_i = A_t S_y$) of 80 kN. Elasticities are such that $k_c = 3k_b$. Figure P10.16 shows a plot of load vs. time which covers initial tightening, light load operation, heavy load operation, light load operation, and finally, load off.

(a) Copy the figure and complete the curves for bolt force and clamping force.

(b) After the above loading experience, the bolt is removed from the machine. Make a simple drawing showing the shape of the residual stress curve in a threaded section of the bolt.

10.17 A grade 5, 1-in. UNF bolt has rolled threads, and is used to clamp two rigid members together such that $k_c = 4k_b$. In keeping with the fatigue treatment of ductile members given in Chapter 8, assume that the material has an idealized stress-strain curve, with "square corner" at the yield strength (not the proof strength). There is a force tending to separate the members that fluctuates rapidly between 0 and 20,000 lb. The construc-

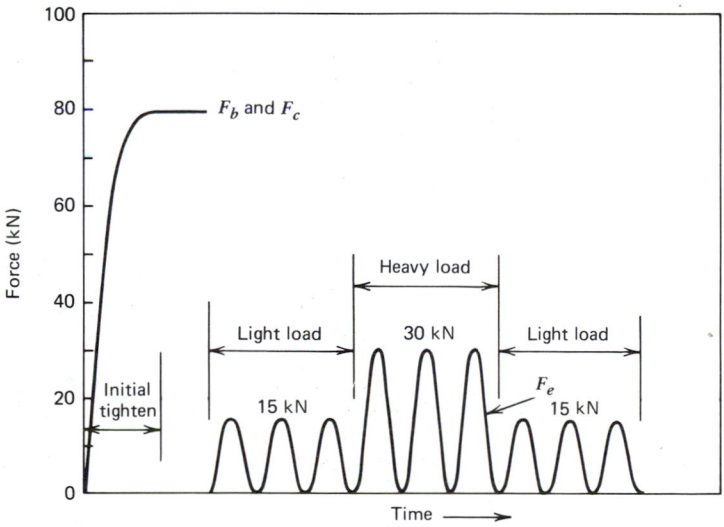

FIGURE P10.16

tion and precision of manufacture are such that there is a possibility of slight bolt bending. Two values of initial tension are to be investigated: (1) the normal value that a mechanic might tend to apply [Eq. (e) in Section 10.7], and (2) the theoretical limiting value of $A_t S_y$.

(a) Make a force vs. time plot (like Fig. P10.16) showing the fluctuating external load, bolt force, and fluctuating clamping force for each of the two values of initial tension. Begin the plot with initial tightening, show about three load cycles, and conclude with load off.

(b) Estimate the safety factor corresponding to each of the values of initial tension, where failure is considered to be *either* eventual fatigue fracture *or* joint opening (clamping force dropping to zero).

[Ans.: (b) Approximately 1.0 for initial tension (1); 2.0 for initial tension (2)]

10.18 A grade 7, ¾-in. UNF bolt with rolled threads is used in a joint having a soft gasket such that the clamped member stiffness is only half the bolt stiffness. The bolt initial tension corresponds to Eq. (e) in Section 10.7. During operation, there is an external separating force that fluctuates between 0 and P. This application involves negligible bending of the bolts.

(a) Estimate the maximum value of P that would not cause eventual fatigue failure.

(b) Estimate the maximum value of P that would not cause joint separation.

[Ans.: (a) Approximately 5600 lb, (b) 36,000 lb]

10.19 The cap of an automotive connecting rod is secured by two class 10.9, M8 x 1.25 bolts with rolled threads. The grip and unthreaded length can both be taken as 16 mm. The connecting rod cap (the clamped member) has an effective cross-section area of 250 mm^2 per bolt. Each bolt is tightened to an initial tension of 22 kN. The maximum external load, divided equally between the bolts, is 18 kN.

(a) Estimate the bolt tightening torque required.

(b) What is the maximum total load per bolt during operation?

(c) Figure P10.19 shows free-body diagrams for zero load. Make a similar pair of

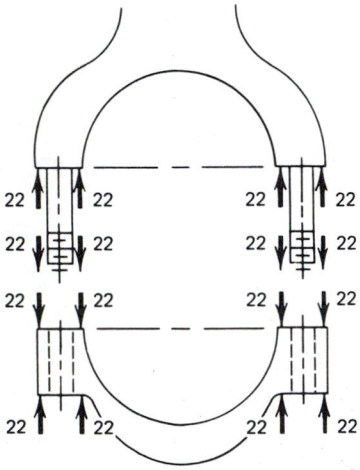

FIGURE P10.19

free-body diagrams when the maximum load of 18 kN is pushing downward on the center of the cap.

(d) What is the safety factor (factor by which the 18 kN load could be multiplied) for fatigue. Assume an idealized stress-curve based on S_y.

[Ans.: (a) 35 N·m, (b) 23.5 kN, (d) approximately 1.0]

10.20 Two grade 8.8 bolts with M20 x 2.5 rolled threads are used to attach a pillow block similar to the one shown in Fig. 10.30. The bolts are initially tightened in accordance with Eq. (10.11a). Joint stiffness is estimated to be three times bolt stiffness. Bolt bending may be assumed negligible. During operation the external load tending to separate the pillow block from its support varies rapidly between 0 and P.

(a) Estimate the maximum value of P that would not cause eventual fatigue failure of the bolts. (Assume an idealized stress-strain curve based on S_y.)

(b) Show on a mean stress-alternating stress diagram points representing thread-root stresses: (1) just after initial tightening, (2) during operation with the load fluctuating between 0 and $P/2$ (i.e., using your value of P and a safety factor of 2), and (3) with the machine shut down after operating with the 0 to $P/2$ load.

[Ans.: (a) Approximately 183 kN]

10.21 Two aluminum plates, which are part of an aircraft structure, are held together by a grade 7, ½-in. UNF bolt. The effective area of the aluminum plates in compression is estimated to be 12 times the cross-sectional area of the steel bolt. The bolt is initially tightened to 90% of its proof strength. Gust loads, varying from zero to P, tend to pull the plates apart (this results in negligible bolt bending). With a safety factor of 1.3, what is the maximum value of P that will not cause eventual bolt fatigue failure? What clamping force will remain when this value of P acts?

Section 10.12

10.22 Review your solutions to the problems listed below and, in the light of information in Section 10.12 and Table 10.7, comment on the likely accuracy of the fatigue results. If previous designs had been made based on these earlier results, would it now appear important to specify that the bolt threads be rolled after heat-treatment?

(a) Problem 10.17.
(b) Problem 10.18.
(c) Problem 10.19.
(d) Problem 10.20.
(e) Problem 10.21.

10.23 A critical application requires the smallest possible bolt for resisting a dynamic separating force varying from 0 to 100 kN. It is estimated that by using an extra high strength bolt steel with $S_p = 1200$ MPa, and using special equipment to control initial tightening to the full A_tS_p, a stiffness ratio of $k_c/k_b = 6$ can be realized. Also, any of the bolt threads and finishes listed in Table 10.7 may be selected. With a safety factor of 1.3 with respect to eventual fatigue failure, what is the smallest size metric bolt that can be used? State the thread and finish selected. With this bolt tightened as specified, what clamping force will remain (at least initially) when the 100-kN load is applied?

CHAPTER 11 – RIVETS, WELDING, AND BONDING

Section 11.4

11.1 Two ⅜-in. steel plates with S_y = 50 ksi are butt welded (as in Fig. 11.6). E60 series welding rods are used, and good welding practice is followed. With a safety factor of 3, what tensile load can be applied to the plates per inch of welded plate width?
(Ans.: 6000 lb)

11.2 Two 15-mm steel plates with S_y = 400 MPa are butt welded, using E70 series welding rods and good welding practice. The weld length is 100 mm. What is the maximum tensile load that can be applied to the joint with a safety factor of 3?

11.3 Two steel plates with S_y = 50 ksi are attached by ⅜-in. parallel-loaded fillet welds, as shown in Fig. 11.7*c*. E60 series welding rods are used, and good welding practice is followed. Each of the welds is 3 in. long. With a safety factor of 3, what maximum tensile load can be applied?
(Ans.: 15,400 lb)

11.4 Two steel plates with S_y = 400 MPa are attached by transverse-loaded fillet welds, as shown in Fig. 11.7*d*. Each weld is 100 mm long. E70 series welding rods are used, and good welding practice is followed. What minimum leg length must be used if a force of 150 kN is to be applied with a safety factor of 3?

Section 11.5

11.5 The bracket shown in Fig. P11.5 is to support a total load (equally divided between the two sides) of 60 kN. Using E60 series welding rods and a safety factor of 3.5, what size weld should be specified?

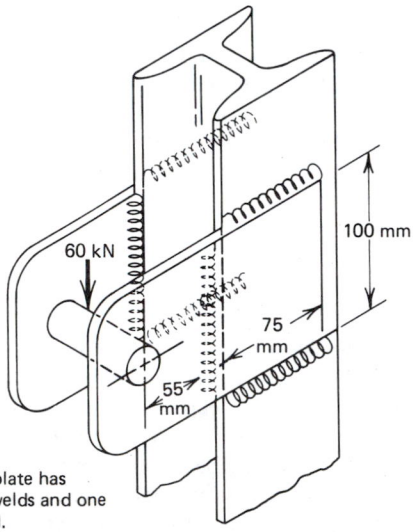

Note: Each plate has two 75 mm welds and one 100 mm weld.

FIGURE P11.5

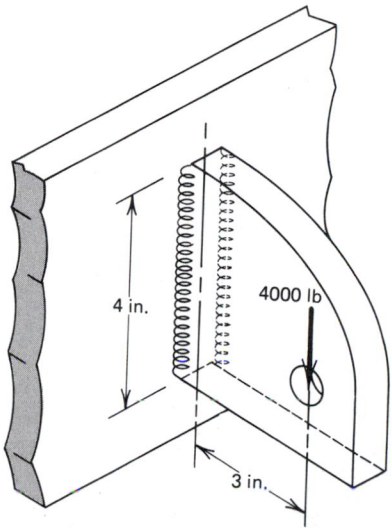

FIGURE P11.6

11.6 The bracket shown in Fig. P11.6 supports a 4000 lb load. The fillet weld extends for the full 4 in. length on both sides. What weld size is required to give a safety factor of 3.5 if Series E60 welding rod is used.

11.7 In Sample Problem 11.4, determine the weld size required if only the top (AB) is welded.

Section 11.6

11.8 Two 20-mm steel plates are butt welded together. Both the plate and the welding electrode materials correspond to strength properties of $S_u = 500$ MPa and $S_y = 400$ MPa. The imposed loading fluctuates rapidly between -20 kN and $+60$ kN. Estimate the length of weld required in order to provide a safety factor of 2.5:

(a) If the weld "reinforcement" is not removed.
(b) If the excess weld material is carefully ground off so as to provide smooth, continuous surfaces.

CHAPTER 12 – SPRINGS

Section 12.2

12.1 Figure P12.1 shows a fully opened trap door covering a stairwell. The door weighs 50 lb, with center of gravity 2 ft from the hinge. A torsion bar spring, extending along the hinge axis, serves as a counterbalance. Determine the length and diameter of a solid steel torsion bar that would counterbalance 80% of the door weight when closed, and provide a 5 lb-ft torque holding the door against the stop shown. Use a maximum allow-

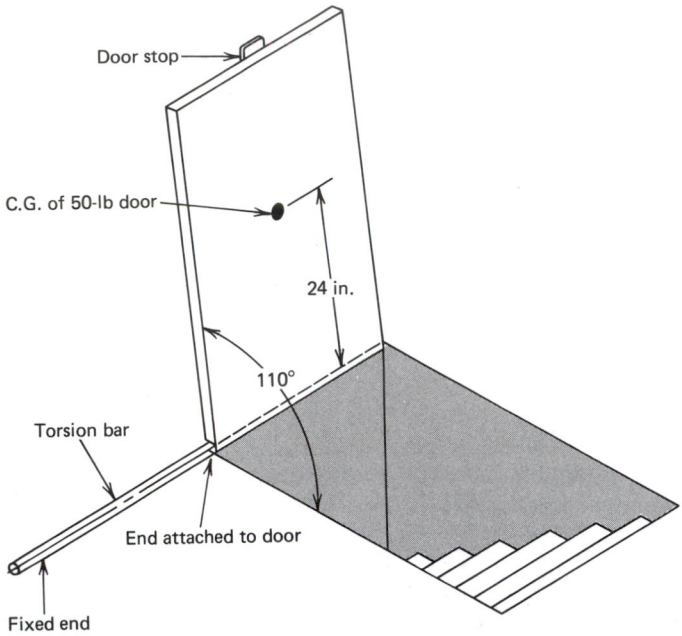

FIGURE P12.1

able torsional stress of 50 ksi. Make a graph showing gravity torque, spring torque, and net torque all plotted against door opening angle.
(Ans.: 100 in., 0.46 in.)

12.2 Suppose that the trap door in Problem 12.1 must be designed to open only 60° (due to clearance problems). Determine the length and diameter of torsion bar that would counterbalance 80% of the door weight when closed, and provide a *net* torque of 10 lb-ft to hold the door against the 60° stop when open. Make a graph showing gravity torque, spring torque, and net torque all plotted against door opening angle.
(Ans.: 221 in., 0.46 in.)

12.3 Repeat Problems 12.1 and 12.2 using a door weight of 200 N with center of gravity 600 mm from the hinge. Spring torque for door open position corresponding to Problem 12.1 is to be 7 N·m; net torque for door open position corresponding to Problem 12.2 is to be 14 N·m. Maximum allowable torsional stress is 350 MPa.

Sections 12.3-12.7

12.4 A helical coil spring with $D = 48$ mm and $d = 9.5$ mm is wound with a pitch (distance between corresponding points of adjacent coils) of 14 mm. The material is ASTM A227 cold drawn carbon steel. If the spring is compressed solid, would you expect it to return to its original free-length when the force is removed?

12.5 Figure P12.5 shows a coiled compression spring that has been loaded against a support by means of a bolt and nut. After the nut has been tightened to the position shown, an

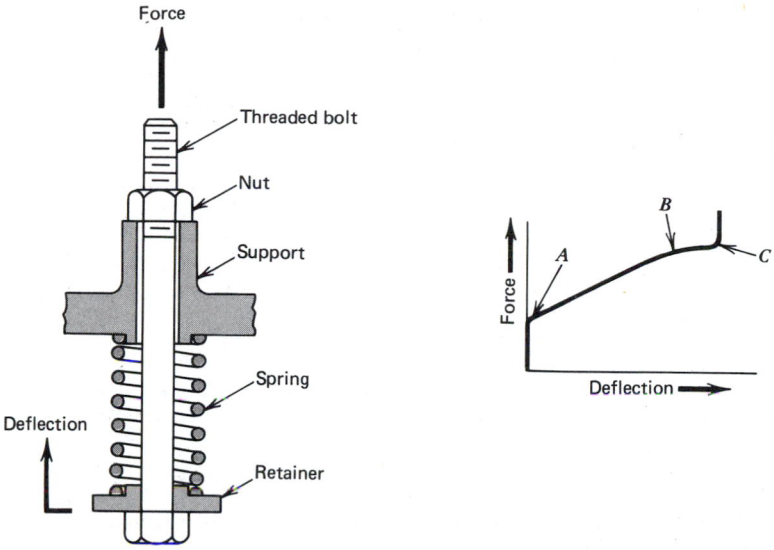

FIGURE P12.5

external force F is applied to the bolt as indicated, and the deflection of the spring is measured as F is increased. A plot of the resulting force-deflection curve is shown. Briefly but clearly state the reasons the curve changes at points A, B, and C.

12.6 A conical compression coil spring, as illustrated in Fig. 12.6, is made of 4-mm-dia. steel wire and has an active coil diameter that varies from 23 mm at the top to 50 mm at the bottom. The pitch (axial spacing between corresponding points on adjacent coils) is 7 mm throughout. There are four active coils. A force is applied to compress the spring, and the stresses always remain in the elastic range.

(a) As the force is increased, which coil deflects to zero pitch first?
(b) Calculate the magnitude of force required to cause the deflection in (a).
(c) Suppose the force is increased until the entire spring is compressed solid. Neatly sketch a force vs. deflection force for the spring.

12.7 A machine uses a pair of concentric helical compression springs to support an essentially static load of 2.25 kN. Both springs are made of steel and have the same length when loaded and when unloaded. The outer spring has $D = 45$ mm, $d = 8$ mm, $N = 5$; the inner spring $D = 25$ mm, $d = 5$ mm, and $N = 10$. Calculate the deflection, and also the maximum stress in each spring.

12.8 A helical compression spring is to be made with a wire diameter of 2 mm, an outside diameter of 20 mm, 10 active coils, and ends closed and ground. The least expensive steel spring wire is to be used, and presetting will not be used.

(a) Estimate the maximum static load that can be applied without encountering more than 2% long-term set.
(b) What is the spring rate?
(c) What should be the free-length of the spring in order that it becomes solid when the load determined in (a) is applied?
(d) Would buckling problems be encountered if one end plate is free to tilt?

(Ans.: 115 N, 2.71 N/mm, 66.4 mm, no)

12.9 A helical compression spring with squared and ground ends is required to exert a maximum working force of 1100 N, and a force of 600 N when the spring is 60 mm longer. Sufficiently few load cycles will be involved to justify designing the spring on the basis of static loading. It is desired to use an existing supply of 5-mm-dia. music wire. Presetting is to be used. Determine appropriate values for D, N, and L_f. Check for possible buckling.

12.10 A helical compression spring with squared and ground ends is to be made of steel, and presetting is to be used. The loading can be considered static. Maximum working force is 100 lb. A force of 50 lb is required when the spring is 1.5 in. longer. Use recommended clash allowance and a steel having $S_u = 200$ ksi. For a ratio of $C = 8$, determine appropriate values for D, d, N, and L_f.
(Ans.: $D = 1.08$ in., $d = 0.135$ in., $N = 11.38$, $L_f = 5.11$ in.)

12.11 A helical compression spring used for essentially static loading has $d = 0.100$ in., $D = 0.500$ in., $N = 8$, and squared and ground ends. It is made of ASTM A227 cold drawn steel wire.

(a) Compute the spring rate and the solid height.
(b) Estimate the greatest load that can be applied without causing long-term permanent set in excess of 2%.
(c) What spring free-length will result in the load determined in (b) just causing the spring to become solid?

[Ans.: (a) 144 lb/in., 1.00 in., (b) 69.3 lb, (c) 1.48 in.]

12.12 A particular machine requires a helical compression spring, having ends squared and ground, to support an essentially static load of 500 lb. The spring constant is to be 200 lb/in., and the stress at the design load is to be 80 ksi. The clash allowance is to be 0.10 in. The dimensions of related parts establishes that D should be 3 in. Determine N, d, and L_f.
(Ans.: 5.28, 0.375 in., 5.33 in.)

Section 12.8

12.13 An automatic production machine requires a compression coil spring to keep a follower in contact with a cam that rotates at speeds up to 2000 rpm. When installed, the spring force is to vary between 140 and 590 N while the spring height varies over a range of 10 mm. An available supply of 4.5-mm-dia. shot-peened wire is to be used, for which Fig. 12.16 applies. Clash allowance is to be 2.5 mm. The design is to be tentatively based on limiting the stress to 800 MPa when the spring is closed solid. Ends are to be squared and ground, and presetting is *not* to be used.

(a) Determine appropriate values for D, N, L_s, and L_f.
(b) Determine the likelihood of buckling, the likelihood of encountering spring surge problems, and the approximate safety factor during normal operation.

[Ans.: (a) $D = 33.75$ mm, $N = 2.35$, $L_s = 19.58$ mm, $L_f = 35.2$ mm; (b) no buckling or surge problems are indicated; safety factor ≈ 1.1]

12.14 What would be the approximate safety factor of the spring in Problem 12.13 (with respect to fatigue failure) if presetting were used, resulting in a residual torsional stress of 100 MPa?

12.15 A coil spring with squared and ground ends is to operate with a load that fluctuates between 100 and 200 lb, during which the deflection is to vary by 1 in. We wish to use an available supply of steel spring wire having $d = 0.200$ in. and fatigue strength properties as shown on Fig. 12.16 for shot-peened wire. Presetting, and a clash allowance of ¼ in. are to be used. As a safety factor, the Fig. 12.16 curve for *non*-shot-peened wire is to be used for design stresses, and residual stresses due to presetting are not to be taken into account. This is considered to provide a sufficient safety factor that no consideration need be given to possible higher stresses due to the spring closing solid. Determine appropriate values for N, D, and L_f.
(Ans.: $N = 14.7$, $D = 1.16$ in., $L_f = 5.59$ in.)

12.16 A coil spring with squared and ground ends is to operate with a load which fluctuates between 50 and 100 lb, during which the deflection is to vary by ½ in. Due to space limitations, a mean coil diameter of 2 in. has been selected. Steel spring wire corresponding to shot-peened wire in Fig. 12.16 is to be used. To provide an appropriate safety factor, design stresses are to be taken from the non-shot-peened curve, and the beneficial effect of the specified presetting is not to be taken into account in the calculations. It is considered that this will provide sufficient safety factor to account for the possibility of higher stresses, should the spring ever close solid. Choose an appropriate clash allowance, and determine appropriate values for N, d, and L_f.
(Partial ans.: $N = 2.02$, $d = 0.183$ in., $L_f = 1.83$ in.)

12.17 A helical compression spring is subjected to a load fluctuating between 90 and 225 N. Fatigue properties of the spring wire correspond to the curve for shot-peened wire given in Fig. 12.16. The spring failed in service after about 10^5cycles. A replacement spring was found which was identical in all respects except that its free-length was slightly shorter. To correct for this, the spring was stretched slightly to increase its free-length to exactly that of the original spring. Show, by means of a τ_{max}-τ_{min} plot, whether you would expect the life of the replacement spring to be the same as, less than, or greater than that of the original.

12.18 An automotive-type disc clutch similar to the one shown in Fig. 18.2 has an axial clamping force provided by six identical helical compression springs. When the clutch is engaged, the springs must provide a clamping force of 1080 lb (180 lb per spring). When the clutch is disengaged the springs are all 0.10 in. shorter, and this should cause the force of each spring to increase as little as reasonably possible, but a 25 lb increase is considered satisfactory. The clash allowance is to be 0.050 in. On the basis of preliminary calculations, a wire size of 0.192 in. has been selected. The material is to be shot-peened wire having fatigue properties as indicated in Fig. 12.16. Use a safety factor for eventual fatigue failure of 1.3. In addition, presetting is to be used but not taken into account in the calculations. Determine a suitable combination of D, N, L_s, and L_f.
(Ans.: $D = 1.08$ in., $N = 6.3$, $L_s = 1.59$ in., $L_f = 2.46$ in.)

12.19 A force of 4.45 kN is required to engage a clutch similar to the one shown in Fig. 18.2. This force is to be provided by nine identical springs equally spaced around the pressure plate of the clutch. Because of space limitations the outside diameter of the coils can be no more than 40 mm, and the length of the springs when the clutch is engaged cannot exceed 52 mm. The pressure plate must move 3 mm to disengage the friction surfaces, and the lowest spring rate reasonably possible is desired. Design the springs, determining a satisfactory combination of D, d, N, wire material, type of ends, L_s, and L_f.

12.20 A helical compression spring is to be designed for infinite life when subjected to a load which fluctuates between 50 and 100 lb. Steel spring wire having $S_u = 180$ ksi, $S_{us} = 144$ ksi, $S_y = 170$ ksi, $S_{ys} = 99$ ksi and a zero-to-maximum torsional endurance limit of 80 ksi is to be used. (These values are assumed to apply to the size range and surface finish involved.) For a value of $C = 7$, determine the wire diameter theoretically required (safety factor = 1), (a) if presetting is not used, and (b) if presetting is used to maximum advantage.

12.21 A coil spring is to be designed for use in the suspension of an experimental trailer. Specifications are as follows: static load per spring = 3500 N, rate per spring = 45 N/mm, jounce = 150 mm (i.e., compression beyond static load position is limited to 150 mm by a rubber bumper), rebound = 50 mm (i.e., extension beyond static load position is limited to 50 mm by a rubber bumper). Infinite fatigue life is required, using a safety factor of 1.3 applied to the maximum load only (safety factor of 1.0 applied to minimum load). Spring is to be preset, with a favorable torsional residual stress of 100 MPa being obtained. A metallurgist from the spring manufacturer has advised that for the size range and surface that will be involved, fatigue strength can be represented by a line between $\tau_{max} = 600$, $\tau_{min} = 0$ and $\tau_{max} = \tau_{min} = 900$ MPa. Determine a suitable combination of d, D, and N.

12.22 An automotive engine requires a valve spring to control the motion of a valve subjected to the accelerations shown in Fig. P12.22. (Note: A spring is required to hold the follower in contact with the cam during *negative accelerations only*.) The critical point for the spring is the "acceleration reversal point," corresponding in this case to a valve lift of 0.201 in. A larger spring force is required at maximum valve lift (0.384 in.) but this is easily satisfied because the spring is further compressed. In fact, the problem will be to make the spring adequately stiff to give a high enough natural frequency without making it so stiff that the spring force at full valve lift causes objectionably high contact stresses when the engine is running slowly. The valve spring is to satisfy the following specifications:

1. Spring length when valve is closed: not over 1.50 in. (due to space limitations).
2. Spring force when valve is closed: at least 45 lb.
3. Spring force when valve lift is 0.201 in. ("reversal point"): at least 70 lb.
4. Spring force at maximum valve lift of 0.384 in.: at least 86 lb but not over 90 lb (to prevent excessive cam nose contact stresses).
5. Spring outside diameter: not over 1.65 in. (due to space limitations).

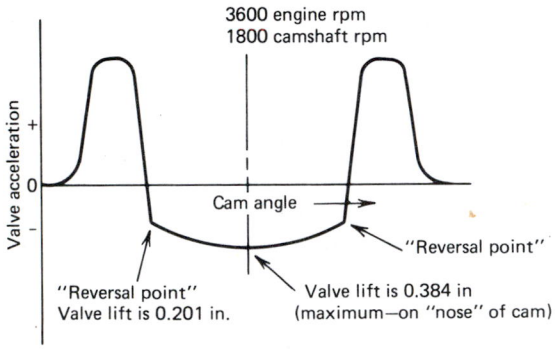

FIGURE P12.22

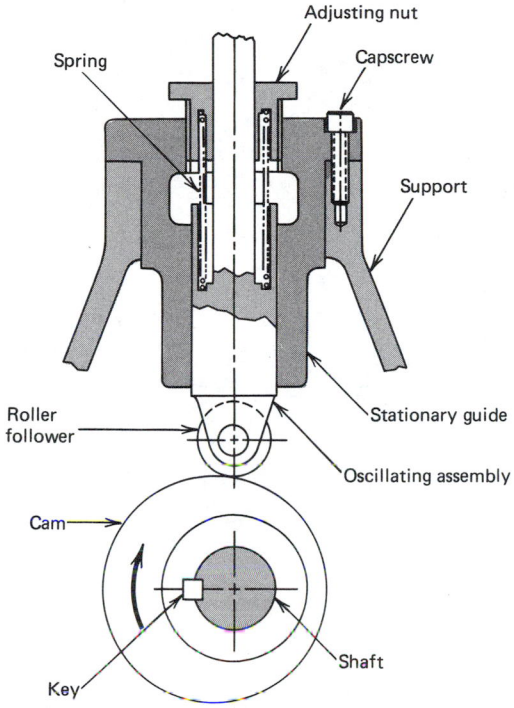

FIGURE P12.23

6. Clash allowance: 0.094 in.
7. Natural frequency: at least as high as the 13th harmonic at 1800 camshaft rpm (i.e., at least 390 Hz).

High-quality valve spring wire is to be used, and full advantage taken of both shot-peening and presetting. Because of this, you can assume in this case that fatigue failure will not occur if the calculated stress with spring solid is limited to 800 MPa. Ends are to be closed and ground. Determine a suitable combination of d, D, N, and L_f.

12.23 A cam and follower similar to that shown in Fig. P12.23 is to rotate at 10 Hz and impart a harmonic or sinusoidal oscillation to the follower. Total follower lift is to be 20 mm, and the weight of the oscillating parts is estimated at 90 N. The function of the spring is to overcome inertia forces and keep the roller follower in contact with the cam. To fit the available space, the spring inside diameter must be at least 25 mm and the outside diameter no more than 50 mm. Determine a satisfactory combination of D, d, N, material, L_s, L_f. Determine the natural frequency of the spring proposed.

Section 12.9

12.24 A helical tension spring, as shown in Fig. 12.19, is used as a component of a mass-produced machine. No difficulties are encountered until a new batch of springs is delivered that conforms to all specifications, except that the hooks are improperly formed,

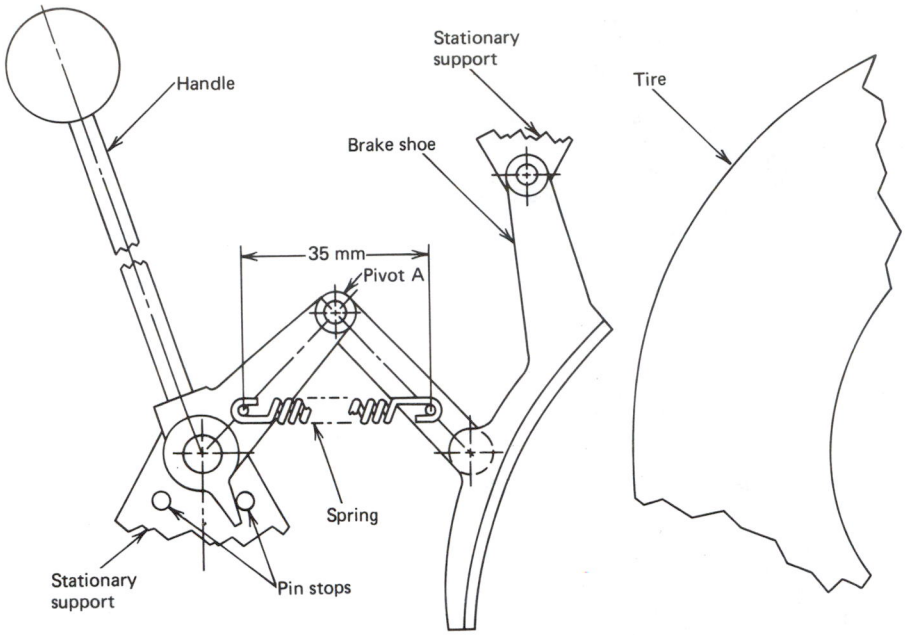

FIGURE P12.26

causing the load to be applied closer to the outside of the spring than to its geometric axis. Upon attempting to use these springs, it is noted that they deform permanently as soon as the maximum normal load is applied. Explain why, briefly.

12.25 A tension spring used in a machine must exert an essentially static force of 135 N. It is to be wound with an initial tension of 45 N, and have a spring rate of 10.5 kN/m. Determine a satisfactory combination of D, d, N, and wire material. What is the length of the coiled section when unloaded and when the 135 N force is applied?

12.26 Figure P12.26 shows a wheel chair brake in its retracted position, with a tension spring exerting a force of 4 N, which holds the handle against the stop-pin. When the handle is moved clockwise, pivot A drops below the axis of the spring (thereby going "over-center"), and the spring acts to hold the brake shoe against the tire. Available space limits the outside diameter of the spring to 10 mm. Estimate dimensions needed by scaling the drawing. Determine a satisfactory combination of D, d, N, wire material, and free-length of coiled section.

Section 12.10

12.27 Experience with the spring designed in Sample Problems 12.4 and 12.5 indicates that a "softer" spring would be desired. Repeat these sample problems for a spring rate of 20 N/mm.

12.28 A semi-elliptic leaf spring similar to the one in Fig. 12.24 has four leaves, each made from 0.1 in. x 2 in. steel having properties of S_u = 180 ksi, S_y = 160 ksi, and S_n = 80 ksi. The latter figure includes appropriate corrections for size and surface. K_f (due to clips and center hole) is 1.3. Use the simplified "triangular plate" model.

(a) What total spring length is needed to give a spring rate of 80 lb/in. (i.e., 80 lb applied at the center causes a 1-in. deflection at the center)?

(b) In service, the spring will carry a static load (applied to the center) of P, plus a superimposed dynamic load which varies from $+P/2$ to $-P/2$. What is the highest value of P that will give infinite life, with a safety factor of 1.3?

(Ans.: 20 in., 130 lb)

12.29 A semi-elliptic leaf spring is to be designed so that it will just have infinite fatigue life at an overload (applied to the center of the spring) that varies between 500 and 1500 N. The total length of the spring is to be 1 m. The steel to be used has properties of $S_u = 1200$ MPa, $S_y = 1030$ MPa, and $S_n = 500$ MPa. The latter figure applies to reversed bending loads, and is corrected for size and surface. Stress concentration at the center hole is such that $K_f = 1.25$.

(a) Estimate the appropriate values of h and b for a spring of proportion $b = 50h$.

(b) Based on a trapezoidal approximation similar to the one in Fig. 12.26b, what will be the deflection at the center of the spring when a 1500 N load is applied?

12.30 A full-elliptic leaf spring operates normally with a load which fluctuates between 100 and 200 lb, but is to be designed for an overload that fluctuates between 100 and 300 lb. Total spring length is 20 in., $h = 0.1$ in., $K_f = 1.3$, and the steel to be used has $S_u = 180$ ksi, $S_y = 160$ ksi, and $S_n = 80$ ksi (this figure pertains to the actual size and surface).

(a) Determine the total width, b, required.

(b) Show, on a σ_m-σ_a diagram the "operating points" for (1) machine turned off and spring supporting a static load of 100 lb only, (2) normal loads applied, and (3) design overloads applied.

(c) Determine the spring rate.

12.31 A semi-elliptic leaf spring for use in a light trailer is to be made of steel having $S_u = 1200$ MPa, $S_y = 1080$ MPa, and a fully corrected endurance limit of 550 MPa. The spring is 1 m long, and has five leaves of 5 mm thickness and 100 mm width. $K_f = 1.4$. When the trailer is fully loaded, the static load applied to the center of the spring is 3500 N.

(a) Estimate the alternating load (due to driving over a rough road) that, when superimposed onto the fully loaded spring, would be just on the verge of causing eventual fatigue failure.

(b) What will be the maximum deflection of the spring when loaded as determined in (a)?

(c) How much energy is absorbed by the spring in going from minimum load to maximum load when loaded as determined in (a)?

(d) To what value could the alternating load be increased if only 10^4 cycles of life are required?

Section 12.11

12.32 Figure P12.32 shows a symmetrically matched pair of steel torsion springs in a conventional arrangement for supporting the weight of a domestic overhead garage door. The

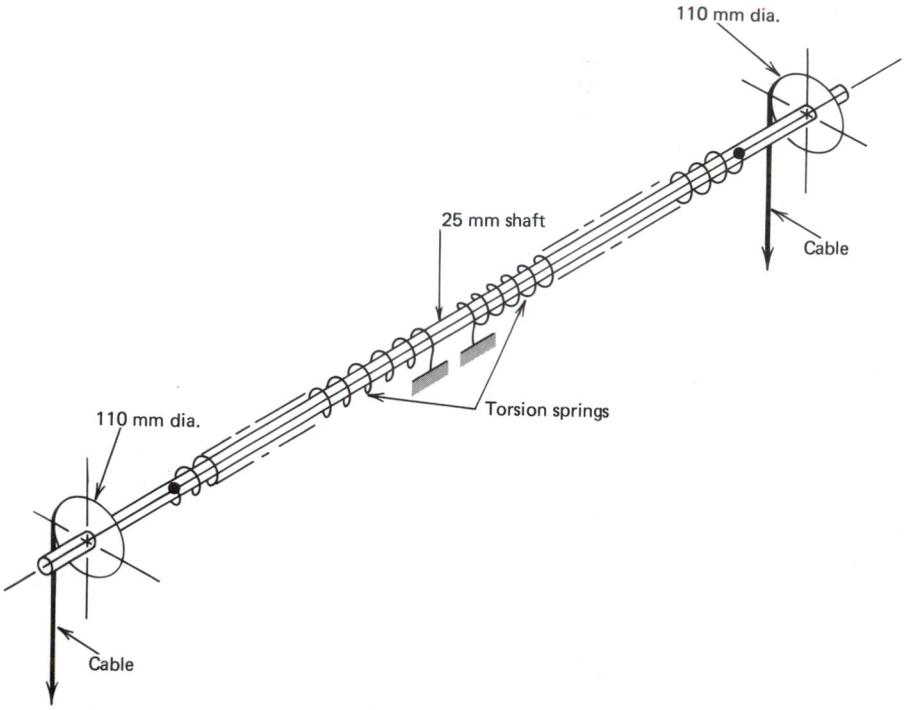

FIGURE P12.32

25 mm shaft is supported to the fixed structure by bearings (not shown) at each end and in the center. A cable wrapped around each of the pulleys supports the weight of the door. (The 110 mm diameter is measured to the axis of the wrapped cable.) Each spring has $D = 45$ mm, $d = 6$ mm, and 140 turns. The total extension of the cables from spring-unloaded to door-closed positions is 2.1 m.

(a) Calculate the nominal bending stress in the springs in the door-closed position.
(b) List important factors that would tend to make the actual maximum stress different than the nominal value.
(c) What supporting force is provided by each cable in the door-closed position?

12.33 Solve Problem 12.32 except use data obtained from your own measurements of a similar overhead garage door support.

12.34 Ordinary window shade rollers have long torsion springs that work on the same principle as the overhead garage door support of Problem 12.32. For a roller of 33 mm diameter, how many turns of 1.2 mm square steel wire with $D = 19$ mm would be required to give a shade pull force of 14 N with the spring wound up 16 turns? What would be the value of maximum bending stress in the spring?

12.35 The torsion bar in Problem 12.1 is objectionably long, and is to be replaced with a torsion spring made of steel having the same physical properties. Space is available for a spring of outside diameter up to 5 in. Determine one appropriate combination of values of d, D, and N. What is the overall length of the coiled section of the spring proposed? How does the weight of the torsion spring compare with that of the torsion bar?

12.36 Repeat Problem 12.35 for a torsion spring replacement of the torsion bar in Problem 12.2.

12.37 Compare the energy absorbing capacity per unit weight or volume of steel springs made in the form of torsion bars, helical compression springs, helical tension springs, rectangular plate cantilever springs, triangular plate cantilever springs, and torsion springs. Assume solid round wire for all except the cantilever springs.

CHAPTER 13 – LUBRICANTS AND SLIDING BEARINGS

Section 13.7

13.1 A journal bearing 120 mm in diameter and 60 mm long has a diametral clearance of 0.20 mm. The journal rotates 3000 rpm and is lubricated with SAE 20 oil at an average temperature of 70 °C. Using Petroff's equation, estimate the power loss and the friction torque.
(Ans.: 1.01 kW, 3.2 N·m)

13.2 An automobile engine has five main bearings, each 2.5 in. in diameter and 1 in. long. The diametral clearance is 0.0015 in. Using Petroff's equation, estimate the power loss per bearing at 3600 rpm if SAE 30 oil is used, with an average oil film temperature of 180 °F.
(Ans.: 0.66 hp)

Sections 13.8 and 13.9

13.3 Repeat Problem 13.1 using the charts, rather than Petroff's equation, and also determine the minimum oil film thickness, h_0:

(a) For a bearing load of 500 N.
(b) For a bearing load of 5000 N.

Note the comparison with answers obtained to Problem 13.1.
[Ans.: (a) 1.02 kW, 3.25 N·m, 0.086 mm, (b) 1.26 kW, 4.0 N·m, 0.041 mm]

13.4 A journal bearing of $L = 1$ in., $D = 2$ in., and a diametral clearance of 0.002 in. supports a 400 lb load while rotating 1800 rpm. SAE 10 oil is used, with an average film temperature of 130 °F. Determine (a) minimum oil film thickness, (b) coefficient of friction, (c) maximum film pressure, (d) angle between load direction and minimum film thickness, (e) angle between load direction and termination of film, (f) angle between load direction and maximum film pressure, (g) total circumferential oil flow rate, and (h) side or leakage flow rate. Would you recommend that a little smaller or a little larger clearance be used? Why?

13.5 A bearing of $L = 100$ mm and $D = 200$ mm carries a 33.4 kN radial load. It has a radial clearance of 0.100 mm and rotates at 900 rpm. Make a plot of friction power and h_0 vs. viscosity, using SAE 10, 20, 30, and 40 oil, all at 71 °C.

13.6 Repeat Problem 13.5 except use only SAE 40 oil and plot friction power and h vs. radial clearance, using values of 0.050, 0.100, and 0.150 mm. In addition, include points corresponding to clearances giving boundaries of the "optimum band" in Fig. 13.11.

13.7 A journal bearing of $D = 1$ in. and $L = 1$ in. is to be used to support a turbine rotor that rotates 12,000 rpm. SAE 10 oil is to be used, and the average lubricant temperature in the bearings is estimated to be about 148°F. Excellent oil filtration is provided, and the journal roughness is less than 32 micro-inches rms. Because of these factors, the minimum oil film thickness can be as little as 0.0003 in.

(a) What diametral clearance will give the bearing its greatest load carrying capacity?
(b) Using this clearance, what maximum load can be carried?
(c) With this clearance and load, what would be the friction power lost in the bearing?

[Ans.: (a) 0.0011 in., (b) 1360 lb, (c) 0.70 hp]

13.8 A bearing of $L = 50$ mm and $D = 50$ mm is to support a radial load of 5 kN. Rotating speed is to be 1200 rpm, and minimum film thickness, h_0, is to be 0.025 mm. For bearing operation at the minimum friction edge of the optimum band, what radial clearance and oil viscosity are required? Using these values, what would be the coefficient of friction and friction power loss?

Sections 13.10–13.13

13.9 A journal bearing is to be designed for a gear-reducer shaft that rotates 1200 rpm and applies a 4.45 kN load to the bearing. Proportions of $L/D = 1$ are desired. SAE 20 oil is to be used, and the average film temperature is expected to be 60°C. The bearing is to be as small as is consistent with reported current practice.

(a) Determine appropriate values of L and D.
(b) Determine values of c corresponding to the two edges of the optimum zone in Fig. 13.11.
(c) Does the value of c for minimum friction satisfy Trumpler's criterion for minimum acceptable film thickness?

[Ans.: (a) 55 mm, (b) 0.029 mm and 0.047 mm]

13.10 A journal bearing is to be designed for a gear reducer shaft which rotates 290 rpm and applies a radial load of 953 lb to the bearing.

(a) Determine an appropriate combination of L, D, material, lubricant, and average oil temperature (assume that this can be controlled to a reasonable specified temperature by an external oil cooler).
(b) Plot f, h_0, ΔT, Q, and Q_s for a range of radial clearances extending somewhat beyond both sides of the optimum range of Fig. 13.11. From this, suggest an appropriate range of production clearances if the tolerances are such that the maximum clearance is 0.001 in. greater than the minimum clearance.

13.11 A ring-oiled bearing is to support a steady load of 4.5 kN when the shaft rotates 660 rpm. Obviously, many designs are possible, but suggest one reasonable combination of D, L, c, bearing material, surface finish, and lubricating oil.

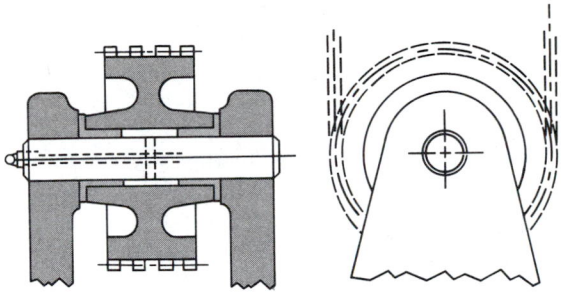

FIGURE P13.12

Section 13.14

13.12 Figure P13.12, which is not drawn to scale, shows a four-strand roller-chain idler sprocket that rotates on a stationary shaft, supported by two sliding-surface bearings inside the sprocket. The chain transmits 3.7 kW at a chain velocity of 4 m/s, and the sprocket diameter is 122.3 mm. The distance between the inside edges of the sprocket shaft support is 115 mm.

(a) Determine a satisfactory combination of bearing length, diameter, and material.
(b) What considerations other than those taken into account in (a) might influence the final decision of bearing diameter and length?

CHAPTER 14 – ROLLING ELEMENT BEARINGS

Section 14.7

14.1 Some bearing manufacturers rate their bearings on the basis of a life of 10^6 revolutions. *If* all other factors are the same, by what value should these ratings be multiplied when comparing them with the ratings in Table 14.2?
(Ans.: 0.259)

14.2 A No. 204 radial contact ball bearing is used in a gearing application considered to be average with respect to shock loading. The shaft rotates 3500 rpm and the bearing is subjected to a radial load of 1000 N and a thrust load of 250 N. Estimate the bearing life in hours for 90% reliability.
(Ans.: 15,000 hr)

14.3 A particular bearing application involves life for 90% reliability of 5000 hr. What are the corresponding lives for 50% reliability and 99% reliability?

14.4 A No. 211 radial ball bearing has a B-10 life of 5000 hr. Estimate the B-10 life that would be obtained if this bearing were replaced by (a) an L11 radial ball bearing, (b) a 311 radial ball bearing, and (c) a 1211 roller bearing.

14.5 A No. 207 radial contact ball bearing supports a shaft that rotates 1000 rpm. A radial load varies such that 50%, 30%, and 20% of the time the load is 3 kN, 5 kN, and 7 kN,

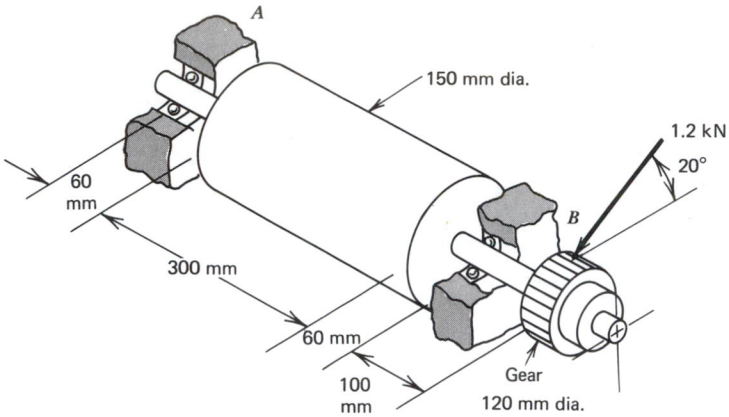

FIGURE P14.10

respectively. The loads are uniform, so that $K_a = 1$. Estimate the B-10 life and the median life of the bearing.
(Ans.: 8700 hr, 44,000 hr)

14.6 A No. 312 radial contact ball bearing is loaded uniformly (without shock) as follows: 55% of the time, 7 kN and 1800 rpm; 25% of the time, 14 kN and 1200 rpm; 20% of the time, 18 kN and 800 rpm. Estimate the bearing life for 90% reliability.
(Ans.: 6400 hr)

14.7 In a given application, a No. 212 radial ball bearing has a B-10 life of 6000 hr. What would be the expected life of the next larger sizes (No. 213 and No. 312) used in the same application?
(Ans.: 10,300 hr, 21,700 hr)

14.8 A bearing supporting a 1000 rpm shaft carries a 3-kN radial load and 1-kN thrust load. The shaft is part of a machine for which the loading involved is borderline between "light" and "moderate" impact. The required life is 5000 hr with only 2% probability of failure. Select a 200 series ball bearing for this application:

(a) Using a radial contact bearing.
(b) Using an angular contact bearing.
[Ans.: (a) No. 208, (b) No. 208]

14.9 Repeat Problem 14.8 with the thrust load increased to:

(a) 1.5 kN.
(b) 3.0 kN.

14.10 Figure P14.10 shows a printing roll driven by the gear to which the 1.2-kN force is applied. The bottom surface of the roll is in contact with a similar roll that applies a uniform (upward) loading of 4 N/mm. Select identical 200 series ball bearings for A and B, if the shaft rotates 350 rpm.

CHAPTER 15 – SPUR GEARS

Sections 15.2 and 15.3

15.1 An 18-tooth pinion with a diametral pitch of 6 rotates 1800 rpm and drives a gear at 900 rpm. What is the number of teeth in the gear, the theoretical center distance, and the circular pitch?
(Ans.: 36 teeth, 4.5 in., $\pi/6$ in.)

15.2 A 3:1 speed ratio is to be provided by a pair of spur gears on 6 in. centers. Using a diametral pitch of 8, what are the numbers of teeth and pitch diameters of the two gears?

15.3 A 24-tooth pinion has a module of 2 mm, rotates 3600 rpm, and drives a 900-rpm gear. Determine the number of teeth on the gear, the circular pitch, and the theoretical center distance.

15.4 Begin a full-size layout of a pair of spur gears having a 4:1 velocity ratio, 7½ in. center distance, a diametral pitch of 3, and standard 20° full-depth teeth. Include only the following items on the drawing, and label each clearly:

(a) Pitch circles (partial).
(b) Base circles (partial).
(c) Pressure angle.
(d) Addendum (for both pinion and gear).
(e) Dedendum (for pinion only).

Show whether interference exists and, if so, indicate the preferred modification for eliminating it.

15.5 The diametral pitch of a pair of standard 20° full-depth spur gears is 2. The pinion has 12 teeth and rotates clockwise. The velocity ratio is 1.5. Make a full-size layout of the gears in the region of tooth contact.

(a) Label pitch circle, addendum circle, dedendum circle, and base circle of the *gear*.
(b) If interference exists, indicate the amount that one or both of the addenda would have to be reduced to eliminate it.
(c) Indicate the path of contact by a heavy solid line that extends exactly the length of the path of contact and no farther. (Assume that any interference has been eliminated by necessary addenda reduction.)
(d) Make a careful sketch of a pair of mating tooth profiles at the end of contact. Label angles of recess for pinion and gear.

15.6 A pair of involute gears have base circle diameters of 50 mm and 150 mm.

(a) If the center distance is 110 mm, what is the pressure angle?
(b) If the center distance is increased to 125 mm, what is the pressure angle?
(c) What is the ratio of the two pitch diameters for each of the two center distances?
[Ans.: (a) 0.4297 rad, (b) 0.6435 rad]

15.7 A pair of mating spur gears with 6-mm module and 0.35-radian pressure angle have 20 and 42 teeth.

(a) Make a full-size drawing of the tooth-contact region, showing (and labeling) (1) both pitch circles, (2) both base circles, (3) both root circles, (4) both outside circles, (5) pressure angle, (6) length of path of contact, (7) both angles of approach, and (8) both angles of recess.

(b) Using values scaled from your drawing, state or calculate numerical values for (1) length of path of contact, (2) angles of approach, (3) angles of recess, and (4) contact ratio.

15.8 A pair of standard 20° spur gears with 7.5-in. center distance have a velocity ratio of 4.0. The pinion has 18 teeth.

(a) Determine P, p, and p_b.

(b) Begin a full-size layout showing partial pitch circles, partial base circles, pressure angle, addendum, and dedendum. Label each of these on your drawing.

(c) Show on your drawing the interference-limiting maximum addendum radii, $r_{ag,max}$, and $r_{ap,max}$. Scale their numerical values from the drawing. Will interference be encountered with teeth of standard proportions?

(d) Measure on the drawing the length of the path of contact for standard tooth proportions and, from this, compute the contact ratio.

15.9 Using the equations in Section 15.3, compute $r_{ag,max}$, $r_{ap,max}$, and the contact ratio for the gears in Problem 15.8. Compare the results with values obtained graphically in Problem 15.8.

Section 15.4

15.10 Figure P15.10 shows a two-stage gear reducer. Identical pairs of gears are used. (This enables input shaft a and output shaft c to be collinear, which facilitates machining of

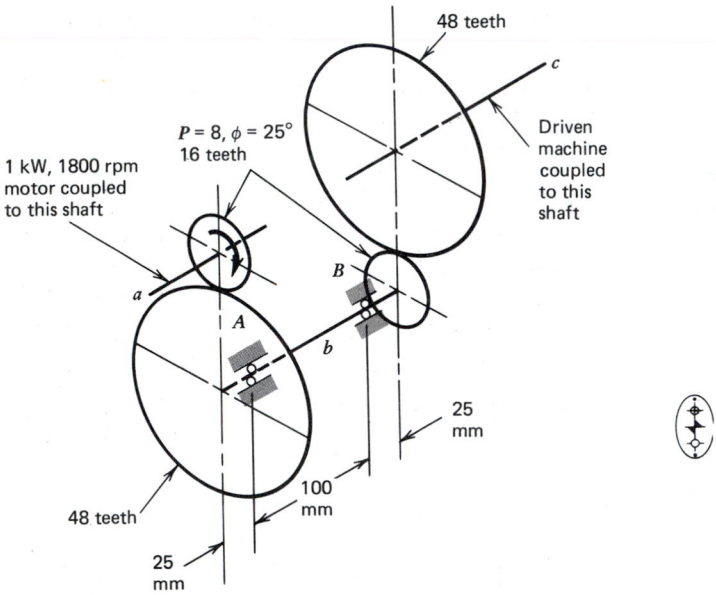

FIGURE P15.10

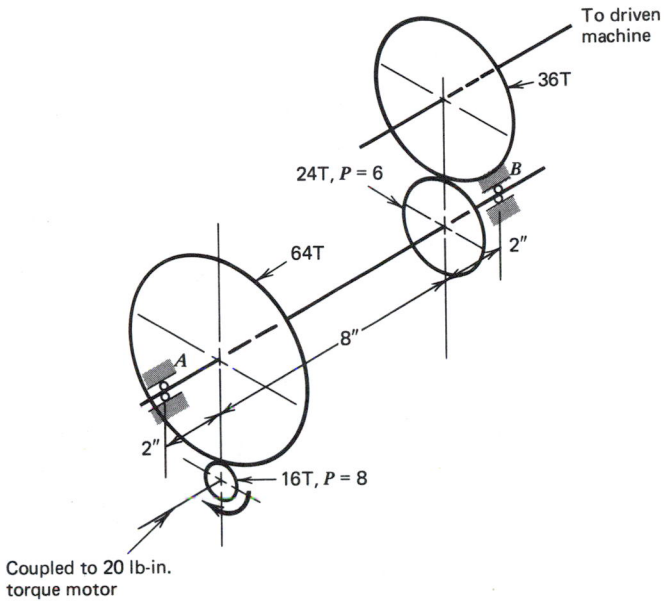

FIGURE P15.11

the housing.) Shaft b, called the countershaft, turns freely in bearings A and B, except for the gear-tooth forces.

(a) Determine the rpm of shafts b and c, the pitch diameters of the pinion and gear, and the circular pitch.

(b) Determine the torque carried by each of the shafts a, b, and c: (1) assuming 100% gear efficiency, and (2) assuming 97% efficiency of each gear pair.

(c) For the 100% gear efficiency case, determine the radial loads applied to bearings A and B, and sketch the countershaft as a free-body in equilibrium.

(*Note:* This problem illustrates a machine designed using SI units except for the gear teeth, which are dimensioned in inches.)

[Partial ans.: (b) $T_b = 15.93$ N·m, $T_c = 47.79$ N·m for 100% efficiency; $T_b = 15.45$ N·m, $T_c = 44.97$ N·m for 97% efficiency]

15.11 The 16-tooth pinion in Fig. P15.11 is driven 1000 rpm by a motor that delivers 20 lb-in. of torque. The gears provide a double reduction in speed, with output taken from the 36-tooth gear. Both the 6-pitch and 8-pitch gears have a 20° pressure angle. Neglecting the small friction loss in the gears and bearings, determine the radial loads applied to countershaft bearings A and B. Sketch the countershaft as a free-body in equilibrium.

(Partial ans.: Radial load on bearing B is 38.3 lb)

15.12 Figure P15.12 shows a machine being gear-driven by a motor. (Use of the idler gear causes the input and output shafts to rotate in the same direction, and increases the space between them.) For a 20° gear pressure angle, what is the relative loading on the six bearings shown?

15.13 Figure P15.13 shows an electric motor driving a machine by means of three straight-

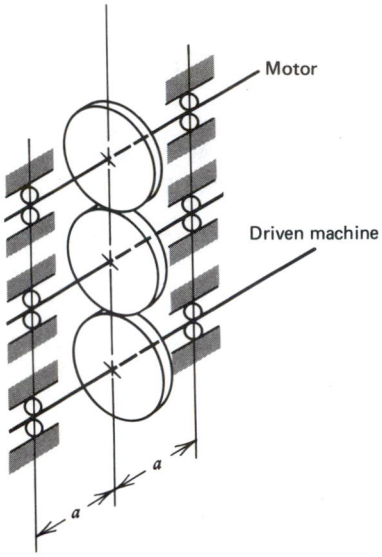

FIGURE P15.12

tooth spur gears having 16, 32, and 24 teeth. The gears have $P = 8$ and $\phi = 20°$. The idler shaft is supported by bearings A and B.

(a) For the direction of motor rotation shown, determine the radial load carried by each bearing.

(b) Determine the bearing loads for the opposite direction of motor rotation.

(c) Explain briefly why the answers to (a) and (b) are different.

[Partial ans.: (a) 289 lb and 96 lb on A and B, respectively]

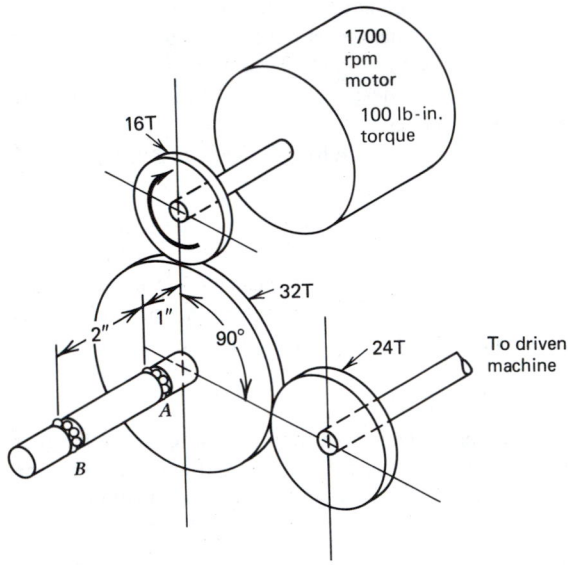

FIGURE P15.13

Sections 15.5-15.8

15.14 Suppose that the gears in Fig. P15.12 are identical, each having a 4-in. diameter, 48 teeth, and 25° pressure angle. Assume that their finish and accuracy correspond to about the best that is commercially obtainable with a shaping cutter.

(a) Which of the gears is the most vulnerable to tooth bending fatigue failure? Why?
(b) If the motor rotates 2000 rpm, determine proper values for V, P, p, K_v, and J.

15.15 A pair of mating gears have 10-pitch, 20° full-depth teeth. Both gear and pinion are made of steel heat-treated to 350 Bhn, and both have a face width of 1¼ in. The teeth are cut with a top-quality hobbing operation. The pinion has 16 teeth and rotates 1720 rpm. It is mounted outboard on the shaft of an electric motor, and drives a 32-tooth gear that is positioned inboard on an accurately mounted blower shaft. The design life corresponds to 5 years of 60 hr/wk, 50 wk/yr operation. Using a reliability of 99% and a safety factor of 1.5, estimate the horsepower that can be transmitted, based only on bending fatigue.
(Ans.: Approximately 11.2 hp)

15.16 To what value could the gear hardness be reduced in Problem 15.15 without making the gear teeth weaker than the pinion teeth in bending fatigue?
(Ans.: 272 Bhn)

15.17 A spur gear reducer has a 16-tooth pinion driven 1800 rpm by an electric motor, and a 48-tooth gear that drives a load involving "moderate shock." A life of 10^6 pinion revolutions is required, and the transmitted load, F_t, is 80 lb (this figure includes a safety factor of 2). Conditions are such that $K_m = 1.6$, and $K_T = C_T = 1$. It is proposed that standard 20° full depth gears be used, with both pinion and gear teeth being cut with a low-cost, average-quality form cutting process from steel of 180 Bhn for the gear and 210 Bhn for the pinion. Diametral pitch is to be 8, and face width 1.2 in. Estimate the reliability with respect to bending fatigue failure.

15.18 The accessory drive unit of a large aircraft engine involves an identical pair of 20° full-depth spur gears. The gears have 50 teeth, and rotate together at 5000 rpm. The diametral pitch is 10 and the face width 0.8 in. The gears are made of alloy steel, case hardened to 62 R_c (680 Bhn). Although the case extends throughout the fillet, it is conservatively decided to use the core hardness of 500 Bhn for bending fatigue strength calculations. The tooth profiles are finished with a fine-grinding operation that gives a degree of precision justifying the use of curve A of Fig. 15.23 and the assumption that teeth share the load. The loading involves only very mild shock, justifying the use of $K_0 = 1.1$. Estimate the horsepower that can be transmitted with 99% reliability, based on tooth bending fatigue.

Sections 15.9 and 15.10

15.19 The 20-tooth pinion of Problem 15.7 rotates 210 rpm. Determine graphically the sliding velocity between the teeth (a) at start of contact, (b) at the pitch point, and (c) at the end of contact. Make vectors large enough to be scaled accurately.

15.20 For the gears in Problem 15.15, estimate the horsepower that can be transmitted, based on surface durability.
(Ans.: Approximately 2½ hp)

15.21 For the gears in Problem 15.17, estimate the reliability with respect to surface durability.

15.22 For the gears in Problem 15.18, estimate the horsepower that can be transmitted for 10^8 cycles with 90% reliability, based on surface fatigue.

15.23 A two-stage gear reducer, similar to the one in Fig. P15.10, uses two identical pairs of gears, each of which has a speed ratio of 2. The pitch-line velocity of the high-speed pair is 2000 ft/min. Compare the relative strengths of the gears for serving in the high-speed and low-speed positions. Consider both bending fatigue and surface durability.

Section 15.11

15.24 A two-stage gear reducer using identical pinion and gear sets for the high-speed and low-speed stages (similar to Fig. P15.10) is to be used between an 8 hp, 3600 rpm motor and a 400 rpm load. The center distance is to be 6 in. The motor involves "light shock" and the driven machine "moderate shock." The shafts and their mountings correspond to ordinary good industrial practice, but not "high-precision." Design the gears for a life of 10^7 cycles with 99% reliability and a safety factor of 1.5. Determine an appropriate combination of diametral pitch, face width, level of manufacturing precision, and material.

15.25 Design a pair of standard spur gears for transmitting 50 hp from a 5100-rpm motor to a 1700-rpm machine. A safety factor of 1.3 should be used, together with 99% reliability and a life of 10^7 pinion revolutions at full load (with the remaining lower-load cycles being negligible). It is important to minimize size and weight. Use case-hardened alloy steel gears of 660 Bhn (pinion) and 600 Bhn (gear), and use Table 15.5 to estimate surface fatigue strength. Specify an appropriate combination of diametral pitch, numbers of teeth, center distance, face width, and level of manufacturing accuracy.

Section 15.13

15.26 The simple planetary train shown in Figs. 15.28–15.30 is used in automotive overdrive units. When the overdrive is not engaged, the entire train rotates together as one solid unit, providing a 1:1 ratio. With overdrive engaged, the sun is fixed, the arm is the input, and the ring is the output (i.e., input and output members are reversed from that shown in Figs. 15.29 and 15.30). One specific overdrive design provides a speed increase (and torque reduction) of 1.42, and uses planets with 18 teeth. (a) How many teeth are in the other gears? (b) Could four equally spaced planets be used? (c) Could three equally spaced planets be used?

15.27 An engineer unfamiliar with the workings of planetary gears is having great difficulty determining a combination of numbers of teeth that will enable a unit like the one in Figs. 15.29 and 15.30 to give a ratio of 2.0. Explain briefly why this is theoretically impossible.

15.28 Figure P15.28 is a schematic representation of a conventional three-speed bicycle hub. Sprocket 1 is driven by the chain. Member 2 always rotates with the sprocket, and slides axially to one of three positions (low, neutral, and high) when the gear shift control is actuated. Sun gear 6 is permanently fixed, and serves as the reaction member. Free-

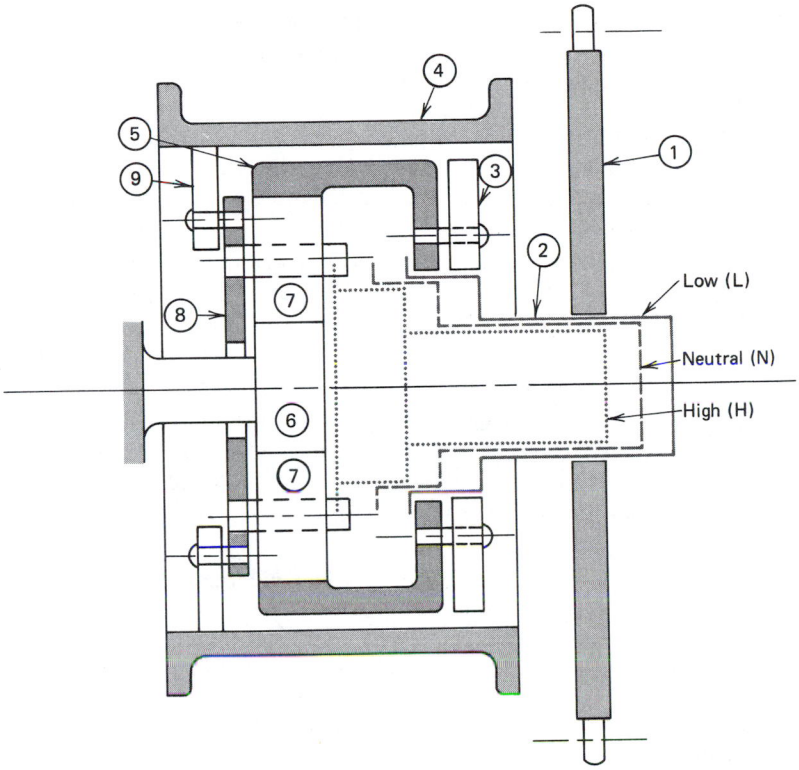

(*a*)

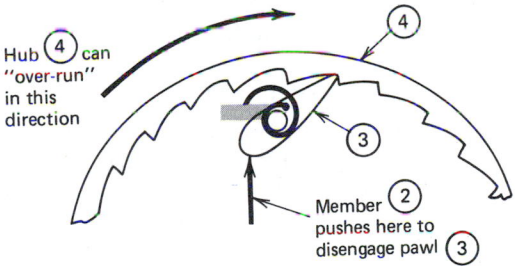

(*b*)

FIGURE P15.28

wheeling pawls 3 (attached to annulus 5) and 9 (attached to arm 8) each represent a set of circumferentially arranged members which function as shown in Fig. 15.28(*b*). Light torsion springs hold the pawl tips against the notched inner surface of hub 4 (the member to which the wheel spokes are attached). The pawls permit hub 4 to rotate faster (in the clockwise or forward direction) than the member to which they are attached, but not slower; thus, they constitute *one-way clutches*. The tips of pawls 9 always bear on the inner hub surface, but in low gear sliding member 2 disengages pawls 3 by pushing against them as shown in Fig. 15.28(*b*). Operation in the three gears is as follows.

Low gear. Member 2 is in position L, where it (1) disengages pawls 3 and (2) drives annulus 5. The output member is the arm, which drives the hub through pawls 9.
Neutral gear. Member 2 is in position N, where it continues to drive the annulus, but releases pawls 3 which drive the hub at sprocket speed. (Pawls 9 remain engaged, but "over-run," causing a "clicking" sound as the pawls go over the notches in the hub).
High gear. Member 2 moves out of engagement with the annulus and into engagement with the projecting planet pins, which are part of arm 8. Pawls 3 drive the hub at annulus speed, and pawls 9 "over-run."

Now for the problem: (a) The sun and each of the four planets have a ⅝ in. pitch diameter and 20 teeth. How many teeth are in the annulus, and what is the diametral pitch of the gears? (b) What is the ratio of wheel rpm to sprocket rpm for each of the three speeds? Determine these ratios using at least two of the three methods given in the text. (c) Explain briefly what happens when the bicycle coasts in any gear.

15.29 A planetary train similar to Fig. 15.30 has a band brake that holds the ring fixed. The sun is driven clockwise at 1000 rpm with a torque of 20 N·m. The arm drives a machine. The gears have a module, m, of 2.5 (mm/tooth) and $\phi = 20°$. The ring has 70 teeth. Only two planets are used, each having 20 teeth.

(a) What is the circular pitch, p?
(b) Sketch each member of the train as a free body in equilibrium (neglect gravity loads).
(c) What is the output torque?
(d) What is the arm rpm? Does it rotate clockwise or counterclockwise?
(e) What pitch-line velocity should be used in determining the velocity factor for each of the gears? (*Hint:* The velocity must be determined with respect to the proper member. For example, the appropriate pitch-line velocity for the pinion and gear of a simple aircraft engine reduction gear when the plane is doing a "barrel roll" is the velocity with respect to the frame, not with respect to the ground.)
(f) What are the nominal radial loads imposed on the bearings supporting each of the gears?
(g) What torque must the brake provide in order to hold the ring fixed?

15.30 Figure P15.30 shows a planetary train with double planets, two suns, and no ring gear. Planets $P1$ and $P2$ are made together from the same piece of metal, and have 30 and

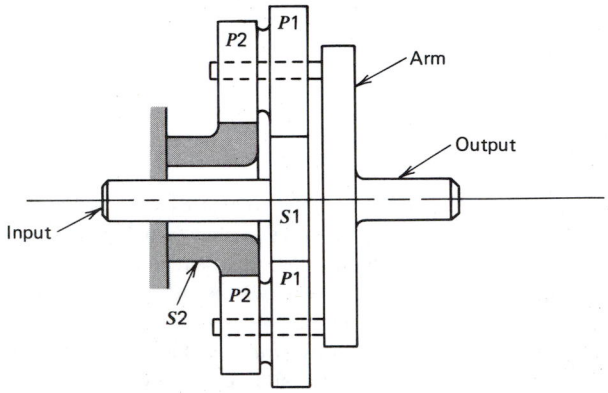

FIGURE P15.30

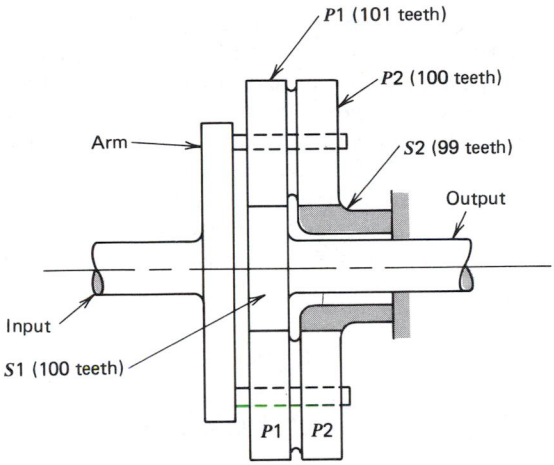

FIGURE P15.31

24 teeth, respectively. Sun $S1$ is the input member, and has 28 teeth. Sun $S2$ is fixed. All gears have the same pitch. For each clockwise revolution of $S1$, what is the motion of the arm?

15.31 Figure P15.31 represents an ingenious gear train for obtaining a tremendous reduction ratio in a small space. It is similar to Figure P15.30 except that the arm is the driver and the diameters of the two suns and the diameters of the two planets (which are made together as a single piece) are very nearly the same. First, notice that if these pairs of gears had exactly the same diameters, the output velocity would be zero. By making the pitch of $S1$ and $P1$ just slightly greater than the pitch of $S2$ and $P2$, the center distances will be the same for the numbers of teeth shown. Determine the magnitude and sign of the gear train speed ratio.
(Ans.: $+0.0001$)

15.32 Figure P15.32 is a diagrammatic representation of the transmission used in the Model

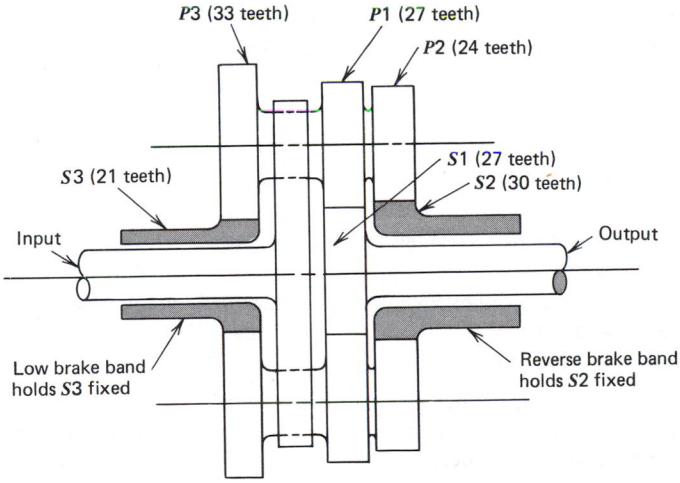

FIGURE P15.32

T Ford. In common with modern automatic transmissions, planetary gears were used. Control was by means of two foot pedals. The low gear pedal applied a band brake to hold sun $S3$ fixed. The reverse pedal applied a band brake holding sun $S2$ fixed. Releasing both pedals applied a direct drive clutch for high gear. (Partial depression of the low gear pedal gave neutral. Applying the hand brake also depressed the low gear pedal to the neutral position—a feature appreciated when starting the engine by hand-cranking!) Determine the transmission ratios for low and reverse gears. (Note the substantially greater ratio in reverse. This meant that for the unusual incline that was too steep for low gear, the car could always be turned around and driven up the incline in reverse!).
(Ans.: +2.75, −4.00)

CHAPTER 16 – HELICAL, BEVEL, AND WORM GEARS

Section 16.2

16.1 Section 15.2 gave the normal range of spur gear face widths as 9/P to 14/P, and Section 16.2 stated that it is usually desirable to make $b \geqq 2.0\ p_a$. Accordingly, it is desired to make a helical gear with face width equal to 12/P and also equal to 2.1 p_a. What helix angle is required to do this? How does this compare with the commonly used range of helix angles cited in Section 16.2?
(Ans.: 28.8°)

16.2 A 20-tooth helical spur gear with $\psi = 30°$ has a 20° pressure angle in the plane of rotation. What is the pressure angle in the normal plane, ϕ_n, and the equivalent number of teeth, N_e. What pressure angle and number of teeth would a straight spur gear tooth of comparable bending strength have?

16.3 An industrial machine utilizes a simple planetary train with 20 teeth in the sun and 120 in the ring. The teeth have a 4-mm module in the normal plane, and the helix angle is 0.42 rad. The manufacturer wishes to make available optional gearing using the same arm, a 24 tooth sun and a 114 tooth ring. If teeth of the same module are used, what must be the helix angle?
(Ans.: 0.4509 rad)

16.4 Figure P16.4 shows a double reduction gear arrangement used in an industrial machine. Modules in the normal plane are 3 mm and 4 mm for the high and low speed gears, respectively. The helix angle of the high speed gears is 0.44 rad:

(a) What is the total speed reduction provided by the four gears?
(b) What must be the helix angle of the low speed gears?
(c) If the low speed gears are replaced by 23 and 37 tooth gears of the same modulus, what helix angle must they have?

(Ans.: 5.1063, 0.3532 rad, 0.4655 rad)

16.5 A capstan winch incorporates a pair of spur gears with 22 and 66 teeth, $b = 45$ mm, and $m = 3.5$ mm. In order to provide quieter operation, these are to be replaced by helical gears. For strength reasons, it is desired to use gears with $m = 3.5$ mm in the normal plane. Determine a suitable combination of numbers of teeth and helix angle.

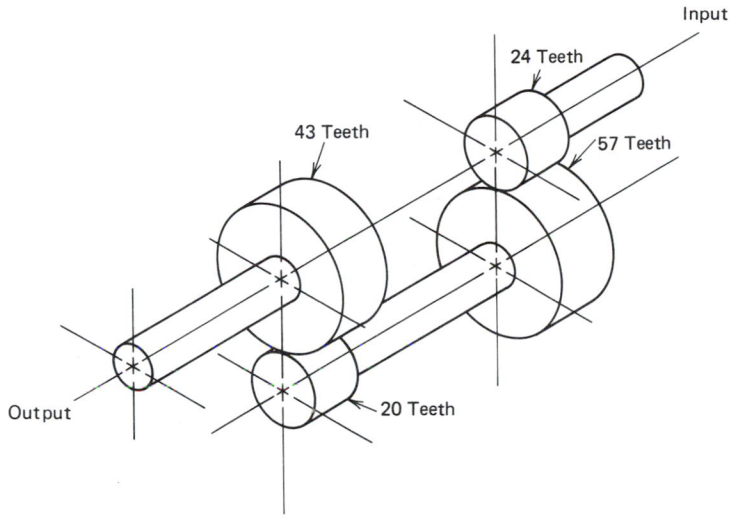

FIGURE P16.4

Section 16.3

16.6 Repeat Sample Problem 16.1 with the following changes: motor hp = 1, motor rpm = 2400, 19 pinion teeth, $P_n = 12$, $\psi = 25°$, with right-hand helix on pinion, gear rpm = 1200.

16.7 A helical gear reducer similar to the one in Fig. 16.7 transmits 1 hp with a speed reduction ratio of 4.0. The gear has 72 teeth and rotates 290 rpm. $\phi_n = 20°$, $\psi_g = 25°$ (right hand), $P_n = 12$, and $b = 1$ in. Determine N_p, ψ_p (and hand of helix), ϕ, P, d_p, d_g, V, ratio of b/p_a, and forces F_t, F_r, and F_a. Make a sketch like Fig. 16.7*b* showing the directions of the forces acting on the pinion, if the pinion rotates opposite the direction shown in Fig. 16.7*b*.
(Ans., in order: 18 teeth, 25° left-hand, 21.88°, 10.88 teeth/in., 1.65 in., 6.62 in., 502.6 ft/min, 1.61, 65.66 lb, 26.37 lb, 30.62 lb)

16.8 The helical gear reducer represented in Fig. P16.8 is driven at 1140 rpm by a motor developing 14.9 kW. The teeth have $\psi = 0.52$ rad and $\phi_n = 0.35$ rad. Pitch diameters are 70 mm and 280 mm for pinion and gear. Determine the magnitude and direction of the three components of the gear-tooth force. Make a sketch like the one in the figure except with the shafts separated vertically and show all force components acting on both gears.
(Ans.: $F_t = 3566$ N, $F_r = 1500$ N, $F_a = 2042$ N)

16.9 The four helical gears shown in Fig. P16.9 have a module in the normal plane of 4 mm and a pressure angle in the normal plane of 0.35 rad. The motor shaft rotates 600 rpm and transmits 22 kW. Other data are on the drawing.

(a) What is the speed ratio between the motor (input) and output shafts?
(b) Determine all force components that the 24 tooth pinion applies to the 54 tooth gear. Make a sketch showing these forces applied to the gear.

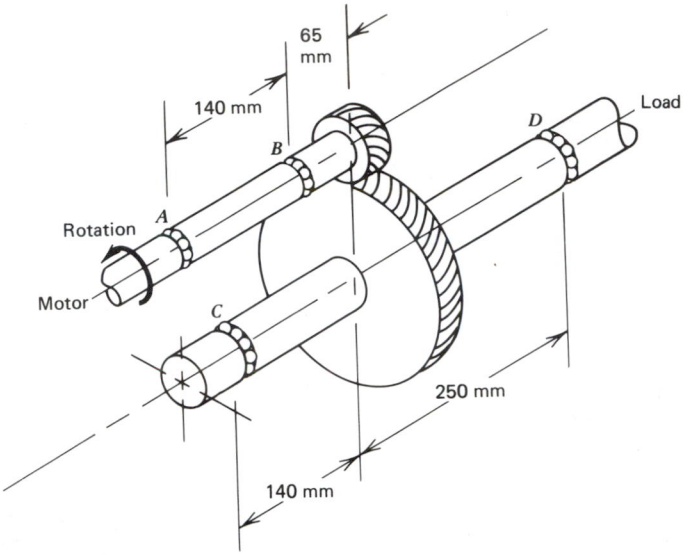

FIGURE P16.8

(c) The same as (b) but for the force components that the 50 tooth gear exerts on the 22 tooth pinion.

Section 16.4

16.10 A helical gear reducer is to be used with an 1800 rpm electric motor and a 450 rpm load which involves a moderate amount of shock. The 24-tooth pinion has $P_n = 8$, $b = 1.5$ in., $\phi_n = 20°$, and $\psi = 26°$. Both gears are made of AISI 8620 steel, with surfaces

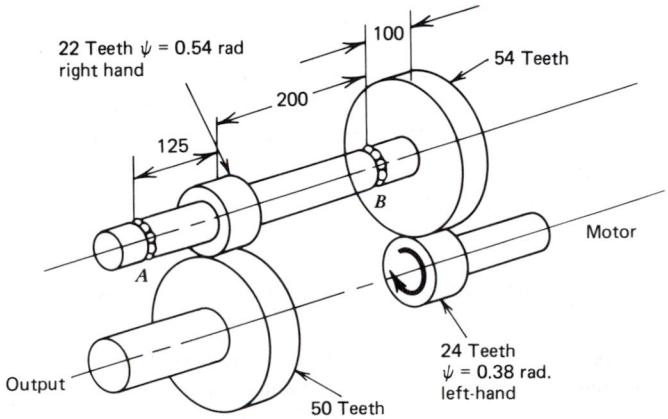

FIGURE P16.9

carburized to give properties listed in Appendix C-7. Manufacturing precision corresponds to curve B of Fig. 15.23. Estimate the horsepower that can be transmitted for 10^7 pinion revolutions with 99% reliability and a safety factor of 2. Check for both bending and surface fatigue.

16.11 A pair of helical gears mounted on parallel shafts have $P_n = 4$, $\phi_n = 20°$, and $b = 3$ in. The 32-tooth pinion and 44-tooth gear are made of steel having 400 Bhn and 350 Bhn, respectively. Manufacturing precision corresponds to curve C of Fig. 15.23. Center distance is 10 in.

(a) What helix angle is required?

(b) Estimate the horsepower that can be transmitted for 10^7 pinion revolutions with 99% reliability and a safety factor of 2.5 if the driving motor rotates 860 rpm and involves light shock, and the driven load involves medium shock. Check for both bending and surface fatigue.

16.12 Suggest a suitable design for a pair of helical gears to transmit 110 hp between a 2400 rpm electric motor and an 800 rpm load that is essentially free of shock. Forty-hour-per-week operation is anticipated. State a satisfactory combination of pitch, numbers of teeth, helix angle, pressure angle, face width, manufacturing accuracy, and material and hardness.

Section 16.7

16.13 A pair of straight-tooth bevel gears mounted on perpendicular shafts transmits 30 hp at 900 rpm of the 36-tooth pinion. The gear turns 360 rpm. Face width is 2 in., $P = 6$, and $\phi = 20°$. Make a sketch of the pinion showing (a) an assumed direction of rotation, (b) the direction and magnitude of torque applied to the pinion by its shaft, and (c) the direction and magnitude of the three components of force applied to a pinion tooth by a gear tooth. Make a corresponding drawing of the gear and the loads applied to it. (Partial ans.: for pinion, $F_t = 799$ lb, $F_a = 108$ lb, $F_r = 270$ lb)

16.14 A pair of straight-tooth bevel gears mounted on perpendicular shafts transmits 60 hp at 1500 rpm of the 30-tooth pinion. The gear has 60 teeth. Face width is 2.5 in., $P = 5$, and $\phi = 20°$. The gear is mounted in the center of a simply supported shaft with 6 in. span between bearings. Thrust is taken by the bearing having the lighter radial load. Output torque is taken by a flexible coupling connected to a driven machine. Make a sketch of the gear and shaft assembly as a free-body in equilibrium.

Section 16.8

16.15 An 1160 rpm electric motor drives a belt conveyor (which imposes moderate shock loading to the drive train) through a straight-tooth bevel gear reduction unit. $N_p = 19$, $N_g = 61$, $P = 8$, $b = 1$ in., and $\phi = 20°$. Both gears are made of steel having a hardness of 300 Bhn. The gear is straddle mounted and the pinion is overhung, resulting in a mounting considered to be reasonably good from the standpoint of rigidity. The gears are manufactured to standards corresponding to curve B of Fig. 15.23. The design life

is 7 years of 2000 hr/yr operation. Estimate the horsepower that can be transmitted at 99% gear reliability.
(Ans.: about 2 hp)

16.16 Determine an appropriate combination of materials, hardnesses, and manufacturing precision for the gears in Problem 16.13. (State any necessary assumptions.)

16.17 Determine an appropriate combination of materials, hardnesses, and manufacturing precision for the gears in Problem 16.14. (State any necessary assumptions.)

Section 16.9

16.18 An automobile with a standard differential is stuck on an icy road so that the car does not move. The frustrated (and not very intelligent!) driver floors the accelerator and notes that the speedometer reads 75 mph. To what normal car speed does the rotation of the spinning wheel correspond? (This illustrates the kind of "foreseeable abuse" that the engineer must be aware of.)

16.19 An automobile with a standard differential turns sharply to the left. The left driving wheel turns on a 15m radius. Distance between right and left wheels is 1.5 m. What are the rotating speeds of each driving wheel as fractions of the drive-shaft speed?

16.20 In Fig. 16.18, call the radius from the axis of the axle shafts to the resultant gear-tooth forces "r," and call the torque applied to the arm "T." Make a sketch, showing as free-bodies in equilibrium (with all load directions and magnitudes shown):

(a) The assembly composed of the arm, two pinions P, and their shafts.
(b) The portion shown of the right axle, with gear R attached.
(c) The portion shown of the left axle, with gear S attached.

Section 16.10

16.21 A worm gear with 60 teeth and $P = 8$ mates with a double-threaded worm. Determine (a) the gear ratio, (b) diameter of gear, (c) lead of worm, (d) minimum normally recommended diameter of shell worm, (e) corresponding lead angle of worm, and (f) corresponding shaft center distance.
[Ans.: (a) 30:1, (b) 7.5 in., (c) 0.7854 in., (d) 2.042 in., (e) 6.98°, (f) 4.771 in.]

16.22 A worm gear with 40 teeth and a single-threaded worm are to be mounted with a 6-in. center distance. The worm diameter is to be as small as Eq. (16.28) will permit and still use an integral gear diametral pitch. Determine P, d_g, d_w, and λ.

Section 16.11

16.23 A worm gear reducer driven by an 1160 rpm motor has a chill-cast bronze gear and a hardened steel worm. $N_w = 3$, $N_g = 42$, $p = ½$ in., $c = 4.25$ in., $b = 0.75$ in., and $\phi_n = 20°$.

(a) Determine d_g, d_w, and L.
(b) Does the worm correspond to recommended proportions for maximum power transmitting capacity? Is the worm large enough to be bored and mounted over a separate shaft?
(c) Estimate the efficiency, using Eq. (16.34) and check the result with Fig. 10.8. Is the unit self-locking?

[Ans.: (a) 6.685 in., 1.815 in., 1.50 in., (b) yes, no, (c) about 89%, no (the unit is reversible)]

16.24 An 1160 rpm motor delivers 2 hp to the worm in Problem 16.23. The coefficient of friction is estimated to be 0.029. Determine the magnitude of all gear-tooth force components. Show these acting on both members for the case of a counterclockwise-driven, right-hand worm. Use a sketch similar to Fig. 16.20. Compute the efficiency of the unit from the relative rotating speeds and the torques indicated by your free-body sketches.

16.25 A worm gear speed reducer has a right-hand triple-threaded worm, 15:1 velocity ratio, $p = 0.25$ in., $\phi_n = 20°$, and $c = 2.375$ in. The worm is driven by a 900 rpm motor delivering 1/2 hp. The worm is hardened steel and the gear chill-cast bronze; b = 0.5″.

(a) Determine d_w, d_g, N_g, and λ. Compare d_w with recommended values for maximum power transmitting capacity.
(b) Estimate the coefficient of friction and the efficiency of the gears.
(c) Based on your estimated coefficient of friction, determine all force components applied to the worm and to the gear. Show these on a sketch similar to Fig. 16.20 for the case of clockwise motor rotation, as viewed from the worm.
(d) Compute the worm torque, gear torque, and gear power output. From the latter, check the previously determined value of efficiency.

Sections 16.12 and 16.13

16.26 Estimate the steady-state power input and power output capacity of the reducer in Problems 16.23 and 16.24 (with worm driven by an 1160 rpm motor), based on bending and surface fatigue considerations. What, if any, special cooling provisions would be needed for operation at this capacity?
(Ans.: about 3.7-hp input, 3.3-hp output; fan, or housing fins desired)

16.27 Estimate the safety factor with respect to bending fatigue and with respect to surface fatigue for the worm gear reducer operation described in Problem 16.25. Estimate the steady-state oil temperature if the ambient air temperature is 100°F and the reducer is equipped with an integral fan on the end of the worm shaft.

CHAPTER 17 – SHAFTS AND ASSOCIATED PARTS

Section 17.5

17.1 The six shafts represented in Fig. P17.1 carry various combinations of static and alternating bending, axial, and torsional loads. State the loading involved for each of the shafts, and give a short one-sentence explanation of the cause of the loading.

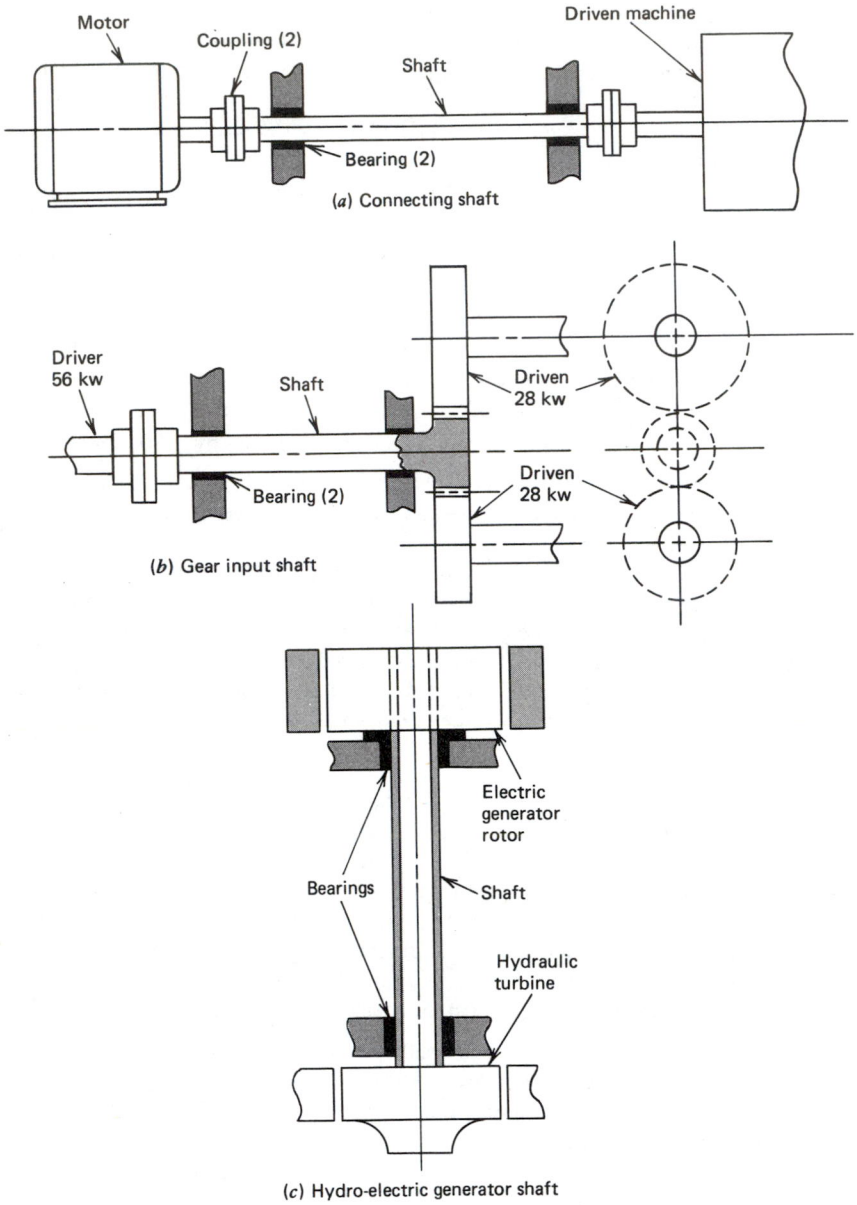

FIGURE P71.1

17.2 Figure P17.2 shows the load components acting on a helical gear mounted on a simply supported shaft. Bearing B takes thrust. A flexible coupling for transmitting torque attaches to the right end of the shaft. The left end is free.

(a) Draw load, shear force, and bending moment diagrams for the shaft, in both the horizontal and vertical planes. Also draw diagrams showing the torsional loading and the axial force loading. (The desired diagrams include the seven shown in Fig.

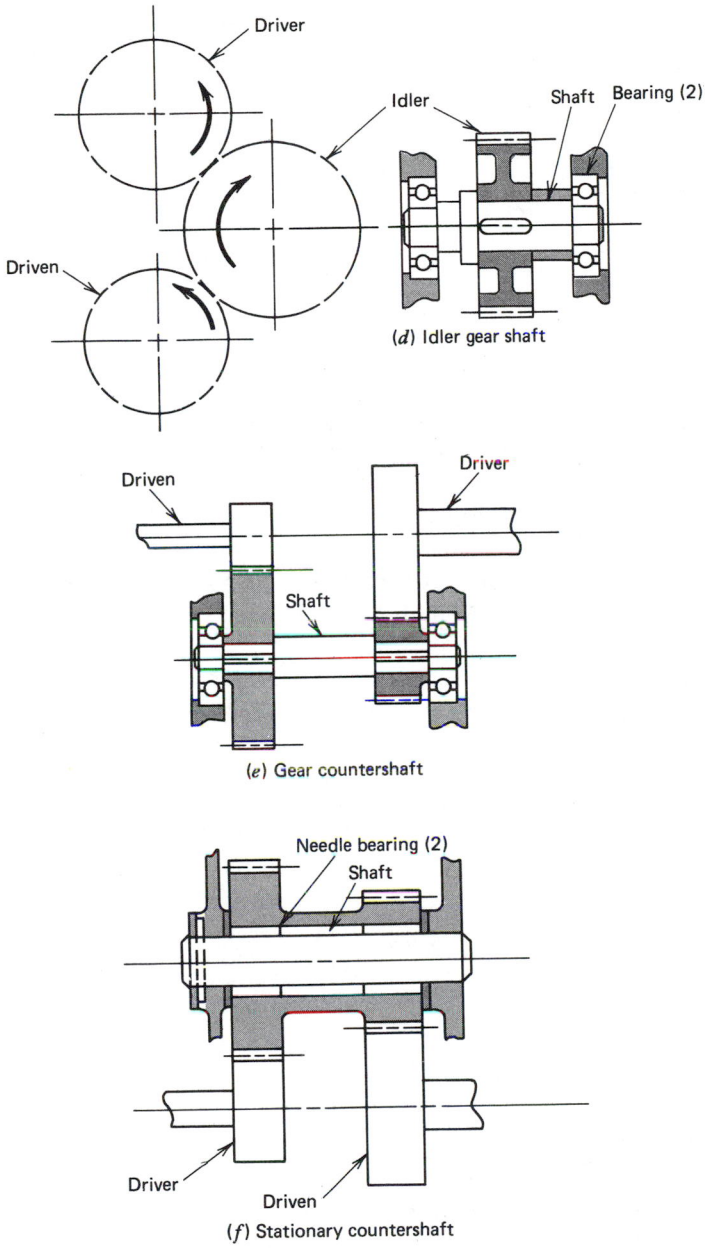

FIGURE P17.1 (*continued*)

17.6*c*. Add a similar diagram showing axial load, with tension plotted positive and compression negative.)

(b) What radial and thrust loads are applied to the bearings?

(c) Identify the most critically loaded shaft cross section, and for this location determine the diameter theoretically required for infinite life. Assume that the shaft will

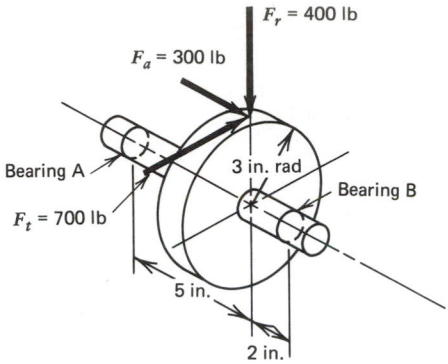

FIGURE P17.2

be machined from steel having S_u = 150 ksi and S_y = 120 ksi, and that K_f = 2.0, 1.5, and 2.0 will apply to bending, torsional, and axial loading respectively at the critical location.

[Ans.: (b) 200 lb radial to A; 649 lb radial and 300 lb thrust to B, (c) just to the right of the gear; approximately 0.88 in.]

17.3 A bevel pinion and shaft are shown in Fig. P17.3. Bearing A takes thrust. The left end of the shaft is coupled to an electric motor, and the right end is free. Load components applied by the mating bevel gear are shown.

(a) Draw load, shear force, and bending moment diagrams for the shaft in both horizontal and vertical planes, plus torsional load and axial load diagrams. (The desired diagrams include the seven shown in Fig. 17.6*c* plus a similar diagram for axial loads, with tension plotted as positive.)

(b) Determine the radial and thrust loads applied to the two bearings.

(c) Identify the critically loaded shaft cross section, and estimate the safety factor with

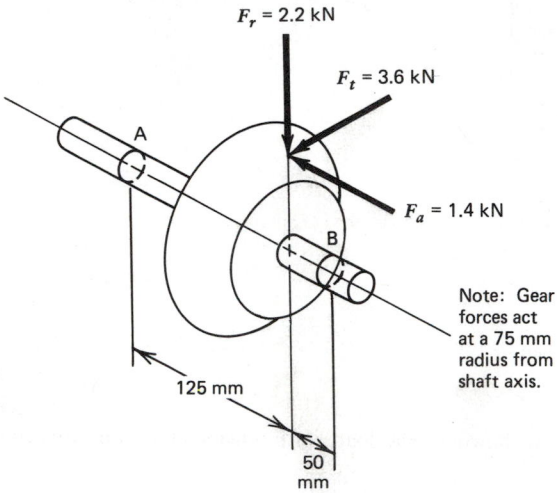

FIGURE P17.3

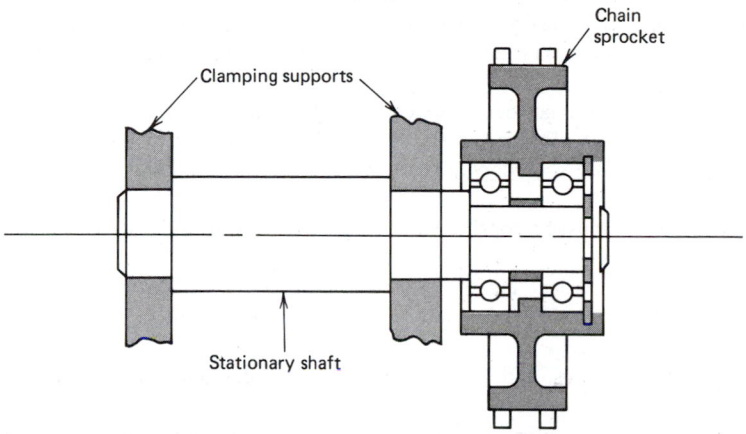

(*a*)

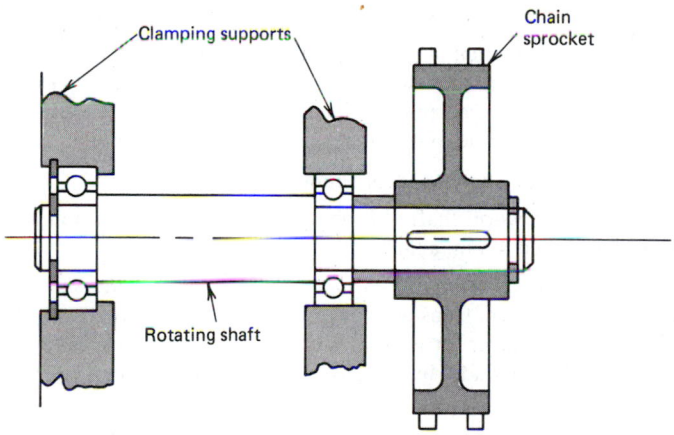

(*b*)

FIGURE P17.4

respect to eventual fatigue failure using the following data: shaft diameter = 33 mm, K_f = 1.3, 1.2, and 1.3 for bending, torsional, and axial loading respectively, material is steel with S_u = 900 MPa and S_y = 700 MPa, and critical surfaces have a ground finish.

17.4 Figure P17.4 shows two alternative approaches to the problem of supporting an overhung or cantilevered chain idler sprocket (or spur gear or belt sheave). What fundamental differences are there between the two with respect to shaft loading and bearing loading? How would the comparison change if a bevel gear were substituted for the chain sprocket?

17.5 The pinion shaft involved in Problem 16.8 is to be designed:

(a) Draw load, shear force, and bending moment diagrams for the shaft, in both the horizontal and vertical planes. Also draw diagrams showing the shaft torsional and axial force loadings. (The desired diagrams include the seven shown in Fig. 17.6*c* plus a similar diagram for axial load, with tension plotted as positive and compression negative.)

(b) What are the radial and axial loads applied to the bearings?

(c) Make a preliminary scale drawing of the shaft, with "reasonable-looking" diameters and with appropriate shoulders and other provisions for the axial retention of parts.

(d) What primary factors must be considered when determining appropriate numerical values of the diameters?

(e) Determine, at fatigue-critical location(s), an appropriate combination of design detail (such as fillet radius), shaft diameter, material and hardness, and surface finish.

(f) Select appropriate bearings, and add dimensions for all diameters shown in your drawing in (c).

(g) Using an appropriate computer program (such as the one given in Section 5.7) or other means, determine the linear deflection at the gear and the angular deflection at the bearings.

[Answer to (d): (1) Bearing considerations—sufficient diameter, standard sizes, and possible use of bearings already on hand for other applications. (2) Fatigue strength. (3) Deflections—lateral deflection of gear should not generally exceed 0.05–0.10 mm or 0.002–0.004 in., and angular deflection at bearings should not normally exceed about 0.03. In some applications, lateral and torsional shaft stiffnesses must be checked to ensure keeping natural frequencies out of the operating speed range.]

17.6 Repeat Problem 17.5 for the countershaft of Problem 16.9.

17.7 Repeat Problem 17.5 for the worm shaft of Problem 16.25, with the worm mounted midway between bearings spaced 4.5 in. apart.

17.8 Problems 17.5–17.7 dealt with motor-driven input shafts and a countershaft for which the loads were essentially known. Consider now the more involved problem of designing the output shaft of a gear reducer for general use. The driven machine may be coupled directly to the reducer output shaft, but it may also be driven through gears, belts, or chains.

(a) In designing the reducer output shaft and bearings for the most severe reasonable application, would you consider the largest or the smallest gear, sprocket, or sheave that might be mounted on this shaft? Explain, briefly.

(b) Other factors being equal, would the most severe loading be applied with a gear, belt, or chain drive? Explain, briefly.

Section 17.6

17.9 Estimate the length of flat key required to transmit a torque equal to the elastic torque capacity of a round shaft of diameter d. Assume that the key and shaft are made of the same ductile material and that the key is tightly fitted at its top and bottom. Compare this result with the length of square key required, and suggest a possible reason why a flat key might be preferred in some cases.
(Ans.: $L = 2.4\,d$)

CHAPTER 18 – CLUTCHES AND BRAKES

Section 18.2

18.1 A multiple-disk clutch is to operate in oil, and be able to transmit a design overload torque of 650 N·m. The disks are alternately high carbon steel and molded asbestos, with inside and outside diameters of 90 and 150 mm respectively. Design values based on test experience for this application are p_{max} = 1000 kPa and f = 0.10. What total number of disks is required?
[Ans.: 15 (7 disks in one set and 8 in the other, giving 14 friction interfaces)]

18.2 A multiple-disk wet clutch is required to provide a torque capacity of 35 lb-ft. Design values of p_{max} = 200 psi and f = 0.1 are to be used. Disk inside and outside diameters are to be 3 and 4 in., respectively.

(a) What total number of disks should be used? Make a simple sketch showing how these are arranged with respect to the input and output members.
(b) Using your answer from (a), what is the smallest value of axial clamping force that would provide the necessary torque capacity?
(c) Assuming that your solution was (appropriately) based on a uniform rate of wear across the friction surfaces, explain briefly what will happen when the clutch is new and the pressure distribution is uniform.

18.3 A multiple-disk dry clutch for an industrial application must transmit 4 hp at 100 rpm. Based on space limitations, disk inside and outside diameters have been set at 4 and 6 in., respectively. Materials are hardened steel and sintered bronze. Tests have indicated that design values of p_{max} = 225 psi and f = 0.20 are appropriate.

(a) For a safety factor of 1.5 with respect to clutch slippage, what total number of disks is needed?
(b) Using this number of disks, what is the least clamping force that will provide the desired torque capacity?
(c) Using this clamping force, what is the interface pressure at the inside and outside contact radii (assume that "initial wear" has taken place)?

[Ans.: (a) 5, (b) 1890 lb, (c) 151 and 100 psi, respectively]

18.4 An automotive-type clutch, as shown in Fig. 18.2, has inside and outside diameters of 150 and 225 mm, respectively. Clamping force is provided by nine springs, each compressed 6.5 mm to give a force of 625 N when the clutch is new. The molded friction material provides a conservatively estimated coefficient of friction of 0.35 when in contact with the flywheel and pressure plate. Maximum engine torque is 280 N·m.

(a) What is the safety factor with respect to slippage of a brand new clutch?
(b) What is the safety factor after "initial wear" has occurred?
(c) How much wear of the friction material can take place before the clutch will slip (assuming no change in coefficient of friction)?

[Ans.: (a) 1.34, (b) 1.32 (assuming negligible spring force change during initial wear), (c) 1.57 mm]

18.5 An automotive clutch, similar to the one shown in Fig. 18.2, is to be designed for use with an engine with a maximum torque of 275 N·m. A friction material is to be used for which design values are f = 0.35 and p_{max} = 350 kPa. A safety factor of 1.3 should be used with respect to slippage at full engine torque, and it is desired to have as small an outside diameter as possible. Determine appropriate values of r_0, r_i, and F.

18.6 When in use, the motor and flywheel of a punch press run continuously. A compressed-air-actuated, multiple-disk clutch connects the flywheel to the countershaft each time the press is to punch or form metal. The clutch torque required is 525 N·m. Alternate sintered metal and cast-iron disks are to be used. For design purposes it is decided to use 75% of the average value of f, and ⅓ of the average value of p given in Table 18.1. In addition, a safety factor of 1.15 is to be used with respect to torque capacity. Space requirements limit the disk outside diameter to 220 mm.

(a) Determine the minimum satisfactory total number of disks.

(b) Using this number of disks, determine the minimum area of compressed-air actuating surface that must be provided if air will always be available at a pressure of at least 0.40 MPa.

Section 18.3

18.7 The wheels of a standard adult bicycle have a rolling radius of approximately 13.5 in. and a radius to the center of the caliper disk brake pads of 12.5 in. The combined weight of bike plus rider is 200 lb, equally distributed between the two wheels. If the coefficient of friction between the tires and road surface is twice that between the brake pads and the metal wheel rim, what clamping force must be exerted at the caliper in order to slide the wheels?
(Ans.: 108 lb)

18.8 A disk brake similar to the one shown in Fig. 18.4 uses a double caliper. Each half has a 50 mm diameter round pad on each side of the disk. The center of contact of each of the four pads is at a radius of 125 mm. Outside diameter of the disk is 300 mm. The pads have woven linings that provide a coefficient of friction of approximately 0.35. The average pressure on the pads is to be limited to 500 kPa.

(a) What clamping force must be provided in order to develop the limiting pad pressure?

(b) With this clamping force, what approximate brake torque is obtained?

[Ans.: (a) 982 N/pad or 1964 N total, (b) 172 N·m]

Section 18.4

18.9 The brake in Problem 18.8 is used to bring a rotating assembly with mass-moment of inertia = 6.5 N·m·s^2 to a stop from a rotating speed of 600 rpm.

(a) Assuming instantaneous full application of the brake and a constant coefficient of friction, how long does this take? Solve the problem and check your solution, using two alternative approaches: (1) by relating kinetic energy in the system to the energy absorbed by the brake per revolution, and (2) by determining the rate of negative acceleration produced by the brake torque.

(b) How does the average value during stopping of the product pV compare with representative values listed in Table 18.3?

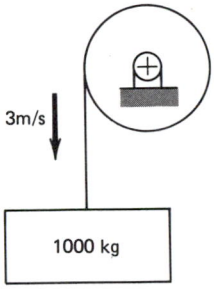

FIGURE P18.10

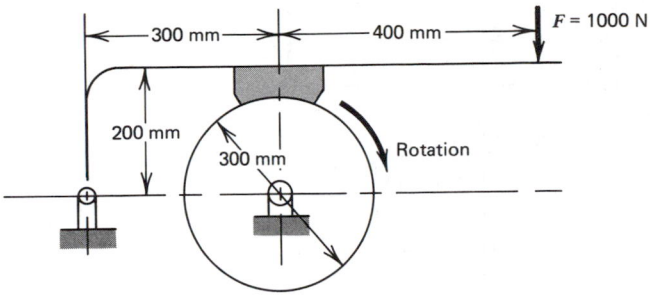

FIGURE P18.13

18.10 Figure P18.10 shows a 1000 kg mass being lowered at a uniform rate of 3 m/s from a drum of 600 mm diameter weighing 2 kN and with 250-mm radius of gyration.

(a) What is the kinetic energy in the system?

(b) The uniform rate of descent is maintained by a brake on the drum which applies the necessary torque of 300 kg·m. What additional brake torque is required to bring the system to rest in 0.50 s?

[Ans.: (a) 5137 J, (b) 2054 N·m]

18.11 The automobile in Sample Problem 2.2 (Section 2.2) accelerates to 90 mph and then makes a panic stop. If the brakes are applied so as to take full advantage of a friction coefficient of 0.8 between tires and pavement, what is the average horsepower absorbed by the brakes during deceleration from 90 to 80 mph?
(Ans.: 584 hp)

18.12 An industrial diesel engine is attached to a 3:1 ratio gear reducer, the output of which is coupled by means of a friction clutch to a machine having a mass moment of inertia of 15 $N \cdot m \cdot s^2$. Assume that the clutch is controlled so that during its engagement the engine operates continuously at 2100 rpm, delivering a torque of 170 N·m:

(a) What is the approximate time required for the clutch to accelerate the driven machine from rest to 700 rpm?

(b) How much heat energy is generated in the clutch during this engagement?

(c) How much energy is delivered to the driven machine in bringing it up to 700 rpm?

Section 18.6

18.13 Figure P18.13 shows a brake with only one shoe, being applied by a 1 kN force. (The complete brake would normally have a second shoe in order to balance the forces, but only one shoe is considered here to keep the problem short.) Four seconds after force F is applied, the drum comes to a stop. During this time the drum makes 100 revolutions. Use the "short shoe" approximation and an estimated coefficient of friction of ⅓.

(a) Draw the brake shoe and arm assembly as a free-body in equilibrium.

(b) Is the brake self-energizing, or de-energizing, for the direction of drum rotation involved?

(c) What is the magnitude of the torque developed by the brake?

(d) How much work does the brake do in bringing the drum to a stop?

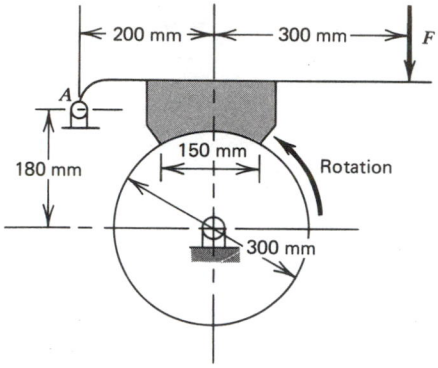

FIGURE P18.14

(e) What is the average braking power during the four-second interval?
(f) How far below the drum center would the arm pivot need to be to make the brake self-locking for $f = 1/3$?

[Ans.: (b) self-energizing, (c) 140 N·m, (d) 88.0 kJ, (e) 22.0 kW, (f) 750 mm]

18.14 Figure P18.14 shows a brake with one shoe (a second shoe would normally be used to balance the forces, but only one shoe is shown here to keep the problem short). The width of shoe contact with the drum is 40 mm. The friction material provides a coefficient of friction of 0.3, and permits an average pressure based on *projected* area of contact of 500 kPa. Use the "short-shoe" approximate relationships.

(a) What value of force F can be applied without exceeding the allowable contact pressure?
(b) What is the resulting brake torque?
(c) Is the brake self-energizing, or de-energizing for the direction of rotation involved?
(d) What radial force is applied to the pivot bearing at point "A"?
(e) If full brake application brings the drum from 240 rpm to a stop in 6 seconds, how much heat is generated?
(f) What is the average power developed by the brake during the stop?

18.15 The brake shown in Fig. P18.15 is applied by the spring and released by a hydraulic

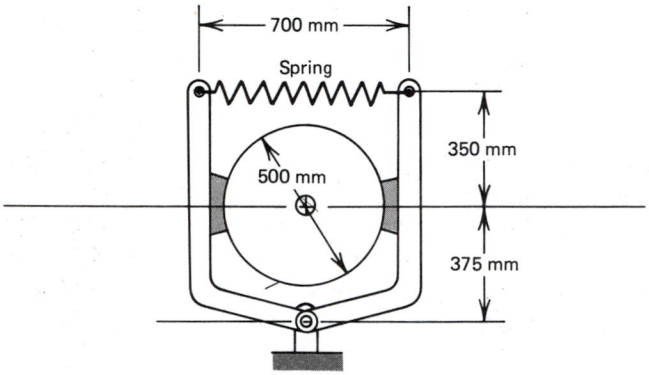

FIGURE P18.15

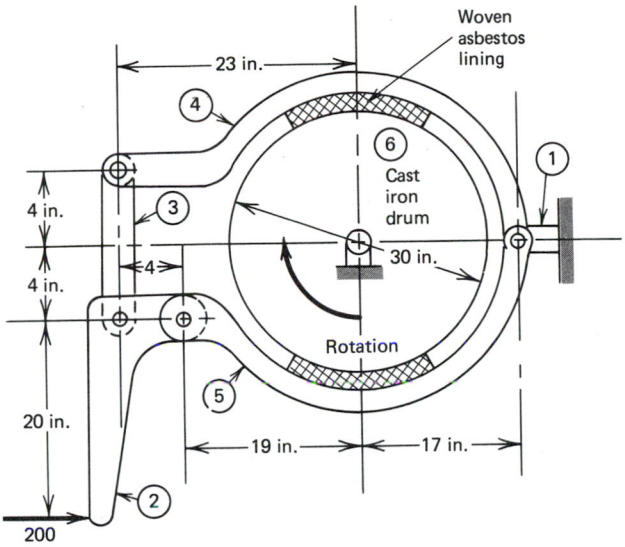

FIGURE P18.16

cylinder (not shown). Use the "short-shoe" equations and an assumed coefficient of friction of 0.3.

(a) Draw as free-bodies in equilibrium each of the brake shoe and arm assemblies, the spring, and the drum. Show forces in terms of spring force, F_s.

(b) What spring force is required to produce a braking torque of 1500 N·m?

[Ans.: (b) 4960 N]

18.16 The shoes of the drum brake shown in Fig. P18.16 can be assumed to be "short." A 200-lb force is applied as shown. Use the minimum value of f (dry) from Table 18.1:

(a) Draw each of the five numbered members as a free body in equilibrium.

(b) What torque is developed by the brake?

Section 18.7

18.17 If the shoe in Problem 18.13 extends 45° on each side of the centerline:

(a) Estimate the brake torque using "long-shoe" equations.

(b) Determine the shoe width required for an allowable pressure equal to the average value listed for woven asbestos on cast iron in Table 18.1.

18.18 Estimate the brake torque in Problem 18.14 using "long-shoe" equations.

18.19 If the shoes in Problem 18.15 extend 45° on each side of the centerline:

(a) Determine the spring force required, using "long-shoe" equations.

(b) Determine the shoe width required for an allowable pressure equal to the minimum value listed for woven asbestos on cast iron in Table 18.1.

18.20 If the shoes in Problem 18.16 extend 60° on each side of the centerline:

(a) Determine the brake torque using "long-shoe" equations.

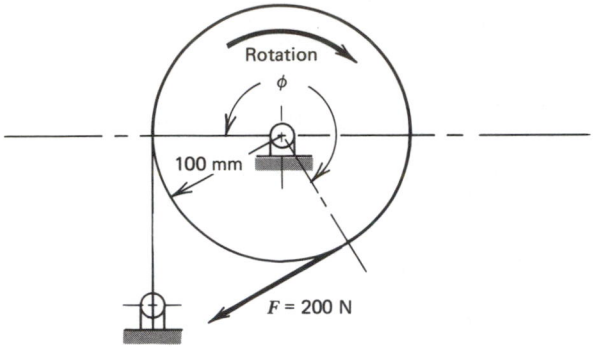

FIGURE P18.21

(b) Determine the shoe width required for an allowable pressure equal to the maximum value listed for woven asbestos on cast iron in Table 18.1.

Section 18.9

18.21 Figure P18.21 shows a simple band brake operated by an air cylinder that applies a force F of 200 N. The drum radius is 100 mm. The band is 30 mm wide and is lined with a woven material that provides a coefficient of friction of 0.4.

(a) What angle of wrap, ϕ, is necessary to obtain a brake torque of 100 N·m?
(b) What is the corresponding maximum lining pressure?
[Ans.: (a) 257°, (b) 400 kPa]

18.22 A differential band brake similar to the one shown in Fig. 18.17 is being considered for use with a hoist drum. With the hoist operating at rated load, a brake torque of 5000 N·m is required to hold the drum from slipping. The brake is to be designed to slip at a 20% overload. Available space limits the cast iron brake drum diameter to 600 mm, with brake band contact extending over an angle of 250°. Design values for the brake lining selected are $f = 0.35$, $p_{max} = 1.1$ MPa.

(a) Determine forces P_1 and P_2 corresponding to maximum brake torque, and select an appropriate value of brake band width.
(b) If distance a (Fig. 18.17) is 100 mm, what value of distance s would enable the brake to be operated with a force F of 200 N at the end of a lever of length c = 650 mm?
(c) Using the linkage dimensions from (b), what value of coefficient of friction would make the brake self-locking?

18.23 Figure P18.23 shows a band brake used with a punch press like the one involved in Problem 18.6. When in use, the clutch (Problem 18.6) is released when the crank is 120° past bottom dead center. The brake is to be engaged at this point, and bring the crank to rest at top dead center. The crank assembly has a mass moment of inertia of approximately 12 N·m·s^2 and is rotating at the rate of 40 rpm when the brake is engaged. The brake will be used about 3 times per minute, so the maximum pressure on the band lining should be limited to about 0.18 MPa for long life. Coefficient of friction can be taken as 0.35.

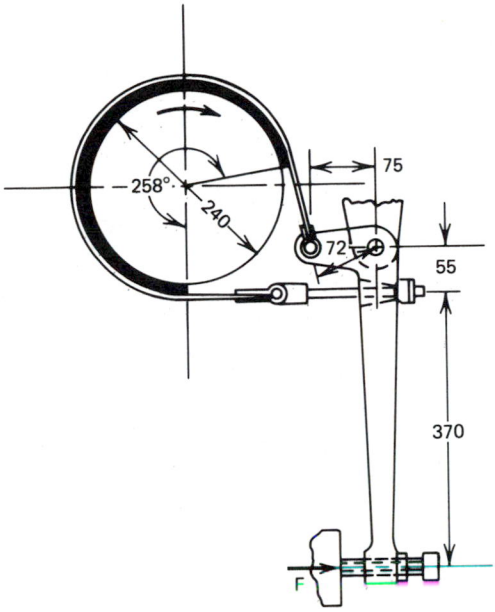

FIGURE P18.23

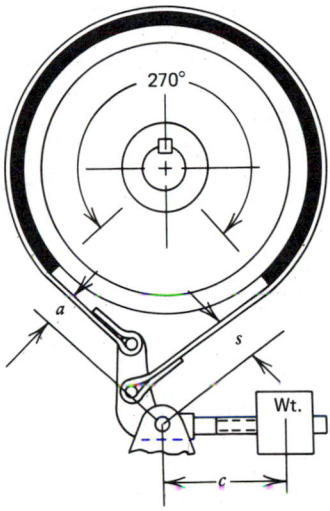

FIGURE P18.24

(a) Determine the required band width.
(b) Determine the required force, F.
(c) Would any combination of direction of rotation and value of coefficient of friction make the brake self-locking? Explain, briefly.

[Ans.: (a) 49 mm, (b) 174 N]

18.24 Figure P18.24 shows a differential band brake with the linkage modified from Fig. 18.17 so as to permit a greater angle of band contact. This particular brake is to be self-locking for counterclockwise rotation. To accomplish this, the adjustable weight is provided at the end of the lever. Its function is merely to ensure that the band is in contact with the drum (any excess weight increases the drag torque during clockwise drum rotation). If self-locking action is to be obtained with coefficients of friction as low as 0.20, what relationship must exist between dimensions a and s?

CHAPTER 19 – MISCELLANEOUS POWER TRANSMISSION COMPONENTS

Section 19.2

19.1 A belt drive, as shown in Fig. 19.1a, has an angle of wrap on the small pulley of 150°. Adding an idler, as in Fig. 19.1c, increases this to 200°. If the slack-side tension is the same in the two cases and if centrifugal force is negligible, by what percentage is the capacity of the belt drive increased by adding the idler if $f = 0.3$?
(Ans.: 55%)

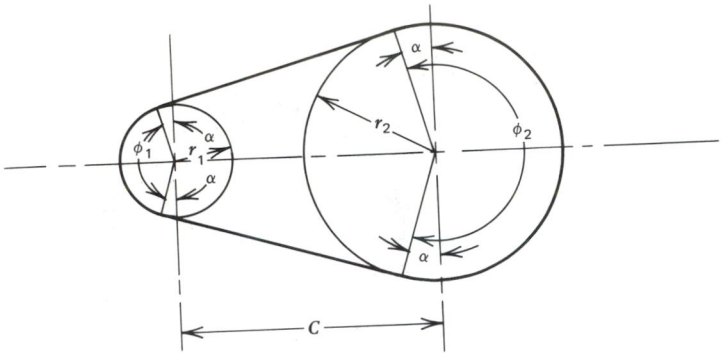

FIGURE P19.2

19.2 Figure P19.2 shows that angles of wrap ϕ_1 and ϕ_2 are equal to $\pi - 2\alpha$ and $\pi + 2\alpha$, respectively. Derive an equation relating α to r_1, r_2, and c.
[Ans.: $\sin \alpha = (r_2 - r_1)/c$]

Section 19.3

19.3 A 30 hp, 1750 rpm electric motor drives a machine through a multiple V-belt. The pulley on the motor shaft has a 3.7 in. pitch diameter (a standard size), and the geometry is such that the angle of wrap is 165°. The size 5V belts used have an angle, β, of 18° and a unit weight of 0.012 lb/in. It is conservatively assumed that the maximum belt tension should be limited to 150 lb, and that the coefficient of friction will be at least 0.20. How many belts are required?
(Ans.: 5)

19.4 A standard V-belt with $\beta = 18°$ is used with a 40 mm pitch diameter motor pulley and a smooth cylindrical drum of 120 mm diameter. Center distance is 120 mm. What is the approximate relationship between the slip-limited power transmission capacities at the pulley and at the drum?

19.5 Find the maximum power that can be transmitted by the smaller pulley of a V-belt drive under the following conditions: pulley speed = 4000 rpm, r = 100 mm, $\beta = 18°$, $\phi = 170°$, $f = 0.20$, belt maximum tension = 1300 N, and belt unit weight = 1.75 N/m.

19.6 What power could be transmitted by the pulley in Problem 19.5 if (a) 2 belts were used, each identical to the belt in Problem 19.5, and (b) a single belt is used, with twice the cross-section of the belt in Problem 19.5, and hence twice the maximum tension?

19.7 A single V-belt with $\beta = 18°$ and unit weight of 2.2 N/m is to be used to transmit 12 kW from a 1750 rpm, 180-mm diameter driving pulley to a 1050-rpm driven pulley. Center distance is 400 mm.

(a) Based on a coefficient of friction of 0.20 and an initial belt tension marginally adequate to prevent slippage, determine the values of P_1 and P_2.

(b) Determine the torsional and resultant radial loads applied by the belt to each shaft.

(c) Determine the initial belt tension when the drive is not operating.
(d) Determine values of P_1 and P_2 when the drive is operating at normal speed but transmitting only 6 kW.

[Ans.: (a) 926 N, 199 N, (b) 65.5 N·m and 109.2 N·m torque to the driving and driven shafts; 1118 N radial load to each shaft, (c) 562.5 N, (d) 744 N, 381 N]

19.8 A single V-belt with $\beta = 18°$ and unit weight of 0.012 lb/in. is to transmit 12 hp from a 1750 rpm, 6 in. diameter driving pulley to a 12 in. diameter driven pulley. Center distance is to be 20 in.

(a) Based on a coefficient of friction of 0.20 and on initial belt tension just adequate to prevent slippage, determine values of P_1 and P_2.
(b) Determine the torsional and resultant radial loads applied by the belt to each of the shafts.
(c) Determine values of P_1 and P_2 when the drive is operating at normal speed but transmitting only 3 hp.

Section 19.8

19.9 A fluid coupling of the type shown in Fig. 19.10 has its input and output shafts directly coupled to an electric motor and driven machine, respectively. The motor and coupling characteristics conform to the performance curves in Fig. 19.11.

(a) During normal operation the motor rotates 1780 rpm and drives the machine with 55% of its rated power. What is the rotating speed of the machine input shaft? What percent of the motor torque reaches the machine? What percent of the power output of the motor is converted into heat in the fluid coupling?
(b) How much does the coupling slip during overload operation requiring full rated motor power?
(c) If the machine is overloaded to the point that the motor cannot drive it and the machine stops, what is the motor rpm? What fraction of normal rated motor power is converted into heat in the fluid coupling under these conditions?

19.10 Assume that the motor and coupling characteristics represented in Fig. 19.11 correspond to a well-engineered installation incorporating a 10 kW, 1750 rpm motor. A similar but smaller installation is being considered, for which the power requirements are 50% less. Accordingly, a 5 kW, 1750 rpm motor and a smaller, but geometrically similar, fluid coupling will be used. By what factor could the diameter of the fluid coupling be reduced? If the 5-kW power source used in the new installation rotates 3500 rpm, by what factor could the coupling diameter be reduced?

Section 19.9

19.11 A hydrodynamic torque converter like the one represented in Fig. 19.13 provides a 2.4 torque multiplication when operating with a driving motor providing a torque of 100 N·m. What torque is applied to the one-way clutch?

Index

ABEC (Annular Bearing Engineers' Committee), 441
Abrasive wear, 255-256
  of gear teeth, 485
Acme thread, 275, 278, 286-287
Actuator, linear (rolling-element screw), 284-285
Adhesives, 348-350
  use with rolling-element bearings, 411
Adhesive wear, 253-255
  of gear teeth, 484-485
AFBMA (Anti-Friction Bearing Manufacturers Association), 441
AGMA (American Gear Manufacturers Association), 457
Air (pneumatic) springs, 351
AISI (American Iron and Steel Institute), 331, 624, 631-633
Aluminum alloys, 61
  fatigue strength, 207, 208
  properties of (tables), 619, 620, 634-635
  temper designations, 636
ANSI (American National Standards Institute), 274, 275, 394, 441, 622
Area ratio, R (index of ductility), 49
Assembly fits and tolerances, 652-653
ASME (American Society of Mechanical Engineers), 331
ASTM (American Society for Testing Materials), 59, 62, 337, 394, 621, 622, 641, 643
Automatic transmission ("Hydra-Matic"), 598-610
  see also Fluid couplings; Torque converters
Automobile:
  free-body analysis of power train, 28-31
  performance and economy analysis, 21-23, 26-28
AWS (American Welding Society), 337
Axial loading fatigue strength, 209
Axial stress, limitation of "P/A" formula, 69-72
Axles, 535

Babbitt, 415
Backlash in gears, 466
Ball bearings, *see* Bearings, ball and roller
Ball-bearing screws, 284, 285
Ball bushings, 436, 440
Base circle of involute gears, 459
Basic hole system of dimensioning, 652-653
"Beach marks" in fatigue fracture, 200
Beams, 126-127
  computer analysis, 128-131
  curved, bending, 77-82
  deflection and slope, 127-132, 646-651
  as springs, 377-382
  straight, bending of, 76-77
  superposition method of deflection determination, 128
  transverse shear in, 82-88
Bearing, general term, 387
Bearing loading:
  in journal bearings, 391
  in pinned joints, 39-40
  in screw threads, 289
Bearings, ball and roller:
  design, 437-441
  dimensions, 443-447
  equivalent load, 451
  fatigue life, 449-450
  fitting and installation, 441
  history, 429
  lubrication, 427
  manufacturing tolerances, 441

materials used, 437-439
mounting, for taking thrust, 454-456
needle (roller), 430, 433, 436, 437
rated capacities, 448
reliability, 450-451
selection, 442-452
series, 430
thrust, 430, 431, 433, 434, 437
types of, 427-429, 431-437
Bearings, journal:
bearing characteristic number, 402
boundary lubricated, 422-424
design, 415-424
dynamic stability of ("whip"; "whirl"), 409
friction and heat losses in, 391, 413-414
lubricant supply to, 410-413
materials, 414-415, 423, 424
partial, 388
thrust, 424, 425
types, 388
see also Lubrication
Belleville washer springs, 384-385
Belts, 580-586
flat, 581-583
toothed, 586
"V," 583-586
used in adjustable speed drives, 585
Bending of:
beams, curved, 77-82
sign convention for bending of, 79
beams, straight, 76, 77
bolts, 310, 312
gear teeth, 474-484, 510, 516-518, 528-529
shafts, 203, 230-232, 544
welded joints, 340-345
Bevel gears, 501-503, 512-521
Bolts and screws, 273-327
design for maximum fatigue strength, 327-328
fatigue strength and selection, 318-328
grade (class) designation, 297-298
initial tightening, 299-302
load analysis (effect of elasticities), 305-310
load distribution among threads, 288-291
materials and manufacture, 296-299
power screws, *see* Screws, power
static strength and selection, 310-318
stress concentration factors, 318-319
stress calculations, 288-291
thread fits and tolerances, 275, 277
thread geometry:
special, 290-291, 325, 327-328
standard, 274-278
thread loosening and thread locking, 303-306
types, 294-296
see also Screws, power; Screw threads
Boundary lubrication, 389, 422-424
Boyd and Raimondi lubrication charts, 402-410, 425
Brakes:
band, 575-577
caliper (disk), 559-560
disk, 559-560
drum, 563-575
force analysis:
band, 575-577
external shoe drum, 563-573
internal shoe drum, 573-575
heat considerations, 560-561
materials used in, 558, 559, 579-580
self-energizing action, 565
Brazing, 347
Brittle fracture, 200
Brittle materials, fatigue strength of, 210
Buckingham, Earle, 486
Buckling:
column, 145-154, 292-293
end fixity conditions, 148-149
Euler equation, 145-148, 150-152, 156
of helical compression springs, 363-364
Johnson parabola, 149-152
local "wrinkling," 154-155
of power screws, 292
secant formula, 152-153
Bushings, 389

Cadmium, 268
Cams (load, energy, power analysis), 17-19
Carburizing, influence on fatigue strength, 239
Cast iron, 57-58
fatigue strength, 207
properties (table), 619, 621-623
Castigliano's theorem:
for deflection determination, 132-145
for redundant reaction determination, 140-145
Cavitation damage to surfaces, 252
Chain couplings, 551
Chains, 580-581, 586-590
beaded, 580-581
chordal action, 587-588
inverted tooth ("silent"), 589-590
roller, 586-589
Clash allowance of spring, 361
Clutches:
cone, 561-563
disk, 553-559
free-wheeling (over-running or one-way), 726, 727
materials used in, 558, 559, 578-579

multiple-disk, 554, 556-559
Coefficients of friction, *see* Friction, coefficients of
Coil springs, *see* Springs, coil conical and helical
Coining, effect on fatigue strength, 238
Cold-rolling, effect on fatigue strength, 238
Cold-working:
effect on fatigue strength, 238-239
effect on static strength, 50
Columns, *see* Buckling
Combined stresses, 91-94
in bolts and power screws, 299-301
in shafts (disk sander example), 230-232
Composite materials (fiber reinforced plastics), 66
Computer use:
in shaft deflection determination, 128, 130-131
in spring design, 368
Conjugate action of gear teeth, 458-459
Contact stresses, 259-263
in gear teeth, 486-491
in rolling element bearings, 437
Copper alloys, 61-62, 415
fatigue strength of, 207
properties of (table), 619, 620, 637
Corrosion, 241-252
fatigue, 251-252
fretting, 256, 257
galvanic series, 243
stress cracking, 249-251
Coulomb-Mohr theory of failure, 169-170
Couplings:
flexible, 550
fluid, 591-595, 599
rigid, 549
universal joints, 550-552
Crankshaft, fatigue failure photograph of, 201
Critical speed of shafts, 536, 539-542
Cumulative damage (Palmgren-Miner rule), 232
Cycloidal gear teeth, 459
Cylinders, properties of, 617

Datsko, Joseph, 53, 54, 56, 620
Davis, G. H. B., 396
Dean, E. W., 396
Decarburization, effect on fatigue strength, 236
Deflection:
by Castigliano's method, 132-145
of beams, 127-132, 646-651
Density of common metal (table), 619
Design considerations (general), 13-14
Differential, automotive, 520-521
Disk clutches, *see* Clutches
Distortion energy theory of failure, 167-169
Distribution (statistical):
normal (Gaussian), 177-180
Weibull, 450
Ductile iron, *see* Nodular iron

Eccentric loading:
of axially loaded members (in fatigue), 209
of bolted joints, 315-318
of bolts, 310, 312
of columns, 152-153
of welded joints, 340-345
Efficiency:
of fluid couplings, 595, 597
of hydrodynamic torque converters, 597
of power screws, 283-287
of welded joints, 337
of worm gears, 523-528
Elastic limit of material, 48
Elastic modulus of material, 48
values of, for common metals (table), 619
Elastic stability, *see* Buckling
Elastohydrodynamic lubrication, 425, 485
Elongation at fracture of a material, 49
values of (tables), 622-642
Embeddability of bearing materials, 414-415
Endurance limit, *see* Fatigue, strength
Energy:
absorbing capacity of materials, 54-54, 189-190
associated with brakes and clutches, 560-561
associated with impact loading, 183, 185-198
methods for:
deflection analysis (Castigliano's theorem), 132-145
impact analysis, 186-188
storage in flywheels, 19-21
work, power basic relationships, 17-24
Engineering stress-strain curve, 47-50
Epicyclic (planetary) gear trains, 496-499, 599-606
Equivalent bending stress for biaxial stress fatigue, 211
Euler column buckling, 145-148, 150-152, 156
Expansion, thermal coefficient of (table), 619

Face width, *b*, of:
bevel gears, 513, 514
helical gears, 505, 506
spur gears, 463, 464-465
worm gears, 521, 522
Factor of safety, *see* Safety factor
Failure theories for static loading, 164-172
Fasteners, *see* Bolts and screws; Rivets
Fatigue:
crack propagation, 202
failure:
basic concepts, 201-203
description, 200

history, 200
illustration, 201
of surfaces due to contact stresses, 259, 263-267
strength:
of aluminum alloys, 207-208
biaxial stress, effect of, 210, 211, 217
of cast iron, 207
constant life diagrams, 216-221
corrosion, effect of, 212
decarburization, effect of, 236
equivalent stress approach, factors influencing, 211, 214
Goodman line, 216
laboratory test of R. R. Moore, 203-204
load-type (C_L) effect of, 207-210, 214
mean stress, effect of, 215-222
of non-ferrous alloys, 207
number of load cycles, effect of, 203-205
of *S-N* curves, 203-205
statistical variation, 205
of steel, 204-206
stress concentration, effect of, 222-232
for keyways in steel shafts, 548
for steel screw threads, 318
stress gradient (C_G), effect of, 209, 213, 214
stress level variations, effect of, 232-235
surface finish (C_S), effect of, 211, 212, 214, 267
surface treatments, effect of, 236-239
Finger (human), free-body analysis of internal forces, 34, 35
Fits, 652-653
Flame hardening (influence on fatigue strength), 239
Fluid couplings, 591-595, 599
Flywheels, 19-21
Force analysis, *see* Free-body analysis of loads; Load analysis, applied to
Force-flow concept, applied to:
location of critical sections, 37-40
redundant ductile structures, 43-45
stress distribution within axially loaded members, 70-71
Fracture:
brittle, 160-164, 200
mechanics, 160-164, 200, 202
Free-body analysis of loads, 25-37
Free-wheeling clutch, 596, 597, 726, 727
Fretting (fretting corrosion), 256-257
influence on fatigue failure, 256
Friction, coefficients of:
in belts, 582
in brakes and clutches, 558, 559
in journal bearings, 389, 398
in rolling-element bearings, 429
in screws, 281, 282
in worm gears, 526
Friction materials for brakes and clutches, 578-579

Galling, 254
Galvanic series, 243
Gaskets, influence of, in bolted connections, 307-310
Gaussian distribution, 177-180
Gear:
housing, heat dissipation, 530-531
lubrication, 484
manufacture, 466-467
materials, 488, 489, 495
mounting, 471, 481, 518
teeth:
backlash, 466
bending stresses and strength, 473-484, 510, 516-518, 528-529
conjugate action, 458-459
contact ratios, 466-470, 476-477
cycloidal form, 459
force analyses, 470-473, 507-510, 514-516, 523-527
geometry, 458-466
factors, 479
interference (involute), 466-470
involute form, 459-461, 465-466
Lewis equation and form factor, 475, 476
load sharing, 469, 476-477, 479, 480
nomenclature, 459-466, 505-507, 512-514, 521-523
pitch, *see* Pitch of gears
standard proportions, 464
stress concentration, 473, 480
surface stresses and strength, 484-491, 510-512, 518, 528-529
velocity factors, 480
trains (ratios), 495-499
epicyclic (cycloidal), 496-499, 520-521, 599-610
Gears:
bevel, 501-503, 512-521
differential, 520-521
mounting, 518
spiral, 514, 516, 517, 519
Zerol ®, 514-515
crossed-axis helical, 512
geometric (theoretical) stress concentration factor, K_t, 98-105
helical, 501, 505-512
herringbone, 501
history of, 457

internal, 465, 495
spur, 457-499
design procedure, 491-495
worm, 503-504, 521-534
Goodman line (fatigue strength criterion), 216
Gradient factor, C_G (for fatigue), 209, 213, 214
Graphical integration, 131-132
Grease lubrication, 387-388, 394
Gyration, radius of, 615, 616

Handbook material properties data, limitations of, 56
Hand of helical gears, 505
Hand of screws and worms, 274
Hardenability, 59
Jominy end-quench test for, 59
Hardness (from penetration hardness tests), 54
number scales, relationships between, 55
relationship to static and fatigue strengths, 54, 56, 205, 206
of various metals (tables), 621, 622, 624, 625, 631, 641
Harmonics of cam motion, 371-373
Heat balance of bearings, 413-414, 421
Heating and heat dissipation:
of bearings, 413-414, 421
of brakes and clutches, 560-561
of worm gear housings, 523-525, 530-534
Helical gears, 501, 505-512
Helical springs, *see* Springs, coil, helical
Herringbone gears, 501
Hertz contact stresses, 259-263
in gear teeth, 486-487
in rolling-element bearings, 437
High strength, low alloy steel, 60
Holmes, Oliver Wendell, 157
Hooke's joint, 550
Horsepower, definition and formula, 18
Hydra-Matic transmission, 598-610
Hydraulic drives, *see* Fluid couplings; Torque converters, hydronamic
Hydrodynamic lubrication, 389-392, 398-410, 415-422
Hydrogen embrittlement, 237
Hydrostatic lubrication, 389-390
Hydrostatic stress, 168
Hypoid gears, 502, 503

Idealized stress-strain curve, 105
use of in fatigue calculations, 225-227
Impact:
and shock loading, 182-198
in bolts, 196-198
definition of, 183
in gears, 476, 477, 480
in rolling-element bearings, 452
strength:
effect of strain rate on, 183-184
effect of stress raisers on, 194-198
torsional, 192-194
Indeterminate loads, *see* Redundant reactions, determination
Induced stresses, 88-91
Initial tension:
in belts, 581, 583, 586
in bolts, 299-302, 324-325
in tension springs, 375-376
Initially curved beams, stresses in, 77-82
Instability, *see* Buckling
Interference fits, 538, 652-653
Interference of involute gear teeth, 466-470
Interference theory of reliability prediction, 179-181
Internal friction theory of failure, 170
Internal gears, 465, 495
International system of units (SI) 14, 612-614
Integration, graphical, 131-132
Inverted-tooth ("silent") chains, 589-590
Involute gear teeth, 459-462
Involute splines, 540
Iron, cast, *see* Cast iron
ISO (International Standards Organization) standards:
for screw threads, 274, 277
for viscosity, 394

Johnson's column formula, 149-152
Joints:
bolted, 305-310
bonded (with adhesives), 348-350
brazed, 347
gasketed, 305-310, 702
riveted, 330-334
soldered, 347
stiffness of bolted, 305-310
welded, 336-347
Jominy, Walter, 59
Jominy end-quench hardenability test, 59
Journal bearings, *see* Bearings, journal

Keys (shaft), 536-537, 546-549
Keyways, fatigue stress concentration factors for, 548
Kinematic viscosity, *see* Viscosity
Kinetic energy, 21, 187, 561

Lead of screw thread, 274
Lead of worm gear, 521, 522
Leaf spring, 377-382

Lewis, Wilfred, 474-475
Lewis equation for gear tooth stresses, 475
Lewis form factor for gear teeth, 475, 476
Life considerations:
  in fatigue loading:
    basic *S-N* curve, 203-207
    contact stress *S-N* curves, 267
  in gear teeth (surface fatigue), 489
  in roller chains, 588
  in rolling-element bearings, 449-450
Line of action of gears, 462
Linear actuators, *see* Screws, power
Load analysis, applied to:
  beams, 34-37
  bearing loading, *see* Bearing loading
  free bodies (in general), 25-34
  gears:
    bevel, 514-516
    helical, 507-510
    spur, 470-473
    worm, 523-528
  impact loading, 184-187
Load factor, C_L (for fatigue), 209
Lock nuts and lock washers, 303-305
Lubricant:
  additives, 387
  definition, 387
  gas (as air), 389
  greases, 387-388
  solid types, 387
  types, 387-388
  viscosity, *see* Viscosity
  viscosity index, 396
Lubrication:
  of chains, 587, 589
  of gears, 486
  of journal bearings, 389-392, 402-412, 422-424
  of rolling-element bearings, 441, 449
  types of:
    boundary (thin-film), 389, 422-424
    elastohydrodynamic, 425, 485
    hydrodynamic (full-film, thick-film), 389-392, 398-410, 415-422
    hydrostatic, 389-390
    mixed film, 389
    squeeze-film, 392, 409

Machinability, 57
Magnesium alloys, 62
  fatigue strength, 207, 208
  properties of (tables), 619, 620, 638
  temper designations, 636
Malleable cast iron, 58
  properties and uses of (table), 622
Materials:
  for bolts and screws, 296-299
  for clutches and brakes, 558, 559, 578-579
  for gears, 495
  for journal bearings, 414-415
  mechanical properties and uses (tables), 620-644
  physical properties of metals (table), 621
  for rivets, 332
  for rolling-element bearings, 437-439
  for springs, 359, 360
  tensile properties of metals (table), 620
  see also specific materials
Maximum-distortion-energy theory of failure, 167-169
Maximum-normal-stress theory of failure, 165-166
Maximum-shear-stress theory of failure, 166, 167
Mean value, 179
Miner-Palmgren cumulative damage rule, 232-235
Mixed-film lubrication, 389, 422-424
Modified Mohr theory of failure, 169, 170
Module (of gear teeth), 463
Modulus:
  of elasticity, 48
    values of, for common metals (table), 619
  of resilience, 53
  of rigidity (shear modulus, torsional modulus), 56
    values of, for common metals (table), 619
  of toughness, 53
Mohr circle:
  for strains, 118-125
  for stresses, 88-89
  three-dimensional aspect of, 96-98, 124-125
Mohr, Otto, 91
Mohr theory of failure, 169-171
Molybdenum disulphide, 387
Moment diagram, 34-37
  sign convention, 36
Moment of inertia:
  of areas (table), 616
  of flywheels, 20-21
  of solids (table), 617
  of welds, 341-345
Moments and deflections in beams (tables), 646-651
Monel metal, 63, 639
Moore's fatigue test, 203-204
Multiple disc clutch, 554-559
Multiple roller-chain drive, 588, 589
Multiple V-belt drive, 583-585
Music wire, 360

Natural frequency of springs, 372
Natural frequency of vibration, 183
Needle bearings, 430, 433, 437, 439
Neg'ator springs, 384-385

Newton's law of motion, 15
Newton's law of viscous fluid flow, 392
Nickel alloys, 62, 63
  fatigue strength, 207
  properties (tables), 619, 639
Nitriding, influence on fatigue strength, 239
Nodular iron, 58
  properties (table), 623
Normal distribution, 177, 180
Normal pitch, plane, pressure angle (gears), 525
Notch sensitivity, 223, 224

Octahedral plane and stresses, 168
Octahedral shear stress theory of failure, 167-169
Ocvirk's short bearing analysis, 402
Oil, *see* Lubricant; Viscosity
Oil grooves in bearings, 411, 412
Oldham couplings, 550, 552
"One-Hoss Shay" (Oliver Wendell Holmes), 158-159
One-way (free-wheeling, overrunning) clutches, 596, 597, 726, 727
O-ring seals, 308
Overhauling of power screws, 282-283
Overhauling of worm gearing, 523, 526-527

Palmgren-Miner cumulative damage rule, 232-235
Palmgren on rolling-element bearing life, 449
Partial bearing, 388
Peening, fatigue strengthening influence of, 238-239
Penetration hardness tests, *see* Hardness, from penetration hardness tests
Percentage elongation, *see* Elongation at fracture of material
Petroff's equation for journal bearing friction, 397-398
Photoelastic representation of gear tooth stresses, 473
Physical properties of common metals (table), 619
Pinion, definition of, 460
Pins, for use with shafts, 536, 538
Pipe thread, 278
Pitch of chains (p), 587, 590
Pitch circle of gears, 460
Pitch cone of bevel gears, 512, 513
Pitch diameter:
  of gears, 460
  of screw threads, 275
Pitch of gears:
  axial, 506, 521
  base, 468
  circular, 463
  diametral, 463
Pitch of helical springs, 357, 358
Pitch line velocity of gears, 470
Pitch point of gears, 459
Pitch of screw threads, 274, 275
Pitch of worms, 521
Pitting (of surfaces), 263-267
  of gear teeth, 484, 486
Pivoted shoe brake, 570
Plain bearing, 388
Plain strain in fracture mechanics, 161
Plain stress in fracture mechanics, 161
Planetary (epicyclic) gear trains, 496-499, 499-606
Plastics (materials), 63-68
  properties of (tables), 642-644
Plate clutches, *see* Clutches
Plating, influence of on fatigue strength, 237
Poise (unit of viscosity), 393
Poisson's ratio (table of values), 619
Porous metal bearings, 423
Power:
  loss in bearings, 398, 421
  screws, *see* Screws, power
  work, energy basic relationships, 17-24
Preloading of rolling-element bearings, 429
Presetting of springs, 361, 362
Press fits, *see* Interference fits
Pressure:
  angle, in gears, 461, 505
  distribution in journal bearings, 408
  lubricated journal bearings, 412-413
Principal stresses, planes, directions, 90, 95, 96
Principal strains, 119
Probability, *see* Reliability
Progressive fracture, *see* Fatigue, failure
Proof load of bolts, 299
Proof strength of bolt materials, 297, 298
Proportional limit, 48
Punch press, load-energy-power analysis of, 19-21
Pure bending of standard fatigue test specimens, 203

Quality control, *see* Reliability

R. R. Moore fatigue test, 203-204
Rack, 465
Rack cutter for gears, 467
Radius of gyration of sections (table), 616
Raimondi and Boyd lubrication charts, 402-410, 425
Randomly varying fatigue loading, 232
Rated capacities of rolling-element bearings, 442, 448
Rate of springs, 125-127, 357
Ratio of gear trains, 495-499, 599-604
Reduction in area (index of ductility), 49

values of (tables), 624-625, 627-629, 631
Redundant ductile structure strength, 43-45, 71-72
Redundant reactions, determination, 40-42, 140-142
Redundant support, definition, 40
Reliability, 177-181
factor in fatigue, 482
Relaxation of bolt initial tension, 302
Relaxation of springs (i.e., "set"), 359
Residual stress:
due to rolling of screw threads, 299
due to thermal gradients, 112-113
due to yielding, 106-111
influence of, 113, 114
on fatigue strength, 236, 238-239, 265-267
on fretting fatigue, 257
in springs, 361-362, 370-371
Resilience, 53
Resistance welding, 335
Retaining rings, 537-539
Reversed axial loading fatigue strength, 209
Reversed bending fatigue strength, 207-209
Reversed biaxial loading fatigue strength, 210-211
Reversed torsion fatigue strength, 209-210
Reyn (unit of viscosity), 392-393
Reynolds, Osbourne, 393
Reynolds' equation:
for one-dimensional flow, 398-402
for two-dimensional flow, 402
Ring-oiled bearing, 410
Rings, "snap" or retaining, 537-539
Rivets, 330-334
Riveted joint (triple-riveted butt), 43-45
Rockwell hardness, 54-55
Rolled threads, 299
Roller bearings, *see* Bearings, ball and roller
Roller chains, 586-589
Rotating bending fatigue strength, 203-207
Rubber bearings, 415

SAE (Society of Automotive Engineers) specifications, 59, 394-395
Safety considerations, 5-8
Safety factor, 173-176
representative values, 176
Saybolt viscosity, 393-396
Scoring, 254
of gear teeth, 484, 485
Screws, *see* Bolts and screws
Screws, power, 278-293
coefficient of friction, 281-282
collar friction, 280-282
column loading, 292-293
efficiency, 283-287
load analysis, 279-284, 286-287
overhauling, 282, 283
rolling-contact type, 284-285
self-locking, 282, 283
thread design, 275-279
Screw threads, 275-288
inch series fastener thread sizes, 277
inch series power screw thread sizes, 279
metric series fastener thread sizes, 277
see also Bolts and screws
Scuffing, 254
Seals, bearing and shaft, 432-433
Secant formula for column buckling, 152-153
Section modulus, 77, 616
Sections, properties of, 615-616
Self-aligning bearings, 431, 432
Self-energizing brakes and clutches, 565
Self-locking brakes and clutches, 565
Self-locking power screws, 282-283
Self-locking worm gears, 523, 526-527
Sems, 294
Set screws, 536-537
Set of springs, 359
Shaft:
critical speeds, 536, 539-542
couplings, *see* Couplings
deflections, 126-132
computer determination, 130-131
design:
basic procedure, 542-546
criteria for, 535-536, 542
fatigue strength (sample problem), 543-546
keys and keyways, 536-537, 546-549
support by bearings, 536
Shear:
deflection:
by Castigliano's method, 132-137
due to torsion, 125, 127
direct, 72-73
loading diagram in beams, 34-36
sign convention, 36
pure, 73
stress sign convention, 74
stress, transverse, distribution of, 82-88
stresses:
in beams, 82-88
due to torsion, 73-76
Shock loading, *see* Impact, and shock loading
Shot peening, 238
applied to springs, 368, 371
effect on fatigue strength, 238
Size, influence on fatigue strength, 210, 214
Sleeve bearings, *see* Bearings, journal

Slenderness ratio, 147
Snap (retaining) rings, 537, 539
Societal values and considerations, 10-13
Soldering, 347
Sommerfeld variable (lubrication), 402
S-N curves, 203-205, 207
  "knee" of, 205, 207
Spalling of surfaces, 263-267, 486
Speed ratios in gear trains, 495-499
Spindles, 535
Splines, 538, 540, 549
Spring index of coil springs, 354
Spring rate (spring constant, spring scale), 125-126, 357
  influence of in bolt loading, 309
Springs:
  beam, 377-382
  Belleville washer, 384-385
  coil, conical, 358
  coil, helical, 353-377
    active coils in, 356-357
    compression:
      buckling, 363-364
      clash allowance, 361
      deflection, 356
      design procedure, 364-375
      pitch, 357
      presetting, 361-362, 370-371
      solid height, 362
      types of ends, 362-363
  constant force ("Neg'ator"), 385-386
  extension, 375, 376
    end hook stresses and deflections in, 376, 377
    deflection, 376
    initial tension, 375-376
  garter, 385
  index, 354
  leaf, 377-382
  material specifications and costs, 359-360, 370
  natural frequency ("spring surge") of, 371-372
  Neg'ator (constant force), 385-386
  rate (constant, scale), 357
    in bolted joints, 309
  residual stresses, 370-371
  safety aspect, compression vs. extension, 375
  shot peening effect, 368, 371
  stresses, 353-357, 359-362
  torsion, 383-384
    bar, 352-353
  volute, 385
  Wahl factor, 354-357
  washer, 384-385
  wire form, 385
Spring steels, strength and cost, 359-360, 370
Sprocket, 586-587
Spurgears, *see* Gears, spur
Squeeze-film lubrication effect, 392, 409, 415
Standard deviation, 178, 179
  of endurance limit, 205
Statistical approach to fatigue strength, 205
Statistical aspect of reliability, 177-181
Steel, 58-61
  high-strength low alloy (HSLA), 60
  properties of (tables), 619, 620, 624-632
  stainless, 61, 620, 632
Stiffness, *see* Spring rate
Strain, 47, 51, 116-124
  gage rosette analysis, 119-123
  gages, 118
  true, 51
Strain-strengthening exponent (m), 52
  values of (table), 620
Strength properties of materials (tables), 620-625, 627-635, 637-643
Stress:
  corrosion cracking, 249-251
  engineering, 47
  gradient factor for factor C_G, 209, 213, 214
  residual, *see* Residual stress
  thermal, 111-113
  three-dimensional aspect of, 96-97
  true, 51
Stress concentration factors:
  fatigue (K_f), 222-225
    for keyways, 548
    for screw threads, 318-319
    for welded joints, 346
  influence for fatigue loading, 222-228
  influence for static loading, 102-105
  theoretical (or geometric) K_t, 98-105
Stress-strain curve:
  engineering, 47-50
  true, 51-53
Superalloys, iron-based, 61
  properties of (tabular), 633
Superposition, principle of for deflection determination, 128
Surface:
  coatings, 267-268
  fatigue failure, 259, 263-267
    of gear teeth, 484-491
    of rolling-element bearings, 442
  influence on fatigue strength, 211, 212, 214, 267
  treatments for fatigue-strengthening, 238-239

Tapered roller bearings, 433, 435, 436
Teflon bearings, 387
Tension, *see* Axial loading fatigue

strength, Axial stress, limitation of PIA formula
Tensile test, 47
Theoretical stress concentration factors K_t, 98-105
Theories of failure for static loading, 164-172
Threaded fasteners, *see* Bolts and screws
Threads, *see* Screws, power; Screw threads
Three-dimensional stresses, 96-98
Thermal capacity of worm gears, 530-534
Thermal expansion coefficient, values of (table), 619
Thermal stresses, 111-113
Thrust bearings, 424, 425
Thrust collar, 280
Timing belts, 586
Titanium alloys, 63
  fatigue strength of, 207
  properties of (tables), 619, 640
Tolerances, 652-653
Toothed belts, 586
Torque converters, (hydronamic), 596-597
Torsional deflection, 127
Torsional fatigue strength, 209-210
Torsional impact, 192-194
Torsional stresses, 73-76
  in shafts, 228-232, 544
  in welded joints, 340-344
Torsion bar springs, 352-353
Torsion springs, 383-384
Toughness, 53
  modulus of, 53
Tower, Beauchamp (lubrication experiment), 398-399
Train value, 499
Transmission, automotive automatic, 598-610
Transverse shear stress distribution, 82-88
Tredgold's approximation, 513
Triaxial (three-dimensional) stresses, 96-98
True stress, 51
True stress-strain curve, 51-53
True strain, 51

Ultimate strength, 47
  correlation with Brinnel hardness number, 54
Ultrasonic welding, 335
Undercutting of gear teeth, 467
Unground ball bearings, 433, 439
Unified thread series, 274, 275
Units:
  basic, 15
  conversion factors between SI and English, 612
  derived, 15
  systems, 14-17
Universal joints, 550-552
  constant-velocity type, 552
  cross-type (Cardan and Hooke's joint), 550-552

Variable speed V-belt drive, 584-585
V-belts, 583-585
Velocity ratio of gear trains, 495-499
Vibration, natural period of, 183
  in coil springs, 371-372
  in journal bearings, 409
Virtual number of teeth, 506, 513
Viscosity, 390-396
  index, 396
  kinematic, 394
  temperature effect on, 396
  pressure effect on, 396
Volute spring, 385

Wahl, A. M., 354
Wahl factor in coil springs, 354
Warping in heat-treatment, 59
Wear, 252-259
  abrasive, 255-256
  adhesive, 253-255
  analytical approach to, 257, 259
  in clutches and brakes, 557, 578
  coefficient, 257-259
  fretting, 256-257
  in gear teeth (surface deterioration), 484-487
  types, 252-253
  in worm gears (surface deterioration), 528-530
Weibull, W., 450
Weibull distribution (statistical), 450
Weight, unit, of common metals (table), 619
Weldability of common metals, 336, 634
Welded joints, 336-347
  fatigue strength, 346-347
  static strength, 336-345
Welding:
  electrodes, 337
  fatigue considerations, 346-347
  processes:
    arc, 333-335
    gas, 335
    resistance, 335
    special fusion types, 335
  of thermoplastics, 335-336
Welds, fillet, 337
  strength for parallel loading, 338-339
  strength for transverse loading, 340
"Whip" and "whirl" in journal bearings, 409
Woodruff key, 537
Work-energy-power basic relationships, 17-24

Work hardening, *see* Strain strengthening exponent (m)
Worm gears, 503-504, 521-534

Yield strength, 48
Yoke connection, stresses in, 38-40
Young's modulus, 48

Zerol bevel gears, 514, 515
Zinc alloys, 63, 619, 641

PROPERTIES OF SECTIONS

A = area, in.^{2}

I = moment of inertia, in.^{4}

J = polar moment of inertia, in.^{4}

Z = section modulus, in.^{3}

ρ = radius of gyration, in.

$\bar{y}$ = centroidal distance, in.

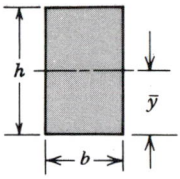

Rectangle

$$A = bh \qquad I = \frac{bh^3}{12} \qquad Z = \frac{bh^2}{6}$$

$$\rho = 0.289h \qquad \bar{y} = \frac{h}{2}$$

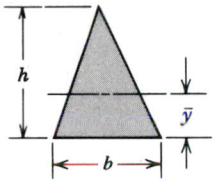

Triangle

$$A = \frac{bh}{2} \qquad I = \frac{bh^3}{36} \qquad Z = \frac{bh^2}{24}$$

$$\rho = 0.236h \qquad \bar{y} = \frac{h}{3}$$

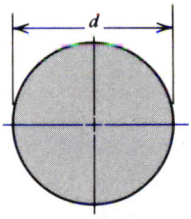

Circle

$$A = \frac{\pi d^2}{4} \qquad I = \frac{\pi d^4}{64} \qquad Z = \frac{\pi d^3}{32}$$

$$J = \frac{\pi d^4}{32} \qquad \rho = \frac{d}{4}$$

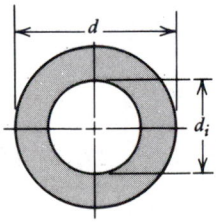

Hollow circle

$$A = \frac{\pi}{4}(d^2 - d_i^2) \qquad I = \frac{\pi}{64}(d^4 - d_i^4) \qquad Z = \frac{\pi}{32d}(d^4 - d_i^4)$$

$$J = \frac{\pi}{32}(d^4 - d_i^4) \qquad \rho = \sqrt{\frac{d^2 + d_i^2}{16}}$$

MASS AND MASS MOMENTS OF INERTIA OF HOMOGENEOUS SOLIDS

ρ = mass density

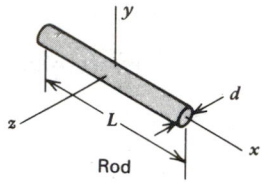

Rod

$$m = \frac{\pi d^2 L \rho}{4}$$

$$I_y = I_z = \frac{mL^2}{12}$$

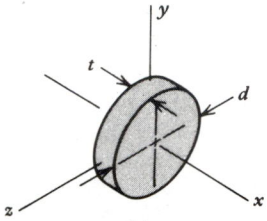

Disk

$$m = \frac{\pi d^2 t \rho}{4}$$

$$I_x = \frac{md^2}{8}$$

$$I_y = I_z = \frac{md^2}{16}$$

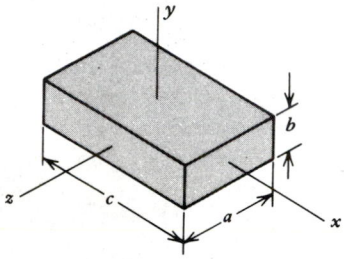

Rectangular prism

$$m = abc\rho$$

$$I_x = \frac{m}{12}(a^2 + b^2)$$

$$I_y = \frac{m}{12}(a^2 + c^2)$$

$$I_z = \frac{m}{12}(b^2 + c^2)$$

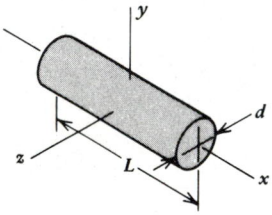

Cylinder

$$m = \frac{\pi d^2 L \rho}{4}$$

$$I_x = \frac{md^2}{8}$$

$$I_y = I_z = \frac{m}{48}(3d^2 + 4L^2)$$

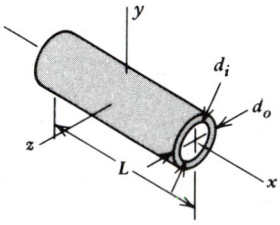

Hollow cylinder

$$m = \frac{\pi L \rho}{4}(d_o^2 - d_i^2)$$

$$I_x = \frac{m}{8}(d_o^2 - d_i^2)$$

$$I_y = I_z = \frac{m}{48}(3d_o^2 + 3d_i^2 + 4L^2)$$